T0145387

Studies in Computational Intelligence

Volume 730

Series editor

Janusz Kacprzyk, Polish Academy of Sciences, Warsaw, Poland
e-mail: kacprzyk@ibspan.waw.pl

The series "Studies in Computational Intelligence" (SCI) publishes new developments and advances in the various areas of computational intelligence—quickly and with a high quality. The intent is to cover the theory, applications, and design methods of computational intelligence, as embedded in the fields of engineering, computer science, physics and life sciences, as well as the methodologies behind them. The series contains monographs, lecture notes and edited volumes in computational intelligence spanning the areas of neural networks, connectionist systems, genetic algorithms, evolutionary computation, artificial intelligence, cellular automata, self-organizing systems, soft computing, fuzzy systems, and hybrid intelligent systems. Of particular value to both the contributors and the readership are the short publication timeframe and the world-wide distribution, which enable both wide and rapid dissemination of research output.

More information about this series at http://www.springer.com/series/7092

Aboul Ella Hassanien · Diego Alberto Oliva
Editors

Advances in Soft Computing and Machine Learning in Image Processing

Editors
Aboul Ella Hassanien
Faculty of Computers and Information, Information Technology Department
Cairo University
Giza
Egypt

and

Scientific Research Group in Egypt (SRGE)
Cairo
Egypt

Diego Alberto Oliva
CUCEI, Departamento de Ciencias Computacionales
Universidad de Guadalajara
Guadalajara
Mexico

and

Scientific Research Group in Egypt (SRGE)
Cairo
Egypt

ISSN 1860-949X ISSN 1860-9503 (electronic)
Studies in Computational Intelligence
ISBN 978-3-319-87627-6 ISBN 978-3-319-63754-9 (eBook)
https://doi.org/10.1007/978-3-319-63754-9

Softcover reprint of the hardcover 1st edition 2017

Printed on acid-free paper

This Springer imprint is published by Springer Nature
The registered company is Springer International Publishing AG
The registered company address is: Gewerbestrasse 11, 6330 Cham, Switzerland

Foreword

In this book, the best chapters that explore the combination of machine learning techniques with image processing methods are selected. Nowadays, these topics are important not only in the scientific community, but also for other areas such as engineering and medicine. The problems described in each chapter are relevant for the area of image processing and computer vision. Moreover, the use of soft computing and machine learning techniques is also demonstrated.

This book brings together and explores possibilities for combining image processing and artificial intelligence, both focused on machine learning and soft computing that are two relevant areas and fields in Computer Science. The selected chapters include topics from different areas and implementations. The editor classifies them accordingly with the importance and usefulness. In Part I, chapters related with image segmentation are included. This problem is very important because it is considered as a preprocessing step in most of the image processing systems. For example, one of the chapters presents a survey of the advantages and disadvantages of using different colors spaces in segmentation using clustering. Another work is focused in the use of multi-objective optimization for thresholding. Moreover, other interesting chapters provide an overview of the use of evolutionary approaches for the segmentation of thermal images. On the other hand, Part II addressed the implementation of both soft computing and machine learning in medical applications of image processing. Chapters for liver tumor recognition based on different classifiers, a method for the detection of lesions of coronary disease, glaucoma monitoring to mention some are included.

Images and videos are extensively used in security, Part III is dedicated to the use of machine learning and soft computing algorithm for security and biometric applications. The chapters contained in this part consider biometric problems as a signature recognition based on neural networks or finger print identification using different topological structures. Meanwhile for security methods such as the use of support vector machines for detecting violent activities of humans in videos is presented.

The aim of image processing and computer vision is detection, recognition, and analysis of the objects contained in the scenes. Part IV of this book is dedicated to new approaches for this task. To mention some interesting approaches, chapters such as object recognition for robots, distance measurements for geometric figures, and image reconstruction using Fourier transform are included.

The book was designed for graduate and postgraduate education, where students can find support for reinforcing or as the basis for their consolidation or deepening of knowledge, researchers. Also teachers can find support for the teaching process in areas involving machine vision or as examples related to the main techniques addressed. Additionally, professionals who want to learn and explore the advances on concepts and implementation of optimization and learning-based algorithms applied in image processing can be found in this book, which is an excellent guide for such purpose.

The book which is concise and comprehensive on the topics addressed makes this work an important reference in image processing, which is an important area where a significant number of technologies are continuously emerging and sometimes untenable and scattered along the literature. Therefore, congratulations to the authors for their diligence, oversight, and dedication for assembling the topics addressed in the book. The computer vision community will be very grateful for this well-done work.

May 2017

Marco Pérez-Cisneros
Universidad de Guadalajara
Guadalajara, Mexico

Preface

Several automatic systems require cameras to analyze the scenes and perform the desired task. Images and videos are taken from the environment, and after that, some processing algorithms should be used to analyze the object contained in the frames. On the other hand, in the past two decades, the amount of users of cameras has been increased exponentially. Cameras are present in smartphones, computers, cars, gadgets, and many other apparatuses used every day. Based on such facts, it is necessary to generate robust algorithms that permit the analysis of all this information. These algorithms are used to extract the features that permit the identification of the objects contained in the image. To achieve this task, it is necessary to introduce computational tools from artificial intelligence. The tendency is to have automatic applications that can analyze the images obtained with the cameras. Such applications involve the use of image processing algorithms combined with soft computing and machine learning methods. This book presents a study of the use new methods in image and video processing. The selected chapters explore areas from the theory of image segmentation until the detection of complex objects in medical images. The implementation concepts from machine learning, soft computing, and optimization are analyzed to provide an overview of the application of this tools in image processing.

The aim of this book is to present a study of the use of new tendencies to solve image processing problems. We decide to edit this book based on the fact that researchers from different parts of the world are working in this field. However, such investigations are published in different journals, and it is hard to find a compendium of them. The reader could see that our goal is to show the link that exists between intelligent systems and image processing. Moreover, we include some interesting applications in areas like medicine or security that are very important nowadays.

The content is divided into four parts; Part I includes the methods involved with theory and applications of image segmentation. For example, the use of multiobjective approaches or different color spaces. Part II includes the applications of machine learning and soft computing algorithms for medical purposes, for example, glaucoma or coronary diseases. Meanwhile, in Part III approaches for security and biometry are included . Some of them are related to finger print identification or the analysis of videos for activity recognition. Finally, Part IV contains 11 chapters about object recognition and analysis in the scenes.

Editing this book was a very rewarding experience, where many people were involved. We acknowledge to all the authors for their contributions. We express our gratitude to Prof. Janusz Kacprzyk, who warmly sustained this project. We also acknowledge to Dr. Thomas Ditzinger, who kindly agreed to its appearance.

Finally, it necessary to mention that this book is just a small piece in the puzzles of image processing and intelligence. We would like to encourage the reader to explore and expand the knowledge in order to create their implementations according to their necessities.

Cairo, Egypt — Aboul Ella Hassanien
Guadalajara, Mexico — Diego Alberto Oliva
May 2017

Contents

Part I
Image Segmentation

Color Spaces Advantages and Disadvantages in Image Color Clustering Segmentation

Edgar Chavolla, Daniel Zaldivar, Erik Cuevas and Marco A. Perez

Abstract Machine learning has been widely used in image analysis and processing for the purpose of letting the computer recognize specifics aspects like color, shape, textures, size, and position. Such procedures allow to have algorithms capable of identifying objects, find relationships, and perform tracking. The present chapter executes an analysis of the one image attribute that is listed among the basic aspects of an image and the color. The color is chosen due to the importance given by humans to this attribute. Also the color is used main filter criteria in object recognition and tracking. In this section the effect of the color is studied from the point of view of the most popular color spaces available in image processing. It will be tested the effect of the selection of a given color space one of the most common clustering machine learning algorithms. The chosen algorithm is K-means ++, a variation of the popular K-means, which allows a more fair evaluation since it mitigates some of the randomness in the final cluster configuration. Every advantage or weakness will be exposed, so it can be known what color spaces are the right choice depending on the desired objective in the image processing.

E. Chavolla (✉) · D. Zaldivar · E. Cuevas · M.A. Perez
Universidad de Guadalajara, CUCEI, Guadalajara, Jalisco, Mexico
e-mail: chavolla@gmail.com

D. Zaldivar
e-mail: daniel.zaldivar@cucei.udg.mx

E. Cuevas
e-mail: erik.cuevas@cucei.udg.mx

M.A. Perez
e-mail: marco.perez@cucei.udg.mx

A.E. Hassanien and D.A. Oliva (eds.), *Advances in Soft Computing and Machine Learning in Image Processing*, Studies in Computational Intelligence 730,
https://doi.org/10.1007/978-3-319-63754-9_1

1 Colors Models and Spaces

Color is one of the most important aspects in vision, since it is used generally to discriminate and recognize information. From the Physics view, the color is caused by the reflection of a portion of light in an object. Basically, it is just an electromagnetic wave in a range of frequencies. The longest wavelength produces the red and the shortest blue-violet (Fig. 1).

The part of the electromagnetic spectrum analyzed is the one called the visible spectrum, which contains all the possible colors. Other colors existing outside the visible spectrum range are not contemplated in this section (Infrared and ultraviolet) [15].

Since representing the colors as a specific wave frequency with a certain amplitude and longitude is not trivial or natural to human beings, it created systems that help to describe the colors in a more natural and ordered way. The descriptions systems are known as color models.

A color model is actually a set of equations and/or rules used to calculate a given color and are capable of describing a wide range of tones from the visible spectrum of light.

A color space is defined as all the possible tones generated by a color model; in this work the colors are managed from the color space perspective.

1.1 Additive Color Spaces

Additive color spaces are those set of colors that are created by mixing primary colors. The most used primary colors are the red, blue, and green. Originally, these colors were formed by adding two or more rays of light into an obscure surface. The created color space using this technique is named RGB [23].

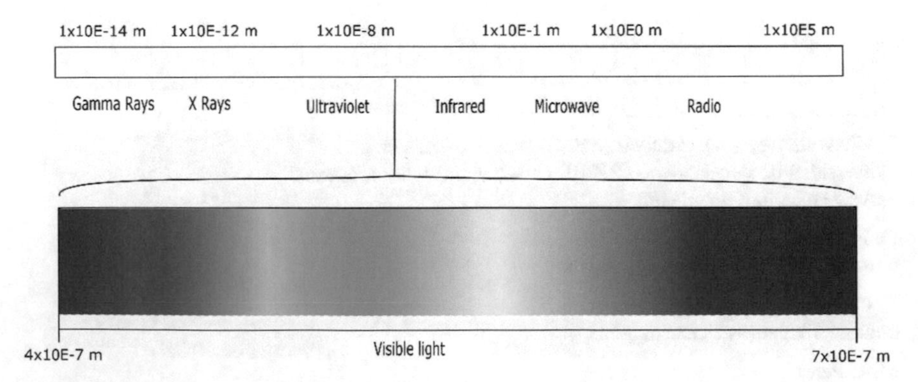

Fig. 1 Visible light spectrum. Wavelengths are in meters

A variant of the additive color spaces is the subtractive color space, which instead of adding rays of light, it blocks or subtracts certain frequencies from a ray of light containing the complete frequency range from the visible light spectrum.

This section will use only the additive color space in form of the RGB model, since it is the most frequent color space used in computer images.

RGB This model is most widely used in computer graphics. Many works use this model as a base for color filtering, segmentation, and analysis in image processing. The ease of use and intuitive model for color creation and manipulation makes it an ideal choice for drawing and coloring (Fig. 2).

Unfortunately, it has some disadvantages when it is used in image processing. The model produces a nonlinear and discontinues space, which makes the changes in color hue hard to follow due to discontinuities. Another issue found in RGB model is the fact that the color hue is easily affected by illumination changes, making the color tracking and analysis a nontrivial task.

Despite the issues described for the RGB model, the produced space is preferred in most of the computational tasks. One of the reasons is that most of the computer hardware and software is developed under this model, so it can be used directly without additional work.

Normalized RGB This model is a variation from the RGB, and basically is created under the premise of protecting the color model from the illumination changes. The main idea behind this variation is that the colors are formed using a certain proportion of three primary colors from the model, not a defined amount of each one. So the sum of all the colors proportions must equal a 100%. The process of normalization is straightforward and can be easily computed (Eqs. 1–3). Basically, each of the equations used to transform RGB to normalized RGB obtains the proportion of each color. Another advantage of this model is that any color can be described just using two colors instead of three. This is possible due to the fact that the third color will always be the difference between one and the addition of the other two colors:

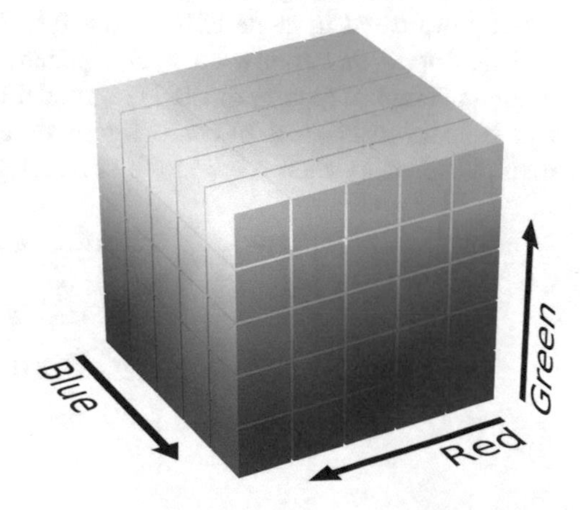

Fig. 2 RGB color space (Wikimedia commons RGB)

$$r = \frac{R}{R+G+B} \tag{1}$$

$$g = \frac{G}{R+G+B} \tag{2}$$

$$b = \frac{B}{R+G+B}. \tag{3}$$

This variant certainly reduces the negative effect of the illumination changes like shadows or shines, but in exchange it reduces object detection capability. This is due to the loss of contrast that the same illumination provides [5].

This model is used in machine learning with some degree of success, and it will part of the comparison performed in this work

1.2 Perceptual Color Spaces

Perceptual color spaces are created from those models that treat the color in a more human intuitive form. In order to achieve this objective, any color is represented by a specific tone or hue, a level of saturation for the hue, and the amount of light available or illumination.

Inside this category, it can be found Hue Saturation Value model (HSV), Hue Saturation Lightness (HSL), and Hue Saturation Intensity (HSI). All of these color models have a similar description about the color hue but differ in the saturation and illumination definition.

HSL This color model can be computed from the RGB standard model by using a set of equations (Eqs. 4–6). The major advantage of using this model is the fact that it presents immunity to illumination changes, since the illumination is enclosed in the lightness component of the model. The other feature of this model can be found in the color hue changes; they are continuous and linear. The hue component is usually expressed in angle terms, from 0 to 359°, since every tone has a previous and next tone that follows a cyclic pattern and can be visualized in a circular diagram. The last feature of the HSL model is the geometric representation which can be expressed as a bi-conic figure or a cylinder (Fig. 3), which allows to manipulate easily each color hue by changing its illumination and saturation [7]:

$$\mathrm{H} = \begin{cases} 60 * \left(\frac{G-B}{max(R,G,B)-min(R,G,B)}\right) & R = max(R,G,B) \\ 60 * \left(\frac{2+(B-R)}{max(R,G,B)-min(R,G,B)}\right) & G = max(R,G,B) \\ 60 * \left(\frac{4+(R-G)}{max(R,G,B)-min(R,G,B)}\right) & B = max(R,G,B) \end{cases} \tag{4}$$

$$S = \frac{max(R,G,B) - min(R,G,B)}{1 - |(max(R,G,B) + min(R,G,B) - 1|} \tag{5}$$

$$L = \frac{max(R,G,B) + min(R,G,B)}{2}. \tag{6}$$

Even though the HSL model could seem ideal, it has some issues; the first occurs when the maximum and minimum values for RGB are the same, which corresponds to the gray tones. In this scenario there is an undefined value for hue, which usually the existing software libraries set to zero (Red color). This causes some incorrect color interpretations. The other visible issues occur when the maximum and minimum sum two; in this scenario the saturation is undefined.

HSV This model is similar to HSL, and share the same definition for the hue component (Eq. 4), but differs in the way it interprets the color saturation (Eq. 7) and the illumination (Value component) (Eq. 8):

$$S = \frac{max(R,G,B) - min(R,G,B)}{max(R,G,B)} \tag{7}$$

$$V = max(R,G,B). \tag{8}$$

The issues found in HSL can be found as well in HSV; there is the same undefined scenario for hue and regarding saturation, the issue arises when the maximum value for RGB is zero (black color).

Sometimes, HSV is preferred due to the geometric representation, which is usually more natural than HSL, which allows a better color hue manipulation (Fig. 4).

Some works use variations from the HSV and HSL models, which does not involve extra computer work. This variation is just to ignore the illumination component, which theoretically would eliminate the effect produced by shadow or shines [3, 10, 24]. Other works go beyond and also ignore the saturation component, leaving only the hue component as the only criteria for color manipulation and analysis [11, 8].

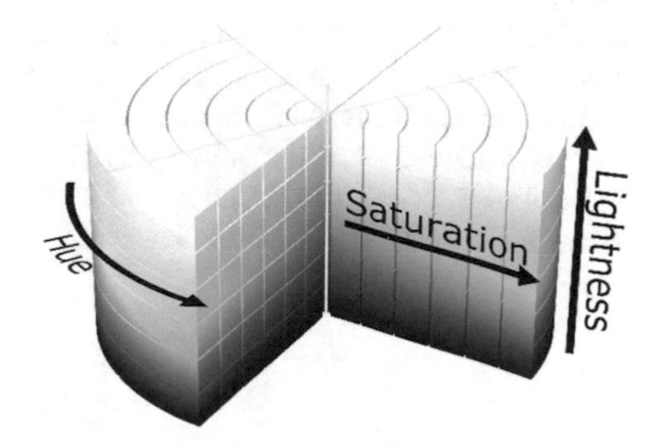

Fig. 3 HSL color space (Wikimedia commons HSL)

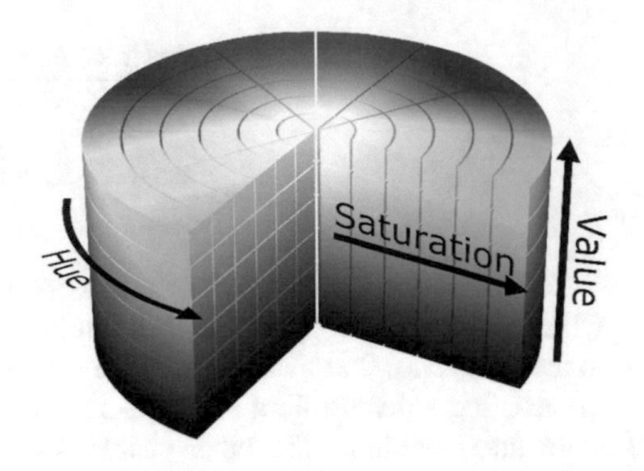

Fig. 4 HSV color space (Wikimedia commons HSV)

A common problem found in the perceptual color spaces is the hue representation. Usually, it is represented as the angle of a circle. It means that one hue is at the restarting point of the circle (where the angle changes from 359° to 0°). This causes a discontinuity in the color hue. It is usually fixed by using two ranges for the hue at this position. In this case the affected hue corresponds to the red color.

1.3 CIE Color Space

The international Commission on Illumination (CIE) is worldwide nonprofit organization, created with the purpose to gather and share information related to the science and art of light, color, vision, photobiology, and image technology (CIE, [4].

The CIE created the first color models based primarily on the physic aspect of the light. Among the most notable models, the CIE RGB, CIE XYZ, and CIELab can be found. All of them describe how all the colors can be formed on a plane with a given illumination level.

The CIE XYZ and CIE RGB are calculated by using the light wavelength from the physic representation of the color, while CIELab is indirectly obtained from the CIE XYZ.

Lab (also known as L * a * b*) This is one of the most used models from the CIE family. The lab model is formed by a Lightness component and two chromatic or color components (a and b). This color model can express a wider color range or gamut than the RGB, and usually, is used to enhance color images. The lab model can be obtained from the primary CIE XYZ model by using Eqs. 9–11, where $X_n = 95.047$, $Y_n = 100.0$, and $Z_n = 108.883$ are the tristimulus reference value for white for the CIE XYZ model; basically, the values are pseudo-normalized:

$$L^* = 116f\left(\frac{Y}{Y_n}\right) - 16 \tag{9}$$

$$a^* = 500\left(f\left(\frac{X}{X_n}\right) - f\left(\frac{Y}{Y_n}\right)\right) \tag{10}$$

$$b^* = 200\left(f\left(\frac{Y}{Y_n}\right) - f\left(\frac{Z}{Z_n}\right)\right). \tag{11}$$

The CIELab model can represent colors that are not handled by other models; theoretically, it could represent an infinite number of chromatic combinations. Unfortunately, its creation is a complex process, and the color space produced is not natural to humans as in RGB or perceptual color spaces (Fig. 5).

1.4 Other Color Spaces

Additional color spaces and models exist, but those are beyond the scope of this work, sincc in this chapter it will be covered only the most commonly used color spaces in computer image processing.

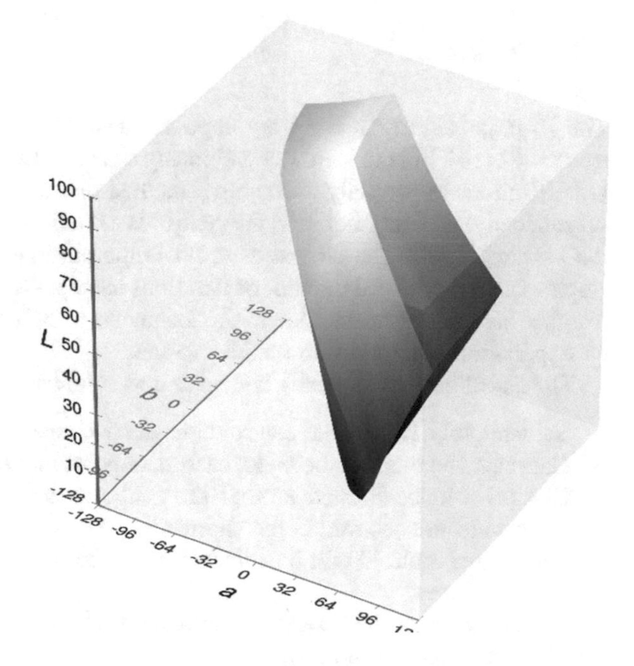

Fig. 5 RGB space represented using lab color model. (Wikimedia commons CIELab)

2 Clustering Algorithms and Machine Learning

Clustering algorithms refer to those algorithms capable of separate and classify a group of observations. Inside machine learning there are some specialized algorithms dedicated to perform clustering and classification tasks. The most interesting algorithms are those referred as unsupervised learning algorithms.

2.1 Unsupervised Learning Algorithms

Most of the algorithms in machine learning are required to go through a training phase in order to calibrate inner values. As a result of the training the algorithm is able to perform classification or to respond in a specific manner to certain entry data.

But in unsupervised learning, algorithms never require a training phase; instead, these algorithms are capable of discovering by themselves some interesting structures found in the input data [16].

The most common algorithms are clustering methods. Among them, it can be found K-means, Gaussian mixture, Hidden Markov models, and Hierarchical or Tree models. Those algorithms are capable to segment data into confined spaces, so existing and new data can be classified.

2.2 K-Means

The K-means algorithm is a very popular method for clustering. The term comes from 1967 [14], but the idea was previously stated in 1957, but disclosed until 1982 [12]. K-means is basically an iterative method that starts with k clusters centroids set randomly. At every iteration the centroids are adjusted, so that the distance from the existing data to the closest centroid is the minimum possible. The algorithm stops when a predefined number of iterations have passed or a desired data-centroid distance has been reached. Due to this behavior, usually the K-means is referred as an expectation maximization method variant.

The algorithm for K-means is depicted as follows:

- Set randomly K possible centroids in the data space;
- Calculate the distance between each data point and each centroid;
- Clusters will be formed around the centroids using those data points which distance to the centroid is the shortest;
- The cluster centroid will be adjusted using the data points inside the previously created cluster;
- If the distance or the iteration criteria are reached, the algorithm stops, otherwise go back to the second step.

Based on the K-means method some other algorithms have been proposed, among the most popular are the Fuzzy C-means, K-means ++, etc.

2.2.1 K-Means ++

K-means usually generates acceptable results, but there is a couple of known issues that affect its performance.

- If two centroids are too close to each other, they could be trying to classify the same cluster. In this case, the real cluster would be split among two K-means clusters.
- The time required to reach the optimum centroid for the existing clusters could vary depending on the initial centroids setup. If the initial random centroid is set close to the real centroid of the real clusters, the algorithm will take little time or iteration to find the optimums centroids. On the other hand, if the initial random centroids are set far from the real centroids, they could take significantly more time.
- Even more, the resulting clustering could be made using the wrong data points. This last one is caused due to the fact the K-means method only looks for minimizing the distance between the centroid and data point. At some point of the iterations, some data points could be changing clusters, and if the algorithm reaches the iteration number criteria, the data points stay in the current cluster. No validation is made to corroborate if each data point is assigned to the right cluster. Sometimes, this issues could create quite different clusters every time the K-means algorithm is executed.

Using the previous statements, it can be concluded that K-means algorithm performance depends on the initial centroids setting. The K-means ++ algorithm is created as a response to the initialization problem. The main idea behind the K-means ++ algorithm is to create a method to optimize the initial setup for the centroids in the K-means algorithm. Actually, K-means ++ algorithm is identical to K-means algorithm, but differs in the first step (The initial centroids setup). Instead of choosing arbitrary random values for all the centroids, K-means ++ uses an algorithm to select those centroids that can speed up and optimize the cluster search [1]. K-means ++ uses a simple probabilistic approach to generate the initial cluster setup, calculating the probability of how well a given point is doing acting as a possible centroid.

The K-means algorithm gets the K centroids (c) using a set X of data points as follows:

- Sample the first centroid $c(1)$ using a uniform random distribution over X.
- Iterate from $k = 2$ until K.

- Sample the k_{th} center $c[k]$ from a multinomial over X where a point x has probability θ_x, defined by Eq. 12, where $D(x)$ is defined as the distance to the closest existing centroid:

$$\theta_x = \frac{D(x)^2}{\sum_{x' \in X} D(x')^2} \alpha D(x)^2. \qquad (12)$$

K-means ++ allows a better cluster initialization, but it has a high computational cost. The method uses an iterative procedure to calculate the centroids with the best probability to be the centroids. The amount of iteration and calculations involves all the points in the data space, causing a significant work overhead.

The previous drawback can be alleviated by the fact that the rest of the K-mean ++ algorithm will find the cluster configuration faster and the cluster configuration will usually be the most convenient for the given problem.

As a result of the previously stated, it can be expected that K-means ++ would achieve to get the best clustering on each of the color spaces that will be tested.

2.2.2 K-Nearest Neighbor

A usual application for K-means is the nearest neighbors ' selection. This can be achieved by considering the data points from a given resulting cluster to be the closest neighbors from the resulting centroid. This process is used to categorize information by the closeness of a given known point.

3 Image Segmentation by Machine Learning

The idea of performing image segmentation and image analysis by using some of the algorithms covered by machine learning is widely spread across several works and literature. Among some recent examples of segmentation using machine learning algorithm are neural networks [2, 17–19], Gaussian mixture model [21, 22], support vector machine [9, 13, 6], and support vector machine with K-means family-based training [20, 25].

From the selection, recent works can be found that there is no standard color space regarding color feature extraction for the segmentation or classification. The most used models are RGB, nRGB, HSI, HSV, CIELab, Gray scales, and YCbCr. This scenario is the motivation for the present chapter. It exposes and explains valuable information about the most used color spaces regarding its properties. The properties should be considered in the election of a given color space. Just a few works expose the reason behind a specific color selection but never explain the possible issues that can arise from the mentioned selection.

4 Testing and Experimentation

The main purpose of the present tests is to demonstrate the benefits and issue coming from selecting a given color space. For the testing scenario, it was selected some of the most popular color spaces used in the image processing field:

- RGB
- nRGB
- HIS
- HSV
- HS
- H
- CIELab

The HS and H are not properly color spaces themselves, but variations from the HSV color space. It was decided to be included in the testing since these variations are attractive for the exclusion of the illumination component. A small explanation about the mentioned color spaces was already stated in the color section in this chapter.

For the targeted machine learning algorithm, it was selected the K-means algorithm. Even though it is an old algorithm, it is still widely used and still performs in an acceptable level against some more recent algorithms. The K-means ++ variant was selected, since it is helpful alleviating the randomness in the resulting classification clusters.

A set of three images presenting a solid background and the three primary colors (red, green, and blue) is selected:

- A synthetic image (made up image with solid colors);
- An image with defines colors, but some noise is present;
- A natural image with shades and color not so well defines (boundaries are close).

The second and third images are taken from the freeimages database (http://www.freeimages.co.uk).

4.1 *Synthetic Image Testing*

The first image is a synthetic image consisting of four solid pure colors (Red, Blue, Green, and White) (Fig. 6). The image is composed by four rectangles: upper left is red, upper right is white, lower left is blue, and lower right is green.

The K-means algorithm is expected to generate four clusters, one for each color available in the figure. One consideration made for the testing regarding the test for H only component is that using this model is impossible to distinguish red from white. The previous is one of the present issues coming from most of the perceptual

Fig. 6 Base synthetic image with *red, blue, green,* and *white* colors

color models, where white and gray tones are undefined for the hue (H) component (Eq. 4). Usually in the case of undefined value for the H component, a zero is used instead. For the H component, zero value corresponds to red hue.

The test for the synthetic image shows no major problem for the selected color spaces, just the already mentioned issue for the H component color space. In Fig. 7f it could be noted how the white and red are merged together, since there is no way to differentiate between both hues. For the white color the hue is zero, while for the pure red the hue is also zero.

4.2 Testing Image with Some Noise

The following image to be tested is an image containing the alphabet letters alternating the hues red, blue, and green. The background for the image is set to yellow; it is expected that for the perceptual color spaces the effect of issue for having white or gray tones is diminished, since the only gray source is coming from the noise (Fig. 8).

The present noise found in the image in Fig. 8 is expected to confuse the clustering algorithm, by producing some pixels that are located close to the boundaries of other colors. Depending on the properties in each of the testing color spaces, the clustering algorithm will be able to overcome the noise.

From the result clusters in Fig. 9, it can be stated some interesting aspects:

RGB (Fig. 9a) mixed the red and green letters, this mostly due to the noise getting the boundaries close and adding additional gray tones.

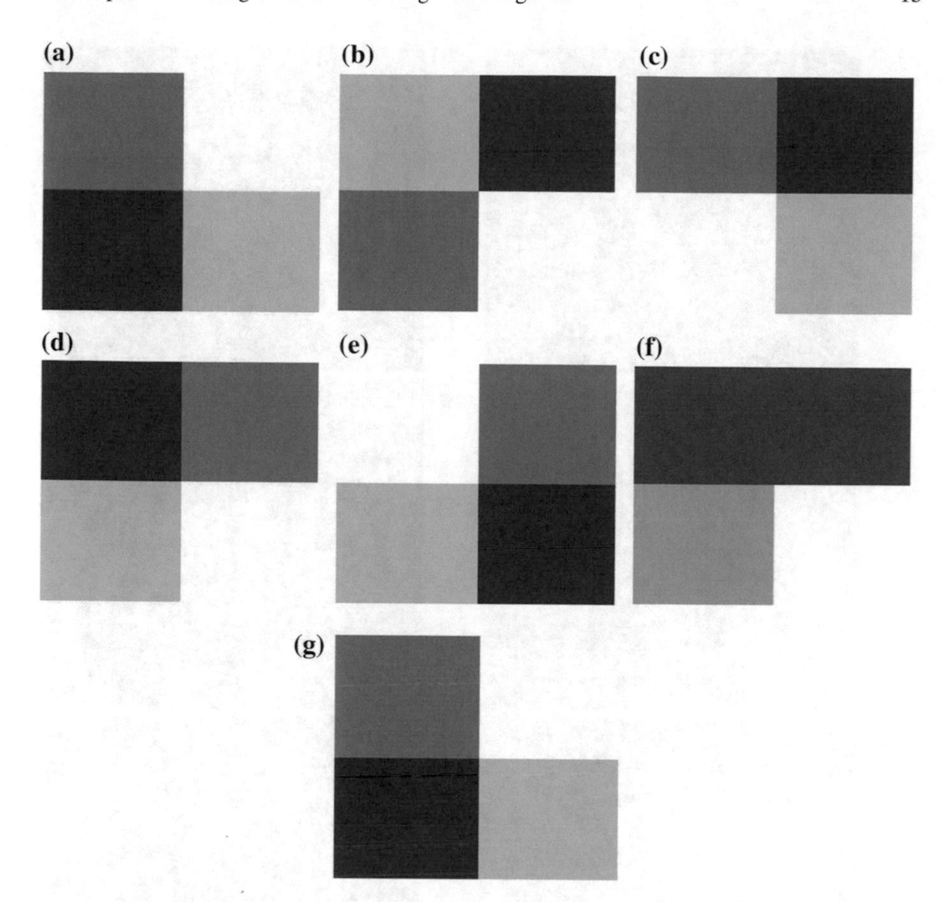

Fig. 7 **a** Clusters resulting from RGB. **b** Clusters resulting from HSV. **c** clusters coming from HSL. **d** Clusters coming from CIELab. **e** Cluster coming from nRGB. **f** Clusters coming from H. **g** Clusters coming from HS

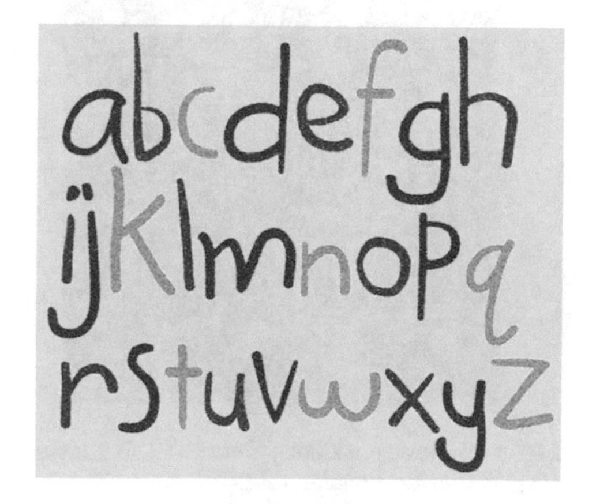

Fig. 8 Colored image with some noise. First-row sequence *red–blue–green*, second-row sequence *blue–red–green*, third-row sequence *blue–red–green*. Edited image from http://www.freeimages.co.uk/

Fig. 9 **a** Clusters resulting from RGB. **b** Clusters resulting from HSV. **c** clusters coming from HSL. **d** Clusters coming from CIELab. **e** Cluster coming from nRGB. **f** Clusters coming from H. **g** Clusters coming from a fixed H component. **h** Cluster coming from HS

HSV (Fig. 9b) was capable to select green and blue in an acceptable level; it creates another cluster for the background part of the noise and gray tones were classified as one cluster, but in the red letters, the clustering was a mixture of all other clusters.

HSL (Fig. 9c) in this case the background and blue letters were acceptable, but the red and green letters were incorrectly classified, using a combination of other clusters.

CIELab (Fig. 9d) for this color space all the letters and background presented an acceptable classification. Just a few pixels erroneous sorted inside the letters and surrounding most of the letters.

nRGB (Fig. 9e) the normalized RGB presents the best performance, even showing better results than CIELab. The previous can be stated from the fact that erroneous pixels inside each letter are less than CIELab.

H (Fig. 9f) using just the hue from the perceptual color spaces performs an acceptable clustering for the background, blue letters, and green letters. It can be noted that the noise inside these letters is not present in the final segmentation. It exhibits little noise around some letters. Regarding the red letters, it was able to create a cluster for the color, but the cluster is heavily affected by the discontinuity in the hue definition. This issue is evident while classifying the red hue, since it is defined in two ranges at the low values of hue and the highest values of hue in the H component.

H-Fixed (Fig. 9g) in an effort to demonstrate how the red hues are affected by the discontinuity in the H component, a special scenario, is created. In this scenario the two red hue ranges in the H component are merged. This can be done by appending the first values (0–20) at the end of the scale, then performing subtraction of 20 units on each value. This procedure is similar to circular shift. After performing the join of the two ranges, it can be seen an almost perfect clustering, just a few misclassified pixels.

HS (Fig. 9h) selecting the hue and saturation produces a really noisy clustering for the red and green letters, while regarding the blue letters, the clustering shows a better result, but still with some noise.

From the result over the tested image, normalized RGB presented the best results regarding clustering. But using just the H component from the perceptual color spaces resulted in having a more noise immune classification. This was notable when the discontinuity in the red hue was solved. Unfortunately, the shift procedure is not always a possibility in real scenarios. The colors that will be in the processing images are not usually known, so performing the shift could move the discontinuity to another color. So the shift procedure is only worthy when can assure that the red color is the image and it should be included in a cluster. Also it is needed to assure that pure white and gray tones are not in the image, so they will not be misclassified as red hues (white and gray tones are undefined and usually given an H value of zero).

4.3 Natural Image

The next scenario is a photograph of three colored cubes. This will test the capability from each color space to add significant information when shades and lights are present in the image (Fig. 10). An additional error source observed from the target image is the fact that the hues from the colors are not well defined, especially regarding the blue cube (upper cube), which seems to have a close boundary with the green hue.

This scenario is set to reveal some of the common issues that each color space brings while trying to segment an image using its color as the main feature. The shades and lights produce a wide distribution in the range of possible values for each of the selected primary colors. As the background of the image is white it is expected some confusion between the background and the lighter parts of the cubes. Also from the shades is expected to create some possible misclassification issues.

A detailed analysis of the resulting clusters exhibits the benefits and issues coming from each color space:

RGB (Fig. 11a) Errors in all the clusters can be noticed. The red cube is grouped along with the shadow, while the part of the blue cube is erroneously classified with the back ground (the lighter section) and finally part of the green cube is classified in the blue cluster. This is caused by the illumination issue present in the RGB color space.

HSV (Fig. 11b) In this case the background cluster is acceptable, just presenting some noise. Regarding the cubes, the clusters are mixed. A point in favor for the model is the fact that it was not affected by the shadow below the blue (upper) cube.

Fig. 10 Colored cubes. *Upper* cube is *Blue*, *left lower* is red and *right lower* is green. Image from http://www.freeimages.co.uk/

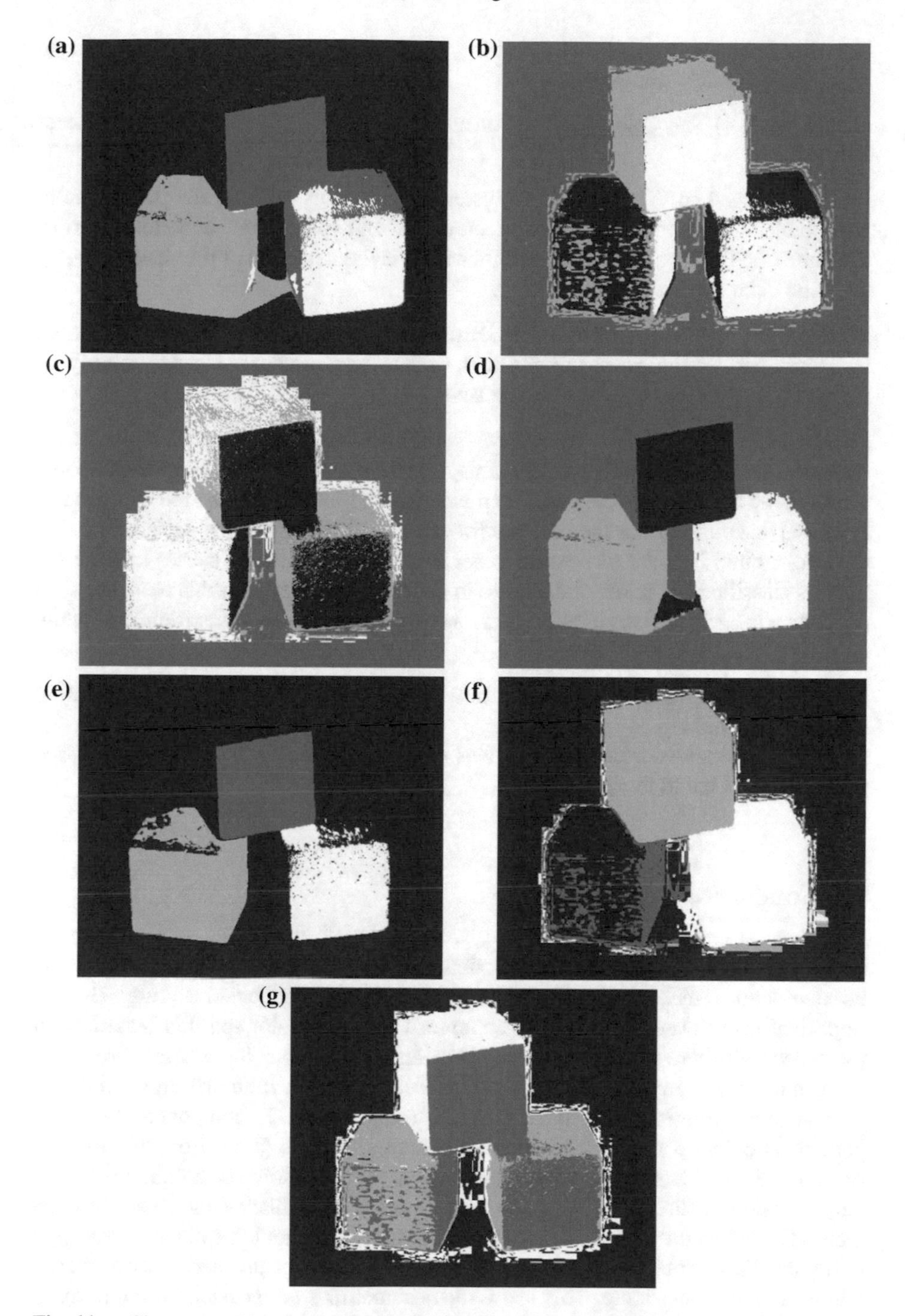

Fig. 11 **a** Clusters resulting from RGB. **b** Clusters resulting from HSV. **c** clusters coming from HSL. **d** Clusters coming from CIELab. **e**.Cluster coming from nRGB. **f** Clusters coming from H. **g** Clusters coming from HS

This last behavior comes from the ability to decouple the illumination component from the chromatic component.

HSL (Fig. 11c) This color space performs in a similar way to HSV, but with noisier clustering.

CIELab (Fig. 11d) The resulting clustering using this color space present solid acceptable clusters for the red and green cubes. But for the blue cube loses part of the cube (the lighter sides). Also, another issue is the detection of the shadow below the blue cube as part of the red cube.

nRGB (Fig. 11e) The normalized RGB presents a similar behavior as the CIELab, but loses part of the green cube. On the other hand demonstrate its property of immunity to shadows by avoiding the selection of the shadow below the blue cube.

H (Fig. 9f) In this scenario this approach performs an acceptable clustering in the blue and green cubes, and a really noisy in the red cube. Also, it presents some errors in the back ground cluster. The misclassified pixels inside the red cube comes from the discontinuity in the red hue for the H component, and also from the fact the background is white (hue value is set to zero). The previous causes that the red cube is classified in the red cluster and in of the white cluster. In this case, the shift procedure used in the previous image is not applicable due to the existence of white color.

HS (Fig. 9g) This selection behaves in a similar way as the HSV, just presenting slightly more noise.

From the previous exercise, there was not a clear winner, since ever color space presented noticeable issues.

5 Conclusions

From all the performed scenarios it is demonstrated that every color space presents its own issues that could affect the segmentation done by clustering. But an important result observed from the behavior from each color space is regarding to the aspect of failure in different areas. It means that some color spaces were better performing the segmentation of a given color, or a given illumination condition.

The most promising approach was the usage of the H component from the perceptual color spaces, as long as the discontinuity issue (jump from 360 to 0) is fixed and the white and gray values are avoided. There is some technique that could help to improve the robustness for the H component, like using other distance method based on the angle distance (cosine distance). It should be noted that despite fixing the discontinuity issue in the perceptual models, it is also necessary to fix the white and gray tones issue. This last issue is sometimes not possible or not a trivial task.

An approach to follow it is to use a given color space depending on the conditions of the scenario where it should be implementing. So the features of a given color space can really help to solve the required clustering. Unfortunately, this approach is not possible in most of the scenarios, where the machine learning algorithm must be prepared for unknown scenarios.

From the previous can be stated one of the most common issues in machine learning algorithms, that most of the classifiers are weak classifiers. In the case scenarios presented in this chapter, it can also use the same solution as in any machine learning procedure; it is the boosting approach. Boosting can be coupled weak classifiers in order to create a stronger classifier. Each configuration presented in the chapter is really a weak classifier. Mixing some of them can improve the final clustering result. The boosting using different color spaces is out of the scope of the chapter, but a given path for further development.

The chapter does not go deep in searching for techniques to correct each possible failure in the color spaces, since its main purpose is to remark the possible benefits and issues coming from each color space.

References

1. Arthur, D., Vassilvitskii, S.: k-means ++: the advantages of careful seeding. In: Proceedings of the Eighteenth Annual ACM-SIAM Symposium on Discrete Algorithms. Society for Industrial and Applied Mathematics pp. 1027–1035 (2007)
2. Arumugadevi, S., Seenivasagam, V.: Color image segmentation using feedforward neural networks with FCM. Int. J. Autom. Comput. **13**(5), 491–500 (2016)
3. Blanco, E., Mazo, M., Bergasa, L.M., Palazuelos, S.: A method to increase class separation in the HS plane for color segmentation applications. In: 2007 IEEE International Symposium on Intelligent Signal Processing, WISP (2007)
4. CIE: Commission internationale de l 'Eclairage proceedings, 1931. Cambridge University Press, Cambridge (1932)
5. Finlayson, G., Xu, R.: Illuminant and gamma comprehensive normalisation in logRGB space. Pattern Recogn. Lett. **24**(11), 1679–1690 (2003)
6. Gong, M., Qian, Y., Cheng, L.: Integrated foreground segmentation and boundary matting for live videos. IEEE Trans. Image Process. **24**(4), 1356–1370 (2015)
7. Hanbury, A., Serra, J.: A 3D-polar coordinate colour representation suitable for image analysis. Pattern Recognition and Image Processing Group Technical Report 77. Vienna, Austria, Vienna University of Technology (2003)
8. Khaled, S., Islam, S., Rabbani, G., Tabassum, M., Gias, A., Kamal, M., Muctadir, H., Shakir, A., Imran, A.: Combinatorial color space models for skin detection in sub-continental human images. In Visual Ibnformatics, First International Visual Informatics Conference, IVIC, pp. 532–542 (2009)
9. Kim, K., Oh, C., Sohn, K.: Non-parametric human segmentation using support vector machine. IEEE Trans. Consum. Electron. **62**(2), 150–158 (2016)
10. Kuremoto, T., Kinoshita, Y., Feng, L., Watanabe, S., Kobayashi, K., Obayashi, M.: A gesture recognition system with retina-V1 model and one-pass dynamic programming. Neurocomputing **116**, 291–300 (2013)

11. Liu, W., Wang, L., Yang, Z.: Application of self-adapts to RGB threshold value for robot soccer. In: International Conference on Machine Learning and Cybernetics (ICMLC), vol. 2 (2010)
12. Lloyd., S.P.: Least squares quantization in PCM. IEEE Trans. Inf. Theory **28**, 129–137 (1982)
13. Lucchi, A., et al.: Learning structured models for segmentation of 2-D and 3-D imagery. IEEE Trans. Med. Imaging **34**(5), 1096–1110 (2015)
14. MacQueen, J.B.: Some methods for classification and analysis of multivariate observations. In: Proceedings of 5th Berkeley Symposium on Mathematical Statistics and Probability, pp. 281–297 (1967)
15. Malacara, D.: Color Vision and Colorimetry: Theory and Applications. SPIE Press (2011)
16. Murphy, K.: Machine Learning —A probabilistic Perspective. MIT Press, Cambridge Massachusetts (2012)
17. Oh, S., Kim, S., Approaching the computational color constancy as a classification problem through deep learning. Pattern Recognit. **61**, 405–416 (2017)
18. Pan, C., Park, D.S., Lu, H. et al.: Color image segmentation by fixation-based active learning with ELM. Soft Comput. **16**(9), 1569–1584 (2012)
19. Pan, C., Park, D.S., Yang, Y., et al.: Leukocyte image segmentation by visual attention and extreme learning machine. Neural Comput. Appl. **21**(6), 1217–1227 (2012)
20. Pratondo, A., Chui, C., Ong, S.: Integrating machine learning with region-based active contour models in medical image segmentation. J. Vis. Commun. Image Represent. (2016)
21. Sang, Q., Lin, Z., Acton, S.: Learning automata for image segmentation. Pattern Recognit. Lett. **74**, 46–52 (2015)
22. Sridharan, M., Stone, P.: Structure-based color learning on a mobile robot under changing illumination. Auton. Robots **23**(3), 161–182 (2007)
23. Velho, L., Frery, A., Gomes, J.: Image Processing for Computer Graphics and Vision, 2nd edn. Springer (2009)
24. Yanga, H., Wanga, Y., Wanga, Q., Zhanga, X.: LS-SVM based image segmentation using color and texture information. J. Vis. Commun. Image Represent. **23**(7), 1095–1112 (2012)
25. Wang, X., Wang, Q., Yang, H., Bu, J.: Color image segmentation using automatic pixel classification with support vector machine. Neurocomputing **74**(18), 3898–3911 (2011)
26. Wikimedia commons CIELab, SRGB gamut within CIELAB color space isosurface.png. https://commons.wikimedia.org
27. Wikimedia commons, HSL color solid cylinder alpha lowgamma.png. https://commons.wikimedia.org
28. Wikimedia commons, HSV color solid cylinder alpha lowgamma.png, https://commons.wikimedia.org
29. Wikimedia commons, RGB color solid cube.png. https://commons.wikimedia.org

Multi-objective Whale Optimization Algorithm for Multilevel Thresholding Segmentation

Mohamed Abd El Aziz, Ahmed A. Ewees, Aboul Ella Hassanien, Mohammed Mudhsh and Shengwu Xiong

Abstract This chapter proposes a new method for determining the multilevel thresholding values for image segmentation. The proposed method considers the multilevel threshold as multi-objective function problem and used the whale optimization algorithm (WOA) to solve this problem. The fitness functions which used are the maximum between class variance criterion (Otsu) and the Kapur's Entropy. The proposed method uses the whale algorithm to optimize threshold, and then uses this thresholding value to split the image. The experimental results showed the better performance of the proposed method to solving the multilevel thresholding problem for image segmentation and provided faster convergence with a relatively lower processing time.

Keywords Multi-objective · Swarms optimization · Whale optimization algorithm · Multilevel thresholding · Image segmentation

1 Introduction

In recent years, the intelligent systems that depend on machine learning and pattern recognition are widely used in numerous fields. These include the application

M.A. El Aziz
Faculty of Science, Department of Mathematics, Zagazig University, Zagazig, Egypt
e-mail: abd_el_aziz_m@yahoo.com

A.A. Ewees (✉)
Department of Computer, Damietta University, Damietta, Egypt
e-mail: a.ewees@hotmail.com; ewees@du.edu.eg

A.E. Hassanien
Faculty of Computers and Information, Information Technology Department, Cairo University, Giza, Egypt
e-mail: boitcairo@gmail.com

M. Mudhsh · S. Xiong
School of Computer Science and Technology, Wuhan University of Technology, Wuhan, China
e-mail: xiongsw@whut.edu.cn

A.E. Hassanien and D.A. Oliva (eds.), *Advances in Soft Computing and Machine Learning in Image Processing*, Studies in Computational Intelligence 730,
https://doi.org/10.1007/978-3-319-63754-9_2

of face and voice recognition, objects identification, computer vision, and so on. Nevertheless, the researchers are still working to improve the accuracy of these systems, especially when they are used in real-time environments. When these systems acquire their data from images, they should use image processing techniques to prepare and process the images to be able to identify and recognize the objects on them. Image segmentation is an essential phase in this stage. It works for splitting an image into segments with similar features (i.e., color, contrast, brightness, texture, and gray level) based on a predefined criterion [1]. Image segmentation has been applied in several applications such as medical diagnosis [2], satellite image [3], and optical character recognition [4]. However, it could be a complex process if the images are corrupted by noises from environments or equipment. There are many methods for applying image segmentation, such as edge detection [5], region extraction [6], histogram thresholding, and clustering algorithms [7]; as well as, threshold segmentation [8], it is one of the popular methods for performing this task to locate the best threshold value [9, 10]; it can be divided into two types: bi-level which can be used to produce two groups of objects and multilevel that used to segment complex images and separate pixels into multiple homogeneous classes (regions) based on intensity [1, 11]. Bi-level thresholding method can produce adequate outcomes in cases where the image includes two levels only, however, if it has been used with multilevel the computational time will be often high [12]. On the other hand, the results of bi-level thresholding are not suitable to real application images; so, there is a wide requirement to use multilevel thresholding [11]. There are two methods to determine the thresholds, namely, a global and local level. In a local level, thresholds are determined for each portion of the image; on the other hand, at a global level, one threshold is taken to the whole image [13]. So, by using the image histogram, the global thresholding can be determined. Several thresholding methods explore for the thresholds by optimizing some fitness functions that are defined from images and they handle the determined thresholds as parameters. So, the determination of optimal thresholds in multilevel thresholding is an NP-hard problem [14]. Many methods analyze the image histogram to determine the optimal thresholds, by either minimizing or maximizing a fitness function with consideration of the values of threshold.

When the number of thresholds is small, classical methods are acceptable; but if there are several threshold numbers, it is a best practice to perform a swarm intelligence (SI) technique to optimize this task, such as, genetic algorithm (GA), particle swarm optimization (PSO), firefly optimization (FFO), and bat algorithm.

Jie et al. (2013) [15] introduced a multi-threshold segmentation method that utilized k-means and firefly optimization algorithm (FA). The results showed that the proposed method obtained a low run-time and higher performance than the classical fast FCM and PSO-FFCM models. In the same effort, Chaojie et al. (2013) [16] proposed a method based on FA that outperformed GA algorithm.

Vishwakarma et al. (2014) [17] compared their proposed model that based on FA with the classical K-means clustering algorithm and the model achieved the best results. Sarkar (2011) [18] presented a technique based on differential evolution for multilevel thresholding using minimum cross entropy thresholding (MCET). It was applied to some of the real images and the results showed high efficiency than PSO and GA. Moreover, Fayad et al. [19] proposed a segmentation model based on ACO algorithm. It achieved good results and small errors in comparison to the ground truth. On the other hand, Abd ElAziz et al. [20] introduced a hybrid model that combined SSO and FA (FASSO) for image segmentation. It showed faster convergence and lower preprocessing time. The PSO and its edition [21–26] are implemented in image segmentation to locate the multilevel thresholding. Moreover, there are several swarm techniques that applied for segmentation including honey bee mating optimization (HBMO) [27], harmony search (HS) algorithm [28], cuckoo search (CS) [29], and artificial bee colony (ABC) [30, 31]. However, most of these techniques are either trapped on local optima or predefined control parameters such as GA, PSO, CS, and HS algorithms.

In this chapter, we present a new multilevel thresholding method for image segmentation method. The multilevel thresholding is considered as multi-objective optimization problem, in which the popular two image segmentation functions namely, Otsu's and entropy are used as the fitness function which optimized by the whale optimization algorithm. The properties of these two functions are used to improve the accuracy of image segmentation via multilevel thresholding. The characteristics of the WOA are the ability of fast convergence. The rest of this chapter is organized as follows: Sect. 2 presents the materials and methods. Section 3 introduces the proposed method. Section 4 illustrates the experiments and discussions. The conclusion and future work are given in Sect. 5.

2 Materials and Methods

2.1 Problem Formulation

In this section, the multilevel thresholding problem definition is introduced, by considering an gray level image I contains $K+1$ groups. Therefore, the $t_k, k = 1, \ldots, K$ thresholds are needed to split I to subgroups C_k as in the following equation:

$$\begin{aligned} C_0 &= \{I(i,j) \in I \mid 0 \le I(i,j) \le t_1 - 1\} \\ C_1 &= \{I(i,j) \in I \mid t_1 \le I(i,j) \le t_2 - 1\}, \\ &\ldots \\ C_K &= \{I(i,j) \in I \mid t_k \le I(i,j) \le L - 1\}, \end{aligned} \tag{1}$$

where $I(i,j)$ is (i,j)th pixel value and L is the gray levels of $I \in [0, L-1]$.

The aim of the multilevel thresholding is to find the threshold values construct these groups C_k, which can be determined by maximizing the following equation:

$$t_1^*, t_2^*, \ldots, t_K^* = \max_{t_1,\ldots,t_K} F(t_1, \ldots, t_K), \tag{2}$$

where $F(t_1, \ldots, t_K)$ may be Kapur's entropy or the Otsu's function.

- Otsu's function:

This function is defined mathematically as

$$F_{Ots} = \sum_{i=0}^{K} A_i(\eta_i - \eta_1)^2, \tag{3}$$

$$A_i = \sum_{j=t_i}^{t_{i+1}-1} P_j, \tag{4}$$

$$\eta_i = \sum_{j=t_i}^{t_{i+1}-1} i\frac{P_j}{A_j}, \quad where \ \ P_i = h_i/N, \tag{5}$$

where η_1 is the mean intensity of I with $t_0 = 0$ and $t_{K+1} = L$. The h_i and P_i are the frequency and the probability of the ith gray level, respectively.

- Kapur's Entropy:

The Kapur's entropy function determines the optimal threshold values through maximizing the overall entropy [32] that is defined as:

$$F_{Kap} = \sum_{i=0}^{K}(-\sum_{i=t_i}^{t_{i+1}-1} \frac{P_j}{A_j} ln(\frac{P_j}{A_j})). \tag{6}$$

2.2 Whale Optimization Algorithm (WOA)

The whale optimization algorithm (WOA) is a new meta-heuristic technique that mimics the Humpback whales [33]. In this technique, the optimization begins by producing a random population of whales. These whales search for the prey's (optimum) location, then attach (optimize) them by one of these methods encircling or bubble-net.

In the encircling method [33] the Humpback whales improve their location based on the best location as follows:

$$\mathbf{D} = |\mathbf{C} \odot \mathbf{X}^*(t) - \mathbf{X(t)}| \tag{7}$$

$$\mathbf{X}(t+1) = |\mathbf{X}^*(t) - \mathbf{A} \odot \mathbf{D}|, \tag{8}$$

where **D** describes the distance between the position vector of both the prey $\mathbf{X(t)}^*$ and a whale **X(t)**, and t denotes the current iteration number. **A** and **C** are coefficient vectors, and defined as follows:

$$\mathbf{A} = 2\mathbf{a} \odot \mathbf{r} - \mathbf{a} \tag{9}$$

$$\mathbf{C} = 2\mathbf{r}, \tag{10}$$

where r is a random vector $\in [0, 1]$, and the value of $\mathbf{a}$ is linearly decreased from 2 to 0 as iterations proceed.

Whereas the bubble-net method can be performed by two approaches. The first is the shrinking encircling that given by reducing the value of $\mathbf{a}$ in equation (9), also, $\mathbf{A}$ is reduced. The last is the spiral updating position. This method is applied to mimic the helix-shaped movement of Humpback whales around prey:

$$\mathbf{X}(t+1) = \mathbf{D}' \odot e^{bl} \odot cos(2\pi l) + \mathbf{X}^*(t), \tag{11}$$

where $\mathbf{D}' = |\mathbf{X}^*(t) - \mathbf{X}(t)|$ is the distance between the whale and prey, b is a constant for determining the shape of the logarithmic spiral, $\odot$ is an element-by-element multiplication, and l is a random value in $[-1, 1]$.

The whales can swim around the victim through a shrinking circle and along a spiral-shaped path concurrently:

$$\mathbf{X}(t+1) = \begin{cases} \mathbf{X}^*(t) - \mathbf{A} \odot \mathbf{D} & if \;\; p \geq 0.5 \\ \mathbf{D}' \odot e^{bl} \odot cos(2\pi l) + \mathbf{X}^*(\mathbf{t}) & if \;\; p < 0.5 \end{cases} \tag{12}$$

where $p \in [0, 1]$ is a random value which describes the probability of choosing either the shrinking encircling method or the spiral model to adjust the position of whales.

In exploration phase, the Humpback whales search randomly for prey. The position of a whale is adjusted by determining a random search agent rather than the best search agent as follows:

$$\mathbf{D} = |\mathbf{C} \odot \mathbf{X}_{rand} - \mathbf{X}(\mathbf{t})| \tag{13}$$

$$\mathbf{X}(t+1) = |\mathbf{X}_{rand} - \mathbf{A} \odot \mathbf{D}|, \tag{14}$$

where $\mathbf{X}_{rand}$ is a random position determined from the current population. Algorithm 1 illustrates the whole structure of the WOA.

3 The Proposed Method

In this section the proposed method for determining the multilevel thresholding values is introduced. In the first the fitness function is defined, based on the combination of the Otsu's and Kapur entropy functions, as

$$Fit = \alpha F_{Ots} + \beta F_{Kap}, \tag{15}$$

where α and β are random values in the range [0, 1] and the parameters represent the balance between the two fitness functions.

The input to the proposed method is the image histogram, the number of whales N and the dimension of each whale position is the threshold level *dim*. The WOA starting by generating a random population of N solutions in the search domain $[0, L]$ (here $L = 265$), for each position the fitness function Fit_i is computed using equation (15). Then the fitness function F_{best} and its corresponding best whale position x_{best} are determined. Based on each value of decrease the parameter a from 2 to 0, the values of two parameters A and C are computed, then the position of each whale is updated based on the value of the parameter p as illustrated in Sect. 2.2. The previous steps are repeated until the stop criteria are satisfied, and the proposed method is shown in Algorithm 1.

Algorithm 1 Whale Optimization Algorithm (WOA)

1: Input: *dim* dimension of each whale, N: number of whales, t_{max}: maximum number of iterations.

2: Output: x_best Threshold values.
3: Generate a population of N whales $\mathbf{x}_i$, $i = 1, 2, \ldots, N$
4: $t = 1$
5: **for** all $\mathbf{x}_i$ do // parallel techniques **do**
6: Calculate the fitness function Fit_i for $\mathbf{x}_i$.
7: **end for**
8: Determine the best fitness function F_{best} and its position whale x_{best}.
9: **repeat**
10: **for** For Each value of a decrease from 2 to 0 **do**
11: **for** $i = 1 : N$ **do**
12: Calculate C and A using (10) and (9) respectively.
13: $p = rand$
14: **if** $p \geq 0.5$ **then**
15: Update the solution using (11)
16: **else**
17: **if** $| A | \geq 0.5$ **then**
18: Update the solution using (14)
19: **else**
20: Update the solution using (7)
21: **end if**
22: **end if**
23: **end for**
24: **end for**
25: $t = t + 1$
26: **until** $G < t_{max}$

4 Experiments and Discussion

In this section, the experimental environment for the proposed method is introduced. The image description is illustrated in the first, then the setting of the parameters for each algorithm and the measurements used to evaluate the quality of segmentation image is discussed.

Fig. 1 Samples of the tested images, from *left* TestE1, TestE2, TestE3, and TestE7

4.1 Benchmark Images

The proposed methods used in this chapter are tested on four common grayscale images from the database of Berkeley University [34]. These images are called TestE1, TestE2, TestE3, and TestE7 as illustrated in Fig. 1.

4.2 Experimental Settings

The proposed method results are compared with four algorithms, namely, WOA, SSO, FA, and FASSO; these algorithms are previously proposed for multilevel image segmentation and introduced good results. To make the comparison process fair, the population size is 25, the dimension of each agent is the number of thresholds (m) and the same stopping criteria (maximum number of iterations is 100, with a total of 35 runs per algorithm). The parameters of each algorithm used in this paper are illustrated in Table 1.

The experiments were computed on using the following threshold numbers: 2, 3, 4, and 5. All of the methods are programmed in "Matlab 2014" and implemented on "Windows 64bit" environment on a computer having "Intel Core2Duo (1.66 GHz)" processor and 2 GB memory.

Table 1 The parameters setting of each algorithm

Algorithm	Parameters	Value
WOA	a	[0, 2]
	b	1
	l	$[-1, 1]$
SSO	Probabilities of attraction or repulsion (pm)	0.7
	Lower female percent	65
	Upper female percent	90
FASSO	γ_{FA}	0.7
	β_{FA}	1.0
	α_{FA}	0.8
	Probabilities of attraction or repulsion (pm)	0.7
	Lower female percent	65
	Upper female percent	90
FA	γ_{FA}	0.7
	β_{FA}	1.0
	α_{FA}	0.8

4.3 Segmented Image Quality Metrics

The accuracy of the segmented image is evaluated based on fitness function, time, peak signal-to-noise ratio (PSNR), and the structural similarity index (SSIM), where PSNR is defined as

$$PSNR = 20log_{10}(\frac{255}{RMSE}), \quad RMSE = \sqrt{\frac{\sum_{i=1}^{N}\sum_{j=1}^{M}(I(i,j) - \hat{I}(i,j))^2}{N.M}}, \tag{16}$$

where I and $\hat{I}$ are original and segmented images of size $M \times N$, respectively. The high value of PSNR refers to the high performance of segmentation algorithm. The SSIM is defined as

$$SSIM(I, \hat{I}) = \frac{(2\mu_I\mu_{\hat{I}} + c_1)(2\sigma_{I,\hat{I}} + c_2)}{(\mu_I^2 + \mu_{\hat{I}}^2 + c_1)(\sigma_I^1 + \sigma_{\hat{I}}^2 + c_2)}, \tag{17}$$

where μ_I ($\mu_{\hat{I}}$) and σ_I ($\sigma_{\hat{I}}$) are the mean intensity and the standard deviation of the image I ($\hat{I}$), respectively. The $\sigma_{I,\hat{I}}$ is the covariance of I and $\hat{I}$ and $c_1 = 6.5025$ and $c_2 = 58.52252$ are two constants [35]. The highest value of SSIM and PSNR indicates better performance (Figs. 2, 3, 4 and 5).

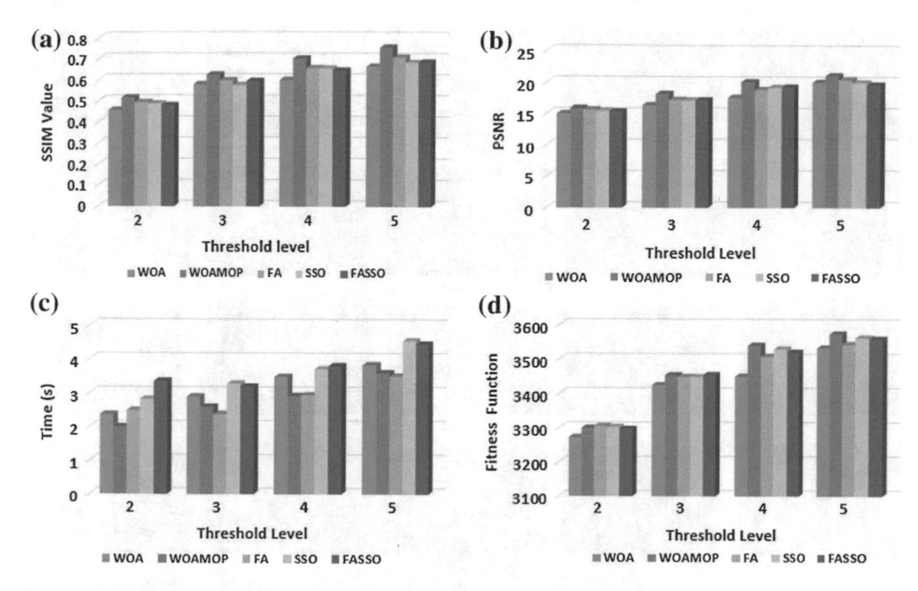

Fig. 2 The average of results of measures overall the testing images

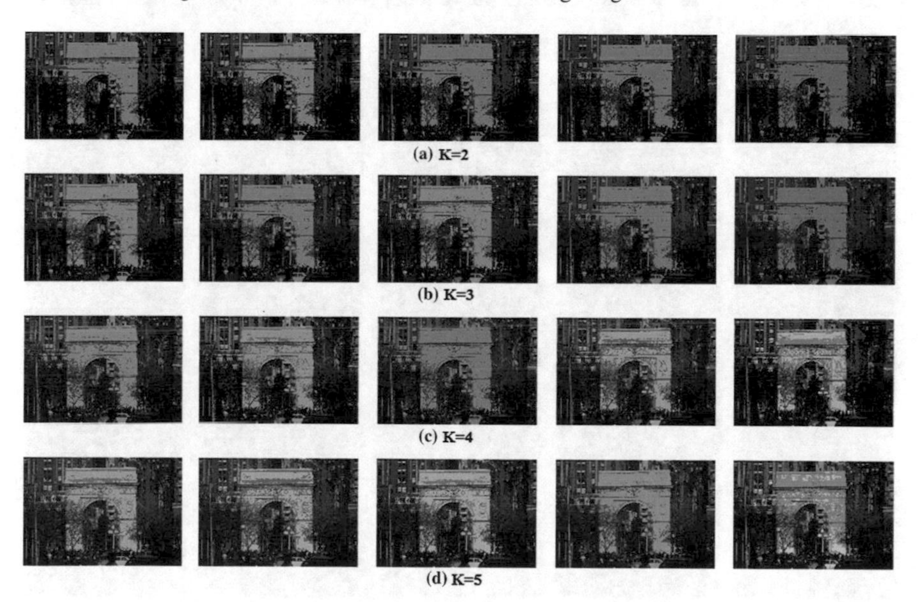

Fig. 3 The result of segmentation TestE1 image using (from *left* to *right*) SSO, FASSO, FA, WOAMOP, and WOA

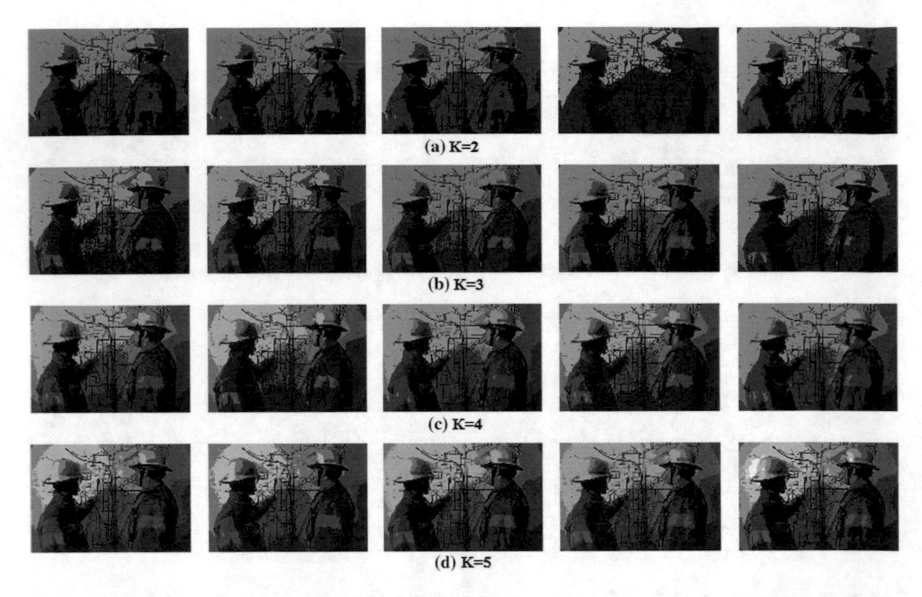

Fig. 4 The result of segmentation TestE2 image using (from *left* to *right*) SSO, FASSO, FA, WOAMOP, and WOA

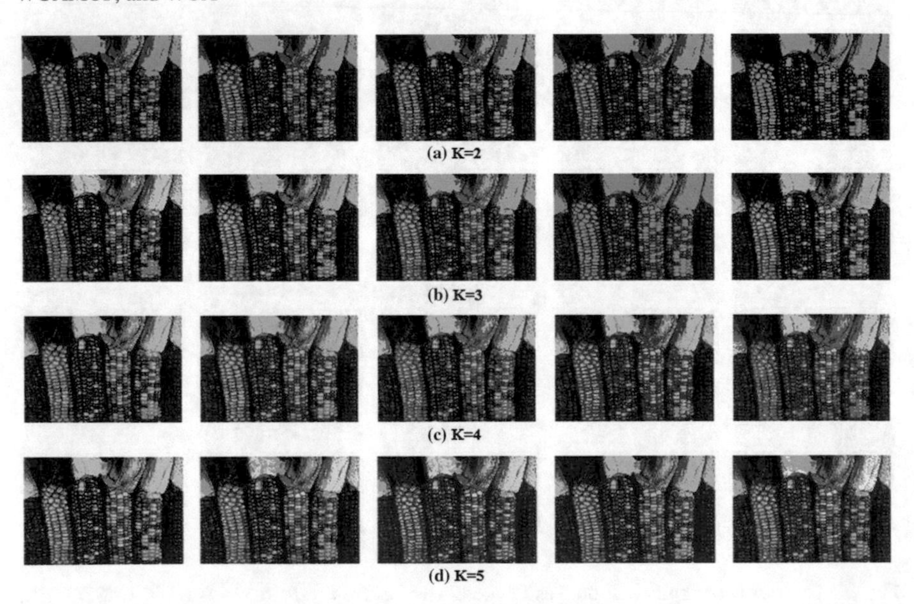

Fig. 5 The result of segmentation TestE3 image using (from *left* to *right*) SSO, FASSO, FA, WOAMOP, and WOA

Table 2 The average of fitness function values and CPU process time of different segmentation techniques

Fitness function							Time (s)				
Images	K	WOA	WOAMOP	FA	SSO	FASSO	WOA	WOAMOP	FA	SSO	FASSO
TestE1	2	3451.335	3473.352	3474.179	3468.534	3460.652	2.3367	2.2682	1.9391	2.5095	2.6167
	3	3556.136	3605.717	3596.408	3615.769	3620.536	2.7800	2.7063	2.3811	3.1488	3.1537
	4	3565.543	3697.336	3673.351	3691.681	3679.444	3.2220	3.1448	3.1551	3.8013	4.1316
	5	3702.948	3744.108	3716.655	3742.238	3730.328	4.0102	3.5838	2.9527	4.0258	4.0410
TestE3	2	3460.841	3535.650	3556.535	3560.862	3550.880	2.3861	1.5464	2.4646	2.5000	4.0586
	3	3711.475	3705.523	3708.666	3695.661	3719.520	2.7932	2.1946	2.5323	3.3532	3.1982
	4	3714.290	3787.483	3757.122	3774.966	3777.062	3.2639	2.1506	2.8329	3.6672	3.7002
	5	3795.133	3832.902	3782.561	3810.940	3821.005	3.7084	3.5551	4.4066	5.3610	5.5121
TestE2	2	4079.675	4100.774	4095.949	4099.082	4081.655	2.3500	2.0995	3.5750	3.6269	3.7653
	3	4243.405	4288.824	4281.223	4273.422	4287.896	3.1398	2.7166	2.2690	3.3195	2.9403
	4	4328.520	4409.415	4370.016	4396.643	4369.280	3.7066	3.1582	2.8544	3.5133	3.5661
	5	4361.881	4443.372	4408.191	4446.256	4411.608	3.7597	3.5995	3.2680	4.2011	4.0670
TestE7	2	2097.983	2088.951	2093.031	2078.644	2090.938	2.3472	2.0487	1.9413	2.6079	2.9397
	3	2185.131	2210.436	2206.062	2205.575	2186.233	2.8317	2.7107	2.3150	3.2536	3.4244
	4	2189.891	2265.325	2229.118	2252.467	2252.220	3.6996	3.1712	2.8970	3.8260	3.7835
	5	2270.616	2273.391	2265.752	2249.944	2272.461	3.8285	3.6192	3.3637	4.5745	4.1576

Table 3 Selected thresholds of techniques

Images	K	WOA	WOAMOP	FA	SSO	FASSO
TestE1	2	63 129	59 141	69 144	77 148	74 155
	3	32 67 133	44 81 144	48 113 161	52 109 159	43 94 151
	4	76 80 146 193	34 73 115 177	39 60 101 148	31 62 105 159	53 101 125 170
	5	54 100 128 157 208	35 67 100 148 165	44 76 133 178 187	31 60 80 129 178	30 58 78 123 184
TestE3	2	115 164	79 167	92 175	88 168	84 169
	3	77 140 192	59 122 159	66 114 182	84 138 209	77 132 189
	4	49 92 123 207	43 103 161 185	91 123 197 82	122 154 200 59	50 112 155 193
	5	47 83 121 184 242	27 72 93 148 195	37 66 110 179 219	72 114 132 178 204	67 95 120 172 225
TestE2	2	72 157	58 144	62 139	62 141	69 139
	3	44 115 150	68 95 151	42 94 142	26 92 161	37 84 164
	4	49 65 117 156	26 83 122 177	41 72 129 167	48 83 125 175	31 97 130 187
	5	63 103 159 198 247	25 67 101 140 179	54 92 137 177 187	42 68 93 149 196	22 72 86 138 199
TestE7	2	55 134	80 151	80 144	71 128	62 130
	3	48 133 176	82 130 175	53 128 159	54 124 155	57 134 180
	4	32 69 132 191	66 108 138 174	69 112 141 190	50 94 141 171	63 99 133 165
	5	37 82 131 149 163	53 92 144 158 170	27 101 121 139 167	53 96 132 173 243	41 107 137 156 197

Table 4 The average PSNR and SSIM of different segmentation techniques

SSIM							PSNR				
Images	K	WOA	WOAMOP	FA	SSO	FASSO	WOA	WOAMOP	FA	SSO	FASSO
TestE1	2	0.4630	0.4997	0.4568	0.4386	0.4445	15.4602	15.7190	15.7193	15.5803	15.6746
	3	0.6428	0.6263	0.5882	0.5819	0.6263	17.1204	18.2775	17.9933	18.0240	18.2690
	4	0.4867	0.7132	0.6846	0.7270	0.6149	16.6475	20.0431	19.1112	19.9239	19.0666
	5	0.6153	0.7250	0.6826	0.7623	0.7579	19.3225	21.0601	20.2145	21.2919	20.8062
TestE3	2	0.3271	0.4083	0.3915	0.4037	0.4092	12.7984	14.1648	13.8011	13.9347	14.0133
	3	0.4888	0.5886	0.5317	0.4583	0.4934	15.4088	16.8358	16.0666	14.9637	15.4732
	4	0.6373	0.6727	0.5579	0.5035	0.6047	17.8221	18.9062	16.6948	15.6900	17.5301
	5	0.6967	0.7857	0.7661	0.5532	0.5773	19.2657	19.7803	20.0627	16.6521	17.0998
TestE2	2	0.4988	0.5558	0.5397	0.5406	0.5084	15.6136	15.4336	15.2995	15.3727	15.1418
	3	0.5950	0.6093	0.6327	0.6055	0.6457	16.4259	17.8354	16.7074	16.9031	17.6821
	4	0.6476	0.6791	0.6729	0.6649	0.6158	18.1127	19.8322	18.9827	19.7417	18.3249
	5	0.5912	0.7168	0.6510	0.6959	0.6675	19.4139	20.8858	20.3041	21.1293	19.4707
TestE7	2	0.5375	0.5947	0.5818	0.5624	0.5512	16.0103	18.0508	17.7184	16.7979	16.4532
	3	0.5896	0.6698	0.6375	0.6495	0.6084	16.4043	19.5508	18.0620	18.6063	17.0253
	4	0.6241	0.7370	0.7040	0.7190	0.7399	17.7795	21.4809	20.5230	21.2755	21.9815
	5	0.7486	0.7876	0.7284	0.7141	0.7313	22.0388	22.5572	21.1294	20.8070	21.0581

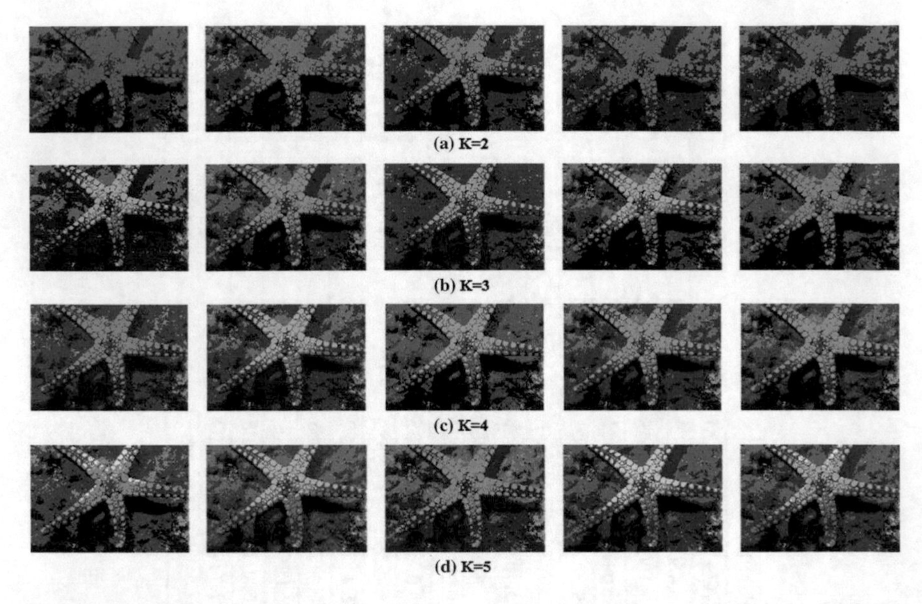

Fig. 6 The result of segmentation TestE7 image using (from *left* to *right*) SSO, FASSO, FA, WOAMOP, and WOA

4.4 The Results and Discussions

The results of comparison between the proposed algorithm and other algorithms are illustrated in Tables 2, 3, 4 and Fig. 6.

In Table 2, the average results of fitness values and time(s) are computed at thresholds 2, 3, and 4. From this table and Fig. 6d we can conclude that, based on the fitness function (as a measure), in general the WOAMOP is the better algorithm than the SSO is in the second rank followed by the FASSO, FA, and WOA. However, at threshold level equal to the FA and SSO give results better than that obtained by WOAMOP, also at level three of segmentation, the FASSO is outperformed WOAMOP (very small difference). At the high-level thresholding (4 and 5) the WOAMOP is better than all other algorithms followed by SSO in the second rank. Also from this table and Fig. 6c the best algorithm based on the time elapsed is the proposed algorithm followed by FA (however, this very small difference).

Table 4 and Fig. 6a–b show the SSIM and PSNR values. From this table and 6b we can observe that, at $K = 2, 3, 4$, and 5 the WOAMOP is better than all other algorithms (however, at $k = 2$ the difference between the algorithm is small). Also, the FA is in the second rank followed by SSO, FASSO, and WOA.

From all previous discussion we can conclude that the proposed method gives better performance based on the quality measures that used (PSNR, SSIM, time, and fitness function).

5 Conclusion and Future Work

Image recognition applications use image processing methods to prepare and process the images to be able to identify and recognize the objects on them. So, image segmentation techniques is an essential preprocessing step in several applications; it divides an image into segments with similar features based on a predefined criterion. In this chapter, a new multi-objective whale optimization algorithm (WOAMOP) was proposed for multi-thresholding image segmentation. The proposed method used the hybrid between the Kapur's entropy and the Otsu's function as a fitness function. The WOAMOP applied to determine the best solution (threshold values) and then used this thresholding values to divide the image. The experiment results of the proposed method were compared with four algorithms, namely, original WOA, FA, SSO, and FASSO. The WOAMOP achieved better results than all algorithms, and also it provides a faster convergence with relatively lower processing time. In future, the WOAMOP can be applied to other complex image segmentation problems such as color images.

References

1. Sarkar, S., Sen, N., Kundu, A., Das, S., Chaudhuri, S.S.: A differential evolutionary multilevel segmentation of near infra-red images using Renyis entropy. In: Proceedings of the International Conference on Frontiers of Intelligent Computing: Theory and Applications (FICTA), Chicago, pp. 699-706. Springer, Heidelberg (2013)
2. Zhao, F., Xie, X.: An overview of interactive medical image segmentation. Annals of the BMVA **7**, 1–22 (2013)
3. Pare, S., Bhandari, A.K., Kumar, A., Singh, G.K., Khare, S.: Satellite image segmentation based on different objective functions using genetic algorithm: a comparative study. In: 2015 IEEE International Conference on Digital Signal Processing (DSP), pp. 730-734. IEEE (2015)
4. Kim, S.H., An, K.J., Jang, S.W., Kim, G.Y.: Texture feature-based text region segmentation in social multimedia data. Multimedia Tools Appl., 1–15 (2016)
5. Ju, Z., Zhou, J., Wang, X., Shu, Q.: Image segmentation based on adaptive threshold edge detection and mean shift. In: 2013 4th IEEE International Conference on Software Engineering and Service Science (ICSESS), pp. 385–388. IEEE (2013)
6. Li, Z., Liu, C.: Gray level difference-based transition region extraction and thresholding. Comput. Electr. Eng. **35**(5), 696–704 (2009)
7. Tan, K.S., Isa, N.A.M.: Color image segmentation using histogram thresholding fuzzy c-means hybrid approach. Pattern Recogn. **44**(1), 1–15 (2011)
8. Zhou, C., Tian, L., Zhao, H., Zhao, K.: A method of two-dimensional Otsu image threshold segmentation based on improved firefly algorithm. In: Proceeding of IEEE International Conference on Cyber Technology in Automation, Control, and Intelligent Systems 2015, Shenyang, pp. 1420–1424 (2015)
9. Guo, C., Li, H.: Multilevel thresholding method for image segmentation based on an adaptive particle swarm optimization algorithm. In: AI 2007: Advances in Artificial Intelligence, pp. 654–658. Springer, Heidelberg (2007)
10. Zhang, Y., Lenan, W.: Optimal multi-level thresholding based on maximum Tsallis entropy via an artificial bee colony approach. Entropy **13**(4), 841–859 (2011)

11. Bhandari, A.K., Singh, V.K., Kumar, A., Singh, G.K.: Cuckoo search algorithm and wind driven optimization based study of satellite image segmentation for multilevel thresholding using Kapurs entropy. Expert Syst. Appl. **41**(7), 3538–3560 (2014)
12. Dirami, A., Hammouche, K., Diaf, M., Siarry, P.: Fast multilevel thresholding for image segmentation through a multiphase level set method. Signal Process. **93**(1), 139–153 (2013)
13. Akay, B.: A study on particle swarm optimization and artificial bee colony algorithms for multilevel thresholding. Appl. Soft Comput. **13**(6), 3066–3091 (2013)
14. Marciniak, A., Kowal, M., Filipczuk, P., Korbicz, J.: Swarm intelligence algorithms for multilevel image thresholding. In: Intelligent Systems in Technical and Medical Diagnostics, pp. 301–311. Springer, Heidelberg (2014)
15. Jie, Y., Yang, Y., Weiyu, Y., Jiuchao, F.: Multi-threshold image segmentation based on K-means and firefly algorithm, pp. 134–142. Atlantis Press (2013)
16. Yu, C., Jin, B., Lu, Y., Chen, X., et al.: Multi-threshold image segmentation based on firefly algorithm. In: Proceedings of Ninth International Conference on IIH-MSP 2013, Beijing, pp. 415–419 (2013)
17. Vishwakarma, B., Yerpude, A.: A meta-heuristic approach for image segmentation using firefly algorithm. Int. J. Comput. Trends Technol. (IJCTT) **11**(2), 69–73 (2014)
18. Sarkar, S., Ranjan, G.P., Das, S.: A differential evolution based approach for multilevel image segmentation using minimum cross entropy thresholding. In: International Conference on Swarm, Evolutionary, and Memetic Computing, pp. 51–58. Springer, Heidelberg (2011)
19. Fayad, H., Hatt, M., Visvikis, D.: PET functional volume delineation using an ant colony segmentation approach. J. Nucl. Med. **56**(supplement 3), 1745–1745 (2015)
20. El Aziz, M.A., Ewees, A.A., Hassanien, A.E.: Hybrid swarms optimization based image segmentation. In: Hybrid Soft Computing for Image Segmentation, pp. 1–21. Springer International Publishing (2016)
21. Djerou, L., Khelil, N., Dehimi, H.E., Batouche, M.: Automatic multilevel thresholding using binary particle swarm optimization for image segmentation. In: International Conference of Soft Computing and Pattern Recognition, 2009. SOCPAR'09, pp. 66–71. IEEE (2009)
22. Ghamisi, P., Couceiro, M.S., Benediktsson, J.A., Ferreira, N.M.: An efficient method for segmentation of images based on fractional calculus and natural selection. Expert Syst. Appl. **39**(16), 12407–12417 (2012)
23. Nakib, A., Roman, S., Oulhadj, H., Siarry, P.: Fast brain MRI segmentation based on two-dimensional survival exponential entropy and particle swarm optimization. In: 29th Annual International Conference of the IEEE in Engineering in Medicine and Biology Society, 2007. EMBS 2007, pp. 5563–5566 (2007)
24. Wei, C., Kangling, F.: Multilevel thresholding algorithm based on particle swarm optimization for image segmentation. In: 27th Chinese Conference in Control, 2008. CCC 2008, pp. 348–351. IEEE (2008)
25. Yin, P.Y.: Multilevel minimum cross entropy threshold selection based on particle swarm optimization. Appl. Math. Comput. **184**(2), 503–513 (2007)
26. Zhiwei, Y., Zhengbing, H., Huamin, W., Hongwei, C.: Automatic threshold selection based on artificial bee colony algorithm. In: The 3rd International Workshop on Intelligent Systems and Applications (ISA), 2011, pp. 1–4 (2011)
27. Horng, M.-H.: Multilevel minimum cross entropy threshold selection based on the honey bee mating optimization. Expert Syst. Appl. **37**(6), 4580–4592 (2010)
28. Oliva, D., Cuevas, E., Pajares, G., Zaldivar, D., Perez-Cisneros, M.: Multilevel thresholding segmentation based on harmony search optimization. J. Appl. Math. 2013 (2013)
29. Agrawal, S., Panda, R., Bhuyan, S., Panigrahi, B.K.: Tsallis entropy based optimal multilevel thresholding using cuckoo search algorithm. Swarm Evolut. Comput. **11**, 16–30 (2013)
30. Akay, B.: A study on particle swarm optimization and artificial bee colony algorithms for multilevel thresholding. Appl. Soft Comput. **13**(6), 3066–3091 (2013)
31. Bhandari, A.K., Kumar, A., Singh, G.K.: Modified artificial bee colony based computationally efficient multilevel thresholding for satellite image segmentation using Kapurs, Otsu and Tsallis functions. Expert Syst. Appl. **42**(3), 1573–1601 (2015)

32. Kapur, J.N., Sahoo P.K., Wong, A.K.C.: A new method for gray-level picture thresholding using the entropy of the histogram. Comput. Vis. Graphics Image Process. **29**(3), 273–285 (1985)
33. Mirjalili, S., Lewis, A.: The whale optimization algorithm. Adv. Eng. Softw. **95**, 51–67 (2016)
34. Martin, D., Fowlkes, C., Tal, D., Malik, J.: A database of human segmented natural images and its application to evaluating segmentation algorithms and measuring ecological statistics. In: Eighth IEEE International Conference on Computer Vision, 2001. ICCV 2001. Proceedings, vol. 2, pp. 416–423. IEEE (2001)
35. Wang, Z., Simoncelli, E.P., Bovik, A.C.: Multiscale structural similarity for image quality assessment. In: Conference Record of the Thirty-Seventh Asilomar Conference on Signals, Systems and Computers, 2004, vol. 2. IEEE (2003)

Evaluating Swarm Optimization Algorithms for Segmentation of Liver Images

Abdalla Mostafa, Essam H. Houssein, Mohamed Houseni, Aboul Ella Hassanien and Hesham Hefny

Abstract There is a remarkable increase in the popularity of swarms inspired algorithms in the last decade. It offers a kind of flexibility and efficiency in their applications in different fields. These algorithms are inspired by the behaviour of various swarms as birds, fish and animals. This chapter presents an overview of some algorithms as grey wolf optimization (GWO), artificial bee colony (ABC) and antlion optimization (ALO). It proposed swarm optimization approaches for liver segmentation based on these algorithms in CT and MRI images. The experimental results of these algorithms show that they are powerful and can get remarkable results when applied to segment liver medical images. It is evidently proved from the experimental results that ALO, GWO and ABC have obtained 94.49%, 94.08% and 93.73%, respectively, in terms of overall accuracy using similarity index measure.

Keywords Artificial bee colony · Grey wolf · Antlion and segmentation

---

A. Mostafa (✉) · H. Hefny
Scientific Research Group in Egypt (SRGE), Institute of Statistical Studies and Research, Cairo University, Giza, Egypt
e-mail: Abdalla_mostafa75@yahoo.com

E.H. Houssein
Scientific Research Group in Egypt (SRGE), Faculty of Computers and Information, Minia University, Minia, Egypt

M. Houseni
National Liver Institute, Radiology Department, Menofia University, Menofia, Egypt

A.E. Hassanien
Faculty of Computers and Information, Information Technology Department, Cairo University, Giza, Egypt

A.E. Hassanien and D.A. Oliva (eds.), *Advances in Soft Computing and Machine Learning in Image Processing*, Studies in Computational Intelligence 730,
https://doi.org/10.1007/978-3-319-63754-9_3

1 Introduction

There is a kind of complexity in handling liver segmentation in different modalities of abdominal imaging. This complexity motivated many researchers to find out new efficient methods for better segmentation. Nature and swarm observation inspired them to create different meta-heuristic algorithms. Even though any swarm has a simple individual behaviour, but its coordination presents an amazing social organization. Swarm algorithms focus on the collective behaviour and coordination between members in hunting, finding food sources, travelling, and mating. Researchers try to mimic the steps of swarms to develop new methods to solve the different sophisticated optimization problems. In this chapter, some swarm algorithms are discussed to manipulate liver segmentation images using two imaging modalities represented in CT and MRI images.

There are different medical imaging modalities, depending on the usage of radiology of X-ray gamma rays, radio frequencies pulses and ultrasound. It is used in a wide range of medical diagnosis in different branches as hepatology, oncology, cardiology, etc. The most common modalities are computed tomography (CT) and magnetic resonance imaging (MRI) which are used in this discussion. Liver segmentation is the backbone of the discussed usage of the swarms optimization algorithms. At the time being, many researchers mimic the behaviour of different kinds of swarms, aiming to create new algorithms to solve the complex optimization problems in different fields. These algorithms include particle swarm optimization (PSO), firefly optimization algorithm (FOA), glowworm swarm optimization (GSO), artificial bee colony (ABC), fish swarm optimization (FSO), ant colony optimization (ACO) and bacterial foraging optimization algorithm (BFOA).

Sathya et al. [4] used particle swarm optimization (PSO) to divide the image into multilevel thresholds for the purpose of segmentation. The objective functions of Kapur and Otsu methods are maximized using Particle swarm optimization. Jagadeesan [5] combined Fuzzy C-means with firefly algorithm to separate brain tumour in MRI images. The membership function of Fuzzy C-means is optimized using firefly algorithm to guarantee a better brain tumour segmentation. Liang et al. [8] combined the non-parametric ant colony optimization (ACO) algorithm and Otsu with the parametric expectation and maximization (EM) algorithm. Since Em is sensitive to the initial solution, ACO is used to get this one, then EM selects multilevel threshold for objects segmentation. The artificial bee colony (ABC) optimization algorithm is used by Cuevas et al. [3], to compute image thresholds for image segmentation. A. Mostafa et al. [11] clustered the image of liver using ABC algorithm to get the initial segmented liver. Then, simple region growing method is used to segment the whole liver. Sankari [18] could manage the local maxima in expectation–maximization (EM) algorithm by combining glowworm swarm optimization (GSO) algorithm with EM algorithm. GSO clusters the image to find the initial seed points. These seed points are passed to EM algorithm for segmentation. Jindal [6] tried to solve the problem of computational complexity and time. This paper presented an algorithm called bacterial foraging optimization algorithm (BFOA). It is inspired by Escherichia coli bacteria. Needing no thresholding in image segmentation is its main advantage. Sivaramakrishnan et al. [19] used ABC optimization algorithm, combined with Fish swarm algorithm, to diagnose breast tumors in mammogram images. Alomoush et al. [1] built a system that detects the spots of brain tumors in MRI images using a hybrid firefly and Fuzzy C-means.

This chapter focuses on three bio-inspired optimization algorithms: Grey Wolf, Artificial Bee Colony, and Antlion optimization algorithms.

The remainder of this chapter is ordered as follows. Section 2 gives an overview of the nature-inspired optimization algorithms, including Grey Wolf Algorithm in Sect. 2.1, artificial bee colony algorithm (ABC) in Sect. 2.2 and antlion optimization (ALO) in Sect. 2.3. Section 3 presents the implementation of the approaches using GWO, ABC and ALO. This section describes the phases of preprocessing, swarm algorithms followed by the experimental results of the three approaches. Finally, conclusions and future work are discussed in Sect. 4.

2 Swarm Optimization Algorithms

The social behaviour of the animals, birds, fish and insects is the main base for all swarm or nature-inspired algorithms. It tries to mimic this behaviour and organization in hunting, food searching and mating. The following subsections present three swarm algorithms as follows.

2.1 Grey Wolf Optimization

Grey wolf optimizer (GWO) is considered a population-based meta-heuristic algorithm. It mimics the hierarchy of leadership and mechanism of hunting performed by grey wolves in nature. This algorithm was proposed by Mirjalili et al. in 2014 [10]. The following subsections will present an overview of the main concepts of the algorithm and its structure as follows.

Main Concepts and Inspiration

Grey wolves always live in groups, consists of 5–12 members. Each member in the group has their own job in the strict social dominant hierarchy. This hierarchy defines the leadership and coordination. The hierarchy also defines the way of gathering information about the prey for hunting and finding the place to stay in. Figure 1 shows the hierarchy of grey wolves group.

The social hierarchy of grey wolves group consists of four levels as follows.

1. **Alpha wolves** (α)
   The alpha wolves are the leaders of the pack. They are a male and a female,

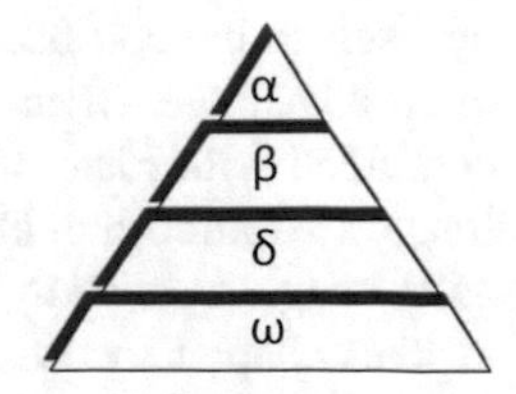

Fig. 1 Social hierarchy of grey wolves' pack

responsible for making the major decisions about hunting, moving, direction and stay. The pack members have to obey their orders and show their acceptance by holding their tails down. They are the brain and the ruler of the pack and their orders should be followed by all other levels in the pack.

2. **Beta wolves** (β)
   The betas are subordinate wolves, that aid the alpha in making decisions, transferring the commands of the alpha through the pack and giving feedback. It could be male or female and it is the best candidate to be alpha when the current alpha passes away or gets old.
3. **Delta wolves** (δ)
   The delta wolves are not alpha nor beta wolves and have to submit to them and follow their orders. But they dominate the omega (the lowest level in wolves social hierarchy). There are different categories of delta wolves as follows.

   - **Scouts**
     The scout wolves' responsibility is to watch the boundaries of the pack against enemies and warn the pack for danger. Also, they are responsible for telling about the candidate prey.
   - **Sentinels**
     They are the pack's guards. The sentinel wolves protect the packs from any danger.
   - **Elders**
     The elder wolves are the wise and experienced ex alpha or beta who left their position because of age.
   - **Hunters**
     The hunters help the alpha and beta wolves in the process of hunting to provide food for the pack.
   - **Caretakers**
     The caretakers take care of the weak, ill, young and wounded wolves in the pack.

4. **Omega** (ω) **(lowest level)**
   The omega wolves are all other members in the pack that have to submit to all the other up levels of the dominant wolves. They are not important members and they are the last to eat.

The following subsection presents the mathematical models of the packs' social hierarchy and other operations of tracking, encircling and attacking prey.

Social Hierarchy

In grey wolf optimizer algorithm, the alpha α are considered the fittest solution, while beta β and delta δ are the second and the third fittest solutions. Any other solutions are considered omega ω. The α, β and δ solutions guide the hunting operation, and the ω solutions follow these three wolves.

Encircling Prey

When the grey wolves start hunting, they encircle prey. The encircling behaviour is presented as a mathematical model in the following equations:

$$\boldsymbol{D} = |\boldsymbol{C} \cdot \boldsymbol{X}_p(t) - \boldsymbol{A} \cdot \boldsymbol{X}(t)| \tag{1}$$

$$\boldsymbol{X}(t+1) = \boldsymbol{X}_p(t) - \boldsymbol{A} \cdot \boldsymbol{D}, \tag{2}$$

where t is the current iteration, $\boldsymbol{A}$ and $\boldsymbol{C}$ are coefficient vectors, $\boldsymbol{X}_p$ represents the prey's position vector, and $\boldsymbol{X}$ indicates the a grey wolf's position vector.

The coefficient vectors $\boldsymbol{A}$ and $\boldsymbol{C}$ are calculated as follows:

$$\boldsymbol{A} = 2\boldsymbol{a} \cdot \boldsymbol{r}_1 - \boldsymbol{a} \tag{3}$$

$$\boldsymbol{C} = 2 \cdot \boldsymbol{r}_2, \tag{4}$$

where components of $\boldsymbol{a}$ are linearly reduced from 2 to 0 through the running iterations and $\boldsymbol{r}_1$, $\boldsymbol{r}_2$ are random vectors in the range [0, 1].

Hunting

The alpha α guides the hunting operation, and both beta β and delta δ might participate hunting. The mathematical model of hunting behaviour assumes that alpha α, beta β and delta δ know the potential location of prey better than the others. They represent the first three best solutions. These solutions, forcing the other agents to update their positions according to the best solution, are shown in the following equations:

$$\begin{aligned} \boldsymbol{D}_\alpha &= |\boldsymbol{C}_1.\boldsymbol{X}_\alpha - \boldsymbol{X}|, \\ \boldsymbol{D}_\beta &= |\boldsymbol{C}_2.\boldsymbol{X}_\beta - \boldsymbol{X}|, \\ \boldsymbol{D}_\delta &= |\boldsymbol{C}_3.\boldsymbol{X}_\delta - \boldsymbol{X}| \end{aligned} \tag{5}$$

$$\begin{aligned} \boldsymbol{X}_1 &= \boldsymbol{X}_\alpha - \boldsymbol{A}_1 \cdot (\boldsymbol{D}_\alpha), \\ \boldsymbol{X}_2 &= \boldsymbol{X}_\beta - \boldsymbol{A}_2 \cdot (\boldsymbol{D}_\beta), \\ \boldsymbol{X}_3 &= \boldsymbol{X}_\delta - \boldsymbol{A}_3 \cdot (\boldsymbol{D}_\delta), \end{aligned} \tag{6}$$

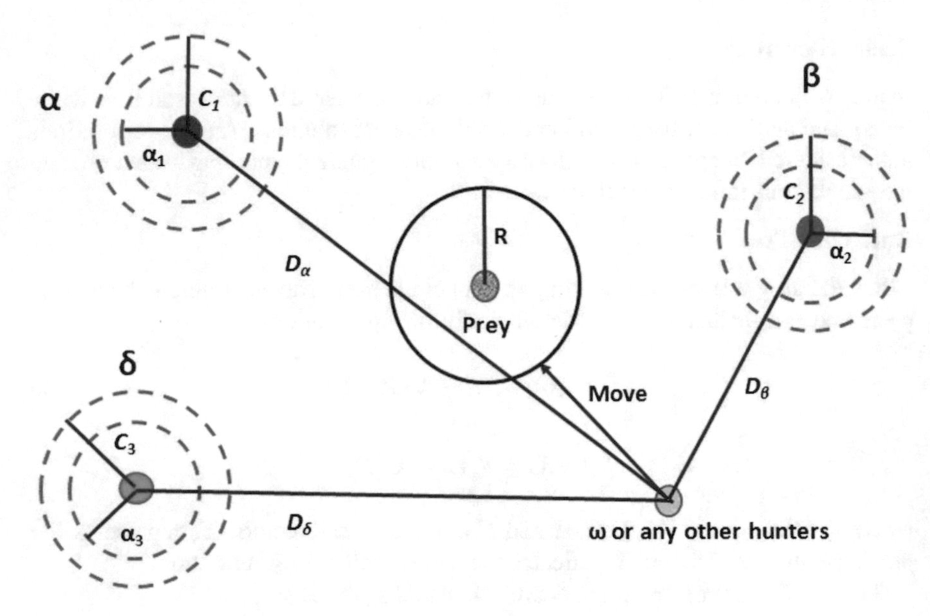

Fig. 2 Position updating in GWO

$$X(t+1) = \frac{X_1 + X_2 + X_3}{3}, \tag{7}$$

Figure 2 shows the search agent position updating process.

Searching for Prey (Exploration)

GWO algorithm presents the exploration process in the change of positions of α, β and δ. They diverge from each other when they search for prey and converge to attack the targeted prey. The exploration process is mathematically modelled by using $\boldsymbol{A}$ with random values less than -1 or greater than 1. This random value forces the search agent to diverge from the prey. When $|A| > 1$, the wolves moves away from the prey to search for a fitter prey.

Attacking Prey (Exploitation)

GWO algorithm finishes hunting when the prey stops moving and is attacked. The vector $\boldsymbol{A}$ is a random value in interval $[-2a, 2a]$, where a is reduced from 2 to 0 during the algorithm iterations. When $|A| < 1$, the wolves move towards the prey to attack. This represents the exploitation process.

In the following subsection, the GWO algorithm is described as follows.

GWO Algorithm

Algorithm 1 describes the steps of GWO algorithm in detail.

Algorithm 1 Grey wolf optimizer algorithm

1: Set the initial values of different parameters including the population size n, the maximum number of iterations, Max_{itr},parameter a and coefficient vectors $\boldsymbol{A}$, $\boldsymbol{C}$.
2: Set $t := 0$.
{**Initialize counter.**}
3: **for** $(i = 1 : i \leq n)$ **do**
4: Generate random values of the initial population $\boldsymbol{X}_i(\boldsymbol{t})$.
5: Calculate the value the fitness function for each search agent (solution) $f(\boldsymbol{X}_i)$ in the population.
6: **end for**
7: Assign the values of the best three solutions to $\boldsymbol{X}_\alpha$, $\boldsymbol{X}_\beta$ and $\boldsymbol{X}_\delta$, respectively.
8: **repeat**
9: **for** $(i = 1 : i \leq n)$ **do**
10: Update the value of each search agent in the population as shown in Eq. 7.
11: Decrease the value of parameter a from 2 to 0.
12: Update the coefficients $\boldsymbol{A}$ and $\boldsymbol{C}$ according to Eqs. 3, 4, respectively.
13: Evaluate the fitness function value for each search agent (vector) $f(\boldsymbol{X}_i)$.
14: **end for**
15: Update the vectors $\boldsymbol{X}_\alpha$, $\boldsymbol{X}_\beta$ and $\boldsymbol{X}_\delta$.
16: Increase the iteration number, $t = t + 1$.
17: **until** $(t < Max_{itr})$.
{**Termination criteria satisfied**}.
18: Produce the best solution $\boldsymbol{X}_\alpha$.

2.2 *Artificial Bee Colony Algorithm*

This section highlights the main concepts and structure of the artificial bee colony algorithm as follows.

Main Concepts

While searching for food, there is a kind of communication and cooperation between the bees in the swarm. They communicate using dancing to share information about the food source. They also share, store and memorize the information they obtained according to the changes in the surrounding environment. Each bee can update its position using the shared information. The different bees behaviour can be summarized as follows.

Food Sources

For every bee, a flower is considered a food source. The bee collects the information of the amount of nectar in the flower, its distance and direction from the beehive. The bee memorizes and shares this information with other bees in the swarm.

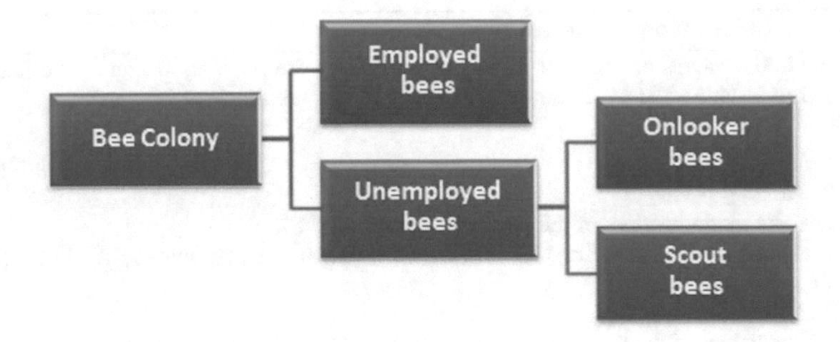

Fig. 3 Bee colony types

Employed and Unemployed Bees

There are two kinds of working bees inside the beehive, employed and unemployed bees as shown in Fig. 3.

The employed bees are responsible for the exploitation of the food source and for watching the amount of the remaining nectar in the associated food source. The other kind in the bee colony is the unemployed. They share information with the employed bees to select a food source. They are divided into two categories, the onlooker and the scout bees. The onlooker bees are responsible for collecting the information from the employed bees to be able to select a proper food source for themselves. The scout bees have a different job. They search for new food sources when the current food sources are exhausted and the nectar amount is dramatically decreased. Usually in the colony, the employed bees represent half of the swarm, while the unemployed bees represent the other half. The scout bees normally represent 10% of the total number of bees.

Foraging Behaviour

In foraging process, a bee starts to search for the food to get the nectar from it. The amount of nectar that it can extract depends on the richness of the food source and its distance from the hive. The enzymes in the bee stomach are used to transfer the nectar into honey.

Dancing

The communication between the bee to share information with others is done by dancing in different forms. The bee shares the food source information using three dancing forms. Two dances are used to express the distance from the hive, and the third is for expressing the profitability of the food source.

- **Round dance**. The bee uses this dances when the food source is close to hive.
- **Waggle dance**. The employed bees dance the waggle dance to show that the food source is far from the hive. The speed of the waggle dance performed by the bee is proportional to the distance from the hive towards the food source.

- **Tremble dance**. This type of dance means that the bee does not know about the amount of nectar and the profitability of its food source.

ABC Algorithm

The artificial bee colony (ABC) algorithm was proposed by Dervis Karaboga in 2005 [2, 7]. It is a population-based meta-heuristics algorithm using the foraging behaviour of honey bee colonies. The algorithm has four phases, described as follows.

The Initial Population Phase

The initial population in ABC algorithm is generated randomly. The population contains a number of solutions (*NS*), where each solution x_i is a D dimensional vector. The x_i represents the ith food source. Each solution is generated as in Eq. 8:

$$x_{ij} = x_{Lj} + r(x_{jU} - x_{jL}), \;\; j = 1, 2, \ldots, D, \tag{8}$$

where L and U are bounds of x_i in jth direction and r is a random number and $r \in [0, 1]$. The ABC algorithm has three main phases, presented in details in Algorithm 2.

The Employed Bees Phase

The employed bees define the nectar amount and modify the value of the current solution according to the fitness values of each solution. The position of the bee is updated towards the better food source if the fitness value of the old food source is less than the new one. The bee position is updated according to Eq. 9:

$$v_{ij} = x_{ij} + \phi_{ij}(x_{ij} - x_{kj}), \tag{9}$$

where $\phi_{ij}(x_{ij} - x_{kj})$ is the step size, k and j are randomly selected indices, $k \in 1, 2, \ldots, NS$, $j \in 1, 2, \ldots, D$ and ϕ_{ij} are random numbers and $\phi_{ij}(x_{ij} - x_{kj}) \in [-1, 1]$.

Onlooker Bees Phase

The employed bees share the information of the nectar amount with the onlooker bees. This information represents the fitness values of the food source. The onlooker calculates the probability of the values compared to the total fitness values for other search agents. The onlooker selects the highest probability p_i as the best solution. The probability is calculated as follows in Eq. 10:

$$p_i = \frac{f_i}{\sum_{j=1}^{NS} f_j}, \tag{10}$$

where f_i represents the fitness value of the ith solution.

Scout Bees Phase

When the food source is not updated for a number of iterations. This means that the bee cannot determine the amount of nectar and abandon food source with tremble dance. This bee becomes a scout bee and generates a new solution (food source). The new solution is randomly generated according to Eq. 11:

$$x_{ij} = x_{Lj} + r(x_{jU} - x_{jL}), \quad j = 1, 2, \ldots, D, \tag{11}$$

where L and U are lower and upper bounds of x_i in jth direction and r is a random number, $r \in [0, 1]$.

Algorithm (2) describes the implementation of ABC algorithm.

Algorithm 2 ABC algorithm

1: Set the parameters of population size, number of iterations and the dimension (clusters) of the values for search agents.
2: Randomly, generate the initial population values x_i, where $i = \{1, \ldots, NS\}$.
{**Initialization**}
3: Calculate the fitness function $f(x_i)$ of all search agents (solutions) in the population.
4: Keep the value of the best solution as x_{best}.
{**Memorize the best solution**}
5: Set iteration=1
6: **repeat**
7: Generate a new solution v_i from the old solution x_i using the next equation.

$$v_{ij} = x_{ij} + \phi_{ij}(x_{ij} - x_{kj})$$

where

$$\phi_{ij} \in [-1, 1], k \in \{1, 2, \ldots, NS\},$$
$$j \in \{1, 2, \ldots, D\} \text{ and } i \neq k$$

{**Employed bees**}
8: Evaluate the value of fitness function $f(v_{ij})$ for all solutions in the population.
9: Compare between the best current solution and candidate solutions, and keep the best one as new best solution.
{**Greedy selection**}
10: Calculate the probability p_i, for the solutions x_i, where $p_i = \frac{f_i}{\sum_{j=1}^{NS} f_j}$.
11: Generate the new solution v_i from the selected solutions depending on its p_i.
{**Onlooker bees**}
12: Calculate the fitness function f_i for all solutions in the population.
13: Keep the best solution between current and candidate solutions.
{**Greedy selection**}
14: Find out the abandoned solution and replace it with a newly random generated solution x_i.
{**Scout bee**}
15: Keep the best solution (x_{best}) found so far in the population
16: $iteration = iteration + 1$
17: **until** iteration $\leq MCN$

2.3 Antlion Optimization Algorithm

Antlion optimization (ALO) algorithm is a new nature-inspired optimization algorithm proposed by Mirjalili. It simulates the hunting mechanism of antlion insect.

The antlion digs a conic hole in the sands and hides at the bottom, waiting for trapping and hunting insects, especially ants [9]. In the ALO algorithm, the prey moves with random walks inside the search space. Its movement is affected by the closest trap. The traps are built according to the fitness function and the antlion ability to build larger holes. The larger hole increases the probability of catching ants. The ant is considered caught when it is fitter than the antlion [9]. The main parameters in the algorithm include the size of search space (antlions and ants), the number of iterations and the number of clusters.

Operators of ALO Algorithm

The ALO algorithm simulates the ants' actual movement in its real environment and mimics its way of building traps for hunting [9]. This subsection will clear the main concepts of the antlion behaviour.

- **Building trap**
  A roulette wheel selection is used to model the antlion capabilities for insects hunting. The roulette wheel is originally a genetic operator used in genetic algorithms to select the potential useful solutions. Each ant is supposed to be trapped by only one antlion. So, the roulette wheel operator can select the candidate antlion according to its proper fitness value. If the antlion has a higher fitness, it has a higher chance to catch ants.
- **Random walks of the ants**
  ALO algorithm assumes that every ant is walking randomly within the search space. The following equation rules the random movement of the ant:

$$X_i^t = \frac{(X_i^t - a_i) \times (d_i - C_i^t)}{(d_i^t - a_i)} + C_i, \tag{12}$$

  where C_i^t represents the least value of ith variable at *tth* iteration, a_i is the least value of random walk of ith variable, and d_i^t is the maximum of ith variable at *tth* iteration. The ant's move must be kept inside the search space. If it walks outside the space, its random walk is normalized to minimum or maximum value of the search space.
- **Trapping in antlions pits**
  The ant's random walk is affected by the existing traps close to its position. The following equations show that the ant's random walks are controlled by a range of values (c, d), representing the minimum of all variables for i*th* ant and the maximum of all variables for *ith* ant, respectively.

$$c_i^t = Antlion_j^t + C^t \tag{13}$$

$$d_i^t = Antlion_j^t + d^t. \tag{14}$$

- **Sliding ants towards antlion**
  When an ant enters the conic hole, the antlion starts to shoot sands to force it to move towards the bottom or centre of the trap, where it resides. The following equation is used to decreases the radius of the ants random walks:

$$c^t = \frac{c^t}{I} \tag{15}$$

$$d^t = \frac{d^t}{I}, \tag{16}$$

  where c^t represents the minimum value of all variables at *tth* iteration, and d^t is the vector including the maximum of all variables at *tth* iteration. Also, I is calculated as follows:

$$I = 10^w \frac{t}{T}, \tag{17}$$

  where T is the maximum number of iterations and t is the current iteration.
- **Prey catching and trap re-building**
  At the point that the ant becomes fitter than the antlion, the antlion starts to catch and eat the captured ant. Then it updates its position to the prey's position, to increase its opportunity for catching a new hunt. The following equation is used for changing position:

$$Antliont_j^t = Ant_i^t \quad if \;\; f(Ant_i^t) > f(Antlion_j^t). \tag{18}$$

- **Elit-Antlion:**
  Through the iterations of the algorithm, it keeps best found values according to the fitness function' value. The elit-antlion is the best solution of all antlions in all iterations. The best fitness value of antlions is compared to the supreme solution (Elit-antlion) in every iteration. It should be noticed that the elit-antlion solution affects the random walks through the following equation. The equation uses the roulette random walk of ants around the elit-antlion (R_E^t) with their random walk around the antlion as follows:

$$Ant_i^t = \frac{R_A^t + R_E^t}{2}. \tag{19}$$

ALO Algorithm

Algorithm (3) describes the implementation of ABC algorithm.

Algorithm 3 ALO algorithm

1: Randomly, initialize the ants and antlions population.
2: Evaluate the fitness value of ants and antlions.
3: Set the best antlions fitness value as the elite.
4: **for** Iteration number **do**
5:     **for** Every ant **do**
6:         Select an antlion using Roulette wheel.
7:         Update c and d using equations Eqs. 2.3 and 2.3.
8:         Create a random walk for ant and normalize it inside the search space.
9:         Update the position of ant using Eqs. 15 and 16.
10:     **end for**
11:     Calculate the fitness of all ants.
12:     Replace an antlion with its corresponding ant it if becomes fitter (Eq. 18).
13:     Update elite if an antlion becomes fitter than the elite.
14: **end for**
15: Return elite.

3 Proposed Approaches Using Swarm Optimization

The swarm optimization based segmentation approaches for liver segmentation consist of three main phases, including preprocessing, swarm clustering and post-processing. These phases are described in detail in the following section, along with the involved steps and the characteristics of each phase.

3.1 Preprocessing Phase

The usage of swarm-inspired based algorithms makes the phase of preprocessing simpler. It depends on image cleaning and ribs connection. The main purpose of using filters as mean, median and gabor is to deepen the boundaries between the organs and clean the noise. There is no need for implementing any filters or morphological operations for deepening the boundaries or smoothing the image. Because clustering the image results in some holes inside the liver which is filled easily in the last phase. But the only used filter is contrast stretching. It is used to stress white ribs to make it easier to connect ribs. The morphological operations are used to remove patients' information and machine bed [20]. There is an advantage of using swarms algorithms in liver segmentation regarding the noise. Segmentation is not affected by noise at all. Since noise is normally small pixels, which is filled by morphological operations step, a statistical image can be used to define all the possible occurrence of liver in the abdominal image. The statistical image is created by summing all binary manual segmented liver. The dataset of available manual segmented images is converted into binary images. Every binary image is summed in one binary image that represents all possible occurrences of liver for any patient. The resulting binary image is the statistical image.

3.2 Swarm Algorithms Phase

In the second phase, the swarm algorithms are used to cluster the image. The swarm algorithms extract a predefined number of clusters that represent the centroids of the intensity values in the image. Handling clusters can be in different ways. (1) The first way is to separate every cluster in a binary image and use morphological operations for everyone to remove small objects and keep the large objects that which represents organs. Then the binary clustered images are gathered in one binary image and multiplied with the original image. This results in an initial segmented liver. (2) The second way of handling clusters is to remove the lowest and highest clusters which represent the black background and the white bones and apply the morphological operations on other clustered binary images. Then the resulting binary images are gathered in one binary image multiplied by the original image to get the initial segmented liver. (3) The third way is to pick up some points by the user that represents the clusters inside the liver. These clustered images are picked up in one binary image and multiplied by the original image to get the initial segmented liver.

In the next subsections, different swarm-based algorithms are used to get the initial segmented liver. The different algorithms have to set up some values for its parameters. The proposed approaches are described in detail for the three main algorithms, including grey wolf, artificial bee colony and antlion optimization.

Grey Wolf Phase

Algorithm (4) shows the steps of using GWO to get the initial segmented liver.

Algorithm 4 Grey Wolf based approach for liver segmentation

1: Set the different values for the parameters that includes number of search agents, clusters and iterations.
2: Apply GWO algorithm on the image to obtain the clusters' centroids.
3: Calculate the distance between each pixel's intensity value and the different clusters. Find the least distance and assign the cluster number to the pixel.
4: Give different colours to each cluster and display the clustered image.
5: Let the user to pick up some points that represents the liver clusters.
6: Get the picked up clusters in one binary image and fill the holes inside the liver.
7: Finally, multiply the original image by the resulting binary image to get the initial segmented liver.

Artificial Bee Colony Phase

Algorithm (5) shows the steps of using ABC as a clustering technique to get the initial segmented liver.

Algorithm 5 The Artificial bee colony based liver segmentation approach

1: Set a value for the parameters of the number of colony bees, clusters and the iterations to start ABC algorithm.
2: Apply ABC algorithm on the CT image.
3: Get the global values, which is the centroids of the clusters.
4: Sort the obtained clusters.
5: Get the different clusters in binary images.
6: Use the morphological operations to fill the holes remove small objects in the clustered images.
7: Exclude two clustered images representing the lowest and highest clusters values.
8: Sum all remaining clustered images together in one binary image.
9: Multiply the original image by the resulting binary image to get the initial segmented liver.

3.2.1 Antlion Optimization Phase

Algorithm (6) describes the steps of the ALO approach to segment the initial liver.

Algorithm 6 The Antlion based liver segmentation approach

1: Prepare the statistical image as mentioned before.
2: Use the morphological operations to clean the image annotations and connect ribs.
3: Use ALO to get the clusters of the abdominal image.
4: Multiply the resulting clustered image by the prepared binary statistical image.
5: Pick up manually the required clusters to get the initial segmented liver.
6: Use the morphological operations to enhance the segmented image by removing small objects.

3.3 Postprocessing Phase

In the third phase, the resulting initial liver is enhanced using one of two methods. The first is to use region growing technique to get the final segmented liver. The second is to use the morphological operations to remove small objects and enhance the boundaries of the segmented liver. The final segmented liver image is validated using similarity index measure to calculate the accuracy.

3.4 Experimental Results and Discussion

3.4.1 Datasets

The datasets of CT or MRI images, used to test the approaches, are described as follows. Every dataset includes a number of original abdominal images and the ground truth of these images. The ground truth is approved by a radiological specialist.

- **Grey Wolf:** A set of 38 CT images, taken in pre-contrast phase, was used for liver segmentation and testing the proposed approach.
- **Artificial bee colony:** A set of 38 CT images were used to experiment the proposed approach. The used images were taken in the first phase of CT scan before the patient is injected with contrast materials.
- **Antlion optimizer:** A set of 70 MRI images were used to experiment the proposed approach in pre-contrast phase and dual phase.

Parameter Settings

All swarm-based algorithms have their predefined parameters. These parameters affect the efficiency of the result of the algorithm. To test the best parameter setting, a number of five images, taken randomly from the tested dataset, are tested using 10 different values for each parameter. Hence, each parameter is tested 50 times to get the best average setting.

Table 1 describes the successful parameter settings when applying the GWO proposed algorithm (Tables 1 and 2).

Table 2 describes the successful parameter settings when applying the ABC proposed algorithm.

Table 3 describes the successful parameter settings when applying the ALO proposed algorithm.

Experimental Results

Evaluation of the approaches is performed using similarity index (SI), defined using the following equation:

Table 1 Parameters of GWO approach

Ser.	Parameter	Setting
1	Population size	10
2	Maximum iterations	20
3	Number of clusters	7

Table 2 Parameters of ABC approach

Ser.	Parameter	Setting
1	Population size	50
2	Food	25
3	Number of solutions	50
4	Maximum iterations	30
5	Number of clusters	6

Table 3 Parameters of ALO approach

Ser.	Parameter	Setting
1	Search Agents size	10
2	Maximum iterations	10
3	Number of clusters	7

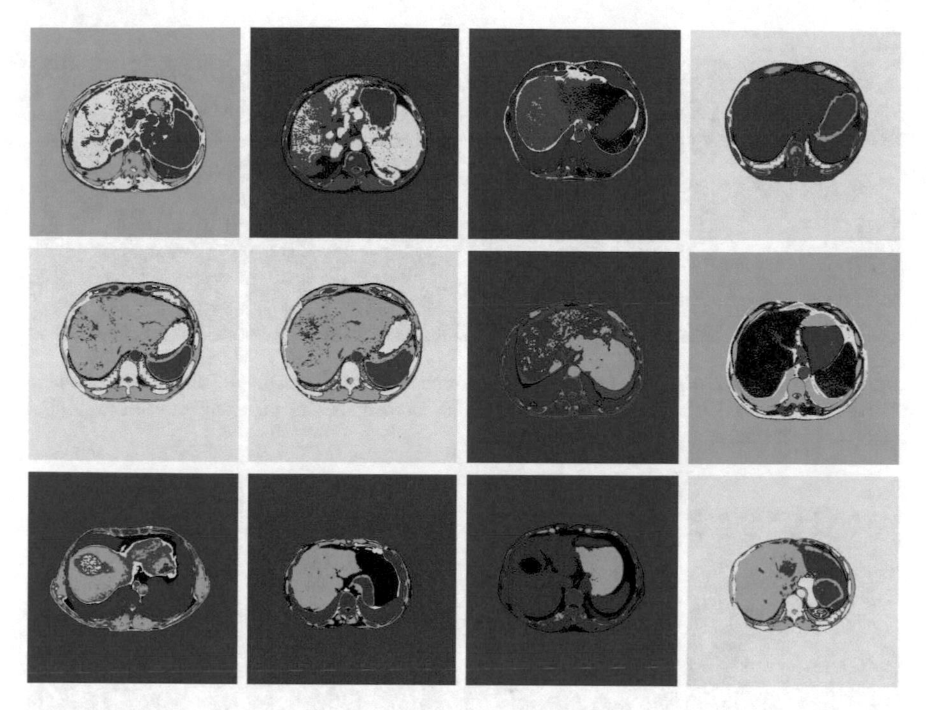

Fig. 4 The effect of applying GWO on different CT images

$$SI(I_{auto}, I_{man}) = \frac{I_{auto} \bigcap I_{man}}{I_{auto} \bigcup I_{man}}, \tag{20}$$

where SI is the similarity index, I_{auto} is the binary automated segmented image, resulting from the phase of final segmentation of the whole liver in the used approach and I_{man} is the binary manual segmented image by a radiology specialist.

GWO experimental results

Figure 4 shows the results of using Grey Wolf Optimization algorithm on different CT images.

Figure 5 shows the resulting image when the required clusters are picked up to represent the liver. The picked up clusters are multiplied by the original image. Obviously, there is no need to pick up the cluster of the small regions of lesions inside the liver. They might be holes, which can be filled in the extracted liver.

Finally, the region growing technique is used to improve the image resulting from applying GWO and statistical image. Figure 6 shows the difference when comparing the segmented image through the proposed approach with the manual image.

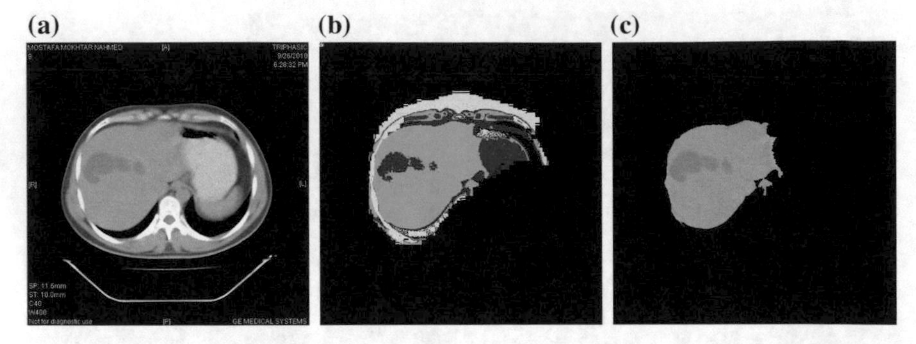

Fig. 5 Picking up the required clusters from the image resulting from applying the statistical image: **a** Original image **b** GWO image multiplied by statistical image, **c** The picked up clusters multiplied by the original image

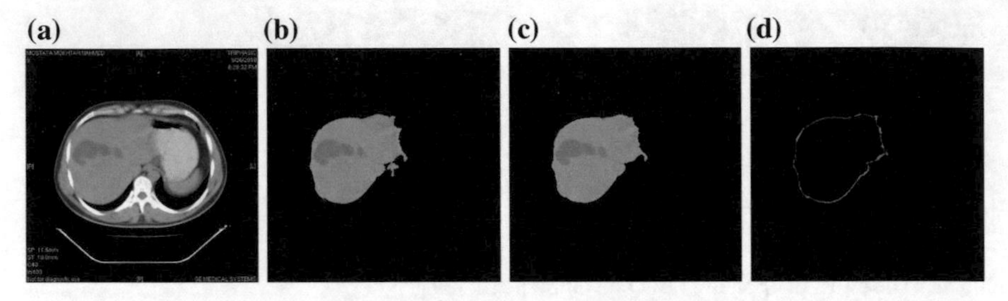

Fig. 6 Enhanced RG segmented image: **a** Original image **b** Picked up clusters image, **c** Enhanced image by region growing, **d** Difference image

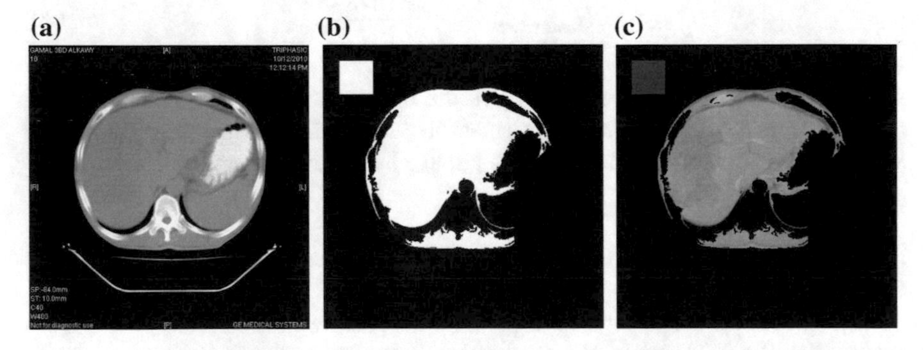

Fig. 7 ABC liver segmentation, **a** original image **b** ABC binary image, **c** ABC segmented image

ABC experimental results

Figure 7 shows the binary image of ABC and the liver segmented images.

Finally, the ABC segmented image is enhanced using simple region growing technique. Figure 8 shows the difference between the segmented image and annotated one.

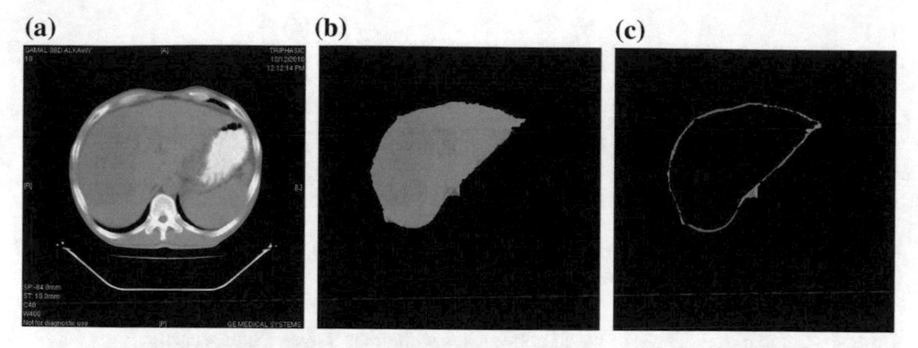

Fig. 8 Final liver segmented image using ABC, compared to the annotated image: **a** Original image, **b** Segmented image, **c** Difference image

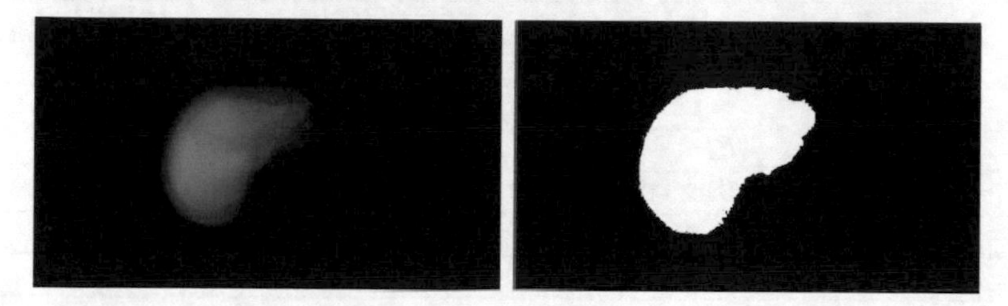

Fig. 9 Statistical occurrence image

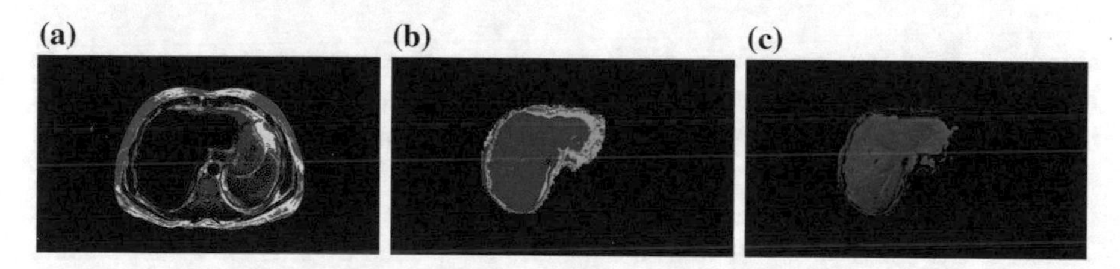

Fig. 10 Picking up the required clusters: **a** ALO image **b** Statistical occurrence image, **c** The picked up clusters multiplied by the original image

It shows that the average performance of liver images segmentation is improved using the proposed approach. Segmentation using region growing has an average result of SI = 84.82%. This result is improved using the proposed approach with SI = 93.73.

Antlion experimental results

Figure 9 shows the statistical image, which is a result of collecting all liver occurrence in different images of the used MRI dataset.

Figure 10 shows ALO image and the resulting image from applying the removal of the right part close to the liver, and the multiplication of all possible statistical occurrence on the abdominal image. This process excludes a great part of the unrequired organs from the image, especially the organs of stomach and spleen. It also

Fig. 11 Enhancing segmented image: **a** Original image **b** Picked up clusters image, **c** Enhanced image by morphological operations

Table 4 Comparison of swarm-based approaches with other traditional approaches

Ser.	Approach		Result
1	Region growing (RG)	[20]	84.82
2	Wolf local thresholding + RG	[12]	91.17
3	Morphological operations + RG	[13]	91.20
4	Level set	[20]	92.10
5	K-means + RG	[14]	92.38
6	Proposed Artificial Bee Colony (ABC)	[15]	93.73
7	Proposed Grey Wolf (GWO)	[16]	94.08
8	Proposed Antlion (ALO)	[17]	94.49

shows the resulting image from picking up the required clusters that represent the liver. The chosen clusters are multiplied by the original image. The user does not have to choose the cluster of the small regions of lesions inside the liver. The small fragments representing the un-chosen clusters of lesion might be holes inside the liver. When the liver is extracted, these holes can be filled easily.

The last process of segmentation enhances the picked up clustered image using morphological operations. It erodes and removes the small objects in the image. Figure 11 shows the result of the enhanced image and segmented ROIs.

Table 4 compares the results (using similarity index) of proposed approaches with other approaches applied on the same dataset. The compared approaches are region growing, wolf local thresholding, morphological operations, artificial bee colony and K-means.

4 Conclusion and Future Work

In this chapter, the main concepts of some of nature-inspired algorithms were presented such as grey wolf, artificial bee colony and antlion optimization. The liver segmentation problem was highlighted and solved using the nature-inspired algorithms. Also, we described the use of grey wolf, ABC and ALO optimization algorithms to solve the CT and MRI liver segmentation problems with different techniques. The experimental results showed the efficiency of the three algorithms. In the future work, we will apply more nature-inspired algorithms with different medical imaging applications.

References

1. Alomoush, W., Sheikh Abdullah, S.N., Sahran, S., Hussain, R.Q.: MRI l. J. Theor. Appl. Inf. Technol. **61** (2014)
2. Basturk, B., Karaboga, D.: An Artificial bee colony (ABC) algorithm for numeric function optimization. In: IEEE Swarm Intelligence Symposium 2006, Indianapolis, Indiana, USA, May 12-14, 2006
3. Cuevas, E., Sencin, F., Zaldivar, D., Prez-Cisneros, M., Sossa, H.: Applied Intelligence (2012). doi:10.1007/s10489-011-0330-z
4. Duraisamy, S.P., Kayalvizhi, R.: A new multilevel thresholding method using swarm intelligence algorithm for image segmentation. J. Intell. Learn. Syst. Appl. **2**, 126–138 (2010)
5. Jagadeesan, R.: An artificial fish swarm optimized fuzzy mri image segmentation approach for improving identification of brain tumour. Int. J. Comput. Sci. Eng. (IJCSE) **5**(7) (2013)
6. Jindal, S.: A systematic way for image segmentation based on bacteria foraging optimization technique (Its implementation and analysis for image segmentation). Int. J. Comput. Sci. Inf. Technol. **5**(1), 130–133 (2014)
7. Karaboga, D.: An Idea Based On Honey Bee Swarm For Numerical Optimization, Technical Report-TR06. Erciyes University, Engineering Faculty, Computer Engineering Department (2005)
8. Liang, Y., Yin, Y.: A new multilevel thresholding approach based on the ant colony system and the EM algorithm. Int. J. Innov. Comput. Inf. Control **9**(1) (2013)
9. Mirjalili, S.: The ant lion optimizer, advances in engineering software, pp. 80–98 (2015). doi:10.1016/j.advengsoft.2015.01.010
10. Mirjalili, S., Mirjalili, S.M., Lewis, A.: Grey wolf optimizer. Adv. Eng. Softw. **69**, 46–61 (2014)
11. Mostafaa, A., Fouad, A., Abd Elfattah, M., Hassanien, A.E., Hefny, H., Zhu, S.Y., Schaefer, G.: CT liver segmentation using artificial bee colony optimisation. In: 19th International Conference on Knowledge Based and Intelligent Information and Engineering Systems, Procedia Computer Science, vol. 60, pp. 1622–1630 (2015)
12. Mostafa, A., AbdElfattah, M., Fouad, A., Hassanien, A., Hefny, H.: Wolf local thresholding approach for liver image segmentation in ct images. In: International Afro-European Conference for Industrial Advancement AECIA, Addis Ababa, Ethiopia (2015)
13. Mostafa, A., Abd Elfattah, A., Fouad, A., Hassanien, A., Hefny, H.: Enhanced region growing segmentation for CT liver images. In: The 1st International Conference on Advanced Intelligent System and Informatics (AISI2015), Beni Suef, Egypt (2015)
14. Mostafa, A., Abd Elfattah, M., Fouad, A., Hassanien, A., Kim, T.: Region growing segmentation with iterative K-means for CT liver images. In: International Conference on Advanced Information Technology and Sensor Application (AITS), China (2015)
15. Mostafa, A., Fouad, A., Abd Elfattah, M., Ella Hassanien, A., Hefny, H., Zhue, S.Y., Schaeferf, G.: CT liver segmentation using artificial bee colony optimisation. In: 19th International Conference on Knowledge Based and Intelligent Information and Engineering Systems, Procedia Computer Science 60, Singapore, pp. 1622–1630 (2015)
16. Fouad, A.A., Mostafa, A., Ismail, S.G., Abd, E.M., Hassanien, A.: Nature Inspired Optimization Algorithms for CT Liver Segmentation. Medical Imaging in Clinical Applications:- Algorithmic and Computer-Based Approaches (2016). doi:10.1007/978-3-319-33793-7_19

17. Mostafa, A., Houseni, M., Allam, N., Hassanien, A.E., Hefny, H., Tsai, P.-W.: Antlion Optimization Based Segmentation for MRI Liver Images, International Conference on Genetic and Evolutionary Computing (ICGEC 2016), November 7–9, 2016, Fuzhou City, Fujian Province, China, pp. 265–272 (2016). doi:10.1007/978-3-319-48490-7.31
18. Sankari, L.: Image segmentation using glowworm swarm optimization for finding initial seed. Int. J. Sci. Res. (IJSR) **3**
19. Sivaramakrishnan, A., Karnan, M.: Medical image segmentation using firefly algorithm and enhanced bee colony optimization. In: International Conference on Information and Image Processing (ICIIP-2014), 316–321. Proceedings of the IEEE International Conference on Control and Automation, pp. 166–170 (2007)
20. Zidan, A., Ghali, N.I., Hassanien, A., Hefny, H.: Level set-based CT liver computer aided diagnosis system. Int. J. Imaging Robot. **9** (2013)

Thermal Image Segmentation Using Evolutionary Computation Techniques

Salvador Hinojosa, Gonzalo Pajares, Erik Cuevas and Noé Ortega-Sanchez

Abstract This chapter analyzes the performance of selected evolutionary computation techniques (ECT) applied to the segmentation of forward looking infrared (FLIR) images. FLIR images arise challenges for classical image processing techniques since the capture devices usually generate low-resolution images prone to noise and blurry outlines. Traditional ECTs such as artificial bee colony (ABC), differential evolution (DE), harmony search (HS), and the recently published flower pollination algorithm (FPA) are implemented and evaluated using as objective function the between-class variance (Otsu's method) and the Kapur's entropy. The comparison pays particular attention to the quality of the segmented image by evaluating three specific metrics named peak-to-signal noise ratio (PSNR), structural similarity index (SSIM), and feature similarity index (FSIM).

1 Introduction

The image processing community has focused its attention on the development of automatic vision systems considering the visible light spectrum. Nevertheless, there are some disadvantages when using this kind of images. The visibilities of the scene and colors are dependent on a light source. With every change of the light source, the collected image differs from the original scene leading to challenging compu-

S. Hinojosa (✉) · G. Pajares
Dpto. Ingeniería del Software e Inteligencia Artificial, Facultad Informática,
Universidad Complutense de Madrid, Madrid, Spain
e-mail: salvahin@ucm.es

G. Pajares
e-mail: pajares@ucm.es

E. Cuevas · N. Ortega-Sanchez
Departamento de Electrónica, Universidad de Guadalajara, CUCEI, Jalisco, Mexico
e-mail: erik.cuevas@cucei.udg.mx

N. Ortega-Sanchez
e-mail: noe.ortega@academicos.udg.mx

A.E. Hassanien and D.A. Oliva (eds.), *Advances in Soft Computing and Machine Learning in Image Processing*, Studies in Computational Intelligence 730,
https://doi.org/10.1007/978-3-319-63754-9_4

tational scenarios where the image processing might become arduous. In the worst case, the absence of a light source makes impossible to capture an image. A wider range of the electromagnetic spectrum can be exploited to overcome this issue. Specifically, infrared images are processed with classical computer vision techniques providing good results on applications of many fields such as agriculture [1], building inspection [2], detection and tracking of pedestrians [3], security [4], electronic system validation [5, 6], and health care [7, 8].

All objects with a temperature above absolute zero emit infrared radiation. This thermal radiation can be captured with specialized photon or thermal detectors. Objects emit radiation in the long wavelength and mid-wavelength of the infrared spectrum [9]. This radiation variates on the dominating wavelength and intensity according to the temperature. Thermal cameras take advantage of this property by generating images even with the lack of an illumination source [10]. A grayscale representation of the thermal image encodes high temperatures as bright pixels while objects with low temperature are represented as dark pixels.

Forward looking infrared (FLIR) images show bright objects where the thermal radiation is higher. From the computational point of view, FLIR images are identical to grayscale pictures. However, FLIR images present additional challenges since the resolution of the capturing sensors is usually lower than common cameras. Moreover, FLIR images typically present blurry edges as the heat radiates. The hazy contours difficult simple task like segmentation. Particular attention must be paid to the segmentation as it is a fairly common preprocessing technique.

The objective of image segmentation is to partition the image into homogeneous classes. Each class contains shared properties such as intensity or texture. The simplest case of image segmentation is the image thresholding (TH), where the intensity values of the histogram of the image are analyzed to determinate the limits of each class. On bi-level thresholding, a threshold value is calculated in order to partition the histogram into two classes. This technique was developed to extract an object from its background. For a more complex approach, multilevel thresholding (MTH) can identify a finite number of classes from the image. Segmentation methods based on threshold can be divided into parametric and nonparametric [11]. Parametric approaches estimate parameters of a probability density function to describe each class, but this approach is computationally expensive. By contrast, nonparametric approaches use criteria such as between-class variance, entropy, and error rate [12–14]. These criteria are optimized to find the optimal threshold value providing robust and accurate methods [15].

To reduce the computational time required by nonparametric techniques, multilevel thresholding is computed with evolutionary computation techniques (ECT). ECT comprises algorithms based on the interaction of collective behaviors, usually depending on a population of particles interchanging information to reach the global optima. These mechanisms provide excellent properties such as a gradient-free search and can avoid stagnation on local suboptimal.

The methodology analyzed in this chapter works through the utilization of several evolutionary computation techniques intended to optimize a particular criterion. Many approaches have been utilized such as [16, 17]. Every ECT is designed to work with a particular problem under certain circumstances. Thus, no single algorithm can solve all problems competitively [18]. Due to this fact, a proper performance comparison of ECT applied to FLIR image thresholding becomes relevant.

This paper is devoted to analyze the performance of image thresholding performed by ECTs applied to FLIR images. For this purpose, two segmentation criteria are selected: Otsu's [14] and Kapur's [12] methods. These methodologies are used as objective functions on selected ECTs. The comparison pays special attention to classic approaches such as artificial bee colony (ABC) [19], differential evolution (DE) [20], harmony search (HS) [21], and the recently proposed flower pollination algorithm (FPA) [22].

The remaining part of this chapter consists of the following: Sect. 2 describes the two thresholding techniques considered in this study; Sect. 3 briefly discusses the evolutionary computation techniques used for the implementation; Sect. 4 describes the experimental process and details implementation characteristics; Sect. 5 presents the results and provides information about the comparison. Finally, Sect. 6 summarizes and concludes the chapter.

2 Thresholding Techniques

In this work, several MTH algorithms are analyzed to determinate which technique is more suitable for segmenting FLIR images. The most common approaches will be shortly discussed in this section.

2.1 Otsu

The most popular thresholding technique was proposed by Otsu [14]. This unsupervised technique segments the image by maximizing the difference between various classes. The intensity value of each pixel of the FLIR image generates a probability distribution according to the following equation:

$$Ph_i = \frac{h_i}{NP}, \quad \sum_{i=1}^{NP} Ph_i = 1, \tag{1}$$

where i represents the intensity of such pixel $(0 \le i \le L-1)$, NP is the total amount of pixels present in the image, h is the histogram, and h_i denotes the number of occurrences of the intensity i. The histogram is associated with a probability distribution Ph_i. On the simplest scenario, the bi-level segmentation can be expressed as

$$C_1 = \frac{Ph_1}{\omega_0(th)}, \ldots, \frac{Ph_{th}}{\omega_0(th)} \quad \text{and} \quad C_2 = \frac{Ph_{th+1}}{\omega_1(th)}, \ldots, \frac{Ph_L}{\omega_1(th)}, \tag{2}$$

where $\omega_0(th)$ and $\omega_1(th)$ are probabilities distributions for C_1 and C_2:

$$\omega_0(th) = \sum_{i=1}^{th} Ph_i, \quad \omega_1(th) = \sum_{i=th+1}^{L} Ph_i. \tag{3}$$

The following classes denoted by μ_0 and μ_1 are calculated in Eq. (4):

$$\mu_0 = \sum_{i=1}^{th} \frac{iPh_i}{\omega_0(th)}, \quad \mu_1 = \sum_{i=th+1}^{L} \frac{iPh_i}{\omega_1(th)}. \tag{4}$$

Next, the variances σ_1 and σ_2 of C_1 and C_2, respectively, are calculated via

$$\sigma_1 = \omega_0(\mu_0 + \mu_T)^2, \quad \sigma_2 = \omega_1(\mu_1 + \mu_T)^2, \tag{5}$$

where $\mu_T = \omega_0\mu_0 + \omega_1\mu_1$ and $\omega_0 + \omega_1 = 1$. Finally, the Otsu's variance operator σ^2 can be calculated using Eq. (6). It must be noted that the number two is part of the operator and it is not an exponent:

$$\sigma^2 = \sigma_1 + \sigma_2. \tag{6}$$

From the optimization's perspective an objective function $f_{Otsu}(th)$ can be generated to maximize the Otsu's variance, where $\sigma^2(th)$ is the Otsu's variance for a given *th* value:

$$f_{Otsu}(th) = \max\left(\sigma^2(th)\right), \quad 0 \le th \le L-1. \tag{7}$$

For the multilevel approach, *nt* thresholds are necessary to divide the original image into *nt + 1* classes. This scenario involves the calculation of *nt* variances and the necessary elements. Thus, the objective function $f_{Otsu}(th)$ is rewritten considering various thresholds as

$$f_{Otsu}(\mathbf{th}) = \max(\sigma^2(\mathbf{th})), \quad 0 \le th_i \le L-1, \quad i = 1, 2, \ldots, nt \tag{8}$$

where $\mathbf{th} = [th_1, th_2, \ldots, th_{nt}]$ is a vector with the thresholds values. The variance equation is updated to work with several threshold values as follows:

$$\sigma^2 = \sum_{j=1}^{nt} \sigma_j = \sum_{j=1}^{nt} \omega_j\left(\mu_j - \mu_T\right)^2, \tag{9}$$

where j is the index of each of the classes to be thresholded, ω_i and μ_j are, respectively, the probability of occurrence and the mean of a class. The values ω_i can be computed for multilevel thresholding as follows:

$$\begin{aligned} \omega_0(th) &= \sum_{i=1}^{th_1} Ph_i \\ \omega_1(th) &= \sum_{i=th_1+1}^{th_2} Ph_i \\ &\vdots \\ \omega_{k-1}(th) &= \sum_{i=th_{nt}+1}^{L} Ph_i, \end{aligned} \tag{10}$$

while the mean equations are also updated to

$$\begin{aligned} \mu_0 &= \sum_{i=1}^{th_1} \frac{iPh_i}{\omega_0(th_1)}\mu_1 \\ \mu_1 &= \sum_{i=th_1+1}^{th_2} \frac{iPh_i}{\omega_0(th_2)} \\ &\vdots \\ \mu_{k-1} &= \sum_{i=th_{nt}+1}^{nt} \frac{iPh_i}{\omega_1(th_{nt})}. \end{aligned} \tag{11}$$

2.2 *Kapur*

An entropy-based method used to find optimal threshold values is the one presented by Kapur [12]. The method is based on the probability distribution of the image histogram and the entropy. Kapur's method searches for the optimal th that maximizes the overall entropy. In this case, when the optimal th value separates the classes, the entropy has the maximum value. For a bi-level example, an objective function can be defined as

$$f_{Kapur}(th) = H_1 + H_2, \tag{12}$$

where the entropies H_1 and H_2 are calculated as

$$H_1 = \sum_{i=1}^{th} \frac{Ph_i}{\omega_0} \ln\left(\frac{Ph_i}{\omega_0}\right), \quad H_2 = \sum_{i=th+1}^{L} \frac{Ph_i}{\omega_1} \ln\left(\frac{Ph_i}{\omega_0}\right). \tag{13}$$

The probability distribution Ph_i of the intensity levels is computed using

$$Ph_i = \frac{h_i}{NP}, \quad \sum_{i=1}^{NP} Ph_i = 1. \tag{14}$$

Also, the probability distribution of each class is determined by

$$\omega_0(th) = \sum_{i=1}^{th} Ph_i \quad \omega_1(th) = \sum_{i=th+1}^{th} Ph_i. \tag{15}$$

The ECT determines the optimal threshold th^* by maximizing the Kapur's entropy,

$$th^* = \arg \max_{th} f_{Kapur}(th). \tag{16}$$

Similarly to the Otsu's method, Kapur's approach can be extended to multiple threshold values. In such case, the image is divided into nt classes. With such conditions, the objective function is defined as th

$$f_{Kapur}(\mathbf{th}) = \sum_{i=1}^{nt} H_i, \tag{17}$$

where $\mathbf{th} = [th_1, th_2, \ldots, th_{nt}]$ is a vector containing multiple thresholds. Each entropy value is computed separately with its respective th value. Thus, Eq. 9 is rewritten for nt entropies:

$$\begin{aligned} H_1 &= \sum_{i=1}^{th_1} \frac{Ph_i}{\omega_0} \ln\left(\frac{Ph_i}{\omega_0}\right) \\ H_2 &= \sum_{i=th_1+1}^{th_2} \frac{Ph_i}{\omega_1} \ln\left(\frac{Ph_i}{\omega_1}\right) \\ &\vdots \qquad \vdots \\ H_k &= \sum_{i=th_k+1}^{th_2} \frac{Ph_i}{\omega_{nt-1}} \ln\left(\frac{Ph_i}{\omega_{nt-1}}\right). \end{aligned} \tag{18}$$

3 Evolutionary Computation Techniques

3.1 Artificial Bee Colony (ABC)

Proposed by Karaboga [19], the artificial bee colony (ABC) has become one of the most known ECT on the community. ABC is inspired by the intelligent foraging behavior of the honeybee swarm. In this technique, a population $\mathbf{L}^k(\{\mathbf{l}_1^k, \mathbf{l}_1^k, \ldots, \mathbf{l}_N^k\})$ of N food locations (particles) evolves from an initial point

($k = 0$) to a total number of generations ($k = gen$). Every location $\mathbf{l}_i^k$ encodes design variables of the optimization problem as a d-dimensional vector $\{l_{i,1}^k, l_{i,2}^k, \ldots, l_{i,d}^k\}$. The population is first initialized and then an objective function is evaluated to determine the fitness of each location. The optimization process continues following the movement rules (operators) of the honeybee types. The first operator generates a new food source $\mathbf{t}_i$ from the neighborhood of a food location $\mathbf{l}_i^k$ as

$$\mathbf{t}_i = \mathbf{l}_i^k + \phi\left(\mathbf{l}_i^k - \mathbf{l}_r^k\right), \quad i, r \in (1, 2, \ldots, N), \tag{19}$$

where $\mathbf{l}_i^k$ is selected randomly ($r \neq i$), and a scaling factor ϕ is drawn from a uniform distribution between [−1, 1]. After the generation of $\mathbf{t}_i$, it is evaluated to determine its fitness value fit ($\mathbf{l}_i^k$). The fitness value of a minimization problem is expressed as

$$\mathrm{fit}\left(\mathbf{l}_i^k\right) = \begin{cases} \frac{1}{1+f\left(\mathbf{l}_i^k\right)} & \text{if } f\left(\mathbf{l}_i^k\right) \geq 0 \\ \frac{1}{1+\mathrm{abs}\left(f\left(\mathbf{l}_i^k\right)\right)} & \text{if } f\left(\mathbf{l}_i^k\right) < 0 \end{cases} \tag{20}$$

where the objective function to be minimized is $f(\cdot)$. Following the calculation of the fitness values, the greedy selection operator determinates whether or not $\mathrm{fit}\left(\mathbf{t}_i^k\right)$ is better than $\mathrm{fit}\left(\mathbf{l}_i^k\right)$. If so, the food location $\mathbf{l}_i^k$ is discarded and replaced with the new food source $\mathbf{t}_i$.

3.2 *Differential Evolution (DE)*

Storn and Price [20] developed a stochastic vector-based ECT called differential evolution (DE). In this technique, a population is initialized containing vectors with feasible solutions. Each particle is defined as

$$\mathbf{x}_i^k = (x_{1,i}^k, x_{2,i}^k, \ldots, x_{N,i}^k), \; i = 1, 2, \ldots, N \tag{21}$$

where $\mathbf{x}_i^k$ is the i-th vector at the generation t. The algorithm involves three main steps: mutation, crossover, and selection.

The mutation process occurs to a given vector $\mathbf{x}_i^k$ at the k generation. For this purpose, three vectors are randomly selected from the population named $\mathbf{x}_p$, $\mathbf{x}_q$, and $\mathbf{x}_r$. The $\mathbf{x}_i^k$ vector is mutated using the following equation, leading to the creation of a new vector:

$$\mathbf{v}_i^{k+1} = \mathbf{x}_p^k + F(\mathbf{x}_q^k - \mathbf{x}_r^k), \tag{22}$$

where $F \in [0, 1]$ is a parameter called differential weight. The second operator called crossover is used over the population according to the crossover rate $C_r \in [0, 1]$.

This crossover starts with a uniformly distributed random number $r_i \in [0, 1]$ and the j-th component of $\mathbf{v}_i$ is manipulated as

$$\mathbf{u}_{j,i}^{k+1} = \begin{cases} \mathbf{v}_{j,i} & \text{if} \quad r_i \leq C_r \\ \mathbf{x}_{j,i}^{k} & \text{otherwise} \end{cases}, \quad j = 1, 2, \ldots, d, \quad i = 1, 2, \ldots, n. \tag{23}$$

Finally, the selection step is performed by comparing the fitness value of the candidate vector against the original:

$$\mathbf{x}_i^{t+1} = \begin{cases} \mathbf{u}_i^{t+1} & \text{if } f(\mathbf{u}_i^{t+1}) \leq f(\mathbf{x}_i^t) \\ \mathbf{x}_i^t & \text{otherwise} \end{cases}. \tag{24}$$

3.3 Harmony Search (HS)

The harmony search algorithm (HS) was introduced by Geem [21]. This particular ECT is inspired by the improvisation process of Jazz players. The population is called harmony memory $\mathbf{HM}^k(\{\mathbf{H}_1^k, \mathbf{H}_2^k, \ldots, \mathbf{H}_N^k\})$ of N particles (harmonies). The optimization process evolves the memory $\mathbf{HM}^k$ from a starting point of ($k = 0$) to the total of iterations ($k = gen$). Each harmony $\mathbf{H}_1^k$ represents a d-dimensional vector of decision variables of the problem to be optimized $\{H_{i,1}^k, H_{i,2}^k, \ldots, H_{i,N}^k\}$. HS works by generating new harmonies considering the harmony memory $\mathbf{HM}^k$. First, the initial memory is randomly generated. Then, a new candidate solution is generated using memory consideration with a pitch adjustment or a random re-initialization. The generation of a new harmony is called improvisation, and it relays on predefined parameters such as the harmony memory consideration rate (HMCR). Such parameter if is set too low only a few members of the memory will be considered in the exploration phase. In this step, the value of the first decision variable for the new harmony $H_{new,1}$ is selected randomly from the values of any harmony for the same decision variable present on the harmony memory. This process continues for all decision variables:

$$H_{new} = \begin{cases} H_j \in \{x_{1,j}, x_{2,j}, \ldots, x_{HMS,j}\} & \text{with probability HMCR} \\ \text{randomly generated} & \text{with probability } (1 - \text{PAR}) \end{cases}. \tag{25}$$

All the considered values of $H_{new,1}$ are perturbed with an operation called pitch adjustment. In this phase, pitch adjustment rate (PAR) is the frequency of the adjustment, and also a bandwidth factor (BW) controls how strong the perturbation is applied:

$$H_{new} = \begin{cases} H_{new} \pm \text{rand}(0,1) \cdot BW & \text{with probability PAR} \\ H_{new} & \text{with probability } (1 - \text{PAR}) \end{cases} \quad (26)$$

Finally, the $\mathbf{HM}^k$ is updated if the new candidate solution outperforms the worst harmony stored in the memory.

3.4 Flower Pollination Algorithm (FPA)

The flower pollination algorithm (FPA) proposed by Yang [22] is an ECT inspired by the pollination process of flowers. In FPA, individuals emulate a set of flowers or pollen gametes which behave based on biological laws of the pollination process. From a computational point of view, in the FPA operation, the population $\mathbf{F}^k(\{\mathbf{f}_1^k, \mathbf{f}_2^k, \ldots, \mathbf{f}_N^k\})$ of N individuals (flower positions) is evolved from the initial point ($k = 0$) to a total gen number of iterations ($k = \text{gen}$). Each flower $\mathbf{f}_i^k (i \in [1, \ldots, N])$ represents a d-dimensional vector $\{\mathbf{f}_{i,1}^k, \mathbf{f}_{i,2}^k, \ldots, \mathbf{f}_{i,N}^k\}$ where each dimension corresponds to a decision variable of the optimization problem to be solved. In FPA, a new population $\mathbf{F}^{k+1}$ is produced by considering two operators: local and global pollination. A probabilistic global pollination factor p is associated with such operators. To select which operator will be applied to each current flower position $\mathbf{f}_i^k$, a uniform random number r_p is generated within the range [0, 1]. If r_p is greater than p, the local pollination operator is applied to $\mathbf{f}_i^k$. Otherwise, the global pollination operator is considered.

Global Pollination Operator. Under this operator, the original position $\mathbf{f}_i^k$ is displaced to a new position $\mathbf{f}_i^{k+1}$ according to the following model:

$$\mathbf{f}_i^{k+1} = \mathbf{f}_i^k + s_i \cdot (\mathbf{f}_i^k - \mathbf{g}), \quad (27)$$

where $\mathbf{g}$ is the global best position seen so far, whereas s_i controls the length of the displacement. A symmetric Lévy distribution generates the value s_i according to Mantegna's algorithm [23]

$$s_i = \frac{\mathbf{u}}{|\mathbf{v}|^{1/\beta}}, \quad (28)$$

where $\mathbf{u}(\{u_1, \ldots, u_d\})$ and $\mathbf{v}(\{v_1, \ldots, v_d\})$ are n-dimensional vectors and $\beta = 3/2$. Each element of $\mathbf{u}$ and $\mathbf{v}$ is calculated by considering the following normal distributions:

$$u \sim N(0, \sigma_u^2), v \sim N(0, \sigma_v^2) \quad (29)$$

$$\sigma_u = \left(\frac{\Gamma(1+\beta) \cdot \sin(\pi \cdot \beta/2)}{\Gamma((1+\beta)/2) \cdot \beta \cdot 2^{(\beta-1)/2}} \right)^{1/\beta}, \quad \sigma_v = 1 \tag{30}$$

where $\Gamma(\cdot)$ represents the Gamma distribution.

Local Pollination Operator. In the local pollination operator, the current position $\mathbf{f}_i^k$ is perturbed to a new position $\mathbf{f}_i^{k+1}$ as follows:

$$\mathbf{f}_i^{k+1} = \mathbf{f}_i^k + \varepsilon \cdot \left(\mathbf{f}_j^k - \mathbf{f}_h^k \right); \quad i, j, h \in (1, 2, \ldots, N) \tag{31}$$

where $\mathbf{f}_j^k$ and $\mathbf{f}_h^k$ are two randomly chosen flower positions, satisfying the condition $i \neq j \neq h$. The scale factor ε is a random number between [−1, 1].

4 Implementation

4.1 Multilevel Thresholding

This chapter analyzes the performance of several ECT applied to the image thresholding problem, specifically to the segmentation of FLIR images. The following experiments are designed to maximize the objective function either of Otsu's method (Eq. 8)

$$\begin{array}{l} \arg\ \max_{\mathbf{th}} f_{Otsu}(\mathbf{th}) \\ \text{subject to } \mathbf{th} \in \mathbf{X} \end{array} \tag{32}$$

or Kapur's entropy (Eq. 18)

$$\begin{array}{l} \arg\ \max_{\mathbf{th}} f_{Kapur}(\mathbf{th}) \\ \text{subject to } \mathbf{th} \in \mathbf{X} \end{array}, \tag{33}$$

where $\mathbf{X} = \{th \in \mathbb{R}^{nt} | 0 \leq th_i \leq 255, i = 1, 2, \ldots, nt\}$ stands for the restrictions that bound the feasible region. Such restrictions are originated due to the characteristics of the problem since the intensity of the pixels lies in the interval 0–255.

The candidate thresholds *th* at each generation are encoded as decision variables on every element of the population as follows:

$$\mathbf{Sp}_t = [\mathbf{th}_1, \mathbf{th}_2, \ldots, \mathbf{th}_N], \mathbf{th}_i = [th_1, th_2, \ldots, th_{nt}]^T, \tag{34}$$

where t is the iteration number, N is the size of the population, T refers to the transpose operator, and nt is the number of thresholds applied to the image.

After the conclusion of the optimization process, the best threshold values found by the ECT are used to segment the image pixels. This chapter follows the traditional rule for segmenting:

$$\mathbf{I}_S(r,c) = \begin{cases} \mathbf{I}_{Gr}(r,c) & \text{if} \quad \mathbf{I}_{Gr}(r,c) \leq th_1 \\ th_1 & \text{if} \quad th_1 < \mathbf{I}_{Gr}(r,c) < th_2 \\ \mathbf{I}_{Gr}(r,c) & \text{if} \quad \mathbf{I}_{Gr}(r,c) > th_2 \end{cases}. \tag{35}$$

4.2 *Experimental Setup*

Since the objective of this chapter is to analyze the performance of MTH with ECT on FLIR images, a diverse benchmark set is assembled with images of different scenes and fields of study. All images are 480 × 480 JPEG and their specific properties are described in Table 1. The experiments were performed using Matlab 8.3 on an i5-4210 CPU @ 2.3Ghz with 6 GB of RAM.

For every evaluation of the algorithm, a 3000 generation stop criteria is established. The quality of the segmentation is evaluated using standard statistical methods and signal quality metrics. The specific parameters of each ECT are described on Table 2.

Table 1 Properties of benchmark image set

Image	Capture device	Source	Image
Cars	FLIR One	Captured on site	
Circuit	FLIR SC660	Multispectral image database [24]	
Crop	FLIR One	Captured on site	

(continued)

Table 1 (continued)

Image	Capture device	Source	Image
Medical	FLIR SC-620	Database for mastology research [25]	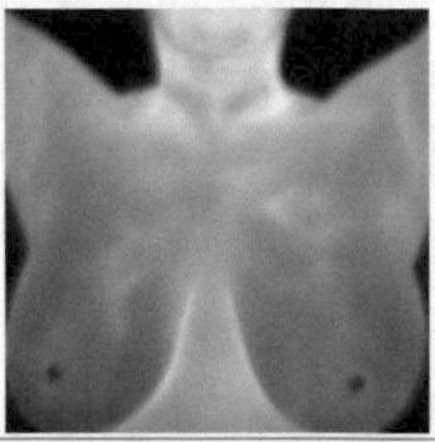
Office	FLIR One	Captured on site	
Tools	FLIR One	Captured on site	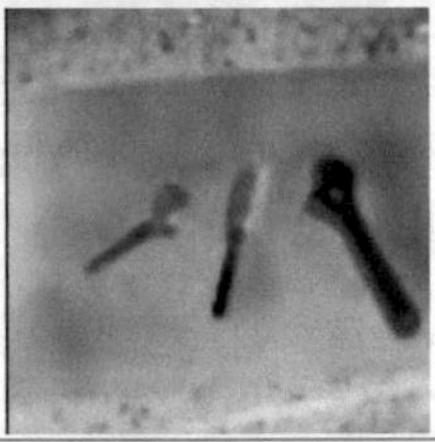
Pedestrian	Raytheon 300 D	OTCBVS Benchmark Dataset 01 [26]	

Table 2 Implementation parameters

Algorithm	Parameters
ABC	The algorithm has been implemented using the guidelines provided by its reference [19]
DE	The variant implemented is DE/rand/bin where $cr = 0.5$ and *differential weight* = 0.2 [20]
HS	The algorithm was implemented using the guidelines provided by its reference [21] with $HMCR = 0.95$ and $PAR = 0.3$
FPA	The algorithm remains as described in its proposal [22]

The first metric considered is the standard deviation (STD). This metric focuses on the stability of the algorithm; if the STD value decreases the algorithm becomes stable [27]:

$$STD = \sqrt{\sum_{i=1}^{Iter_{\max}} \frac{(\sigma_i - \mu)}{Ru}}. \quad (36)$$

The peak-to-signal ratio (PSNR) compares the similarity of the original image against the segmented. This metric is based on the mean square error (RMSE) of each pixel. Both PSNR and RMSE are defined as

$$PSNR = 20 \log_{10}\left(\frac{255}{RMSE}\right) \mathrm{dB} \quad (37)$$

$$RMSE = \sqrt{\frac{\sum_{i=1}^{ro} \sum_{j=1}^{co} \left(I_{Gr}(i,j) - I_{th}(i,j)\right)}{ro \times co}}, \quad (38)$$

where I_{Gr} is the original image and I_{th} is the segmented image; the total number of rows is ro, and the total number of columns is co.

The structure similarity index (SSIM) is used to compare the structures of the original image against the thresholded result [28], and it is defined in Eq. 39. A better segmentation performance produces a higher SSIM:

$$SSIM(I_{Gr}, I_{th}) = \frac{\left(2\mu_{I_{Gr}}\mu_{I_{th}} + C1\right)\left(2\sigma_{I_{Gr}I_{th}} + C2\right)}{\left(\mu_{I_{Gr}}^2 + \mu_{I_{th}}^2 + C1\right)\left(\sigma_{I_{Gr}}^2 + \sigma_{I_{th}}^2 + C1\right)} \quad (39)$$

$$\sigma_{I_0 I_{Gr}} = \frac{1}{N-1} \sum_{i=1}^{N} \left(I_{Gr_i} + \mu_{I_{Gr}}\right)\left(I_{th_i} + \mu_{I_{th}}\right), \quad (40)$$

where $\mu_{I_{Gr}}$ and $\mu_{I_{th}}$ are the mean value of the original and the thresholded image, respectively; for each image the values of $\sigma_{I_{Gr}}$ and $\sigma_{I_{th}}$ correspond to the standard deviation. C1 and C2 are constants used to avoid the instability when $\mu_{I_{Gr}}^2 + \mu_{I_{th}}^2 \approx 0$; experimentally in [29] both values are C1 = C2 = 0.065.

One last method used to quantify the quality of the thresholded image is the feature similarity index (FSIM) [30]. FSIM establishes the similarity between two images; in this case, the original gray scale image and the segmented image (Eq. 41). As on PSNR and SSIM, a higher value indicates a better performance of the evaluated methodology.

$$FSIM = \frac{\sum_{w \in \Omega} S_L(w) PC_m(w)}{\sum_{w \in \Omega} PC_m(w)}, \quad (41)$$

where Ω represents the entire domain of the image:

$$S_L(w) = S_{PC}(w)S_G(w) \tag{42}$$

$$S_{PC}(w) = \frac{2PC_1(w)PC_2(w) + T_1}{PC_1^2(w) + PC_2^2(w) + T_1} \tag{43}$$

$$S_G(w) = \frac{2G_1(w)G_2(w) + T_2}{G_1^2(w) + G_2^2(w) + T_2}. \tag{44}$$

G is the gradient magnitude (GM) of an image and is defined as

$$G = \sqrt{G_x^2 + G_y^2}. \tag{45}$$

PC is the phase congruence:

$$PC(w) = \frac{E(w)}{\left(\varepsilon + \sum_n A_n(w)\right)}. \tag{46}$$

The magnitude of the response vector in w on n is $E(w)$ and $A_n(w)$ is the local amplitude of scale n. ε is a small positive number and $PC_m(w) = \max(PC_1(w), PC_2(w))$.

5 Results

The values reported on each experiment from Tables 3, 4, 5, 6, 7, and 8 are the averaged results of 35 evaluations of each objective function with its respective ECT at every image and number of thresholds *nt*. Tables 3 and 6 report statistical data of the Otsu's method and Kapur's entropy, respectively. Tables 4 and 5 are partitioned for readability purposes. The same occurs to Tables 7 and 8. The best value of each experiment on every table is bolded.

5.1 Otsu's Results

Since the Otsu's method is a maximization process, the objective function is expected to be as high as possible. Table 3 shows the mean value and STD of the Otsu's objective function for all the analyzed algorithms. FPA outperforms ABC, DE, and HS in both metrics.

Table 3 Statistical results of ABC, DE, HS, and FPA using Otsu's method as objective function

Image	*nt*	ABC		DE		HS		FPA	
		Mean	Std	Mean	Std	Mean	Std	Mean	Std
Cars	2	1190.9202	0.3017	1191.2043	0.0707	1191.2496	0.0169	**1191.2578**	**0.0049**
	3	1614.8457	0.5955	1614.4202	0.9947	1615.4343	0.3444	**1615.8415**	**0.0339**
	4	1714.3754	0.9779	1713.7907	3.5836	1715.9867	1.0195	**1717.1537**	**0.0666**
	5	1766.6552	1.3444	1765.8426	2.6367	1768.0400	1.8868	**1771.1984**	**0.1200**
Circuit	2	1968.1636	0.1580	1968.4058	0.1277	1968.4245	0.0686	**1968.4741**	**0.0053**
	3	2258.5519	0.4958	2258.2495	0.4141	2258.8803	0.2944	**2259.1170**	**0.0330**
	4	2327.1105	6.1012	2343.6385	0.9983	2345.1314	0.5103	**2345.7375**	**0.0505**
	5	2386.8003	5.6086	2401.2700	1.7315	2404.0981	0.4784	**2405.0080**	**0.0945**
Crop	2	1538.5842	0.1754	1538.5873	0.2875	1538.6882	0.0420	**1538.7057**	**0.0047**
	3	1841.6442	0.4689	1841.4914	0.7492	1842.1634	0.3402	**1842.4936**	**0.0276**
	4	1961.1340	0.7381	1959.8387	1.5208	1962.0959	0.6112	**1962.8465**	**0.0572**
	5	2018.9002	1.3835	2017.1197	2.7514	2020.7496	1.2512	**2022.3693**	**0.0960**
Medical	2	1588.5637	0.0943	1588.7291	0.0383	1588.7470	0.0278	**1588.7605**	**0.0084**
	3	2073.2428	0.8432	2073.3506	0.7975	2073.9934	0.2724	**2074.3334**	**0.0294**
	4	2197.6308	2.2985	2199.1006	1.8829	2200.2549	1.4825	**2202.0307**	**0.0942**
	5	2247.3319	1.9340	2247.4678	1.8325	2247.7920	2.9289	**2251.6854**	**0.0826**
Office	2	1392.2126	0.3049	1392.6247	0.1388	1392.6854	0.0681	**1392.7315**	**0.0028**
	3	1673.6714	3.4990	1679.5073	0.6906	1680.1208	0.2712	**1680.4159**	**0.0302**
	4	1859.8373	1.1902	1859.4691	1.9370	1861.6973	1.0111	**1862.8328**	**0.0850**
	5	1933.3062	1.2189	1932.5815	3.5319	1936.4498	0.9864	**1938.0710**	**0.0878**
Tool	2	674.3096	0.2216	674.5791	0.1487	674.5937	0.2727	**674.7168**	**0.0071**
	3	1005.9789	0.8258	1006.1544	1.0812	1006.8794	0.2997	**1007.1833**	**0.0328**
	4	1102.8980	1.4988	1104.1747	1.0409	1104.7636	1.4774	**1106.5584**	**0.0471**
	5	1160.2180	1.7741	1158.2793	3.2629	1161.6290	1.5385	**1163.8411**	**0.1139**

(continued)

Table 3 (continued)

Image	*nt*	ABC		DE		HS		FPA	
		Mean	Std	Mean	Std	Mean	Std	Mean	Std
Pedestrian	2	111.9835	0.1632	112.1431	0.0789	112.1686	0.0735	**112.2064**	**0.0024**
	3	158.7331	0.6780	159.6061	0.2010	159.7849	0.1075	**159.8552**	**0.0058**
	4	188.4953	0.5837	188.5496	0.3837	189.0816	0.1678	**189.3136**	**0.0246**
	5	200.6837	0.7315	201.0701	0.6576	202.0394	0.1989	**202.3881**	**0.0285**

Table 4 Quality results of the segmented image for ABC and DE using Otsu's method

Image	*nt*	ABC			DE		
		PSNR	SSIM	FSIM	PSNR	SSIM	FSIM
Cars	2	**10.2954**	**0.2538**	0.6705	9.4639	0.1638	0.6662
	3	13.9399	0.4901	0.7046	**14.6155**	0.4963	**0.7144**
	4	15.2883	0.5149	0.7401	15.5772	0.5254	0.7493
	5	16.9750	0.5824	0.7495	17.3467	0.5761	0.7662
Circuit	2	13.6374	0.4528	0.7370	14.6589	0.4956	0.7367
	3	17.1852	0.5556	0.7701	18.0840	0.5839	**0.7769**
	4	18.8839	0.6250	0.7919	20.7318	0.6653	0.8036
	5	20.5025	0.6627	0.8051	20.8779	0.6671	0.8087
Crop	2	**12.0396**	**0.2526**	**0.5341**	11.6380	0.2201	0.5291
	3	14.2759	**0.3547**	0.6136	14.4982	0.3498	0.6286
	4	15.6913	0.4288	0.6673	16.7028	0.4539	0.7026
	5	17.9979	0.5247	0.7306	18.9015	0.5538	0.7669
Medical	2	13.3951	0.7284	0.8128	13.5220	0.7489	0.8186
	3	15.3872	0.7514	0.8157	16.2999	0.7680	0.8214
	4	16.2494	0.7524	0.8212	17.7156	0.7697	0.8211
	5	18.5999	0.7773	0.8211	19.8676	0.7771	0.8167
Office	2	10.7762	**0.2864**	**0.7355**	**10.8032**	0.2626	0.7332
	3	13.8166	0.4906	0.7400	**15.5397**	0.5631	**0.7348**
	4	15.9192	**0.6302**	**0.7559**	16.4900	0.5975	0.7466
	5	19.0038	0.7143	0.7791	20.2068	0.7352	0.7819
Tool	2	**9.9161**	**0.4581**	**0.6241**	9.4845	0.4369	0.6090
	3	15.0970	0.6379	0.6890	16.3156	0.6705	0.6899
	4	17.6627	0.6601	0.7173	18.5422	0.6669	0.7237
	5	20.4009	0.7310	0.7827	**21.7611**	0.7301	0.7953
Pedestrian	2	**13.5776**	**0.2426**	0.5700	13.2081	0.2350	**0.5772**
	3	14.0902	0.2733	0.5933	14.2089	0.2862	0.6031
	4	**18.1878**	**0.4948**	0.6706	17.4529	0.4682	0.6699
	5	21.0671	0.6264	0.7408	21.7515	0.6710	0.7406

Even though FPA surpasses other algorithms maximizing the objective function and providing consistent results, the quality indicators of the segmented image do not reflect this behavior. Tables 4 and 5 show the PSNR FSIM and SSIM quality metrics. These metrics show a disperse behavior; HS is the algorithm with the higher amount of best experiment values and it is closely followed by FPA. The ECT which accumulated less best experiment values is DE.

Table 5 Quality results of the segmented image for HS and FPA using Otsu's method

Image	*nt*	HS			FPA		
		PSNR	SSIM	FSIM	PSNR	SSIM	FSIM
Cars	2	9.5091	0.1663	0.6705	9.4615	0.1648	**0.6673**
	3	14.6123	0.4951	0.7130	14.6126	**0.4949**	0.7126
	4	**15.6444**	**0.5252**	0.7502	15.6022	0.5253	**0.7511**
	5	**17.8162**	**0.5930**	0.7631	17.3501	0.5765	**0.7667**
Circuit	2	14.9888	0.5036	0.7464	**15.0464**	**0.5097**	**0.7507**
	3	**18.1943**	**0.5848**	0.7782	18.1592	0.5825	0.7761
	4	**20.7454**	**0.6671**	**0.8049**	20.6473	0.6636	0.8016
	5	**20.9371**	**0.6672**	0.8103	20.8335	0.6661	0.8065
Crop	2	11.7740	0.2242	0.5403	11.6126	0.2191	0.5266
	3	14.4809	0.3491	0.6259	**14.5093**	0.3516	**0.6296**
	4	16.7014	0.4547	0.7024	**16.7147**	**0.4559**	**0.7034**
	5	**19.0409**	**0.5598**	**0.7695**	18.9596	0.5569	0.7690
Medical	2	14.3094	**0.7585**	0.8188	**14.4381**	0.7582	**0.8189**
	3	16.6016	0.7645	0.8197	**16.3592**	**0.7688**	**0.8216**
	4	**17.7807**	0.7685	**0.8215**	17.7072	**0.7704**	**0.8215**
	5	**20.1848**	**0.7818**	**0.8202**	19.7926	0.7773	0.8171
Office	2	10.7984	0.2623	0.7327	10.7272	0.2631	0.7329
	3	15.5333	0.5630	0.7339	15.4713	**0.5634**	0.7347
	4	**16.4959**	0.5971	0.7463	16.4532	0.5975	0.7452
	5	**20.3213**	**0.7387**	**0.7836**	20.1876	0.7382	0.7820
Tool	2	9.5028	0.4400	0.6114	9.4829	0.4384	0.6095
	3	**16.3974**	0.6691	**0.6916**	16.3915	**0.6698**	**0.6916**
	4	**18.5818**	0.6650	**0.7253**	18.5346	**0.6680**	0.7229
	5	21.7453	**0.7307**	**0.7959**	21.7444	**0.7307**	0.7936
Pedestrian	2	13.2043	0.2339	0.5761	13.2042	0.2341	0.5766
	3	**14.2107**	**0.2864**	**0.6038**	14.2103	0.2863	0.6037
	4	17.4041	0.4660	**0.6702**	17.4016	0.4658	0.6694
	5	**21.8061**	**0.6727**	**0.7442**	21.6591	0.6675	0.7385

5.2 Kapur's Entropy Results

Table 6 presents statistical data of the Kapur's method applied to the selected ECT. The objective function also maximizes the Kapur's entropy of each image for every experiment. As a result, a higher value of the mean of the objective function indicates a better segmentation. Contrary to Otsu's method, ABC outperforms at most of the experiments.

Table 6 Statistical results of ABC, DE, HS, and FPA using Kapur's entropy as objective function

Image	*nt*	ABC		DE		HS		FPA	
		Mean	Std	Mean	Std	Mean	Std	Mean	Std
Cars	2	**17.6758**	**0.0014**	17.6628	0.0087	17.6644	0.0149	17.6194	0.0346
	3	**22.1336**	**0.0056**	22.0928	0.0171	22.1090	0.0270	22.0250	0.0568
	4	**26.1868**	0.0827	25.9645	0.0740	26.0083	**0.0549**	25.9566	0.1526
	5	**30.0658**	0.0442	29.8316	0.1262	29.8338	**0.0856**	29.7889	0.1707
Circuit	2	**17.3273**	**0.0018**	17.2930	0.0137	17.3012	0.0183	17.2077	0.0587
	3	**21.6443**	**0.0298**	21.5546	0.0600	21.5241	0.0475	21.4583	0.1298
	4	**25.8571**	**0.0185**	25.7014	0.0641	25.7563	0.0677	25.6115	0.1012
	5	**29.6390**	**0.0270**	29.4349	0.0628	29.5821	0.0446	29.2699	0.2038
Crop	2	**17.8285**	**0.0012**	17.8174	0.0068	17.8133	0.0091	17.7597	0.0532
	3	**22.2342**	**0.0036**	22.2152	0.0099	22.2112	0.0197	22.1498	0.0468
	4	**26.3697**	**0.0137**	26.3130	0.0212	26.3055	0.0459	26.2505	0.0729
	5	**30.4170**	**0.0697**	30.1759	0.0888	30.2176	0.0449	30.1592	0.1782
Medical	2	**17.5238**	**0.0019**	17.5033	0.0087	17.4839	0.0304	17.4490	0.0478
	3	**21.8562**	**0.0023**	21.8220	0.0156	21.8314	0.0146	21.7630	0.0612
	4	**26.2693**	**0.0063**	26.1912	0.0293	26.2363	0.0201	26.0938	0.0811
	5	**30.2272**	**0.0210**	30.0929	0.0569	30.1590	0.0939	30.0474	0.0720
Office	2	**17.3833**	0.0020	17.3661	0.0085	17.3682	0.0160	17.2859	0.0652
	3	**21.7850**	0.0038	21.7344	0.0176	21.7633	0.0194	21.6500	0.0687
	4	**25.9259**	0.0900	25.6474	0.1458	25.6545	0.1159	25.5330	0.1824
	5	**29.7348**	0.0513	29.4132	0.1638	29.4566	0.1260	29.3374	0.1791

(continued)

Table 6 (continued)

Image	*nt*	ABC		DE		HS		FPA	
		Mean	Std	Mean	Std	Mean	Std	Mean	Std
Tool	2	**17.3930**	**0.0033**	17.3808	0.0075	17.3600	0.0464	17.3274	0.0447
	3	**21.7232**	**0.0095**	21.6974	0.0134	21.6217	0.0766	21.6505	0.0592
	4	**25.8280**	**0.0278**	25.7369	0.0270	25.7745	0.0338	25.6829	0.0759
	5	**29.8829**	**0.0537**	29.6592	0.0945	29.6438	0.1062	29.5880	0.1327
Pedestrian	2	**16.8579**	**0.0052**	16.8318	0.0138	16.8102	0.0257	16.7485	0.0539
	3	**21.4753**	**0.0060**	21.3942	0.0315	21.4241	0.0248	21.2526	0.1067
	4	**25.6262**	**0.0172**	25.4739	0.0451	25.4842	0.0806	25.3198	0.1225
	5	**29.2976**	**0.0168**	29.0905	0.0555	29.1945	0.0738	28.9653	0.1283

Table 7 Quality results of the segmented image for ABC and DE using Kapur's entropy

Image	*nt*	ABC			DE		
		PSNR	SSIM	FSIM	PSNR	SSIM	FSIM
Cars	2	13.2778	0.4617	0.6900	14.1686	0.4736	0.7007
	3	14.6202	**0.5192**	0.7156	**15.1789**	0.5082	**0.7323**
	4	15.9592	0.5863	0.7220	**17.6729**	**0.5936**	**0.7527**
	5	17.7815	**0.6542**	0.7437	**18.6284**	0.6230	**0.7658**
Circuit	2	**14.8876**	**0.5145**	0.7488	14.6412	0.4987	**0.7505**
	3	15.0393	**0.5502**	0.7581	15.5329	0.5251	0.7627
	4	17.5823	**0.6173**	0.7825	17.8950	0.5958	0.7859
	5	19.0199	0.6498	0.7942	**20.8381**	0.6563	0.8144
Crop	2	12.6244	**0.2927**	0.5547	13.1440	0.2785	**0.5822**
	3	**15.8271**	**0.4413**	0.6381	15.3553	0.3836	**0.6638**
	4	15.7842	0.4608	0.6398	17.9957	0.5012	0.7209
	5	18.2075	0.5553	0.7090	**20.0144**	**0.5871**	**0.7646**
Medical	2	11.2497	0.6328	**0.8083**	**14.4628**	**0.7255**	0.7849
	3	14.8435	0.7613	**0.8220**	**15.4723**	0.7694	0.8175
	4	17.2505	0.7833	0.8285	19.2433	**0.7979**	0.8458
	5	17.8286	0.7910	0.8352	**20.7799**	0.8104	0.8513
Office	2	**12.3584**	**0.4383**	**0.7375**	12.3439	0.3739	0.7204
	3	15.6265	0.6365	0.7568	17.8563	0.7161	0.7620
	4	16.8774	0.6918	**0.7681**	**18.3948**	0.7223	0.7677
	5	17.6706	0.7116	0.7756	19.6462	0.7351	0.7774
Tool	2	14.1762	0.6414	0.6905	**16.5409**	**0.6671**	**0.6814**
	3	15.8012	0.6834	0.7086	17.9838	0.6990	0.7114
	4	17.6140	0.7002	0.7299	**18.7458**	**0.7173**	**0.7433**
	5	16.4082	0.6975	0.7340	**20.0630**	0.7224	**0.7641**
Pedestrian	2	**14.7716**	**0.2837**	**0.5838**	10.1687	0.0149	0.5534
	3	17.3160	0.4410	0.6171	**17.5064**	**0.4588**	**0.6222**
	4	**18.3683**	**0.5001**	0.6617	17.5323	0.4609	0.6271
	5	18.1330	0.4842	0.6761	**19.9934**	**0.5994**	**0.7582**

Besides, Tables 7 and 8 report quality metrics with no clear winner. The algorithm that tops most of the experiments is HS. ABC and DE show a similar amount of best values and FPA has the least number of best values.

Table 9 shows a qualitative comparison of the proposed approaches with three threshold values ($nt = 3$) as an example. The difference on the presented images is not easily spotted by the naked eye. All four approaches present similar results that can only be quantified by the corresponding metrics.

Table 8 Quality results of the segmented image for HS and FPA using Kapur's entropy

Image	*nt*	HS			FPA		
		PSNR	SSIM	FSIM	PSNR	SSIM	FSIM
Cars	2	**14.1711**	**0.4738**	**0.7008**	14.1686	0.4736	0.7007
	3	15.1759	0.5080	0.7322	15.1772	0.5080	0.7323
	4	16.3196	0.5408	0.7385	15.6756	0.5284	0.7321
	5	18.0457	0.6088	0.7602	17.5270	0.5848	0.7559
Circuit	2	14.6538	0.4981	0.7501	14.6429	0.4988	**0.7505**
	3	**16.2929**	0.5359	**0.7651**	14.6147	0.4998	0.7562
	4	17.8598	0.5978	0.7850	**17.9018**	0.5976	0.7854
	5	20.6348	**0.6642**	0.8163	20.6868	0.6635	0.8154
Crop	2	**13.1881**	0.2808	0.5819	13.1440	0.2785	0.5822
	3	15.8010	0.4031	0.6625	15.3606	0.3839	**0.6638**
	4	**18.5330**	**0.5258**	**0.7276**	17.3676	0.4727	0.7061
	5	18.0291	0.5098	0.7422	18.5830	0.5293	0.7338
Medical	2	**14.4628**	**0.7255**	0.7849	**14.4628**	**0.7255**	0.7849
	3	15.3062	**0.7702**	0.8183	15.2861	0.7701	0.8184
	4	19.2404	**0.7979**	0.8458	**19.2474**	0.7982	0.8457
	5	20.5062	**0.8150**	**0.8518**	20.6473	0.8114	**0.8518**
Office	2	12.3439	0.3739	0.7204	12.3439	0.3739	0.7204
	3	17.8720	**0.7230**	**0.7638**	**17.8736**	0.7178	0.7624
	4	18.1053	**0.7312**	0.7678	17.9242	0.7171	0.7652
	5	**19.6923**	**0.7717**	**0.7830**	18.9943	0.7316	0.7722
Tool	2	16.5230	0.6875	0.6958	**16.5409**	**0.6671**	**0.6814**
	3	**18.0815**	**0.7011**	**0.7162**	17.9719	0.6992	0.7114
	4	18.7427	0.7132	0.7371	18.6826	0.7083	0.7313
	5	19.8386	**0.7241**	0.7627	19.8999	0.7196	0.7553
Pedestrian	2	10.1686	0.0149	0.5534	10.1687	0.0149	0.5534
	3	17.5060	0.4587	**0.6222**	17.2622	0.4440	0.6202
	4	17.5389	0.4609	0.6268	17.5639	0.4618	0.6273
	5	19.8776	0.5932	0.7526	19.9250	0.5950	0.7540

Table 9 Qualitative using Kapur's entropy

ABC	DE	HS	FPA

6 Summary

This chapter analyzes the thresholding problem on FLIR images. For this purpose, two classical thresholding techniques named Otsu's method and Kapur's entropy are implemented. Both methodologies are used as objective functions on selected ECTs. ABC, DE, HS, and FPA are used to maximize each of the objective functions to partition the image into various classes. The performance comparison includes two approaches; the first compares the fitness values to determine which algorithm is best fitted to optimize the analyzed objective function by comparing average fitness and average stability, while the second approach focuses on the segmented image quality.

Considering only the fitness values obtained for Otsu's method, FPA proves to be the best-suited option from the algorithms considered in this evaluation. For Kapur's entropy, ABC outperforms the other algorithms. Nevertheless, the results indicate that a good fitness function value does not necessarily reflect on a high quality of the segmented image according to the PSNR FSIM and SSIM metrics. This gap between both results might be a consequence of the noisy nature of FLIR images and the smooth transition between the intensity values of the classes. Such conditions make FLIR image processing a challenging task.

According to the quality metrics of the segmented images, HS showed better performance in the majority of experiments for both Otsu and Kapur. The HS algorithm provided excellent results regarding quality and required less computational time to perform the optimization than ABC and FPA making it an attractive alternative for FLIR image thresholding.

Despite that the results did not show an absolute winner for all tests and images, this chapter provides useful information about the behavior of the analyzed ECT applied to Otsu and Kapur for the MTH problem with the challenging characteristics of FLIR images.

References

1. Vadivambal, R., Jayas, D.S.: Applications of Thermal Imaging in Agriculture and Food Industry-A Review. Food Bioprocess Technol. **4**, 186–199 (2011). doi:10.1007/s11947-010-0333-5
2. Al-Kassir, A.R., Fernandez, J., Tinaut, F.V., Castro, F.: Thermographic study of energetic installations. Appl. Therm. Eng. **25**, 183–190 (2005). doi:10.1016/j.applthermaleng.2004.06.013
3. Leykin, A., Hammoud, R.: Pedestrian tracking by fusion of thermal-visible surveillance videos. Mach. Vis. Appl. **21**, 587–595 (2010). doi:10.1007/s00138-008-0176-5
4. Wei, W., Xia, R., Xiang, W., Hui, B., Chang, Z., Liu, Y., Zhang, Y.: Recognition of Airport Runways in FLIR Images Based on Knowledge. IEEE Geosci. Remote Sens. Lett. **11**, 1534–1538 (2014). doi:10.1109/LGRS.2014.2299898
5. Al-Obaidy, F., Yazdani, F., Mohammadi, F.A.: Intelligent testing for Arduino UNO based on thermal image. Comput. Electr. Eng. **58**, 88–100 (2017). doi:10.1016/j.compeleceng.2017.01.014

6. Pitarma, R., Crisóstomo, J., Jorge, L.: Analysis of Materials Emissivity Based on Image Software. Springer, Cham, pp. 749–757 (2016). doi:10.1007/978-3-319-31232-3_70
7. Ring, E.F.J., Ammer, K., Ring, E.F.J.: Infrared thermal imaging in medicine. Physiol. Meas. **33**, R33 (2012). doi:10.1088/0967-3334/33/3/R33
8. Mehra, M., Bagri, A., Jiang, X., Ortiz, J.: Image analysis for identifying mosquito breeding grounds. In: 2016 IEEE International Conference on Sensing Communication and Networking SECON Work. IEEE, pp. 1–6 (2016). doi:10.1109/SECONW.2016.7746808
9. Gade, R., Moeslund, T.B.: Thermal cameras and applications: A survey. Mach. Vis. Appl. **25**, 245–262 (2014). doi:10.1007/s00138-013-0570-5
10. Vollmer, M., Möllmann, K.-P.: Wiley InterScience (Online service), Infrared Thermal Imaging : Fundamentals, Research and Applications. Wiley-VCH (2010)
11. Akay, B.: A study on particle swarm optimization and artificial bee colony algorithms for multilevel thresholding. Appl. Soft Comput. **13**, 3066–3091 (2013). doi:10.1016/j.asoc.2012.03.072
12. Kapur, J.N., Sahoo, P.K., Wong, A.K.C.: A new method for gray-level picture thresholding using the entropy of the histogram. Comput. Vis. Gr. Image Process. **29**, 273–285 (1985). doi:10.1016/0734-189X(85)90125-2
13. Kittler, J., Illingworth, J.: Minimum error thresholding. Pattern Recognit. **19**, 41–47 (1986). doi:10.1016/0031-3203(86)90030-0
14. Otsu, N.: Threshold selection method from gray-level histograms. IEEE Trans. Syst. Man Cybern. **SMC-9**, 62–66 (1979). http://www.scopus.com/inward/record.url?eid=2-s2.0-0018306059&partnerID=tZOtx3y1
15. Sankur, B.: Survey over image thresholding techniques and quantitative performance evaluation. J. Electron. Imaging **13**, 146 (2004). doi:10.1117/1.1631315
16. Yin, P.-Y.: Multilevel minimum cross entropy threshold selection based on particle swarm optimization. Appl. Math. Comput. **184**, 503–513 (2007). doi:10.1016/j.amc.2006.06.057
17. Horng, M.-H., Liou, R.-J.: Multilevel minimum cross entropy threshold selection based on the firefly algorithm. Expert Syst. Appl. **38**, 14805–14811 (2011). doi:10.1016/j.eswa.2011.05.069
18. Wolpert, D.H., Macready, W.G.: No free lunch theorems for optimization. IEEE Trans. Evol. Comput. **1**, 67–82 (1997). doi:10.1109/4235.585893
19. Karaboga, D., Basturk, B.: A powerful and efficient algorithm for numerical function optimization: artificial bee colony (ABC) algorithm. J. Glob. Optim. **39**, 459–471 (2007). doi:10.1007/s10898-007-9149-x
20. Storn, R., Price, K.: Differential evolution—a simple and efficient heuristic for global optimization over continuous spaces. J. Glob. Optim. **11**(n.d.), 341–359. doi:10.1023/A:1008202821328
21. Loganathan, G.V.V., Geem, Z.W., Kim, J.H., Loganathan, G.V.V.: A new heuristic optimization algorithm: harmony search. Simulation. **76**, 60–68 (2001). doi:10.1177/003754970107600201
22. Yang, X.S.: Flower pollination algorithm for global optimization. Lecture Notes Computer Science (Including Subser. Lecture Notes Artificial Intelligence Lecture Notes Bioinformatics), vol. 7445 pp. 240–249, LNCS (2012). doi:10.1007/978-3-642-32894-7_27
23. Mantegna, R.: Fast, accurate algorithm for numerical simulation of Levy stable stochastic processes. Phys. Rev. E. (1994). http://journals.aps.org/pre/abstract/10.1103/PhysRevE.49.4677 20 Oct 2015
24. Zukal, M., Mekyska, J., Cika, P., Smekal, Z.: Interest points as a focus measure in multi-spectral imaging. Radioengineering. **22**, 68–81 (2013). doi:10.1109/TSP.2012.6256402
25. Silva, A., Saade, L.F., Sequeiros, D.C.M., Silva, G.O., Paiva, A.C., Bravo, A.C., Conci, R.S.: A new database for breast research with infrared image. J. Med. Imaging Heal. Inform. **4**, 92–100 (2014)
26. Davis, J., Keck, M.: A two-stage approach to person detection in thermal imagery. IEEE Work. Appl. Comput. Vis. (2005)

27. Ghamisi, P., Couceiro, M.S., Benediktsson, J.A., Ferreira, N.M.F.: An efficient method for segmentation of images based on fractional calculus and natural selection. Expert Syst. Appl. **39**, 12407–12417 (2012). doi:10.1016/j.eswa.2012.04.078
28. Wang, Z., Bovik, A.C., Sheikh, H.R., Simoncelli, E.P.: Image quality assessment: from error visibility to structural similarity. IEEE Trans. Image Process. **13**, 600–612 (2004). doi:10.1109/TIP.2003.819861
29. Agrawal, S., Panda, R., Bhuyan, S., Panigrahi, B.K.: Tsallis entropy based optimal multilevel thresholding using cuckoo search algorithm. Swarm Evol. Comput. **11**, 16–30 (2013). doi:10.1016/j.swevo.2013.02.001
30. Zhang, D.: FSIM: a feature similarity index for image quality assessment. IEEE Trans. Image Process. **20**, 2378–2386 (2011). doi:10.1109/TIP.2011.2109730

News Videos Segmentation Using Dominant Colors Representation

Ibrahim A. Zedan, Khaled M. Elsayed and Eid Emary

Abstract In this chapter, we propose a new representation of images. We called that representation as "Dominant Colors". We defined the dissimilarity of two images as a vector contains the difference in order of each dominant color between the two image representations. Our new image representation and dissimilarity measure are utilized to segment the news videos by detecting the abrupt cuts. A neural network trained with our new dissimilarity measure to classify between two classes of news videos frames: cut frames and non-cut frames. Our proposed system tested in real news videos from different TV channels. Experimental results show the effectiveness of our new image representation and dissimilarity measure to describe the images and segment the news videos.

Keywords News videos • Video segmentation • Abrupt cut • Dominant colors

1 Introduction

The amazing increase in video data on the internet leads to an urgent need of research work to develop efficient procedures for summarization, indexing, and retrieval of this video data. Earlier video indexing methodologies utilize people to manually annotate videos with textual keywords. This methodology is tedious, impeded by a lot of human subjectivity [1]. As shown in Fig. 1, content-based video indexing methodologies are categorized into two approaches: the query by example approach and the query by keywords approach. In the query by example approach, the video is segmented into shots and key frames are extracted for each shot. The user input is a

I.A. Zedan (✉) · K.M. Elsayed · E. Emary
Faculty of Computers & Information, Cairo University, Cairo, Egypt
e-mail: i.zedan@fci-cu.edu.eg

K.M. Elsayed
e-mail: k.mostafa@fci-cu.edu.eg

E. Emary
e-mail: e.emary@fci-cu.edu.eg

A.E. Hassanien and D.A. Oliva (eds.), *Advances in Soft Computing and Machine Learning in Image Processing*, Studies in Computational Intelligence 730,
https://doi.org/10.1007/978-3-319-63754-9_5

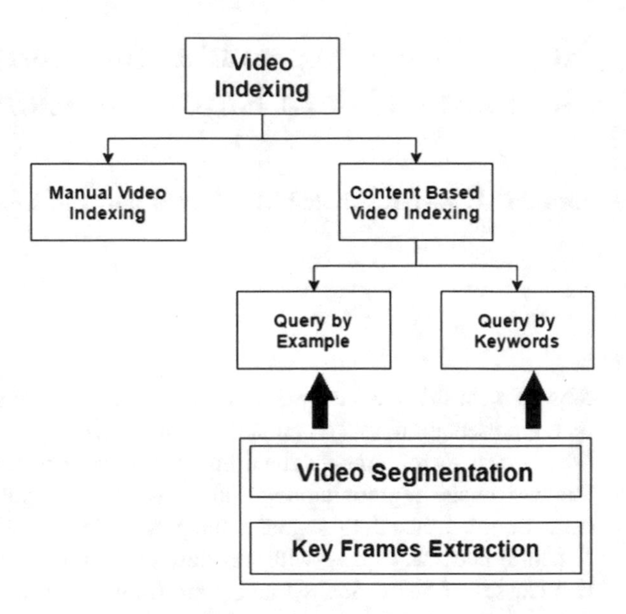

Fig. 1 Video indexing methodologies

query image and the retrieval system searches for similar key frames. In the query by keywords approach, the video is segmented into shots and each shot is assigned to some keywords. The user presents a semantic query and the retrieval system returns the shots that match the query keywords [2]. From this point of view, we can see that video segmentation and key frame extraction are essential steps for the organization of huge video data. Video segmentation segments a video to shots by identifying the boundaries between camera shots. Key frame extraction has been perceived as an important research issue in video information retrieval [1].

Among all video data the news videos have gained a special importance as many individuals are concerned about news video. News videos are closely related to us as it gives us the main news in the world recently. News video is a special type of video that contains a set of semantically independent stories. News story is defined as an arrangement of shots with a coherent focus which contain no less than two independent declarative clauses. The appearance of the anchor person is an example of the declarative clauses that surround and define the news story. Also, the news separator is another example of the declarative clauses. Story segmentation in news videos is usually based on video shot segmentation that followed by a shots clustering process. It is highly needed to extract key frames of news video in order to know the main content of news and avoid spending a lot of time to watch every detail of the video. News video should be segmented into shots in order to extract the key frames [3].

News video has relatively fixed structure. As shown in Fig. 2, the shots of news video are categorized into several classes [2, 3]. The introduction and the ending sequences are usually contain graphics. Separator is a sequence of frames inserted to separate between different news stories and its pattern is similar to the beginning

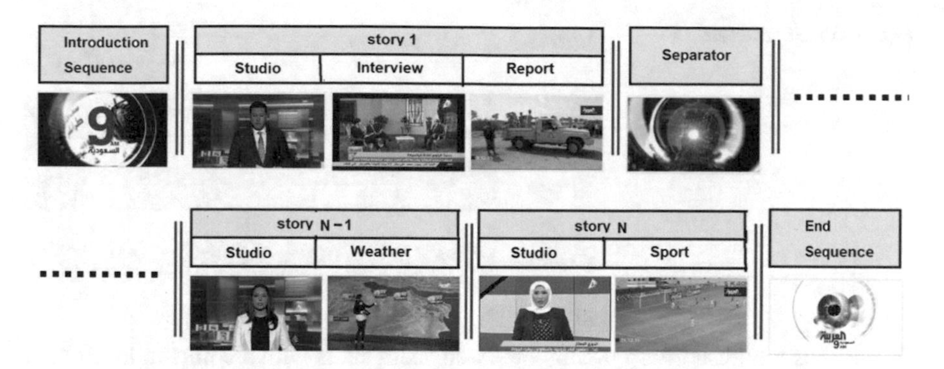

Fig. 2 News videos structure

and the ending sequences. The appearance of the anchor person shots is similar among a large portion of the TV channels. Interview shots are a special type of studio shots where an interview between the anchor person and the interviewed person is taking place. Weather forecast shots are studio shots showing a weather map that generally contains animated symbols that represent the weather conditions. Sports shots contain a lot of motion-like football matches. Report shots are recorded outside the studio.

In this chapter we introduce a new image representation and dissimilarity measure. To prove the effectiveness of our new image descriptor, we applied it to the segmentation problem in news videos as video segmentation is normally the first and essential step for content-based video summarization and retrieval. We focused on the detection of abrupt cuts as it is the most common shot boundaries.

The rest of the chapter is organized as follows; an overview of the video segmentation foundation, challenges, and methods is given in Sect. 2. A detailed description of the proposed system is given in Sect. 3. Experimental results are shown in Sect. 4. Section 5 presents our conclusion and future work.

2 Video Segmentation

Video segmentation is the initial step of video content analysis, indexing, and retrieval. Video segmentation is the method of automatically identifying the shots transitions inside a video [4]. Video segmentation is a vital step for video annotation [5] as the common procedure of the video annotation is to temporally segment the video into shots, extracting the key frames from the segmented video shots, and the textual annotations are formed by analyzing these key frames. Video shot is a series of consecutive interrelated frames representing continuous activities in time and space and recorded contiguously by a single camera [6, 7]. As shown in Fig. 3, video is segmented into scenes and each scene can be segmented into shots [8].

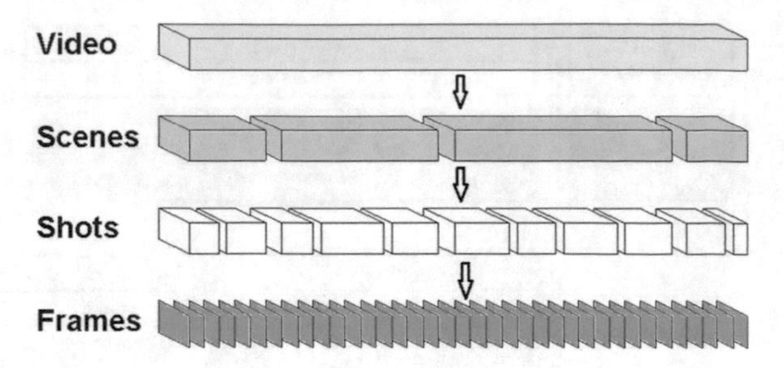

Fig. 3 Video structure

Scene is a logical group of temporally adjacent shots with a common location, a common object of interest, or a common thematic concept. Video scenes are independent in nature but we cannot say that video shots are independent. Sometimes, a transition from one video shot to another video shot is just a change in camera angle or switch from different cameras that photo the same scene [5] as shown in Fig. 4. Scene segmentation is usually based on video shot segmentation that followed by a shots clustering process.

2.1 Types of Shot Transitions

As shown in Fig. 5, two shots can be combined with a gradual transition or an abrupt cut. The abrupt cut is the simplest transition between the shots and is defined as an instantaneous transition from one shot to the next shot [6]. Figure 6 shows an example of abrupt cut. Detection of cuts is very useful as it is significantly more common than gradual transitions. In [9] detection of cuts over ranges of frames is utilized to select key frames for the summarization purpose.

The gradual transition is a more complex transition that artificially combines two shots by using special effects like fade-in/out, dissolve, and wipe [10, 11]. Fade-in

Fig. 4 Video scene

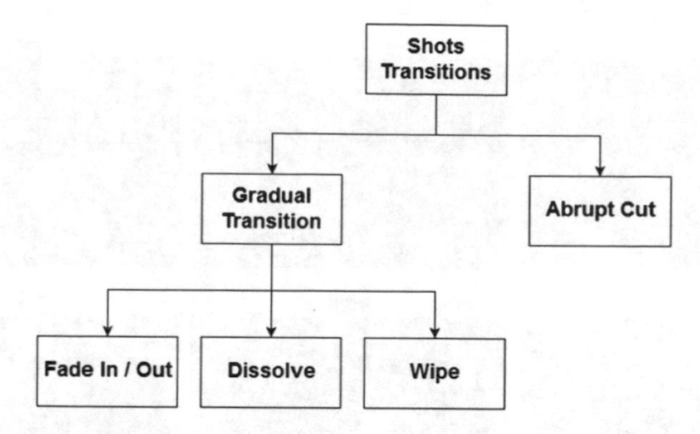

Fig. 5 Types of shot transitions

Fig. 6 Abrupt cut

is a progressive transition in which shot begins with a single color frame and develops slowly till it illuminates to its full force [6, 7] as shown in Fig. 7a. Fade-out is the converse of a fade-in [6, 7]. Figure 7b shows an example of fade-out.

Dissolve is the interfering of two shots where both shots are slightly apparent [6, 7]. Figure 8 shows an example of short dissolve that happened in a single frame. Also, dissolved effect can be spread over many frames as shown in Fig. 9. In the transition begin, the previous shot share is high. During the transition the previous shot share gradually decreases while the next shot share increases. In the transition end, the previous shot completely disappeared [4].

Wipe transition occurs as a pattern moves across the screen, progressively replacing the previous shot with the next shot [1]. Figure 10 shows a wipe example where a horizontal line moves from the left of the screen to its right progressively replacing the previous shot with the next shot.

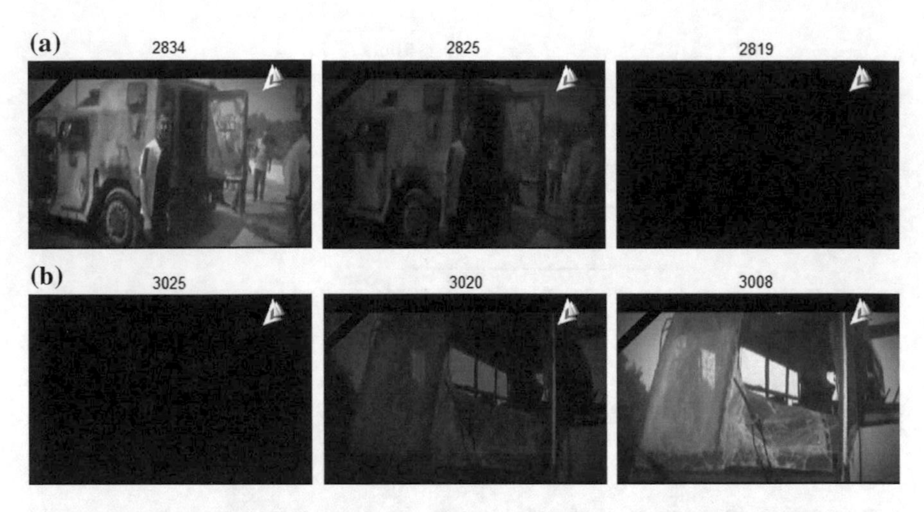

Fig. 7 **a** Fade-in **b** Fade-out

Fig. 8 Short dissolve

2.2 Video Segmentation Foundation and Challenges

The foundation of video segmentation is detecting the frames discontinuities. As shown in Fig. 11, the dissimilarity (distance) between adjacent frames is measured and compared with a threshold to detect the significant changes that correspond to shot boundaries. The video segmentation performance relies on the suitable choice of the similarity/dissimilarity measure that used to detect shot boundaries. In video segmentation algorithms, threshold selection is an important issue. The utilization of a dynamic adaptive threshold is more favored than global thresholding as the threshold used to detect a shot boundary changes from one shot to another and should be based on the distribution of the video inter-frames differences [4, 11].

Video segmentation faces many challenges such as detection of gradual transitions, flash lights, and object/camera motion. Many examples of camera motion exist such as panning, tilting, and zooming [11]. Detection of gradual transitions is the most challenging problem in video segmentation as the editing effects used to

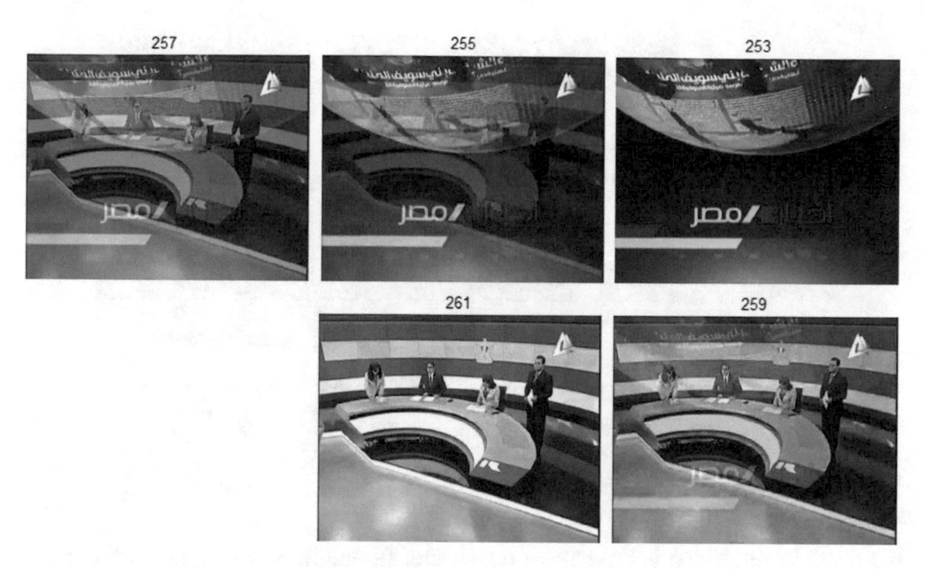

Fig. 9 Long dissolve

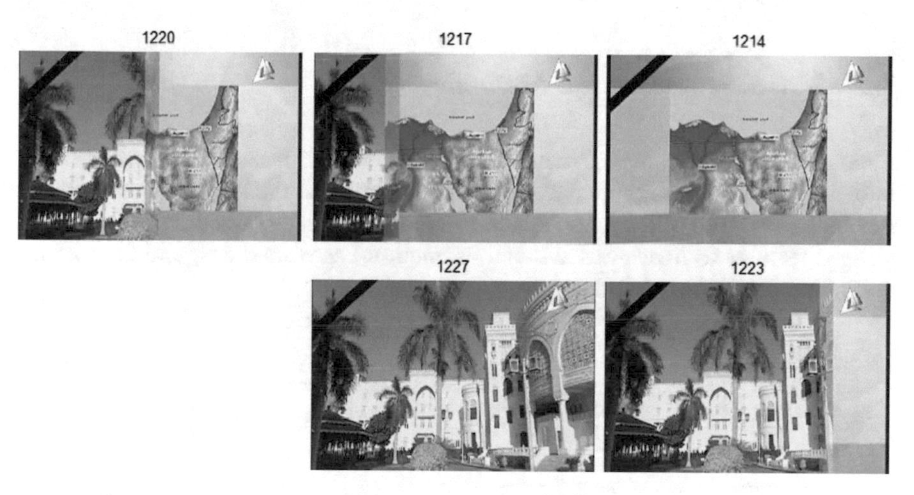

Fig. 10 Wipe

create a gradual transition are very comparable to the patterns of object/camera motion. It is hard to distinguish between shot boundaries, and object/camera motion as quick motion causes content change similar to cuts, and slow motion causes content change similar to gradual transitions. The vast majority of the video content representations utilize color features. Color is the essential component of video content. Illumination changes and especially the abrupt illumination changes such as flashlights might be recognized as a shot boundary by a large portion of the video segmentation tools [1]. Another example of illumination changes that happen

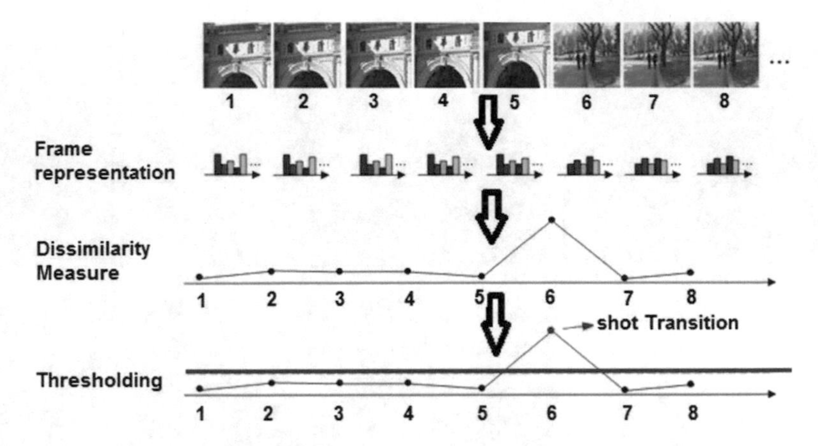

Fig. 11 Video segmentation foundation

naturally in the video is the smoke or the fire. In addition, screen split or video in the video is a challenging issue in news videos segmentation.

2.3 Video Segmentation Methods

As shown in Fig. 12, diverse methodologies have been proposed to segment video into shots such as the compressed domain methods and the uncompressed domain methods [10]. The video segmentation methods are categorized according to the utilized features to image-based methods, motion-based methods, and the methods that utilize audio-based features.

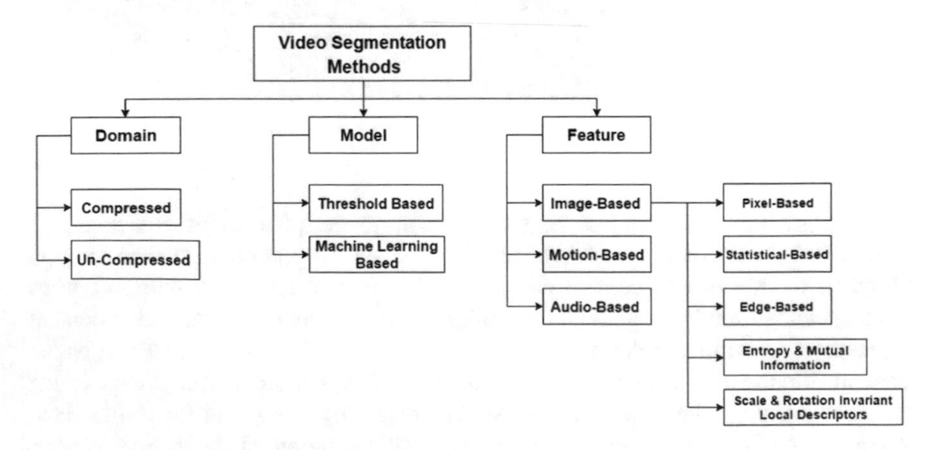

Fig. 12 Video segmentation methods

Image-Based Methods In the uncompressed domain, video segmentation methods are mainly based on computing image-based features. Image features can be based on pixel differences, statistical-based methods, edge differences, or their combination [4]. Methods based on pixel differences are easy to implement, but it is very sensitive to motion and noise [3].

Statistical-based methods include histogram differences which can be calculated on the entire frame level or at a block level [12]. In [1] each video frame is partitioned into blocks. For each successive frame pair, the histogram matching difference between the corresponding blocks is computed. Video frame is assumed as a shot boundary if it has histogram matching difference greater than a threshold. Threshold is computed as a function of the mean and standard variance of all histogram differences over the entire video. In [4], R, G, and B pixels are quantized and only one color index is computed. JND histogram is computed at the entire frame level. In JND histograms, each bin contains visually similar colors and visually different from other bins. For every two consecutive frames, histogram intersection is computed. For each video frame, a sliding window is centered. A frame is considered as a cut if its similarity degree is the window minimum and much less than the window mean and the second minimum. A frame is considered a part of gradual transition if it is less than a threshold related to the window similarity mean. In [13] histogram intersection in HSI color space is used as a similarity measure between frames. A placement algorithm up to eight frames is applied to convert gradual transition to abrupt cut. Shot boundaries are extracted by thresholding. Generally, spatial information is missed in the statistical-based methods. As a result, two images with comparable histograms may have totally different contents and away to enhance results is to work on a block level rather than the entire image level. In [14] histogram of gradient is combined with HSV color histogram to enhance results. In [15] fuzzy logic and genetic algorithm are utilized for shot boundary detection. For each consecutive frame, the difference of the normalized color histogram is calculated. Genetic algorithm is utilized to calculate the membership values of the fuzzy system. The fuzzy system classifies the types of shot boundaries into gradual transitions and abrupt cuts.

Edge-based video segmentation methods identified two sorts of edge pixels which are the entering and exiting edges. Entering edges are new edges showed up far from the locations of the previous shot edges. Exiting edges are the old shot edges that vanish. The main disadvantage of the edge-based methods is its execution time [1].

Several video segmentation algorithms that depend on the scale and rotation invariant local descriptors were proposed as in [3, 12]. In [12] each video frame is modeled by its HSV histogram and the SURF descriptors. For every pair of frames, a similarity score is calculated as a function of the normalized histogram correlation, and the ratio of the matched SURF descriptors. Cut is detected if the similarity score is less than a threshold. For all video frames, the moving similarity average is calculated. A frame that corresponds to a local minimum of the moving similarity average is considered a part of gradual transition. The detected shot boundary is considered as false alarm and removed if the similarity between it and the two surrounding local maxima is higher than a threshold. In [3] only abrupt cuts are detected in news video. Frames are converted to HSV color space and divided into uneven blocks with different weighting coefficients. A sliding window is utilized. Window is divided into right and left sub-windows. The difference between sub-windows is calculated as weighted frame difference. If the difference is large, an adaptive binary search process is continued on the sub-window whose larger discontinuity until obtaining a window size of 2 frames represents the candidate cut. Final cuts are assumed if the SURF points matching ratio are below a threshold. In [16] a combination of local and global descriptors is integrated to represent the video frames. Dissimilarity scores between the frames are calculated and shot boundaries are extracted utilizing adaptive thresholding.

Many works utilizing high-level features such as the entropy and the mutual information as in [17]. In [17] a temporal window is centered in each frame index and the cumulative mutual information is calculated. The cumulative mutual information combines the mutual information between multiple pairs inside the window. Shot boundaries are detected by identifying the local minimums of the cumulative mutual information curve. False shot boundaries caused by motion are eliminated by identifying local minimums that are below a threshold, crosses zero, and have positive value of the second derivative. Flashes effect is minimized as the cumulative mutual information combines multiple pairs of frames.

Motion-based Methods Motion-based algorithms are not favored in the uncompressed domain as the estimation of motion vectors expends huge computational power [1]. In [10] a method for abrupt cut detection based on motion estimation is introduced. Video frame is divided into blocks. Block matching algorithm is applied to calculate the blocks motion vectors. Average magnitude of the motion vectors is extracted to reflect the motion intensity. The quantitative angle histogram entropy of the motion vectors in each frame is extracted to reflect the motion irregularity degree. Candidate abrupt cut is identified when both the average motion magnitude and the quantitative angle histogram entropy are greater than specified thresholds as the camera motion cannot destroy the motion regularity. False cuts caused by gradual transitions and flashes are eliminated using a temporal window.

Utilizing Audio Features Audio features have been utilized to enhance the video segmentation results as in [2, 11]. In [2] a scene boundary detection and scene classification for news video are combined in one process using HMM. For each video frame, a feature vector that combines several features is derived. Average color components and the luminance histogram difference are used as image features. For adjacent video frames, some features derived from the difference image are computed as motion indicators. The logarithmic energy and the cepstral vector of the audio samples of each video frame are used as audio features. In [11] a technique to segment video into shots is presented. Multiple independent features are combined within an HMM framework. The absolute grayscale histogram difference between adjacent frames is used as an image feature. The acoustic difference in long intervals before and after the adjacent frames is used as an audio feature to reflect the audio types. Video frames are divided into blocks. A block matching algorithm is applied between adjacent frames to extract the blocks motion vectors. For the blocks motion vectors, the magnitude of the average and the average of the magnitude are used as motion features to detect camera movements.

Several machine learning techniques such as Support Vector Machines (SVM) are used for video segmentation by classifying the video frames into boundary frames and non-boundary frames [12].

3 Proposed System

In our previous works [18, 19] we introduced a method for caption detection and localization in news videos. We showed the patterns of caption insertion and removal. Actually, the process of caption insertion or removal takes a range of frames. Caption insertion and removal introduce many changes and can happen during the same video shot. Figure 13 shows sample frames associated with its frame index to illustrate the caption insertion pattern. By observing Fig. 13, it is clear that the caption background appears smoothly and text inserted to that background in steps. Also, caption may be static and shared in several consecutive shots. For all these reasons it is clear that captions in news videos disrupt the video segmentation process.

Our idea for detecting shots boundaries is based on the elimination of the candidate caption area and a new feature for describing an image we called it as "dominant colors". The proposed system is shown in Fig. 14.

Fig. 13 Caption insertion pattern

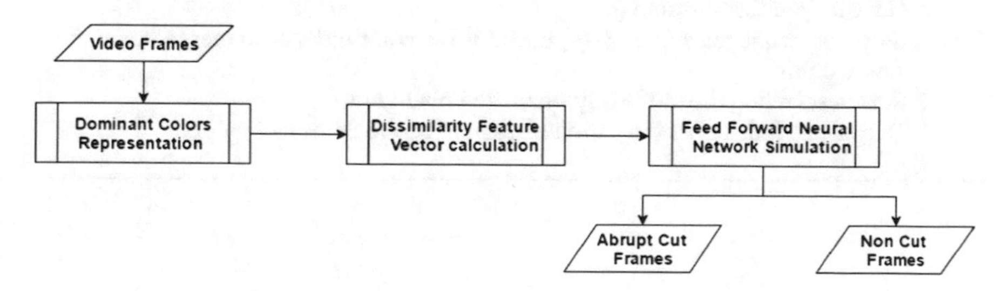

Fig. 14 The block diagram of the proposed system

3.1 Dominant Colors Representation

We represent the image by the order of its colors sorted from the highest frequency to the lowest frequency. As described in Algorithm 1 for each video frame, we extract the R, G, and B components. At first we truncate the bottom 25% of the three color components as it is the candidate caption area and actually the most important part of image is always at the frame center. Then we generate the dominant colors of each color component as described in Algorithm 2. The RGB representation of the video frame is the concatenation of the dominant colors of the three components.

Algorithm 1. RGB Frame Representation Extraction

1: Extract R, G, and B color components.
2: Truncate the bottom 25% from the color components.
3: Get the dominant colors of the R component as Dom_R.
4: Get the dominant colors of the G component as Dom_G.
5: Get the dominant colors of the B component as Dom_B.
6: Output the frame representation as the concatenation of Dom_R, Dom_G, and Dom_B.

Colors quantization is involved in the process of dominant colors extraction. The main purpose of the colors quantization is to absorb the lighting conditions that result in variation of the pixels color intensities of the same object. We carried out several experiments with different quantization steps and traced the accuracy results to conclude the colors quantization impact and select the best model. Also we carried out several experiments based on the gray representation of video frames. Algorithm 3 explains the gray representation extraction of a video frame. For each video frame, we transform it from RGB to HSI. We work only on the intensity component. Dominant gray intensities are extracted with several quantization steps.

Algorithm 2. Dominant Colors Extraction

1: Get the frame histogram.
2: Get the quantization step (s).
3: Sum the frequencies of each (s) consecutive color indexes to merge them into single color.
4: Sort descending the uniformly quantized histogram.
5: Output the indexes of the sorted histogram as the dominant colors.

Algorithm 3. Gray Representation Extraction of a Video Frame

1: Convert the RGB colored image to gray scale luminance/intensity image.
2: Truncate the bottom 25% of the luminance image.
3: Output the frame representation as the dominant colors of the luminance image.

3.2 *Dissimilarity Feature Vector Calculation*

We defined the dissimilarity between two video frames as a vector contains the difference in order for each color. If two images are similar so its order of dominant colors representations will be similar. Our idea for detecting the shots boundaries is based on the fact that if a shot boundary exists so a large difference of the dominant colors order will occur. To calculate the dissimilarity vector (dist) between frame i (F_i) and frame j (F_j), first we get its dominant colors representation Rep_i and Rep_j. The algorithm of calculating the dissimilarity vector is described in Algorithm 4. An example to calculate the dissimilarity of two frames represented by its eight levels dominant colors is shown in Fig. 15.

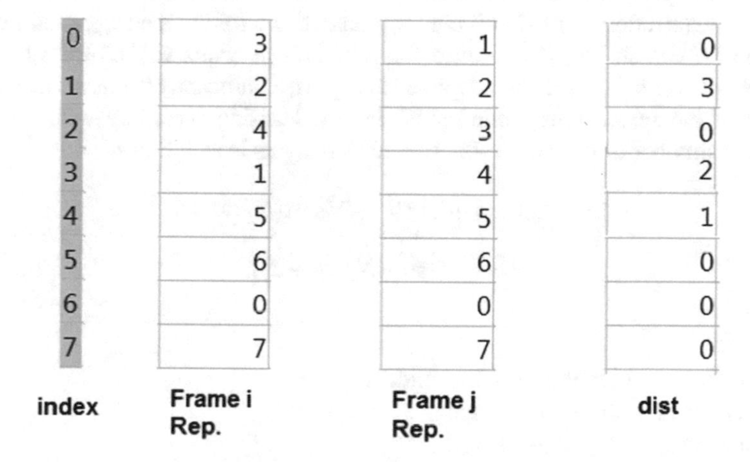

Fig. 15 An example of the dissimilarity vector calculation

Algorithm 4. Dissimilarity Feature Vector Calculation

1: Get Rep_i, Rep_j
2: **For** k = 0 to Rep Length -1 **do**
3: ind1 = the index of k in Rep_i
4: ind2 = the index of k in Rep_j
5: dist(k) = abs (ind1 - ind2)
6: **End For**

3.3 Feed-Forward Neural Network Training

We trained a feed-forward neural network to classify the video frames to abrupt cut frames or non-cut frames. The neural network is trained by the feature vectors of the consecutive frames dissimilarity after excluding the frames of gradual transitions. Also the frames of beginning, ending, and separators are excluded as it represents graphics. The samples of similar frames are selected as consecutive frames belong to the same video shot. The samples of different frames are selected as consecutive frames belong to different shots or in other words consecutive frames that correspond to abrupt cut frames. We carried out several experiments considering the grayscale dominant colors representation with different quantization steps and also the RGB representation with different quantization steps. In all our experiments we follow the suggested network structure in [20]. In [20] it is proved that two hidden layers of feed-forward neural network with the below structure depicted in Fig. 16 significantly decrease the required number of hidden neurons while preserving the learning capability with any small error. The number of neurons in the two hidden layers is as follows:

$$L1 = \text{sqrt}((M+2)*N) + 2*\text{sqrt}(N/(M+2)) \quad (1)$$

$$L2 = M*\text{sqrt}(N/(M+2)), \quad (2)$$

where

- $L1$ Number of neurons in the first hidden layer.
- $L2$ Number of neurons in the second hidden layer.
- M Number of output classes = 2.
- N The length of dissimilarity feature vector that varied according to the used representation (gray or RGB) and the used quantization step.

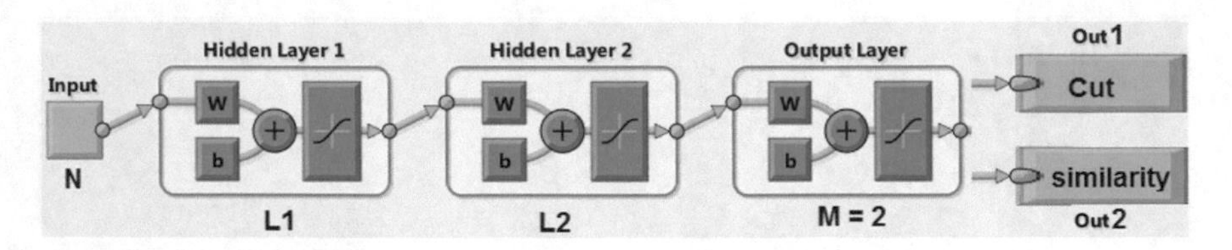

Fig. 16 The feed-forward neural network structure

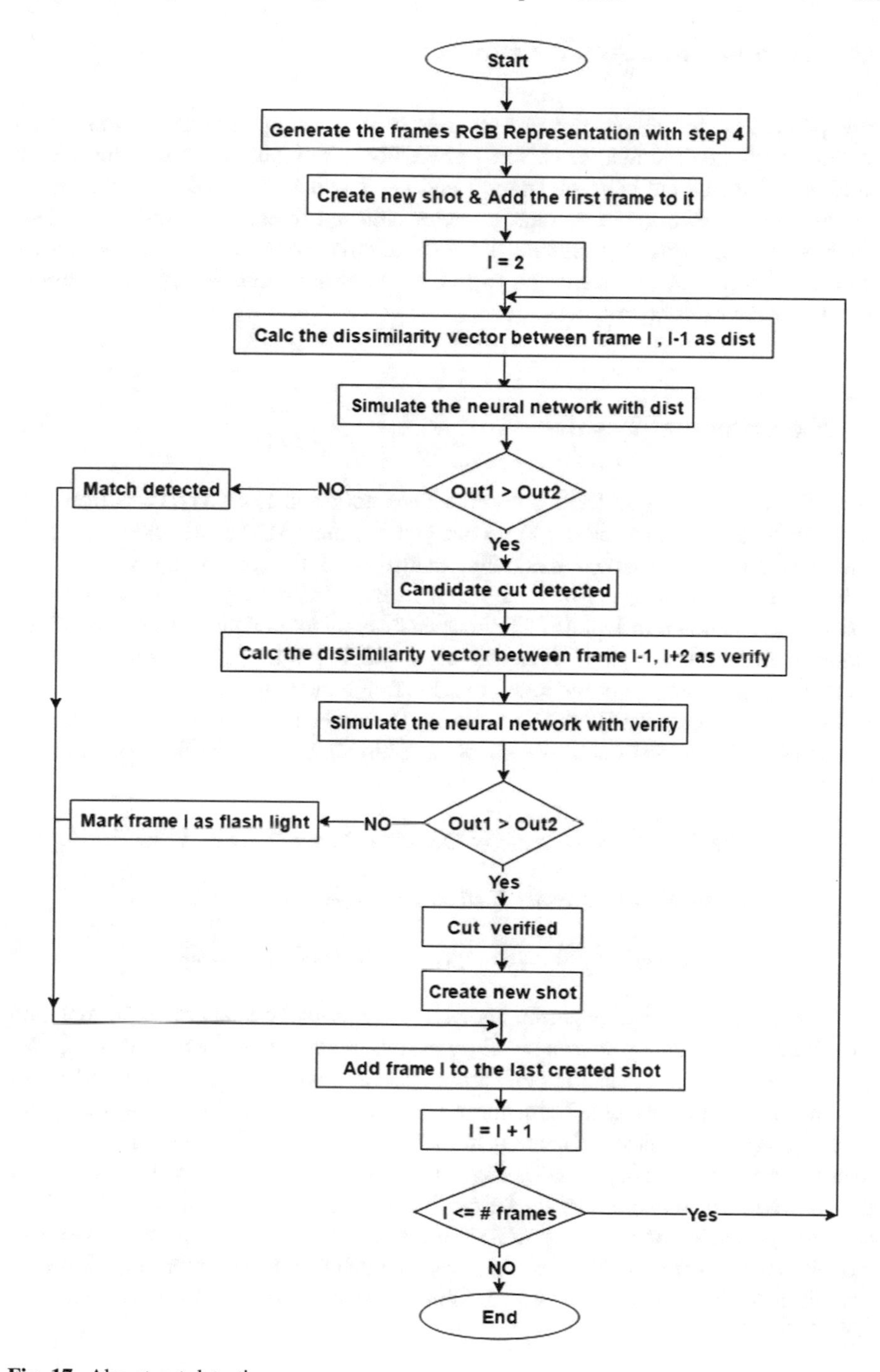

Fig. 17 Abrupt cut detection process

3.4 Abrupt Cut Detection Process

The process of the abrupt cut detection is carried out by simulating the neural network with the dissimilarity vectors between two consecutive video frames. Each candidate detected cut between frame i and i − 1 will be verified by checking the neural network with the dissimilarity vector between frame i − 1 and i + 2. This verification step aims to eliminate the false alarms come from flashes as flashes effects are introduced in maximum in two successive frames. Figure 17 illustrates the abrupt cut detection process.

4 Experimental Results

We build our data set as RGB colored videos collected from YouTube from different news TV channels as the Egyptian first channel, AlArabiya, and Al Jazeera. Shot boundaries are marked manually. Statistics about our data set is shown in Table 1. We developed a MATLAB tool to help us to manually mark the shot boundaries as shown in Fig. 18. The tool enables us to navigate through the video frames with different steps of 1, 6 and jump to specific frame index. Ability to select a shot type from specific frame index to other frame index is available. Tool enables us to select the shot transition type. Figure 19 shows a sample file from our shot boundary detection data set. We used the following measures for assessing the system results:

$$\text{Precision}\,(P) = \#\text{Correctly Detected Cuts}/\#\text{Detected Cuts} \quad (3)$$

$$\text{Recall}\,(R) = \#\text{Correctly Detected Cuts}/\#\text{Ground Truth Cuts} \quad (4)$$

$$F\,\text{Measure} = 2*\text{Precision}*\text{Recall}/(\text{Precision} + \text{Recall}). \quad (5)$$

We carried out several experiments with the dominant colors gray representation and RGB representation with several quantization steps as shown in Table 2. We used the F measure alone for evaluation as it combines the precision and recall together with equal weight. From the results, it is clear that increasing the quantization step will result in enhancement in the results up to limit and after that will result in negative feedback that is logical as loss of information will occur. RGB representation results are slightly better than the gray representation and the best accuracy is reached when using RGB representation with quantization step = 4 so we will use this network. The test data results when using RGB representation with quantization step = 4 are detailed in Table 3 without flash light elimination.

Table 1 Data set statistics

1040	# of abrupt cuts	53	# of videos
268	# of gradual transitions	85	Total duration in minutes

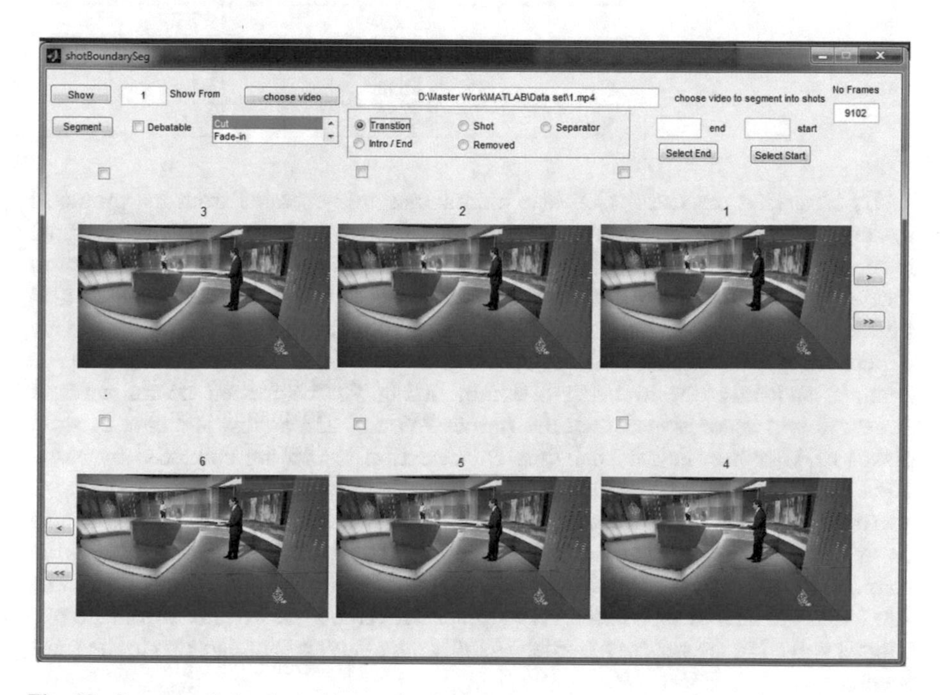

Fig. 18 Our manual shot boundary marker tool

```
Segment No 1 from 1 to 82 Introduction Sure
from 82 to 83 Transition Cut Sure
Segment No 2 from 83 to 124 Shot Sport Sure
from 124 to 125 Transition Cut Sure
Segment No 3 from 125 to 267 Shot Sport Sure
from 267 to 268 Transition Cut Sure
Segment No 4 from 268 to 338 Shot Sport Sure
from 338 to 339 Transition Cut Sure
Segment No 5 from 339 to 372 Shot Separator Sure
from 372 to 373 Transition Cut Sure
Segment No 6 from 373 to 539 Shot Sport Sure
```

Fig. 19 Sample file from our shot boundary detection data set

Table 2 Gray and RGB representations results on test data without flash light elimination

Quantization step	Gray representation	RGB representation
	F measure (%)	F measure (%)
2	91.00	91.56
4	91.08	93.40
8	92.50	93.30
16	91.36	91.40
32	86.83	89.49
64	63.00	82.05

By observing the reasons of false alarms that are generated from the proposed system, we will found mainly four reasons for false alarms: flash light, rapid motion, natural challenges, and dissolves especially short dissolves. As shown in Fig. 20a, the reason for the false alarm at frame 495 is flashlight. As shown in Fig. 20b, the reason for false alarm at frame 4370 is the rapid zoom on the money paper. In Fig. 20c false alarm is generated at frame 1499 because the video frame is changed suddenly due to firing a bomb. In Fig. 21 the trained neural network generates two successive cuts at the frames 257 and 258 as this is a case of short dissolve. After handling flash lights, cut detection results are enhanced by about 1.6% as detailed in Table 4.

Also by observing the miss cases as shown in Fig. 22, the system fails to detect its cuts. In Fig. 22a the system misses a cut at frame 1851, in (b) misses a cut at frame 4512, in (c) misses a cut at frame 1151, and in (d) misses a cut at frame 2362. We found the main reason is the very similar background or in other words the two shots involved in the cut are recorded in the same place and corresponds to the same scene.

Table 3 Test data results using RGB dominate colors representation with quantization step = 4 without flash light elimination

Video	Ground truth	Correct	Miss	False alarms	P (%)	R (%)	F (%)
Vid1	68	63	5	9	87.50	92.65	90.00
Vid2	27	26	1	2	92.86	96.30	94.55
Vid22	23	23	0	0	100.00	100.0	100.0
Vid30	26	25	1	5	83.33	96.15	89.29
Vid43	25	25	0	4	86.21	100.0	92.60
Vid9	29	29	0	0	100.00	100.0	100.0
Overall	198	191	7	20	90.52	96.46	93.40

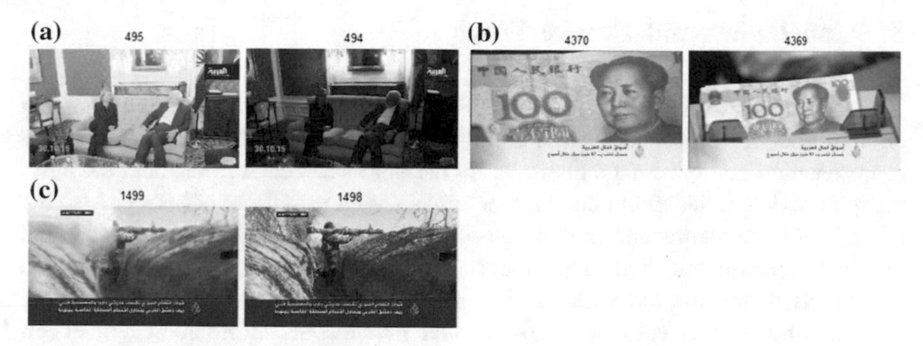

Fig. 20 False alarms observation

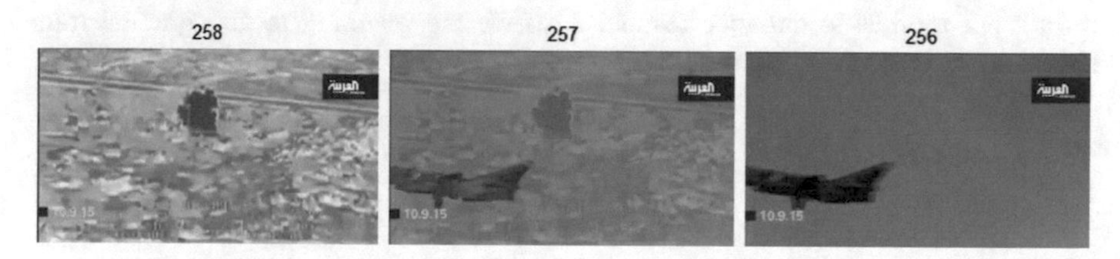

Fig. 21 Short dissolve

Table 4 Test data results using RGB dominate colors representation with quantization step = 4 utilizing flash light elimination

Video	Ground truth	Correct	Miss	False alarms	P (%)	R (%)	F (%)
Vid1	68	62	6	8	88.57	91.17	89.86
Vid2	27	26	1	2	92.86	96.30	94.55
Vid22	23	23	0	0	100.00	100.0	100.0
Vid30	26	25	1	2	92.59	96.15	94.34
Vid43	25	25	0	0	100.00	100.0	100.0
Vid9	29	29	0	0	100.00	100.0	100.0
Overall	198	190	8	12	94.06	95.96	95.00

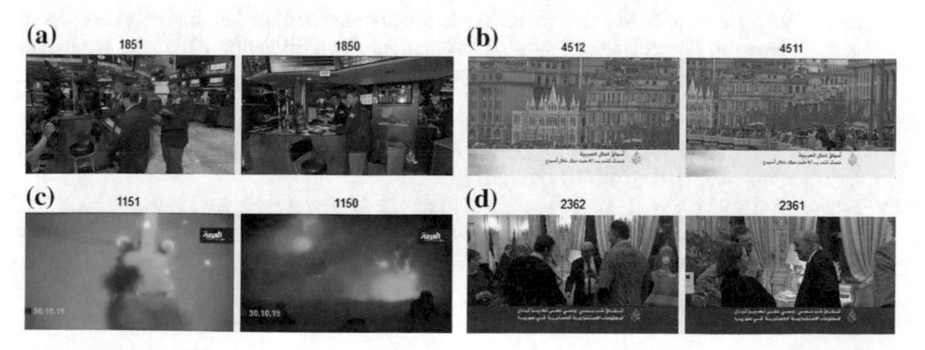

Fig. 22 Misses observation

5 Conclusion and Future Work

There are several highlights in our work. Although the training data is not big enough and the proposed system is tested on challenging data, the new image representation and our image dissimilarity measure succeeded to segment the news videos and detect the abrupt cuts in news videos with promising accuracy. Also the reasons of false alarms and misses caused by the proposed system are logical and justified. The quantization step is important in the dominant colors extraction process as it absorbs the variation in video frame pixels values due to lighting condition but this is valid up to limit, after that loss of information will occur. The RGB dominant color representations have slightly better results than the grayscale representation. We are planning to extend our data set, publish it, and retrain our models to enhance results. Also we are planning to add gradual transition detector to our system.

References

1. Dhagdi, S.T., Deshmukh, P.R.: Keyframe based video summarization using automatic threshold & edge matching rate. Int. J. Sci. Res. Publ. **2**(7), 1–12 (2012)
2. Eickeler, S., Muller, S.: Content-based video indexing of tv broadcast news using hidden markov models. In: International Conference on Acoustics, Speech, and Signal Processing, vol. 6, pp. 2997–3000, Phoenix, AZ, 15–19 Mar 1999
3. Jiang, M., Huang, J., Wang, X., Tang, J., Wu, C.: Shot boundary detection method for news video. J. of Comput. **8**(12), 3034–3038 (2013)
4. Janwe, N.J., Bhoyar, K.K.: Video shot boundary detection based on JND color histogram. In: IEEE Second International Conference on Image Information Processing, pp. 476–480, Shimla, 9–11 Dec 2013
5. Vora, C., Yadav, B.K., Sengupta, S.: Comprehensive survey on shot boundary detection techniques. Int. J. Comput. Appl. **140**(11), 24–30 (2016)
6. Singh, R.D., Aggarwal, N.: Novel research in the field of shot boundary detection—a survey. Adv. Intell. Inform. **320**, 457–469 (2015)
7. Thounaojam, D.M., Trivedi, A., Singh, K.M., Roy, S.: A survey on video segmentation. Intell. Comput. Netw. Inform. **243**, 903–912 (2014)
8. Rajendra, S.P., Keshaveni, N.: A survey of automatic video summarization techniques. Int. J. Electron. Electr. Comput. Syst. **3**(1), 1–6 (2014)
9. Zedan, I.A., Elsayed, K.M., Emary, E.: An innovative method for key frames extraction in news videos. In: Proceedings of the 2nd International Conference on Advanced Intelligent Systems and Informatics (AISI 2016), Advances in Intelligent Systems and Computing, vol. 533, pp. 383–394, Cairo, Egypt, 24–26 Oct 2016
10. Wang, C., Sun, Z., Jia, K.: Abrupt cut detection based on motion information. In: International Conference on Intelligent Information Hiding and Multimedia Signal Processing, pp. 344–347, Dalian, 14–16 Oct (2011)
11. Boreczky, J.S., Wilcox, L.D.: A hidden Markov model framework for video segmentation using audio and image features. In: IEEE International Conference on Acoustics, Speech and Signal Processing, vol. 6, pp. 3741–3744, Seattle, WA, 12–15 May 1998

12. Apostolidis, E., Mezaris, V.: Fast shot segmentation combining global and local visual descriptors. In: IEEE International Conference on Acoustics, Speech and Signal Processing, pp. 6583–6587, Florence, 4–9 May 2014
13. El-bendary, N., Zawbaa, H.M., Hassanien, A.E., Snasel, V.: PCA-based home videos annotation system. Int. J. Reason.-Based Intell. Syst. **3**(2), 71–79 (2011)
14. Shao, H., Qu, Y., Cui, W.: Shot boundary detection algorithm based on HSV histogram and HOG feature. In: 5th International Conference on Advanced Engineering Materials and Technology, pp. 951–957 (2015)
15. Thounaojam, D.M., Khelchandra, T., Singh, K.M., Roy, S.: A Genetic algorithm and fuzzy logic approach for video shot boundary detection. Comput. Intell. Neurosci. **16**(1), 1–11 (2016)
16. Tippaya, S., Sitjongsataporn, S., Tan, T., Chamnongthai, K., Khan, M.: Video shot boundary detection based on candidate segment selection and transition pattern analysis. In: IEEE International Conference on Digital Signal Processing, pp. 1025–1029, Singapore, 21–24 July 2015
17. Cernekova, Z., Nikolaidis, N., Pitas, I.: Temporal video segmentation by graph partitioning. In: IEEE International Conference on Acoustics Speech and Signal Processing Proceedings, vol. 2, pp. 209–212, Toulouse, 14–19 May 2006
18. Zedan, I.A., Elsayed, K.M., Emary, E.: Caption detection, localization and type recognition in Arabic news video. In: Proceedings of the 10th International Conference on Informatics and Systems (INFOS 2016), pp. 114–120, Cairo, Egypt, 9–11 May 2016
19. Zedan, I.A., Elsayed, K.M., Emary, E.: Abrupt cut detection in news videos using dominant colors representation. In: Proceedings of the 2nd International Conference on Advanced Intelligent Systems and Informatics (AISI 2016), Advances in Intelligent Systems and Computing, vol. 533, pp. 320–331, Cairo, Egypt, 24–26 Oct 2016
20. Huang, G.: Learning capability and storage capacity of two-hidden-layer feed forward networks. IEEE Trans. Neural Netw. **14**(2), 274–281 (2003)

Part II
Applications of Image Processing in Medicine

Normalized Multiple Features Fusion Based on PCA and Multiple Classifiers Voting in CT Liver Tumor Recognition

Ahmed M. Anter and Aboul Ella Hassenian

Abstract Liver cancer is a serious disease and is the third commonest cancer followed by stomach and lung cancer. The most effective way to reduce deaths due to liver cancer is to detect and diagnosis in the early stages. In this paper, a fast and accurate automatic Computer-Aided Diagnosis (CAD) system to diagnose liver tumors is proposed. First, texture features are extracted from liver tumors using multiple texture analysis methods and fused feature is applied to overcome the limitation of feature extraction in single scale and to increase the efficiency and stability of liver tumor diagnosis. Classification-based texture features is applied to discriminate between benign and malignant liver tumors using multiple classifier voting. We review different methods for liver tumors characterization. An attempt was made to combine the individual scores from different techniques in order to compensate their individual weaknesses and to preserve their strength. The experimental results show that the overall accuracy obtained is 100% of automatic agreement classification. The proposed system is robust and can help doctor for further treatment.

Keywords Classification · Fusion · Feature extraction · SVM · Feature reduction · CAD

A.M. Anter (✉)
Faculty of Computers and Information, Beni-Suef University, Beni Suef, Egypt
e-mail: sw_anter@yahoo.com
URL: http://www.egyptscience.net

A.E. Hassenian
Faculty of Computers and Information, Information Technology Department, Cairo University, Giza, Egypt
e-mail: aboitcairo@gmail.com

A.M. Anter · A.E. Hassenian
Scientific Research Group in Egypt (SRGE), Cairo, Egypt

A.E. Hassanien and D.A. Oliva (eds.), *Advances in Soft Computing and Machine Learning in Image Processing*, Studies in Computational Intelligence 730,
https://doi.org/10.1007/978-3-319-63754-9_6

1 Introduction

Medical image analysis becomes more and more popular in recent years due to the advances of the imaging techniques, including Magnetic Resonance Imaging (MRI), Computer Tomography (CT), Mammography, Positron emission tomography (PET), X-ray, and Ultrasound or Doppler Ultrasound. Computed tomography has been identified as an accurate and robust imaging modality in the diagnosis of the liver cancer. Liver cancer is one of the most common internal malignancies worldwide and also one of the leading death causes. Early detection and diagnosis of liver cancer is an important issue in practical radiology. Liver cancer in men is the fifth most frequently diagnosed cancer worldwide but the second most frequent cause of cancer death based on global cancer statistics. In women, it is the seventh most commonly diagnosed cancer and the sixth leading cause of cancer death. According to global cancer statistics in 2008 [1], there were estimated to be 748 300 new liver cancer cases in the world, and 695 900 people died from liver cancer. The incidence and mortality rate of primary liver cancers were increasing across many parts of the world. In this context, the earliest possible detection of such a disease becomes critical to successful treatment [2].

The most effective way to reduce deaths due to liver cancer is to detect and diagnosis in the early stages. One of the most common and robust imaging techniques for the diagnosis of liver cancer is Computed Tomography (CT). Visual analysis of liver CT scans, performed by an experienced radiologist, often is not sufficient to correctly recognize the type of pathology. Due to the fact that those performing the analysis are able to identify only a small part of information stored in images, invasive techniques (such as a needle biopsy) still remain a gold standard for a definitive diagnosis of hepatic disorders. The use of invasive procedures could be avoided, if doctors had the appropriate tools to interpret the image content. The solution could be the image-based Computer-Aided Diagnosis (CAD) systems, which have recently and rapidly become growing interest [3].

Due to the semantic gap problem, which corresponds to the difference between the human image perception and what automatically extracted features convey, an important aspect of the feature extraction task is to obtain a set of features that is able to succinctly and efficiently present Region of Interest (ROI) [4, 5].

The feature extraction stage is one of the important components in any pattern recognition system. The performance of a classifier depends directly on the choice of feature extraction. The feature extraction stage is designed to represent and analysis tumors. These features are used by the classifier to classify tumors into normal, benign, and malignant [6, 7].

The main goal is to develop an efficient system to assist radiologists in categorizing liver into normal and abnormal such as benign and malignant. Normal liver usually have a regular structure, but due to the presence of the abnormal tissues the complexity of abnormal liver increases. Thus, naturally they will have higher fractal dimension. Malignant masses are generally rough and have more irregularity structure, whereas benign masses commonly have smooth, round, oval contours.

In this paper, texture features are extracted from liver tumors using multiple texture analysis: Gray-Level Co-occurrence Matrix (GLCM), Local Binary Pattern (LBP), Segmentation-based Fractal Texture Analysis (SFTA) and First-Order Statistics (FOS). Also, the fused features (FF) is applied to increase the performance of classification rate. Principle component analysis (PCA) is used to reduce the feature dimensionality. Texture feature-based classification system is then applied to discriminate between liver tumor, benign and malignant. The classification system is developed based on multiple classifier voting: K-nearest neighbor (KNN), Artificial Neural Network (ANN), Support Vector Machine (SVM), and Random Forest (RF). The classifiers performances are evaluated using measures: Precision, Recall, and accuracy.

The rest of this paper is ordered as Sect. 2, discusses and demonstrate the previous and related work on liver tumor characterization. Section 3, details of the proposed feature extraction methods are given. The proposed classification methods are presented in details in Sect. 4. The proposed CAD system is presented in Sect. 5. Section 6, shows the experimental results and analysis. Finally, conclusion and future work are discussed in Sect. 7.

2 Related Work

CAD system is used to distinguish between benign and malignant tumors. The features are automatically extracted from each Region of Interest (ROI). The features are normalized, reduced, and merged with linear classifiers or artificial intelligence techniques to further refine the detection and diagnosis (benign vs. malignant) of potential abnormalities. Several researchers have focused their attention on the use of texture features to describe liver tumor.

Gletsos et al. [8], proposed gray-level difference method, FOS, SGLDM, Laws' texture energy features, and fractal dimension (FD) methods to extract features from liver tumors. Neural Network is used to classify between normal and abnormal tissue. The classification accuracy obtained differs from 91 to 100%. Cavouras et al. [9], extracted 20 features from CT density matrix. Twenty cases of benign tumor and thirty-six cases of malignant tumor were used to train a multi-layer perceptron classifier. The performance accuracy obtained is 83%.

Chen et al. [10], proposed a modified probabilistic neural network (MPNN) and feature descriptors to discriminate tumors generated by FD and GLCM. The system was applied on 20 malignant and 10 benign tumors. The classification accuracy achieved is 83%.

Mougiakakou et al. [11], proposed five distinct sets of texture features are extracted from FOS, SGLDM, GLD, Laws' texture energy measures, and FD methods. Two different approaches were constructed and compared. The first one consists of five multi-layer perceptron. The second one comprised from five different primary classifiers: multilayer perceptron, probabilistic NN, and three KNN classifiers. The final decision was extracted using appropriate voting schemes. The fused feature set was obtained. Data set contains 38 normal cases and 59 abnormal cases. The best performance achieved is 84%.

Guozhi et al. [12], proposed Gaussian Mixture model (GMM), and expectation maximization (EM) are used to classify Metastases and Normal cases. The accuracy achieved by similarity index 95.8%. Wang et al. [13], applied texture features methods: FOS, SGLDM, GLRLM, and GLDM and applied SVM Classifier to distinguish between HCC, Hemangioma, and Normal tissue. The accuracy achieved is 97.78%.

Mala et al. [14], applied texture features methods: Wavelet transform and extract Angular Second moment, Energy, Contrast, Homogenity, and Entropy and used Learning vector quantization (LVQ) classifier to distinguish between Hemangioma, Adenoma, HCC, Cholangiocarcinoma.

Kumar et al. [15], proposed Wavelet, and Fast Discrete Curvelet Transform (FDHCC) methods and Feed Forward NN classifier to distinguish between HCC and Hemangioma. The accuracy obtained using Curvelet transform is 93.3%, and wavelet is 88.9%. Gunasundari et al. [16], applied texture features methods: Co-occurance matrix and Fast discrete Curvelet transform and used BPN, PNN, Cascade feed forward BPN (CFBPN) for Classification phase to distinguish between Hemangioma, and HCC. The accuracy achieved by classifier BPN is 96%, PPN is 96%, and CFBPN is 96%.

Mougiakakou et al. [17], proposed CAD system based on texture features and a multiple classification scheme for the characterization of four types of hepatic tissue from CT images: normal liver, cyst, hemangioma, and hepatocellular carcinoma. For each ROI, five distinct sets of texture features are extracted using the following methods: first-order statistics, spatial gray-level dependence matrix, gray-level difference method, Laws texture energy measures, and fractal dimension measurements. Classification of the ROI is then carried out by a system of five neural networks (NNs). The proposed system has achieved a total classification performance of the order of 97%.

Vijayalakshmi et al. [18], proposed CAD system based on extracted texture feature using local binary pattern and statistical features are extracted by Legendre moments. This communication presents a comparative analysis between these Legendre moments, local binary pattern, and combined features. The classification accuracy obtained is 96.17%.

Chien et al. [19], proposed CAD system for three kinds of liver diseases including cyst, hepatoma, and cavernous hemangioma. The diagnosis scheme includes two steps: features extraction and classification. The features, derived from gray levels, co-occurrence matrix, and shape descriptors, are obtained from the region of interests (ROIs) among the normal and abnormal CT images. Finally, the receiver operating characteristic (ROC) curve is used to evaluate the performance of the diagnosis system. The average accuracy obtained using SVM is 78%.

Pavlopoulos et al. [20], also characterized diffused liver diseases automatically using Fuzzy Neural Network (FNN). Twelve texture features for classification were extracted using FDTA, SGLDM, GLDS, RUNL, and FOP. These features were further reduced to 6 using different feature combinations. Then voronoi diagram of training patterns was constructed, which was used by FNN to generate fuzzy sets and built class boundaries in a statistical manner. For validation, the authors used 150 liver images and showed 82.67% classification accuracy.

Ritu et al. [21], optimal features are extracted and used as input for neural network classifier. Texture features are utilized and extracted from histogram-based features and wavelet transform- based features. This feature set is very large and it may contain redundant features. So, for reducing this feature set, feature selection algorithm is used and optimal features are selected. For optimal feature selection, genetic algorithm is used. Finally, these features are given as input to neural network which is a pixel-based classifier and it classifies pixels into liver and non-liver area. The Average Accuracy obtained from NN and GA is 96.02%.

Kumar and Moni [22], designed a CAD system consisting of liver and tumor segmentation, feature extraction, and classification module is presented characterizing the CT liver tumor as hemangioma and hepatoma. The experiment results show that the classification accuracy of Fast Discrete Curvelet Transform (FDCT)-based feature extraction and classification is higher than the wavelet-based method. FDCT-achieved accuracy is 93.3% and Wavelet-achieved accuracy 88.88%.

Hussein and Amr [23], proposed an automated CAD system consists of three stages; first, automatic liver segmentation and lesions detection. Second, extracting features. Finally, classifying liver lesions into benign and malignant by using contrasting feature difference approach. Several types of intensity, texture features are extracted from both; the lesion area and its surrounding normal liver tissue. Machine learning classifiers are then trained on the new descriptors to automatically classify liver lesions into benign or malignant. With accuracy reached 98.3%.

The proposed research work addresses these limitations and designed CAD system that can automatically diagnose liver tumors in large scale of dataset. To overcome the limitation of feature extraction in a single scale, we used multiple methods and fused features to increase the efficiency and stability of liver tumor diagnosis. Also multiple classifier are used to increase the efficiency of classification rate with high accuracy.

3 ROI Feature Extraction Methods

In this paper, different kinds of features extraction are tested and evaluated, LBP, GLCM, SFTA, FOS, and fused feature (FF) between them.

3.1 Gray-Level Co-occurrence Matrix (GLCM)

GLCM is widely used to discriminate texture images, it is second-order statistical approach that measures the relationship and dependency between intensities and take in consideration distance and angle between intensity. The method involves statistically sampling the way certain gray levels occur in relation to other gray levels.

This method obtains the GLCM of specified texture, which further gives various descriptors by measuring texture properties. A set of features derived from GLCM were used as texture features. In this paper, we used four angles 0°, 45°, 90°, and 135°, and one displacement. The following statistics were used to describe texture: Energy, Entropy, homogeneity, Contrast, Dissimilarity, Angular Second Moment (ASM), Correlation, Variance, Maximum Probability (MP), Cluster Tendency, Cluster Shade, Cluster prominence, Sum Entropy, Sum Average, Entropy Difference, Sum variance, and Difference Variance. Each vector consists of 68 values. [24, 25].

3.2 First-Order Statistics (FOS)

FOS is a first-order statistical approach used to describe texture and various moments based on gray level histogram computed from a digital image can be used to describe statistical properties such as mean, variance, skewness and Kurtosis. The first four moments are easier to describe intuitively. The first moment is the mean intensity; the second central moment is the variance, describing how similar the intensities are within the region. The third central moment skewness, describes how symmetric the intensity distribution is about the mean and fourth central moment kurtosis, describes how flat the distribution is. Let random variable I represents the gray levels of image region. The first-order histogram $P(I)$ is defined as

$$P(I) = \frac{\textit{number of pixels with gray level } I}{\textit{total number of pixels in the region}} \tag{1}$$

3.3 Local Binary Pattern (LBP)

LBP was first introduced by Ojala as a shift invariant complementary measure for local image contrast. LBP works with the 4-neighbors of the pixels and using value of the center pixel as a threshold. The LBP code computation for a 33 neighborhood is produced by first thresholding the neighbor values with the center pixel; such that the pixel greater than or equal to the center pixel will be set to 1, otherwise it will be set to 0. After thresholding the resulted 3×3 window will be multiplied by weighted window. Then, the central pixel will be replaced by the summation of the window thresholded. Then describtors are extracted from GLCM (Mean, Variance, Standard Division, Skewness, and Kurtosis) [25].

3.4 Segmentation-Based Fractal Texture Analysis (SFTA)

Geometric primitives that are self-similar and irregular in nature are termed as fractals. Fractal Geometry was introduced to the world of research in 1982 by Mandelbrot

[25]. FD measurements can be used to estimate and quantify the complexity of the shape or texture of objects. Liver tumor texture is a combination of repeated patterns with regular/irregular frequency. Tumor structure exhibit similar behavior, it has maximum disparity in intensity texture inside and along boundary which serves as a major problem in its segmentation and classification. Fractal dimension reflects the measure of complexity of a surface and the scaling properties of the fractal, i.e., how its structure changes when it is magnified. Thus, fractal dimension gives a measure of the irregularity of a structure. In fact, the concept of fractal dimension can be used in a large number of applications, such as shape analysis and image segmentation.

Segmentation-based Fractal Texture Analysis (SFTA) algorithm introduced by Costa et al. [26] consists of decomposing the input image into a set of binary images from which the fractal dimensions of the resulting regions are computed in order to describe segmented texture patterns. In order to decompose the input image, a new algorithm, named two-threshold binary decomposition (TTBD) is used. Then SFTA feature vector is constructed as the resulting binary images size, mean gray level and boundary represent fractal dimension. The fractal measurements are employed to describe the boundary complexity of objects and structures segmented in the input image using box counting algorithm.

3.5 Shape Feature Analysis

It is important to classify masses whether it is malignant or benign based on shape morphology. Several shape morphologies are used as follows:

3.5.1 Circularity

Circularity feature is used to measure the degree of tumor. The roundness tumor is one of the criteria of benign; the probability of masses as being benign is higher when circularity is higher. The circularity values is between 0 and 1. If it is equal to 1, this means that the tumor is circular and matches benign tumor. On the other hand, if the circularity is much smaller than 1, this implies the tumor is far from a circle and it is malignant.

Circularity $= \frac{4\pi A}{p2}$

where A is the area of the tumor and P is the tumor perimeter.

Area: is the actual number of pixels in the tumor region.

Perimeter: is the distance around the boundary of the regular/irregular tumor region.

3.5.2 Solidity

Solidity feature used to measure the density of the objects. If the solidity approximately is equal to 1, which means the object is completely solid. If solidity <1, the object is irregular.

$$\text{Solidity} = \text{Area}/\text{Convex hull}$$

3.5.3 Compactness

Area = |O|
Perimeter = length (E)

where O is a set of pixel that belongs to the abnormality segmented and E edge pixels (subset of O).

$$\text{Compactness} = \frac{\text{Perimeter}^2}{\text{Area}}$$

3.5.4 Radial Angle

The speculation is an indicator of the malignant masses. The Radial Angle is used to differentiate the shape of edges of the tumor as speculated or as round and smooth. When tumors tend to be more round, its Radial Angles tend to be near 180 and the average of the Radial Angles tends to be larger. On the other hand, if the tumor with a speculated edge will have a smaller averaged Radial Angle.

3.5.5 Circular Equivalent Diameter

Equivalent diameter is the scalar value that specifies the diameter of a circle with the same area as the region.

$$\text{EquivDiameter} = \sqrt{\frac{4A}{\pi}}$$

4 Liver Tumors Diagnosis-Based Classification System

The classification stage is one of the characterization techniques that is used to predict and describe data classes into benign and malignant.

4.1 *Random Forests (RF)*

RF algorithm is considering one of the best known classification algorithms, which is able to classify big data in short time with high accuracy. RF starts with a standard decision tree machine learning technique. The input is entered at the top and as it traverses down the tree, the data gets bucketed into smaller and smaller sets. The ensemble of simple trees vote for the most popular class. The main principle behind ensemble methods is that a group of weak learners can come together to form a strong learner [27].

4.2 *Artificial Neural Network (ANN)*

ANN has been developed as generalizations of mathematical models of biological nervous systems. Simple cells called neurons form the brain, and each neuron is a specialized cell that processes the incoming information and conducts it to the next neuron. The learning capability of an artificial neuron is achieved by adjusting the weights in accordance to the chosen learning algorithm. ANN is composed of an input layer, an output layer and one or more hidden layers. Each layer is composed of neurons. Several ANN classifiers are used for medical image analysis [28, 29]. In this paper, backpropagation neural network is constructed with 7 hidden, 3 output Layers, and and number of training epochs 1000.

4.3 *K-Nearest Neighbor (KNN)*

KNN is a very simple and nonparametric classifier based on the nearest neighbor. A new sample is classified based on majority of k nearest neighbor. In this classifier, the Euclidean distance is used. It simply works based on a minimum distance from the searching query to the training one to determine the k-nearest neighbors. KNN is simple, high variance, and nonlinear and the disadvantage is it is expensive at test time. After specifying the k-nearest neighbor classes, the new sample follows (predicts) the major class of KNN [30].

4.4 *Support Vector Machine (SVM)*

SVM is first introduced by Vapnik (1995) and widely used in machine learning. SVM modeling algorithm computes the optimal separating hyperplane with the maximum margin by separating a positive class from a negative class. It is directly motivated by Kernels. Kernel functions are used to map the input data into a higher dimension space, where the data are supposed to have a better distribution and then an optimal separating hyperplane in the high-dimensional feature space is chosen. In this paper, the linear kernal function is used for simplicity [31].

5 Proposed Liver Tumor Classification System

The proposed CAD system consists of four main phases to classify liver CT abnormality into benign, malignant, and normal as shown in Fig. 1.

Feature extraction phase: In this phase, we used feature extraction methods such as GLCM, FOS, LBP, SFTA, Shape features, and Feature Fusion to discriminate between benign and malignant tissues.

Normalization phase: We applied data normalization technique to decrease the gap between features and to reach all features near to others. All values exclusive between [0, 1] are used to increase the classification rate especially in the features with high values.

Feature reduction phase: The main purpose of feature reduction is to determine a small set of features from whole features in the problem. The features extracted have irrelevant, redundant, misleading, and noisy features. Remove these data that affects the prediction and classifiers accuracy can be useful. Principle Component Analysis (PCA) is used to reduce dimensionality between the extracted features.

Classification phase: The classifiers KNN, ANN, SVM and RF are used to classify abnormality into three classes (normal, benign, and malignant). We used voting system to vote for high accuracy selected from these classifiers to increase the efficiency of the proposed system.

Analysis and evaluation: Evaluation criteria for classifiers performance using confusion matrix, ROC and these parameters "TP, FP, TN, FN, Precision, Recall, Accuracy, and Over-all accuracy."

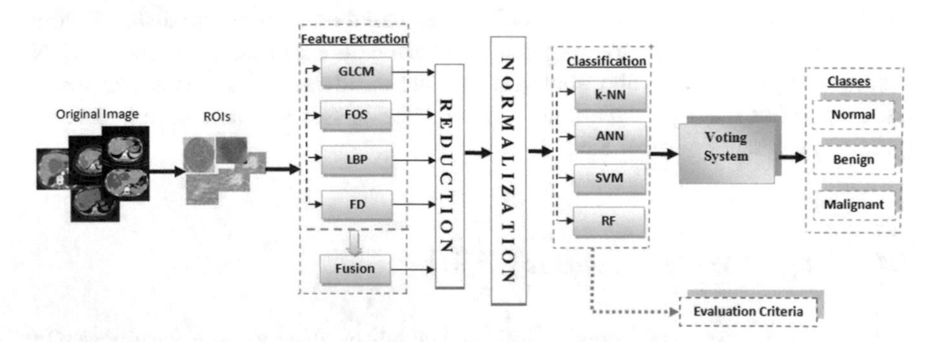

Fig. 1 The proposed CAD system for liver tumor diagnosis

6 Experimental Results and Discussion

In this section, the performance of the proposed diagnosis system is presented. The used programming tool is MATLAB V. 7.9 and the perfromance was measured using DELL with Intel, Core i7, CPU 2.2 GHz, memory 8GB, and operating system MS Windows 7.

6.1 CT Data Set Description

CT scanning is a diagnostic imaging procedure that uses X-rays. The proposed system will work on a difficult dataset with different tumors. The dataset is divided into three categories depending on the tumor type and abnormality: Normal, Benign, and Malignant. 62 image in normal cases, 392 in benign cases, and 308 in malignant cases. All these ground truth tumor images were croped by two expertise. Each image has resolution dimensions 64×64 pixel [3].

6.2 Performance Analysis

The performance of liver tumor classification is analyzed with the following parameters: ***Accuracy*** is a statistical measure used to measure the proportion of true cases among the total number of cases examined, can be calculated using the equation:

$$\text{Accuracy} = \frac{(\text{TP} + \text{TN})}{(\text{TP} + \text{TN} + \text{FP} + \text{FN})} \tag{2}$$

Recall is the proportion of positive cases that were correctly identified, calculated using the following equation:

$$\text{Recall} = \frac{\text{TP}}{(\text{TP} + \text{FN})} \tag{3}$$

Precision is the proportion of the predicted positive cases that were correct, as calculated using the equation:

$$\text{Precision} = \frac{\text{TP}}{(\text{TP} + \text{FP})} \tag{4}$$

where TP, FP, TN, FN being the number of true positives, false positives, true negatives, and false negatives, respectively.

Imag.	Area	Perimet er	Compact	Circulari ty	Solidity	Diame ter	Orienta tion
	2429	401	66.20	0.18973	0.8	55.63	-1.7
	375	154.02	63.26	0.19854	0.5752	21.86	-52.28
	2468	687.6	191.57	0.06556	0.7	56.07	57.1
	7302	679.1	63.16	0.19887	0.9	96.45	-32.4
	326	213.97	140.43	0.08944	0.6468	20.38	0.095

Fig. 2 Samples of shape features results for different tumors

6.3 Feature Extraction and Classification Accuracy

The proposed CAD system consists of four main phases, in the feature extraction phase five sets of texture features are calculated for each ROI. These features are obtained from Gray-Level co-occurrence matrices, which are constructed to extract feature vector with 68 values to represent each ROI. Seventeen features were used to represent ROI: Energy, Entropy, homogeneity, Contrast, Dissimilarity, Angular Second Moment (ASM), Correlation, Variance, Maximum Probability (MP), Cluster Tendency, Cluster Shade, Cluster prominence, Sum Entropy, Sum Average, Entropy Difference, Sum variance, and Difference Variance with four directions (0°, 45°, 90°, 135°) and one displacement equal to 1 is used. LBP is constructed to extract feature vector using mean, variance, skewness, and kurtosis to represent each ROI. FOS is used to construct feature vector for each ROI composed from (mean, variance, skewness, and kurtosis). SFTA is constructed to obtain feature vector for each ROI with 36 values. The SFTA features correspond to the number of binary images obtained by TTBD multiplied by 3. Then these measures are calculated from each binary image: fractal dimension, mean gray level and size. Five shape features are extracted to represent each ROI: Circularity, Solidity, Compactness, Radial Angle, and Circular Equivalent Diameter as shown in Fig. 2.

Finally, fused features were adopted to address some of the limitations of the feature extraction methods utilities. It is a process of combining feature vectors from different sources with the aim of maximizing the useful information content. It improves the reliability or discrimination capability and offers the opportunity to minimize the data retained. The fact of combining multiple features derived from multiple methods does improve the correct classification rate. The features extracted from GLCM, LBP, FOS, SFTA, and shape features are combined in one feature vector (V) with 117 values to improve the performance of correct classification rate.

Normalization function is applied to decrease gap between features and normalizing all features between [0, 1]. Also, features dimensionality are reduced using PCA to increase the classifier performance with high prediction and decrease time computation.

Table 1 Precision, Recall, and Accuracy for three classes Normal (N.), Benign (B.), and Malignant (M.) using four feature methods (GLCM, LBP, FOS, SFTA, Feature Fusion) and four classifiers (KNN, RF, ANN, SVM)

Meth	Class.	Precision			Recall			Accuracy			Time/S
		B.	M.	N.	B.	M.	N.	B.	M.	N.	
GLCM (68)	KNN	100	100	100	100	100	100	100	100	100	**1.19**
	RF	100	100	100	100	100	100	100	100	100	**3.07**
	ANN	100	100	100	100	100	100	100	100	100	**7.17**
	SVM	100	100	100	100	100	100	100	100	100	**0.87**
LBP											
LBP (4)	KNN	88.3	76.98	83.33	80.94	87.17	71.11	84.44	81.72	76.67	**0.78**
	RF	97.39	91	80	90.32	97.57	87.78	93.18	93.63	83.68	**3.90**
	ANN	98.65	95.63	97.12	97.02	98.65	92.89	97.83	97.12	95.43	**7.23**
	SVM	79.55	76.8	84	93.59	77.17	72.22	81.86	86.76	84.78	**0.84**
FOS											
FOS (4)	KNN	92	88.44	80	94.02	82.6	83.33	90.9	83.52	80	**0.8**
	RF	88.28	83.7	75.43	96.58	83.7	77.78	92.24	83.7	80	**4.37**
	ANN	89.26	80	100	92.31	86.96	83.33	90.76	83.33	90	**2.48**
	SVM	77.33	84	100	99.15	78.48	71.11	86.89	75.45	70	**0.94**
SFTA											
SFTA (36)	KNN	96.46	91.66	100	93.16	95.65	100	94.78	93.62	100	**0.98**
	RF	95.73	92.55	100	95.73	94.57	88.89	95.73	93.55	94.12	**4.36**
	ANN	100	100	100	100	100	100	100	100	100	**5.63**
	SVM	99.15	96.81	100	100	98.91	83.33	99.57	97.85	90.91	**1.13**
Fusion											
Fusion (117)	KNN	92.6	90.74	87.33	90.5	98	85	92.54	94.23	88.57	**1.78**
	RF	100	100	100	100	100	100	100	100	100	**2.66**
	ANN	90.16	88.57	88.57	89.94	96	90	92	93.73	89.27	**19.7**
	SVM	88.46	87.39	100	89.84	97	100	89.15	91.94	100	**1.30**

Finally, for the classification experiment and evaluations, we used four different classifiers to classify feature vectors which represent abnormality ROI. These four classification methods are KNN, ANN, RF, and SVM. Each classifier has pros and cons in terms of time execution and accuracy as seen in Table 1. Cross validation is used for training and testing cases. 70% used for training cases and 30% for testing.

Table 1 shows the precision, recall, and accuracy for the three classes Normal, Benign, and Malignant using four feature extraction and fused future (GLCM, LBP, FOS, SFTA) and four classifiers (KNN, RF, ANN, SVM). The final decision is dependent on multiple classifier voting system based on weighted or a majority of

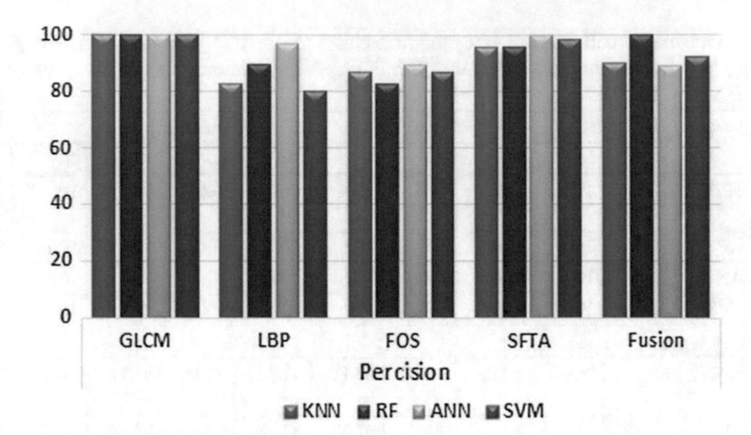

Fig. 3 The precision average for each classifier and feature extraction method

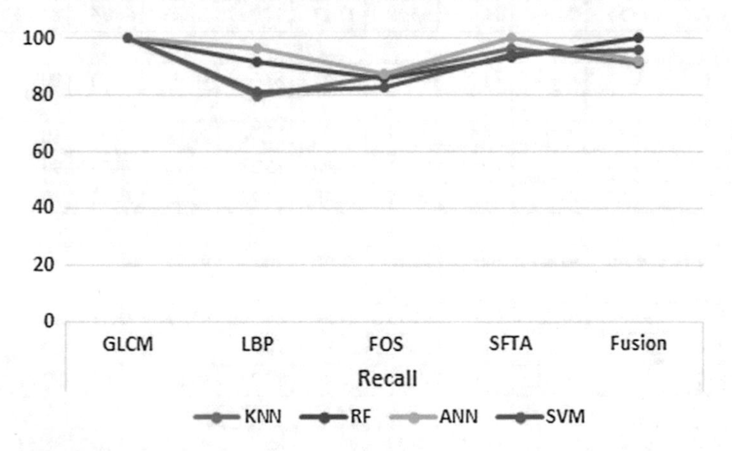

Fig. 4 The recall average for each classifier and feature extraction method

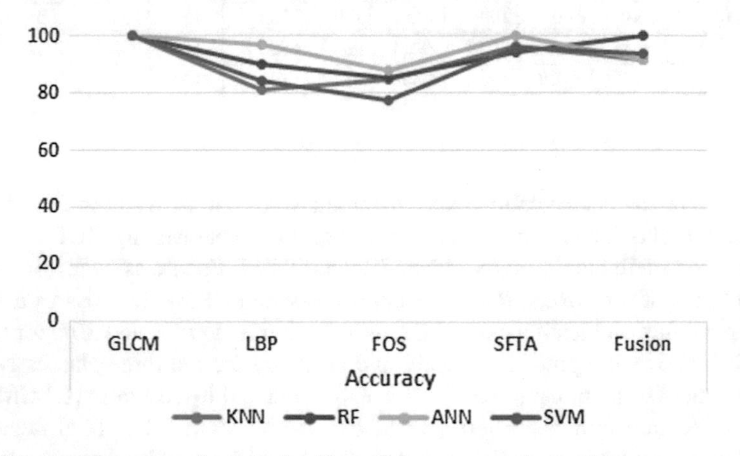

Fig. 5 The overall accuracy for each classifier and feature extraction method

voting system. From the results, the GLCM feature extraction method with all classifier has high accuracy with 100% for precision, recall, and accuracy Followed by SFTA method, the artificial neural network with SFTA feature extraction has high accuracy 100%. As we can see SVM classifier has very small time consumption in all feature extraction methods in comparison with ANN and RF classifiers. Figures 3, 4, and 5 showing the over-all precision, recall and accuracy for normal, benign, and malignant classes. As seen the high accuracy is obtained from GLCM.

7 Conclusion and Future Work

A CAD system is proposed to distinguish between benign and malignant tumors using normalized texture and shape features which extracted from multiple feature extraction methods (GLCM, LBP, SFTA, and FOS) to overcome the limitation of single-scale feature extraction, also fused features is used to increase the efficiency and stability of liver tumor diagnosis. Principle component analysis is used to reduce the feature dimensionality and decrease the homogeneity between features. In classification phase, multiple classifiers are used (KNN, SVM, ANN, and RF) with voting system to increase the efficiency of classification rate with high accuracy results. The proposed CAD system achieved a total classification performance of 100%. The proposed system is robust and can help doctors for further treatment. In the future, we plan to search for the features that are independent on the resolution (or on the reconstruction diameter) and take more in the consideration tumor shape, location, and size.

References

1. Jemal, A., Bray, F., Ferlay, M.M.J., Ward, E., Forman, D.: Global cancer statistics. CA: A Cancer J. clin. **61**(2), 69–90 (2011)
2. Duda, D., Kretowski, M., Bezy-Wendling, J.: Computer-aided diagnosis of liver tumors based on multi-image texture analysis of contrast-enhanced CT. Selection of the most appropriate texture features. Stud. Logic Gramm. Rhetor. **35**(1), 49–70 (2003)
3. Anter, A., Hassanien, A., Schaefer, G.: Automatic segmentation and classification of liver abnormalities using fractal dimension. In: Pattern Recognition (ACPR), 2nd IAPR Asian Conference, IEEE, pp. 937–941 (2013)
4. ElSoud, M.A., Anter, A.M.: Computational intelligence optimization algorithm based on meta-heuristic social-spider: case study on CT liver tumor diagnosis. Int. J. Adv. Comput. Sci. Appl. **1**(7), 466–475 (2016)
5. Anter, A.M., Hassanien, A.E., ElSoud, M.A., Kim, T.H.: Feature selection approach based on social spider algorithm: case study on abdominal CT liver tumor. In: 2015 Seventh International Conference on Advanced Communication and Networking (ACN), pp. 89–94. IEEE, July 2015
6. Gunasundari, S., Janakiraman, S.: A study of textural analysis methods for the diagnosis of liver diseases from abdominal computed tomography. Int. J. Comput. Appl. **74**(11), 59–67 (2013)

7. Anter, A.M., Hassenian, A.E.: Computer aided diagnosis system for mammogram abnormality, pp. 175–191. In Medical Imaging in Clinical Applications, Springer International Publishing (2016)
8. Gletsos, M., Mougiakakou, S., Matsopoulos, G., Nikita, S., Nikita, A., Kelekis, D.: A computer-aided diagnostic system to characterize CT focal liver lesions, design and optimization of a neural network classifier. IEEE Trans. Inf. Technol. Biomed. **7**(3), (2003)
9. Cavouras, D., Prassopoulos, P., Karangellis, G., Raissaki, M., Kostaridou, L., Panayiotakis, G.: Application of a neural network and four statistical classifiers in characterizing small focal liver lesions on CT. In: 18th Annual International Conference of the IEEE EMBS, Amsterdam (1996)
10. Chen, E., Chung, P., Tsai, H., Chang, C.: An automatic diagnostic system for CT liver image classification. IEEE Trans. Biomed. Eng. (1998)
11. Mougiakakou, S., Valavanis, G., Nikita, I., Nikita, A.: Differential diagnosis of CT focal liver lesions using texture features, feature selection and ensemble driven classifiers. Artif. Intell. Med. (2007)
12. Tao, G., Singh, A., Bidaut, L.: Liver segmentation from registered multiphase CT data sets with EM clustering and GVF level set. Int. Soc. Opt. Photonics SPIE Med. Imaging **7623**, 1–9 (2010)
13. Wang, L., Zhang, Z., Liu, J., Jiang, B., Duan, X., Xie, Q., Li, Z.: Classification of hepatic tissues from CT images based on texture features and multiclass support vector machines. In: Advances in Neural Networks, pp. 374–381, Springer, Berlin, Heidelberg (2009). ISN 2009
14. Mala, K., Sadasivam, V.: Wavelet based texture analysis of liver tumor from computed tomography images for characterization using linear vector quantization neural network. In: International Conference on IEEE in Advanced Computing and Communications (ADCOM2006), pp. 267–270 (2006)
15. Gunasundari, S., Ananthi, M.: Comparison and evaluation of methods for liver tumor classification from CT datasets. Int. J. Comput. Appl. **39**(18), 46–51 (2012)
16. Kumar, S., Moni, R.: Diagnosis of liver tumor from CT images using curvelet transform. Int. J. Comput. Sci. Eng. **2**(4), 1173–1178 (2010)
17. Mougiakakou, S., Valavanis, I., Nikita, K., Nikita, A., Kelekis, D.: Characterization of CT liver lesions based on texture features and a multiple neural network classification scheme. In: Engineering in Medicine and Biology Society, 2003. Proceedings of the 25th Annual International Conference of the IEEE, vol. 2, pp. 1287–1290. IEEE, Sept 2003
18. Vijayalakshmi, B., Bharathi, V.: Classification of CT liver images using local binary pattern with Legendre moments. Curr. Sci. **110**(4), 687–691 (2016)
19. Lee, C., Chen, S., Chiang, Y.: Automatic liver diseases diagnosis for CT images using kernel-based classifiers. In 2006 World Automation Congress pp. 1–5. IEEE, July 2006
20. Pavlopoulos, S., Kyriacou, E., Koutsouris, D., Blekas, K., Stafylopatis, A., Zoumpoulis, P.: Fuzzy neural network-based texture analysis of ultrasonic images. IEEE Eng. Med. Biol. Mag. **19**(1), 39–47 (2000)
21. Punia, R., Singh, S.: Automatic detection of liver in CT images using optimal feature based neural network. Int. J. Comput. Appl. **76**(15), 2013
22. Kumar, S., Moni, R.: Diagnosis of liver tumor from CT images using curvelet transform. Int. J. Comput. Sci. Eng. **2**(4), 1173–1178 (2010)
23. Alahmer, H., Ahmed, A.: Computer-aided classification of liver lesions using contrasting features difference. In: ICMISC 2015: 17th International Conference on Medical Image and Signal Computing, London, 27–28 Nov 2015
24. Namita, A., Agrawal, R.: First and second order statistics features for classification of magnetic resonance brain images. J. Signal Inf. Process. **3**, 146–153 (2012)
25. Anter, A., El Souod, M., Azar, A., Hassanien, A.: A hybrid approach to diagnosis of hepatic tumors in computed tomography images. Int. J. Rough Sets Data Anal. (IJRSDA) **1**(2), 31–48 (2014)
26. Costa, A., Humpire, F., Traina, G.: An efficient algorithm for fractal analysis of textures. In: IEEE Conference on Graphics, Patterns and Images (SIBGRAPI), pp. 39–46 (2012)

27. Zlem, A., Oguz, G.: Classification of multispectral images using random forest algorithm. J. Geodesy Geoinf. **1**(2) (2012)
28. Yu, B., Zhu, D.: Automatic thesaurus construction for spam filtering using revised: back propagation neural network. J. Expert Syst. Appl. **37**(1), 24–30 (2010)
29. Bishop, C.: Neural Networks for Pattern Recognition. Oxford University Press (1995)
30. Anter, A.M., ElSoud, M.A., Hassanien, A.E.: Automatic mammographic parenchyma classification according to BIRADS dictionary. Comput. Vis. Image Process. Intell. Syst. Multimed. Technol. IGI Glob. 22–37, 2014
31. Wu, Q., Zhou, D.: Analysis of support vector machine classification. Int. J. Comput. Anal. Appl. **8**, 99–119 (2006)

Computer-Aided Acute Lymphoblastic Leukemia Diagnosis System Based on Image Analysis

Ahmed M. Abdeldaim, Ahmed T. Sahlol, Mohamed Elhoseny and Aboul Ella Hassanien

Abstract Leukemia is a kind of cancer that basically begins in the bone marrow. It is caused by excessive production of leukocytes that replace normal blood cells. This chapter presents Computer-Aided Acute Lymphoblastic Leukemia (ALL) diagnosis system based on image analysis. It presented to identify the cells ALL by segmenting each cell in the microscopic images, and then classify each segmented cell to be normal or affected. A well-known dataset was used in this chapter (ALL-IDB2). The dataset contains 260 cell images: 130 normal and 130 affected by ALL. The proposed system starts by segmenting the white blood cells. This process includes sub-processes such as conversion from RGB to CMYK color model, histogram equalization, thresholding by Zack technique, and background removal operation. Then some features were extracted from each cell, each of them represents aspects of a cell. The extracted features include color, texture, and shape features. Then each feature set was exposed to three data normalization techniques z-score, min-max, and grey-scaling to narrow down the gap between the features values. Finally, different classifiers were used to validate the proposed system. The proposed diagnosing system achieved acceptable accuracies when tested by well-known classifiers; however, K-NN achieved the best classification accuracy.

A.M. Abdeldaim
Culture & Science City, 6th of October 15525, Egypt
e-mail: a7medabdeldaim@gmail.com

A.T. Sahlol (✉)
Damietta University, Damietta 34517, Egypt
e-mail: atsegypt@du.edu.eg

M. Elhoseny
Mansoura University, Mansoura, Egypt
e-mail: mohamed.elhoseny@unt.edu

A.E. Hassanien
Faculty of Computers and Information, Information Technology Department, Cairo University, Giza, Egypt
e-mail: aboitcairo@gmail.com

A.M. Abdeldaim · A.T. Sahlol · M. Elhoseny · A.E. Hassanien
Scientific Research Group in Egypt (SRGE), Cairo, Egypt

A.E. Hassanien and D.A. Oliva (eds.), *Advances in Soft Computing and Machine Learning in Image Processing*, Studies in Computational Intelligence 730,
https://doi.org/10.1007/978-3-319-63754-9_7

Keywords Leukemia • Acute lymphoblastic leukemia (ALL) • Image analysis and segmentation • Data normalization

1 Introduction

Microscopic blood cell images enable doctors for diagnosing several diseases. Leukaemia can be considered as a type of blood cancer, it can be detected via the medical analysis of white blood cells or leucocytes. French–American–British (FAB) determined two types of leukaemia; Chronic, which are subclassified into Chronic Myelogenous Leukaemia (CML) and Chronic Lymphocytic Leukaemia (CLL). Acute, which is subclassified into: Acute Lymphoblastic Leukaemia (ALL) and Acute Myelogenous Leukaemia (AML) [1].

Acute lymphoblastic leukemia can be diagnosed by the morphological identification of lymphoblasts by microscopy, the immunophenotypic assessment of lineage commitment, and developmental stage by flow cytometry [2]. Blood samples can be observed and diagnosed for different diseases by doctors. Any human-based diagnosing suffers from nonstandard precision as it basically depends on doctor's skill; also it is unreliable in a statistical point of view. There are currently various systems that can count the number of blood cells based on measuring the physical and chemical properties of blood cells using a light detector that uses fluorescence or electrical impedance to identify cell types [3].

Although the quantification results are precise, it costs much money, also it does not detect the morphological abnormalities of the cells; therefore, a complementary blood analysis based on microscopic image is required. Image processing and their subsections like image segmentation can provide solutions for counting the number of cells in the blood, and accordingly, it can provide valuable information about cells morphology. Automated diagnosing systems are more accurate and not temperamental like human-based systems. Also, they are statistically reliable and can be generalized. So, the white blood cell (WBS) affected by acute lymphoblastic leukemia will be counted and classified.

There are three main components of blood image: red blood cells, platelets, and leucocytes. The red blood cells transport oxygen from the hurt to all organs and living tissues and in the same time, they carry away carbon dioxide. They are present at a percentage up to 50% of the total blood volume. RBCs' diameter is 6–8 μm. The WBCs play an important role in the body's immune system as they defend the body against infection and diseases. Therefore, analysis and classification of WBC are essential.

WBCs can be classified into two groups (by the presence of granules in the cytoplasm). The first group is Granulocytes, which includes basophil, eosinophil, and neutrophil. Basophil is responsible for allergic reaction and antigen, basophil's granules are of irregular distribution, they represent only 0–1% of all lymphocytes in human blood [4]. Eosinophil plays an important role in killing parasites, they present at 1–5% in human blood [5]. Neutrophil is most abundant in the blood stream.

Table 1 Normal and affected Lymphocyte cells

Type	Lymphocyte	
WBC	Normal cells	Affected cells
Samples from dataset [9]	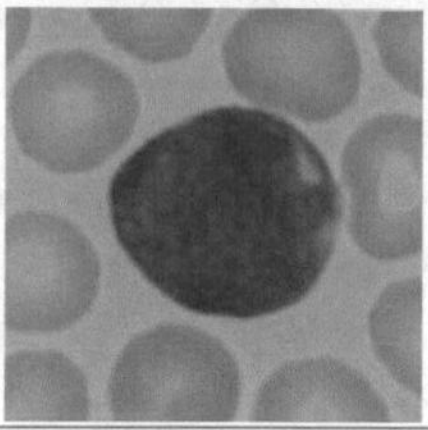	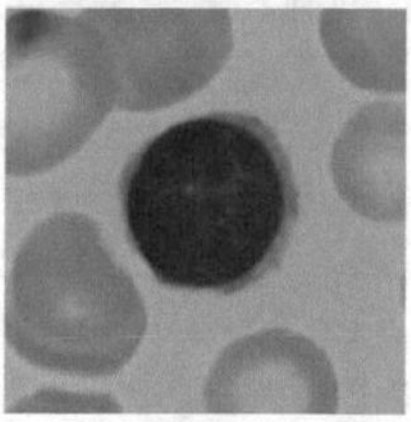

It has multiple-lobed nuclei, they present in human blood at a percentage ranging between 50 and 70% [6]. The second group is agranulocytes, it includes lymphocyte and monocyte. Lymphocyte is very common in human blood, with a percentage of 20–45% [7]. Monocytes are the largest WBCs; they represent 3–9% of circulating leucocytes [8].

Lymphocytes are regularly shaped and have a compact nucleus with regular and continuous edges. On the contrary, lymphocytes that are suffering from ALL are called lymphoblast. They are irregularly shaped and contain small cavities in the cytoplasm (termed vacuoles) and spherical particles within the nucleus (termed nucleoli), the more morphological changes increasing the more severity of the disease indicated [9]. Healthy and acute lymphoblastic leukemia lymphocytes are shown in Table 1.

Although, leukocytes can be easily identified (as they appear darker "purple" than the background), but due to wide variations in their shapes, dimensions, and edges, the analysis and the processing become very complicated. The generic term leucocyte refers to a set of cells that are quite different from each other. Thus, these cells can be distinguished according to their shape or size.

Image analysis techniques were adopted in several fields; medical [10–12] and others [13, 14]. Mohapatrain [15] used an ensemble classifier system for the early diagnosis of acute lymphoblastic leukemia in blood microscopic images. The identification and segmentation of WBCs were done followed by extracting different types of features. Features covered most aspects of WBC such as shape, contour, fractal, texture, color, and Fourier descriptors. Finally, an ensemble of classifiers is trained to recognize acute lymphoblastic leukemia. The results of this method were good, however, the reproducibility of the experiment and comparisons with other methods are not possible as the dataset was not available for public. In [16], it was proposed a technique to segment the acute leukemia cell images by transforming the RGB color space to C-Y color space in the C-Y color space, the luminance component is used to segment (ALL). The proposed algorithm runs on 100 microscopic ALL images and the experimental result shows that it achieved a good segmentation of ALL from its complicated background, as the segmentation accuracy reached 98.38%.

While in [17], two sets of blood WBC images were used in this study's experiments. The first contains 555 images with 601 white blood cells. They were collected from Rangsit University. The second contains 477 cropped WBC images which were downloaded from CellaVision.com. The proposed system comprises a preprocessing step, nucleus segmentation, cell segmentation, feature extraction, feature selection, and classification. Naïve Bayes classifiers were applied for performance comparison. It was found that the proposed method is consistent and coherent in both datasets, with dice similarity of 98.9 and 91.6% for average segmented nucleus and cell regions, respectively. Furthermore, the overall correction rate in the classification phase is about 98 and 94% for linear and naïve Bayes models, respectively.

In [18], an automated blast counting method to detect acute leukemia in blood microscopic images was proposed to identify the WBCs through a thresholding operation performed on the HSV color space S component. Morphological erosion for image segmentation was done. The results of this study were promising; however, no features or classifiers were presented. Also, there is no method to determine the optimum threshold for segmentation. Also in [3], the whole leucocyte was isolated and then the nucleus and cytoplasm were separated, then different features, such as shape, color, and texture are extracted. The feature set was used to train different classification. 245 of 267 total leucocytes were properly identified (92% segmentation accuracy). Different classification models were tested; however, the support vector machine with a Gaussian radial basis kernel has the highest performance for the identification of acute lymphoblastic leukemia by achieving a 93% of accuracy and a 98% of sensitivity.

In this chapter, it is focused only on the acute lymphoblastic leukemia (ALL), which is well known as it affects basically children and older people. Children under 5 years and older people over 50 years are at higher risk of acute lymphoblastic leukemia, also, it can be fatal if it is not treated earlier as it is rapidly spread into some vital organs and the bloodstream too [19]. Therefore, earlier diagnosis greatly aids in providing the appropriate treatment for ALL and is developing a fully automated computer-aided diagnosis system for detection, segmentation, feature extraction, and classification of WBCs affected by acute lymphoblastic leukemia. The system inputs are microscopic images (some affected by ALL while others not), and the decision is taken by the model is the identifying whether each image is suffering from acute lymphoblastic leukemia or not.

The remainder of this chapter is presented as follows. Section 1 gives an overview of the related works; Sect. 2 describes in details the proposed system. The results and discussions are presented in Sect. 3. Finally, the conclusions and future chapter are discussed in Sect. 4. The proposed computer-aided acute lymphoblastic leukemia diagnosis system.

2 The Proposed System

The proposed computer-aided acute lymphoblastic leukemia diagnosis system aims to optimally select the most powerful features that can be used in the lymphoblastic leukemia diagnosis system. The proposed lymphoblastic leukemia diagnosis system consists of three basic phases: Image segmentation, feature extraction, and classification. The overall architecture of the proposed system and its phases are summarized in Fig. 1.

2.1 *Cell Segmentation Phase*

Unlike many methods in the literature, the proposed system detects the nuclei and the entire membrane at the same time.

Each image in the dataset contains only one cell. The images are in RGB color space which is difficult to be segmented. So, the images were converted to CMYK color space (Eqs. 1–4). In fact, leukocytes are more contrasted in the Y component of CMYK color model because the yellow color is presented in all elements of the image; then Y was used as the whole cell and M as the nucleus.

$$K = \min(255 - r, \min(255 - G, 255 - B)) \tag{1}$$

$$C = \frac{255 - R - K}{255 - K} \tag{2}$$

$$M = \frac{255 - G - K}{255 - K} \tag{3}$$

$$Y = \frac{255 - B - K}{255 - K} \tag{4}$$

For Y component, segmentation is a bit easier because the whole cell exists in pure black color; so the non-black elements were ignored in the image. However, for M component, it was a bit challenging because of the variation of magenta color levels in the image. Therefore, as in [3], the segmentation was performed by Zack algorithm [20] which was used to determine the thresholding value. Before it is applied, a histogram for the grayscale cell image has to be produced then the intensity values of it are exposed to the Zack's. The thresholding value estimation by Zak's can be calculated by constructing a line between the maximum (a) and the minimum (b) of the histogram grey level axis, where the minimum (b) is larger than 0. The distance (L) which is the length between the line and the histogram is computed for all values between (b) and (a). The thresholding value is the value of the grey intensity axis when the distance (L) reaches its maximum value. This algorithm is particularly effective when the object pixels produce a weak peak in the histogram. Figure 2 shows how threshold value can be calculated by Zack algorithm.

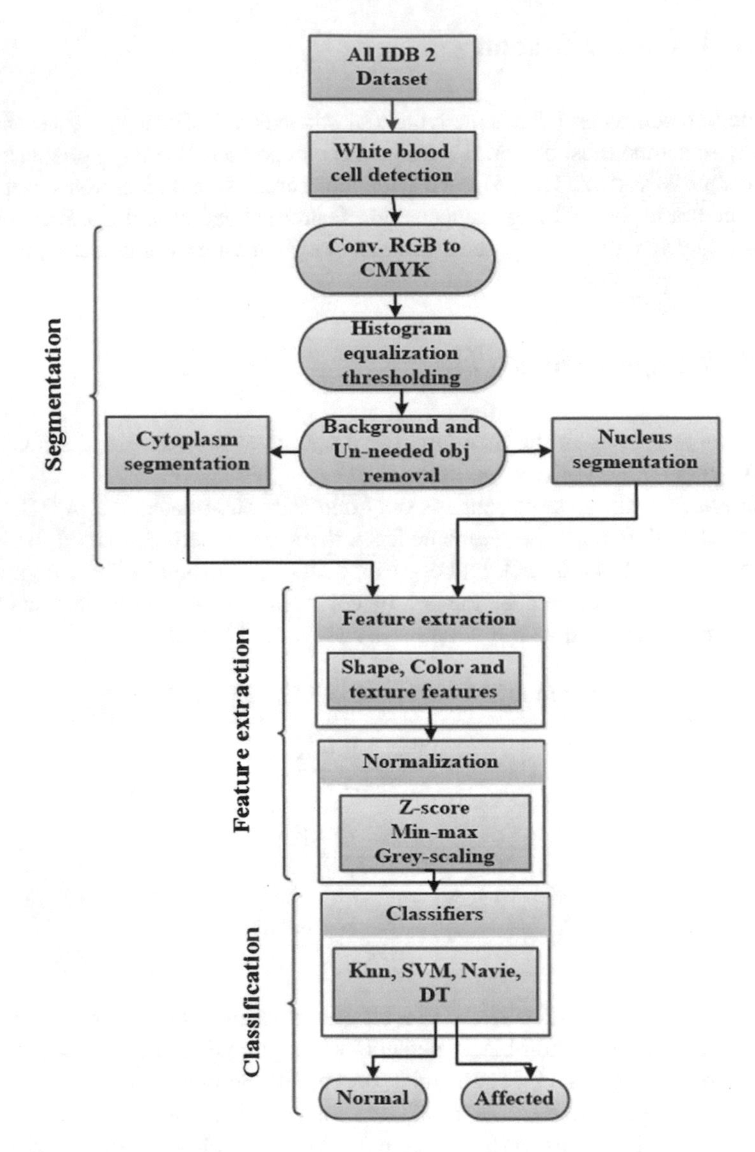

Fig. 1 The proposed lymphoblastic leukemia diagnosis System

Cleaning the image edge is a simple task. In this chapter, in order to remove the nonlymphocyte cells, the roundness equation (Eq. 5) was used, as the lymphocyte cell has a unique circular shape, so it measures how closely the morphology of a shape to be a circle is. Roundness is dominated by the shape's gross features rather

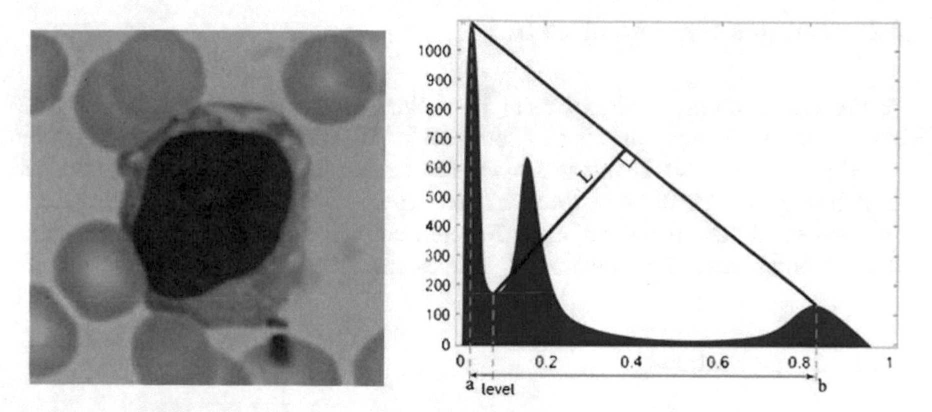

Fig. 2 *First* Image from the data set. *Second* threshold calculated by Zack's

than the definition of its edges and corners, or the surface roughness of a manufactured object:

$$\text{Roundness} = \frac{4 \times \pi \times \text{Area}}{\text{Convex Perimeter}} \tag{5}$$

If roundness is equal to 1 then the shape is circular, the closer the roundness to 1 the closer the shape has a circular shape. Roundness is relatively insensitive to irregular boundaries. After several trials on the images, it was assumed that the best minimum roundness value for a Lymphocyte cell is 0.52. Beside roundness, the solidity of the object, solidity (Eq. 6) was calculated. It can be defined as the ratio of the area of an object to the area of a convex hull of an object.

$$\text{Solidity} = \frac{\text{Area}}{\text{Convex Area}} \tag{6}$$

A solidity value of 1 signifies a solid object, and a value less than 1 represents that an object has an irregular boundary (or containing holes). After several trials, it was considered the minimum value of solidity for a lymphocyte cell is to be 0.98.

So, if the roundness and the solidity of an object are more than the assumed values then this object classified is as a lymphocyte cell and it proceeds to the final step, otherwise, it does nothing. Filling hole method [20] was used to recover images from distortion of important pixels resulted after the previous preprocessing operations.

2.2 Feature Extraction Phase

Feature extraction in image processing is a technique of transforming the input data into the set of features. In order to construct an effective feature set, several published articles were studied, it was noted that certain features were widely used as they gave good classification results. Three types of features were extracted from the segmented cells including shape features, color features, and texture features. Table 2 summarizes the all extracted features and their numbers.

2.2.1 Shape Features

Starting from binary images of the nucleus and the nucleus with cytoplasm, the extracted shape features include: Area, Perimeter, Major Axis Length, Minor Axis Length, Convex Area, Convex Perimeter, Orientation, Roundness, Solidity, Elongation, Eccentricity, Rectangularity, Compactness, Convexity, and the Area Ratio. All these features were calculated for two types of images (nucleus and nucleus with cytoplasm). In this chapter, 30 shape features were extracted.

2.2.2 Color Features

The color features are the most discriminatory features of blood cells. The mean, standard deviation, skewness, kurtosis, and entropy were calculated from the three channels (red, green, blue) of each image. The extracted color features included only 15 features.

2.2.3 Texture Features

The texture features were extracted from each cell image in the grey level format. The following descriptors were evaluated: Autocorrelation, Contrast, Correlation, Cluster Prominence, Cluster Shade, Dissimilarity, Energy, Homogeneity 1, Homogeneity 2, Maximum Probability, Sum of Squares Variance, Sum Average, Sum Variance, Sum Entropy, Difference Variance, Difference Entropy, Information Measure of Correlation 1, Information Measure of Correlation 2, Inverse Difference is Homom, Inverse Difference Normalized, and Inverse Difference Moment Normalized. These features

Table 2 Types of the extracted features

Feature type	Numbers
Color features	15
Shape features	30
Texture feature	84

were calculated for the 0, 45, 90, and 135 angles. The extracted texture feature set contains 84 features.

Unfortunately, the shape features cannot only be trusted because they are sensitive to segmentation errors. Thus, these features were combined together with regional features, which are less susceptible to errors. We have one more feature set that was hybridized by color and texture features. The combination of shape, color, and texture features produces a feature set of 129 features.

2.3 *Feature Normalization*

To narrow down the gap between the highest and the lowest value of extracted features and to improve the classification results. Three different normalization techniques were applied; they include grey-scaling, min-max and Z-score techniques.

2.3.1 Grey-Scaling

It is an image normalization technique used to convert a matrix to a greyscale image. This can be performed by scaling the entire image to the range of brightness values from 0 to 1. It works by normalizing each individual columns or rows to a range of brightness values from 0 to 1.

2.3.2 Min-Max

Data are scaled to a fixed range usually 0–1. The cost of having this bounded range in contrast to standardization is that it will be ended up with smaller standard deviations, which can suppress the effect of outliers. Min-max scaling (Eq. 7) is typically done via thc following equation:

$$X_i = \frac{X_i - X_{\min}}{X_{\max}}, \tag{7}$$

where X_i is the original feature vector, $X_{\min}$ is the minimum value of that feature vector, and $X_{\max}$ is the maximum value of it.

2.3.3 Z-Score

In this normalization method, the mean and the standard deviation of each feature are calculated (Eq. 8). Next, the mean was subtracted from each feature. Finally, the product values were divided by the standard deviation.

$$Z_i = \frac{X_i - \overline{X}}{\sigma}, \tag{8}$$

where X_i is the original feature vector, X is the mean of that feature vector, and σ is its standard deviation.

2.4 Classification Phase

The proposed system can be considered a binary classification problem because there are only two classes (normal or affected cell) as outputs and number of variables (features) as inputs. In this chapter, as there are different number of features (starting from 5 to 127), several classifiers were used. The extracted feature sets were tested by k-Nearest Neighbor (k-NN) [21] using the Euclidean distance measure with different values of k, Naive Bayes (NB) [22, 23], by a Gaussian (G) and kernel data distribution (K), support vector machines [24] with different kernels and Decision Trees [25]. The chosen classifiers proved to achieve acceptable results with other pattern recognition problems [26, 27]. The performance of the models was evaluated using a k-fold, cross validation. Considering k = 10, the whole dataset is randomly divided into tenfolds.

3 Experimental Results and Discussions

The proposed system was implemented by "MATLAB 2016b" on "Windows 7 (64 bit). The dataset was provided by Department of Information Technology – Università degli Studi di Milano [9]. The dataset was captured with an optical laboratory microscope coupled with a Canon PowerShot G5 camera. All images are in JPG format with 24-bit color depth. The images were taken with a different magnification of microscope ranging from 300 to 500. The ALL-IDB database has two distinct versions (ALL-IDB1 and ALL-IDB2). The proposed system worked on the ALL-IDB2 version which has been designed for testing the performances of classification systems. The ALL-IDB2 is a collection of cropped area of interest of normal and blast cells that belongs to the ALL-IDB1 dataset. It contains 260 images, half of them lymphoblast's. ALL-IDB2 images have similar grey level properties to the images of the ALL-IDB1, except the image dimensions.

The accuracy (Eq. 9) was used to test the classification performance of the proposed system:

$$\text{Accuracy} = \frac{\text{TP} + \text{TN}}{\text{TP} + \text{TN} + \text{FP} + \text{FN}}, \tag{9}$$

where

- TP (True positives) is the number of elements that are correctly classified as positive by the test.
- TN (True negatives) is the number of elements that are correctly classified as negative by the test.
- FP (False positive) is also known as type I error, it is the number of elements that are classified as positive by the test, but they are not.
- FN (True positive) is also known as type II error, it is the number of elements that are classified as negative by the test, but they are not.

In order to investigate the performance of the segmentation, the output of the segmentation process was compared to that was provided by the data set truth table. Figures 3, 4 and 5 show the segmentation results of some images from the data set.

The segmentation accuracy was tested using (Eq. 9) and the achieved accuracy reached 99.2%, represents 189 true positives, 69 true negatives, 2 false positives, and there were no false negative cell images out of 260 cell images. As mentioned, the proposed system produced several feature sets (six feature sets). Each was exposed to three data normalization techniques, so 18 feature sets were produced. Tables 3, 4, 5, 6, 7, and 8) represent the classification accuracy of each of the extracted feature sets when tested by k-nearest neighbor, Naive Bayes by a Gaussian (G), and kernel data distribution (K), support vector machines and decision trees classifiers.

In order to investigate the performance of the used classifiers, the best classification accuracy achieved by each of them was presented, as seen in Fig. 7.

From Fig. 7, it is obvious that the classification accuracy did not go lower than 86%; however, the best classifier was K-NN. This matches with our previous work [28], that K-NN has several parameters like the distance function as well as the number of neighbors that can adapt the learning mechanism toward improving the classification performance. As three normalization techniques were adopted in this

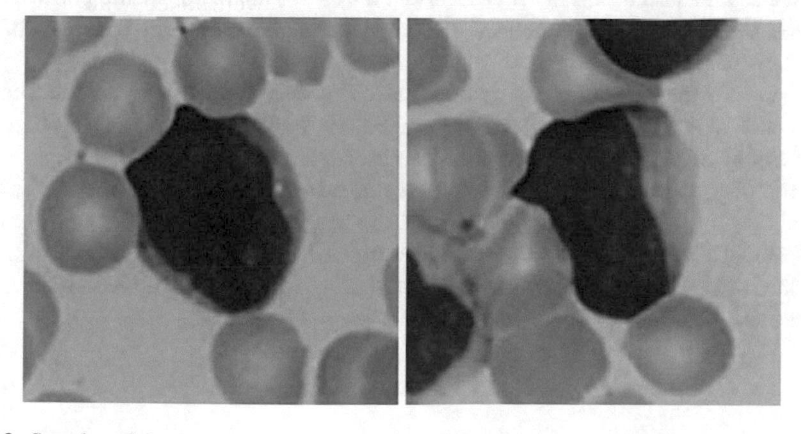

Fig. 3 Samples of the data sets

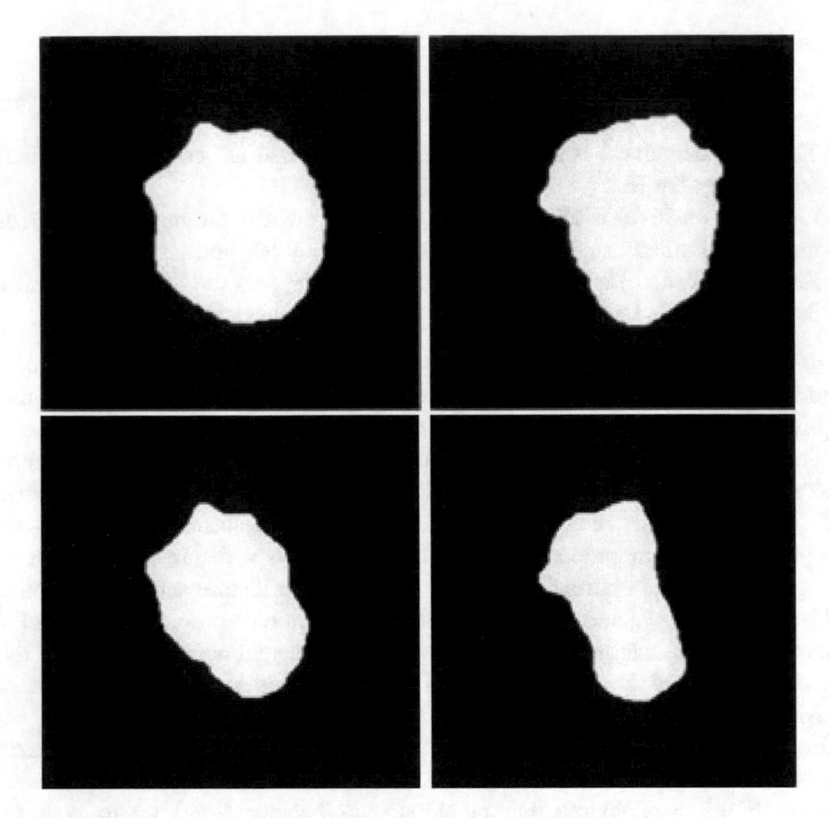

Fig. 4 Leucocyte segmentation. *Up* whole leucocyte images. *Down* nucleus images

chapter, so a comparison between the best and the worst classification accuracy for each of them is shown in Fig. 8.

Figure 8 shows the three used normalization techniques and the best and the worst classification accuracy. It is seen that they have almost the same performance for the best experiment; however, the grey-scaling has the best accuracy of the worst experiments.

The performed experiments included also the relationship between the number of features and the classification accuracy. Figure 9 shows the relationship between the number of features and the achieved accuracy.

From Fig. 9, it is noted that the accuracy gets better the when number of features increased for all the used classifiers. Also, the worst accuracy did not go lower than 80% except for the Naive Bayes. It is also noticed that the gap between the lowest and the highest number of features in terms of accuracy is not high as it does not exceed 5%.

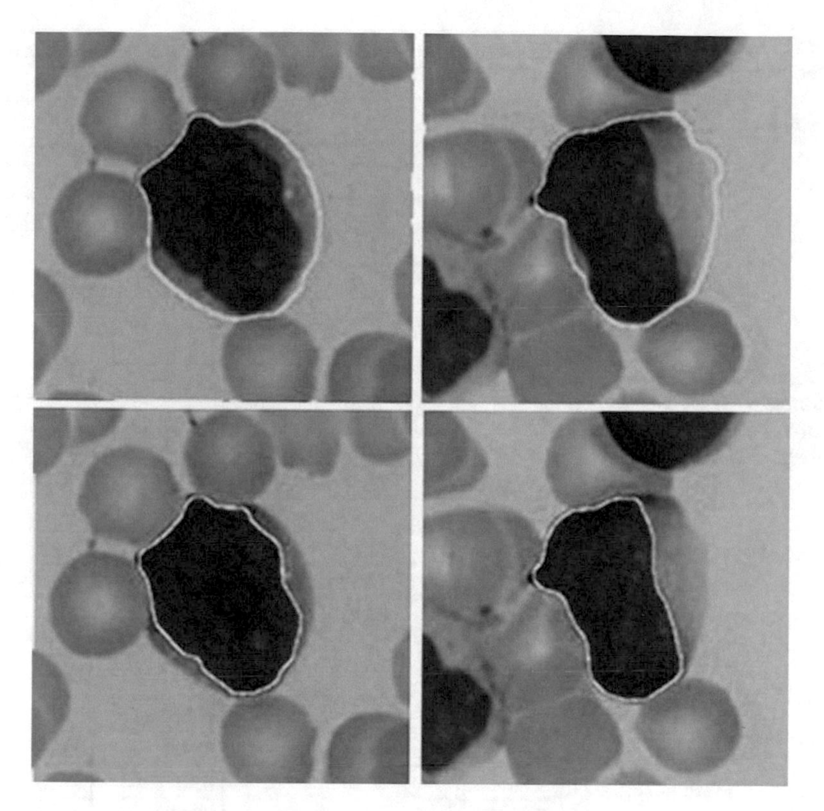

Fig. 5 Leucocyte identification (border highlighted). *Up* whole leucocyte cell images. *Down* nucleus cell images

Table 3 Classification performance of some texture features with the three normalization techniques

Classifier	Grey-scaling	Z-Score	Min-max
Knn	**95.78**	88.82	89.25
SVM-RBF	90.87	89.27	83.64
SVM-L	89.07	89.87	68.35
SVM-P	92.18	90.56	88.46
NB-G	83.5	82.12	78.3
NB-K	86.14	86.78	81.94
TREE	88.48	88.59	86.13

Table 4 Classification performance of all texture features with the three normalization techniques

Classifier	Grey-scaling	Z-Score	Min-max
Knn	90.94	**95.99**	92.57
SVM-RBF	91.49	69.77	93.41
SVM-L	90.14	92.55	89.58
SVM-P	88.96	87.83	92.51
NB-G	79.66	84.09	84.4
NB-K	82.7	86.81	86.44
TREE	79.66	89.28	85.47

Table 5 Classification performance of some color features with the three normalization techniques

Classifier	Grey-scaling	Z-Score	Min-max
Knn	**88.95**	88.66	87.96
SVM-RBF	82.38	88.12	77.97
SVM-L	77.7	86.98	69.75
SVM-P	85.4	88.77	86.08
NB-G	80.58	83.14	72.89
NB-K	82.24	82.29	77
TREE	88.02	85.43	85.09

Table 6 Classification performance of all color features with the three normalization techniques

Classifier	Grey-scaling	Z-Score	Min-max
Knn	86.93	89.63	87.57
SVM-RBF	87.53	89.83	88.45
SVM-L	72.43	81.52	73.48
SVM-P	91.62	90.35	**92.52**
NB-G	72.16	77.61	76.89
NB-K	79.52	80.23	79.38
TREE	84.51	85.715	89.41

Table 7 Classification performance of colors and texture features with the three normalization techniques

Classifier	Grey-scaling	Z-Score	Min-max
Knn	**96.42**	90.69	93.63
SVM-RBF	94.41	65.3	93.57
SVM-L	92.51	93.73	95.34
SVM-P	94.06	88.59	94.87
NB-G	87.85	82.83	90.34
NB-K	85.73	83.3	89.45
TREE	84.59	82.15	90.93

Table 8 Classification performance of colors, shape and texture features with the three normalization techniques

Classifier	Grey-scaling	Z-Score	Min-max
Knn	**96.01**	84.69	81.38
SVM-RBF	92.80	65.98	82.68
SVM-L	93.43	91.54	89.05
SVM-P	93.89	89.01	91.15
NB-G	89.97	82.81	83.89
NB-K	86.02	81.99	82.62
TREE	86.81	82.29	81.55

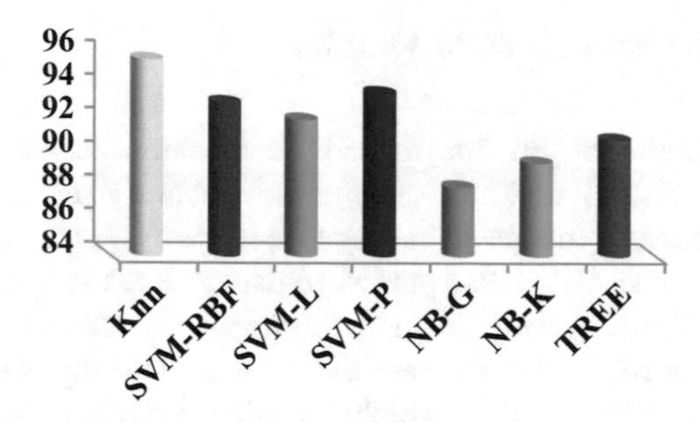

Fig. 7 Different accuracies based on different classifiers

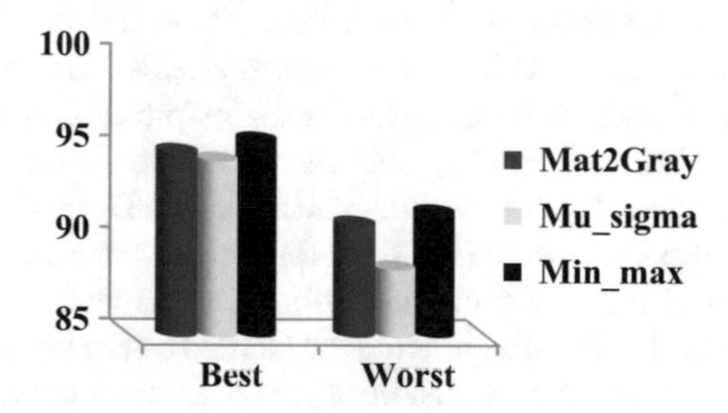

Fig. 8 Different accuracies based on different normalization methods

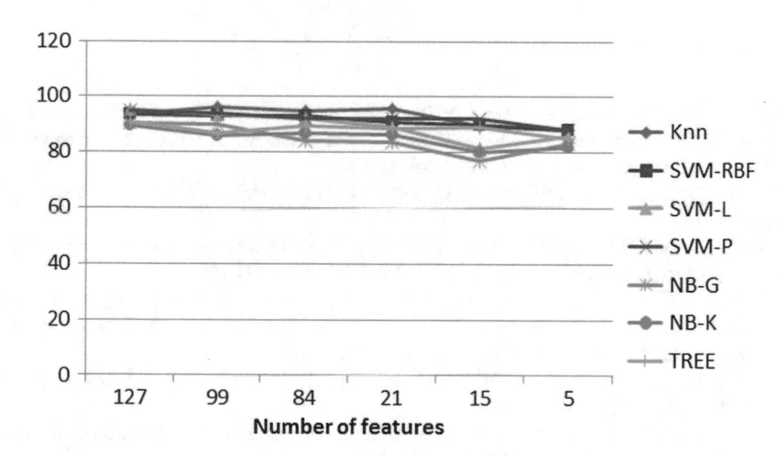

Fig. 9 Different accuracies based on different number of features

4 Conclusions and Future Works

This chapter focused on lymphocyte cells which are affected by lymphoblastic cancer. The main objective of this chapter is to identify the lymphocyte by segmenting the microscopic images then diagnose (classify) each segmented cell to be normal or affected. The ALL-IDB2 published dataset was used in this chapter to test the efficiency of the proposed diagnosis system. The dataset contains 260 cell images; 50% are normal, and the rest 50% are affected by ALL. The proposed system starts with segmentation which included the conversion from RGB to CMYK color model then contrast stretching by histogram equalization then thresholding and noise removal including background removal operation. Second, some features were extracted from each cell. They can be classified as color, texture, and shape features. Third, three data normalization techniques (z-score, min-max and grey-scaling) were applied to each extracted features to improve the classification performance. Finally, different classifiers were used (K-Nearest Neighbor, Naive Bayes, Support vector machines and Decision Trees) to test the efficiency of the proposed system. The performance of the proposed system was acceptable in terms of the segmentation performance as well as the accuracy of all classifiers, especially K-NN which achieved the best classification accuracy. The future work might include the extension of this chapter to cover the other kind of white blood cancers.

References

1. Bennett, J.M., Catovsky, D., Daniel, M.T., Flandrin, G., Galton, D.A., Gralnick, H.R., et al.: Proposals for the classification of the acute leukemias. French–American–British (FAB) co-operative group. Br. J. Hematol. (1976)
2. Inaba, H., Greaves, M., Mullighan, C.G.: Acute lymphoblastic leukaemia. Lancet **381**, 1943–1955 (2013)
3. Putzu, L., Caocci, G., Di Ruberto, C.: Leucocyte classification for leukaemia detection using image processing techniques. Artif. Intell. Med. **62**, 179–191 (2014)
4. Qiu, H.N., Wong, C.K., Chu, I.M., Hu, S., Lam, C.W.: Muramyl dipeptide mediated activation of human bronchial epithelial cells interacting with basophils: a novel mechanism of airway inflammation. Clin. Exp. Immunol. **172**, 81–94 (2013)
5. Meeusen, E.N., Balic, A.: Do eosinophils have a role in the killing of helminth parasites?. Parasitol. Today **16**, 95–101 (2000)
6. Kolaczkowska, E., Kubes, P.: Neutrophil recruitment and function in health and inflammation. Nat. Rev. Immunol. **13**, 159–175 (2013)
7. Thompson, S.C., Bowen, K.M., Burton, R.C.: Sequential monitoring of peripheral blood lymphocyte subsets in rats. Cytometry **7**, 184–193 (1986)
8. Brown, A.L., Zhu, X., Rong, S., Shewale, S., Seo, J., et al.: Omega-3 fatty acids ameliorate atherosclerosis by favorably altering monocyte subsets and limiting monocyte recruitment to aortic lesions. Arterioscler. Thromb. Vasc. Biol. **32**, 2122–2130 (2012)
9. DonidaLabati, R., Piuri, V., Scotti, F.: ALL-IDB: The acute lymphoblastic leukemia image data base for image processing. In: The 18th IEEE International Conference on Image Processing (ICIP), pp. 2045–2048 (2011)

10. Mostafa, A., Fouad, A., Elfattah, M.A., Hassanien, A.E., Hefny, H., Zhu, S.Y., Schaefer, G.: CT liver segmentation using artificial bee colony optimisation. Procedia Comput. Sci. **60**, 1622–1630 (2015)
11. Zidan, A., Ghali, N.I., Hassanien, A.E., Hefny, H., Hemanth, J.: Level set-based CT liver computer aided diagnosis system. J. Intell. Robot. Syst. **7**, Number S13 (2012)
12. Anter, A.M., Hassenian, A.E., ElSoud, M.A., Tolba, M.F.: Neutrosophic sets and fuzzy C-means clustering for improving CT liver image segmentation. In: The 5th International Conference on Innovations in Bio-Inspired Computing and Applications, Ostrava, Czech Republic, 22–24 June 2014
13. Sahlol, A.T., Suen, C.Y., Zawbaa, H.M., Hassanien, A.E., Elfattah, M.A.: Bio-inspired BAT optimization algorithm for handwritten arabic characters recognition. In: 2016 IEEE Congress on Evolutionary Computation (WCCI-2016), Vancouver, Canada, pp. 1749–1756, 22–24 July 2016
14. Elhoseny, M., Farouk, A., Batle, J., Abouhawwash, M., Hassanien, A.E.: Secure image processing and transmission schema in cluster-based wireless sensor network. In: Handbook of Research on Machine Learning Innovations and Trends (2017)
15. Mohapatra, S., Patra, D., Satpathy, S.: An ensemble classifier system for early diagnosis of acute lymphoblastic leukemia in blood microscopic images. J. Neural Comput. Appl. **24**(7–8), 1887–1904 (2014)
16. Mohammed, R., Nomir, O., Khalifa, I.: Segmentation of acute lymphoblastic leukemia using C-Y color space. Int. J. Adv. Comput. Sci. Appl. (IJACSA) **5**(11), 99–101 (2014)
17. Prinyakupt, J., Pluempitiwiriyawej, C.: Segmentation of white blood cells and comparison of cell morphology by linear and naíve Bayes classifiers. Biomed. Eng. OnLine **14**(1), 63 (2015)
18. Halim, N.H.A., Mashor, M.Y., Hassan, R.: Automatic blasts counting for acute leukemia based on blood samples. Int. J. Res. Rev. Comput. Sci. **2**(4), 971–976 (2011)
19. Biondi, A., Cimino, G., Pieters, R., Pui, C.H.: Biological and therapeutic aspects of infant leukemia. Blood **96**, 24–33 (2000)
20. Zack, G., Rogers, W., Latt, S.: Automatic measurement of sister chromatid exchange frequency. J Histochem. Cytochem. **25**(7), 741–753 (1977)
21. Cover, T.M., Hart, P.E.: Nearest neighbor pattern classification. IEEE Trans. Inf. Theory **13**, 21–27 (1967)
22. Duda, R.O., Hart, P.E.: Pattern Classification and Scene Analysis. Wiley, New York (1973)
23. Langley, P., Iba, W., Thompson, K.: An analysis of Bayesian classifiers. In: Rosenbloom, Paul, Szolovits, Peter (eds.) Proceedings of the Tenth National Conference on Artificial Intelligence, pp. 223–228. The AAAI Press, Menlo Park, CA (1992)
24. Vapnik, V.: Statistical Learning Theory. Wiley (1998)
25. Quinlan, J.R.: Induction of decision trees. Mach. Learn. (1986)
26. Sahlol, A.T., Suen, C.Y., Zawbaa, H.M., Hassanien, A.A., AbdElfattah, M.: Bio-inspired BAT optimization technique for handwritten Arabic characters recognition. In: IEEE Congress on Evolutionary Computation (WCCI-2016), Vancouver, Canada, pp. 1749–1756 (2016)
27. Sahlol, A.T., AbdElfattah, M., Suen, C.Y., Hassanien, A.A.: Particle swarm optimization with random forests for handwritten Arabic recognition system. In: Proceedings of the International Conference on Advanced Intelligent Systems and Informatics (AISI 2016), Cairo, Egypt, pp. 437–446 (2016)
28. Sahlol, A.T., Suen, C.Y., Elbasyoni, M.R., Sallam, A.A.: Investigating of preprocessing techniques and novel features in recognition of handwritten Arabic characters. In: Artificial Neural Networks in Pattern Recognition, pp. 264–276. Springer International Publishing (2014)
29. Soille, P.: Morphological Image Analysis: Principles and Applications, pp. 170–171. Springer (1999)

Telemammography: A Novel Approach for Early Detection of Breast Cancer Through Wavelets Based Image Processing and Machine Learning Techniques

Liyakathunisa Syed, Saima Jabeen and S. Manimala

Abstract Telehealth monitoring is an innovative process of synergising the benefits of information and communication technologies (ICT) and Internet of Things (IoT) to deliver healthcare services to remote, distant, and underserved regions. The objective of this study is to deliver healthcare services to patients outside the conventional settings by connecting the patient and healthcare providers with technology. As technologies for telehealth monitoring have become more advanced, they have become fully integrated into delivery of healthcare service. One of the most publicized telehealth services is the use of telemammography in the early diagnosis of breast cancer from remote and rural locations. Automated detection and classification of tumor in telemammographic images is of high importance for physicians for accurate prediction of the diseases. This study presents advances in telehealth services and also proposes novel telemammography system for early detection of breast cancer from remote and underserved areas. In this study, we have used efficient wavelet-based image processing techniques for preprocessing, detection, and enhancing the resolution of mammographic images. A detailed comparative analysis is performed to select the best classification model using different classification algorithms. We have used Multi-Layer Perceptron Neural Networks, J48 decision trees, Random Forest, and K-Nearest Neighbor classifier for classifying the tumor into three categories namely: benign, malignant, and normal. The classification is based on the area, volume, and boundaries of tumor masses. All the tumor features and

L. Syed (✉)
Department of Computer Science, Prince Sultan University, Riyadh, Saudi Arabia
e-mail: lsyed@psu.edu.sa

S. Jabeen
Department of Computer Science, COMSATS Institute of Information Technology (CIIT), Abbottabad, Pakistan
e-mail: uop.saima@gmail.com

S. Manimala
Department of Computer Science, Sri Jayachamarajendra College of Engineering, Mysore, India
e-mail: malasjce@gmail.com

A.E. Hassanien and D.A. Oliva (eds.), *Advances in Soft Computing and Machine Learning in Image Processing*, Studies in Computational Intelligence 730,
https://doi.org/10.1007/978-3-319-63754-9_8

classification methods are compared using Accuracy, Sensitivity, Specificity, Precision, and Mean Square Error. Experimental results on the Mammographic Image Analysis Society (MIAS) database are found to give the best results when neural network classifier is used for classification of mammographic images.

1 Introduction

Health care is one of the worlds biggest economic challenges, it is estimated that over 200 million people in the world suffer from chronic diseases like Alzheimer's Disease, Arthritis, Asthma, Cancer, Chronic Obstructive Pulmonary Disease (COPD), Cystic Fibrosis, Diabetes, and Heart diseases. The aging population will further place significantly higher burden on the healthcare systems. In 2010, there were 207 million people over 75 years of age worldwide. This figure is estimated to rise to 265 million by 2020 [1]. All of these challenges drive for an immense need for telehealth services for remote monitoring of patients for improved patient health care and reduce cost. Telehealth can be defined broadly as the use of electronic information and telecommunications technologies to provide access to health assessment, diagnosis, intervention, consultation, supervision information, and education across distance [2].

The use of Information and Communication Technology (ICT) and Internet of Things (IoT) has dramatically revolutionized health care by improving the quality of patientcare and lowering costs. Telehealth innovations are considered to have immense potential in health care which will be very helpful to resolve significant issues. The potential benefits include continuous monitoring of patient health, regardless of patients location, enhanced quality of care, enhanced accessibility to health care, and reduced cost of care. Telehealth and telemammography emerged with aspire of providing healthcare services; former by remotely monitoring a patients condition such as blood pressure, heart rate and measures of health status while latter is involved with the secure transfer of mammographic images from one place to another. Telehealth services comprise of teleconsultation, remote patient monitoring, and Intraoperative Monitoring (IOM). Teleconsultation is associated with sharing of medical information such as CT, MRI or ultrasound scans taken by local physician who incorporates these images into an electronic medical record and forwards to an expert for analysis and treatment recommendations. In Remote Patient Monitoring (RPM), sensors placed on a device store and transmit physiological data of a patient with the help of wireless solutions for review by a health professional. In result of happy marriage of software and hardware technology under the umbrella of Internet of Things (IoT), where IoT connects healthcare devices like MRI, CT scanner or other lab test equipment to Internet, trillions of data are being produced through connecting multiple devices and sensors with cloud and making sense of data with intelligent tools. This data prompts the need of complex algorithm to analyze it in order to assist medical professionals for their appropriate health recommendations; e.g., in diabetes management, real-time transmission of blood glucose and blood pressure readings enables immediate alerts for patients and healthcare providers to intervene when needed. Analyzing data has significant importance among all these stages, i.e., from health data collection and transmission, its evaluation, to notify, and to intervene. Hence, there is a need of exploiting soft computing and machine learning techniques on top of traditional approaches which lack efficiency in accessing, processing, and analyzing the complex-natured medical data collected in the form of images.

This chapter discusses analysis of telemammogram images where tumors from mammogram images are aimed to be mined, then these extracted tumors are classified as benign, malignant or normal by means of soft computing techniques. Section 2 presents some related work in telehealth and telemammography by means of soft computing techniques. Section 3 presents challenges in telehealth monitoring. Then, machine learning and image processing techniques to medical image analysis are discussed in Sects. 4 and 5. Section 6 presents the proposed framework for telemammography based on image processing and machine learning approach. Result analysis and discussion are provided in Sect. 7, while Sect. sec008 presents the conclusion.

2 Related Work in Telehealth and Telemammograpy Using Soft Computing Techniques

A large amount of work has been done in the past several years on image processing and medical imaging following different research directions but the work done in the field of telehealth and telemammography is not yet satisfyingly enough.

The use of various techniques such as Bayesian classifier, Artificial Neural Network (ANN), wavelet transform, fuzzy c-means clustering, fuzzy index, connected component analysis, local density measures, stochastic resonance noise, geometric linear discriminant analysis and support vector machines, etc., has been made to work with abnormalities in mammograms and ultrasound images.

Li and Chen in 2016 [3] presented a review on computer-aided detection (CAD) methods, served as a second view for radiologists in the early detection of breast cancer, with mammography in order to detect and classify abnormalities. Image processing techniques and pattern recognition are typical ways to detect and classify abnormalities in mammograms. It is reported that there is still need to improve abnormalities classification whereas recent research trend of using deep learning methods for classification is encouraged. Su in 2011 [4] used texture and morphological features to automatically detect, segment and classify breast tumors in Ultrasonic images. Neural network was used to detect region of interest (ROI) by considering local characteristics of texture and position. Segmentation was performed by a graph theory-based clustering algorithm, modified normalized cut approach and regional-fitting active contour model. For the last step of classification, eight features (three textures and five morphologic) are extracted from each breast tumor. An unsupervised approach was adopted using affinity propagation clustering to classify malignant and benign tumors by making use of the extracted texture and morphological features.

Based on some previous results, Nascimento et al. in 2016 [5] stated that morphological features perform better than texture features in order to classify breast tumors and also reduced number of features give better performance than large number of features. The scalar selection technique with correlation and Fishers Discriminant Ratio (FDR) were used for identifying most important features. K-fold cross validation was performed to evaluate classifiers where neural networks showed better performance over Support Vector Machines (SVM) to classify breast lumps. In another work presented by Wahdan et al. in 2016 [6], a system is proposed to detect breast tumors from ultrasound images which are considered less expensive and less invasive than x-rays in mammography and computerized tomography. The proposed system consists of preprocessing, feature extraction and classification. Gaussian blurring, anisotropic diffusion, and histogram equalization are used to reduce noise and to enhance the image quality in preprocessing step. Features are extracted and reduced using Principle Component Analysis (PCA) in the second step. In the third step, images are classified as with/without tumors by employing two different classification methods; i.e., SVM and ensemble classifier using bagging technique which were implemented using Weka 3.6.9. bagging ensemble classifier have shown better performance over SVM.

A system is proposed in [7] to segment and classify the breast cancer in ultrasound images. ROIs are identified using marker controlled watershed transformation technique. Statistical and texture features are extracted by applying wavelet transform. Then identified ROIs are differentiated as normal or focal lesion and focal lesion as malignant or benign using SVM, K-NN, and CART. Study demonstrates that the SVM and CART showed best results using texture features to differentiate between normal and focal lesion and CART gave best result in classifying malignant and benign by considering both statistical and texture features.

3 Telehealth and TeleMammography

There are people all over the world, suffering from various chronic diseases, as they do not have ready access to efficient health monitoring systems. Telehomecare also known as remote patient monitoring focuses on the delivery of health care to patients outside of conventional care settings (e.g., a patient's home), made possible by connecting the patient and healthcare provider through technology [8]. Small powerful wireless solutions connected through IoT are now making it possible for monitoring to come to these patients instead of vice versa [9]. The use of IoT in health care has significantly revolutionized by lowering costs and improving quality. With the advent of IoT and wireless sensors, patient health data can be elicited from a variety of sensors, image, and data analytics algorithms can be applied to analyze the data, generate reports, and share it with medical professionals through wireless connectivity for further diagnosis and intervention purpose, which was never before available for access from remote patients and underserved areas. Thus, IoT-based solutions are making it possible to improve health and reduced cost by increasing the quality care with availability of experts' specialist intervention.

As technology for collecting, analyzing, and transmitting data in IoT continues to mature, more and more exciting new IoT-driven healthcare applications and systems emerge [9]. A variety of wireless health monitoring systems such as IoT wearable healthcare gadgets are used to capture patient health data, also smartwatches for heart disease, Google smart contact lenses for diabetes patients, Leaf healthcare ulcer sensor, wearable sensors and monitoring patches, FDA-cleared Body Guardian which integrates ECG, heart rate, respiration rate, physical activity data, and Agamatrix IBGStar blood glucose monitoring system are used for healthcare monitoring (Fig. 1).

Patient data collected from different sources is sent through cloud and IoT to the central processing unit. Analysis is performed on this unstructured and semistructured data by applying complex image processing and machine learning algorithms in order to provide clinical decision support system, evaluate and generate reports, share it through wireless connectivity with medical professionals. If there are any areas of concern, notification alerts will be sent to patient care givers and specialist intervention will be provided in case of emergencies.

Thus Tele-health care may be defined as the use of information technology to provide healthcare services at a distance. It includes anything like medical services at the inpatient or at the outpatient stage [10]. A much earlier intervention can be provided to patients with efficient data transmission and accurate analysis of the patient data.

Telemedicine can be extremely advantageous for people living in isolated communities and remote regions. In recent years, it has been widely applied in almost all medical domains. With Advances in IoT and remote patient monitoring systems, patients living in underserved areas with limited medical expertise can be examined by a doctor or a specialist, who can provide a precise and complete examination. The patient does not need to travel the normal distances, like the earlier conventional hospitals approaches [11] (Fig. 2).

Telemedicine is widely used in most of the healthcare sectors, few of the advances and implications of telemedicine in healthcare are as below:

Fig. 1 Remote patient monitoring process

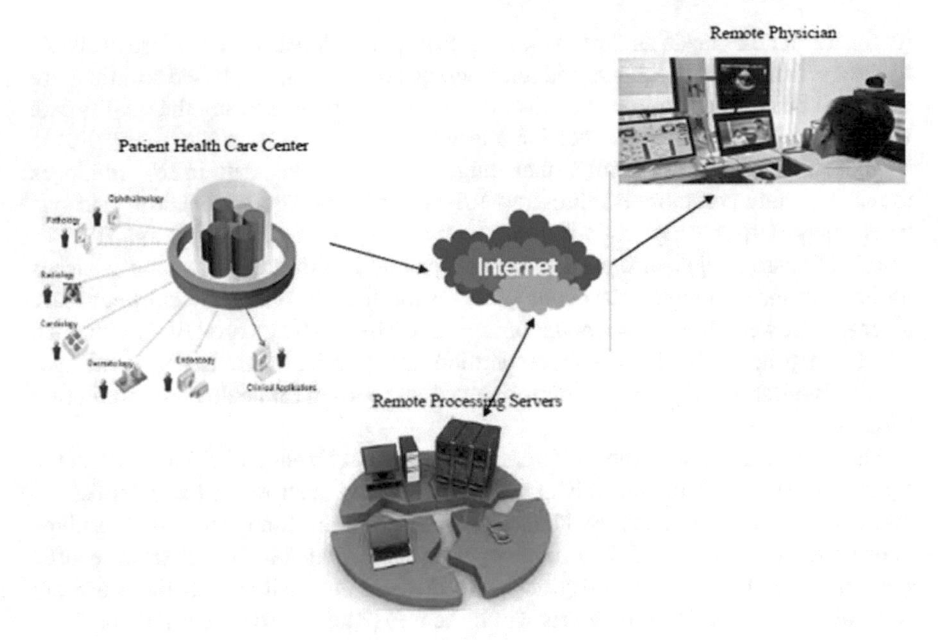

Fig. 2 Telemedicine system

Telecolposcopy: It involves a distant expert colposcopists evaluation of women with potential lower genital tract neoplasia. Existing telemedicine network and computer systems provide an audiovisual interface between local colposcopists and expert colposcopists at other locations [12].
Telepsychiatry: Is the application of telemedicine to the specialty field of psychiatry. The term typically describes the delivery of psychiatric assessment and care through videoconferencing [12].
Telemedicine in pregnancy: One area of women's health care where telemedicine has offered some of the greatest opportunities is in pregnancy and prenatal care. There are many examples of the use of telemedicine in pregnancy, one being the remote and distant monitoring of blood glucose in diabetic pregnancies [11, 12]. Telemedicine in fetal monitoring in high-risk pregnancy has shown high efficacy and less cost effective.
TeleMammography: is one of the areas in which telemedicines is showing major growth.

3.1 TeleMammography

Breast cancer is the most common cause of cancer death in women between age of 40 and 45 years. It is one of the leading causes of mortality in women. The World

Health Organizations International Agency for Research on Cancer in Lyon, France, estimates that more than 1,50,000 women worldwide die of breast cancer each year [13]. It is expected that 89,000 new cases of breast cancer will be found each year. It is estimated that in 2012, 1.7 million new cases and 5,21,900 deaths from this cause were detected worldwide [14]. One out of every 15 newly born girls is expected to develop breast cancer. In addition, 80% of cancer cases detected in low-income countries are at a stage where cure is impossible and palliation is the only possible treatment. For breast cancer in high-resource countries, the fatality rate is 23.9, while it is 56.3 for low-income countries [15, 16]. Mammography, Xeroradiography, and Thermography are used for the detection of cancer [16].

A cancerous tumor in the breast is a mass of breast tissue that is growing in an abnormal, uncontrolled way. The primary signatures of this disease are masses and micro-calcifications. Masses are space occupying lesions, described by their shapes, margins, and dense properties. Micro-calcification is tiny deposits of calcium that appear as small bright spots in the mammogram. Small clusters of micro-calcification appearing as collection of white spots on mammograms show an early warning of breast cancer (Fig. 3).

Tumors will appear as shades of gray, while calcifications (micro-calcification) are white. Benign lesions are most often circumscribed and in regular contours, while tumors often appear as spiculated masses (needle-like shapes).

Benign and malignant calcifications:

The term benign implies a mild and nonprogressive disease, and indeed, many kinds of benign tumor are harmless to the health. Malignant is a medical term used to describe a severe and progressively worsening disease. Malignant tumor is synonymous with cancer. Cancer is a class of disease in which a group of cells displays uncontrolled growth (division beyond normal limits), invasion (intrusion on and destruction of adjacent tissues), and sometimes metastatic (spread to other locations in the body via lymph or blood). These three malignant properties of cancers differ-

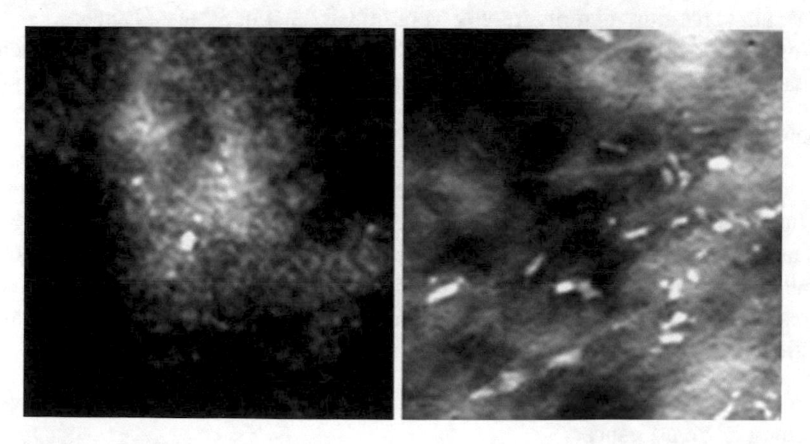

Fig. 3 Micro-calcification

entiate them from benign tumors, which are self-limited, do not invade or metastasize [17].

Primary prevention seems impossible, since the cause of this disease still remains unknown. An improvement of early diagnostic technique is critical for women's quality of life [18]. Early detection of cancer does the treatment easier, reducing risks and the percentage of mortality upto 25% [19].

There is a clear evidence which shows that early diagnosis and treatment of breast cancer can significantly increase the chance of survival for patients [20, 21]. The early detection of cancer may lead to proper treatment. Mammography is the main test used for screening and early diagnosis. Mammograms are good at catching infiltrating breast cancer. The current clinical procedure is difficult, time consuming, and demands great concentration during reading. Due to large number of normal patients in the screening programs, there is a risk that radiologists may miss some subtle abnormalities [22].

Mammography test can be performed either through film mammography or digital mammography. Although both film and digital mammography use X-rays to produce an image of the breast. The actual procedure of positioning and compressing the breast for examination are identical. There are several differences between these two types of mammography that should be noted.

Film Mammography: This has been successfully used as a screening tool for breast cancer for 35 years, which uses film to produce an image of the breast. Even though film mammography is considered a very good screening tool for detecting irregular breast characteristics, studies have revealed that film mammography is less sensitive for women who have dense breasts. Perhaps the most limiting issue with film mammography is the film itself. Once an image of the breast has been created on the film, it cannot be significantly altered, magnified, or brightened, which would allow a better assessment of the image [23].

False-negative results:

- A false-negative mammogram looks normal even though breast cancer is present. Overall, screening mammograms do not find about 1 in 5 breast cancers.
- Women with dense breasts have more false-negative results. Breasts often become less dense as women age, so false negatives are more common in younger women.

False-positive results:

- A false-positive mammogram looks abnormal but no cancer is actually present. Abnormal mammograms require extra testing (diagnostic mammograms, ultrasound, and sometimes MRI or even biopsy) to find out if the change is cancer.
- False-positive results are more common in women who are younger, have dense breasts, have had breast biopsies, have breast cancer in the family, or are taking estrogen.

In order to elevate the above limitations, digital mammography is being used for screening of breast cancer.

Digital Mammography: Digital mammography is one of the several new breast-imaging technologies developed to improve the detection of breast cancer in its earliest stages. Approved by the Food and Drug Administration in 2000, digital mammography is done the same way traditional mammography is done but, instead of the X-rays being detected on a photographic film, the X-rays are converted to digital images read by a computer [24].

Digital mammography is more advanced to the conventional film-screen mammography for cancer detection in women. Digital mammography is most efficient as it includes improved image archiving over hard-copy storage [24, 25]. This electronic method allows images to be stored and shared more easily than film mammography. Images created by digital mammography can be modified, for example, images can be enhanced or brightened for further evaluation [24, 25]. Digital mammography uses less radiation when compared to film mammography by lowering the dose of radiation exposure in women.

Image processing algorithms can be applied to the mammogram images to detect abnormalities. Digital mammography images can be transmitted, archived, and interpreted offsite, which allows for increased access to high-quality imaging. Digital mammography supports Picture Archiving and Communication System (PACS), and hence postprocessing of image is possible unfortunately, most of rural women do not undergo routine annual mammography because of nonavailability of mammography services and high cost [25]. With the advancement in ICT and the use of IoT technologies, women from rural and underserved areas can be well connected to experts and thereby reducing the mortality rate caused by breast cancer.

A telemammography service offers low-cost healthcare services at the doorstep of rural breast cancer victim [26]. The quality of health care will be improved faster resulting in a better diagnosis, as remote experts can be consulted for complicated cases of abnormal findings, by providing better access control to medical facilities for women at remote locations for consultation, early screening and education [25, 27, 28].

Multiple international organizations in Europe, America, and Japan collaborated in the development of the Digital Imaging and Communications in Medicine (DICOM) standards for the transmission of radiographic images [29]. These DICOM standards were adopted in Europe under the name MEDICON and this standardization paved the way for all teleradiography and specifically telemammography for women [11]. Telemammography for women is rapidly becoming a standard in todays world.

3.2 Technical Challenges in Telehealth and Telemammography

The rapid increase in cases is due to several factors, according to the world health organization, including the asymptomatic nature of the disease in its early stage, the high dropout rate of patients due to lack of critical information. Barriers to improved telemammography surveillance also include socioeconomic factors, geographic challenges, and lack of patient and physician awareness regarding the importance of annual mammographic examinations, as well as demographic and cultural barriers among minorities.

Although not unique to telemammography, some of the common telehealth challenges are security and resolution [30]. Security services can be broadly classified as confidentiality, availability, and integrity of patients health data. The Health Insurance Portability and Accountability Act (HIPAA) was enacted in 1996 (Pub. L 104–191) [31]. Congress sought to streamline electronic health record systems while protecting patients, improving healthcare efficiency, and reducing fraud and abuse. As per the HIPAA, the requirements of security services such as confidentiality, availability, and integrity of patient health records have to be met in any tele-healthcare Network [32].

Security: Robust and highly efficient security measures are required to ensure that health information is disclosed only to authorized individuals. Access to patient information should be granted only to authorized medical practitioner and patient care. In order to ensure integrity of data only accurate information should be transmitted electronically, with the use of IoT and wireless transmission patient information should be readily available to specialist for intervention in case of areas of serious concerns.

Image Resolution: High-resolution, quality videos, and images require a significant amount of network bandwidth (the maximum capacity for data transfer available to an information system or organization). Low-resolution videos and images require less bandwidth, but they provide poorer quality images that may be blurry. Although high-resolution technologies are often necessary for providers to accurately interpret the results of an image or scan, in some instances low-resolution images have been shown to provide enough detail to permit clinical Diagnosis [30, 33]. Another challenging issue at the image processing server is failures to obtain the images of adequate quality for grading such as masses and micro-calcifications is not visible, small vessels are blurred major arcade vessels are just blurred, significant blurring of major arcade vessels in more than one-third of the image. These images have to be preprocessed by applying appropriate denoising and deblurring filters for accurate grading. Image restoration and enhancement techniques can be used to improve the visibility of mammography details from photography.

4 Machine Learning Techniques in Medical Image Analysis

4.1 Overview of ML Techniques

Machine learning explores the study and construction of algorithms that can learn from and make predictions on data [34]. These techniques are employed in various applications such as optical character recognition (OCR), computer vision, search engines, spam filtering, detection of network intruders, etc. In data analytics, machine learning is a way to develop complex models and algorithms that can be used for prediction, also known as predictive analytics. Based on past relationships and trends in the data, these analytical models allow researchers and scientists to generate reliable, repeatable decisions, and results [35]. Some commonly used machine learning algorithms are linear regression, logistic regression, decision tree, SVM, Naive Bayes, KNN, K-means, Random Forest, Gradient Boost and Adaboost, and Dimensionality reduction algorithms.

Machine learning algorithms are categorized into three broad classes:

- **Supervised Learning**: Here, a given set of independent variables (also known as predictors) are used to predict a dependent variable as a target/outcome variable. A function based on these set of variables is produced to map inputs to desired outputs. The process of training the model continues until it gives satisfactory level of accuracy on the training data. Decision Trees, Nave Bayes Classification, Regression, Logistic Regression, SVM, KNN, Random Forest, and Ensemble methods are some examples of supervised learning.
- **Unsupervised Learning**: Here, no target/outcome variable to predict is used. Supervised learning is used to form clusters in different groups. Some examples of unsupervised learning are Apriori algorithm, K-means, Principle Component Analysis (PCA), and Singular Value Decomposition (SVD).
- **Reinforcement Learning**: Machine is trained to make specific decisions in this algorithm where machine trains itself continuously using trial and error by learning from past experience with the aim to capture best possible knowledge to make accurate business decisions. Markov Decision Process is an example of reinforcement learning.

There are a number of research efforts employing machine learning techniques in the field of medical image analysis.

4.2 Feature Extractions

Features extraction and selection of relevant features have main importance and crucial task in breast cancer depiction and classification. The performance and execution time of the classifier depends on the suitable features and their dimension. Classifiers exhibit poor accuracy and complications if encountered with too many irrelevant features. Improved accuracy is achievable by selecting appropriate features. Literature shows that malignant and benign lesions are differentiated using features related with morphology and texture of the tumors. Some studies in the literature found the better performance of morphological features set than texture features. In [36], ultrasound images are preprocessed by wavelet filters to reduce speckle noise then 57 texture and shape features were extracted to classify breast tumors. Local features of texture and position are incorporated in [4] to generate region of interest (ROI) by using a self-organizing map neural network then ROI were partitioned into clusters using a modified normalized cut approach. Three textures and five morphologic features are extracted from each breast tumor to provide a basis for classification of malignant and benign tumors by using affinity propagation clustering scheme. [6] used PCA to reduce the dimensions of the feature vector for the step of feature extraction and dimensionality reduction.

4.3 Classification Techniques

There are a number of techniques used for classification in medical image analysis such as decision trees, neural networks, etc. These classifiers use training data to construct the learning model in order to predict the class of new instances. Following is a brief overview of some of the classification techniques which have widely been used in medical image analysis:

4.3.1 Decision Tree Classifier

A decision tree is generated from a training dataset. The leaf nodes represent the class labels while the intermediate nodes split the dataset into subsets. The division of the training set continues till all the records in the training set are covered by the decision tree. The main problem in this classifier is that there could be more than one decision trees for a single dataset. To choose the best decision tree amongst the candidate decision trees, a number of quality matrices can be used; e.g., gini index, entropy, and misclassification error. Decision tree is well suited for categorical features/attributes while continuous attributes are discretized before the construction of the decision tree. Being popular classifiers, decision trees are used in different domains such as activity recognition, expert systems, medical diagnosis, etc. ID3, C4.5, C5, and CART are some of the examples of decision tree algorithms.

4.3.2 K-Nearest Neighbor (KNN) Classifier

This is one of the popular classification technique in which unknown instances are classified based upon their similarity/distance with the records in the training set. Either similarity or distance of a test record is computed with every record of the training set. Then, all the records are sorted according to the proximity function (similarity/distance). Afterwards, top-k records are chosen to be the k-nearest neighbors of the test record. If k-nearest neighbors of a record contain instances of more than one class then the decision is made according to the majority vote. The main issue of this algorithm is selecting value for k. In case of smaller value of k, KNN is sensitive to noise while for a very big value of k, the decision of classifier may be hijacked by the majority ive classes in the neighbor. Selecting appropriate proximity function depends on the nature of the underlying dataset.

4.3.3 Naive Bayes Classifier

It is a probabilistic classifier based on the Bayes theorem. Suppose X is a set of attributes representing a record and y is the class label then according to the Bayes theorem:

$$P(X|y) = \frac{P(X, y)}{P(y)} \Rightarrow \quad P(X, y) = P(X|y) \times P(y) \tag{1}$$

$$P(y|X) = \frac{P(X, y)}{P(X)} \Rightarrow \quad P(X, y) = P(y|X) \times P(X) \tag{2}$$

$$P(y|X) = \frac{(P(X|y) \times P(y))}{P(X)} \tag{3}$$

where $P(X)$ is called evidence and is fixed for a given record therefore, it is not necessary to compute. $P(y)$ is called prior which is simple and straightforward to be computed from the training dataset. The remaining part of the above Eq. 3 is only $P(X|y)$ known as class conditional probability which can be estimated either by Nave method or Bayesian belief network. According to Nave, all the attributes are statistically independent. We can write $P(X|y) = P(X_1, X_2, X_n|y)$. If the attributes are assumed to be independent then

$$P(X_1, X_2, \ldots, X_n|y) = P(X_1|y).P(X_2|y), \ldots, P(X_n|y) = \prod_{i=1}^{n} P(X_i|y) \tag{4}$$

Equation 4 is substituted in Eq. 3 to calculate the posterior P(y|X). The posterior $P(y|X)$ is calculated for each class label and the class label with the highest probability is assigned to the unknown X record.

4.3.4 Random Forest

As the name of this algorithm illustrates, forest is a collection of trees. It comes under the umbrella of ensemble classifier. The training set is divided into a number of random subsamples spaces (sub-training sets). A decision tree is created for each training subset leading to a number of decision trees known as random forest. A test sample is assigned class label by each and every decision tree. The final decision is made on the basis of majority vote of the decision trees. The main issue of this algorithm is to decide about the number of trees in the forest. The time complexity of a single decision tree is O(mn log n) but for M number of trees in the forest, computational complexity becomes O(M(mn log n)) where m is the number of attributes, n is the number of records in the training dataset.

4.3.5 Support Vector Machine

Support vector machine (SVM) is a non-probabilistic binary linear classifier used to analyze data for classification and regression analysis. It belongs to the family of supervised learning models with associated learning algorithms. A model generated by an SVM training algorithm assigns new examples to one category of the two marked categories given a set of training examples. Classification of breast tumor in ultrasound images was made by using support vector machines and neural networks [5]. Neural networks with selected stop criterion performed better than SVM in the study. Two classification approaches; i.e., SVM and ensemble classifier using bagging are employed for classification step in [6] and a comparison is also made between the two techniques. The study shows the better performance of ensemble classifier using bagging technique over the SVM classifier. In [7], SVM, KNN and CART are employed for classification of normal or focal lesion from the identified ROIs and then to classify focal lesion as malignant or benign.

4.4 Application of ML to Medical Image Analysis

The use of machine learning techniques is dramatically increasing in medical image analysis for computer-aided diagnosis in recent years which can help to extract reliable diagnostic cues. ML techniques are heavily being investigated for diagnosing abnormalities in thoracic, abdominal, brain, and retinal imaging. In neurodegenerative diseases, ML-based approaches are using brain MR images to diagnose Alzeimer's disease or other forms of dementia or predict conversion to dementia from mild cognitive impairment (MCI). ML is also being used to detect diabetic retinopathy in retinal fundus photographs [37].

5 Image Processing Techniques for Medical Image Analysis

Medical image analysis and processing has emerged into one of the most significant fields within scientific imaging due to rapid and continuing progress in computerized image analysis and visualization methods, it has become a vital part within health care for early detection, diagnosis, treatment planning, guiding, and monitoring of disease progression [38].The major strength in the application of computers to medical imaging lies in the use of image processing techniques for effective visualization, analysis, and intervention. Medical images are always visual in nature, subjective analysis can be performed by an expert interpretation based on his/her experience and expertise. However with advances in image processing techniques, if analysis is performed with the appropriate logic, it can potentially add objective strength to the interpretation of the expert. Thus, it becomes possible to improve the diagnostic accuracy and confidence of even an expert with many years of experience [39].

A sequence of image processing techniques are required, for the effective analysis, detection, and visualization of medical images,each of these techniques is discussed in the following sections.

5.1 Image Preprocessing

Digital images have some degree of noise and get corrupted with noise during its acquisition from various sensors and transmission by different media. The most commonly used medical images modalities are MRI (Magnetic Resonance Imaging), CT (Computed Tomography), Mask and contrast images (Digital subtraction angiography), and X-ray equipments. Usually, the existence of noise into medical image reduces the visual quality that complicates early detection, analysis, diagnosis, and treatment. The perception of the addition of noise as artifacts can lead to false diagnosis. Generally, MRI images are corrupted with speckle noise and Rician noise. CT images are contaminated with gaussian, salt and pepper, and structural noise [40]. Image denoising techniques are necessary to remove random additive noises while retaining as much as possible the important image features. The main objective of these types of random noise removal is to suppress the noise while preserving the original image details [41, 42].

Spatial images can be filtered using linear or nonlinear filtering techniques. Filtering can be performed in two modes namely correlation and convolution. Depending on the noise in the image one can select either average, disk, Gaussian, Laplacian, or log filters. Median filtering is a nonlinear operation often used in image processing to reduce "salt and pepper" noise. A median filter is more effective than convolution when the goal is to simultaneously reduce noise and as well as preserve edges [43]. Every pixel in denoised image contains the median value in the m-by-n neighborhood around the corresponding pixel in the input image.

Statistical filters like average filter, Wiener filter can be used for removing additive white gaussian, random and salt and pepper noise which has been acquired during image transmission and acquisition, but wavelet based denoising techniques proved better results than these filters [42, 44].

Wavelet transforms compress the essential information in an image into a relatively few, large coefficients which represent image details at different resolution scales. In recent years, there has been a fair amount of research on wavelet thresholding and threshold selection for image denoising [45–47]. In our proposed approach, we have used wavelet-based denoising for removal of noise and blur from the degraded low-resolution medical images.

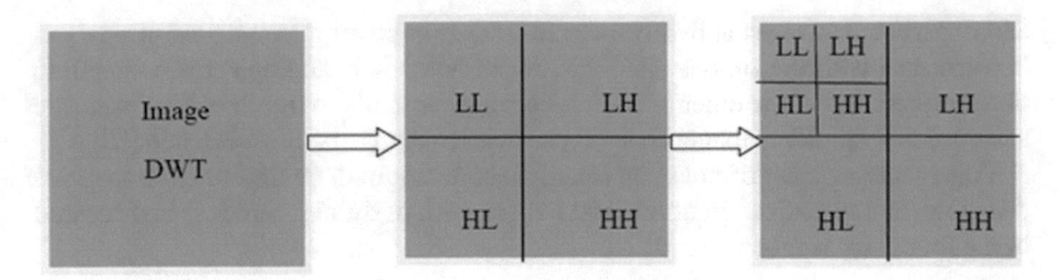

Fig. 4 Image decomposition using DWT

5.1.1 Proposed Wavelet Based Denoising

Wavelet Transforms decompose the image into different frequency subbands, normally labeled as LL(Low-Low), LH(Low-High), HL(High-Low) and HH(High-High) frequencies, as in the schematic, depicted in Fig. 4. The LL subband can be further decomposed into four subbands labeled as LL, LH, HL, and HH.

In wavelet decomposition of an image, we obtain one approximate (LL) and three details (LH, HL and HH) subbands, detail coefficients in the subbands(LH, HL and HH) are dominated by noise while approximate coefficients with large absolute value carry more image information than noise. Replacing noisy coefficients by zeros and an inverse wavelet transform lead to a reconstruction that has lesser noise.

Normally hard thresholding and soft thresholding techniques are used for denoising:

Hard Thresholding

$$\begin{aligned} D(X,T) &= X \; if \; |X| > T \\ &= 0 \; if \; |X| < T \end{aligned} \tag{5}$$

Soft Thresholding

$$D(X,T) = Sign(X) \times max(0, X - T) \tag{6}$$

where X is the input subband, D is the denoised band after thresholding and T is the threshold level. The denoising algorithms, which are based on thresholding suggests that each coefficient of every detail subband is compared to threshold level and is either retained or killed if its magnitude is greater or less respectively.

In our proposed approach, we have adapted soft thresholding technique for denoising of degraded images. Selecting an optimum threshold value (T) for soft thresholding is not an easy task. An optimum threshold value should be selected based on the subband characteristics. In wavelet subbands, as the level increases the coefficient of the subband becomes smoother. For example, when an image is decomposed into two levels DWT using Daubechies 4 tap (DB4) wavelet transform, we get eight subbands as shown in Fig. 4 above. The approximate coefficients(LL) are not submitted in this process. Since, on one the hand, they carry the most important information about the image, on the other hand the noise mostly affects the high frequency subbands. Hence the HH subband contains mainly noise The HH subband of first level contains large amount of noise, hence the noise level is estimated for the LH, HL, and HH subbands using Eq. 7. For estimating the noise level, we use the Median Absolute Deviation (MAD) as proposed by Donoho [48].

$$\sigma = \frac{Median|Y_{ij}|}{0.6745}, Y_{ij} \epsilon LH, HL, HH \tag{7}$$

Once the noise level is estimated, we select the threshold value T. The threshold value T is estimated using Eq. 6.

$$T = \sigma - (|HM - GM|) \tag{8}$$

Here σ is the noise variance of the corrupted image. As given in [41], the Harmonic Mean(HM) and Geometric Mean(GM) are best suited for the removal of Gaussian noise, hence we use the absolute difference of both the Harmonic Mean (HM) and Geometric Mean (GM) or either of the means also can be considered for denoising the image corrupted by Gaussian noise.

The harmonic mean filter is better at removing Gaussian type noise and preserving edge features than the arithmetic mean filter. Hence, we have considered harmonic mean than arithmetic mean. The process is repeated for LH and HL bands and threshold is selected for all the three bands once threshold is estimated, soft thresholding of Eq. 6, is performed to denoise the image.

$$HM = \frac{M^2}{\sum_{i=1}^{M} \sum_{j=1}^{M} \frac{1}{g(i,j)}} \tag{9}$$

$$GM = \left[\prod_{i=1}^{M} \prod_{j=1}^{M} g(i,j) \right]^{\frac{1}{M^2}} \tag{10}$$

Based on understanding, the conceptual theme of HM and GM for selecting the threshold, it can be clearly evident that our proposed approach for denoising the image results in a smoother image.

5.2 Image Enhancement

Digital image processing systems have the ability to preprocess the images with mathematical settings that enhance the quality of images. Image interpolation techniques are used to enhance the quality of digital images, image super resolution techniques have proven to be very successful in obtaining a high-resolution image for analysis of medical images [42]. Super resolution refers to the process of producing a high spatial resolution image than what is afforded by the physical sensor through postprocessing, making use of one or more LR observations [49].

In our proposed approach, wavelet-based interpolation is applied to enhance the quality of image to produce a high-resolution image which can be used for further analysis and diagnosis. In wavelet-based interpolation, the image contains one low-frequency (LL) subband and three high-frequency subbands (HL, LH, and HH) resulting in wavelet coefficient image of size $M \times N$. Each subband contains different information about the image. The HL and LH subbands contain edge information in different direction. Dropping the HH subband does not impact the perceptual quality of the upsampled image. Hence, we drop the HH subband. The next step is to obtain an image of size $2M \times 2N$. The LL subband is multiplied by a scaling factor. We consider the scaling factor as two. The HH subband is a matrix of zeros with dimensions $M \times N$. The new LH and HL subbands are generated by inserting zeros in alternate rows and columns, as shown in Fig. 5. Applying inverse DWT, we obtain an image with double the resolution to that of the original image.

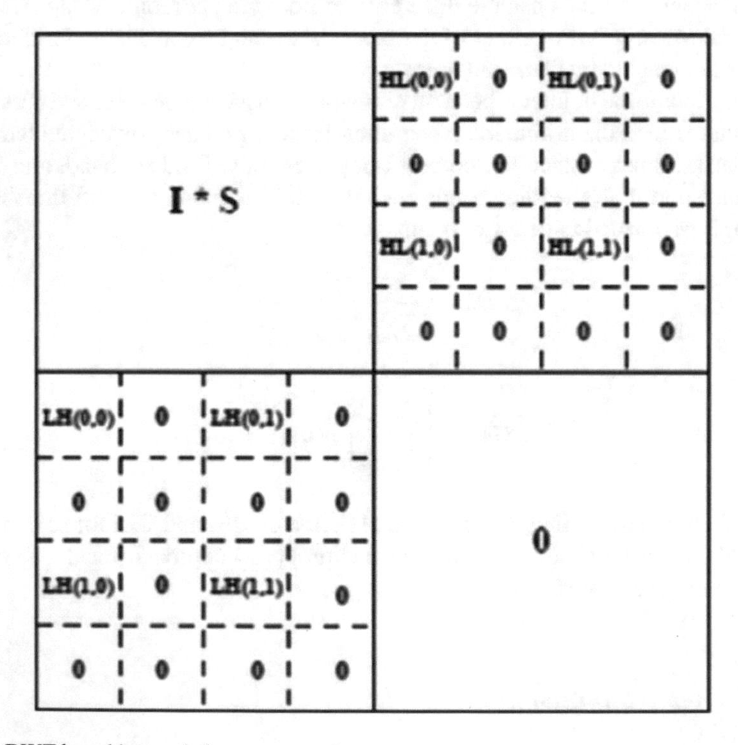

Fig. 5 DWT based interpolation

5.3 Image Segmentation

Segmentation of the image is basically subdividing the image into its constituent components, regions or objects. Segmentation is a middle-level image processing technique, where input is a digital image and output could be extracted attributes like edges, contours, etc.

Image segmentation algorithms can be based on either discontinuity or similarity. Discontinuity algorithms are based on abruptly changing intensity for example points, lines or edges in an image. Similarity based algorithms partition an image into segments or regions based on a set of predefined criteria like thresholding, region growing, region splitting or region merging, etc.

5.3.1 Segmentation Based on Discontinuity

A mask of 3X3 with middle element as 8 and all other elements value as -1 can be used to detect a point in an image. Line can be detected using four masks representing horizontal, vertical, or diagonal. In order to detect an edge in an image either first-order derivative or second-order derivative is used. Many operators are available for detecting edges in literature like Prewitt, Sobel, Laplacian, Robert or Mexican hat, and many more. After obtaining the edge pixels it should properly be linked to get meaningful edges, which can be achieved either by locally processing similar pixels or global processing using Hough transform or graph theoretic techniques [41, 50].

5.3.2 Segmentation Based on Thresholding

Image segmentation based on thresholding can be implemented using single or multiple thresholds. These algorithms are very simple to implement. Depending on the number of class objects present, thresholds are chosen. Consider an image containing three regions or classes, then two thresholds T_1 and T_2 are required.

The algorithms can be implemented as follows:

If $f(x, y) < T1$; Pixel belongs to region 1
If $f(x, y) > T1$ *and* $f(x, y) < T2$; Pixel belongs to region 2
If $f(x, y) > T2$; Pixel belongs to region 3

Depending on how the function $f(x, y)$ is chosen the threshold can be local or global. Basic global thresholding can be performed by using intensity of the image. When converting a grayscale image to binary all the pixels greater than some threshold T are converted as black pixels otherwise white pixels. Clean segmentation is possible only when shadows are eliminated from the image. When thresholding depends on the local characteristics of the image, then it is called local thresholding and it gives better segmentation when compared to global thresholding.

Sometimes, the image may be taken at different illumination conditions. Hence, the image with uneven illumination requires little transformation using thresholding for a part of the image instead of the whole image, i.e., adaptive thresholding. These algorithms need to address two key issues regarding subdividing the image and estimating threshold for each sub-image. Thresholds can be based on many parameters like multispectral, color, three-dimensional histogram, etc.

5.3.3 Region Based Segmentation

Segmentation of image is performed based on regions. Any algorithm designed using the concept of regions should satisfy the following basic requirement: Segmentation must be complete. Points within the region must be connected Regions must be disjoint. All points in a segmented region must have the same intensity.(Intra-class Similarity) Intensity of two regions must be different.(Inter-class dissimilarity)

Segmentation based on regions can be broadly classified into Region Growing and Region Splitting and Merging.

5.3.4 Region Growing

Region growing is a procedure that groups the pixels or sub-regions into a larger region based on predefined criteria. These algorithms start with seed points and from these seed points the regions are grown until the neighboring pixels fails the condition or criteria. The real challenge to region growing algorithms lay in selecting the seed points. Depending on the problem domain, similarity and dissimilarity measure can be selected. Some of the similarity measures could be intensity in case of grayscale images, color in case of color images, texture, etc. [41, 50].

5.3.5 Region Splitting and Merging

In region splitting algorithms, initially the given image is split into several disjoint regions and an attempt is made to either split or merge the regions based on the five rules discussed above [41, 50].

6 Framework for TeleMammography Using Image Processing and Machine Learning

Figure 6 illustrates the proposed framework for automatic detection of Cancerous tumors through Telemammography and Data Analytics.

Data Acquisition

At the primary health care, data will be collected by trained physician which consist of Electronic Medical Records (EMR) and mammogram images of patients. These

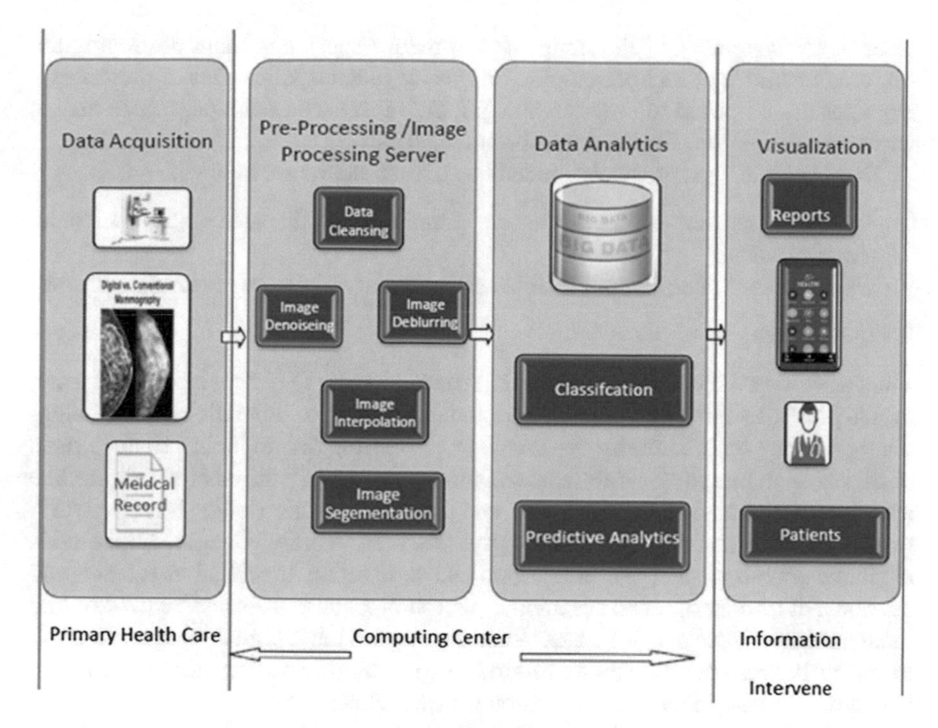

Fig. 6 Proposed framework for detection and classification of telemammography images

images are taken by the local physician are incorporated into an electronic medical record and transmitted through IoT wirelessly to the image processing server. Clinical data such as Name, PatentId, Age, gender, address and nonclinical data such as pathological reports, Mammogram Images, Blood Pressure, Diabetic Status will be collected. The storage format of these data is in diverse formats such as structured, semi structured, or unstructured format. During transmission of the images, DICOM standards have been considered. DICOM supports integration of imaging equipment from different manufacturers to support a range of modalities, as well as computed radiography and digitized film radiographs. DICOM conforms to international standard in the market [51].

Preprocessing

At the image processing server, the data collected includes Electronic medical records of patients which consists of patients examination findings (identification, demographic and medical information) along with mammographic images. Frequently, the information collected will not be in a format ready for analysis [52]. Preprocessing is the primary step in many systems in medical image segmentation and analysis. During image acquisition, different visual angles of patients cause bright speckles to spread over the image i.e. inhomogeneous background. These images are also corrupted during transmission. The task of image preprocessing is to enhance the quality of the image and to reduce speckle without destroying the important features of mammographic images for diagnosis. A series of preprocessing steps are proposed in order to obtain a high-resolution mammographic image, which can be used for further analysis and diagnosis.

Highly efficient preprocessing techniques, listed below are applied.

1. Wavelet-based denoising by fixing an optimum threshold value based on subband characteristics.
2. Wavelet-based interpolation to enhance the quality of the low-resolution images.

Image Segmentation

The preprocessed that is denoised and enhanced images are further segmented using region growing segmentation technique and morphological operations. Morphological operations try to simplify the image by preserving the important characteristic of an image. A morphological transformation would satisfy the basic principles like invariability to translation and scaling, and possess local knowledge. It is also worth to note that the transformation would not cause any abrupt changes. Minkowskis algebraic operations like vector addition and subtraction are called as dilation and erosion. Hit or miss operation is another morphological transformation used to find local patterns of pixels in an image. When an image is transformed using erosion followed by dilation operations, a new morphological transformation known as opening is evolved. Dilation followed by erosion is called closing [53].

Top-hat filtering or transformation is a tool to extract the object from the image even when the background is dark. Bottom hat filter or transformation is applied when the object is dark and the background is bright [53].

Features from the preprocessed mammographic images are extracted by

- Applying Top-Hat morphological operation
- Region Growing based segmentation techniques

The acquired mammographic images are preprocessed using wavelet-based denoising and converted to grayscale. Further, the image is enhanced using our proposed wavelet based interpolation techniques to clearly identify the region of interest using top-hat morphological transformation. The resultant image is segmented using region growing method. The features like area and volume of the tumor is computed from the segmented region mammographic image.

$$Area = [bwarea('segmented\_binary\_image') * 0.035 * 0.034]\,\mathrm{cm}^2 \tag{11}$$

$$Volume = [Area * 0.5]\,\mathrm{cm}^3 \tag{12}$$

The features area and volume are sent to the classification algorithm to classify the sample as benign, malignant or normal.

Measuring a tumor mass is one of the most tedious tasks for most radiologists.

Classification

In this study, several machine learning algorithm such as Neural Networks, J48 Decision Trees, Random Forest and K- Nearest Neighbor are applied on the extracted features by segmentation technique to classifying the tumor into Benign, Normal, and Malignant.

The mammographic clinical descriptors considered for classification are

- Area: the number of pixels in the tumor region
- Volume: For volume computation, the shape of the tumor nodules was approximated to an ellipsoid
- Radius:Approximate radius (in pixels) of a circle enclosing the abnormality
- Background Tissue: Fatty, Fatty-glandular, Dense-glandular
- Centroid: x and y image coordinates of center of abnormality
- Edges or boundaries of abnormality: Calcification, circumscribed masses, Spiculated masses, ill-defined masses, Architectural distortion, Asymmetry, Normal

Nonclinical descriptors such as ImageId and Patient Id are also included.

In total, nine features are considered to measure the severity of abnormality such as Benign, Normal, or Malignancy of Mammographic images, a matrix data is created and fed into the classification algorithms for training and testing.

Data Visualization: Reports are generated by applying data visualization techniques and sent to Web User Interface at the Web Server. People diagnosed with diabetic retinopathy and glaucoma are directed for further treatment through our web-based user interface and mobile app which can be accessed from any part of the world. The report will direct them for further treatment.

Web Server: A Mammographic File Accessing System (MFAS) is designed which is based on Web User Interface, the MFAS consist of two groups of users Group 1 are patients and Group 2 consists of medical experts. In our proposed approach, first the user has to authenticate with username and password. Once authenticated, user will be allowed to enter the patientId and request for information, the request will be transferred to Computing Center which consists of Image processing and Data Analytics server, the request will be processed and response will be returned back to user, users will be able to view patients information such as patient report, which states the number of true positive rate and true negative rate, severity of abnormality such as Benign, Malignant, or Normal and advise them for further treatment. Medical experts will be authorized to view the report as well as the mammographic images, as the resultant image is of much higher resolution with fine details and more accurate because of machine prediction, it will be very helpful for the medical experts for a proper treatment recommendation.

Mobile App: A Mobile App is also designed for the Mammographic File Accessing System, through which patients can be notified by sending notification alerts which will be transmitted wirelessly.

Hence, our proposed telemammography system will be highly efficient in early detection and prevention of breast cancer which is a major cause of mortality in women in rural areas.

7 Implementation and Results

Mammographic Image Analysis Society (MIAS) database has been used in this research. The database contains 322 digitized mammograms. The database has been reduced to a 200 micron pixel edge and padded/clipped so that all the images are 1024×1024 [54].

Our proposed algorithm for detection and classification of severity of tumor in mammographic images consists of following steps.

1. Apply forward Discrete Wavelet Transform (DWT) to the low-resolution mammogram images to a specified number of levels. At each level, one approximation, i.e., LL subband and three detail subbands, i.e., LH, HL and HH coefficients are obtained.
2. The wavelet decomposed image contains LL, LH, HL, and HH subbands.

   (a.) Obtain the noise variance (σ) using Eq. 7, Sect. 5.1 from our proposed wavelet based denoising for LH, HL and HH subbands of level one.

   $$\sigma = \frac{Median|Y_{ij}|}{0.6745}, Y_{ij} \epsilon LH, HL, HH$$

   (b.) Compute Eq. 8, Sect. 5.1 and select the threshold (T) for LH, HL and HH subbands of level one.

   $$T = \sigma - (|HM - GM|)$$

   where

   $$HM = \frac{M^2}{\sum_{i=1}^{M} \sum_{j=1}^{M} \frac{1}{g(i,j)}} \quad (13)$$

   and

   $$GM = \left[\prod_{i=1}^{M} \prod_{j=1}^{M} g(i,j) \right]^{\frac{1}{M^2}} \quad (14)$$

   (c.) Denoise all the detail subband coefficients of level one (except LL) using soft thresholding given in Eq. 6, Sect. 5.1 by substituting the threshold value obtained in step (2b).

   $$D(X,T) = Sign(X) * max(0, X - T) \quad (15)$$

3. Apply inverse DWT to obtain a High-Resolution restored mammographic image.
4. Apply wavelet-based interpolation technique, discussed in Sect. 5.2 to obtain an enhanced quality image.
5. Perform morphological top-hat filtering on the input image.

6. Apply region growing Segmentation technique.
7. Extract the tumor masses.
8. Compute the Area and Volume of the tumor masses.
9. Apply Machine learning algorithm(Neural Networks, Random Forest, Decision Trees and K-Nearest Neighbor) to classify the severity of the tumor into Benign, Normal or Malignant.

7.1 Telemammography Image Results

The results for the three different cases of mammography images are shown below:
Case 1 (Fig. 7).
Case 2 (Fig. 8).
Case 3 (Fig. 9).

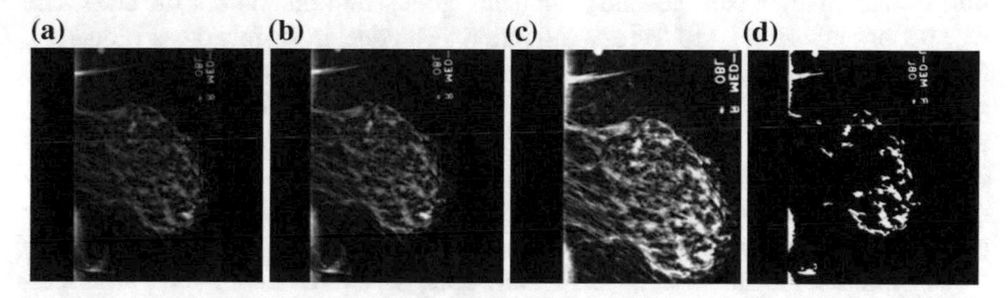

Fig. 7 **a** Mammogram image **b** Preprocessed image **c** Enhanced image **d** Segmented image with malignant tumor

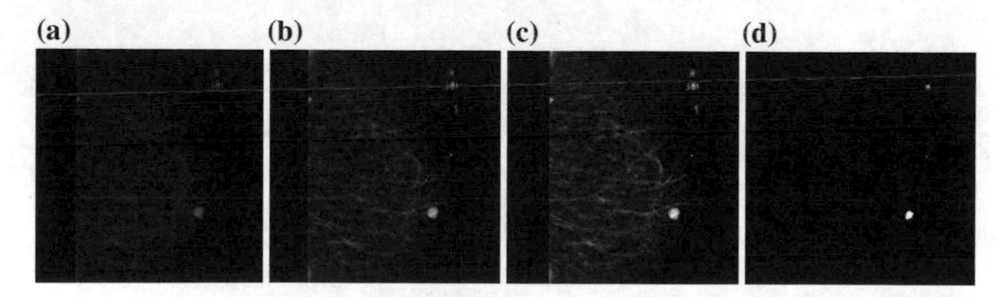

Fig. 8 **a** Mammogram image **b** Preprocessed image **c** Enhanced image **d** Segmented image with malignant tumor

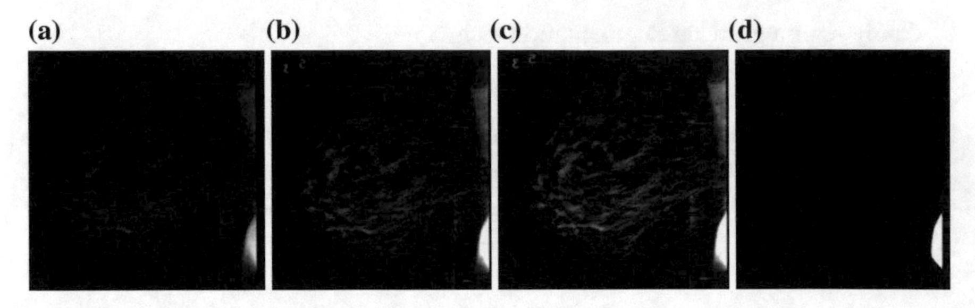

Fig. 9 **a** Mammogram image **b** Preprocessed image **c** Enhanced image **d** Segmented image with benign tumor

7.2 Classification Results

Four different machine learning techniques were used to classify the mammography images into Benign, Normal, and Malignant class. The validation of the classifiers was performed using 10-fold cross-validation technique, in 10-fold cross validation, the original sample is randomly partitioned into 10 equal size subsamples. Of the 10 subsamples, a single subsample is retained for testing the model, and the remaining nine(10 − 1) subsamples are used for training the data. The cross-validation process is then repeated 10 times (the folds), with each of the 10 subsamples used exactly once as the validation data. The 10 results from the folds will then be averaged to produce a single estimation.

Classification Results using Neural Networks, K-NN, Random Forest, and Decision Trees are discussed in sections to follow

(i) Neural Networks based Classification

Table 1 Classification results using multilayer perception neural network (MLP NN) for Mammogram images

Stratified cross validation	
Correctly classified instances	158
Accuracy	96.9325%
Incorrectly classified instances	5
Accuracy	3.0675%
Kappa statistic	0.9422
Mean absolute error	0.0264
Root mean squared error	0.1387
Relative absolute error	7.423%
Root relative squared error	32.9795%
Total number of instances	163

Table 2 Confusion matrix of classification using multi-layer perception neural network (MLP NN) for mammogram images

Benign	Normal	Malignant	¡– classified as
33	0	3	Benign
0	103	0	Normal
2	0	22	Malignant

We considered using only one layer as a hidden layer in this network since almost all applications perform well using single-hidden layer MLPNN classifiers. The number of input and output neurons in input and output layers were set to 9 and 3 with respect to the number of features and the number of result values (class labels). In order to choose the optimal number of hidden neurons, we checked the accuracy and MSE of this classifier for different numbers of hidden neurons for the training set (Tables 1 and 2).

Classification Results:

Time taken to build model: 156.49 s
Test mode: 10-fold cross validation

(ii) K-Nearest Neighbor

In this classifier, the Euclidean function is used as a distance measurement for calculating the distance between a new instance and the training instances. In order to choose an optimal K parameter value for this classifier, the accuracy and the mean square error of K-NN for different K values (the number of neighbors) is calculated. The 10-fold cross-validation technique was used for this calculation. K = 3 is the best choice for the parameter k as accuracy is maximum and mean squared error (MSE)

Table 3 Classification results using K-Nearest Neighbor (K = 3) for Mammogram images

Stratified cross validation	
Correctly classified instances	153
Accuracy	93.865%
Incorrectly classified instances	10
Accuracy	6.135%
Kappa statistic	0.8837
Mean absolute error	0.0508
Root mean squared error	0.1781
Relative absolute error	14.2777%
Root relative squared error	42.3423%
Total number of instances	163

Table 4 Confusion matrix of classification using K-Nearest Neighbor (K = 3) for Mammogram images

Benign	Normal	Malignant	¡– classified as
31	0	5	Benign
0	103	0	Normal
4	1	19	Malignant

is minimum at this point. The accuracy of the K-NN hypothesis was tested with the test set. The results are presented in Tables 3 and 4.

Classification Results:

Time taken to test model on training data: 0.01 s
Test mode: 10-fold cross validation

(iii) Random Forest based Classification

In random forest, n-decision trees were generated for our dataset using random selected attributes. The value of n was fixed after multiple trials and when the accuracy value starts to get stabilized. Random forest classifier was tested using n-trees equal to 100, 200, 300, respectively (Tables 5 and 6).

Classification Results

Bagging with 200 iterations and base learner
Time taken to build model: 0.16 s
Test mode: 10-fold cross validation

Table 5 Classification results using Random Forest- Mammogram images

Stratified cross validation	
Correctly classified instances	154
Accuracy	94.4785%
Incorrectly classified instances	9
Accuracy	5.5215%
Kappa statistic	0.8944
Mean absolute error	0.208
Root mean squared error	0.2584
Relative absolute error	58.5022%
Root relative squared error	61.4298%
Total number of instances	163

Table 6 Confusion matrix of classification using Random Forest for Mammogram images

Benign	Normal	Malignant	¡– classified as
31	1	4	Benign
0	103	0	Normal
2	2	24	Malignant

(iv) J48- Decision Trees based Classification

The Java implementation of the C4.5 decision tree classifier is called J48. Default parameters in WEKA for J48 are used in this evaluation. The results are provided in Tables 7 and 8.

Classification Results:

Time taken to build model: 0.01 s
Test mode: 10-fold cross validation

Table 7 Classification results using Decision Trees—Mammogram images

Stratified cross validation	
Correctly classified instances	156
Accuracy	95.7055%
Incorrectly classified instances	7
Accuracy	4.2945%
Kappa statistic	0.9186
Mean absolute error	0.0403
Root mean squared error	0.1571
Relative absolute error	11.3393%
Root relative squared error	37.3594%
Total number of instances	163

Table 8 Confusion matrix of classification using Decision Trees— Mammogram images

Benign	Normal	Malignant	¡– classified as
32	1	3	Benign
0	103	0	Normal
3	0	21	Malignant

Fig. 10 Confusion matrix

	Predicted Positive (Benign)	Predicted Negative (Normal)	Predicted Negative (Malignant)
Actual Positive (Benign)	TP	FN	FN
Actual Negative (Normal)	FP	TP	TN
Actual Negative (Malignant)	FP	TN	TP

7.3 Performance of Classification Results

Experiments were performed on 163 MIAS database with nine descriptors. Ten-fold cross-validation technique (66% is used for Training and 34% for testing) was performed on different classification algorithms such as Multilayer Perceptron Neural Network (MLP NN), K-Nearest Neighbor, Random Forest and Decision Trees.

After applying different machine learning techniques to the training and testing data using 10-fold cross validation, the performance of different classifiers was evaluated using confusion matrix data. A confusion matrix is a tabular representation that provides classifiers performance based on the correctly and incorrectly predicted benign or malignancy cases shown in Fig. 10.

In General

True positive = correctly identified
False positive = incorrectly identified
True negative = correctly rejected
False negative = incorrectly rejected

Accuracy, Sensitivity, Specificity, and Precision are used in order to measure the performance of each classifier where:

1. **Accuracy**: is the percentage of correct predictions. On the basis of Confusion Matrix, it is calculated by using the below equation

$$Accuracy = \frac{(TP + FP)}{(TP + FP + TN + FN)} \tag{16}$$

2. **Sensitivity (Recall)**: is the ability of a test to correctly identify those with the disease. Sensitivity is measured using the equation

$$Sensitivity = \frac{TP}{(TP + FN)} \tag{17}$$

3. **Specificity**: is the ability of the test to correctly identify those without the disease. Specificity is measured using the equation

$$Specificity = \frac{TN}{(TN + FP)} \tag{18}$$

4. **Precision**: is also called positive predictive value, that is retrieved instances that are relevant of positive predictive value. It is measured using the equation

$$Precision = \frac{TP}{(TP + FP)} \tag{19}$$

7.4 Comparative Analysis of Different Classifiers for Mammogram Images

The performance of the classifiers has been measured by considering Accuracy, Specificity, Sensitivity, Mean Absolute Error, Precision, and Time.

After running the different classifiers on the training and testing dataset, the confusion matrix was generated using cross-validation technique to reflect the TP, TN, FP, and FN. However, the classifiers performance can be measured and evaluated by accuracy, precision, sensitivity and specificity for selected features. The values generated from the confusion matrix of each classifiers are used to calculate the accuracy, precision, sensitivity and specificity.

A summary of classifiers performance in terms of Accuracy, Sensitivity, Specificity, Precision, Mean Square Error and Time is shown in Table 9. Accuracy value represents the total number of correct predictions made by the classifier. According to the analysis, Decision Trees has the highest accuracy when compared to other classifiers. Higher sensitivity represents the higher ability of the corresponding classifier to recognize patients with Malignant tumor which helps the physician in accurate diagnosis of the tumor, whereas higher specificity describes a higher ability of

Table 9 Comparative analysis of different classifiers for mammogram images

Classifier	Accuracy (%)	Sensitivity	Specificity	Mean Square Error	Precision	Time (s)
MLPNN	96.93	98.13%	98.42%	0.1387	98.75%	156.49
K-NN	93.865	96.83	96.85	0.178	97.45	0.01
Random forests	94.47	96.93%	97.72%	0.2584	98.75%	0.16
Decision trees	97.70	88.88%	97.63%	0.1571	91.4%	0.01

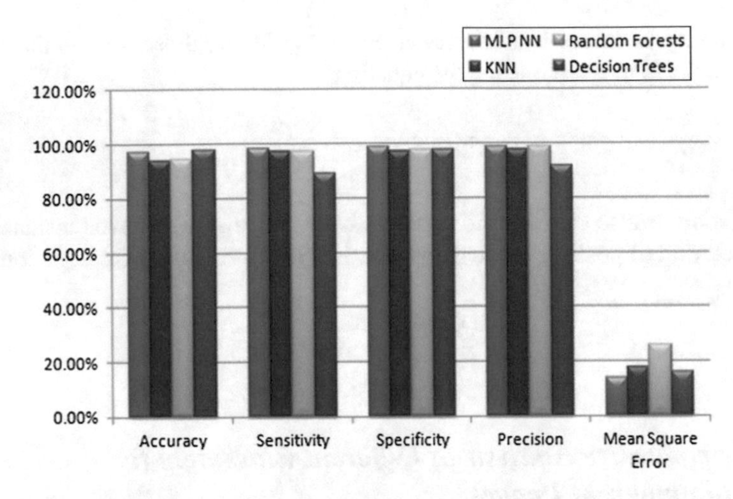

Fig. 11 Comparative analysis of different classifiers

the classifier to identify patients without a disease i.e. benign tumor or normal cases. Multi Layer Perceptron Neural Network has the highest Sensitivity and Specificity i.e. is the ability to correctly identify Malignant and Benign cases. Although in most applications, accuracy is used to evaluate models performance, in medical applications sensitivity and specificity are also more important. The reason behind their importance is their power in detecting how likely will the classifier decide if a patient suffer from benign or malignant tumor.

Precision refers to the level of measurement and exactness of description in mammographic images. Therefore, in our results, the optimal classifier is Multilayer Perceptron Neural Network(MLP NN) then followed by Random Forest and K-NN.

MLP Neural Network also has the Least Mean Square Error compared to Decision trees, K-NN and Random forests.

From the Table 9, it is evident that Multilayer Perceptron Neural Network has the highest Sensitivity, Specificity, and Precision with least MSE compared to other classifiers, Hence MLP NN can be considered as the best classifier for detecting malignant and benign tumors (Fig. 11).

8 Conclusion

This pilot research demonstrates the efficient use of technology in telehealth monitoring and telemammography in a collaborative and community practice, is a feasible solution and will improve access to the quality of healthcare services. Telehealth will reduce emergency departmental visits and hospital stays, and increases patient satisfaction and quality of life. Our studies have focussed on the use of telehealth for telemammography. As a result, a new framework for telemammography is introduced which is based on image processing and machine learning techniques to achieve better performance in terms of accuracy and precision. Furthermore, the proposed approach can be helpful to improve the physician's ability to detect and analyze pathologies leading for more reliable diagnosis and treatment of breast cancer.

References

1. Telemedicine—Remote Patient Monitoring Systems. http://www.aeris.com/for-enterprises/healthcare-remote-patient-monitoring (n.d.)
2. Telehealth and Remote Patient Monitoring for Long-Term and Post-Acute Care: A Primer and Provider Selection Guide 2013 (Rep.) (2013)
3. Li, Y., Chen, H.: A survey of computer-aided detection of breast cancer with mammography. J. Health Med. Inform. **7**(4), (2016). doi:10.4172/2157-7420.1000238
4. Su, Y.: Automatic detection and classification of breast tumors in ultrasonic images using texture and morphological features. Open Med. Inform. J. **5**(1), 26–37 (2011). doi:10.2174/1874431101105010026
5. Nascimento, C.D., Silva, S.D., Silva, T.A., Pereira, W.C., Costa, M.G., Filho, C.F.: Breast tumor classification in ultrasound images using support vector machines and neural networks. Res. Biomed. Eng. **32**(3), 283–292 (2016). doi:10.1590/2446-4740.04915
6. Wahdan, P., Saad, A., Shoukry, A.: Automated breast tumour detection in ultrasound images using support vector machine and ensemble classification. J. Biomed. Eng. Biosci. **3** (2016). ISSN: TBA, DOI: TBA (Avestia Publishing)
7. Abdelwahed, N.M.A., Eltoukhy, M.M., Wahed, M.E.: Computer aided system for breast cancer diagnosis in ultrasound images. J. Ecolog. Health Environ. J. Eco. Heal. Env. **3**(3), 71–76 (2015)
8. Digital Health in Canada—Canada Health Infoway. https://www.infoway-inforoute.ca/en/home/193-consumer-e-services (n.d.)
9. Niewolny, D.: How the Internet of Things Is Revolutionizing Healthcare, White Paper (2013)
10. Sebastian, S., Jacob, N.R., Manmadhan, Y., Anand, V.R., Jayashree, M.J.: Remote Patient Monit. Syst. Int. J. Distrib. Parallel Syst. **3**(5), 99–110 (2012). doi:10.5121/ijdps.2012.3509
11. Telemedicine in women's health care. http://www.physicianspractice.com/blogs/telemedicine-women%E2%80%99s-health-care (n.d.) (2011)
12. Ladyzynski, P., Wojcicki, J.M., Krzymien, J., Blachowicz, J., Jozwicka, E., Czajkowski, K., Janczewska, E., Karnafel, W.: Teletransmission system supporting intensive insulin treatment of out-clinic type 1 diabetic pregnant women. Technical assessment during 3 years application. Int. J. Artif. Organs **24**, 15763 (2001)
13. Breast cancer statisticsl: World Cancer Research Fund International (n.d.)
14. GLOBOCAN Cancer Fact Sheets: Cervical cancer. http://globocan.iarc.fr/old/FactSheets/cancers/cervix-new.asp (n.d.)
15. The state of oncology. Int. Prev. Res. Inst. http://www.i-pri.org/oncology2013 (2013) (n.d.)
16. Liyakathunisa, Kumar, C.N.R.: A novel and efficient lifting scheme based super resolution reconstruction for early detection of cancer in low resolution mammogram images. Int. J. Biometr. Bioinf. (IJBB), **5**(2), 53–75 (2011)
17. Santra, A.K., Singh, W.J., Arul, D.: Pixcals statistical based algorithm to detect microcalcifications on mammograms. Int. J. Comput. Intell. Res. **6**(2), 275–288 (2010)
18. Basha, S.S., Prasad, K.S.: Automatic detection of breast cancer mass in mammograms using morphological operators and fuzzy c-means clustering. J. Theor. Appl. Inf. Technol. (2009)
19. Global cancer rates could increase by 50% to 15 million by 2020. http://www.who.int/mediacentre/news/releases/2003/pr27/en/ (n.d.)
20. Smith R.A.: Epidemiology of breast cancer categorical course in physics. Tech. Aspects Breast Imaging. Radiol. Sco. N. Amer. 21–33 (1993)
21. Shapiro, S., Venet, W., Strax, P., Venet, L., Roester, R.: Ten-to fourteen year effect of screening on breast cancer mortality. JNCL **69**, 349 (1982)

22. Phadke, A.C., Rege, P.P.: Fusion of local and global features for classification of abnormality in mammograms. Sadhana **41**(4), 385395 (2016) (India Academy of Sciences)
23. Faridah, Y.: Digital versus screen film mammography: a clinical comparison. Biomed. Imaging Interv. J. (2008)
24. Patterson, S.K., Roubidoux, M.A.: Update on new technologies in digital mammography. Int. J. Womens Health **6**, 781788 (2014)
25. Patil, K.K., Ahmed, S.T.: Digital telemammography services for rural India, software components and design protocol. In: 2014 International Conference on Advances in Electronics Computers and Communications (2014). doi: 10.1109/icaecc.2014.7002442
26. Sheybani, E., Sankar, R.: Survey of telemedicine teleradiology/telemammography network architectures. SPIE J. Electron. Imaging **2** (2001)
27. Breast Cancer Information and Awareness. http://www.breastcancer.org/ (n.d.)
28. Sheybani, E.: ATMTN: a test-bed for a national telemammography network. Iran. J. Electr. Comput. Eng. **1**(1) (2002) (Winter-Spring)
29. Neri, E., Thiran, J., Caramella, D., Petri, C., Bartolozzi, C., Piscaglia, B., Macq, B., Duprez, T., Cosnard, G., Maldague, B., Pauw, J.D.: Interactive DICOM image transmission and telediagnosis over the European ATM network. IEEE Trans. Inf. Technol. Biomed. **2**(1), 35–38 (1998). doi:10.1109/4233.678534
30. Dixon, B.E., Hook, J.M., McGowan, J.J.: Using Telehealth to Improve Quality and Safety: Findings from the AHRQ Portfolio (Prepared by the AHRQ National Resource Center for Health IT under Contract No. 290-04-0016). AHRQ Publication No. 09-0012-EF. Rockville, MD: Agency for Healthcare Research and Quality (2008)
31. Rafalski, E.M.: Health Insurance Portability and Accountability Act of: HIPAA. Encycl. Health Serv. Res. (1996) (n.d.). doi:10.4135/9781412971942.n180
32. Kumar, M., Wambugu, S.: A Primer on the Security, Privacy, and Confidentiality of Electronic Health Records. MEASURE Evaluation, University of North Carolina, Chapel Hill, NC (2015)
33. Overhage, J.M., Aisen, A., Barnes, M., Tucker, M., McDonald, C.J.: Integration of radiographic images with an electronic medical record. Proc. AMIA Symp. 513–7 (2001)
34. Kohavi, R., Provost, F.: Glossary of terms. Mach. Learn. **30**(2/3), 271–274 (1998). doi:10.1023/a:1017181826899
35. Machine Learning: What it is and why it matters. http://www.sas.com/en_us/insights/analytics/machine-learning.html (n.d.)
36. Singh, B.K., Verma, K., Thoke, A.: Adaptive gradient descent backpropagation for classification of breast tumors in ultrasound imaging. Procedia Comput. Sci. **46**, 1601–1609 (2015). doi:10.1016/j.procs.2015.02.091
37. Bruijne, M.D.: Machine learning approaches in medical image analysis: from detection to diagnosis. Med. Image Anal. **33**, 94–97 (2016). doi:10.1016/j.media.2016.06.032
38. Costin, H., Rotariu, C.: Medical image processing by using soft computing methods and information fusion. In: Recent Researches in Computational Techniques, Non-Linear Systems and Control (2011)
39. Dougherty, G.: Introduction. In: Medical Image Processing Biological and Medical Physics, Biomedical Engineering, vol. 1–4, (2011). doi: 10.1007/978-1-4419-9779-1_1
40. Dogra, A., Goyal, B.: Medical image denoising. Austin J. Radiol. **3**(4), 1059 (2016)
41. Gonzalez, R.C., Woods, R.E.: Digital Image Processing, 3rd edn. Prentice-Hall (2008)
42. Liyakathunisa, Kumar, C.N.R.: A novel and robust wavelet based super resolution reconstruction of low resolution images using efficient denoising and adaptive interpolation. Int. J. Imag. Process. IJIP, CSC J. Publ. **4**(4), 401–420 (2010). ISSN: 1984-2304
43. Filter2: https://www.mathworks.com/help/images/ref/medfilt2.html (n.d.)
44. Liyakathunisa, Kumar, C.N.R.: A novel and efficient lifting scheme based super resolution reconstruction for early detection of cancer in low resolution mammogram images. Int. J. Biometr. Bioinform. (IJBB) **5**(2) (2011)
45. Chang, S.G., Yu, B., Vattereli, M.: Adaptive Wavelet Thresholding for Image denoising and compression. In: Proceedings of IEEE, Transaction on Image Processing, vol. 9, pp. 1532–15460 (2000)

46. Mohiden, S.K., Perumal, S.A., Satik, M.M.: Image Denoising using DWT. IJCSNS Int. J. Comput. Sci. Netw. Secur. **8**(1) (2008)
47. Gnanadurai, D., Sadsivam, V.: An efficient adaptive threshoding technique for wavelet based image denoising. IJSP **2** (2006)
48. Donoho, D.L., Stone, I.M.J.: Adapting to unknown smoothness via wavelet shrinkage. J. Am. Assoc. **90**(432), 1200–1224 (1995)
49. Jiji, C.V., Chaudhuri, S.: Single-frame image super-resolution through contourlet learning. EURASIP J. Adv. Signal Process. **2006**, 1–12 (2006). doi:10.1155/asp/2006/73767
50. Sonka, M., Hlavac, V., Boyle, R.: Digital Image Processing and Computer Vision. Cenage Learning (2008)
51. Patil, K.K., Ahmed, S.T.: Digital telemammography services for rural india, software components and design protocol. In: IEEE International Conference on Advances in Electronics, Computers and Communication (2014)
52. Challenges and Opportunities with Big Data, White paper
53. https://www.quora.com/Why-use-the-top-hat-and-black-hat-morphological-operations-in-image-processing
54. Suckling, J., Parker, J., Dance, D.R.: The mammographic image analysis society digital mammogram database. Exerpta Medica Int. Congr. Ser. **1069**, 375–378 (1994)

Image Processing in Biomedical Science

Lakshay Bajaj, Kannu Gupta and Yasha Hasija

Abstract Images have been of utmost importance in the life of humans as vision is one of the most important sense, therefore, images play a vital role in every individual's perception. As a result, image processing from its very first application in the 1920s to till date has advanced many folds. There are various fields in which image processing flourished but one of the major and upcoming field is Medical Science. There has been a dramatic expansion in Medical Image Processing in last two decades due to its ever-increasing and non-ending applications. The main reason that the field evolved in such short time is because of its interdisciplinary nature, it attracts expertise from different background like Computer Science, Biotechnology, Statistics, Biology, etc. Computerised Tomography (CT), Positron Emission Tomography (PET), Magnetic Resonance Imaging (MRI), X-Ray, Gamma Ray, and Ultrasound are some of the commonly used medical imaging technologies used today. The rush in the development of new technology demands to meet the challenges faced such as—how to improve the quality of an image, how to automate medical imaging and predictions, and how to expand its reach to all medical fields. The sole and only purpose of this chapter is to provide an introduction to medical image application and techniques so that more interest can be developed for further research in the same field.

1 Introduction

The term image processing, in general, refers to the processing of images using mathematical operations where input can be an image, series of images, videos, or even video frames. When we refer to image processing, generally we refer to digital

L. Bajaj · Y. Hasija (✉)
Department of Biotechnology Engineering, Delhi Technological University,
Shahbad Daulatpur, Main Bawana Road, Delhi 110042, India
e-mail: yashahasija@gmail.com

K. Gupta · Y. Hasija
Department of Computer Science Engineering, Delhi Technological University,
Shahbad Daulatpur, Main Bawana Road, Delhi 110042, India

A.E. Hassanien and D.A. Oliva (eds.), *Advances in Soft Computing and Machine Learning in Image Processing*, Studies in Computational Intelligence 730,
https://doi.org/10.1007/978-3-319-63754-9_9

image processing but two other types are also possible, i.e. analog and Optical image processing. Image processing has advanced many folds in the last decade.

1.1 History

Historical roots of the digital signal processing go way back to the twenty-fifth-century BC and are related to the 'Palermo stone' with earliest records of Nile's floods observed on the time base of 12 months. Processing of these records was concentrated on the prediction of floods fundamental for watering fields [1].

The scientific essentials of digital image processing techniques depend on numerical investigation that blocks the innovation of present day computers by numerous hundreds of years utilising works of eminent mathematicians including that of Isaac Newton (1643–1727), Joseph Louis Lagrange (1736–1813), and Leonhard Euler (1707–1783). The framework hypothesis presented amidst the nineteenth century incorporating thoughts of Gottfried Wilhelm Leibnitz (1646–1716), and Carl Friedrich Gauss (1777–1855) organised together are currently one of the fundamental scientific instruments [1].

The maiden utilisation of digital picture started in the daily paper industry, as pictures were sent by an undersea submarine link between London and New York. The Bartlane cable picture transmission framework was presented in the 1920s which diminished the time contrasted with the time required before to transport a photo over the Atlantic from over 7 days to under 3 h. Specific printing equipment coded pictures for connect transmission and after that recreated them at the recipient end. The pictures were transmitted in this way and recreated on a broadcast printer that was fitted with typefaces reproducing a halftone design. Initially, there were some problems that appeared in improving the visual quality of these digital images, some of them being the selection of printing procedures and how the intensity level can be distributed. The methods that were used to print these early images were discarded by the end of 1921 for a technique that was based on photographic projection built from perforated tapes at the telegraphic receiving end. The Bartlane system's competency was increased by 15 times in the year 1929 [2].

The examples mentioned above only involve digital images but are not considered results of digital image processing as computer were not used for the production of these images. Digital Image Processing is solely dependent on the development of computers because of the need of high storage and computational powers. Table 1 shows how the advancement of computer took place over the time. In the late 1960s and early 1970s, the medical field began to use the digital image processing. Johann Karl (1887–1956) was the first researcher who laid the foundation of computer tomography which completely opened the gate of digital image processing for the medical field. In the medical field, images obtained are used for research purposes and for extracting information about the various biological changes taking place below the skin or may be on the skin. Therefore, it helps in determining an unsure disease making diagnoses a bit more uncomplicated or straightforward. The development of new and better medical imaging technologies like Computerised Tomography (CT), Positron Emission Tomography (PET), Magnetic Resonance Imaging (MRI), and much more have certainly revolutionised this field.

2 Applications in Biomedical Science

Medical imaging is widely used in medical science for determination of the internal structures of the human body and diagnosis of diseases. The internal part of the body can be touched without actually having to open the body [3]. Hence, medical imaging is important for the detection and assessment of the treatment of a disease. The various techniques for medical imaging include- X-ray imaging, Gamma ray imaging, Ultrasound, Computes Tomography (CT) Scanner, PET (Positron Emission Tomography) and Magnetic Resonance Imaging (MRI) (Fig. 1) [4].

2.1 X-Ray Imaging

Nowadays, X-ray imaging is the most common and one of the oldest imaging technique used for clinical purposes. X-Rays can penetrate the human body, and so

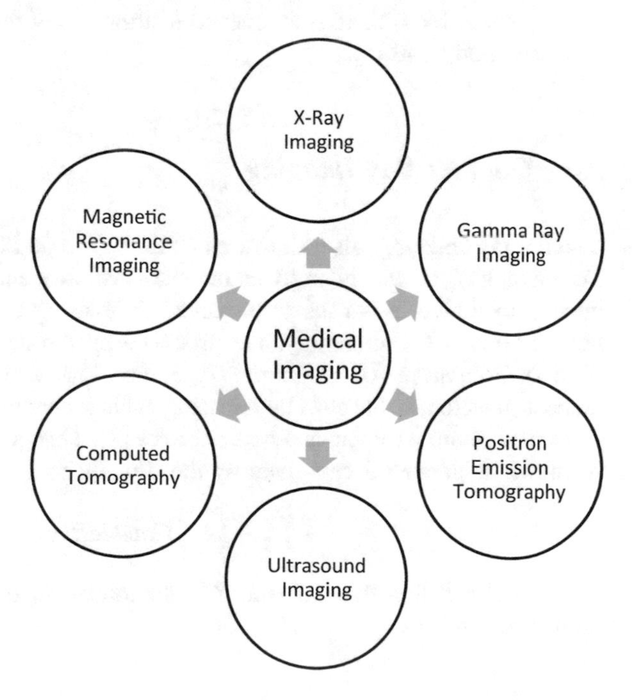

Fig. 1 The various techniques for medical imaging include- X-ray imaging, gamma ray imaging, ultrasound, Computes Tomography (CT) scanner, PET (Positron Emission Tomography) and Magnetic Resonance Imaging (MRI)

are used to detect the anatomical structure of the body. X-Rays are generated when high-speed electrons interact with heavy atoms like tungsten. The resultant loss of energy produces X-ray photons. X-rays are electromagnetic waves whose wavelength varies between 0.01 and 10 nm. Since these rays have a short wavelength, they can easily penetrate and transmit through the human body. Usually, X-rays having wavelength in the range of 0.01 and 0.1 nm are used for diagnostic applications as they have lower energy while the ones with higher energy are used for therapeutic purposes [4]. X-rays are generated in a scattered manner, which are made to pass through a narrow X-ray tube, consisting of an anode and a cathode, to produce a concentrated beam of X-ray radiation. The radiation beam is concentrated over a particular part of the body which needs to be scanned [2]. The X-rays penetrate through the body and undergo absorption and scattering as they pass through the body.

Different tissues of the body absorb different amount of X-rays depending on their density, resulting in a variation in the intensity of the radiation escaping from different locations of the body. The variation is determined by subtracting the emitted radiation from the body with the X-ray radiation passed through the body. This intensity variation is scanned and recorded on a radiographic film to produce an X-ray radiograph (an image that is produced on a fluorescent screen or a photographic film). In this way, the 3D (three-dimensional) internal structure of the human body is recorded on a film as its 2D (two-dimensional) projection [4].

The X-ray imaging technique is generally used for the clinical diagnosis of bones. It is the most commonly used imaging technique as it has a low cost and is easier to implement. It also has certain drawbacks such as the final image would be incomprehensible if there is an uneven distribution of the radiation intensity on the subjected body part [5].

2.2 Gamma Ray Imaging

Gamma ray imaging, also known as SPECT (Single-Photon Emission Computed Tomography), is widely used in nuclear medicine and astronomical studies. In nuclear medicine, it is used to produce 3D images of organs and even the entire human body [6]. Gamma rays are produced by the decay of radioactive isotopes such as Potassium-40, Thallium-201, Iodine-123, and Gallium-68. An unstable nucleus goes to a stable state by releasing nuclear energy and emitting photons such as gamma photons, alpha, and beta particles [7]. This generation of gamma photons is known as gamma decay given by the (Eq. 1).

$$^{X}_{Y}A \rightarrow {}^{X}_{Y}A + GammaRays, \tag{1}$$

where A is a radioactive element, X is the total number of nucleons, and Y is the number of protons.

In this imaging system, the patient is injected with a gamma-emitting radioisotope into its blood stream through the use of radiopharmaceutical drugs. This is absorbed by the tissues and result in the emission of gamma rays. These rays transmit through the body and are then detected by the gamma ray detectors that are surrounding the patient. SPECT imaging is executed by using gamma camera (or scintillation camera), which provides 2D images of a 3D structure [6]. Multiple 2D images or projections are acquired from different gamma ray detectors placed at different angles to the body. A tomographic reconstruction algorithm is then used to produce a 3D view of the body [7].

Gamma ray imaging is a low-cost technique but produces poor quality images because of scattering problems. It is used to detect the location of bone pathologies such as tumours or diseases, cardiovascular problems, and thyroid [2].

2.3 Positron Emission Tomography (PET)

Like gamma ray imaging, PET is also a radionuclide imaging technique. In this imaging method, beta decay (a type of radioactive decay) of radioisotopes is taken into use. Beta decay producсs positrons or beta particles having a positive charge. A significant amount of kinetic energy is also released along with the positron. Examples of positron emitting radioisotopes are Oxygen-15, Nitrogen-13, Carbon-11, and Fluorine-18 [8]. Most of these isotopes have relatively short half-lives due to which the execution of this technique needs to be done as fast as possible. The beta decay is given (Eq. 2).

$$ {}^{X}_{Y}A \rightarrow {}^{X}_{Y}A + Positrons, \tag{2} $$

where A is a radioactive element, X is the total number of nucleons, and Y is the number of protons.

As the positron travels, it loses its kinetic energy and slows down. It ultimately collides with a neighbouring electron. They annihilate each other to produce two gamma ray photons having 511 keV energy (Eq. 3) [7].

$$ \mathrm{e}^{-} + \mathrm{e}^{+} \rightarrow 2\gamma \tag{3} $$

The two gamma photons are emitted in opposite directions, i.e. 180° apart from each other. Hence, it is possible to locate their source through a straight line of coincidence also known as *Line of response*.

PET follows the same procedure as SPECT, with a difference that gamma rays are generated indirectly by positron emitting radioisotopes. The two gamma photons can be detected by scintillation cameras within a small time frame also known as coincidence detection. Therefore, the source location and distribution can be determined through image reconstruction algorithms by detecting multiple coincidences [7].

The main feature of PET is that it can retrieve both the functional and the metabolic information of the tissue which is possible because of the unique interaction of the positron with the tissue's material. PET is well suited for monitoring the body at early stages of a disease [7]. PET imaging has the ability to detect many types of cancer in a single experiment. Also, PET imaging using Fluorine-18 as a radioisotope, which is the most commonly used isotope, has been useful in determining the metabolism and flow of blood in the tissue. This information is very important for detecting the presence of tumours [9].

2.4 Ultrasound Imaging

Ultrasound imaging, which came up in the 1970s and 1980s, make use of ultrasound waves. A sound wave is a mechanical wave which causes back and forth vibration of the particles in the elastic medium in which it is travelling. A sound cannot propagate through vacuum as it does not have any particles [11]. Human ears can detect only those sound waves that have a frequency in the range of 15–20 kHz. Thus, the sound waves with a frequency higher than 20 kHz are not audible to humans. Such waves are known as ultrasound waves. The principle of reflection, refraction, and super-position are followed by light as well as sound waves. Therefore, when a sound wave is incident on a surface that marks the change in medium, a part of it is reflected and the rest is transmitted to the second medium at a different angle. The attenuation of a sound wave depends on the various mechanical properties of the medium such as elasticity, viscosity, density and scattering properties [10].

For generating ultrasound images, ultrasound waves are passed through the patient's body. A high-frequency (1–5 MHz) sound generating system transmits ultrasound waves to the body. The system comprises a computer, a display and a piezoelectric crystal-based transducer consisting of a sender and a receiver. The ultrasound waves sent to the body undergo reflection and refraction when a change of medium takes place between tissues, such as fluid and bone tissues. The reflected waves or echoes are received by the receiver of the probe and are sent to the computer, which calculates the distance of the tissue from the probe by using the speed of sound in the tissue. This information is used to construct a two-dimensional image of the tissue or organ [2].

Ultrasound imaging has most of its applications in medical science. It can be used to obtain images of an unborn child from the mother's uterus, but it can also be misused to determine the sex of the unborn baby. It is also effectively used in the imaging of anatomical structures, characterising tissues and measuring blood flow in the body [10]. The major reasons for its success are features such as low cost, safety and portability. Also, in certain imaging scenarios, it can generate images of comparable quality to CT and MRI [11].

2.5 Computed Tomography (CT)

The term tomography is made up of the Greek words *tomos* (i.e. a cut or a slice) and *graphein* (i.e. to write). During the CT process, each image produced is a section or a slice of an organ, bone or tissue of the body. These slices are the projections or the two-dimensional views of the body generated by X-ray imaging. These 2D images are taken at multiple angles around a single axis of rotation to produce a 3D view of the body. This is a great advancement in the imaging techniques and is very useful for certain scenarios, where a three-dimensional view is necessary such as detection of 3D shape of a tumour for its diagnosis [12].

CT imaging technique follows the same basic principle as that of X-ray imaging. X-rays are sent to the patient's body, which travel through the body and are collected by an array of detectors. The reconstruction algorithms are then used to reconstruct a two-dimensional image of the body. A 3D image is reconstructed by piling the 2D reconstructed images along the z-axis [12].

CT scan can be used to detect changes in the internal structure of lungs, to detect a fracture in bones, and generate images of the heart and brain. It has also proven to be very beneficial for cancer treatments by the following ways:

- Helps in the diagnosis of a tumour
- Determines the shape and size of the tumour
- Provides data about the different stages of cancer
- Helps in planning the appropriate treatment
- Directs where to perform a biopsy procedure
- Detects whether a treatment is effective
- Detects the unusual growth or recurrence of a tumour.

CT imaging enables to acquire images in a short period of time [13]. The images acquired are clear and have a high-contrast resolution. Also, it can capture almost every area of the body, which allows the doctors to get a full body scan of the patient. However, the X-rays generated during a CT scan are a form of ionisation radiation which can be very harmful. These rays can damage the body cells including DNA and cause cancer. Children are more sensitive to these radiations than the adults. Thus, certain established guidelines need to be followed before performing a CT scan [13].

2.6 Magnetic Resonance Imaging (MRI)

Like Computed Tomography, MRI is also a tomography technique that generates a three-dimensional image of the anatomical structure of the body and therefore, it provides both the structural and the functional information about the internal organs and tissues. But, unlike Computed Tomography, it does not produce images based on the transmission of radiations through the body. Instead, it works on the principle of a directional magnetic field associated with moving charged particles. It is based on the nuclear magnetic resonance (NMR) property of some selected nuclei in the matter of the body. Certain nuclei when placed in an external magnetic field, emit and absorb radio frequency energy. Hydrogen atoms are most commonly used in MRI as they are plenty of hydrogen atoms present in most living organisms, particularly in the form of fat and water [12].

MRI produces good quality images that are rich in information. Certain improvements have been made to MRI, by introducing *MR Spectroscopy* and *Functional Magnetic Resonance Imaging* (fMRI). These provide effective means to obtain local characteristic information about the functional behaviour of the organs and tissues. Functional MRI can be used to determine the response of different parts of the brain to a passive activity in an inactive state. For example, Functional MRI can be used to obtain images of the brain through which the changes in the blood flow when an acoustic signal is presented to the subject [12].

MRI is the most widely used imaging technique today for diagnosing brain tumours. A brain tumour is a substance that is either formed due to an uncontrolled growth of cells in the brain or spread from other organs having cancer. Detection of a brain tumour in its early stage helps in determining a proper treatment such as a surgery, chemotherapy, etc. MRI can also provide information about the shape and size of the tumour in the brain [14].

A significant advantage of MRI is that it does not transmit any ionisation radiations through the patient's body, and so does not cause any harmful effect to the patient. Another advantage is that it is fast and produces images with a very high spatial resolution that are in the range of 1–0.01 mm. Therefore, it provides an effective method for imaging the anatomical structures along with their functional characteristics such as oxygenation and blood flow [15]. But, due to heterogeneity in the magnetic field, deviation in temperature or motion of the tissue, noise can be introduced to the MRI images. Thus, de-noising of the images becomes mandatory [14].

3 Techniques Used in Biomedical Science

3.1 Image Segmentation

Image segmentation is considered to be most crucial part of image processing. Segmentation is nothing but the division of something into distinct parts or sections. Therefore, image segmentation can be defined as subdividing an image into several regions or objects until the region of interest is reached [16]. Many segmentation methods have been developed over the time for medical imaging which has tremendously changed the way of medicine, diagnosis and treatment. With the advancement in imaging technique, it has become easier for the biologist to view a variety of biological phenomenon or processes. The main purpose of image segmentation is to provide better visual quality of image so that the process of

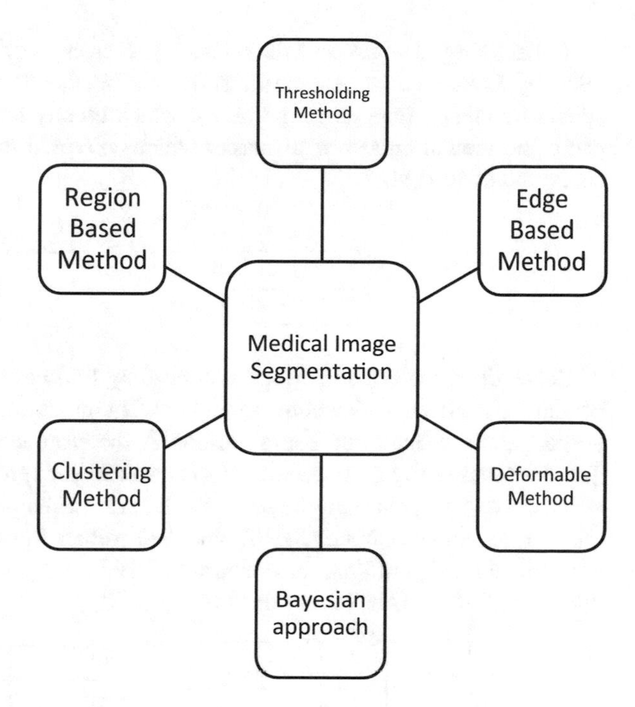

Fig. 2 Medical image segmentation techniques

detection can be more efficient and effective. However, medical image processing may include more reasons such as:

- Disease Diagnosis
- Disease Progression
- Quantification
- Monitoring.

Glancing back at the historical backdrop of strategies and systems proposed with regards to medical image segmentation, we can say that there is an incredible change in such manner. With the progression of time, more viable and proficient procedures have been discovered as appeared in (Fig. 2).

3.1.1 Edge Detection

Edge in an image is defined by pixel because an edge occurs where there is a sharp change in the intensity or value of pixel [18]. For example, an image with black filled circles on white background, an edge will be the periphery of the circle. Edge detection is considered to be a full-fledged field in image processing for that reason it is used as a base for other segmentation techniques [19]. Region boundaries and edges are narrowly related because of which the edges recognised are sometimes nebulous. So, the user needs to define the parameters of the program in order to get the edge detected. Edge detection methods are discussed below.

Robert Edge Detection Filter The first ever edge detection filter introduced in 1965 by Lawrence Roberts [16]. A simple but quick method as it detects high spatial frequency regions in the image which mostly resemble to edges. It makes use of horizontal and vertical masks which convolve the entire image and hence detect the edge [18].

Gx

-1	0
0	+1

Gy

0	-1
+1	0

Sobel Edge Detection Filter Introduced by Irwin Sobel (1970) is considered to be the simplest edge detection filter [16]. It comprises of two masks, one being vertical and the other one being horizontal, therefore contemplated as linear filter. The filter makes use of a 3*3 matrix to calculate an approximate value of first-order x-derivative and y-derivative operators. It leads to the points where the gradient is the highest. Sobel operator has '2' and '−2' values in centre of first, third rows of the vertical mask and first, third column of the horizontal mask which increases the intensity of edge. It is widely used in MRI [21].

Gx

-1	-2	-1
0	0	0
+1	+2	+1

Gy

-1	0	+1
-2	0	+2
-1	0	+1

Canny Edge Filter It is one of the most robust edge detection methods when it comes to detecting edges by removing noises from the image. It is one of the most sophisticated programs and still outperforms a number of other algorithms. The filter approximates the value of the first derivative of 2D Gaussian unlike the Sobel filter which is a plus as it smoothens the images and able to identify weak edges [20]. It is very useful in CT scans [21].

3.1.2 Threshold Method

Threshold method is the most common but powerful method used in image segmentation for analysing the foreground excluding the background of the image [16]. The input image is a grey-scale image and is converted to a white and black image or a binary image in the output. It works on the basis of intensity, dividing the image into two parts with two different intensities. The first part of the image consists of the foreground which has the pixels having intensity greater or equal to the threshold and the second part consists of background having pixels intensities less than the threshold. Its application is found mostly in CT scans where thresholding is used to extract the bone area from the background [22]. There are two types of thresholding methods—(1) Global Thresholding (2) Local Thresholding as given in (Table 1).

Table 1 Types of thresholding methods [20, 22, 25, 26]

Name	Methodology	Advantage	Disadvantage
Global thresholding	Fix a threshold value then pixels larger than that are converted into white pixels and the others are black	Simple and offer a lot of alternatives for determining threshold	Grayscale distributions of bright objects and dark background is assumed to be constant
Local thresholding	Splits the image into various sub-images and by virtue of that calculates the threshold value for each sub-image	When global thresholding lacks, i.e. the constant threshold value is not uniform then local thresholding will resolve the problem by setting a fitting threshold at each pixel	Time required to segment an image is more

3.1.3 Region Growing Method

Region growing is another common and popular image segmentation technique. The technique is simple as it begins by considering each pixel as a segment [23]. Then the region of interest is picked up using predefined condition. Initial point or region of interest can be defined either manually or through edge or intensity detail of image. Those set points are called as seeds. Once the seed or region is selected, the region starts growing on the basis of the homogeneity (e.g. greyscale, colour, texture, shape, etc.) of its neighbouring pixels.

The main application of this method in the medical field is to represent tumour regions [23]. The method has a disadvantage that it significantly depends on initial seed point selection which can be done either by using additional operators or doing it manually [22]. Therefore, it is clear that the program cannot work on its own. Some of the recent region growing methods in medical image segmentation are discussed in brief below.

- It is used in brain abnormality segmentation. The method works on the basis of seed growing region. It takes MR images of different sizes of the brain of both female and male adults. The brain tissue and background are then divided further into different categories so that different sized MR images are received as input [24].
- It is also used in ultrasound image segmentation. The seed points are automatically detected that are based on textual features with help of co-occurrence matrix. This method is faster as it automatically selects the seeds rather than doing it manually [25].

3.1.4 Clustering Based Method

Objects having same characteristics are grouped together, hence are known as clusters. Clustering approach is considered to be an unsupervised method as it does not make use of the training set rather train itself using available data [17, 22]. The inability to learn is balanced by repetitively dividing the image using segmentation process and then demonstrating the divisions. In other words, this method tries to summarise the existing data and tries to train itself using that data [17]. The method is quite appropriate for medical image segmentation as for the anatomy of human, objects with similar attributes are grouped together and different objects in another group [18]. There are two methods of clustering that are most commonly used for medical image segmentation—*K*-means, and Fuzzy *c*-means.

***K*-means** It is the most extensively used unsupervised method. First, the image is divided into *K* clusters which is done by repetitively calculating intensity values of isolated class or *K* cluster of the image. Then the segmentation is done by putting every data point/pixel into a cluster which has the nearest distance to clusters mean [22].

Fuzzy c-means (FCM) It is also an unsupervised method. The difference between the two processes is that *K*-means method categorises the points as a separate class whereas this method allows to connect the points to many classes. It is one of the most appropriate clustering methods for medical image segmentation [17].

Some of the recent works in medical image segmentation are discussed below in brief.

- FCM is used in the segmentation of MR images. A new version FCM has made it possible to automatically determine the number of clusters required for segmentation process. This method uses a statistical histogram to reduce the iteration time and during the iteration process, an ideal number of clusters are detected. The new version came out to be more accurate and faster than the basic version [27].
- LM-K means technique is also used for MR image segmentation. A simple *K*-mean method is applied after pre-processing and converted into a supervised method by the use of *Levenberg–Marquardt optimisation* technique. This method achieved higher precision in comparison to the classic approach [28].

3.1.5 Bayesian Approach

Bayesian approach is mostly used for classification purpose, therefore, allowing the use of prior information which benefits in the image analysis [17]. The method works on the principle of posterior probability, which abridges the level of one's certainty pertaining to a particular situation. In order to derive prior knowledge of a specific object shape, a training set of shape samples is required, and such knowledge can be expressed as a prior distribution, which expounds and represents the local as well as global variations in the training set, for defining a Bayesian estimate

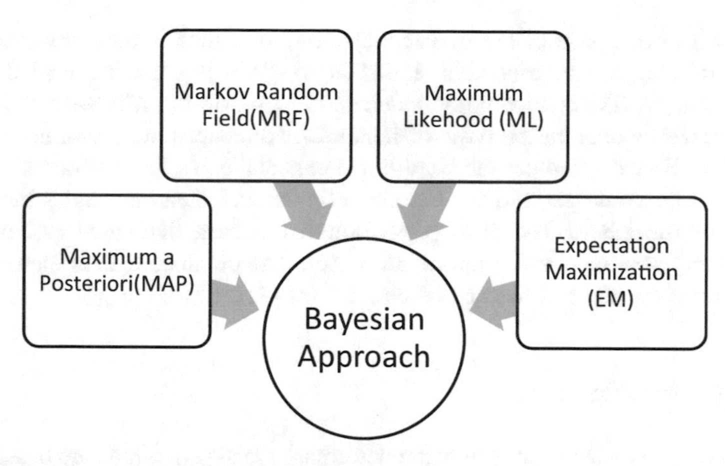

Fig. 3 Bayesian approaches

as the most favourable solution for the object contour [29]. This method is so vast and useful that it can be applied to segment almost any biomedical image (e.g. MRI, CT Scans, and Ultrasound images, etc.). There are four most used Bayesian approaches in image segmentation (Fig. 3) that are discussed below in brief.

Maximum a Posterior (MAP) In Bayesian figures, MAP estimation is an approach to the back assignment. The MAP can be utilised to obtain a summit estimation of an unnoticed measure occurring in the establishment of experimental test data. It is closely interrelated to Fisher's procedure of Maximum probability (ML) technique despite the fact that possesses an increased enhancement reason which coordinates a first portion over the measure one longings to inexact. MAP estimation can be viewed as regularisation of ML assessment [30, 31].

Markov Random Field (MRF) Markov random field (MRF) hypothesis gives a helpful and reliable method for displaying setting context-dependent elements, for example, image pixels and related features highlights. It fundamentally makes utilisation of undirected diagram that decides the Markov estimations of some self-assertive factors contained inside a graphical model. MRF is very much like the Bayesian approach in perspective of illustration. The main distinction is that this approach is undirected though the Bayesian method is included directed graphs [32, 33].

Maximum Likelihood (ML) It is considered as one of the most essential segmentation algorithms that have been generally utilised as a part of numerous applications, including some biomedical image processing problems [35]. In a few situations, it is additionally used to boost the probability work when we are given with settled measure of information together with its factual model from where qualities are chosen of the parameters that complete general employment of amplification [30, 34].

Expectation Maximisation (EM) It is a general procedure for discovering (ML) estimates with incomplete information. In EM, the total information is considered to comprise of the two sections, one of which is only observed while other is missing (or unobservable, and hidden). With just the fragmented data or information, an EM methodology endeavours to tackle the ML estimation issue. This approach works on the basis of iterations. Here steps are performed in alterations; The E-step computes the conditional expectation of the unobservable labels given the observed data and the current estimate and then substitutes the expectations for the labels. The M-step performs maximum likelihood estimation as though there was no missing information [36]. The obtained data is then used for the following E-step and the procedure goes on [31].

3.1.6 Deformable Model

Another recently used technique is the deformable methods, which are based on the boundaries of the objects. Shape, evenness and internal forces along with the external forces on the object are the characteristics that are taken into consideration for the analysis of the image boundaries. This has made this method an appealing approach for image segmentation [37]. The object boundaries are defined using shapes and closed arcs. Initially, a closed curvature or plane is placed close to the selected edge in order to outline the object boundary and then it goes under iterative reduction movement. The internal forces are developed to keep the segmentation process effortless. In order to initiate a plane towards the desired part in the image, the external forces are also developed. This method has significant advantages of piece-wise continuity and noise insensitivity. The deformable methods can be categorised into two main categories—parametric deformable methods (explicit) and nonparametric deformable methods (implicit) as shown in (Fig. 4) [17].

Fig. 4 Deformable methods

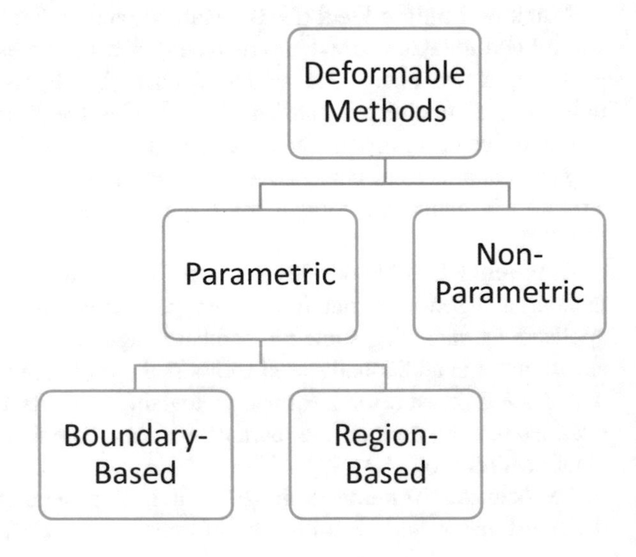

Parametric deformable methods This method, also known as active contours, uses a finite number of parameters. The shape model is characterised by making use of curves produced by the parameters [17]. The model further has two types:

- **Boundary-based method**—The required object boundary is acquired by estimating the object boundary of interest and then imposing feature matching and smoothness constraints on it. This method is very noise sensitive as the boundary information can be easily altered by any sort of noise [38].
- **Region-based method**—The image is partitioned into homogeneous regions by applying smoothness constraints on the image instead of applying on the boundary. Thus, this type of image segmentation method is called region-based method [38].

A major disadvantage of these methods is that it is not easy to manage unknown segmentation of entities [39].

Nonparametric deformable methods Based on the idea of convolution theory, these methods are also known as geometric active contour methods. A level set function is made use of along with the added time feature in order to outline the curve for segmentation [17]. Unlike the parametric deformable methods, these methods do not make use of parameters for the evaluation of the curve. These overcome the drawbacks of the parametric models as it can manage changes in the topological aspects. Edge- and region-based methods can also be used by these methods but the execution is very different from that of parametric deformable methods [40].

Deformable models provide a unique and influential approach for image segmentation as it unites physics, geometry and the approximation theory. These models are used for segmenting, tracking and matching the internal structures of the body by using information about the size, shape and location of the structures. These models have various applications in the analysis of medical images which includes segmentation, matching, motion tracking and shape representation. The variations in the biological structures over time and among different persons can be handled by these models [41].

Table 2 lists all the image segmentation method discussed in this chapter and compares them. The comparison is based on their methodology, advantages, disadvantage and the application in various medical fields. These methods have been developed overtime for the segmentation of both biomedical images and images related to different fields. This will help in choosing the best method for a given condition or as indicated by the required result.

Table 2 Comparison of image segmentation methods [16–20, 22, 23, 26, 28, 29, 37, 41]

Name	Methodology	Advantages	Disadvantages	Application
Edge detection	Discontinuity detection, generally tries to locate sharp change in the intensity or value of pixel	Easy to implement and works well for images with good contrast between regions	Cannot be applied to all sorts of problems	Can be applied in all types of biomedical image segmentation
Threshold	Divides image into two segments on basis of intensity. The first part being foreground and second being background	Does not need prior knowledge of image, simple, fast, and easy to implement	Does not consider spatial information of images which leads to sensitivity to noise and intensity in homogeneities	Applicable to structures that have divided intensity for, e.g. CT scans
Region growing	Considering each pixel as a segment or seed. If the neighbouring regions have same property, combine them	Simple, assure continuity and more immune to noise	Requires at least one seed point, therefore the result significantly depends on the seed point selection	Images having high-contrast boundaries for e.g. MRI
Clustering	Pixels with same properties are grouped together	Simple, require less time	Does not refer to spatial information, difficulty in pixel value similarity	Used for MRI, does not work well for CT scans
Bayesian approach	An ideal classifier which utilises posterior probability distribution as the distinguishable function by estimating statistical properties such as likelihood	Joins the earlier accessible data with the given information and offers a reasonable circumstance for a broad assortment of models	Does not give legitimate ways to make sense of priors and makes utilisation of posterior distributions which are exceedingly subject to priors	These are mainly applicable to any kind of medical image for verification problems
Deformable	Based on the boundaries of the objects; considers features such as smoothness, shape, internal and external forces on the object	Accommodates the variations in biological structures over time; provides sub-pixel sensitivity and is noise insensitive; ensures piece-wise continuity	Slows down the system as it requires the adjusting of parameters	Work best with statistical regional data of the picture. Segmenting, tracking and matching the anatomical structures of the body

3.2 Feature Extraction

Feature extraction is the stage where the desired part of the image is acquired. Feature extraction methods are basically designed with an intent so that different representations of the characters (e.g. solid binary character, character contours, skeleton or grey-level sub-images) of each individual character can be obtained. Therefore, it is very useful in medical modalities as it makes the diagnosis easier. Various feature extraction methods are discussed below.

3.2.1 Low-Level Feature Extraction

The low-level feature may be defined as basic features that can be automatically extracted from the image even without making use of the shape information. Thresholding method is also a form of low-level feature extraction technique. Apparently, high-level feature extraction makes use of low-level feature extraction in order to find shapes in an image [42]. A well-known fact, that people can be recognised from caricaturists portraits, i.e. the first low-level feature extraction called edge detection the aim of which is to produce line drawing. There are some very basic techniques and on the contrary, some advanced ones, thereby we will have a brief look on some of the most popular approaches (Table 3).

3.2.2 High-Level Feature Extraction

High-level feature extraction is mainly concerned with finding shapes. For example, for recognising faces automatically, an approach can be used to extract the components features such as the eye, the ear, and the nose which are major facial

Table 3 Brief description of low-level feature extraction techniques [23, 42, 46]

Name	Methodology	Types
First-order edge detection	Highlights image contrast, which is the difference in intensity. Therefore emphasises change. These are group operators, the aim of which is to deliver an output approximate to result of first-order differentiation	Roberts cross, smoothing, Sobel, Prewitt and Canny
Second-order edge detection	An alternative to first order, f(x) is greatest where the rate of change is greatest and zero when rate of change is constant. At the peak of the first-order derivative, the rate of change is constant. This is where the sign changes in the second-order derivative	Laplacian, and Marr–Hildreth
Image curvature detection	Curvature is considered as the rate of change in edge direction. Therefore, rate of change illustrates the points in a curve; corners are the point where the edge direction changes rapidly, whereas when there is little change it corresponds to straight lines	Planar curvature, curve fitting and Harris corner detection
Optical flow estimation	Used to detect moving objects. The points which move faster are made brighter so that they can be detected easily	Area-based approach, and differential approach

Table 4 A brief description of high-level feature extraction techniques [23, 43]

Name	Methodology	Types
Pixel operation	Shape is obtained from the image on the basis of its pixel. The brightness of the pixel forming shape is obtained and then it is extracted from the background. If background details are known then the background is removed from the images leaving only the pixels forming the shape	Thresholding and background subtraction
Template matching	Shape extraction is done by matching pixels with a predefined model. We just need to match the template with an image, where the template is a sub-image containing shapes that we need to find. The template is centred on an image point and then it is counted how many points match with the image. The point with the max no of match is considered to be the shape	Intensity template, and binary template
Hough transform	Locate shapes in images. A better version of template matching as it produces same results faster. It is achieved by reconstruction of template matching process based on evidence gathering approach	Lines, circle and ellipse detection

features. These features are found by determining their shape as the white part in eyes is ellipsoidal, the mouth can appear as two line upside down and similar for eyebrows two line in opposite orientation. Shape extraction simply implies finding their position, size and orientation. Extraction is more complicated than detection as extraction infers that we have all the information regarding shape such as position and size, detection, on the other hand, infers to having a mere knowledge of the existence of shape in the image [23, 43]. The most commonly used feature extraction techniques have been presented in (Table 4).

3.2.3 Flexible Shape Extraction

The last section discussed on how to find shapes by matching which infers to information of a model (mathematical or template) of the shape required. The shape extracted there was fixed, they are flexible in a way the parameters define the shape, or how the parameter defines the appearance of a template. However, it is not always possible to provide a model with sufficient accuracy or template for the target shape. In that case, we need techniques that may evolve to the target, or adjust the result according to the data which implies the need to use flexible shape extraction, therefore, some of the most popular approaches are presented in (Table 5). The techniques can be distinguished on the basis of matching functional which is used to indicate the extent of match between the shape and data. [44].

Active shape model finds its application in MRI, and face recognition. Recently, a similar approach has also been developed that also incorporates textures, known as active appearance model (AAM). One major difference in these two is that AAM explicitly includes textures and updates the model parameter so that the points can

Table 5 Brief description of flexible shape extraction techniques [23, 44, 47]

Name	Methodology	Types
Deformable template	Aimed to find facial features for the purpose of recognition for example, it may consider an eye to be comprised of an iris that sits within the sclera and which can be modelled as a combination of circles that lies within parabola	Optimisation technique, and gradient technique
Active contour (snakes)	Set of points which aim to enclose target shape. It is basically like a balloon which is placed over the shape and it shrinks until it fits the target shape. In this way set of points are arranged so as to describe a feature enclosed by it	Energy minimisation, greedy algorithm Kass snake
Active shape model	The approach is concerned with a point model of a shape: the variation in these points is called the point distribution model. The aim is to capture all possible variations. The statistics related to the variation of the point's position describes the way in which the shape appears	Active shape model, and active shape appearance

be moved nearer to image points by matching texture. Some of the essential differences between AAM and ASM are mentioned below [47].

- ASM makes use of texture information that is local to a point, whereas AAM makes use of texture information in the whole region,
- ASM tries to minimise the distance between image point and model point on the one hand, AAM tries to minimise the distance between a target image and a synthesised model on the other hand

3.3 Image Classification

Medical image classification can play a vital role in diagnosis and even for teaching purpose. There are many machine learning methods that can be used for classification of images. One way to classify is to find the texture of the image and analyse it. In texture classification, the aim is to allocate an unknown sample image to a known texture class. It has some very important application in computer image analysis as well as in medical imaging. Efficient image texture is required for successful classification and segmentation. Another way is by using Neural Networks Classification. Neural networks have appeared to be an important tool in recent times. It involves a multilayer perception which is used for learning purpose and is considered to be a standard supervised network. Another popular method is *K*-Nearest neighbour which works by placing *k* diverse points in the feature space characterised by clustering objects. After insertion of points, every object is assigned to the cluster that has a contiguous centroid. The position of each *k* point is altered again once the groups have been assigned. This step is iterated until the centroids movement becomes static. This repetition will formulate a metric that needs to be calculated by splitting the objects into different clusters [48]. Some of the popular classification techniques are discussed below.

3.3.1 Texture Classification

Texture classification is an image processing technique that classifies the features of images based on their texture properties. The goal of this technique is to assign a set of known texture classes to a sample image [48]. Image texture classification is an important method in computer vision and image processing related areas. An effective classification method should have an efficient description of the texture of an image. The texture can be described by defining the basic primitive patterns and their replacement rules [49]. Texture classification when done manually becomes variable and depends on the individual visual perception. Thus, tools for automated pattern recognition and image analysis are necessary that provide objective information and help in reducing the variability. Most of the textual features fall under the following three categories—

- **Syntactic** This type of texture analysis analyses the primitive features of the image [48].
- **Statistical** The basic statistical features are mean, standard deviation, energy and variance. More advanced features such as sum average, sum entropy, sum variance, angular second movement, correlation, contrast, etc., also come under this category. The coarseness of a texture in a particular direction can be evaluated by computing run-length features. A grey-level run contains a set of consecutive colinear pixels in a given direction [48].
- **Spectral** The co-occurrence or run-length features may not be able to identify larger scale changes in spatial frequency. This comes under spectral features [48].

Texture analysis methods are of four types—statistical, geometrical, model-based and signal processing [48]. The statistical method derives a set of statistics from the relationship of local features by analysing the spatial distribution of the grey-values in the image. Texture analysis using **Grey Level Co-occurrence Matrix (GLCM)** is a statistical method that considers the spatial relationship of pixels. A GLCM is created by characterising the texture of an image by computing how often pairs of pixels with specific values, and in a specific spatial relationship occur in an image. For example, contrast calculates the variations in the GLCM, correlation calculates the probability of occurrence of the specified pixel pairs, homogeneity evaluates the proximity of the distribution of elements in the GLCM to the GLCM diagonal, and energy gives the sum of squared elements in the GLCM. The matrix provides the statistical information about the texture of an image [49].

Texture classification techniques are used mostly in industrial and medical surface assessment. For example, classification and segmentation of images of satellites, searching defects and diseases, segmentation of textured regions in document analysis, etc. Although it has various useful applications, it is not widely used in industries [48].

3.3.2 Neural-Based Classification

Neural Network in recent times has become one of the most important tools for classification. Neural Networks have shown promise so that it can be used as an alternative to conventional classification methods. There are several advantages of using neural networks, some of these are mentioned below.

- It is a self-adaptive method driven by data, means that they can adjust themselves to data without the need of any explicit specification of functional or distributional of model.
- Considered to be universal functional approximator which can approximate almost any function with arbitrary accuracy.
- Neural networks are nonlinear model which helps them to model complex real-world applications.
- Able to estimate the subsequent probabilities, on the basis of which classification rules are established.

Neural Networks make use of both supervised and unsupervised techniques. With time, medical imaging is gaining importance in the diagnosis of diseases. Methods like Artificial Neural Networks (ANN) have gained considerable attention in recent times for use in medical imaging. We do not aim to go into details of any specific algorithm or illustrate experimental results, rather our aim is to summarise major strengths and weaknesses in medical imaging. Therefore, we now discuss the strengths and weakness of neural networks [48].

Major strengths There are different types of neural networks that are most widely used in medical image processing—Hopfield neural network, Feedforward neural network and Self-organising feature map (SOM).

- The main advantage *Hopfield neural network* provides in medical imaging is that the problem of medical imaging reconstruction can be solved by minimising the energy function so that the network converges to stable state. When compared with conventional techniques, even the idea of optimisation of medical images by Hopfield neural networks makes processing of medical images a lot easier.
- The *feedforward neural network* is also a supervised technique. This type of technique is considered to be one of the best in medical image processing when standards are available. When compared with Hopfield technique or any other convention technique, the method has an edge because of its ability to control

the negotiation between the noise performance and resolution of image reconstruction. The method is easy to implement as compared to others.

- When there are no standards, *Self-organising feature map (SOM)* is an exciting alternative to supervised techniques. It has the ability to learn and differentiate different medical image information.

In general, neural network methods that are employed in medical image processing when compared to other conventional methods, the time was recorded to be negligible small when a trained neural network method is applied to solve medical image problem, although the training time is more for neural networks [50, 51].

Weakness Though, ANN has several applications it also has several limitations:

- The first problem arises that how to choose an efficient neural network method and architecture related to it. However, some work has been done on model selection in recent times but there is still no appropriate data describing what type of network should be built for a particular task. Therefore, networks have to be designed by trial and error, however, this pragmatic approach to network design is difficult to overcome. Furthermore, a danger always resides of over-training the neural networks which may minimise the error measure but may not correspond to searching a well-generalised neural network.
- The second problem arises due to its black box character. A trained neural network always provides a corresponding output for an input but the problem does not explain how it reached to that output and how reliable is that output. In medical imaging, it is certainly problematic which limits the use of neural networks [50, 51].

3.3.3 Support Vector Machine (SVM)

SVM is a binary classifier that tends to generalise better, as it classifies two classes of instances by finding the maximum separating hyper plane between the two [48]. It can be fed the grey-level values of the raw pixels directly with only some pre-processing done like the selection of certain pixels following the configuration of autoregressive features. Thus, SVM includes both feature extraction and classification. The feature extraction is performed implicitly by a kernel, which is the scalar product of two mapped patterns. A kernel can perform the same functions as that of the conventional feature extraction methods such as statistical feature extraction [52].

A binary class regulated classification is done by taking n training samples ($<x_i>$, y_i), where $<x_i> = (x_{i1}, x_{i2}, \ldots, x_{im})$ is an input feature vector and $y_i \in \{-1, +1\}$ is the target label. The job of the classifier is to learn the patterns in the training samples in such a way that it can predict a y_i for an unknown x_i consistently. As SVM includes feature extraction also, nonlinear mapped input patterns can be used as feature vectors. SVM primarily performs binary classification but can be extended to multi-class situations as well by adopting one-against-others decomposition method. This functions by first applying SVM's in order to separate one class from all the other classes, and then deciding between the classes [53].

SVM classifiers have several advantages over other classifiers. These are as follows:

- Solves small sample size class.
- Works effectively in higher dimensional spaces.
- Performs well even when training data is not sufficient [54].
- Mathematically, much less severe.
- Nonlinear mapped input patterns can be used as feature vectors.
- Based on structural risk minimisation.
- Does not degrade due to noisy data.

Thus, SVM's are well suited for image classification [45].

3.3.4 *K*-Nearest Neighbour (KNN)

K-nearest neighbour-based classifier is one of the simplest and most widely used techniques for classification of a multi-class image. The KNN method classifies an input feature vector X by determining the *k*-closest neighbouring classes according to an appropriate distance metric [55]. Each element in the space is defined by position vectors in a multidimensional feature space. Distance is an important element of this method. Thus, the best-fit class for a point can be predicted. If the value of *k* is one, then the method merely becomes the nearest neighbour method and the element is classified as the nearest neighbour class [56].

The distance metric used in the KNN algorithm is the *Euclidean distance*, which is the distance between two points in the Euclidean space. The distance between two points A = (ax, ay) and B = (bx, by) in two-dimensional Euclidean geometry is given in (Eq. 4).

$$d(a,b) = \sqrt{(bx - ax)^2 + (by - ay)^2} \tag{4}$$

Let P and Q be two points in an *n*-dimensional Euclidean space such that P = (p1, p2, p3,...., p*n*) and Q = (q1,q2,q3,...., q*n*), then the Euclidean distance between P and Q is given by (Eqs. 5–6) [43]—

$$d(P,Q) = \sqrt{(p1 - q1)^2 + (p1 - q1)^2 + \cdots + (pn - qn)^2} \tag{5}$$

$$d(P,Q) = \sqrt{\sum_{i=0}^{n} (pi - qi)^2} \tag{6}$$

For an optimal value of *k*, the *K*-nearest neighbour-based classifier yields good performance [55].

4 Future Research

So far, image processing has numerous applications such as image noise reduction, defect detection, food quality evaluation, medical imaging, etc. With the advancements in technology, its use in biomedical science (e.g. X-ray imaging, Gamma ray imaging, PET, CT, and MRI) has increased manifolds. However, not much research has been done on the application of image processing in dermatology. It can be used to detect skin diseases by extracting useful information from the images of the skin. Thus, more research needs to be done in this field.

5 Conclusion

This chapter gives a brief overview of biomedical applications and techniques that come under medical image processing. Image processing majorly comprises of image segmentation, feature extraction and image classification. A number of diseases and problems are found in the various medical modalities that come under the medical field. Thus, this chapter basically examines the various methods that may be implemented in all these modalities in order to help solve a particular issue in the medical field. Each technique has many advantages as well as some disadvantages. Based on the type of application and the resources available, the subsequent methods are selected for performing image processing. In spite of several decades of research up till now, there is no universally accepted method for image processing, as the result of image processing is affected by lots of factors, such as: homogeneity of images, spatial characteristics of the image continuity, texture and image content. Along these lines, there is no single strategy which can be viewed as useful for all kind of images, similarly, not all techniques can be used for a specific sort of image. Therefore, there is still a huge scope for developing more powerful and effective methods.

References

1. Proch´azka A Vyˇsata, O.: History and biomedical application of digital signal and image processing, Paper presented at international workshop on computational intelligence for multimedia understanding, 1–2 Nov 2014
2. Gonzalez, R.C.,Woods, R.E.: Introduction. In: Marcia, J.H. (ed.) Digital Image Processing, 2nd edn., Addison Wesley, USA, 1987 (2004)
3. Elangovan, A., Jeyaseelan, T.: Medical imaging modalities: a survey. Presented at the IEEE international conference on emerging trends in engineering, technology and science, Pudukkottai, 24–26 Feb 2016
4. Dhawan, A.P.: Medical imaging modalities: X-ray imaging. In: Lajoh, H. (ed.) Medical Image Analysis, 2nd edn, pp. 79–98. Wiley, New Jersey (2011)

5. Chou, K.Y., Lin, C.S., Chien, C.H., Chiang, J.S., Hsia, C.H.: Using statistical parametric contour and threshold segmentation technology Applied in X-ray bone images. Paper presented at the international symposium on intelligent signal processing and communication systems. 24–27 Oct 2016
6. Fysikopoulos, E., Kopsinis, Y., Georgiou, M., Loudos, G., et al.: A sub-sampling approach for data acquisition in gamma ray emission tomography **63**(3), 1399–1407 (2016)
7. Dhawan, A.P.: Nuclear medicine imaging modalities. In: Lajoh, H. (ed.) Medical Image Analysis, 2nd edn, pp. 139–156. Wiley, New Jersey (2011)
8. Gambhir, S.S.: Molecular imaging of cancer with positron emission tomography. Nat. Rev. Cancer **2**, 683–693 (2002)
9. Hubner, K.F., McDonald, T.W., Neithammer, J.G., Smith, T.G., Gould, H.R., Buonocore, E., et al.: Assessment of primary and metastatic ovarian cancer by Positron emission tomography (PET) using 2-[18F] deoxyglucose (2-[18F]FDG). J. Gynecol. Oncol. **51**, 197–204 (1993)
10. Dhawan, A.P.: Medical imaging modalities: ultrasound imaging. In: Lajoh, H. (ed.) Medical Image Analysis, 2nd edn, pp. 157–172. Wiley, New Jersey (2011)
11. George, S., Cheng, R., Ignjatovic, Z.: A novel ultrasound imaging technique for portable and high speed imaging. Paper presented at the 13th International on new circuits and systems conference, 7–10 June 2015
12. Dhawan, A.P.: Medical imaging modalities: magnetic resonance imaging. In: Lajoh, H. (ed.) Medical Image Analysis, 2nd edn, pp. 99–138. Wiley, New Jersey (2011)
13. Zhang, H., Han, H., Liang, Z., Hu, Y., Liu, Y., Moore, W., Ma, J., Lu, H., et al.: Extracting information from previous full-dose CT scan for knowledge-based Bayesian reconstruction of current low-dose CT images. Journal. IEEE transactions on medical imaging **35**(3), 860–870 (2016)
14. Tunga, P., Singh, V.: Extraction and description of tumour region from the brain mri image using segmentation techniques. Paper presented at the IEEE international conference on recent trends in electronics information communication technology, 20–21 May 2016
15. Guo, L., Wu, Y., Liu, X., Li, Y., Xu, G., Yan, W.: Threshold optimisation of adaptive template filtering for mri based on intelligent optimisation algorithm. Paper presented at the 28th IEEE annual international conference on engineering in medicine and biology society, New York City, 30 Aug–3 Sept 2006
16. Kaur, A.: A review paper on image segmentation and its various techniques in image processing. Intern. J. Sci. Res. **3**(12), 12–14 (2012)
17. Masood, S., Sharif, M., Masood, A., Yasmin, M., Raza, M., et al.: A survey on medical image segmentation. Curr. Medical Imaging Rev. **11**(1), 3–14 (2015)
18. Thomas, H.M.W., Kumar, S.C.P.: A review of segmentation and edge detection methods for real time image processing used to detect brain tumour. Paper presented at the IEEE international conference computational intelligence and computing research, Madurai, 10–12 Dec 2015
19. Patil, D.D., Deore, S.G.: Medical Image Segmentation: A Review. Intern. J. Comput. Sci. Mob. Comp. **2**(1), 22–27 (2013)
20. Gupta, G., Tiwari, S.: Boundary extraction of biomedical images using edge operators. Intern. J. Innovative Res. Comput. Commun. Eng. **3**(11) (2015)
21. Fabijanska, A., Sankowski, D.: Edge detection in brain images. Paper presented at the IEEE international conference on perspective technologies and methods in mems design, Lviv, 21–24 May 2008
22. Norouzi, A., Rahim, M.S.M., Altameem, A., Saba, T., Rad, A.E., Rehman, A., Uddin, M., et al.: Medical image segmentation methods, algorithms, and applications. J. IETE Tech. Rev. **31**(3), 199–213 (2014)
23. Uchida, S.: Image processing and recognition for biological images. Develop. Growth Differ. (2013)
24. Siddique, I., Bajwa, I.S., Naveed, M.S., Choudhary, M.A.: Automatic functional brain mr image segmentation using region growing and seed pixel. Paper presented at the IEEE 4th international conference on information & communications technology, 10–12 Dec 2006

25. Oghli, M.G., Fallahi, A., Pooyan, M.: Automatic region growing method using gsmap and spatial information on ultrasound images. Paper presented at the IEEE 18th Iranian conference on electrical engineering, 11–13 May 2010
26. Adegoke, B.O., Olawale, B.O., Olabisi, N.I.: Overview of medical image segmentation. Intern. J. Eng. Res. Develop. **8**(9), 13–17 (2013)
27. Poonguzhali, S., Ravindran, G.: A complete automatic region growing method for segmentation of masses on ultrasound images. Paper presented at the IEEE International conference on biomedical and pharmaceutical engineering, 11–14 Dec 2006
28. Kumbhar, A.D., Kulkarni, A.V.: Magnetic resonant image segmentation using trained k-means clustering. Paper presented at the IEEE world congress on information and communication technologies, 11–14 Dec 2011
29. Li, S.Z.: Low-level MRF. In: Singh, S.(ed.) Models Markov Random Field Modeling in Image Analysis, 3rd edn., pp. 49–90. Springer, London, (2009)
30. Chen, S., Cao, L., Wang, Y.: Image segmentation by MAP-ML estimations. IEEE Trans. Image Process. **19**(9), 2254–2264 (2010)
31. Wells, W.M., Grimson, W.E.M., Kikinis, S., Jolesz, F.A.: Adaptive segmentation of MRI data. IEEE Trans. Med. Imaging **15**(4), 429–442 (1996)
32. Bouhlel, N., Sevestre, S., Rajhi, H., Hamzad, R.: A New Markov random field model based on K-distribution for textured ultrasound image. Paper presented at medical imaging 2004: Ultrasonic imaging and signal processing, San Diego, CA, 14 Feb 2004
33. Li, S.Z.: Introduction. In: Singh, S (ed.) Models Markov Random Field Modeling in Image Analysis, 3rd edn., pp. 1–20. Springer, London (2009)
34. Chen, S., Cao, L., Liu, J., Tang, X.: Iterative MAP and ML estimations for image segmentation. Paper presented at IEEE conference on computer vision and pattern recognition, Minneapolis, USA, 17–22 June 2007
35. Liu, X., Yetik, I.S.: A maximum likelihood classification method for image segmentation considering subject variability. Paper presented at 2nd IEEE southwest symposium on image analysis & interpretation (SSIAI), Austin, USA, 23–25 May 2010
36. Li, S.Z.: MRF parameter estimation. In: Singh, S. (ed.) Models Markov Random Field Modeling in Image Analysis, 3rd edn., pp. 183–214 Springer, London (2009)
37. Metaxas, D.N.: Motion-based part segmentation and tracking. Physics-Based Deformable Models: Applications To Computer Vision, Graphics, and Medical Imaging, 1st edn., pp. 149–178. Kluwer Academic Publishers, New York (1997)
38. Gauch, J.M., Pien, H., Shah, J.: Hybrid deformable models for three-dimensional biomedical image segmentation. Paper presented at nuclear science symposium and medical imaging IEEE conference, 30 Oct–5 Nov 1994
39. Hegadi, R., Kop, A., Hangarge, M.: A survey on deformable model and its applications to medical imaging. J. IJCA Spec. Issue Recent Trends Image Proces. Pattern Recogn. **2**(7), 64–75 (2010)
40. Heinz, D.: Hyper markov non-parametric processes for mixture modeling and model selection. Dissertation, Carnegie Mellon University (2010)
41. McInerney, T., Terzopoulos, D.: Deformable models in medical image analysis: a survey. Med. Image Anal. **1**(2), 91–108 (1996)
42. Nixon, M.S., Aguado, A.S.: Low- level feature extraction (including edge Detection). Feature Extraction and Image Processing, 1st edn, pp. 99–160. Reed educational and professional publishing, Woburn (2002)
43. Nixon, M.S., Aguado, A.S.: Feature extraction by shape matching. Feature Extraction and Image Processing, 1st edn, pp. 161–216. Reed educational and professional publishing, Woburn (2002)
44. Nixon, M.S., Aguado, A.S.: Flexible shape extraction (snakes and other techniques). Feature Extraction and Image Processing, 1st edn, pp. 217–246. Reed educational and professional publishing, Woburn (2002)
45. Mandloi, G.: A survey on feature extraction techniques for colour images. Intern. J. Comput. Sci. Info. Technol. **5**(3), 4615–4620 (2014)

46. Beauchemin, S.S., Barron, J.L., et al.: The computation of optical flow. J. ACM Comput. Surv. **27**(3) (1995)
47. Cootes, F.T., Edwards, G., Taylor, C.J.: A comparative evaluation of active appearance model algorithms. Paper presented at the 9th British machine vision conference (1998)
48. Smitha, P., Shaji, L., Mini, M.G.: A review of medical image classification techniques. Paper presented at international conference on VLSI, communication & instrumentation (2011)
49. Song, E., Pan, N., Hung, C.C., Li, X., Jin, L.: Reflection invariant local binary patterns for image texture classification. Paper presented at Racs proceedings of the ACM conference on research in adaptive and convergent systems, 9–12 Dec 2015
50. Shi, Z., He, L., Suzuki, K., Nakamura, T., Itoh, H., et al.: Survey on neural networks used for medical image processing. Int. J. Comput. Sci. **3**(1), 86–100 (2009)
51. Jiang, J., Trundle, P., Ren, J., et al.: Medical image analysis with artificial neural networks. Comput. Med. Imaging Graph. **34**(8), 617–631 (2010)
52. Kim, K.I., Jung, K., Park, S.H., Kim, H.J., et al.: Support vector machines for texture classification. IEEE Trans. Pattern Anal. Mach. Intell. **24**(11), 1542–1550 (2002)
53. Suralkar, S.R., Karode, A.H., Pawade, P.W., et al.: Texture image classification using support vector machine. Int. J. Comp. Tech. Appl **3**(1), 71–75 (2012)
54. Zhang, B.B., Zhou, H.P.: An improved SVM for aliasing dataset: demarcation threshold support vector machine. Paper presented at the IEEE 9th international symposium on computational intelligence and design, 10–11 Dec 2016
55. HemaRajini, N., Bhavani, R. Classification of MRI brain images using k-nearest neighbour and artificial neural network. Paper presented at the IEEE international conference on recent trends in information technology, 3–5 June 2011
56. Patidar, D., Jain, N., Parikh, A. Performance analysis of artificial neural network and K nearest neighbours image classification techniques with wavelet features. Paper presented at the IEEE international conference on computer communication and systems, 20–21 Feb 2014

Automatic Detection and Quantification of Calcium Objects from Clinical Images for Risk Level Assessment of Coronary Disease

R. Priyatharshini and S. Chitrakala

Abstract Medical diagnosis is often challenging, owing to the diversity of medical information sources. Significant advancements in healthcare technologies, potentially improving the benefits of diagnosis, may also result in data overload while the obtained information is being processed. From the beginning of time, humans have been susceptible to a surplus of diseases. Of the innumerable life-threatening diseases around, heart disease has garnered a great deal of consideration from medical researchers. Coronary Heart Disease is indubitably the commonest manifestation of Cardiovascular Disease (CVD), representing some 50% of the whole range of cardiovascular events. Medical imaging plays a key role in modern-day health care. Automatic detection and quantification of lesions from clinical images is quite an active research area where the challenge to obtain high accuracy rates is an ongoing process. This chapter presents an approach for mining the disease patterns from Cardiac CT (Computed Tomography) to assess the risk level of an individual with suspected coronary disease.

Keywords Image segmentation • Active contour model • Coronary disease diagnosis • Calcium object detection • Risk level categorization

1 Introduction

Cardiovascular diseases, the leading reason behind premature death in the world, include heart attacks, strokes, and different vascular diseases. A series of environmental, social, and structural changes that occur over time leads to exposure to risk factors for chronic diseases. Cardiovascular Disease (CVD) is common within

R. Priyatharshini (✉)
Department of Information Technology, Easwari Engineering College, Chennai, India
e-mail: priya.sneham@gmail.com

S. Chitrakala
Department of Computer Science and Engineering, Anna University, Chennai, India
e-mail: au.chitras@gmail.com

A.E. Hassanien and D.A. Oliva (eds.), *Advances in Soft Computing and Machine Learning in Image Processing*, Studies in Computational Intelligence 730,
https://doi.org/10.1007/978-3-319-63754-9_10

the general public, affecting the greater part of adults past the age of 60 years. Whereas a general assessment of the relative risk for CVD is approximated by investigating the number of prior risk factors present in a patient, additional precise estimation of the absolute risk for a primary CVD event is essential when designing treatment recommendations for a selected individual. Cardiovascular disease is instigated by disorders of the heart and blood vessels, and includes coronary heart disease (heart attacks), cerebrovascular disease (stroke), hypertension, peripheral artery disease, congenital heart disease, and heart failure. The major reasons for cardiovascular disease are tobacco use, physical inactivity, an unhealthy diet, and harmful use of alcohol.

Generally speaking, the excess calcium present in the blood stream accumulates as coronary plaque, incorporated in progressive plaque layers, and develops into hardened calcified plaque. A review of the literature shows that the amount of calcium present in the artery is a key indicator for the risk assessment of CHD. Calcium quantification is routinely carried out on low-dose, non-contrast enhanced CT scans, annotating all calcium objects contained in the primary vessels of the heart. Subsequently, on the basis of all selected calcium objects, the quantification is done by using different scoring methods such as Agatston, volume, and mass to assess the risk level of the Individuals. In clinical practices, the scoring of coronary artery calcification is challenging task for the radiologist. Further, higher error percent in calcium score quantification of individual is observed in the manual assessing method. Computer-aided recognition and quantification of calcifications in the arteries deliver better results than the conventional manual methods. An approach to detect and quantify the calcium deposits (lesions) in the coronary artery is required to assess the risk level of an Individual for immediate treatment planning.

2 Research Background

Detecting anatomical structures and locating abnormalities or lesions are the key objectives of medical image analysis. Machine learning approaches are increasingly successful in image-based diagnosis, disease prognosis, and risk assessment [1, 2]. An approach to automatically detect and quantify calcium lesions on non-contrast-enhanced cardiac computed tomographic images has been proposed [3]. First, candidate calcium objects are determined from the CT scan using an atlas-based estimate of the coronary artery locations, which permits the system to assign calcium lesions to the correct coronary arteries. The system uses a machine learning approach to discriminate true calcium objects from all detected candidate objects. A limitation of this system is that the patient scans were acquired on equipment from only one vendor, Siemens. A supervised classification-based approach is proposed to distinguish the coronary calcifications from all the candidate regions [4]. A two-stage, hierarchical classifier is developed for automated coronary calcium detection. At each stage, they learn an ensemble of classifiers where each classifier is a cost-sensitive learner trained on a distinct asymmetrically sampled data subset.

A method for automatic coronary calcium scoring with low-dose, non-contrast-enhanced, non-ECG-synchronized chest CT is developed [5]. First, a probabilistic coronary calcium map was created using multi-atlas segmentation. This map assigned an a priori probability for the presence of coronary calcifications at every location in a scan. Subsequently, a statistical pattern recognition system was designed to identify coronary calcifications by texture, size, and spatial features. The spatial features were computed using the coronary calcium map. The detected calcifications were quantified in terms of volume and Agatston score. To automatically identify and quantify calcifications in the coronary arteries, a supervised machine learning system was developed [6]. First, candidate calcium objects were extracted from the images based on their intensity. Each of these candidate calcifications was described with a set of features, characterizing its size, shape, intensity, and location information. Potential calcifications that the system could not label with high certainty were detected and optionally selected for expert review. Finally, identified calcium objects were quantified with the volume and Agatston score.

A novel distance-weighted lesion-specific Coronary Artery Calcium quantification framework has been developed to predict cardiac events [7]. This framework consists of a novel lesion-specific Coronary Artery Calcium quantification tool that measures each calcific lesion's attenuation, morphologic, and geometric statistics and a distance-weighted event risk model to estimate the risk probability caused by each lesion and a Naive Bayesian-based technique for risk integration. A vessel segmentation method is developed which learns the geometry and appearance of vessels in medical images from annotated data and uses this knowledge to segment vessels in unseen images [8]. Vessels are segmented in a coarse-to-fine fashion. First, the vessel boundaries are estimated with multivariate linear regression using image intensities sampled in a region of interest around an initialization curve. Subsequently, the position of the vessel boundary is refined with a robust nonlinear regression technique using intensity profiles sampled across the boundary of the rough segmentation and using information about plausible cross-sectional vessel shapes. An approach to automatically detect and quantify coronary artery stenosis was developed for diagnosing coronary artery disease [9]. First, centerlines are extracted using a two-point minimum cost path approach and a subsequent refinement step. The resulting centerlines are used as an initialization for lumen segmentation, performed using graph cuts.

All the level set segmentation methods presented above are based on image gradient intensity making them prone to leaking problems in areas with low contrast [10]. Medical images typically suffer from insufficient and spurious edges inherent to physics of acquisition and machine noise from different modalities. Detecting anatomical structures and locating abnormalities or lesions are the key objectives of medical image analysis. The automatic detection and quantification of lesions from clinical images is quite an active research area, where the challenge to obtain high accuracy rates is an ongoing process. To better discriminate lesions from other objects, more optimal features are required [11]. Although disease patterns provide important information about biomarkers from clinical images, it is no easy problem to detect and quantify them to assess the risk level of an individual.

3 Coronary Calcium Scoring for Risk Level Assessment of Coronary Disease

Medical imaging plays a key role in modern-day health care. Given the immense possibilities in high-quality images of anatomical structures in human beings, efficiently analyzing these images can be of immense use to clinicians and medical researchers in monitoring disease. Imaging modalities such as Computed Tomography (CT), Magnetic Resonance Imaging (MRI), and Positron Emission Tomography (PET) are commonly used to detect biomarkers of coronary disease diagnosis. A biomarker is a perceptible substance in the body that stipulates a particular disease state, organ function, or other aspects of health. Physicians and researchers use biomarkers or disease patterns to help predict, diagnose and treat a variety of disease states.

Coronary artery calcium scores have been recognized as independent markers for an adverse prognosis in coronary disease. Assorted methods have been proposed to this end, the most frequently used being the Agatston score. Other methods described include calcium volume and mass scores [12]. A framework for extracting disease patterns is essential, especially in coronary disease diagnosis which upgrades clinical decision making and plays a vital role in Clinical Decision Support (CDS). Hence, algorithms need to be developed that can automatically extract disease patterns from multimodal clinical data, paving the way for early detection of disease. Mining disease patterns finds applications in Image-guided surgery and intervention therapy planning as well as guidance. An approach to segment the coronary artery from CT images in order to detect the calcium objects has been developed for assessing the risk level of an individual with suspected coronary disease.

The overview of the proposed coronary disease diagnosis system is shown in Fig. 1. The processes associated with the given architecture are as follows: Image segmentation, feature selection and risk level categorization. For diagnosing the coronary disease from Cardiac CT, first an approach for segmenting the region of interest from the cardiac CT image is proposed and discussed in Sect. 3.1. An embedded feature selection method for optimal feature set selection is proposed and discussed in Sect. 3.2 to detect calcium objects in the segmented artery. The risk level assessment of an individual is discussed in Sect. 3.3.

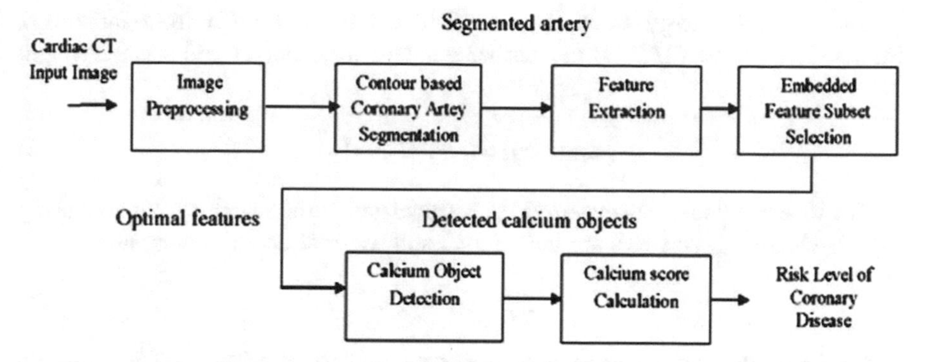

Fig. 1 Overview of proposed system for risk level assessment of coronary disease

3.1 *Cardiac CT Image Segmentation*

A Dual-Phase Dual-Objective approach using the Active Contour Model based region-growing technique (DPDO-ACM) is proposed to segment the coronary artery from the cardiac CT image. The classical active contour model, also called snake, is represented by a parametric curve which can move within the spatial domain of an image where it is assigned. A moving equation is defined to evolve the contours with a set of points $p(s,t) = ((x(s,t), y(s,t))$ on an image parameterized as w.r.t $s \leftarrow [0, 1]$ and t as the time.

Image energy in the proposed method is derived from the robust image gradient characteristic which provides the active contour a global representation in geometric configurations, rendering the method much more robust in handling image noise, weak edges, and initial configurations. The proposed method contains an image attraction force that propagates contours toward artery boundaries, along with a global shape force that deforms the model according to the shape distribution realized from the training set. The image attraction force, derived through the interaction of gradient vectors, differs from traditional image intensity gradient-based approaches, because it makes use of pixel interactions over the image area. A shape distance is described to evaluate dissimilarity among shapes.

The internal energy force E_{int} is calculated from the curvature using Eq. (1) to maintain the search within the spatial image domain using the first and second derivatives of p(s).

$$E_{\text{int}}(p(s,t)) = \frac{1}{2}\left[\alpha(s)\left|\frac{\partial p(s)}{\partial s}\right|^2 + \beta(s)\left|\frac{\partial^2 p(s)}{\partial s^2}\right|^2\right] \quad (1)$$

where α(s) is the curve tension parameter and β(s) the rigidity parameter.

The external energy force E_{ext} is calculated using Eq. (2) from the image information, where $\nabla IP(S)$ is the surface gradient computed at p(s) and γ a weight parameter.

$$E_{ext}(p(s)) = -\gamma|\nabla I(p(s))|^2 \tag{2}$$

The fitness value (image energy) $E_{i,j}$ is calculated in the searching window using Eq. (3) where E_{int} and E_{ext} are the internal and external energy functions.

$$E_{i,j} = E_{\mathrm{int}} + E_{ext} \tag{3}$$

To reduce the extensive computations of the traditional ACM, dual contours are used in our proposed method to steer the parallel evolution of active curves. Dual objectives have been formulated in our proposed method to drive an image attraction force that propagates contours toward artery boundaries, along with a global shape force that deforms the model according to shape characteristics. The curve evolves to minimize the total energy function using Eq. (4).

$$E_{snake} = \arg\min_j((E_{i,j}), K(t)), j \leftarrow W_i \tag{4}$$

The dual objectives formulated are given below:

Objective function 1: To find $X = [x_1, x_2, \ldots x_n]$, which minimizes curvature $K(t)$ where X is an n-dimensional vector for the equation of curve $y = f(X)$.

Objective function 2: To find the control point P_i which minimizes image energy $E_{i,j}$ within its searching window, where $P_i | i = \{1, 2, \ldots n\}$ is the set of control points of the parametric curve.

The DPDO-ACM algorithm is given below.

Declare:

$K(t)$	Curvature
E_{int}	Internal energy
E_{ext}	External energy
$E_{i,j}$	Total image energy
E_{snake}	Level set energy function
$\alpha(s)$	Curve tension parameter
$\beta(s)$	Rigidity parameter
$\nabla IP(S)$	Surface gradient computed at p(s)
γ	Weight parameter

Step 1. Initialize dual contours.
Step 2. For every contour,
Step 3. Create signed distance map from the mask.
Step 4. For every pixel in create signed distance map,

Step 4.1 Calculate curvature $K(t)$ using Eq. (5).

$$K(t) = \frac{X^{\bullet}(t)\,Y^{\bullet\bullet}(t) - X^{\bullet\bullet}(t)y^{\bullet}(t)}{\left[X^{\bullet}(t)^2 + Y^{\bullet}(t)^2\right]^{1.5}} \tag{5}$$

Step 4.2 Calculate internal energy E_{int} using (1).
Step 4.3 Calculate external energy E_{ext} using (2).
Step 4.4 Calculate total image energy $E_{i,j}$ using (3).

Step 5. Evolve the curve to a pixel with minimum curvature and image energy using the following level set re-initialization function with Eq. (4).

A screenshot of the segmented artery using the DPDO-ACM is shown in Fig. 2. The segmented artery from coronary CT is used further for detecting and

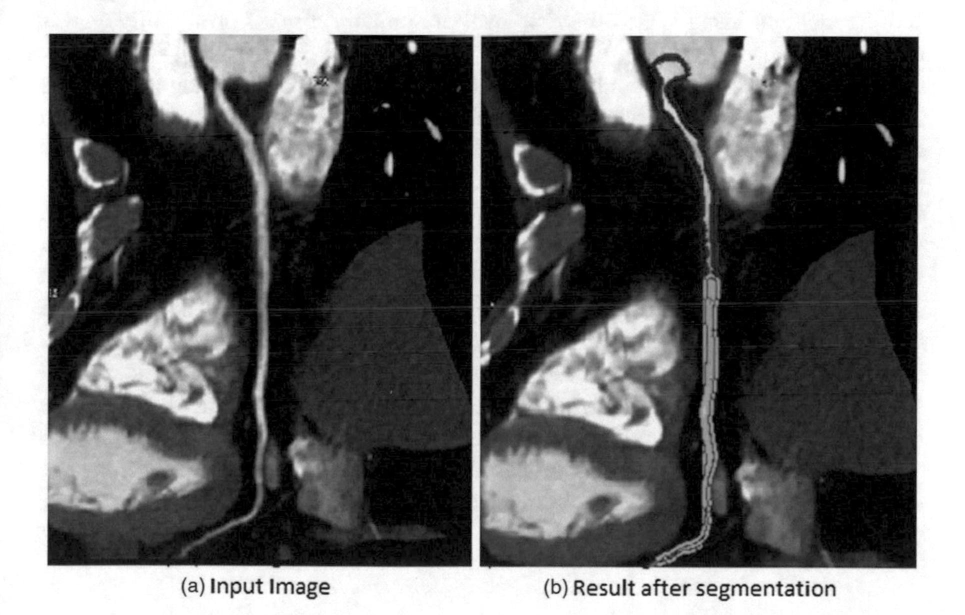

Fig. 2 Results of coronary artery segmentation

quantifying lesion using agatston score in Sect. 3.2 for categorizing the risk level of an individual with suspected coronary disease.

3.2 Detection and Quantification of Calcium Objects

For a CT input image, two fundamental morphological functions, erosion and dilation, are utilized to remove noise and enhance the vessels. Coronary calcifications appear as high-density structures in the coronary calcium CT scan. It is inherently difficult to identify them automatically. It is apparent that the choice of features plays an important role in resolving the problem. Several features—based on the appearance, shape, and size of the calcifications—were used to detect calcium objects. During online processing, when the user supplies an input image, the artery is segmented and candidate calcium objects selected, based on the min HU intensity threshold. From the selected set of candidate objects, selective features are extracted which successfully capture calcium object characteristics and the textural distribution of the pixels that form the feature vector. After the true calcium objects are detected, their quantification is done applying the agatston scoring method, using which the risk level of coronary disease of an individual is assessed.

The Correlation-Based Embedded Feature Selection (CB_EFS) algorithm aims to select the most appropriate subset of features that adequately describe a given classification task. The proposed CB-EFS method starts with a randomly selected subset and generates its successors by adding the remaining features from the feature subset sequentially. In every step, the newly generated subset is evaluated with an independent measure I_m and a learning algorithm A. For the optimal subset of cardinality K, it searches all possible subsets of cardinality $K + 1$. A subset generated at cardinality $K + 1$ is evaluated by an independent criterion I_m and compared with the previous optimal subset. Then a learning algorithm A is applied to the current optimal subset and the performance compared with that of the optimal subset at cardinality K. If the performance of current subset at cardinality $K + 1$ is higher than the previous optimal subset at cardinality K then the current subset is assigned as optimal subset. This process is iteratively done until the stopping the criterion is met and finally it returns the optimal subset. The algorithm of our proposed CB-EFS method is summarized as follows.

Input:

$D = \{X, L\}$ // A training dataset with n number of features where $X = \{f_1, f_{2,} \cdots f_n\}$ and L are labels

Output:

X'_{opt} //An optimal subset

Declare:

X' : Initial feature subset

X_{opt} : Initial optimal subset

θ : Stopping criterion

I_m : Independent evaluation measure

A : Learning algorithm

δ_{opt} : Initial optimal subset evaluated by the learning algorithm

ϕ_{opt} : Initial optimal subset evaluated by the independent evaluation measure

X_g : Subset generated

Step 1. Start

Step 2. Begin

Step2.1 Let the initial feature subset be X' and assign X' to X_{opt}.

Step2.2 Evaluate X' using the independent evaluation measure I_m and assign it to ϕ_{opt}

Step 2.3 Evaluate X' using Mining Algorithm A and assign it to δ_{opt}.

Step 2.4. Calculate the cardinality of X' and assign it to C_0

C_0=C(X');

Step 3: For each K=C_{0+}1 cardinality of X'

Step 4: For each i=0 to n features

Step 4.1 Generate subset X_g for evaluation with cardinality K.

$$X_g = X_{opt} \cup \{f_i\}$$

Step 4.2 Evaluate the current subset using I_m and assign it to ϕ if X_g is optimal.

$\phi = E(X_g, I_M)$;

Step 4.3 IF ($\phi > \phi_{opt}$)

then

$\phi_{opt} = \phi$;

$X'_{opt} = X_g$;

Step4.4 Evaluate X'_{opt} by learning algorithm A and assign it to δ.

$\delta = E(X'_{opt}, A)$;

Step 4.5 Compare δ and δ_{opt} to find the optimal subset.

IF ($\delta > \delta_{opt}$)

$X'_{opt} = X_{opt}$;

$\delta_{opt} = \delta$;

Else

Return X'_{opt} and go to step 3.

Step 5 The most optimal feature subset X'_{opt} is selected.

Step 6. End

Optimal features selected using CB_EFS are used to detect the true calcium objects from candidate calcium objects. The result of detected calcium objects using CB_EFS method is shown in Fig. 3. Quantification of detected calcium objects using Agatston score for assessing the risk level of an individual is discussed in Sect. 3.3.

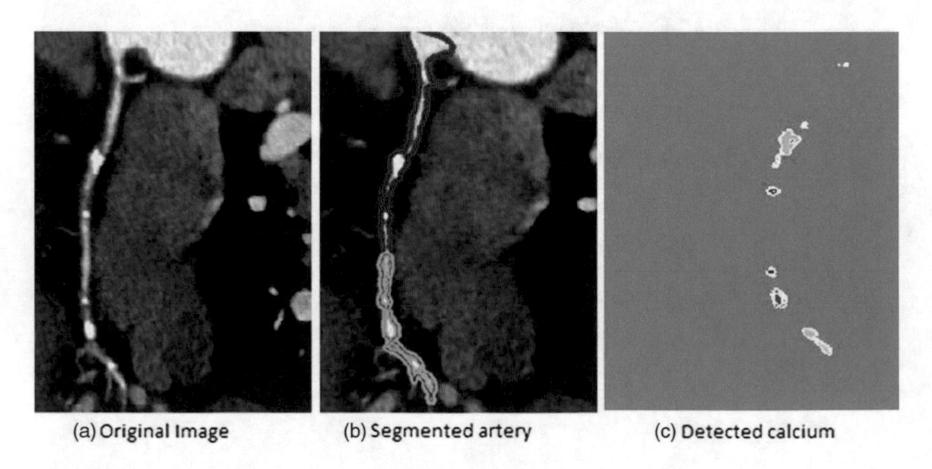

Fig. 3 Results of detected calcium from a coronary CT image

3.3 Risk Level Assessment of Coronary Disease

The coronary CT is a noninvasive CT scan of the heart. It is used to calculate the risk of developing coronary artery disease by measuring the amount of calcified plaque in the coronary arteries. The Coronary Artery Calcium Score measured in Agatston units is a method of measuring calcification in the coronary arteries. It is used to measure the overall coronary calcified plaque burden, thereby delivering prognostic information regarding the occurrence of future cardiovascular events. The Agatston rating was proposed by Agatston and helps physicians identify pre-symptomatic clients at risk for a cardiac malfunction using Eq. (6).

$$AS = A_i * W_i \tag{6}$$

where AS is the Agatston score and A_i the area of the calcified lesion with different HU ranges. The weighting factor W_i is defined using Eq. (7).

$$W_i = \begin{cases} 1, & 130\,HU \le I_i < 200\,HU \\ 2, & 200\,HU \le I_i < 300\,HU \\ 3, & 300\,HU \le I_i < 400\,HU, \\ 4, & 400\,HU \le I_i \end{cases} \tag{7}$$

where I_i is the maximum intensity of the calcification. W_i offers a range from 1 to 4, according to established ranges for a pixel with the largest depth in the plaque. The rating for any plaque is simply the solution of the plaque spot in that slice, as well as a weighting factor. Therefore, an insignificant Hounsfield unit difference yields a major Agatston score difference. The calcium objects detected in the coronary CT with the help of the CB-EFS method are quantified using the Agatston score, and the risk level of an individual is assessed based on the calcium score guidelines given in Table 1.

Table 1 Calcium score guidelines

Total score	Risk category
0	Very low
1–10	Low
11–100	Moderate
101–400	High
>400	Very high

Table 2 Confusion matrix for the two-class problem

	Actually healthy	Actually not healthy
Classified as healthy	TP	FP
Classified as unhealthy	FN	TN

4 Experimental Results

A coronary CT, acquired from a Somatom Definition AS + CT scanner with a slice thickness of 3 mm, is taken for the experiment. The 105 images gathered are divided into a training dataset of 55 images and a testing dataset of 50 images. The proposed risk level categorization approach has been tested over a corpus of 50 patients' coronary CT images and various metrics have been evaluated from the tested results. The proposed risk level categorization system for diagnosing coronary disease has been tested with the coronary CT image dataset. A well-known confusion matrix was attained to calculate sensitivity, specificity and accuracy. The confusion matrix for the two-class problem is represented using a 2×2 matrix, as shown in Table 2. The upper left cell represents the number of samples classified as true while they were true (TP), and the lower right cell represents the number of samples classified as false while they were actually false (TN). The other two cells indicate the number of samples misclassified. Particularly, the lower left cell indicates the number of samples classified as false while they actually were true (FN), and the upper right cell indicates the number of samples classified as true while they actually were false (FP).

For a multi-label classification problem with the $n \times n$ matrix, the diagonal elements represent the TP of the respective classes. The FN for a class is the sum of values in the corresponding row (excluding the TP of that class) and the FP for a class is the sum of values in the corresponding column (excluding the TP of that class). The TN will be the sum of all columns and rows excluding that particular class column and row. The confusion matrix illustrating the performance of the risk level categorization system using our proposed CB-EFS approach is shown in Table 3. Among the 50 patients, five scans were assigned to different risk categories; and it was discovered that one was off by two categories, and the others were off by one category. It is observed from Table 3 that cases close to the boundary can easily move to the neighbouring category.

Table 3 Results of risk level categorization of coronary disease (Confusion matrix representation)

Risk level	Very low	Low	Moderate	High	Very high
Very low	15	1	0	0	0
Low	0	8	0	1	0
Moderate	0	0	11	1	0
Moderate high	0	0	0	7	1
Very high	0	0	1	0	5

5 Conclusion and Future Works

From the clinical imaging modalities (coronary CTs), patients were assigned different risk categories on the basis of the whole-heart Agatston scores. While these approaches show promising results, there are still a number of elements that are to be considered, each of which could lead to a potential rise in accuracy. Algorithms are needed to handle the noise and artifacts that may arise due to thermal, cardiac and breathing motion. Research is needed to explore other features that can efficiently distinguish between calcified and noncalcified regions. Going forward, this system could be extended to diagnose diseases based on multimodal medical data, and the calcium score can be used as a predictor of stenosis and myocardial ischemia.

References

1. de Bruijne, M.: Machine learning approaches in medical image analysis: From detection to diagnosis. Elsevier **33**, 94–97 (2016)
2. Criminisi, A.: Machine learning for medical images analysis. Elsevier **33**, 91–93 (2016)
3. Shahzad, R., van Walsum, T., Schaap, M., Rossi, A., Klein, S., Weustink, A.C., de Feyter, P. J., van Vliet, L.J., Niessen, W.J.: Vessel specific coronary artery calcium scoring: an automatic system. Acad. Radiol **20**(1), 1–9 (2012)
4. Metz, C.T., Schaap, M., Weustink, A.C., Mollet, N.R.A., van Walsum, T., NiessenW, J.: Coronary centerline extraction from CT coronary angiography images using a minimum cost path approach. Med. Phys. **36**(12), 5568–5579 (2009)
5. Shahzad, R., van Walsum, T., Kirisli, H.A., Tang, H., Metz, C.T., Schaap, M., van Vliet, L.J., Niessen, W.J.: Automatic stenosis detection, quantification and lumen segmentation of the coronary arteries using a two point centerline extraction scheme. Int. Conf. Med. Image Comput. Comput. Assist. Interv. (2012)
6. Kurkure, U., Chittajallu, D.R., Brunner, G., Le, Y.H., Kakadiaris, I.A.: A supervised classification-based method for coronary calcium detection in non-contrast CT. Int. J. Cardiovasc Imaging **26**(7), 817–828 (2010)
7. Qian, Z., Marvasty, I., Anderson, H., Rinehart, S., Voros, S.: Leison-specific coronary artery calcium quantification better predicts cardiac events. IEEE Trans. Inf Technol. Biomed. **15**(5), 673–680 (2011)
8. Schaap, M., van Walsum, T., Neefjes, L., Metz, C., Capuano, E., de Bruijne, M., Niessen, W.: Robust shape regression for supervised vessel segmentation and its application to coronary segmentation in CTA. IEEE Trans. Med. Imag. **30**(11), 1974–1986 (2011)
9. Shahzad, R., Kiris li, H., Metz, C., Tang, H., Schaap, M., van Vliet, L., Niessen, W., van Walsum, T.: Automatic segmentation, detection and quantification of coronary artery stenoses on CTA. Int. J. Cardiovasc. Imaging **29**(8), 1847–1859 (2013)
10. Anita, S., Satish, C.: Meta-heuristic approaches for active contour model based medical image segmentation. Int. J. Adv. Soft Comput. Appl. 6(2), 1–22 (2014)
11. Tomar, D., Agarwal, S.: Hybrid feature selection-based weighted least squares twin support vector machine approach for diagnosing breast cancer. Hepatitis Diab. Adv. Artif. Neural Syst. **2015**, 1–10 (2015)
12. Isgum, I.: Computer-aided detection and quantification of arterial calcifications with CT (Ipskamp, Enschede) (2007)

Semi-automated Method for the Glaucoma Monitoring

Nesma Settouti, Mostafa El Habib Daho, Mohammed El Amine Bechar, Mohamed Amine Lazouni and Mohammed Amine Chikh

Abstract The current trend of computer vision and image processing systems in biomedical field is the application of the Computational Intelligence (CI) approaches, which include the use of tools as machine learning and soft computing. The CI approaches bring a new solution to automatic feature extraction for a particular task. Based on that techniques, we have proposed in this work a semi-automated method for the glaucoma monitoring through retinal images. Glaucoma is a disease caused by neuro-degeneration of the optic nerve leading to blindness. It can be assessed by monitoring Intraocular Pressure (IOP), by the visual field and the aspect of the optic disc (ratio cup/disc). Glaucoma increases the rate of cup/disc (CDR), which affects the loss of peripheral vision. In this work, a segmentation method of cups and discs regions is proposed in a semi-supervised pixel-based classification paradigm to automate the cup/disc ratio calculation for the concrete medical supervision of the glaucoma disease. The idea is to canvas the medical expert for labeling the regions of interest (ROI) (three retinal images) and automate the segmentation by intelligent region growing based on machine learning. A comparative study of semi-supervised and supervised methods is carried out in this proposal, by mono approaches (decision tree and *SETRED*) and multi-classifiers (*Random Forest* and *co-Forest*). Our proposition is evaluated on real images of normal and glaucoma cases. The obtained results are very promising and demonstrate the efficacy and potency of segmentation by the multi-classifier systems in semi-automatic segmentation.

N. Settouti (✉) · M. El Habib Daho · M.E.A. Bechar ·
M.A. Lazouni · M.A. Chikh
Biomedical Engineering Laboratory GBM, Tlemcen University, Chetouane, Algeria
e-mail: nesma.settouti@gmail.com

M. El Habib Daho
e-mail: mostafa.elhabibdaho@gmail.com

M.E.A. Bechar
e-mail: am.bechar@gmail.com

M.A. Lazouni
e-mail: aminelazouni@gmail.com

M.A. Chikh
e-mail: mea_chikh@mail.univ-tlemcen.dz

A.E. Hassanien and D.A. Oliva (eds.), *Advances in Soft Computing and Machine Learning in Image Processing*, Studies in Computational Intelligence 730,
https://doi.org/10.1007/978-3-319-63754-9_11

Keywords Pixel-based classification · Semi-supervised learning · Semi-automatic segmentation · Fuzzy C-means · *co-Forest* · Glaucoma monitoring

1 Introduction

The majority of applications in the field of medical diagnostic aid require the acquisition of imaging data of various natures: radiologies, scanner or MRI examinations, ultrasound imaging, video, etc. A fundamental task in the processing of this data is segmentation, i.e., the extraction of structures of interest in images, in 2D or 3D format. This information serves, in particular, as a basis for the visualization of organs, the classification of objects, the generation of simulation models, or surface or volumetric measurements. In this work, we are interested in the annotation of images at the pixel level by a semi-automatic segmentation approach. This method requires more or less important interaction of the expert. This type of tool is useful either to process the data directly or to define a reference result that can be applied for the evaluation of automatic segmentation methods.

Automatic image segmentation aims at the automated extraction of objects characterized by a border (contour). Its purpose is to cluster pixels according to predefined criteria, usually the gray levels or the texture. The pixels are thus grouped into regions, which constitute a partition of the image. Nevertheless, this task remains difficult to achieve especially in the cases where the edges of an object are missing and/or there is a low contrast between the regions of interest (ROI) and the background.

Segmentation of retinal images, mainly the fundus images, is an important step in the medical monitoring of glaucoma. Indeed, the diagnosis of glaucoma is determined by doctors in studying many factors: family history, the intraocular pressure, the thickness of the central cornea, the appearance of the anterior chamber angle, the optic nerve configuration including nerve fiber layer, and the optic nerve function. For now, the diagnosis cannot be based on a single analysis. For example, looking for a high intraocular pressure is the first step to detect glaucoma. However, one-third of patients with glaucoma have normal intraocular pressure. Most glaucoma screening tests are time consuming and require the intervention of glaucoma specialists and the use of diagnostic equipment. Consequently, new automated techniques for diagnosing glaucoma at an early stage, combining precision and speed, are required.

Many approaches have been proposed in the literature for the segmentation of the fundus of the eye images in order to extract the region of the optical disc. These works can be distributed within the families of approaches such as segmentation methods, feature extraction techniques, and classification methods that include supervised and unsupervised algorithms.

Supervised segmentation using methods such as neural networks and support vector machines (SVMs) lead to high precision, but generally, these techniques require a large amount of labeled data for their learning, and the unavailability of labeled data due to the boring task of labeling pixels and images, or the unavailability of experts (e.g., medical imaging), make the task particularly difficult, expensive, and slow to acquire in real applications. On the other hand, unsupervised learning methods such as K-Means and C-Fuzzy Means (FCM) suppress labeling costs but perform less than supervised methods. To solve these problems, we propose a semi-supervised learning (SSL) approach for the segmentation of the cup and disc regions in the objective of medical monitoring by the calculation of the CDR ratio.

The availability of unlabeled data and the difficulty of obtaining labels, make the semi-supervised learning methods gain great importance. With the goal of reducing the amount of supervision required compared to supervised learning, and at the same time improving the results of unsupervised clustering to the expectations of the user. The question that arises is whether the knowledge of points with labels is sufficient to construct a decision function that can correctly predict the labels of unlabeled points. Different approaches propose to deduct unlabeled points of additional information and include them in the learning problem. Several semi-supervised algorithms such as self-training [36], Co-training [6], Expectation Maximization (EM) [27], and in the last few years, the ensemble method co-forest [12, 25] have been developed, but none of them were used for semi-supervised segmentation.

A method of segmentation and automatic recognition of regions cups and discs for measuring the CDR report in a semi-supervised context is proposed here. The intervention of an ophthalmologist expert is important in identifying cups and discs areas in retinal image. For this, the expert will realize a windowing of 5% of the image data set (3 retinal images). This approach helps to automate the segmentation of glaucoma's parts using intelligent techniques.

Thus, a comparative study of several techniques is proposed. The principle is based on a region growing by classifying the neighboring pixels from the pixels of interest of the image using semi-supervised learning. The points of interest are detected by the Fuzzy C-Means (FCM) algorithm. Four classifiers with different principle are applied in this work: Decision Tree (mono-supervised classifier), *Random Forest* (supervised ensemble method) *SETRED* (method of self-learning in SSL), and the algorithm *co-Forest* (Ensemble method in SSL). This study will adapt the best approach to the segmentation of retinal images with minimal intervention of the medical expert.

This work is organized as follows: in Sect. 2, a review of some segmentation and pixel-based classification methods of the retinal images is performed. We explain then in Sect. 3, the general process of our proposed approach and its different steps (characterization, pretreatment and SSL classification methods). After that, we validate our approach and the choices we have made in an experimental phase. Moreover, we show the capacity of our approach to automatic segmentation by applying several methods. Finally, we come to an end with a conclusion that summarizes the contributions made and the tracks defining possible opportunities for future work as well as the difficulties faced with the realization of this work.

2 State of the Art

Glaucoma is an eye disease associated with abnormal increase in the pressure of the ocular fluid. This abnormal pressure leads gradually and most often painless to irreversible visual impairment. It can be evaluated by monitoring the appearance of the optic disc (CDR Disc Cup Report).

The CDR value increases with the increase of neuro-degeneration and retinal vision is lost completely in the CDR value = 0.8. Several methods for extracting features from images of the eye funds are reported in the literature [1, 7, 17, 19, 23, 28]. The techniques described in the literature for the location of the optical disc are generally intended to identify either the approximate center of the optical disc or to place it in a specific region, such as a circle or a square. Lalonde et al., used the Canny detector [23] Ghafar et al. [17], the Hough transform to detect the optical disc (OD). Bock et al. [7], called for the concept of principal component analysis (PCA), bitsplines and Fourier analysis for feature extraction and Support Vector Machines (SVM) as classifier for predicting glaucoma.

The retinal image automatic analysis is becoming an important screening tool for the early detection of eye diseases. The manual review of the Optical Disc (OD) is a standard procedure used to detect glaucoma. The best way to control the glaucoma disease is by using the digital retinal camera. These images are stored in RGB format, which is divided into three channels: red, green, and blue. Other studies have focused on image processing techniques to diagnose glaucoma based on the CDR evaluation of retinal color images.

Madhusudhan et al. [26] have developed a system for processing and automatic image classification based on the usual practice in clinical routine. Therefore, three different image processing techniques namely multi-thresholding segmentation methods based on active contours region are proposed for the detection of glaucoma. Mohammad et al. [29] have presented an approach which includes two main steps. First, a pixel-based classification method for identifying pixels which may belong to the boundary of the optic disc. Second, a match-up of the circular template to estimate the approximation of the circular edge of the optical disc. The characteristics of the used pixels are based on the texture which is calculated from the local picture intensity differences. The Fuzzy C-Means (FCM) and Naive Bayes are used to group and classify image pixels.

Chandrika et al. [11] adopted an automatic identification technique of optical disc retinal images by calculating the ratio CDR. In the first place, a threshold is applied and then the image segmentation is performed using k-means and the Gabor wavelet transform. Second, the contour smoothing of the disc and the optical cup is performed using different morphological characteristics.

More conventionally, Hatanaka et al. [18] proposed a method to measure the ratio cup/disc with a vertical profile on the optical disc. The edge of the optical disc is then detected by using a Canny edge detection filter. The resulting profile is made around the center of the optical disc in the vertical direction. Thereafter, the edge of the vertical area on the cup profile is determined by a thresholding technique.

Joshi et al. [20] implement an automatic technique for the parameterization of the optical disc (OD) according to the segmented cup and disc regions obtained from the monocular retinal images. A new OD segmentation method is proposed, integrating the information of the local image around each point of interest in the multidimensional function space. This is in order to provide robustness against the variations found in and around the OD region. A method of segmentation of the cup is also proposed, it is based on anatomical evidence such as vessel bends at the border of the cup, deemed relevant by experts in glaucoma. A multi-step strategy is used to obtain a reliable subset of vessel bends called r-bends followed by a fitting to derive the desired cup boundary.

Burana-Anusorn et al. [10] have developed an automatic approach for the calculation of CDR ratio from images of the eye funds. The idea is to extract the optical disc by using a contour detection approach and the level set approach by individual variation. The optical cup is then segmented using a color component analysis method and method of thresholding level set. After obtaining the contour, a step of adjusting by ellipse is introduced to smooth the obtained results. The performance of this approach is assessed by comparing the automatically calculated CDR with that calculated manually. The results indicate that the approach of Burana-Anusorn et al., reached a precision of 89% for the analysis of glaucoma. As a consequence, this study has a good potential in automated screening systems for early detection of glaucoma.

Recently, Khalid et al. [22] have proposed the deployment of dilation and erosion with fuzzy c-means (FCM) as an effective technique for segmentation cup and optical disc color images of the eye funds. Previous works have identified the green channel as the most suitable because of its contrast. Hence, at first, the extracted green channel is segmented with FCM. In another test, all the images are pretreated with dilation and erosion to remove the vascular network. Segmentation is assessed on the basis of the labels described by ophthalmologists. The CDR measures are calculated from the diameter ratio of the cup and segmented disc. The assessment shows that the omission of the vernacular area improves the sensitivity, specificity, and accuracy of the segmented result.

Sivaswamy et al. [35] were interested in the problem of segmentation of the optic nerve head (ONH) which is of crucial importance for the assessment of automated glaucoma. The problem of segmentation involves segmenting the optical disc and the cup of ONH region. The authors highlighted the difficulty to evaluate and compare the performances of existing methods due to the lack of a reference data set. On that account, a complete set of retinal image data which include normal and glaucoma eyes by manual segmentations of several experts was implemented. Both evaluation measures based on size and contour are shown to evaluate a method on various aspects of the problem of the evaluation of glaucoma.

From the literature study, several approaches are carried out to determine mostly the cup and have focused on image processing techniques to diagnose glaucoma based on the CDR evaluation of retinal color images. Computational Intelligence (CI) approaches are alternative solutions for traditional automatic computer vision and image processing methods; they include the use of tools as machine learning and soft computing. In this study, we have proposed a new CI approach for the segmentation process by an iterative learning algorithm to achieve a semi-automatic classification procedure that promises to be more accurate than traditional supervised pixel-based methods. A comparative study will be conducted between mono and ensemble classifier to analyze their impact on the accuracy of prediction in pixel classification task.

3 The Proposed Approach

The aim is to automatically recognize cups and discs regions (Fig. 1) that are essential for measuring the progression of glaucoma. To do this, we propose an approach based primarily on a semi-supervised pixel-based classification.

The intervention of an ophthalmologist expert is important in identifying the disc and cup retinal image. We were inspired by the principle proposed by Reza et al. [4] in the step where the expert is appealing to windowing the region of interest on a minimum of images. In our application, we randomly select three images (5% of the image data set), the labeling step is made by tagging three windows (disc, cup, and background). Thereafter, a characterization phase takes part, where each pixel is represented by a color features.

Our contribution is the application of semi-supervised classification techniques on pixels appraised by the doctor to form a robust and reliable hypotheses that allows pixel-based classification. The proposed algorithm is illustrated in Fig. 2. The idea is to realize at first a preprocessing that allows us to score some pixel in cups and discs regions using the fuzzy c-means method. For a better learning, a classification will be applied to the vicinity of each region of interest that was marked in the preprocessing phase. This phase will be conducted by four different classification approaches (Decision trees [9], *Random Forests* [8], *SETRED* self-training algorithm [24], and *Random Forest*s in semi-supervised learning *co-Forest* [25]).

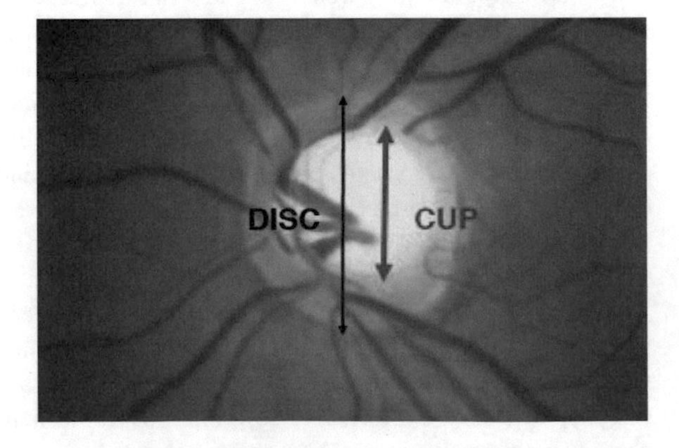

Fig. 1 Glaucomatous papillary excavation (increasing the ratio cup/disc)

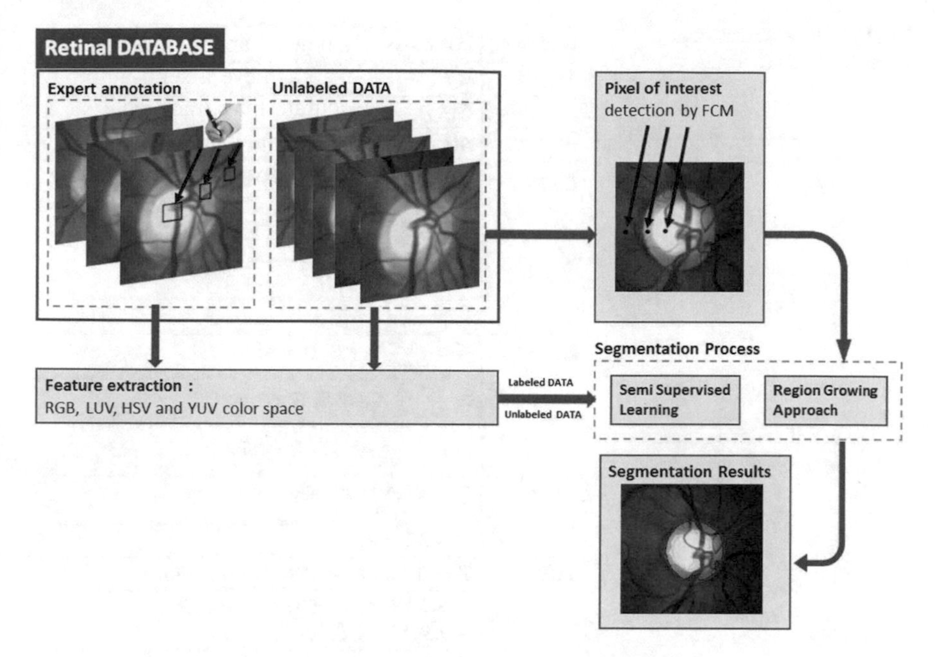

Fig. 2 The proposed approach to semi-automatic segmentation process

3.1 *Feature Extraction Methods*

Color Spaces

Different color spaces have been used in pixel-based classification for segmentation purposes, but many of them share similar characteristics. Therefore, in this work, we are interested in five more representative color spaces that are commonly used in image processing [16]: RGB, LUV, HSV, HSL, YUV.

RGB Color Space

The RGB space is a fundamental and commonly used color space in various computer vision application, it is about describing color by three components: red, green, and blue. Components are combined in various ways to reproduce different colors in the additive model.

HSV Color Space

In the work of [21], they define the Hue as the property of a color that varies in passing from red to green, the Saturation as the property of a color that varies in passing from red to pink, the Value (also called Intensity or Lightness or Brightness) as the property that varies in passing from black to white. The HSV is a linear transformation of RGB to high intensity at white lights, ambient light, and surface orientations relative to the light source.

Table 1 Characterization parameters table

Features	Formulas
RGB	$R(i,j)$
	$G(i,j)$
	$B(i,j)$
LUV	$L = 116(\frac{Y}{Y_n})^{1/3} - 16$ Si $\frac{Y}{Y_n} > 0.008856$
	$= 903.3(\frac{Y}{Y_n})$ Si $\frac{Y}{Y_n} \leq 0.008856$
	$U = 13L(U' - U'_n)$
	$V = 13L(V - V'_n)$
HSV	$\mathrm{H} = \frac{G-B}{(Max-Min)}$ Si R = Max
	$= \frac{B-R}{(Max-Min)} + 2$ Si G = Max
	$= \frac{R-G}{(Max-Min)} + 4$ Si B = Max
	$\mathrm{S} = \frac{Max(R,G,B)-Min(R,G,B)}{Max(R,G,B)}$
	$V = Max(R, G, B)$
YUV	$Y = 0.2989R + 0.5866G + 0.1145B$
	$U = 0.5647(B - Y) = -0.1687R - 0.3312G + 0.5B$
	$V = 0.7132(R - Y) = 0.5R - 0.4183G - 0.0817B$

LUV Color Space

The aim of LUV color space is to produce a more linear color space. Perceptual linear means a variation of the same quantity of color has to produce a variation of which the same visual importance.

YUV Color Space

The YUV space is mainly used for analog video, this representation model is used in the PAL and NTSC video standards. The luminance is represented by Y, whereas the chrominances U and V are derived from the transformation of the RGB space.

Table 1 summarize the characterization parameters of each color space.

3.2 Points of Interest Detection

Considering the time computing, especially in a semi-supervised context, has led us to propose a phase detection process known as points of interest. The objective of this phase is to minimize the computation time and the other to start learning semi-supervised via voltage pixels belonging to our target. We propose the implementation of cluster centers by fuzzy C-means method "FCM".

Fuzzy C-means method (FCM) is an unsupervised fuzzy clustering algorithm. From the C-Means algorithm (C-means), it introduced the concept of fuzzy set in the definition of classes: each point in the data set for each cluster with a certain degree, and all clusters are characterized by their center of gravity. Like any other clustering algorithms, it uses a criterion of minimizing the intra-class distance and maximizing the inter-class distances, but giving a degree of membership in each class for each point. This algorithm requires prior knowledge of the number of clusters and generates classes through an iterative process by minimizing an objective function.

The whole FCM process [5, 14] can be described in the following steps:

Algorithm 1 Fuzzy C-Means Pseudocode

1: **for** $t = 1, 2,$ **do**
2: **Step1** Calculate the cluster center $c_i^{(t)} = \frac{\sum_{j=1}^{N}(\mu_{ij}^{(t-1)})^m x_j}{\sum_{j=1}^{N}(\mu_{ij}^{(t-1)})^m}$
3: **Step2** Calculate the distances D^2_{ijA} avec:
$D^2_{ijA} = (x_j c_i)^T A (x_j c_i), \quad 1 \leq i \leq n_c, 1 \leq j \leq N.$
4: **Step3** Update the Fuzzy partition matrix: $\mu_{ij}^{(t-1)} = \frac{1}{\sum_{k-1}^{n_c}(D_{ijA}/D_{kjA})^{2/(m-1)}}$
5: **end for**
6: **return** $||U^{(t)} - U^{(t-1)}|| < \epsilon$

3.3 Classification Methods

With the availability of unlabeled data and the difficulty of obtaining labels, semi-supervised learning methods have gained great importance. Unlike supervised learning, the semi-supervised learning is the problems with relatively few tagged data and a large amount of unlabeled data. The question is then whether the mere knowledge of the items with labels is sufficient to construct a decision function that can correctly predict the labels of unlabeled points. Different approaches propose to deduct untagged items, additional information and include them in the learning problem.

In this classification part, we focus on improving the performance of supervised classification using unlabeled data (SSL). We set up the first semi-supervised classification under the classification problems, limited to the use of methods of sets in semi-supervised classification. Therefore, we propose in this work to apply the method set type of multi-classifier systems compared to single-supervised learning classifiers methods and semi-supervised: Decision tree [9], *Random Forest*s [8], *SETRED* auto-learning algorithm [24], and Forests in semi-supervised learning *co-Forest* [25].

Decision Trees

Decision trees (DT) represent a very effective method of supervised learning. The goal of DT is to partition a set of data into the most homogeneous groups possible from the point of view of the variable to be predicted. As input, we use a set of data to classify and get a tree, which resembles very much to an orientation diagram, as output. A decision tree consists of a root that is the starting point of the tree, nodes, and branches that connect the root with the nodes, the nodes between them and the nodes with the leaves. There are several algorithms present in the literature, such as: CART [9], ID3 Quinlan86 And C4.5 [32]. In this work, we limit ourselves to the application of the CART algorithm (Classification and Regression Tree).

Self-Training Paradigm

SETRED (self-training with data editing) is the most popular algorithm proposed by Li et al. [24]. Here, the authors studied the potential of data editing techniques as a confidence measure which allows it to reduce the risk of adding mislabeled data to the training set. This paradigm is an iterative mechanism. Its principle is to train a supervised classifier on labeled pixels to predict the labels of unlabeled pixels. Afterward, it iteratively enriches the labeled set by adding newly labeled examples with high confident predictions from the unlabeled data (confidence data). In SETRED, the CEWS (Cut Edge Weight Statistic) [30] rule is applied to measure the confidence level on unlabeled examples. The main steps to calculate the confidence measure are as follows:

The Nearest Neighbor Rule

The nearest neighbor rule (*NNR*) was proposed by Fix and Hodges [15], it is a non-parametric method where the classification is obtained for an unlabeled data taking into account the class of its nearest neighbor in the learning samples. The calculation of the similarity between data is based on distance measurements. Afterward, this rule was developed to $k - NNR$, k represents the size of the neighborhood. The label of a non-classified data is that of the majority class among the labels of its k nearest neighbors.

The Relative Neighborhood Graph

The neighborhood graph is a computational geometry tool that has been exploited in many machine learning applications. By definition, a neighborhood graph $G = (V, E)$ [13] associated with a set of labeled pixel whose vertices S compose the set of edges E. Each pixel in a neighborhood graph is represented by a vertex, existing in the edge between two vertices x_i and x_j if Eq. 1 is verified.

$$(x_i, x_j) \in E \Leftrightarrow dist(x_i, x_j) \leq max\ (dist(x_i, x_k), dist(x_j, x_k)), \forall\ x_k \in TR, k \neq i, j \quad (1)$$

With: $dist(x_i, x_j)$: the distance between x_i et x_j.

The Cut Edge Weight Statistic

Using the previous definition to construct a relative neighborhood graph, Muhlenbach et al. [30] exploited the edge information to calculate a statistical weight in order to cut edges of different classes. The SETRED [24] algorithm follows this principle for the confidence measure.

In the first step, a supervised hypothesis is learned using the labeled pixels. Second, the application of the cut edge weight statistic [30] algorithm, to calculate the ratio R_i by the Eqs. (2, 3, 4, 5). To judge whether the data is well ranked, the ratio R_i must be greater than a threshold that is set by the user. For more information, one may consult [24, 30].

$$R_i = \frac{J_i}{I_i} \tag{2}$$

With:

$$I_i = \sum_{(j \in Neighborhood(x_i))} w_{ij} \tag{3}$$

$$J_i = \sum_{(j \in Neighborhood(x_i), y_j \neq y_i)} w_{ij} \tag{4}$$

$$w_{ij} = \frac{1}{((1 + dist(x_i, x_j)))} \tag{5}$$

Ensemble Method: Random Forests

Random Forest (RF) is a predictor that combines a set of decision trees. In the specific case of CART models (binary tree), Breiman [8] proposes an improvement of bagging with a random forest induction algorithm Forest-RI (Random Input) which uses the "Random Feature Selection" method proposed by Amit and Geman [3]. The induction of the trees is done without pruning and according to the CART algorithm [9], however, at each node, the selection of the best partition based on the Gini index is done only on a subset of attributes (usually equal to the square root of the total number of attributes) selected randomly from the original space of features [34]. The global prediction of the random forest is calculated by taking the majority of votes of each of its trees. This algorithm is defined by Breiman as follows (Algorithm 1) [8]:

Algorithm 2 Pseudocode of the Random Forest algorithm

Input: The Training set T, Number of Random Trees L.
Output: *TreesEnsemble* E
Process: $E = \varnothing$
for $i = 1 \rightarrow L$ **do**
$T^i \leftarrow BootstrapSample(T)$
$C^i \leftarrow ConstructTree(T^i)$ where at each node:

- Random selection of $K = \sqrt{M}$ Variables from the whole attribute space of dimension M
- Select the most informative variable from K using Gini index
- Create children nodes using this variable

$E \leftarrow E \cup \{C^i\}$
end for
Return E

The Random Forest in Semi-supervised Learning "*co-Forest*" Algorithm

co-Forest is an algorithm that extends the paradigm of *co-Training* [6] using *Random Forest* [8]. It was introduced by Li and Zhou [25] in the application to the detection of micro-calcifications in the diagnosis of breast cancer. This ensemble method uses $N \geq 3$ classifiers instead of 3 by *Tri-training* [37]. The $N-1$ classifiers are used to determine confidence examples, called concomitant ensemble $= h_i = H_{N-1}$. Confidence of an unlabeled pixels can be simply estimated by the degree of agreement on the labeling, i.e., the number of classifiers which are agreements assigned by h_i label.

The functioning of *co-Forest* can be summarized in the following steps (Algorithm 2):

Algorithm 3 co-Forest Pseudocode

1: **for** each iteration **do**
2: **Step1** *co-Forest* starts learning H^* on bootstrap of L (labeled pixels).
3: **Step2** all concomitant set examines each sample from U (unlabeled pixels images).
If the number of voters agree on the label of $xu > \theta$,
Then xu is labeled and copied into a new set L'.
4: **Step3** Introduces a weight of the predictive confidence by the concomitant set, due to the situation where $L \geq U$ which will affects the performance of h_i
5: **Step4** Each random tree is refined with newly marked $L \cup L'$ selected by his concomitant set under the following condition examples:

$$e_{i,t}.W_{i,t} < e_{i,t-1}.W_{i,t-1}$$

where $W = \sum w_{ij}$ et w_{ij}: predictive confidence of H_i on x_i in L'
6: **end for**
7: **return**

To insure the success of this ensemble method, two conditions must be satisfied:

- Each individual predictor should be relatively good,
- Each individual predictor should be different from each other.

Even more simple, it is necessary for the individual predictors to be good classifiers, and where a predictor is wrong, the other must take over without making mistakes.

For maintaining the diversity in *co-Forest*, the application of *Random Forest* can inject random learning. To affirm this condition, the authors of *co-Forest* have set a threshold for the labeling of U, where the only U pixels whose total weight is smaller than $e_{i,t-1}.W_{i,t-1}/e_{i,t}$ will be selected.

In summary, the principle of *co-Forest* (Fig. 3) consists of N random trees that are first learned on bootstrap set of L to create a random forest. Then, at each iteration, each random tree will be refined with the newly labeled samples selected by its concomitant set, only when the confidence of the labeled examples exceeds a certain threshold θ. This method will reduce the chances of used biased tree in a *Random Forest* when we use unlabeled data. More details on the *co-Forest* algorithm are in the papers [12, 25, 33].

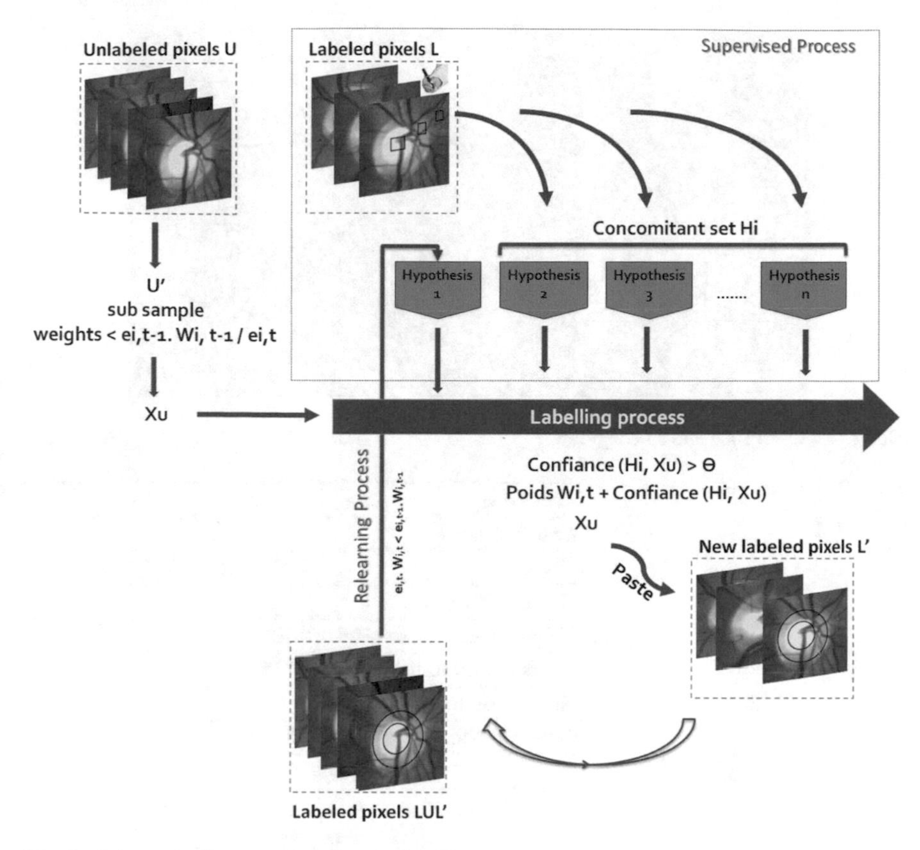

Fig. 3 Schematic diagram of *co-Forest* algorithm

4 Results and Experiments

The retinal image database was constructed from local real images acquired within the eye clinic (clinic LAZOUNI Tlemcen). Eye Backgrounds RTVue XR 100 Avanti Edition of Optovue company provides RGB color images of size 1609 × 1054 pixels. One hundred and three retinal images of the eye funds have been used to test the proposed segmentation algorithm. The ground truth is achieved by using the average segmentation provided by two manual different experts in ophthalmology. We built a learning base, where the expert selects three regions: cup, disc (ROI: regions of interest) and bottom (stopping criterion) (Fig. 4).

In our experiments, we have selected 5% of the database (3 images) to achieve learning. The expert ophthalmologist intervenes in the labeling of these three images by size windowing [576–50466 pixels], allowing a better understanding of areas of interest. In the semi-supervised learning portion, a classification is applied to the vicinity of a degree equals to 50. The application of all methods (*Random Forest* and *co-Forest*) with a number of trees equal to 100 being was chosen. The evaluation

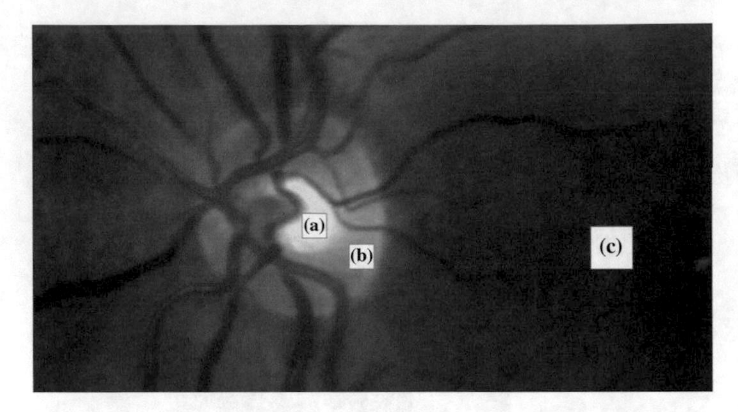

Fig. 4 **a** Cup, **b** Disc and **c** background

Table 2 Classification parameters

Labeled set	3 images
Learning set	33 images
Test set	20 images
Number of clusters	3
Neighborhood of the pixel	50
Confidence level	75%
Number of trees	100
Cross validation	5

is carried out with a cross-validation equals to 5. The details of the experimental parameters are summarized in Table 2.

Table 3 summarizes the classification performance by the four approaches supervised and semi-supervised on 20 test images for the recognition of regions cup and disc. The results achieved by the overall approaches *Random Forest* and *co-Forest* guarantee greater precision of segmentation of the two target regions. However, poor

Table 3 Classification performance by the supervised and semi-supervised techniques with 5% of labeled pictures

Learning type	Techniques	Accuracy performance (%) on 20 images	
		Cup	Disc
Supervised learning	**CART** tree	53.50	72.97
	Random forest	75.60	92.28
Semi-supervised learning	**SETRED**–*CART*	75.56	60.74
	co-Forest	89.00	93.50

results were obtained by the decision tree *CART* but the semi-supervised mode has improved the performance for the method *SETRED-CART*. Our approach by the algorithm *co-Forest* realizes the best segmentation performance for both regions and especially the cup area that is most difficult to extract. These results allow us to consolidate our proposal and affirm its rigor and robustness for-pixel-based classification by region growing task by semi-supervised learning.

4.1 Discussion

For completeness, and to establish a visual assessment of the performance of our approach, we have randomly selected six images of the test basis (Fig. 5) to discuss the performance and quality of segmentation by four single and multi-classifier

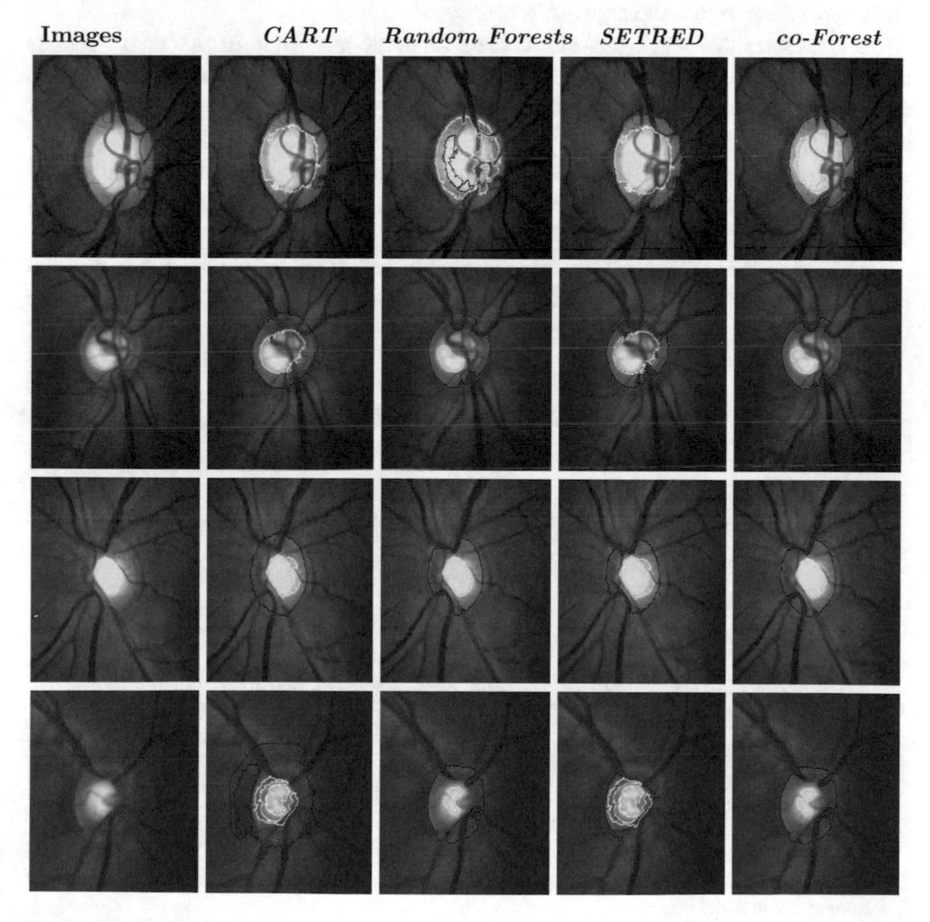

Fig. 5 Examples of automatic segmentation image by the various techniques

approaches supervised mode and semi-supervised namely, respectively, the decision tree and *Random Forest* and the method *SETRED* and *co-Forest*.

Thus, a comparative study of several techniques is proposed. The principle is based on a region growing classifying the neighboring pixels from the pixels of interest of the image semi-supervised learning. In our process, we have used the Fuzzy C-Means (FCM) algorithm to detect the pixels of interest, in order to achieve the segmentation of target regions. The pixels of interest generate by FCM in the pictures N° 2, 3, and 4 (Fig. 5) leads to poor segmentation by the use of mono-classifiers (*CART* and *SETRED*); unlike the multi-classifiers (*Random Forest* and *co-Forest*) was able to successfully separate the cup region.

Another advantage of using the methods together is the network power separation vessel and the bottom region as clearly shown in pictures 2, 3 and 4 (Fig. 5). However, as we can also see in the image 3 (Fig. 5), the contribution of non-labeled pixels to the overall method *co-Forest* has allowed proper identification disc unlike the *Random Forest*. We also notice in the image 1 (Fig. 5) misclassification of pixels of the whole cup approaches except the algorithm *co-Forest*.

As a whole, this work has allowed us to see a multitude of research avenues that are available for automatic segmentation of images. The idea to extrapolate the region segmentation by pixel-based classification semi-supervised learning context by the approach *co-Forest* allowed us to exploit the non-labeled data in the establishment of the set-prediction model. In this sense, non-labeled data has reinforced the recognition of regions of interest for a report cup/disc calculated approximating that of the ophthalmologist.

A comparison of the cup/disc reports performed by an ophthalmologist and the proposed methods on fifteen random images is shown in Fig. 6. Differences in the ophthalmologist's CDR ratios compared to the proposed supervised methods (CART

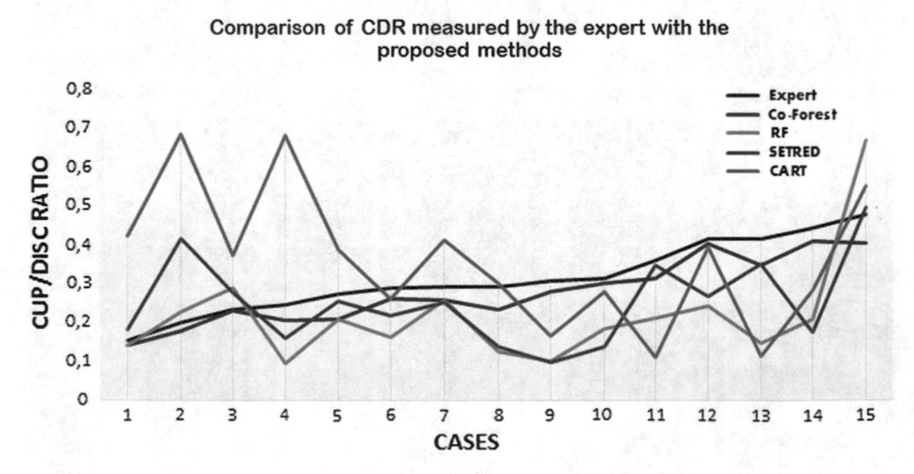

Fig. 6 The cup/disc ratio comparison between the expert and proposed methods calculation. From 1–8 normal, from 9–15 glaucoma

and Random Forest) are widely discarded in cases of glaucoma and normal cases. Even though the CDR ratios calculated by the ensemble approach are closer to the optimal (expert measure).

However, our results for the semi-supervised algorithms *co-Forest* and *SETRED* were slightly different in the value of the ophthalmologist, but our method *co-Forest* tended to show smaller variations in normal cases and glaucoma cases, with almost similar extent to that of the expert.

5 Conclusion

Machine learning (ML) can be used in both image processing and computer vision but it has found more use in computer vision than in image processing. The goal of machine learning is to optimize differentiable parameters so that a certain loss/cost function is minimized. The loss function in ML can have a physical meaning in which case the features learned can be quite informative but this is not necessarily the case for all situations. Computational Intelligence (CI) approaches are alternative solutions for automatic computer vision and image processing systems; they include the use of tools as machine learning and soft computing. The aim is to have an complete autonomy of the computer, but this might not be easily achievable. In this work, we have demonstrated that the semi-supervised learning might be the best solution. Indeed, we have proposed a method of automatic segmentation of disc and cup regions in retinal images by pixel-based classification in semi-supervised learning.

The objective is to involve the expert learning of our model for a better discrimination of regions of interest. Evaluation and segmentation of images tests are performed using the algorithm Fuzzy C-Means. A growth of region is developed by classifying the neighboring pixels by applying four classifiers: *Decision Tree*, *Random Forest*, the method *SETRED* and *co-Forest*.

The results are very convincing and encouraging, showing a great capacity for recognition and segmentation of target regions, this being clearer by applying the *Random Forest* in semi-supervised learning *co-Forest* , heuristics of these ensemble methods allows, using multiple classifiers, to greatly explore the solution space, and by aggregating all predictions, we will take a classifier that considers all this exploration. The contribution of non-labeled data in the establishment of the prediction model can reinforce learning and recognition of relationships between pixels and region.

However, there are some points that deserve some discussion and further development in future works. One must say that the main limitation of our method is the long time processing. Further work to reducing the constraints over time achieving is currently underway to an complete autonomy. One of the proposed solutions to deal with this problem is the parallel programming. Indeed, our algorithm allows us to use a master/slave architecture. In another way, we are currently developing a new semi-automatic segmentation approach based on superpixel-by-superpixel classification [2], where superpixels should both increase the speed and improve the quality of the results.

Acknowledgements The completion of this research could not have been possible without the participation and the support of LIMOS, CNRS, UMR 6158, 63173, Aubiere, France. Their contributions are sincerely appreciated and gratefully acknowledged. However, we would like to express our deep appreciation and indebtedness to the ophthalmic clinic "CLINIQUE LAZOUNI" for providing reel medical database that greatly assisted our work.

References

1. Abdel-Razik Youssif, A.H., Ghalwash, A., Abdel-Rahman Ghoneim, A.: Optic disc detection from normalized digital fundus images by means of a vessels' direction matched filter. IEEE Trans. Med. Imaging **27**(1), 11–18 (2008). doi:10.1109/TMI.2007.900326
2. Achanta, R., Shaji, A., Smith, K., Lucchi, A., Fua, P., Susstrunk, S.: Slic superpixels compared to state-of-the-art superpixel methods. IEEE Trans. Pattern Anal. Mach. Intell. **34**(11), 2274–2282 (2012). doi:10.1109/TPAMI.2012.120
3. Amit, Y., Geman, D.: Shape quantization and recognition with randomized trees. Neural Comput. **9**(7), 1545–1588 (1997)
4. Azmi, R., Norozi, N., Anbiaee, R., Salehi, L., Amirzadi, A.: IMPST: a new interactive self-training approach to segmentation suspicious lesions in breast MRI. J. Med. Signals Sensors **1**(2), 138–148 (2011)
5. Bezdek, J.C.: Pattern Recognition with Fuzzy Objective Function Algorithms. Kluwer Academic Publishers, Norwell (1981)
6. Blum, A., Mitchell, T.: Combining labeled and unlabeled data with co-training. In: Proceedings of the Eleventh Annual Conference on Computational Learning Theory, COLT'98, New York, NY, USA, pp. 92–100 (1998)
7. Bock, R., Meier, J., Nyul, L.G., Hornegger, J., Michelson, G.: Glaucoma risk index: automated glaucoma detection from color fundus images. Med. Image Anal. **14**(3):471–481. doi:10.1016/j.media.2009.12.006. http://www.sciencedirect.com/science/article/pii/S1361841509001509 (2010)
8. Breiman, L.: Random forests. Mach. Learn. **45**, 5–32 (2001)
9. Breiman, L., Friedman, J.H., Olshen, R.A., Stone, C.J.: Classification and Regression Trees. Chapman and Hall, New York (1984)
10. Burana-Anusorn, C., Kongprawechnon, W., Sintuwong, S., Tungpimolrut, K.: Image processing techniques for glaucoma detection using the cup-to-disc ratio. Thammasat Int. J. Sci. Technol. **18**(1) (2013)
11. Chandrika, S., Nirmala, K.: Analysis of CDR detection for glaucoma diagnosis. Int. J. Eng. Res. Appl. (IJERA) NCACCT-19: ISSN: 2248-9622 (2013)
12. Deng, C., Guo, M.: A new co-training-style random forest for computer aided diagnosis. J. Intell. Inf. Syst. **36**(3), 253–281. http://dblp.uni-trier.de/db/journals/jiis/jiis36.html#DengG11 (2011)
13. Devroye, L., Györfi, L., Lugosi, G.: A Probabilistic Theory of Pattern Recognition. Springer, New York (1996)
14. Dunn, J.: A fuzzy relative of the isodata process and its use in detecting compact, well-separated clusters. J. Cybern. **3**, 32–57 (1974)
15. Fix Jr., E.: Discriminatory analysis: nonparametric discrimination: consistency properties. Technical Report Project 21-49-004, Report Number 4, USAF School of Aviation Medicine, Randolf Field, Texas (1951)
16. Foley, J.D., van Dam, A., Feiner, S.K., Hughes, J.F.: Computer Graphics: Principles and Practice, 2nd edn. Addison-Wesley Longman Publishing Co. Inc., Boston (1990)

17. Ghafar, R., Morris, T., Ritchings, T., Wood, I.: Detection and characterization of the optic disc in glaucoma and diabetic retinopathy. In: Medical Image Understand Annual Conference, London, UK, pp. 23–24, Sept 2004
18. Hatanaka, Y., Noudo, A., Muramatsu, C., Sawada, A., Hara, T., Yamamoto, T., Fujita, H.: Automatic measurement of vertical cup-to-disc ratio on retinal fundus images. In: Proceedings of the Second International Conference on Medical Biometrics, ICMB'10, pp. 64–72. Springer, Berlin, Heidelberg (2010). doi:10.1007/978-3-642-13923-9_7
19. Hoover, A., Kouznetsova, V., Goldbaum, M.: Locating blood vessels in retinal images by piecewise threshold probing of a matched filter response. IEEE Trans. Med. Imaging **19**(3), 203–210 (2000). doi:10.1109/42.845178
20. Joshi, G.D., Sivaswamy, J., Krishnadas, S.R.: Optic disk and cup segmentation from monocular color retinal images for glaucoma assessment. IEEE Trans. Med. Imaging **30**(6), 1192–1205 (2011). doi:10.1109/TMI.2011.2106509
21. Kakumanu, P., Makrogiannis, S., Bourbakis, N.: A survey of skin-color modeling and detection methods. Pattern Recogn. **40**(3), 1106–1122 (2007). doi:10.1016/j.patcog.2006.06.010
22. Khalid, N.E.A., Noor, N.M., Ariff, N.M.: Fuzzy c-means (FCM) for optic cup and disc segmentation with morphological operation. Procedia Comput. Sci. **42**(0), 255–262. doi:10.1016/j.procs.2014.11.060. http://www.sciencedirect.com/science/article/pii/S1877050914014987 (2014). Medical and Rehabilitation Robotics and Instrumentation (MRRI2013)
23. Lalonde, M., Beaulieu, M., Gagnon, L.: Fast and robust optic disc detection using pyramidal decomposition and hausdorff-based template matching. IEEE Trans. Med. Imaging **20**(11), 1193–1200 (2001). doi:10.1109/42.963823
24. Li, M., Zhou, Z.H.: SETRED: self-training with editing. In: Ho, T.B., Cheung, D.W.L., Liu, H. (eds.) PAKDD. Lecture Notes in Computer Science, vol. 3518, pp. 611–621. Springer (2005)
25. Li, M., Zhou, Z.H.: Improve computer-aided diagnosis with machine learning techniques using undiagnosed samples. Trans. Syst. Man Cybern. Part A **37**(6), 1088–1098 (2007). doi:10.1109/TSMCA.2007.904745
26. Madhusudhan, M., Malay, N., Nirmala, S., Samerendra, D.: Image processing techniques for glaucoma detection. In: Abraham, A., Mauri, J., Buford, J., Suzuki, J., Thampi, S. (eds.) Advances in Computing and Communications, Communications in Computer and Information Science, vol. 192, pp. 365–373. Springer, Berlin, Heidelberg (2011). doi:10.1007/978-3-642-22720-2_38
27. Maeireizo, B., Litman, D., Hwa, R.: Co-training for predicting emotions with spoken dialogue data. In: Proceedings of the 42th Annual Meeting of the Association for Computational Linguistics (ACL-2004) (2004)
28. Mendonca, A., Campilho, A.: Segmentation of retinal blood vessels by combining the detection of centerlines and morphological reconstruction. IEEE Trans. Med. Imaging **25**(9), 1200–1213 (2006). doi:10.1109/TMI.2006.879955
29. Mohammad, S., Morris, D., Thacker, N.: Texture analysis for the segmentation of optic disc in retinal images. In: 2013 IEEE International Conference on Systems, Man, and Cybernetics (SMC), pp. 4265–4270 (2013). doi:10.1109/SMC.2013.727
30. Muhlenbach, F., Lallich, S., Zighed, D.A.: Identifying and handling mislabelled instances. J. Intell. Inf. Syst. **22**(1), 89–109 (2004)
31. Muramatsu, C., Hatanaka, Y., Ishida, K., Sawada, A., Yamamoto, T., Fujita, H.: Preliminary study on differentiation between glaucomatous and non-glaucomatous eyes on stereo fundus images using cup gradient models. In: Proceedings of SPIE 9035:903,533–903,533–6 (2014). doi:10.1117/12.2043409
32. Quinlan, J.R.: C4.5: Programs for Machine Learning. Morgan Kaufmann (1993)
33. Settouti, N., El Habib Daho, M., El Amine Lazouni, M., Chikh, M.: Random forest in semi-supervised learning (co-forest). In: 2013 8th International Workshop on Systems, Signal Processing and their Applications (WoSSPA), pp. 326–329 (2013). doi:10.1109/WoSSPA.2013.6602385
34. Sirikulviriya, N., Sinthupinyo, S.: Integration of rules from a random forest. In: International Conference on Information and Electronics Engineering IPCSIT, vol. 6. IACSIT Press, Singapore (2011)

35. Sivaswamy, J., Krishnadas, S.R., Joshi, G.D., Jain, M., Tabish, A.U.S.: Drishti-GS: retinal image dataset for optic nerve head(ONH) segmentation. In: IEEE 11th International Symposium on Biomedical Imaging, ISBI 2014, 29 Apr–2 May 2014, Beijing, China, pp. 53–56 (2014). doi:10.1109/ISBI.2014.6867807
36. Yarowsky, D.: Unsupervised word sense disambiguation rivaling supervised methods. In: Proceedings of the 33rd Annual Meeting on Association for Computational Linguistics, Association for Computational Linguistics, ACL'95, Stroudsburg, PA, USA, pp. 189–196 (1995). doi:10.3115/981658.981684
37. Zhou, Z.H., Li, M.: Tri-training: exploiting unlabeled data using three classifiers. IEEE Trans. Knowl. Data Eng. **17**(11), 1529–1541 (2005). doi:10.1109/TKDE.2005.186

Part III
Security and Biometric Applications of Image Processing

Multimodal Biometric Personal Identification and Verification

Mohamed Elhoseny, Ahmed Elkhateb, Ahmed Sahlol and Aboul Ella Hassanien

Abstract Security systems using one identification tool are not ideal. Multisystem security, which using two or more types of security levels like for example using identification password and card, can increase the security of a system, however it is not an ideal security system. Password maybe hacked or forgotten, and Identification card is something we have and could be stolen. This chapter proposes a cascaded multimodal biometric system using fingerprint and iris recognition based on minutiae extraction for fingerprint identification and encoding the log-Gabor filtering for iris recognition. The experiments compare FAR, FRR, and accuracy evaluation metrics for a unimodal biometric system based on either fingerprint or iris and the cascaded multimodal biometric system that sequentially utilizes the fingerprint and iris traits. The proposed system has FAR = 0, FRR = 0.057, and accuracy 99.86%. The results show the superior performance of the proposed multimodal system compared to the unimodal system.

M. Elhoseny (✉) · A. Elkhateb
Faculty of Computers and Information, Mansoura University, Mansoura, Egypt
e-mail: mohamed_elhoseny@mans.edu.eg

A. Elkhateb
e-mail: ahmed_elkhateb@mans.edu.eg

A. Sahlol
Faculty of Specific Education, Damietta University, Damietta, Egypt
e-mail: atsegypt@du.edu.eg

A.E. Hassanien
Faculty of Computers and Information, Information Technology Department, Cairo University, Giza, Egypt
e-mail: aboitcairo@gmail.com

M. Elhoseny · A. Sahlol
Scientific Research Group in Egypt (SRGE), Cairo, Egypt

A.E. Hassanien and D.A. Oliva (eds.), *Advances in Soft Computing and Machine Learning in Image Processing*, Studies in Computational Intelligence 730,
https://doi.org/10.1007/978-3-319-63754-9_12

1 Introduction

Security systems using one identification tool are not ideal. Multisystem security, which using two or more types of security levels like for example using identification password and card, can increase the security of a system, however it is not an ideal security system. Password maybe hacked or forgotten, and Identification card is something we have and could be stolen. Another way for increasing security is using biometrics, which everyone owns unique biometrics data and cannot be forgotten or stolen. Single biometric systems suffer from some problems like noise in sensed data, non-universality, spoof attacks, intra-class variations, and inter-class similarities. Fusion in multimodal biometric systems can be performed using data accessible in any of the modules. Fusion can happen at these levels: (i) sensor level (ii) feature level (iii) score level (iv) rank level and (v) decision level. Different biometric data sources can be utilized as a part of a multimodal biometric system. In view of these sources, multimodal biometric systems could be classified into six distinct classifications: multi sensor, multi algorithm, multi instance, multi sample, multimodal and hybrid. The motivation for working on this chapter is to solve these problems, using an implementation of the multimodal biometric system. Multimodal biometric system is the use of a combination of two or more biometric types to increase the security of a system. In this chapter, a multimodal biometric system using fingerprint and iris recognition system with fusion at cascaded advanced decision level will be introduced.

2 Related Work

Multimodal biometric systems become one of the best security system solutions for most applications in present time. Many researchers have been working on multimodal biometric system. A good survey of the multimodal biometric system was provided by Ross et al. [1]. In this survey, the researchers focused on levels of fusion and score level fusion. Analysis and descriptions on recent multimodal biometric system fusion are contained in the ISO/IEC Technical Report [2]. The report explains requirements supporting multimodal biometric systems. Many research papers explained the types of levels of fusion in multimodal biometric systems. A composite fingerprint image, combining multi part fingerprints, which the user puts finger on a fingerprint sensor surface proposed by Ratha et al. [3]. A face recognition system combining visible and thermal Infrared (IR) images at sensor level was proposed by Singh et al. [4]. Another face recognition system performing a fusion of visual and thermal infrared images without eyeglass at sensor level was proposed by Kong et al. [5]. Fusion of face and iris at feature level was proposed by Son et al. [6]. fusion of hand and face at feature level was performed by Ross et al. [7] and the experiments were performed in three different scenarios. A theoretical framework for combining classifiers was developed by Kittler et al. and various classifier combination strategies were discussed in [8]. Three different classifiers based on the k-nearest-neighbor (k-NN) classifier, logistic regression and decision trees were used to compare the performance at score level fusion by Verlinde et al. [9]. Linear discriminant function and the decision trees at the fusion of match scores were used by Jain et al. [10]. The performance of different fusion methods and normalization techniques in a fusion scenario involving fingerprint, hand geometry and face modalities studied by Jain et al. [11].

Many researchers represented the various types of multimodal biometric frameworks according to the sources of biometric data being fused. A multi-sensor fingerprint system using two sensors (optical and capacitive sensors) was discussed by Marcialis et al. [12]. A multi-algorithm biometric system integrating three different minutiae-based fingerprint matchers was proposed by Jain et al. [13]. Another multi-algorithm gait recognition framework which uses different gait classifiers based on different environmental circumstances was introduced by Han and Bhanu [5]. A multi-instance iris recognition system using a combination of right iris and left iris for the same person is introduced by Wang et al. [14]. A multi-sample system using a composite fingerprint template from multi imprints of the same finger using mosaicking algorithm is proposed by Jain et al. [15]. A hybrid system using multi-sensor and multi-sample of face recognition system is introduced by Bowyer et al. [16]. A multimodal biometric system using face, fingerprint, and voice traits is proposed by Jain et al. [17]. Another multimodal biometric system using face and palmprint is introduced by Yao et al. [18].

Ross and Jain proposed multimodal biometrics system in 2003 [19]. Fusion levels of multimodal biometric systems were introduced in details in Chap. 3. Fingerprint and iris fusion attracted the attention of many researchers. In 2009 Baig et al. [20] presented a multimodal biometric system based on iris and fingerprint using single hamming distance matcher. They used database of WVU containing 400 images and set the threshold equal to EER, the purpose was to improve the percentage of ERR. In 2010 Jagadeesan et al. [21] created a 256-bit cryptographic key using fingerprint and iris based on minutiae extraction and Daugman's approach respectively. Jagadeesan used CASIA database for iris and puplicly available database for fingerprint and make the fusion of the multimodal system at the feature level.

Radha et al. [22] in 2012, proposed a multimodal biometric system using fingerprint and iris at feature extraction fusion level. The proposed system used a combined feature vector from both fingerprint and iris. The proposed system was built using log Gabor filter feature vectors extraction of both modalities.

The final match score generated by Hamming distance. Using database of 50 users the experimental results of FAR, FRR, and execution time were 0%, 4.3, 0.14 s respectively. Abdolahi et al. [23] in 2013 proposed a multimodal biometric framework with two modalities; fingerprint and iris, using fuzzy logic and weighted code. Fusion at decision level combines results after binarizing fingerprint and iris images. Fingerprint and iris codes are weighed as 20% and 80% respectively. The FAR, FRR and accuracy results achieved from this system were 2%, 2%, and 98.3% respectively.

3 Biometric System Process

As an intelligent system [24, 25], the general structure of any biometric framework includes five operations as illustrated in Fig. 1 and listed below.

3.1 *Biometric Data Acquisition*

This process is responsible for collecting a biometric sample from the suitable sensors or devices. A sensor is converting the captured raw signal into a biometric sample, e.g. a unique finger impression picture, iris picture or voice recording.

3.2 *Feature Extraction*

This process is in charge of extracting a set of distinguished features from each biometric sample. These features should be discriminatory enough to represent each individual. The extracted features will be used as a reference during the recognition phase.

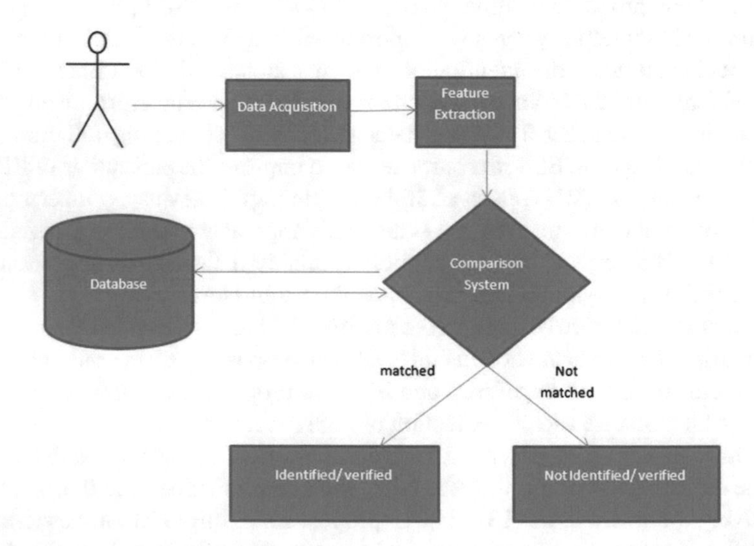

Fig. 1 General structure of biometric system

3.3 Data Storage

All the extracted features during enrollment stage are saved in a data storage system. Along with these features, some other information that related to each individual are also saved such as ID, name, privileges. Biometric and non-biometric data are frequently saved in a distinctive database for security and protection concerns.

3.4 Comparison or Matching Process

In this process, the features are extracted from the input trait are compared to all registered biometric information in the database. There are two main processes: verification or identification. Verification means, an inquiry is addressed "Is this individual who he claims to be?" When doing verification process, a matching score is computed between the input biometric and the corresponding registered biometric information. In an identification process, the inquiry being addressed is "Who is this individual?" So that, the input biometric is compared to all enrolled biometrics for all individual and return the matching scores.

3.5 Decision Subsystem

In light of the matching score(s), the decision process figures out whether the acquired biometric and the enlisted data represent one individual. In verification, the choice is made based on the matching score is either acceptable or not. In the identification scenario, one enrolled identity corresponds to the enrolled biometrics and has the best matching score coincide with the selected choice strategy.

4 Biometric System Errors

The combination of two single biometric qualities is once in a while read precisely the same. This happens because of different reasons, for example, modifications in user's biometric qualities, damaged sensing condition, user's sensor communication and modifications in surrounding conditions. In this way, the result of a biometric system is a matching score calculates the similarity comparing tested template with the stored template.

The biometric system decision relies on a set of threshold t. When the decision score s is greater than the threshold t the test template and stored template are referred to as matched and they both belong to the same user. Otherwise, when decision score s is smaller than threshold t, the test template and stored template are referred to as not matched and the input template does not belong to an authorized user. This can lead us to the definitions of genuine distribution, and imposter distribution. Genuine distribution occurrs when the test template and stored template are matched with matching score s greater than the threshold t. Imposter distribution occurred when the test template is not matched with the stored template or matching score s is smaller than the threshold t (Fig. 2).

False Acceptance Rate (FAR) is the percentage of frauds that were incorrectly recognized over the total number tested. Sometimes it is referred to as False Match Rate FMR.

False Reject Rate (FRR) is the percentage of users that are not recognized falsely to the total numb er tested. Sometimes it is referred to as False Non Match Rate FNMR. Consolidating the FAR and FRR represents the Total Error Rate using the following equation: TER = (Number of False Accepts + Number of False Rejects)/(Total Number of Access).

When increasing the threshold t for high system security, the False Reject Rate FRR increases as well. When decreasing the threshold t to make system tolerant, the False Accept Rate FAR increases as well too. Hence, there is a need to balance between FAR and FRR. The Receiver Operating Characteristics curves (ROC) can be used for measuring the performance of the biometric system. ROC draw the relation between FAR and sensitivity (which equals to 1-FRR) (Fig. 3).

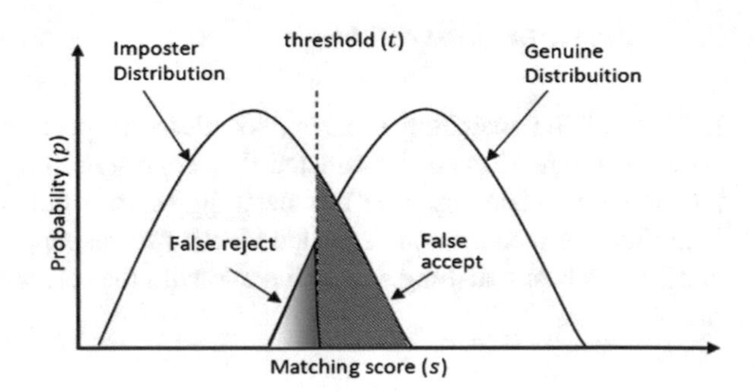

Fig. 2 Biometric system error rates: The *curves* show FAR and FRR for a given threshold t over the genuine and imposter distributions [26]

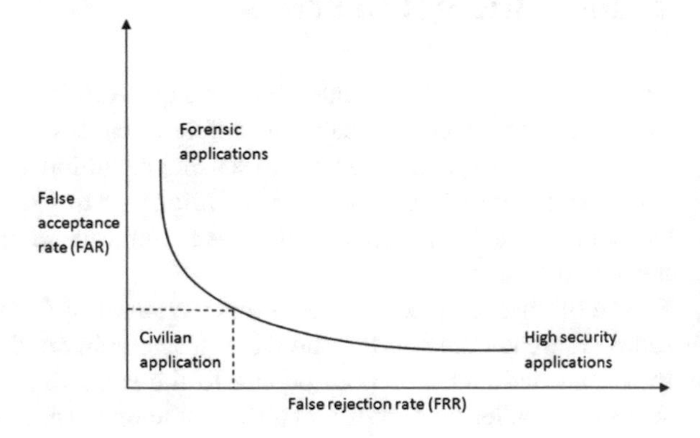

Fig. 3 Receiver Operating Characteristic curve [27]

5 Privacy Issues and Social Acceptance

Acceptability of biometric systems depends on its simplicity and comfort of use. Other factors like cultural, ethnic, and religious also affect the acceptability of the biometric system. As an instance is utilizing biometric traits that do not need contact such as voice, iris, or face reflect high acceptability. Users are more convenient with systems that require less interaction from the user. Biometric traits that acquired with no user interaction may be acquired with no user knowledge and this could be dangerous to privacy of the user. "Privacy is the ability to lead life free of intrusions to remain autonomous, and to control access to ones personal information" [26].

Biometric traits utilization considers some security issues [28, 29] that should be tended to be mentioned. Biometric system users need assurance that their biometric data is secured and protected against misuse and used only for the planned reasons. Most companies using biometric systems store the biometric data in a decentralized encoded database to ensure the security and protection of the biometric data [26].

6 Biometric Systems Challenges

In these days several biometric recognition systems rely only on using one single biometric characteristic to recognize users. Although single biometric systems can offer reliable applications for verification and identification, some limitations and vulnerabilities challenging these single biometric systems like [7]:

Noisy Data The captured biometric data usually contains noise according to, for example, imperfect acquisition conditions or variants in biometric characteristic itself like dirt on fingerprint sensor or a scratch on a fingerprint image. Genuine users always rejected as a result of noise in sensed data.

Non Universality According to some reasons biometric recognition systems sometimes are not capable of capturing perfect biometric data from genuine users resulting in an error named Failure To Enroll (FTE). For instance, drooping eyelids, long eyelashes or certain pathological conditions of eyes may prohibit the iris recognition system from capturing perfect iris data from users.

Spoof Attacks Imposters can try to mimic behavioral biometrics like signature and voice for an enrolled user. Also, creating biometric artifacts can spoof attack physical biometrics such as iris or fingerprint. However, physical traits such as fingerprints and iris are also vulnerable to spoof attacks by creating biometric artifacts. In 2002, Matsumoto et al. [30] explained how fingerprints could be spoof attack as imposters can create gummy fingers using easily obtainable tools and cheap materials. These gummy fingers are granted with high degrees using several fingerprint recognition applications [30]. In 2004, Uludag et al. [31] suggested different ways to protect biometric systems from spoof attacks like liveness detection and detection of known artifacts. Another way by challenging user response like repeating some words "please repeat after beep: 1-5-9-8".

Intra-class variations Some biometric traits change over time like hand geometry and wrong interaction from the user with the sensor represent the main reasons representing intra-class variation. In 2004, Uludag, [31] introduced a solution for intra class variation by storing multiple templates and update these templates periodically over time for each user.

Inter-class similarities Individuals' feature spaces overlapping introduce inter class similarity. The inter-class variation appears in large population identification systems and can result in bigger false acceptance rate. A solution for this problem is to identify the upper bound capacity of users that can be identified effectively using the biometric system.

7 Multibiometric Systems Fusion Levels

Multimodal systems can be constructed in several different ways, based on the biometric information sources and design of the system. Multimodal as usual refers to the system where two or more different biometric sources are in use (such as Iris and fingerprint), however the term multibiometrics is more common. Multibiometric systems include multimodal systems, and also number of different settings.

In the multimodal biometric system, fusion schemes can be performed at any of these different levels; at the sensor level, at the feature-extraction level, at the matching-score level, rank level and at the decision level. Chapter 3 discusses in details the different levels fusion.

8 Multibiometric Systems Evidence Sources

Different biometric data are utilized as part of a multimodal biometric system. In view of these sources, multimodal biometric systems are classified as six distinct classifications [1]: multi-instance, multi-sensor, multi-algorithm, multi-modal systems and hybrid.

Multi-sensor biometric systems acquire the same biometric trait from two or more different sensors. Multi-instance biometric systems use one sensor to capture two or more different instances of the same biometric modality. Multi-algorithm biometric systems process the acquired biometric trait by more than one algorithm. Multimodal biometric systems can use one or more sensors to capture more than one different traits of biometric. Hybrid system means mixing more than one of the above types. Chapter 4 will represent in details sources of evidence in multibiometric systems.

9 Multimodal Biometric Advantages Over Single Biometric

This part represents multimodal biometric systems advantages over single biometric systems [7].

Non-universality is addressed by Multibiometric systems. For instance, when the fingerprints of a user have a poor quality it prevents the user from being enrolled to the system; so usage of other biometric modalities like voice, face, iris, etc. allow systems to utilize other biometric modality and to register an individual to the system.

Multibiometric also addressed the spoof attack as it becomes more difficult for an imposter to spoof multi biometrics of a genuine user at the same time. Using the appropriate fusion technology can possibly help finding if an individual is an imposter or a genuine user. More difficulties could be added to prevent imposters to be enrolled in the system like; the system can ask the user to present modalities in random order, or the system may ask the user to pronounce certain words or numbers to make sure that the user is really alive.

Multibiometric applications report the noisy sensed data effectively. The possibility to depend on data obtained from other traits can come over the noisy sensed data from one biometric trait. Even if many multibiometric systems consider the quality of the sensed data while executing the fusion process and this is a challenging problem itself, multibiometric systems can significantly use these benefits.

Multibiometric systems addressed the problem of fault tolerant by remaining to work even if the information of a certain biometric source is unreliable according to software or sensor faults. In identification systems with large user population usually exist fault tolerance.

The accuracy of biometric systems is improved by fusing evidence from multi biometric sources. Using the suitable sources of evidence and the best fusion technique assures the matching accuracy improvement.

10 Biometric Systems Applications

Applications of Biometrics can be organized as three main groups [26]:

1. Business applications, for example, PC system login, e-trade, Internet access, ATMs or credit cards, physical access control, mobile telephones, personal digital assistant (PDA)s, medical records management, distance learning, and so on.
2. Administration applications, for example, national ID card, driver's license, social security, border control, passport control, welfare-disbursement, and so forth.
3. Legal applications, for example, body identification, criminal examination, terrorist identification, parenthood determination, and so forth.

11 Proposed Multimodal Biometric System Using Fingerprint and IRIS

In multimodal biometric systems, two or more biometrics are employed (e.g. IRIS, fingerprint, face etc.) to enhance system performance and accuracy. The proposed system uses two biometrics; Fingerprint and IRIS. The Proposed system works at two levels; at first level the extracted Fingerprints features are extracted and compared with stored finger prints templates stored in the database, second level the IRIS features are extracted, compared and matched with stored IRIS templates stored in the database. Level-II works only if Level-I is not passed. The fusion is accomplished at cascaded advanced decision level. If Level I is matched, the system avoids for matching IRIS extracted further at level II Fig. 4.

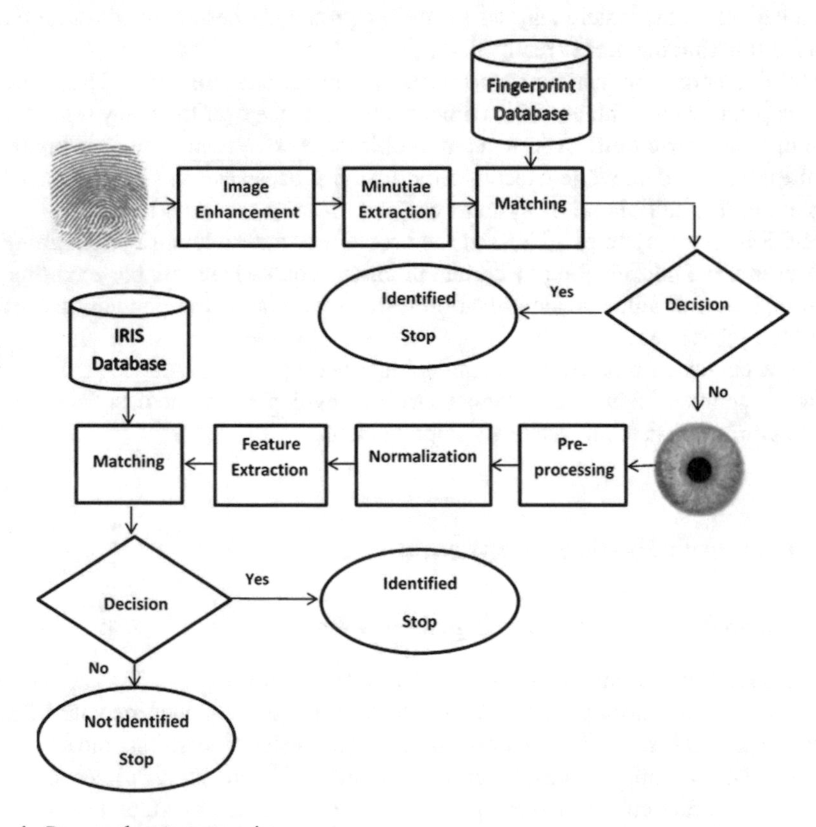

Fig. 4 Proposed system overview

11.1 Level I: Fingerprint

11.2 The Human Fingerprint [27]

Most human body skin is smooth, and contains oil glands and hair, but palm and finger's skin contain no oil glands or hair. Palms and fingers contain a flow pattern of valleys and ridges. Finger ridges (also called friction ridges) help in catching objects, and improve sensing surfaces. Two layers form the friction ridges; Inner layer called dermis, and outer layer called epidermis. Ridges that appear on epidermis enhance the friction between hand and surfaces. Uniqueness of friction ridges even with identical twins helps using it in fingerprint recognition systems for human identification and verification (Fig. 5).

11.3 Fingerprint Recognition

One of the most used single biometrics is fingerprint because it is the most proven modality for user identification. Fingerprint is composed of ridges and valleys found on finger surface. After fingerprint image acquisition, three main steps for fingerprint recognition using minutiae extraction technique which are:

1. Image Enhancement
2. Minutiae Feature Extraction
3. Comparison and Matching

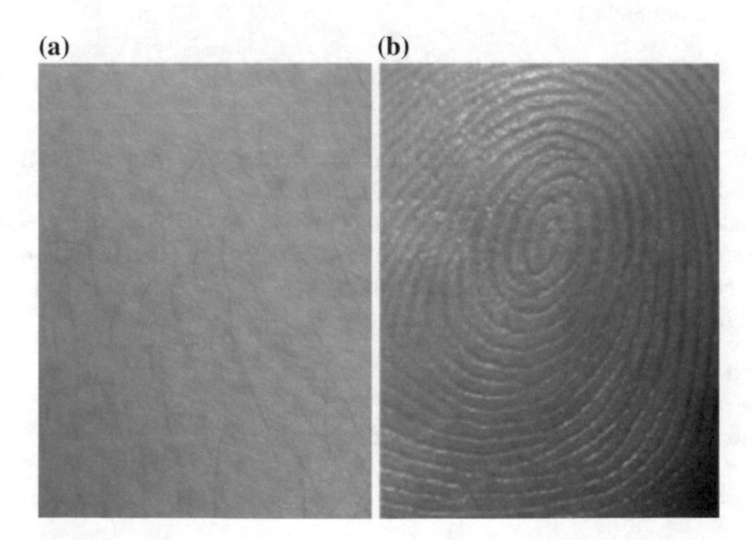

Fig. 5 Two types of skin on the human body: **a** smooth skin and **b** friction ridge skin [27]

11.4 Image Enhancement

Fingerprint image acquisition sometimes results in noisy data like holes, smudges or creases, and this lead to unsuccessful efforts for recovering real valleys or ridges. There is a need for an enhancement algorithm to enhance the structure clarity of valleys and ridges of the image to mask lost regions.

The enhancement process begins with normalizing the input image so that image mean and variance are identified and estimate the image orientation. The resulted image is used to compute a frequency image from which the region mask is acquired using block classification of normalized image. Finally, Gabor filters applied to valley and ridge pixels of normalized image to output the enhanced image of fingerprint. Figure 6b, c shows the mask region and enhanced images of the fingerprint respectively.

11.5 Feature Extraction

Before extracting the minutiae features, a thinning algorithm is applied to the enhanced fingerprint image after being binarized to decrease the thickness of ridges to a single pixel. Minutiae features are the bifurcations and ridge endpoints that are extracted from the resulted skeleton image. Minutiae points' location and orientation are extracted and stored to create a feature set. The crossing number (CN) method uses eight neighborhood connected pixels to extract minutiae points by extracting bifurcations and ridge endings from the enhanced image by testing the nearest pixels to each ridge pixel using 3×3 window. The crossing number (CN) for a given ridge can be defined as:

$$CN = \frac{1}{2} \sum_{k=1}^{8} |V_k - V_{k+1}| \tag{1}$$

where V_i is the pixel value at index i and V9 = V1. According to CN, the ridge pixels can be classified as a ridge ending, bifurcation, or not minutiae point, when CN equals 1, 3, or otherwise respectively. Figure 6d shows the minutiae points on the skeleton image. The feature vector for each detected minutiae point contains its spatial coordinates, and the ridge segment orientation. The following data is stored for each minutiae point extracted:

- The coordinates x and y,
- Ridge segment orientation, and
- Minutiae type (bifurcation, ridge, or ending)

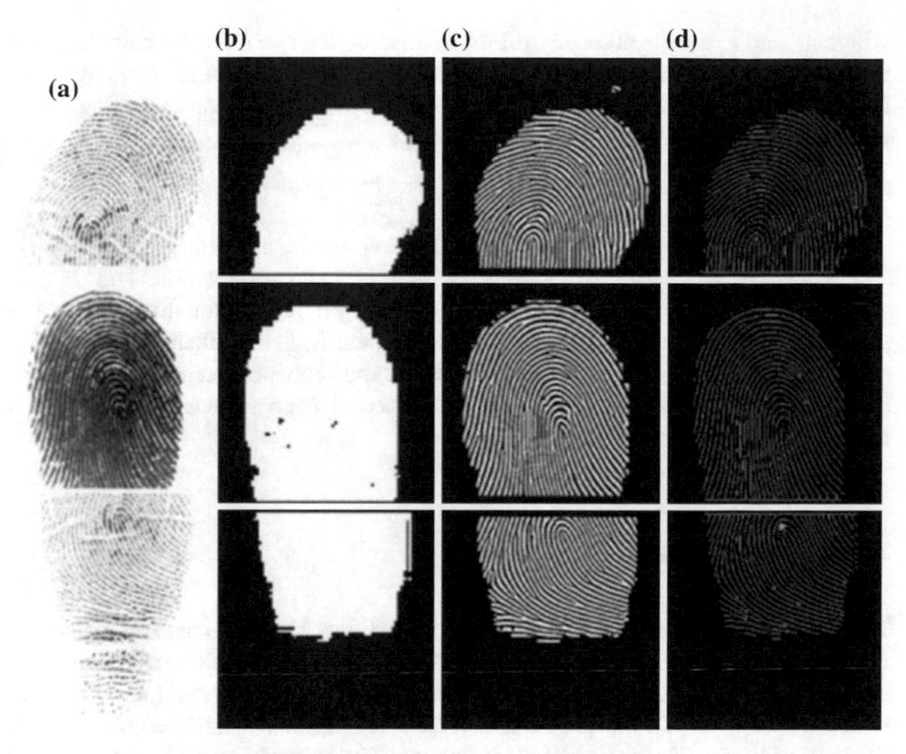

Fig. 6 Fingerprint minutiae feature extraction. **a** original image. **b** mask image. **c** enhanced image. **d** skeleton image with the minutiae points

11.6 Comparison and Matching

Minutiae points extracted from the stored database, and the query fingerprint is presented to the matching algorithm. The matching algorithm finds the association between the input query fingerprint and thc stored template that maximizing the number of minutiae pairings. Consider A = $m_{a1}, \ldots m_{am}$ denotes the set of extracted minutiae points from the template in the stored database, and B = $m_{b1}, \ldots ma_{bm}$ be the extracted minutiae points from the input query fingerprint; where mi = $x; y; \theta$, x and y represents the spatial coordinates of a minutiae point and θ is its orientation. The two minutiae sets are paired if both satisfy the following geometric distance D_s and angle difference D_a constraints:

$$D_s(m_{a_i}, m_{b_j}) = \sqrt{(x_{a_i} - x_{b_i})^2 + (y_{a_i} - y_{b_i})^2} < r_d, \tag{2}$$

$$D_a(m_{a_i}, m_{b_j}) = \min\left(|\theta_{a_i} - \theta_{b_i}|, 360 - |\theta_{a_i} - \theta_{b_i}|\right) < r_a, \tag{3}$$

where r_d, and r_a are the allowed difference between the two minutiae pair. The similarity score is computed based on the number of matching minutiae pairs N_m and a total number of minutia points in the database template N_a and the query fingerprint image N_b.

$$S_{finger} = \sqrt{\frac{N_m^2}{N_a N_b}} \tag{4}$$

The generated similarity score S_{finger} between the tested and stored images is passed to the decision level. In the decision level, the S_{finger} is compared to a decision threshold. If the S_{finger} is greater than or equal to the decision threshold then the user is identified/verified and the system ends, otherwise the system moves to the next level (i.e. iris recognition).

12 Level II: IRIS

The Human Iris IRIS is a thin circular velum, lies between lens and cornea of the eye Fig. 7. IRIS consists of many layers, dense pigmentation cells contained in epithelium; the lowest layer. Above the epithelium layer lies the stromal layer which contains two iris muscles, blood vessels and pigment cells. The color of IRIS determined by density of stromal pigmentation. The externally visible surface of the multi-layered iris contains two zones, which often differ in color [16]. An external ciliary zone and an inward pupillary zone, and these two zones are separated by the collaret-which shows up as a crisscross example.

Arrangement of the iris starts by the third month of embryonic life [16]. The one of a kind example on the surface of the iris is formed the first year of life, and pigmentation of the stroma happens for the initial couple of years. Arrangement of the novel examples of the iris is arbitrary and not identified with any hereditary variables [32]. The main trademark that is subject to hereditary qualities is the pigmentation of the iris, which decides its shading. Because of the epigenetic way of iris examples, according to an individual contain totally autonomous iris examples, and

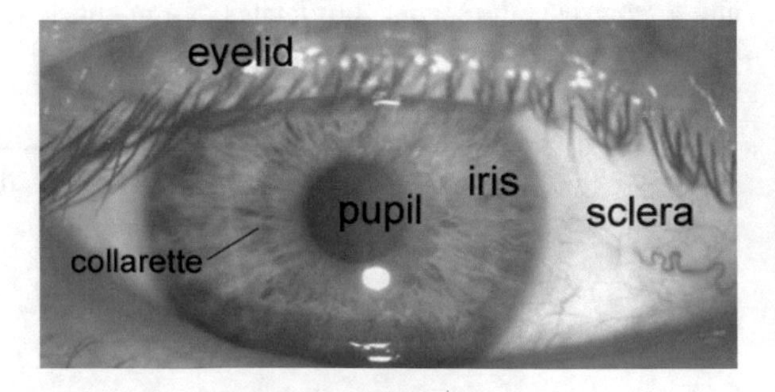

Fig. 7 Human eye view [16]

indistinguishable twins have uncorrelated iris designs. For further points of interest on the life structures of the human eye counsel the book by Wolf [16].

Iris Recognition Each person has a unique iris print which remains stable over his life. Two Circles could estimate The iris region, a circle for the pupil boundary (a central solid black circle of eye) and the other one is for the iris boundary (an annular ring between the pupil boundary and sclera). Pupil size changes according to light; when eye exposed to light the pupil expands, and when dark pupil contracts. Iris is unique for each individual as it contains the unique flowery pattern. The eyelashes and eyelids usually block the lower and upper portions of iris region. Sometimes reflections exist corrupting the iris configuration. A technique is required to locate the circular iris region, and separate and reject these objects. A standard algorithm for detecting Iris boundary is the Hough transform which can be used to derive the radius and center coordinates of the iris and pupil regions. The major steps for iris recognition are:

1. Iris and pupil segmentation
2. Normalization
3. Extracting Features
4. Comparison and Matching

Before applying the four steps mentioned above, iris image need to be captured using a suitable high-quality iris camera because the four steps will depend on image quality.

12.1 *Iris and Pupil Segmentation*

First; we use Canny edge detector to create an edge map. In order to effectively highlight the iris boundary, the gradients were weighted more in the vertical direction. While for pupil detection, the gradients were equally weighted in both directions. Figure 5.5b, c shows the full edge map and vertical edge map obtained by the Canny edge detector. By using the edge map, the circular Hough transform parameters are chosen. These parameters are: radius r, center coordinates x_c and y_c to define the circle according to this equation:

$$x_c^2 + y_c^2 - r^2 = 0 \tag{5}$$

The best circle radius and center coordinates are the maximum points in the Hough space. Note, the Hough transform for the iris is computed firstly. Once the iris region is detected, the second Hough transform is applied within the iris region to detect the pupil. Figure 8d shows the resulted iris and pupil segmentation. To detect the eyelids, linear Hough transform is used to fit a line on the upper and lower eyelid.

When using all gradient data it is found that the eyelids are aligned horizontally, and the eyelid edge interacts with the circular iris outer boundary. When using vertical gradients only to detect iris outer boundary decrease the impact of the eyelids

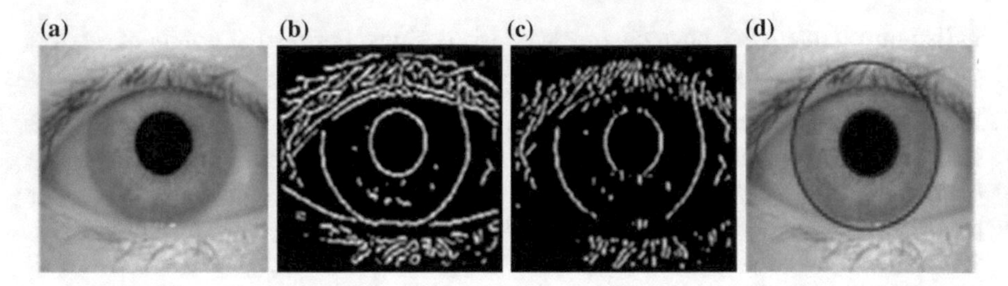

Fig. 8 Iris segmentation. **a** original image. **b** edge map. **c** vertical edge map. **d** segmented iris and pupil boundaries

when performing Hough transform technique, and some edge pixels of the iris circle can be neglected. Using this technique make iris circle localization more accurate and more efficient as it decreases the edge points tested in Hough space.

13 Normalization

The normalization is applied on the segmented iris region to have fixed dimensions of different iris images. The normalization is based on Daugman's rubber sheet model [19]. The iris region I(x; y) is transformed into the strip. The mapping is done by transforming the Cartesian coordinates (x; y) into its polar coordinates $(r; \theta)$ equivalent using

$$I(x(r,\theta), y(r,\theta)) \longrightarrow I(r,\theta) \tag{6}$$

with $x(r;\theta) = (1-r)xp(\theta) + rxi(\theta)$ and $y(r;\theta) = (1-r)yp(\theta) + ryi(\theta)$. Where $x_p; y_p$ and $x_i; y_i$ are the pupil and iris boundaries coordinates along the θ direction. The value of θ and r belongs to $[0; 2\pi]$, and [0; 1] respectively. The center of the coordinate system is at the pupil center. The reflections, eyelashes, and eyelids removed from the normalized image. Here, the polar transformed image has 20×240 dimension represents the radial and angular resolutions as shown in Fig. 9a.

Fig. 9 Iris normalization and feature coding; **a** normalized image. **b** feature codes. **c** mask image

13.1 Extracting Feature

In this stage, the most discriminative characteristics of the iris region are only extracted and encoded in a compact form to improve the accuracy recognition rate. Log-Gabor filter is used to extract iris features. The log-Gabor filter works as a band-pass filter to analysis the texture of the image. The encoding process [33] generates a bitwise template of the iris region by analysis phase information. The filter's phase is categorized into one of four quadrants where each quadrant is represented by two bits.

Here, the total number of bits in the template is 9600. A noise mask is also generated to highlight areas such as eyelids, eyelashes, and reflections identified in the segmentation stage. Figure 9b, c shows the encoding features and the mask region. The result vector is used to make the comparison between the stored iris database and query iris image.

13.2 Comparison and Matching

The matching is accomplished between iris codes Ic generated from iris database images and iris query image using Hamming distance technique. The Hamming distance measures the difference between two bit iris codes using the following equation:

$$S_{Iris} = \frac{\|Ic_A \bigoplus Ic_B \bigcap Im_A \bigcap Im_B\|}{\|Im_A \bigcap Im_B\|} \quad (7)$$

where Ic_A, Ic_B are the iris codes for stored database image and query image, and Im_A, Im_B denotes the noise masks. $\bigoplus, \bigcap$ are the Boolean operators *XOR* and *AND*. This matching score S_{Iris} is used as input to the decision level. so that if the S_{Iris} is smaller than or equal to the decision threshold then the user is identified/verified and the system ends, otherwise the system rejects the user.

14 Experimental Results

MATLAB 7.8.0.347(R2009a) is the programming language used to implement this system. The testing of the performance of the proposed system is applied using the following two databases: 1—CASIA-Iris V1 [34]; It contains 756 images acquired from 108 individuals. 7 images for each eye are captured with an advanced home-made camera for iris. All stored images are formated as BMP with resolution 320 × 280. 2—FVC 2000 ($DB4_B$) and 2002 ($DB1_B, DB2_B, DB3_B$) [35]. Each databases contain 80 fingerprints (80) acquired from ten persons; eight impressions from each person. The FVC database is free downloaded. In this proposed system, the first

Table 1 Confusion matrix

		Predicted template	
Actual		Yes	No
	Yes	TP	FN
Template	No	FP	TN

40 individuals are selected from CASIA Iris V1 and FVC 2000 and 2002 for the experiment; 35 individuals enrolled into system database (4 images for each), and 3 images for each individual is used for testing. Images of individuals from 36 to 40 are not registered in the system but used for testing only. The experiment went through four levels; Fingerprint recognition Level, Iris recognition Level, cascaded multimodal biometric level based on fingerprint and iris recognition, and multimodal biometric level based on Fingerprint and Iris recognition using AND rule at decision level fusion. Biometric applications have a number of performance measures used to characterize the performance of biometric systems. False Acceptance Rate (FAR), False Reject Rate (FRR), system accuracy, and Receiver Operating Characteristics curves (ROC) are the most important performance measures in biometric systems. False Acceptance Rate (FAR) is the percentage of imposters that were incorrectly recognized over the total number of imposters tested. False Reject Rate (FRR) is the percentage of clients that are not recognized falsely to the total number of clients tested. Receiver Operating Characteristics curves (ROC) used for visual comparison of classification models FAR, FRR and Accuracy can be calculated using the confusion matrix shown in Table 1.

True positives (TP): refers to the number of users correctly identified by the system. True negatives (TN): refers to the number of non-users correctly not identified by the system. False positives (FP): refers to the number of non-users were identified by the system. False negatives (FN): the number of users not identified by the system. FAR, FRR, and Accuracy can be calculated according to the following equations using Sensitivity which is true positive rate and specificity which is true negative rate:

$$FAR = \frac{FP}{TN + FP}. \tag{8}$$

$$FRR = \frac{FN}{TP + FN}. \tag{9}$$

$$Acc. = \frac{TP + TN}{TP + TN + FP + FN}. \tag{10}$$

14.1 Lever 1: Fingerprint Recognition Results

In this level, fingerprint Image is captured by a fingerprint scanner device, the captured image is usually corrupted because of some noises like holes, creases, and smudges, so the image needed to be enhanced to improve the quality of fingerprint image using Gabor Filter algorithm and this is the second step. In step 3; the enhanced image is binarized and passed to a thinning algorithm to increase ridge thickness to be one single pixel. The last step is matching in which the input minutiae is compared with stored minutiae templates in database, if matching score is smaller than the given threshold, identification is complete else identification is rejected. In this experiment different thresholds are chosen from 0.25 to 0.7 step 0.05. When using small thresholds the False Accept Rate is increased and False Reject Rate and Accuracy is decreased. With increasing the threshold to 0.6 False Reject Rate is increased slightly, Accuracy also is increased obviously, and False Accept Rate is decreased obviously too. When increasing threshold than 0.6 Accuracy starts to decrease again, and False Reject Rate extremely increased. The experiment results are represented in Table 2.

Figure 10 represents the performance measures; FAR, and FRR curves for the fingerprint recognition system using minutiae extraction, and Fig. 11 represents the accuracy curve.

14.2 Level 2: Iris Recognition Results

In this level Iris Image is captured by a suitable device, then automatic segmentation is applied using Hough transform to generate edge map circles and detect iris

Table 2 Finger print recognition results using minutiae extraction

Finger Threshold	FAR	FRR	Accuracy
0.25	0.8838	0	0.1383
0.3	0.5289	0.0095	0.484
0.35	0.1998	0.0381	0.8043
0.4	0.0488	0.1048	0.9498
0.45	0.012	0.1524	0.9845
0.48	0.0034	0.1714	0.9924
0.5	0.002	0.2095	0.9929
0.55	0	0.2571	0.9936
0.6	0	0.4286	0.9893
0.65	0	0.5143	0.9871
0.7	0	0.6667	0.9833

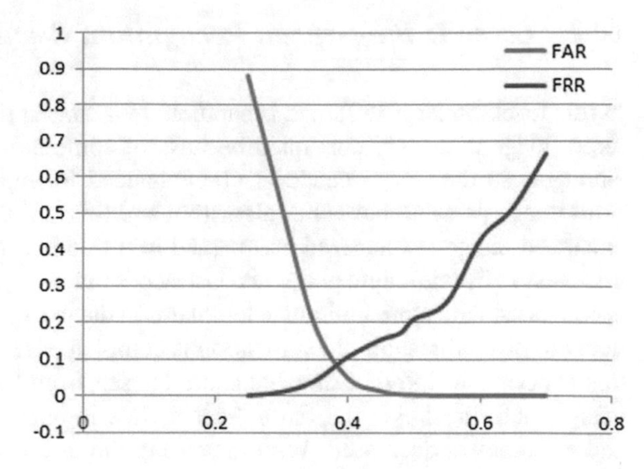

Fig. 10 FAR and FRR curves for fingerprint recognition using minutiae extraction

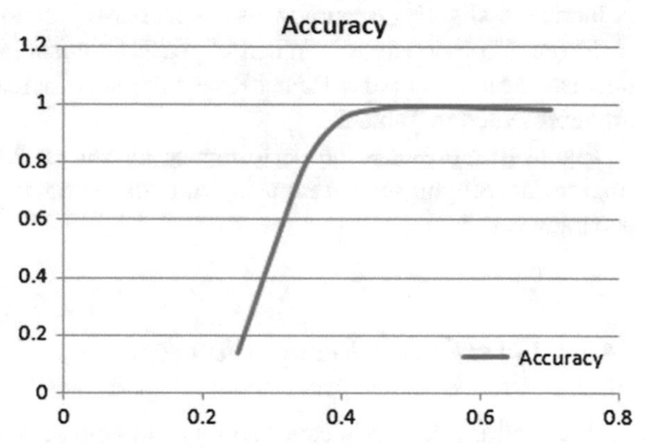

Fig. 11 The accuracy curve for fingerprint recognition using minutiae extraction

boundaries and eyelids, this step is preprocessing. Next step is normalization; in which Iris image is transformed into a strip which involves of points acquired from the outer boundary of iris to the outer boundary of the pupil and normalized to make the strip size constant for different iris images. Features are extracted using Haar wavelet in which Iris image is fragmented into four factors i.e., diagonal, vertical, horizontal, and approximation. The approximation factors are fragmented into four factors. The series of steps are reiterated for five levels and the last level arguments are collected to create a vector. The collected vector is binarized to compare easily between the query image and iris codes stored in the database. The last step is matching in which comparison between query images and iris codes from the stored database is done using hamming distance algorithm. The experiment results are represented in Table 3.

Figure 12 represents the performance measures; FAR, and FRR curves for the Iris recognition system using minutiae extraction, and Fig. 13 represents the accuracy curve.

Table 3 Iris recognition results using Hamming distance

IRIS threshold	FAR	FRR	acc
0.1500	0.000000	1.000000	0.975000
0.1600	0.000000	0.990500	0.975200
0.1700	0.000000	0.990500	0.975200
0.1800	0.000000	0.981000	0.975500
0.1900	0.000000	0.981000	0.975500
0.2000	0.000000	0.961900	0.976000
0.2100	0.000000	0.952400	0.976200
0.2200	0.000000	0.895200	0.977600
0.2300	0.000000	0.847600	0.978800
0.2400	0.000000	0.742900	0.981400
0.2500	0.000000	0.628600	0.984300
0.2600	0.000000	0.466700	0.988300
0.2700	0.000000	0.381000	0.990500
0.2800	0.000000	0.314300	0.992100

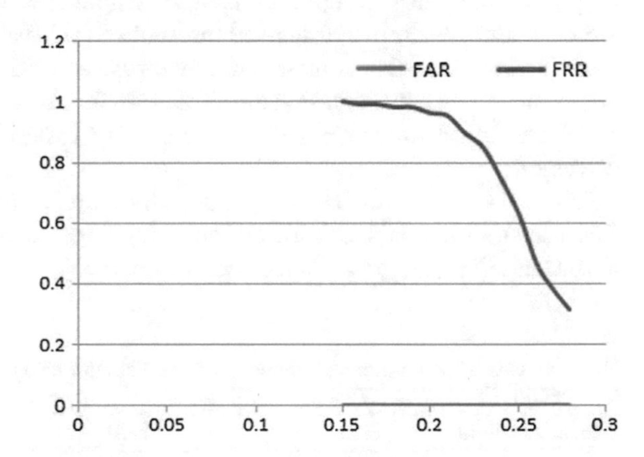

Fig. 12 FAR and FRR curves for Iris recognition using Hamming Distance

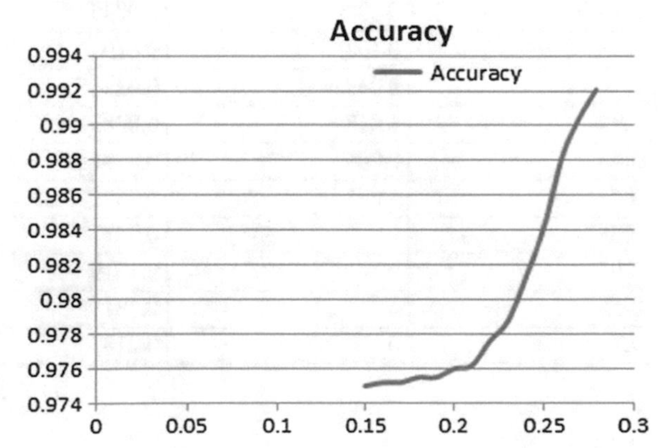

Fig. 13 the accuracy curve for Iris recognition using Hamming Distance

14.3 Level 3: Multimodal Biometric System Results Using Fingerprint and Iris

In this level, the fingerprint image is captured for an individual, enhanced using Gabor Filter algorithm, binarized and passed to thinning algorithm, extract minutiae points, matching with stored templates. If the matching score between the input pattern and the stored template is greater than the given finger threshold then identification/verification is complete else Iris Image is captured for the same individual, automatic segmentation is applied using Hough transform, preprocessing, normalization, features extraction using Haar wavelet, binariztion, matching using hamming distance algorithm.

If the matching score is smaller than the given iris threshold the identification/verification is complete else system stops with no identification/verification. In this experiment different thresholds are chosen from 0.25 to 0.7 step 0.05 for fingerprint and one threshold for Iris 0.28; this threshold is chosen as it is the smallest one achieving the highest accuracy of the tested thresholds with fingerprint thresholds and to reduce the computational complexity. When using small thresholds for fingerprint the False Accept Rate is increased and False Reject Rate and Accuracy are decreased. With increasing the threshold to 0.5 False Reject Rate is increased slightly, Accuracy also is increased obviously, and False Accept Rate is decreased obviously too. When increasing threshold than 0.6 Accuracy starts to decrease again, and False Reject Rate extremely increased. The experiment results are represented in Table 4.

Figure 14 represents the performance measures; FAR, and FRR curves for the Cascaded multimodal biometric system using finger print recognition and Iris recognition, and Fig. 15 represents the accuracy curve.

Table 4 Cascaded multimodal biometric system results using finger print recognition and Iris recognition

Finger threshold	FAR	FRR	Accuracy
0.3	0.5289	0.0095	0.484
0.35	0.1998	0.019	0.8048
0.4	0.0488	0.0286	0.9517
0.45	0.012	0.0381	0.9874
0.48	0.0034	0.0381	0.9957
0.5	0.002	0.0571	0.9967
0.55	0	0.0571	0.9986
0.6	0	0.1619	0.996
0.65	0	0.181	0.9955
0.7	0	0.219	0.9945

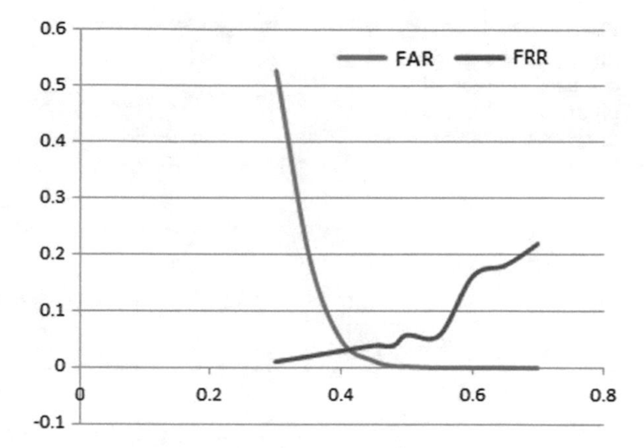

Fig. 14 FAR and FRR curves for the cascaded multimodal biometric system

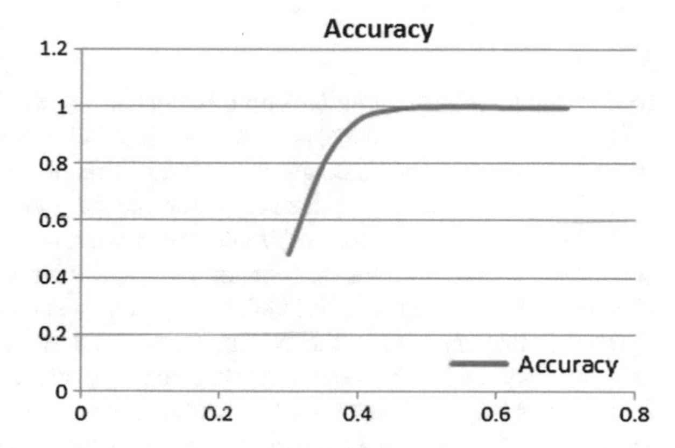

Fig. 15 The accuracy curve for the cascaded multimodal biometric system

The next curve Fig. 16 represents the Receiver Operating Characteristics (ROC) to compare the results of single biometric system using fingerprint and the cascaded multimodal biometric system using fingerprint and Iris recognition. In this curve, the sensitivity (true positive rate) of the multimodal system is farther from the diagonal than the sensitivity of single modal. This means that the multimodal biometric system is more accurate than the single biometric system.

14.4 Level 4: Multimodal Biometric System Results Using Fingerprint and Iris at Decision Level Fusion Using and Rule

In this level, fingerprint and iris images are captured in parallel for the same individual. Each image is processed using the same steps in the previous level. As for fingerprint, the image is enhanced using Gabor Filter algorithm, binarized and passed

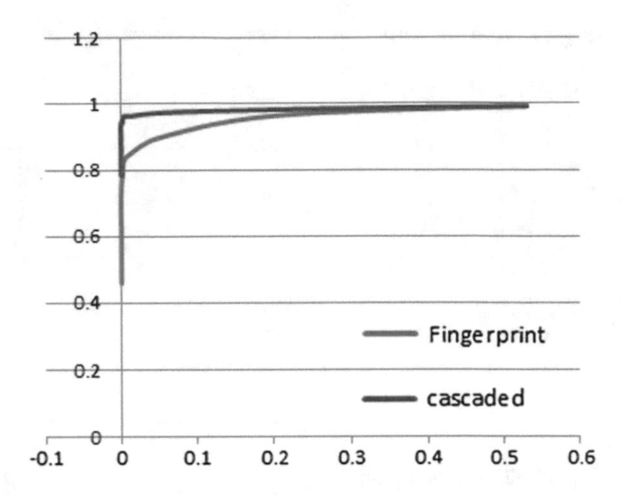

Fig. 16 The ROC curve for fingerprint recognition and cascaded multimodal biometric system using fingerprint and iris

to thinning algorithm, minutiae points extracted, and finally matched with the stored templates. For the iris image, automatic segmentation is applied using Hough transform, preprocessing normalization, Features extraction using Haar wavelet, binariztion, finally using hamming distance algorithm for matching.

If the matching score for both fingerprint and iris are greater than or equal to the pre-specified threshold for both finger and iris the individual is accepted to access the system otherwise the individual is not accepted and have no wright to access the system. In this experiment the system increase the overall security and false accept rate is totally decreased; however, the accuracy is decreased and false reject rate is increased. Also, in this experiment, the time needed for identification or verification is increased as it requires to work on both fingerprint and iris for each user even if one of them is sufficient for identification or verification.

In this experiment, different thresholds are chosen from 0.3 to 0.7 step 0.05 for fingerprint and one threshold for Iris 0.28; this threshold is chosen as it is the one achieving the highest accuracy of the tested thresholds and to reduce the computational complexity.

The experiment results are represented in Table 5.

Figure 17 represents the performance measures; FRR curve for the multimodal biometric system using fingerprint and iris recognition at decision level fusion using the AND rule, and Fig. 18 represents the accuracy curve.

However, using AND rule increase system security as it requires the user to enrol both fingerprint and iris together to the system, but the cascaded system achieved more accuracy than using the AND rule.

Table 5 Multimodal biometric system results using AND rule

Finger Threshold	FAR	FRR	Accuracy
0.3	0	0.314286	0.992143
0.35	0	0.333333	0.991667
0.4	0	0.390476	0.990238
0.45	0	0.428571	0.989286
0.48	0	0.447619	0.98881
0.5	0	0.466667	0.988333
0.55	0	0.514286	0.987143
0.6	0	0.590476	0.985238
0.65	0	0.67619	0.983095
0.7	0	0.790476	0.980238

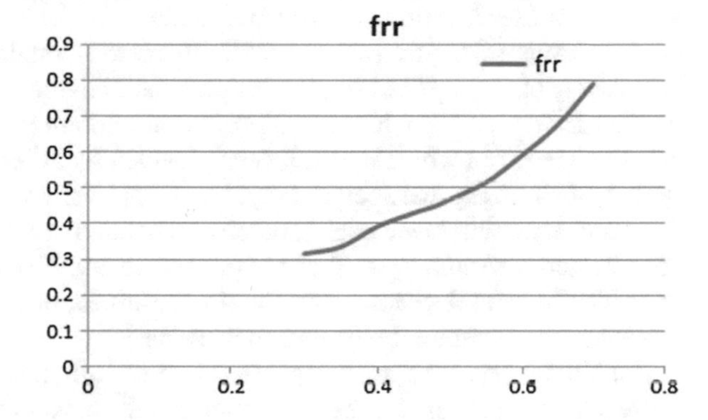

Fig. 17 FAR and FRR curves for the multimodal biometric system using fingerprint and iris at decision level fusion using AND rule

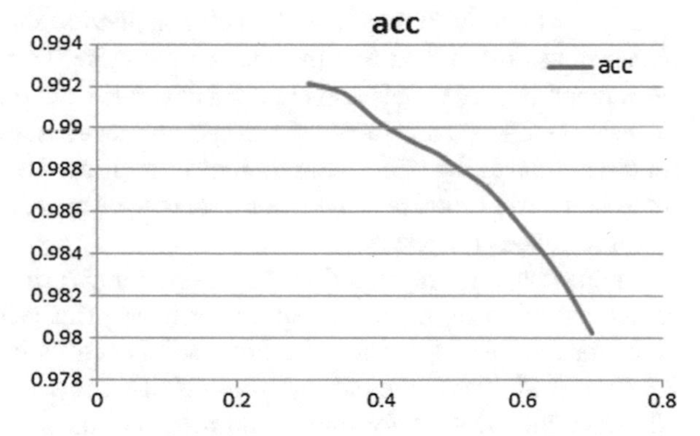

Fig. 18 The accuracy curve for the multimodal biometric system using fingerprint and iris at decision level fusion using AND rule

14.5 Conclusion

Most Security systems can be considered as one of these three types; knowledge based; "What you know" like PIN, passwords, or ID however it may be guessed, forgotten, or shared. Another type is token "What you have" like cards, or key; it may be lost or duplicated and it can be stolen. Last type is the use of biometrics; "What you are" like fingerprint, IRIS, face ..., etc.

Biometric identification systems have the ability to recognize individuals by measuring and analyzing physiological or behavioral characteristics and comparing them against template set stored in the database. Unimodal biometric systems suffer from some problems like noise in sensed data, non-universality, spoof attacks, intra-class variations, and inter-class similarities.

Multimodal biometric system is the use of a combination of two or more biometric types to increase the security of a system (like: Fingerprint and Iris) to increase security for user identification or verification.

Five levels of fusion in multimodal biometric systems: sensor level; in which raw data captured by the sensor are combined, feature level; in this level, features created from each user biometric process are combined to make a single feature set, score level; in which match scores provided by different matchers representing degree of similarity between the input and stored templates, are fused to reach the final decision, rank level; each biometric subsystem assigns a rank to each enrolled identity and the ranks from the subsystems are combined to obtain a new rank for each identity, and decision level; the final result for every biometric subsystem are combined to obtain final recognition decision.

Multibiometric systems categorized into six different types: multi sensor; uses more than one sensor to capture biometric trait to extract various data, multi algorithm; in which more than one algorithm applied to the same biometric data, multi instance; use more than one instance of the same biometric (for example, left and right index fingers or left and right irises), multi sample; more than one sample of the same biometric are captured using the same sensor to acquire a more complete representation of the underlying biometric, multimodal; combine evidence of two or more biometric traits, and hybrid; refers to systems using two or more of the other five mentioned categories.

In this chapter a proposed system using Fingerprint and Iris recognition is presented based on minutiae extraction for fingerprint recognition and hamming distance for IRIS Recognition. The proposed system is implemented with MATLAB 7.8.0.347(R2009a) using dataset from CASIA Iris V1 for Iris recognition and FVC 2000 and 2002 DB1 A for fingerprint recognition.

The experiment results carried on datasets from CASIA Iris V1 for Iris recognition and FVC 2000 and 2002 DB1 A for fingerprint recognition. It compares FAR, FRR, and accuracy metrics for Fingerprint standalone recognition system and the multimodal biometric system based on Fingerprint and Iris and shows that the multimodal system results of FAR and FRR are decreased and accuracy is increased compared to the fingerprint standalone system.

References

1. Ross, A., Nandakumar, K., Jain, A.K.: Handbook of Multibiometrics. Springer- Science + Business Media, LLC (2006)
2. ISO/IEC TR 24722:2007, Information technology biometrics: multimodal and other multibiometric fusion, July 2007
3. Ross A., Govindarajan, R.: Feature level fusion using hand and face biometrics. In: Proceedings of the SPIE Conference Biometric Technology for Human Identification *II*, pp. 196–204, Mar. 2005
4. Singh, S., Gyaourova, A., Bebis, G., Pavlidis, I.: Infrared and visible image fusion for face recognition. In: SPIE Defense and Security Symposium, pp. 585–596 (2004)
5. Heo, J., Kong, S.G., Abidi, B.R., Abidi, M.A.: Fusion of visual and thermal signatures with eyeglass removal for robust face recognition. In: Proceedings of the Joint IEEE Workshop Object Tracking and Classification beyond the Visible Spectrum, June 2004
6. Son, B., Lee, Y.: Biometric authentication system using reduced Joint feature vector of iris and face. In: Lecture Notes in Computer Science vol. 3546, pp. 261–273 (2005)
7. Ross, A., Jain, A.K.: Fusion techniques in multibiometric systems. In: Hammound, R., Abidi, B., Abidi, M. (eds.) Face Biometrics for Personal Identification. Springer, Berlin, Germany (2007)
8. Lanckriet, G.R.G., Ghaoui, L.EI., Bhattacharyya, C., Jordan, M.I.: J. Machine Learning Res. 3, 552 (2002)
9. Verlinde, P., Cholet, G.: Comparing decision fusion paradigms using k-NN based classifiers, decision trees and logistic regression in a multi-modal identity verification application. In: Proceedings of the International Conference Audio and Video-Based Biometric Person Authentication (AVBPA), pp. 188–193, Washington, DC, Mar. 1999
10. Ross, A., Jain, A.K.: Information fusion in biometrics. Pattern Recogn. Lett. **24**(13), 2115–2125 (2003)
11. Jain, A., Hong, L., Kulkarni, Y.: A multimodal biometric system using fingerprint, face and speech. In: Second International Conference on AVBPA, pp. 182–187, Washington, DC, USA (1999)
12. Matsumoto, T., Matsumoto, H., Yamada, K., Hoshino, S.: Impact of artificial gummy fingers on fingerprint systems. In: Proceedings of the SPIE, Optical Security and Counterfeit Deterrence Techniques IV, vol. 4677, pp. 275–289, Jan. 2002
13. Kittler, J., Hatef, M., Duin, R.P.W., Mates, J.: On combining classifiers. IEEE Trans. Pattern Anal. Mach. Intell. **20**(3), 226–239 (1998)
14. Wang, F., Yao, X., Han, J.: Improving iris recognition performance via multi-instance fusion at the score level. Chin. Opt. Lett. **6**(11), 824–826 (2008)
15. Jain, A.K., Nandakumar, K., Ross, A.: Score normalization in multimodal biometric systems. Pattern Recogn. **38**(12), 2270–2285 (2005)
16. Bowyer, K.W., Chang, K.I., Flynn, P.J., Chen, X.: Face recognition using 2-D, 3-D, and infrared: is multimodal better than multisample. Proc. IEEE **94**(11), 2000–2012 (2006)
17. Jain, A.K., Ross, A.: Fingerprint mosaicking. In: Proceedings of the International Conference Acoustic Speech and Signal Processing, vol. 4, pp. 4064–4067 (2002)
18. Yao, Y.-F., Jing, X.-Y., Wong, H.-S.: Face and palmprint feature level fusion for single sample biometrics recognitionion. Neurocomputing **70**, 1582–1586 (2007)
19. Benaliouche, H., Touahria, M.: Comparative study of multimodal biometric recognition by fusion of iris and fingerprint. Sci. World J. **2014**, 113 (2014)
20. Chen, X., Flynn, P.J., Bowyer, K.W.: IR and visible light face recognition. Comput. Vis. Image Underst. **99**(3), 332–358 (2005)
21. Han, J., Bhanu, B.: Gait recognition by combining classifiers based on environmental contexts. In: Lecture Notes in Computer Science, vol. 3546/2005, pp. 113–124 (2005)
22. Ratha, N.K., Connell, J.H., Bolle, R.M.: Image mosaicing for rolled fingerprint construction. In: Proceedings of the International Conference Pattern Recognition, vol. 2, pp. 1651–1653 (1998)

23. Chellappa, R., Wilson, C.L., Sirohey, S.: Human and machine recognition of faces: a survey. Proc. IEEE **83**(5), 705–740 (1995)
24. Metawa, N., Elhoseny, M., Kabir Hassan, M., Hassanien, A.: Loan portfolio optimization using genetic algorithm: a case of credit constraints. In: 12th International Computer Engineering Conference (ICENCO), pp. 59–64. IEEE (2016). doi:10.1109/ICENCO.2016.7856446
25. Metawa, N., Hassan, M.K., Elhoseny, M.: Genetic algorithm based model for optimizing bank lending decisions. Expert Syst. Appl. **80**, 75–82 (2017). ISSN 0957-4174. doi:10.1016/j.eswa.2017.03.021
26. Proenca, H., Alexandre, L.A.: UBIRIS iris image database, Dec. 2006. http://iris.di.ubi.pt
27. Jain, A.K., Ross, A., Prabhakar, S.: An Introduction to Biometric Recognition. IEEE, Biometrics, Vol. 14, No. 1, January 2004
28. Elhoseny, M., Elminir, H., Riad, A., Yuan, X.: A secure data routing schema for WSN using elliptic curve cryptography and homomorphic encryption. J. King Saud Univ.-Comput. Inf. Sci. (2015)
29. Elhoseny, M., Yuan, X., El-Minir, H.K., Riad, A.M.: An energy efficient encryption method for secure dynamic WSN. Secur. Comm. Netw. **9**, 2024–2031 (2016)
30. Poh, N., Bengio, S., Korczak, J.: A multi-sample multi-source model for biometric authentication. In: Proceedings of the IEEE Workshop Neural Networks for, Signal Processing, pp. 375–384 (2002)
31. Uludag, U., Ross, A., Jain, A.K.: Biometric template selection and update: a case study in fingerprints. Pattern Recogn. **37**(7), 1533–1542 (2004)
32. Bromba, M.U.A.: Bioidentification frequently asked questions. http://www.bromba.com/faq/biofaqe.htm
33. Daugman, J.: Recognizing persons by their Iris patterns. In: Jain, A.K., Bolle, R., Pankanti, S. (eds.) Biometrics: Personal Identification in a Networked Society, pp. 103–121. Kluwer, Norwell, MA (1999)
34. Chinese Academy of Sciences: Institute of Automation. http://biometrics.idealtest.org
35. Maio, D., Maltoni, D., Cappelli, R., Wayman, J.L., Jain, A.K.: FVC2000: Fingerprint Verification Competition, Das, R., Signature Recognition. Keesing J. Doc. Identity, **24** (2007)

Suspicious and Violent Activity Detection of Humans Using HOG Features and SVM Classifier in Surveillance Videos

Produte Kumar Roy and Hari Om

Abstract Crimes such as theft, violence against people, damage to property, etc., have become quite common in a society, which a serious concern. The traditional surveillance systems act like post mortem tools in the sense that they can be used for the investigation to detect the person behind the theft, but it is only after the crime has already occurred. In this chapter, we propose a method for automatically detecting the suspicious or violent activities of a person from the surveillance video. We train the SVM classifier with the HOG features extracted from the video frames of two types: frames showing no violent activities and those showing violent activities like kicking, pushing, punching, etc. In the testing phase, the frames from the surveillance video are read and processed in order to classify them as violent or normal frames. If the frames classified as violent frames are detected, an alarm is raised to alert the controller. It can be used to keep track of the time duration for which a person is found loitering at a place being monitored. If the time exceeds a predefined threshold, the alarm is raised to alert about any potential suspicious activity so that it can be checked on time.

1 Introduction

Nowadays, the crime and violence have increased in society in a big way, requiring a person to seek safety and security at each stage of his life. Surveillance camera systems are available to keep an eye on the live happenings in a region of interest. Monitoring such places using cameras are much cheaper than employing human resource who can continuously monitor the live footage for any violent or suspicious action. Even if one manages to hire person(s) to do such task, the efficiency of a person in doing it would be prone to errors as 82% of security observers have multiple

P.K. Roy (✉) · H. Om
Indian Institute of Technology (Indian School of Mines), Dhanbad 826004, India
e-mail: produtekr@gmail.com

H. Om
e-mail: hariom4india@gmail.com

A.E. Hassanien and D.A. Oliva (eds.), *Advances in Soft Computing and Machine Learning in Image Processing*, Studies in Computational Intelligence 730,
https://doi.org/10.1007/978-3-319-63754-9_13

simultaneous duties such as checking-in visitors, attending calls, etc. Thus, there is a need for an intelligent autonomous surveillance system that can detect the suspicious and violent activities of humans in surveillance videos to prevent damage on time. In this chapter, an appropriate algorithm is proposed to meet the above need. Though such systems have been developed, yet it is challenging because there are many issues that need be addressed. One of the issues to be addressed is how intelligent is the system to understand the human behaviour in a region being monitored. Our work focuses on better understanding of human-movement predominantly suspicious behaviour, in order to recognize an intruder using the human movement-detection algorithm. Also, the actions like kicking, pushing, punching, etc., are treated as violent/abnormal activities and an alarm is raised whenever such acts of violence are in process so that no further damage can take place. We analyze, detect and track the motion of a person in a given surveillance video and based upon it we categorize whether there are some kinds of acts of violence, viz. kicking, pushing, punching or a potential suspicious activity like loitering or looking around with malicious intent. To detect acts of violence, a support vector machine (SVM) classifier is pretrained with several examples of acts of kicking, pushing and punching. When the classifier is trained, every fifth frame of the video is provided to the classifier to detect if there is any violence of the above-mentioned categories. If the classifier returns positive result classifying the frame as one in which some act of violence is found, then an alarm is raised to alert the observer. We can use it to compute the time for which a particular person is present in a particular frame loitering there. If the time thus calculated for the person exceeds a pre-determined threshold, then an alarm is raised indicating a potential suspicious activity. Thus, the real-time detection of suspicious human movement can be used to overcome the shortcomings of the existing systems and to provide a proper automated surveillance system.

2 Methodology

The basic steps of a typical video surveillance system are as follows:

- Building a background model
- Foreground pixel extraction
- Object segmentation
- Object classification
- Object tracking
- Action recognition

The very first step is building a background model whose sole purpose is to represent the environment in the absence of foreground objects. In literature, there are many methods for building the background models [2] that differ in the ability to get updated to reflect the alterations in the environment. These methods are broadly classified as adaptive and non-adaptive techniques. The foreground pixel extraction step separates the pixels of an image that are not part of the background model.

This difference forms the basis for further analysis in subsequent steps. Some of the methods for foreground pixel extraction are background subtraction, optical flow, temporal differencing, etc. [4–6]. Object segmentation step groups the similar foreground pixels into homogeneous regions using some similarity metrics. Some of the similarity metrics for object segmentation are location based, colour based, proximity based or hybrid of some or all of the above [7]. The fourth step is object classification. There can be several types of moving objects in the area being monitored and a different tracking method may be used for each type. For example, the moving objects may be humans, vehicles or objects of interest of an investigation application. Thus, the object classification can be referred to as standard pattern recognition task. Two main categories of moving object classification are motion based classification [10] and shape based classification [7]. Different shape information of motion regions like boxes, points, blobs and silhouettes are used for the above classification task. The next step is object tracking that can be defined as a method of locating the object(s) in motion over a time duration, primarily matching or associating the target objects in consecutive frames. Low video frame rate as compared to the rate of object motion poses a great challenge in object tracking. The situation becomes worse when the background has objects similar in features to that of the foreground objects, or there are objects that pose severe occlusion to the object being tracked. The main categories of object tracking methods include region based, contour based, feature based, model based and their hybrid forms. The final step is action recognition, which is roughly a process of identifying the actions in order to understand what is occurring in the environment [10]. In an automatic video surveillance system, there is a great practical importance of suspicious human behaviour [3]. The recognition is achieved through a combination of image processing and artificial intelligence approaches. The former is used to extract low level image features while the later provides expert decisions. There exist several works on low level image features like object detection, recognition and tracking, but few provide reliable classification and analysis of human activities from the video frame sequences. The main steps in our work can be listed as follows:

1. To extract frames from training surveillance video
2. To prepare ground truth
3. To extract HOG (histogram of oriented gradients) features from frames
4. To generate training feature matrix
5. To prepare class label matrix for training feature matrix
6. To train SVM (support vector machine) classifier
7. To detect motion in test (input) video
8. To track and predict the movement of each person in video
9. To send every fifth frame of input video to SVM classifier
10. To raise alarm if test frame is classified as frame with violent activity

We now elaborate these steps.

2.1 To Extract Frames from Training Surveillance Video

We extract the frames of the training surveillance video using *VideoReader* function in MatLab and store them in .jpg format.

2.2 To Prepare Ground Truth

The frames so extracted from the video are analyzed and categorized into frames containing normal activity or any form of violent activity, viz. kicking, pushing, punching, etc.

2.3 To Extract HOG (Histogram of Oriented Gradients) Features from Frames

The histogram of oriented gradient is a feature descriptor used to detect objects and they are mainly suited for detection of humans in images. The HOG descriptor technique counts the occurrence of gradient orientation in localized portions of the image—detection window, or the region of interest. The HOG descriptor is implemented as follows (Figs. 1, 2 and 3):

- The image is divided into small connected regions, referred to as cells. For each cell, a histogram of gradient direction for the pixels is calculated.
- Each cell is discretized into angular bins in accordance to the gradient orientations. There is a contribution of the weighted gradient to the corresponding angular bins for each cell.
- Blocks are considered as spatial regions of groups of adjacent cells. Formation of the blocks is the basis for grouping and normalizing the histograms.
- Block histogram is represented by the normalized group of histograms. The descriptor is in turn represented by the set of these block histograms.

2.4 To Generate Training Feature Matrix

The HOG features are generated for the normal frames as well as the frames showing violent activity to obtain the feature matrix. This feature matrix is used for training the SVM Classifier. The HOG features generated for each frame is of size 1×189036.

Fig. 1 A frame from surveillance video showing violent activity

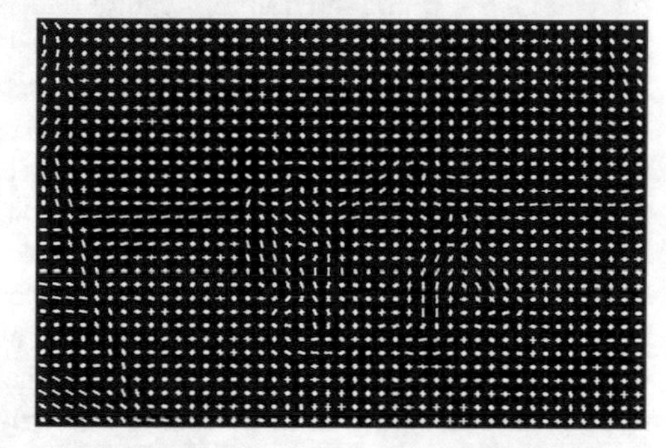

Fig. 2 HOG features visualization of above frame. cell size is 8×8

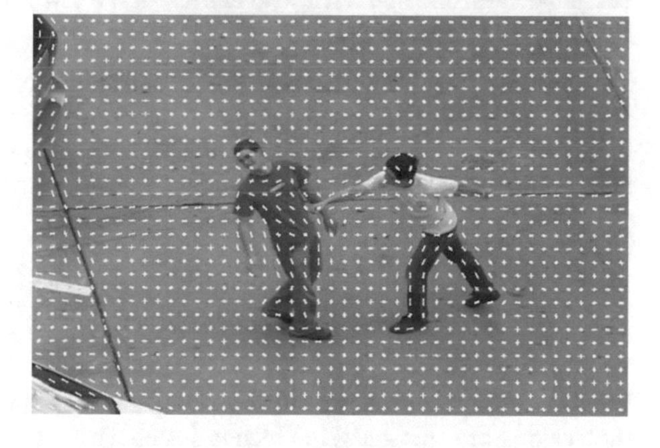

Fig. 3 HOG visualization plotted on original frame

2.5 To Prepare Class Label Matrix for Training Feature Matrix

A class label matrix is created for each kind of frame. The normal frames are labelled as **1** and the frames showing a violent activity are labelled as **2**. The class label matrix is used for training the SVM classifier.

2.6 To Train SVM (Support Vector Machine) Classifier

The SVM classifier was developed by Vapnik (*A training algorithm for optimal margin classifier*). Using the training feature matrix and the class label matrix, we train the SVM classifier that predicts the class label of an input test video frame. Classification is one of the sub-parts of machine learning algorithms. The support vector machine is a supervised learning model that is used for data classification and regression analysis. Consider samples from two classes that are mapped in space. The SVM tries to find a linear decision surface (hyperplane) that can separate the samples of these classes in such a way that the border line samples of each class have the largest distance or gap or margin between them. These border line samples of each class are called the support vectors. In case such linear decision surface does not exist, the data is mapped into a much higher dimensional feature space, where the separating hyperplane can be found. The feature space may be constructed using the mathematical projection, one of them is *kernel trick*. The SVM is important mainly due to the following (Fig. 4):

- It is robust to large number of variables and relatively smaller number of samples.
- It can learn simple to highly complex classification models.
- It can make use of sophisticated principles of mathematics to avoid over fitting.
- It produces superior empirical results.

2.7 To Detect Motion in Test (Input) Video

We input the test video to the main module that reads the video frame by frame and the motion of the person(s) in video is detected by background subtraction using the adaptive Gaussian mixture model. The binary image so obtained is generally noisy. The noise is removed using the morphological operations like erosion and dilation. Finally, the blob analysis detects the groups of connected pixels, which are likely to correspond to moving objects. Figures 5, 6, 7 and 8 show background frame, current frame, background subtracted frame (noisy) and background subtracted frame after noise removal using morphological operations, respectively.

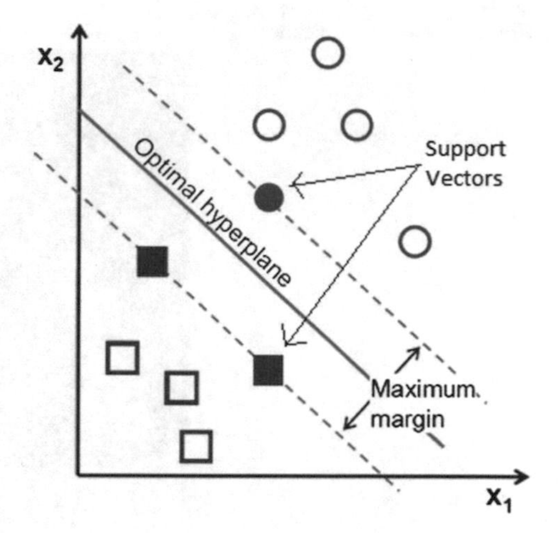

Fig. 4 SVM classifier for two class labels

Background Subtraction: It is a process used to extract the foreground of an image for further analysis. Simple background subtraction techniques such as frame differencing, mean and median filtering are fast, but their global and constant thresholds make them inefficient for the challenging real-world problems. The adaptive background mixture model handles the challenging situations such as bimodal backgrounds, long-term scene changes and repetitive motions in clutter. It can further be improved by incorporating the temporal information, or using some regional background subtraction in conjunction with it.

Morphological Operations: The dilation operator takes two pieces of data as input. The first is the image which is to be dilated and the second is a small set of coordinate points, known as a structuring element. The basic effect of the dilation operator on a binary image is to gradually enlarge the boundaries of regions of foreground pixels (generally white pixels). Thus, the areas of foreground pixels grow in size while the holes within those regions become smaller. In order to compute the dilation of a binary image by a structuring element, we take each of the background pixels in the input image one by one and superimpose the structuring element on the binary image in such a way that the origin of the structuring element coincides with the input pixel position. If at least one pixel in the structuring element coincides with a foreground pixel in the image, then the input pixel is set to the foreground value. If all the corresponding pixels in the image are background, the input pixel is assumed as the background value. The erosion operator takes two pieces of data as input. The first is the image which is to be eroded and the second is a small set of coordinate points, known as a structuring element. The basic effect of the erosion operator on a binary image is to erode the boundaries of regions of foreground pixels (generally white pixels). Thus, the areas of foreground pixels shrink in size and the holes within those areas become larger.

Fig. 5 Background frame

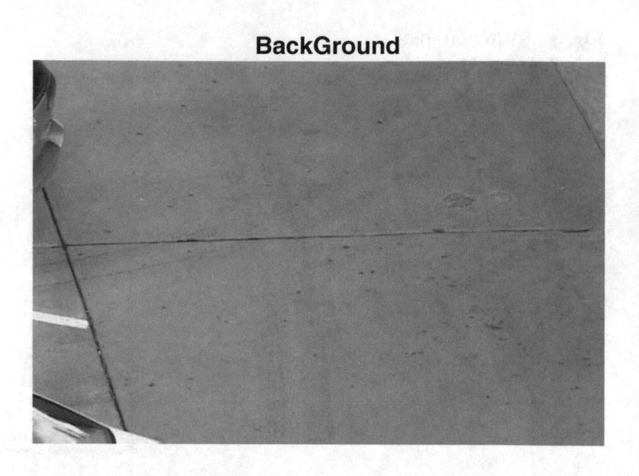

Blob Analysis: This process is used to isolate the blob(s) (objects) in a binary image, calculate its various features like area, centroid, bounding box coordinates, etc., and then it uses them to classify the blobs as per the needs. A blob consists of a group of connected pixels. To find out whether the pixels are connected, we observe their connectivity that tells which pixels are neighbors and which are not. The two most common types of connectivity are: 4-connectivity and 8-connectivity. The 8-connectivity is more accurate than the 4-connectivity, but the 4-connectivity is often applied since it requires fewer computations and hence can process an image faster. There are a number of algorithms for finding the blobs and these algorithms are often called as connected component analysis or connected component labelling. Example of such an algorithm is the Grass-fire algorithm. Area of a blob is the number of pixels covered by the blob and the area feature is used to remove blobs that are too big or too small from the image. Bounding box of a blob is the minimum rectangle which contains the blob. It is defined by going through all pixels for a blob and finding the four pixels with the minimum x-value, maximum x-value, minimum y-value and maximum y-value, respectively. A bounding box can be used as a ROI (Region Of Interest). The bounding box ratio of a blob is defined as the height of the bounding box divided by its width. Compactness of a blob is defined as the ratio of the blob's area to the area of the bounding box. This is used to distinguish compact blobs from non-compact ones. In our work, the blob analysis is used in the process of detection of the moving persons in a surveillance video.

2.8 *Track and Predict the Movement of Each Person in Video*

We use the Kalman filter with constant velocity to track and predict a person in a video. When a person goes out of the frame, its track is deleted from memory. The short lived tracks are ignored and the reliable tracks are dealt with. The Kalman filter is used wherever we have uncertain information about some dynamic system and we

Fig. 6 Current frame

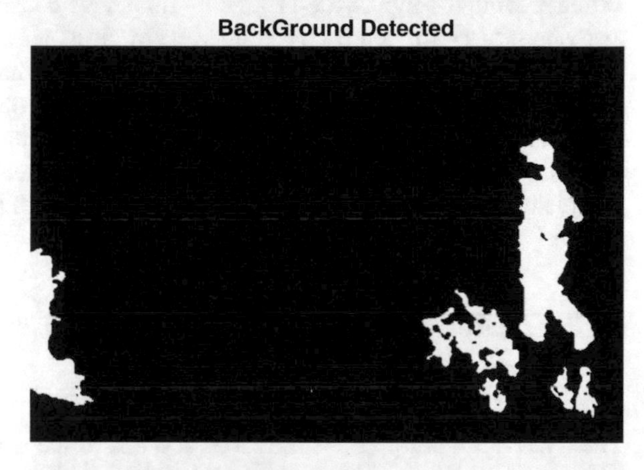

Fig. 7 After background subtraction

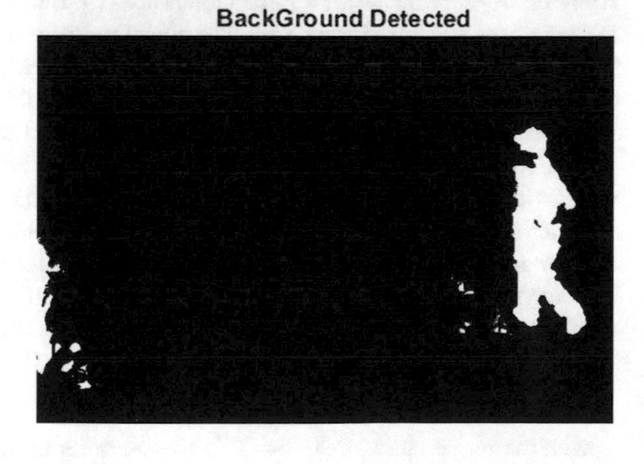

Fig. 8 After removing noise using morphological operations

can make an intelligent guess about what the system is going to do next. Even if there is interference with the clean motion we guessed about, the Kalman filter is good at figuring out what actually happened. It can take advantage of the correlation between phenomena that one would not have generally thought of exploiting. The Kalman filters perform well for dynamic systems. They are light on memory as they do not need to keep any history other than the previous state, and hence are very fast, making them well suited for real time problems and embedded systems.

2.9 To Send Every Fifth Frame of Input Video to SVM Classifier

While reading the input video frame by frame, we consider every fifth frame of the test video for prediction by the trained SVM classifier model so that the frame can be categorized as either normal frame or one which has violent activity in it and accordingly an alarm can be raised to alert the human observer at the other end. We send every fifth frame for prediction instead of all the frames of the test video because checking all the frames would make the detection system lag as prediction by SVM model does take some time for outputting the predicted label for a test frame.

2.10 To Raise Alarm if Test Frame is Classified as Frame with Violent Activity

The SVM classifier predicts the label of a test frame and tells whether it is normal frame or the one which shows some violent activity. If it is of the second type, then the system raises an alarm in the form of a popup message box and some sound to alert the human observer sitting in the control room. There is another alert mechanism for this system that keeps a track of how long a particular person appears in the region being monitored. If the time limit for a person exceeds a certain threshold value (which is pre-determined in our case and can be set to any value based on the application and region being monitored), then the system treats the person loitering and alerts the human observer in the control room about the potential suspicious behaviour of that person by raising an alarm in the form of a popup message box and some sound. The flowchart of the proposed system is shown in Fig. 9. There are primarily two types of learning: supervised and unsupervised learning.

Supervised Learning

- In this case, the patterns are discovered in the data that relate the data attributes with the target class attributes. It is like learning in the presence of an expert or teacher.
- Training data is labelled with a class or value, which means that the expert/teacher provides with correct class of the training data and the system learns from it.

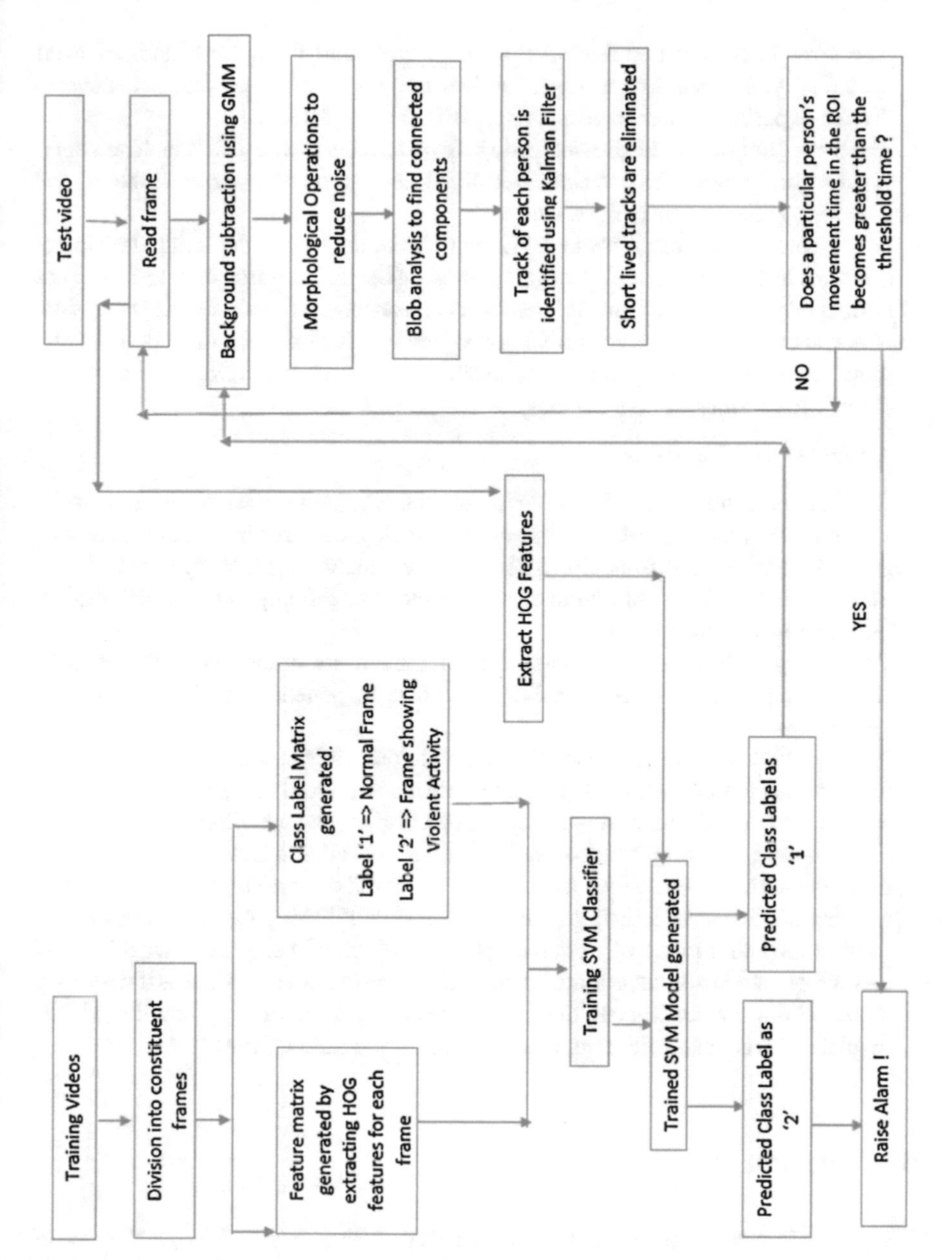

Fig. 9 Flow chart of proposed detection system

Any error is reported as feedback to the system and the system updates itself accordingly. The training is repeated over and over until the system achieves a desired experience level to correctly classify the test data.

- The main goal here is to predict a class or value label of the test data. The supervised learning algorithms include decision trees, support vector machine, neural networks, Bayesian classifiers, etc.
- In our work, the system is trained with samples of the frames depicting the violent activities and then a model is developed that classifies a frame of a test video as normal or violent one. Thus, it is supervised learning in the sense that the system is first shown which one is normal and which one is violent frame. As more and more variety of training frames are available, the system develops the ability to classify the frames more accurately.

Unsupervised Learning

- It is like learning in the absence of a teacher. The system learns on its own by exploring the training data to find some intrinsic structures in it. Thus, it is self-guiding learning algorithm with internal self evaluation against some criteria.
- There is no knowledge of output class or value. The training data is unlabelled or the value is unknown.
- The main goal here is to determine data patterns or grouping. Examples of algorithms using unsupervised learning are: k-means, genetic algorithm, clustering approaches.
- If unsupervised learning is used in our suspicious violent activity detection from the surveillance videos, then the system forms two clusters by itself: one of the normal frames and other one of frames depicting violent activities by observing a large number of training videos and based on the suitable features to distinguish in the clusters formed. When the test frame is input for classification, it may give a probabilistic output indicating the probability of it belonging to each of the two clusters formed, instead of a crisp output in the form of a single cluster it belongs to. Using unsupervised learning, it becomes more efficient over time as it witnesses more training videos. Its one disadvantage is that it needs a lot of training videos to give a correct classification; thus, increasing the training time.

3 Experimental Results

The proposed system can detect the violent activities fairly well and the motion detection is also quite accurate. The person moving is being surrounded by a rectangle along with a tracking number displayed on the top of the bounding box (rectangle). The corresponding background subtracted video is shown along with the original one. We have used a total of 835 video frames from the UT-Interaction dataset, ICPR contest on Semantic Description of Human Activities (SDHA), for training the SVM classifier. Table 1 shows the distribution of frames. The HOG feature vector gener-

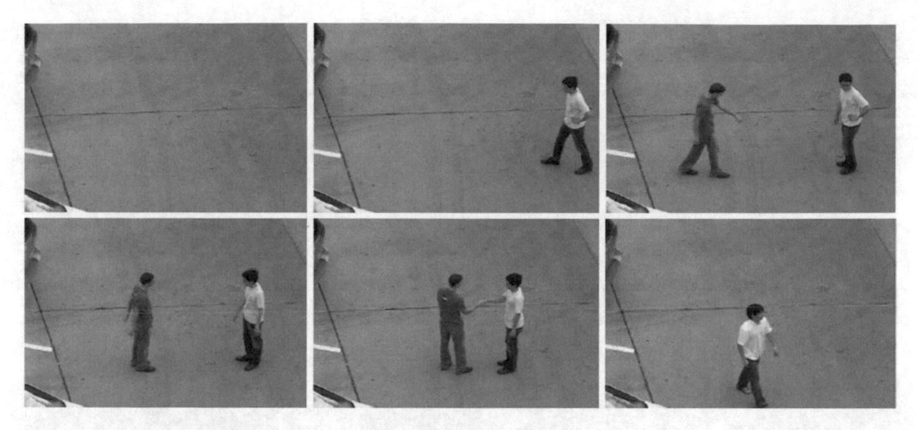

Fig. 10 Normal frames (Frames showing no violent activity)

ated for each frame is of size $\mathbf{1 \times 189036}$ and thus the total training feature matrix is of size $\mathbf{835 \times 189036}$ as there are 835 frames.

For testing purpose, we have input a video to the system and if there is any frame present in the video showing any of the violent activities, the system pauses there to raise an alarm in order to alert the observer about it. The observer decides what to do next whether to take action responding to the alarm raised or to let the surveillance proceed treating the alarm just raised as a false alarm. We have run the system on several input videos and the result so obtained has been recorded in Table 2. Table 3 provides the confusion matrix and the final results are shown in Table 4. In case a particular person has been tracked in a video for a duration greater than a

Table 1 Training set distribution

Training set		
Description	Number of frames	Class label
Normal frames (no violent activity depicted)	576	1
Frames depicting violent activities viz. kicking, pushing or punching	259	2

Table 2 Classification results on testing set

Testing set			
Description	Number of frames	No. of frames correctly classified	No. of frames incorrectly classified
Normal frames (no violent activity)	540	479	61
Frames showing violent activities	454	414	40

Table 3 Confusion matrix of experimental results

	Class 1	Class 2
Class 1	479	61
Class 2	40	414

Table 4 Final experimental results

Precision	Recall	F1 score	Accuracy
0.92	0.89	0.90	0.898

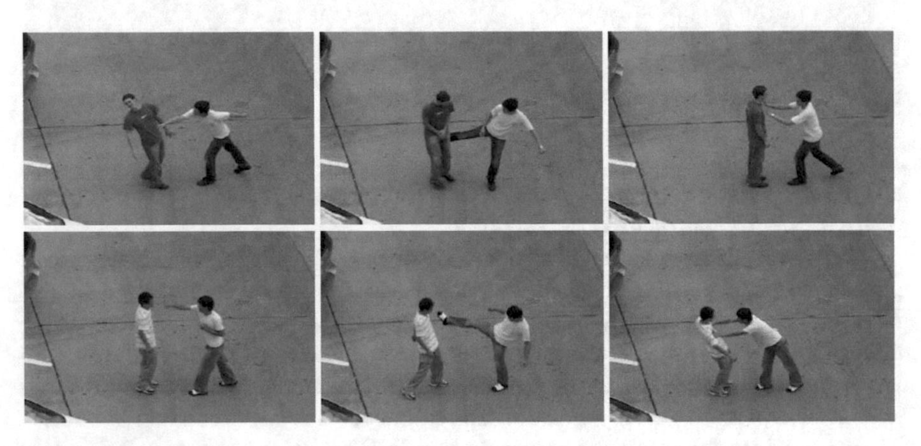

Fig. 11 Violent frames (Frames showing violent activities like kicking, pushing or punching)

pre-determined threshold, an alarm is raised saying the person is loitering and that there may be a potential suspicious activity. This part has been done fairly accurate and the decision lies in the hands of the observer in control room to determine if the loitering is normal or may lead to any suspicious activity (Figs. 10 and 11).

4 Future Work

Although our system performs well and achieves almost 89% accuracy in detecting suspicious and violent activities of humans in surveillance videos; we however like to outline some of the future work that we would undertake in order to improve and make the system more robust.

We would like to use more accurate optical flow features along with background subtraction for motion segmentation. The optical flow features are very common for assessing motion from a set of images, which is an approximation of the two-dimensional flow field from the image intensities. It is the velocity field, which wraps one image into another very similar image.

We have used surveillance videos recorded from the single camera for the system. We would configure the system to provide more accurate and robust results by using videos of the same region of interest captured by multiple cameras kept at different angles. For this, we need to have a synchronization and calibration module which would help these cameras work in coordination to produce more accurate detection of suspicious and violent human activities.

The feature matrix that we have generated to train our SVM classifier is quite large with its dimension being 835×189036 and the memory occupied by it on disc is around 650 MB. So, we would like to use a feature reduction algorithm to minimize the number of relevant features so as to achieve dual advantage viz. smaller feature matrix with no compromise on the training quality. With smaller feature matrix, the system can run much faster with improved accuracy. Apart from HOG features, we would also like to use other features such as shape descriptors so that the human activity is detected more accurately and the number of false alarms is greatly reduced. We would like to train a cascade of different types of detectors/classifiers rather than just using the SVM classifier so that the prediction of the category of a frame is a lot more accurate than in the current system.

Currently, we have just kept track on the amount of time a person is found moving continuously only in the region being monitored by the surveillance camera. If the time exceeds a pre-determined threshold, then the system flags the person as loitering, indicating some potential suspicious act may be performed by him. To add to it, we would like to introduce a mechanism that even if the person goes out of frame for a little while and again returns, he would not be regarded as a new person and the time duration calculation for him would continue from when he left the field of view of the camera. This can be achieved by configuring the system to pause the timer for a particular person for some time even after he has gone out of the frame. The timer resumes if the person returns within the field of view of the camera within the stipulated time duration for which the timer has been paused for a particular person. The duration for which the timer has to be paused would be preset and pre-determined after observing the patterns of several surveillance videos. If the person is out of frame for a longer duration than what was preset, the timer eventually resets for that person. This way, the field of view of the surveillance camera would not be a bar in detection process to some extent.

The likelihood of tracking errors can be reduced by using a more complex motion model such as constant acceleration, or using multiple Kalman filters for every object. In present implementation, it has been assumed that all people move in a straight line with constant velocity. We would like to make our trained classifier model more robust by training it with more varieties of violent activities so that the classification error is reduced.

Besides the existing suspicious and violent activity detection system, an identity detection system (in the form of face cataloger) can also be developed by using the PTZ (pan-tilt-zoom) camera. It can be used to acquire high-resolution images of human faces involved in violent or suspicious activity as detected by the system. There would be an underlying face detection algorithm incorporated in the system. The moment any suspicious or violent activity is detected, the face recognition module will be triggered and the PTZ camera would accordingly be controlled to acquire a zoomed-in image of the human faces involved in the abnormal activity. This data so acquired can later be used for investigation purposes. The functionality is like once the (frontal) face image is detected, the camera gets centred on the face and is zoomed-in. The pan and tilt of the camera gets adjusted based on the relative displacement of the centre of the face with respect to the centre of the image. With the help of intrinsic calibration parameters of the camera and the current zoom level, the relative image coordinate displacements are translated into desired or relative pan and tilt angles.

5 Conclusion

The smart surveillance systems can transform the traditional video surveillance systems from mere data acquisition tool to the information and intelligence acquisition systems. The situational information captured along with real time acquisition of the surveillance data can provide the system an important mechanism to maintain the situational awareness and imparting higher security. In our method, we have used the HOG (histogram of oriented gradients) features of the frames to train the SVM classifier to classify the input test frames to either normal frame or frame showing some act of human violence. If it is of the second type, then an alarm is raised by the system to alert the observer sitting in the control room about the violent activity. Our system can also perform the motion detection of a human using the mixture of Gaussian model, followed by the morphological operations to reduce the noise from the foreground detected binary image, and then blob analysis to find the connected components. The system keeps track of the time duration for which a particular person is found moving in a region being monitored. The system alerts the observer in the control room indicating that suspicious activity may be carried out by the person as he has been found loitering by the system for more than a preset time duration. We have made it to work with real time video acquisition and making it more robust by using more number of relevant features apart from the HOG features to train the SVM classifier and to minimize the number of false alarms. We will try to upgrade the system to work with inputs from the multiple cameras placed at different angles of the same region of interest to make the prediction more accurate. The feature vectors generated need to be optimized to reduce their dimension in order to reduce lagging in the operation.

References

1. Spasic, N.: Anomaly Detection and Prediction of Human Actions in a Video Surveillance Environment, Master Thesis, Computer Science Dept., Cape Town University, South Africa, December (2007)

2. Haque, M., Murshed, M.: Robust background subtraction based on perceptual mixture of gaussians with dynamic adaptation speed. In: IEEE International Conference on Multimedia and Expo Workshops (ICMEW 2012), Melbourne, Australia, pp. 396–401, July 2012
3. Regazzoni, C., Cavallaro, A., Wu, Y., Konrad, J., Hampapur, A.: Video analytics for surveillance: theory and practice. IEEE Signal Process. Mag. **27**(5), 16–17 (2010)
4. Reddy, V., Sanderson, C., Sanin, A., Lovell, B.C.: MRF-based background initialisation for improved foreground detection in cluttered surveillance videos. In: 10th Asian Conference on Computer Vision (ACCV 2010), Queenstown, New Zealand, Vol. Part III, pp. 547–559, November 2010
5. Liu, L., Tao, W., Liu, J., Tian, J.: A variational model and graph cuts optimization for interactive foreground extraction. Signal Process. J. **91**(5) (2011)
6. Suau, X., Casas, J.R., Ruiz-Hidalgo, J.: Multi-resolution illumination compensation for foreground extraction. In: 16th IEEE International Conference on Image Processing (ICIP 2009), pp. 3189–3192, November 2009
7. Loy, C.C.: Activity understanding and unusual event detection in surveillance videos, Ph.D. dissertation, Queen Mary University of London (2010)
8. Cavallaro, A., Steiger, O., Ebrahimi, T.: Tracking video objects in cluttered background. IEEE Trans. Circuits Syst. Video Technol. **15**(4), 575–584 (2005)
9. Karasulu, B.: Review and evaluation of wellknown methods for moving object detection and tracking in videos. J. Aeronaut. Space Technol. **4**(4), 11–22 (2010)
10. Cilla, R., Patricio, M.A., Berlanga, A., Molina, J.M.: Human action recognition with sparse classification and multiple-view learning. Expert Syst. J. (2013). Wiley Publishing Ltd
11. Javed, O., Shah, M.: Tracking and object classification for automated surveillance. In: Proceedings of the 7th European Conference on Computer Vision, Part-IV, pp. 343–357 (2002)
12. Li, T., Chang, H., Wang, M., Ni, B., Hong, R., Yan, S.: Crowded scene analysis: a survey. IEEE Trans. Circuits Syst. Video Technol. **25**(3), 367–386 (2015)
13. Patino, L., Ferryman, J., Beleznai, C.: Abnormal behaviour detection on queue analysis from stereo cameras. In: Advanced Video and Signal Based Surveillance (AVSS), 12th IEEE International Conference, pp. 1–6, August 2015
14. Chan, A.B., Morrow, M., Vasconcelos, N.: Analysis of crowded scenes using holistic properties. In: Performance Evaluation of Tracking and Surveillance workshop at CVPR (2009)
15. Lee, L., Romano, R., Stein, G.: Introduction to the special section on video surveillance. IEEE Trans. Pattern Anal. Mach. Intell. **8**, 740–745 (2000)
16. Zhan, B., Monekosso, D.N., Remagnino, P., Velastin, S.A., Xu, L.Q.: Crowd analysis: a survey. Mach. Vis. Appl. **19**(5–6), 345–357 (2008)
17. Baumann, A., Boltz, M., Ebling, J., Koenig, M., Loos, H., Merkel, M., Niem, W., Warzelhan, J., Yu, J.: A review and comparison of measures for automatic video surveillance systems. EURASIP J. Image Video Process., 1–30 (2008)
18. Cai, Y., de Freitas, N., Little, J.J.: Robust visual tracking for multiple targets. In: European Conference on Computer Vision, LNCS, Vol. 3954, pp. 107–118 (2006)
19. Chang, T., Gong, S., Ong, E.: Tracking multiple people under occlusion using multiple cameras. In: British Machine Vision Conference, pp. 566–575 (2000)
20. Donalek, C.: Supervised and Unsupervised learning (2011)
21. Benezeth, Y., Jodoin, P.M., Saligrama, V.: Abnormality detection using low-level co-occurring events. Pattern Recogn. Lett. **32**(3), 423–431 (2011)
22. Isupova, O., Kuzin, D., Mihaylova, L.: Abnormal behaviour detection in video using topic modeling. In: USES Conference Proceedings. The University of Sheffield, June 2015
23. Cosar, S., Donatiello, G., Bogorny, V., Garate, C., Alvares, L.O., Bremond, F.: Towards abnormal trajectory and event detection in video surveillance
24. Morphology. http://homepages.inf.ed.ac.uk/rbf/HIPR2/morops.htm
25. How a Kalman filter works, in pictures. http://www.bzarg.com/p/how-a-kalman-filter-works-in-pictures/
26. Introduction to support vector machines. http://docs.opencv.org/2.4/doc/tutorials/ml/introduction_to_svm/introduction_to_svm.html

27. Background subtraction for detection of moving objects. https://computation.llnl.gov/casc/sapphire/background/background.html
28. Machine Learning, Part I: Supervised and Unsupervised Learning. http://www.aihorizon.com/essays/generalai/supervised_unsupervised_machine_learning.htm
29. BLOB Analysis (Introduction to Video and Image Processing) Part 1. http://what-when-how.com/introduction-to-video-and-image-processing/blob-analysis-introduction-to-video-and-image-processing-part-1/

Hybrid Rough Neural Network Model for Signature Recognition

Mohamed Elhoseny, Amir Nabil, Aboul Ella Hassanien and Diego Oliva

Abstract This chapter introduces an offline signature recognition technique using rough neural network and rough set. Rough neural network tries to find better recognition performance to classify the input offline signature images. Rough sets have provided an array of tools which turned out to be especially adequate for conceptualization, organization, classification, and analysis of various types of data, when dealing with inexact, uncertain, or vague knowledge. Also, rough sets discover hidden pattern and regularities in application. This new hybrid technique achieves good results, since the short rough neural network algorithm is neglected by the grid features technique, and then the advantages of both techniques are integrated.

Keywords Offline signature · Recognition · Neural network

1 Introduction

Human identification based on biometrics is becoming popular because of ease-of-use and reliability. The use of biometric technologies for human identity verification is necessary for many routine activities such as in government, legal, and commercial transactions, boarding an aircraft, crossing international borders and entering a

M. Elhoseny (✉) · A. Nabil
Faculty of Computers and Information, Mansoura University, Mansoura, Egypt
e-mail: mohamed_elhoseny@mans.edu.eg
URL: http://www.egyptscience.net

A.E. Hassanien
Faculty of Computers and Information, Information Technology Department, Cairo University, Giza, Egypt

M. Elhoseny · A.E. Hassanien · D. Oliva
Scientific Research Group in Egypt (SRGE), Cairo, Egypt

D. Oliva
Departamento de Ciencias Computacionales, Tecnológico de Monterrey, Campus Guadalajara Av. General Ramón Corona, 2514 Jal, Zapopan, Mexico

A.E. Hassanien and D.A. Oliva (eds.), *Advances in Soft Computing and Machine Learning in Image Processing*, Studies in Computational Intelligence 730,
https://doi.org/10.1007/978-3-319-63754-9_14

secure physical location. Biometric systems are based on physiological or behavioral features that are difficult for another individual to reproduce, thereby reducing the possibility of forgery.

A handwritten signature is considered as a behavioral characteristic based biometric trait in the field of security and the prevention of fraud. Handwritten signature is already the most widely accepted biometric for identity verification in our society for years. The long history of trust of signature verification means that people are very willing to accept a signature-based biometric authentication system. It aims to enable computers to verify the identity of a person without human intervention.

Signature identification can be implemented using two approaches: The first is online, where the image is captured directly as handwriting trajectory. The second is offline [14, 15], in which we use a digitizer in order to acquire a digital image. This chapter implements the offline approach that analyzes the static picture of the signature where it is the most commonly used.

Various Handwritten signature recognition algorithms have been proposed to achieve better results. These algorithms include template artificial neural networks, Neurofuzzy, Wavelet Neural, Dynamic time Warping, hidden Markov models, support vector machines (SVM), and Bayesian networks.

Offline Signature Recognition plays an important rule in today's world. It refers to the branch of computer science that involves authenticating the identity of a person by converting signature images into a form that the computer can manipulate and test. Signature identification system can contribute tremendously to the advancement of the automation process of human verification and can reduce the probability of signature forgery.

Over the past decades a considerable amount of time and effort has been expended researching and developing systems of Offline Signature Recognition, but there are still shortcomings in the techniques used to develop these systems. One of the reasons for slow advancements in the offline signature verification systems is that the signatures of a particular person are not exactly the same.

During the application of the recognition system we may require that the signatures should be made carefully but there are always some differences we must deal with. This requires that the identification system should be flexible and allow certain variations within the set of the signatures put down by one person. Much more difficult, and hence more challenging to researchers, is the ability to recognize handwritten characters. The complexity of the problem is greatly increased by the noise problem and by the almost infinite variability of handwriting as a result of the mood of the writer and the nature of the writing.

1.1 Research Objective

Although, handwritten signature recognition has made tremendous achievements in the area of human authentication, it still suffers from different challenges such as the signatures of a particular person are not exactly the same. During the application of the recognition system we may require that the signatures should be made carefully but there are always some differences we must deal with. This requires that the identification system should be flexible and allow certain variations within the set of the signatures put down by one person.

The type of error we want to reduce is the rejection of the genuine signatures. On the other hand, in order to reduce misclassification and improve forgery resistance we must require that certain important features should be exactly recurrent and we must strictly demand their presence. The errors we are trying to minimize in this case are: acceptance of a fake signature and classifying one person's signature as belonging to another one.

Another difficulty, and hence another challenge to researchers, is the ability to automatically recognize handwritten signatures. The complexity of the problem is greatly increased by the noise problem and by the almost infinite variability of handwriting as a result of the mood of the writer and the nature of the writing. These characteristics have made the progress of handwritten signatures more complex and difficult than other biometric verification systems.

Although many efforts and techniques for treating handwritten signatures have been done, these techniques still faceintegrated into the the problems of Lack of semantics, long training time, proneness to overfitting, and large feature vector. These problems result in that the signature recognition systems have low performance level. Hence, an evolutionary rough neuron is needed. In this chapter, the rough neural network and rough sets have been used. Also, a modified rough neural network based on grid features using signature segmentation is introduced.

1.2 Research Goal

According to the previous work, some researchers have reported success using rough neural network techniques for classification purposes. The rough neural networks (RNNs) are the neural networks based on rough set and one kind of hot research in the artificial intelligence in recent years, which synthesize the advantage of rough set to process uncertainly question: attributes reduce by none information losing then extract rule, and the neural networks have the strongly fault tolerance, self organization, massively parallel processing, and self-adapted. So that RNNs can process the massively and uncertainly information, which is widespread applied in our life.

This chapter introduces offline signature recognition using rough neural network. Rough neural network tries to find better recognition performance to classify the input offline signature images. Since offline handwritten signature suffers from confusion and inconsistency, a knowledge representation system that represents the data variation, in addition to an innovative classifier is used. Rough sets have provided an array of tools which turned out to be especially adequate for conceptualization, organization, classification and analysis of various types of data, when dealing with inexact, uncertain, or vague knowledge. Also, rough sets discover hidden pattern and regularities in application. Thus, the main issue tackled in this paper is auto-adaptation occurred in the model of rough neural networks.

Rough sets and rough neuron are integrated in the structure of neural network model. Rough sets are used in the preprocessing step to optimum the input features for the network. Rough sets are used for discovering the suitable neural networks structure and removing noise appeared in our picture. Also, by setting the neuron attribute in the form on knowledge representation table, rough set can determine the need to reduce the network structure by deleting one or more neurons. Using this pruning technique, network with smaller number of neurons and a higher accuracy rate is obtained. Finally, classification is immediately implemented using rough neural network.

Although rough neuron achieves good results, it still suffers from the dependability of the classification process on irrelevant and large set of attributes that are not needed in decision making, and thus cause long training time. Thus, a sampling technique is needed in the preprocessing stage.

2 Related Work

A handwritten signature is considered as a behavioral characteristic based biometric trait in the field of security and the prevention of fraud. Signature verification presents three likely advantages over other biometric techniques from the point of view of adoption in the market place. First, It is the mean accepted method to declare someone's identity in our society for years. The long history of trust of signature verification means that people are very willing to accept a signature-based biometric authentication system [1].

Second, most of the new generation of portable computers and personal digital assistants (PDAs) use handwriting as the main input channel. Third, a signature may be changed by the user, similarly to a password, while it is not possible to change fingerprints, iris, or retina patterns.

The image of the signature is a special type of object when treated as the subject of the recognition process. One of the problems which are likely to arise is that the signatures of a particular person are not exactly the same. Of course, during the application of the recognition system we may require that the signatures should be made carefully but there are always some differences we must deal with.

This requires that the identification system should be flexible and allow certain variations within the set of the signatures put down by one person. The type of error we want to reduce at this moment is the rejection of the genuine signatures. On the other hand, in order to reduce misclassification and improve forgery resistance we must require that certain important features should be exactly recurrent and we must strictly demand their presence. The errors we are trying to minimize in this case are: acceptance of a fake signature and classifying one person's signature as belonging to another one.

Automatic signature verification systems, like all other biometric verification systems, involve two processing modes: training and testing. In the training mode, the user provides signature samples that are used to construct a model or prototype representing some distinctive characteristics of his signature. In the testing mode, the user provides a new signature, along with the alleged identity, and the system judges the likely authenticity of the presented sample with respect to the alleged class model.

Signature identification system has specific phases. First capturing the signature using sensors. Capturing a signature is the process of obtaining information about the signature, which can include both spatial and dynamic information. This information is turned into a form that can be interpreted and processed by a computer. The spatial information is the visual aspects of a signature, such as its shape and size, while the dynamic information relates to the writing process of the signature and defines aspects such as the speed and pressure used. Next phase the core feature extraction is processed where the most significant features are extracted. Finally, a good classifier is needed.

There are two data acquisition methods used for capturing written signatures and the data that is extracted from them; these are the dynamic (online) and static (offline) methods. This paper implements the offline approach that analyzes the static picture of the signature where it is the most commonly used.

Dynamic signatures capture both the spatial and the dynamic information of a signature. The dynamic features that are captured include one or more of the following attributes as the signature is written: acceleration and velocity; the position of the pen; the pressure that is applied; and the pen inclination [2, 3].

Dynamic information is not limited to these factors as it can include any information that is relevant to the writing process. The capture of a signature's dynamic information requires the use of specialized hardware such as a digitizing tablet, or an electronic pen, that captures the written attributes and converts them into processable data.

The advantage of using dynamic signatures for verification is that they produce a greater number of defining features than their static counterparts, making forged signatures easier to detect. The added advantage of digitally capturing this data is that there is typically less preprocessing required than static signatures, making feature extraction easier.

Using this method to capture the writing process also has its associated disadvantages. The main disadvantage is that these tools are not very common and as such, cannot be utilized for every situation where signature verification is required. As well as this, their use is unnatural for the user, which could negatively affect the writing of the signature.

More recent approaches exploit the use of a video camera which is focused on the writing of the signature and can be carried out using ordinary pen and paper. This allows the handwriting to be recovered from its spatiotemporal representation, which is given by the sequence of images that are produced [3].

A static signature is the visual representation of a signature that has been written out in its entirety. Because of this, a static signature is only captured once the writing process has been completed, allowing for visual aspects such as the size, slope, and curvature to be extracted. It is these visual aspects and their derivatives that allow similar signatures to be grouped together and differentiated from others that do not belong in the same group.

A signature is generally written on paper using a standard pen, meaning that the signature is not initially in a digital format. Because of this, the signature has to be captured and transformed into a format that can be processed by a computer. The conversion of a signature to a digital format is carried out either by taking a photograph or scanning the signature.

Both of these methods produce a digital image from which defining features of the signature can then be extracted. In society, a signed hardcopy of a document is the general requirement for binding that document to the signatory. For this purpose, a digitally captured signature.

Many approaches have been developed to counterfeit the signature identification problem such as distance classifiers, artificial neural networks, Neurofuzzy, Wavelet Neural, Dynamic time Warping, hidden Markov models, support vector machines (SVM), and Bayesian networks. But still there is a scope of work since these techniques have weakness such as lack of semantics, long training time, proneness to overfitting, and large feature vector. Distance classifiers were one of the first classification techniques used in the offline signature.

Verification a simple distance classifier is a statistical technique that usually represents a pattern class with a Gaussian probability density function (PDF). Each PDF is uniquely defined by the mean vector and covariance matrix of the feature vectors belonging to a particular class. When the full covariance matrix is estimated for each class, the classification is based on Mahalanobis distance. On the other hand, when only the mean vector is estimated, classification is based on Euclidean distance [4].

Fang et al. extracted a set of peripheral features in order to describe internal and the external structures of the signatures. To discriminate between genuine signatures and skilled forgeries, they used a Mahalanobis distance classifier together with the leave-one-out, cross validation method. The obtained AERs were in the range of 15.6% (without artificially generated samples) and 11.4% (with artificially generated samples) [5].

Another technique is the artificial neural network (ANN) is a massively parallel distributed system composed of processing units capable of storing knowledge learned from experience (examples) and using it to solve complex problems. Multilayer perceptron (MLP) trained with the error backpropagation algorithm has so far the most frequently ANN architecture used in pattern recognition. Mighell et al. were the first ones to apply ANNs for offline signature verification [5].

In a recent work, Armand et al. proposed the combination of the modified direction feature (MDF), extracted from the signature's contour, with a set of geometric features. In the experiments, they compared RBF and resilient backpropagation (RBP) neural network performances. Both networks performed writer-independent verification and contained 40 classes—39 corresponding to each writer and one corresponding to the forgeries. In this case, skilled forgeries were used in the training phase. The best classification rates obtained were 91.21% and 88.0%, using RBF and RBP, respectively [5].

Another signature identification technique is Hidden Markov models that are invented by Rabiner in 1989. They are finite stochastic automata used to model sequences of observations. Although this technique is more suitable to model dynamic data (e.g., as speech and online signatures), it has also been applied in segmented offline signatures. Generally, HMMs are used to perform writer-dependent verification by modeling only the genuine signatures of a writer. In this case, the forgeries are detected by thresholding [12].

Coetzer et al. used HMMs and discrete random transforms to detect simple and skilled forgeries. In this work, some strategies were proposed in order to obtain noise, shift, rotation, and scale invariances. Using a left-to-right ring model and the Viterbi algorithm, EERs of 4.5% and 18% were achieved for simple and skilled forgeries, respectively [5].

Another technique is dynamic time warping (DTW) that is a template matching technique used for measuring similarity between two sequences of observations. The primary objective of DTW is to nonlinearly align the sequences before they are compared [12]. Despite being more suitable to model data that may vary in time or speed, dynamic time warping has been used in offline signature verification. As usually occurs in HMM-based approaches, a test signature is compared to the genuine ones of a writer (writer-dependent verification), and a forgery is detected by thresholding.

Wilkinson and Goodman used DTW to discriminate between genuine signatures and simple forgeries. Assuming that curvature, total length, and slant angle are constant among different signatures of the same writer, they used a slope histogram to represent each sample. In the experiments, they obtained an EER of 7%. Increases in the error rates were observed when the forgers had some a priori knowledge about the signatures.

Another technique is Support vector machines (SVMs) is a kernel-based learning technique that has shown successful results in applications of various domains (e.g., pattern recognition, regression estimation, density estimation, novelty detection, etc.). Signature verification systems that use SVMs as classifiers are designed in a similar way to those that use neural networks. That is, in a writer-dependent approach, there is one class for the genuine signatures and another class for the forgeries. In addiction, using one-class SVMs, it is possible to perform training by using only genuine signatures [5, 12].

Ozgunduz et al. used support vector machines in order to detect random and skilled forgeries. To represent the signatures, they extracted global geometric features, direction features, and grid features. In the experiments, a comparison between SVM and ANN was performed. Using an SVM with RBF kernel, an FRR of 0.02% and an FAR of 0.11% were obtained, whereas the ANN, trained with the backpropagation algorithm, provided an FRR of 0.22% and an FAR of 0.16%. In both experiments, skilled forgeries were used to train the classifier.

Md. Itrat Bin Shams proposed a method of identifying handwritten signature by a good algorithm capable of identifying a signature with high accuracy. First a signature image is segmented and then data is extracted from individual blocks. After data is collected then accuracy is measured from both numbers of segments along with regular line matching coefficient. A mathematical formula is used to perform the accuracy measurement. As the level of accuracy is determined from both number of segments and similarity of lines in these segments, reliability of this method is very high. It has the capability to identify even a skilled forgery. this technique achieves a matching percentage of 84.5% [5].

Daramola Samuel and Ibiyemi Samuel proposed an offline signature verification system that incorporates a novel feature extraction technique. Three new features are extracted from a static image of signatures using this technique. they are image cell size, image center angle relative to the lower cell corner and pixel normalized angles relative to the lower cell corner. From the experimental results, the new features proved to be more robust than other related features used in the earlier systems. The proposed system has 1% error in rejecting skilled forgeries and 0.5% error in accepting genuine signatures [5, 12].

Marianela Parodi proposed feature extraction approach for offline signature verification based on a circular grid. Graphometric features used in the rectangular grid segmentation approach are adapted to this new grid geometry. A Support Vector Machine (SVM) based classifier scheme is used for classification tasks. The FRR is 18.75 and FAR is 2.1 [5, 12].

3 Rough Set

The rough set theory is an extension of set theory for the study of intelligent system with indiscernible, ambiguous, and imperfect data information. Rough set theory appeared two decades ago. Its main thrust is used in attribute reduction, rule generation, and prediction. The rough set concept was introduced by Pawlak [17]. This theory became very popular among scientists around the world in the last decade, and it is now one of the most developed AI methods. The primary goal of this new emerging theory is to classify the information from incomplete information. Its successful application to a variety of problems has shown its usefulness and versatility. The decision table in a rough set scheme plays an important role in clarifying a discretized feature values.

The Rough Sets approach provides mathematical techniques for discovering regularities in data. The principle notion of Rough Sets is that lowering the principle in data representation makes it possible to uncover patterns in the data, which may otherwise be obscured by too many details. At the basis of Rough Sets theory is the analysis of the limits of discernibility of subsets X of objects from the universe of discourse U. Let U be a set of objects (universe of discourse), A be a set of attributes, an information system is a pair $S = < U, A >$. An attribute $a \in A$ can be regarded as a function from the domain U to some value set V_a. A decision system is any information system of the form $\alpha = (U, A \cup \{d\})$, where $d \notin A$ is the decision attribute. The elements of A are called conditions attributes.

An information system [18] may be represented as an attribute value table, in which objects of the universe and columns label rows by the attribute. Similarly, a decision table may represent the decision system.

With every subset of attribute $B \subseteq A$ one can easily associate an equivalence relation I_B on U: $I_B = \{(x, y) \in U :$ for every $a \in B, a(x) = a(y)\}$. Then $I_B = \cap_{a \in B} I_a$. If

$X \subseteq U$, the sets $\{x \in U : [x]_B \subseteq X\}$ and $\{x \in U : [x]_B \cap X \neq \varphi\}$ where $[x]_B$ denotes the equivalence class of the object $x \in U$ relative to I_B, are called the B-lower and B-upper approximation of X in S and denoted by $\underline{B}X$, $\overline{B}X$. It may be regarded that $\underline{B}X$ is the greatest B-definable set contained in X, and $\overline{B}X$ is the smallest B-definable set containing X. This definition is clearly depicted in Fig. 2.

We now define the notions relevant to knowledge reduction. The aim is to obtain irreducible but essential parts of the knowledge encoded by the given information system; these would constitute reducts of the system. So one is, in effect, looking for maximal sets of attributes taken from the initial set (A, say) which induce the same partition on the domain as A. in other words, the essence of the information remains intact, and superfluous attributes are removed. Reducts have nicely characterized in [2–4] by discernibility matrices and discernibility functions. Consider $U = \{x_1, x_2, ..., x_n\}$ and $A = \{a_1, a_2, ..., a_m\}$ in the information system$S = < U, A >$. But the discernibility matrix $\Theta(S)$, of S is meant an $n \times n$ matrix such that

$$c_{ij} = \{a \in A : a(x_i) \neq a(x_j)\} \tag{1}$$

A discernibility function f_S is a function of m Boolean variables $\overline{a}_1, ..., \overline{a}_m$ corresponding to the attributes $a_1, ..., a_m$, respectively, and defined as follows:

$$f_S(\bar{a}_1, ..., \bar{a}_m) = \wedge\{\vee(c_{ij}) : 1 \leq i, j \leq n, j < i, c_{ij} \neq \varphi\} \tag{2}$$

where $\vee(c_{ij})$ is disjunction of all variables $\bar{a}$ with $a \in c_{ij}$. It is seen in [5] that $a_{i_1}, ..., a_{i_p}$ is a reduct in S if and only if $a_{i_1} \wedge ... \wedge a_{i_p}$ is a prime implicit (constituent of the disjunctive normal form) of f_S.

A principle task is the method of rule generation is to compute relative to a particular kind of information system, the decision system. Gelatinized versions of these matrices and function shall be the basic tool used in the computation. R-reducts and d-discernibility matrices are used for this purpose [9]. The methodology is described below.

Let $S = < U, A >$ be a decision table, with C and $D = \{d_1, ..., d_l\}$ its sets of condition and decision attributes, respectively. Divide the decision table $S = < U, A >$ into l tables $S_i = < U_i, A_i >, i = 1, ..., l$ corresponding to the l decision attributes $d_1, ..., d_l$ where $U = U_1 \cup ... \cup U_l$ and $A_i = C \cup \{d_i\}$. Let $\{x_{i1}, ...x_{ip}\}$ be the set of those objects of U_i that occur in $S_i, i = 1, ..., l$. Now for each d_i-reduct $B = \{b_1, ..., b_k\}$, a discernibility matrix is defined as follows

$$c_{ij} = \{a \in B : a(x_i) \neq a(x_j)\},\quad i, j = 1, ..., n \tag{3}$$

for each object $x_j \in x_{i1}, ..., x_{ip}$ the discernibility function

$$f_{d_i}^{x_j} = \wedge\{\vee(c_{ij}) : 1 \leq i, j \leq n, j < i, c_{ij} \neq \varphi\} \tag{4}$$

where $\vee(c_{ij})$ is disjunction of all members of c_{ij}. Then $f_{d_i}^{x_j}$ is brought to its conjunctive normal form (c.n.f). One of thus obtains as a dependency rule r_i. $P_i \rightarrow d_i$ where P_i is disjunctive normal form (d.n.f) of $f_{d_i}^{x_j}$, $j \in i_1, \ldots, i_p$. The dependency factor df_i for r_i is given by

$$df_i = \frac{card(POS_i(d_i))}{card(U_i)}, \tag{5}$$

where $POS_i(d_i) = \cup_{X \in Id_i} l_i(X)$ and $l_i(X)$ is the lower approximation of X with respect to I_i. The nation of the Core is:

$$Core(C) = \{a \in C : c_{ij} = (a), \text{for some } i, j\} \tag{6}$$

or

$$Core(C) = \bigcap \text{Re } duct(C) \tag{7}$$

where Re $duct(C)$ is the family of all reducts for the set of condition attributes C. The Core corresponding to this part of information cannot be removed without loss in the knowledge that can be derived from it [9].

3.1 Rough Neuron

Rough Neuron is first proposed by Lingras in 1996 [16], the definition of rough neuron has some relation with rough pattern; in the pattern each characteristic variable contains the upper boundary and the lower boundary. These characteristic variables cannot describe with accuracy count, but two boundary values or some change sector describe are more appropriate. So the rough neuron is more suitable for this information.

The rough neuron is constituted by traditional neuron and rough set. When the input of neural network is not a single but a range, such as the temperature of climate (daily maximum temperature, daily minimum temperature), rainfall (yearly maximum rainfall, yearly minimum rainfall), if we use traditional neural network, we cannot get good results and the error is relatively large. But if we use the neural network based on rough neuron, we can solve these problems well. Experiment [8], showed that the neural network based on rough neuron can improve the network performance and reduce errors.

Wei Wang proposed the model of the rough neural network and compared it with the traditional neural network. Rough neuron is used for classification. Rough Neural Networks (RNN) consists of rough neurons. Rough neuron defined by representing interval of values where both the upper and lower bounds are used in computations. Each rough neuron is made up of a combination of two individual neurons, namely lower boundary neuron and upper boundary neuron [4, 13]. The lower bound neuron, deals only with the definite or certain part of the input data and generates its output signal called as the lower boundary signal. The second neuron called the upper boundary neuron processes only that part of the input data which lies in the upper boundary region evaluated based on the concepts of rough sets and generates the output called upper boundary Signal.

This interpretation of upper and lower boundary regions is limited only to the learning or training stage of the neural network.

Rough neuron (r) is connected to another rough neuron (s) through two or four connections. Fully connected connection where upper and lower neuron of (r) connected to both upper and lower neuron (s).

Excitatory partial connection where upper neuron of (r) connected to upper neuron of (s), lower neuron of (r) connected to lower neuron of (s). Inhibitory partial connection where upper neuron of (r) connected to lower neuron of (s), lower neuron of (r) connected to upper neuron of (s). The lower boundary and the upper boundary neurons have a generalized sigmoid transfer function [2–10].

$$F(x) = {}^{\alpha}/_{(1+e^{\beta . x+\zeta})} \tag{8}$$

The Input for both the upper and lower neurons will be:

$$(input) = \sum (x - weight)^2 \tag{9}$$

The output of the lower and upper are attained as shown below:

$$(output)_{lower} = \min\left(F\left(input_{lower}\right), F\left(input^{upper}\right)\right) \tag{10}$$

$$(output)^{upper} = \max\left(F\left(input_{lower}\right), F\left(input^{upper}\right)\right) \tag{11}$$

The combined output of the rough neuron is:

$$output = (output)_{lower} + (output)^{upper} \tag{12}$$

Training in rough neural are similar to conventional neural network [3]. During training the network uses inductive learning principle to learn from the training set. In supervised training the desired output from output neurons in the training set is known, the weight is modified using learning equation. Neural network uses back-propagation technique for training, Training using rough Back propagation perform gradient descent in weight space on an error function.

3.2 *Hybrid Model of Rough Neural*

This section presents an implementation for offline signature recognition system using evolutionary Model. The approach used in this system involves two main stages:

- Rough Sets in Preprocessing
- Rough Neurons in Post Processing

This approach concern on the recognition stage of offline signatures images, it uses Rough Neurons as a powerful classifier technique The following algorithm summarize the behavior of Rough neural network Identification system.

Algorithm 1: Rough neural Network Identification

Input: a set of signature images for the trained persons
Output: classification of the signature images for each person
Begin
Step 1: preprocessing phase (Image processing)
Step 1.1: get bound box of image
Step 1.2: convert images to black and white
Step 2: determine structure of RNN (Rough set)
Step 3: classification phase (rough backpropagation)
Input:
Set of input features of signature images to be classified
Processing:
Step-3.1 for each attribute in the attribute set Do
Step-3.2 Compute the upper and lower rough neuron
Step-3.3 Build rough neural Networks
Step-3.4 Compute the relative error
Step-3.5 Calibrate the rough neural Network
Step-3.6 Repeat 4 and 5 until the error become minimum
Step-3.7 Return Class with minimum error.
Output: The final classification output.
End

3.3 *Rough Sets in Preprocessing*

In this section, the main problem of classifying node pixel is totally described. A Knowledge representation system is constructed in which neither condition nor decision attribute is distinguished. Therefore we are basically not interested in dependencies among attributes, but in description of some objects in terms of available attributes, in order to find essential differences between objects of interest. The problem of differentiation of varies options is often crucial importance in pattern identification.

For any signature input pattern U of size M N consists of a set of pixels u_{ij}. The knowledge representation table will be represented as characterization of handwritten signature figures. Each row $u_{i.}$ in the table represents object of the universe where $u_{i.} = \{u_{i1}, u_{i2}, u_{i3} \ldots, u_{iN}\}$ where N is the number of pixels in each row in each signature images. Each column $u_{.j}$ represent the attributes of that object where

$u_{.j} = \{u_{1j}, u_{2j}, u_{3j} \ldots, u_{Kj}\}$. Where k equals Z x M where Z is the number of training signature images for each person and M represent the numbers of pixels in each column in each signature images for that person.

Rough sets will be used for data analysis and feature selection with neural networks. Rough sets theory provides tools for expressing inexact dependencies within data. A minimum description length principle (MDL principle) gives us the reason of why we will use rough sets to reduce the input feature of the data, since MDL is defined to be the minimum number of rules that describe rough neurons which represent variations in data. According to the rough reduct, the date variation can be characterized and then the difference between crisp and rough neurons can be established. A reduct is a subset of attributes such that it is enough to consider only the features that belong to this subset and still have the same amount of information. If the decision table S has Z signature images for a particular person so by $\Theta(S)$ we denote an Z x Z matrix(c_{ij}) called the discernibility matrix of S as shown in Eq. 2.1.

Such that i, j = 1, 2, 3 … Z , Where each element in $\Theta(S)$ represent pixels that are different from their corresponding in other signature images for that person, this set of pixels represented by rough neurons which represent reduct.

The main idea of the reduct algorithm is that if a set of attributes satisfies the consistency criterion (i.e. be sufficient to discern all the required objects), it must have a non-empty intersection with non-empty elements of the discernibility matrix. One can prove that $a \in CORE(C)$ if and only if there exist two objects, which have the same value for each attribute from C except a, this statement may be expressed by mean of matrix elements (c_{ij}) as shown Eq. 2.1.

The input data to the model will be quantized first, i.e. the feature defining the problem should be identified and labeled. If the input data are given as real numbers in the preprocessing stage, one has to divide the data into distinct sets and introduce logical input variable $u_{i.}$ such that

$$IF(x_i \in \mathrm{X}_j) \text{ then } (u_{i.} = \mathrm{label}(\mathrm{x}_i) = true \tag{13}$$

The algorithm to compute reduct of S using indiscernibility matrix is reported as "Min" performs an operation that is analogical to checking for prime implicates of a Boolean function. The returned value is true if the argument R doesn't contain redundant attributes.

With the help of the reduct concept, Rough sets are used in the preprocessing phase. It is an efficient way for data reduction that decreases size of input pattern. Reducing the pattern size allow more speed of classification process, i.e. increase the performance of the signature recognition system. This is done by using Rough sets to determine the structure of the rough neural network. Since neurons in the input layer can be crisp neurons or rough neurons, Rough sets determine the rough neurons depending on the concept of reduct to overcome variability problem which means the vagueness in the human signature identification as follow: if the corresponding pixels in the signatures of one person have the same value then it is crisp neuron otherwise it is considered to be rough neuron [9].

Algorithm 2: Reduct Algorithm

Input: decision table S of signature images in binary form for particular person with set of rules describing rough neuron
Output: S with minimal set of rules describing rough neuron such that the superfluous rules is considered to be conventional neuron (distinguish rough and crisp neurons)
Processing:
Step 1: Compute indiscernibility matrix $\Theta(S) = (c_{ij})$using Equation 2.1
Step 2: Eliminate any empty or non-minimal elements of $\Theta(S)$ and create a discernibility list, $k = (k_1, k_2, ..., K_l)e$,
where e is the number of any non empty element in Θ(S)
Step 3: Build families of sets $R_0, R_1, ..., R_e$ in the following way

$$\circ \text{Set} R_0 = \varphi, i = 1 \tag{14}$$

$$\circ \textit{while } \text{i} \leq \text{e do} \tag{15}$$

$$\bullet \, \text{S}_\text{i} = \left\{ R \in R_{i-1} : R \cap k_i \neq \varphi \right\} \tag{16}$$

$$\bullet \, T_i = \bigcup_{k \in k_i} \bigcup_{R \in R_{i-1} : R \cap k_i \neq \varphi} \{R \cup \{k\}\} \tag{17}$$

$$\bullet \, M_i = \left\{ R \in T_i : Min(R, k, i) = true \right\} \tag{18}$$

$$\bullet \, R_i = S_i \cup M_i \tag{19}$$

$$\bullet \, i = i + 1 \tag{20}$$

The advantage of this approach becomes clear when the pixel corresponding to the crisp neuron mismatch the correct one in some decision making or classification process, i.e. different signature, then the tolerance will be exponentials increase. On the other hand, pixels corresponding to the rough neuron will reduce the associated tolerance or vagueness for human of the correct signature.

Also the limit occurred by our method can be defined by determining the perturbation in the output depending on the changes in the feed forward network. Since the output function is linearly piecewise increasing the function of the weigh link. Then the perturbation in the output is equivalent to the perturbation the weigh link. we can determine this perturbation if we prove that the transient equation is well posed.

The well-posedness of the transient equation, that is considered to be initial value problem, can be proved using the existence and uniqueness theorem by proving that the output function satisfies the Lipschitz condition that states if f(x) is continuous function, it satisfies the Eq. 2.4.

$$\left|f(x_1) - f(x_2)\right| \leq L\left|x_1 - x_2\right. \tag{21}$$

where x1 and x2 defined in the function domain. From the mean value theorem

$$f'(y(\zeta w)) = \frac{f(y(w)) - f(y(0))}{y(w) - y(0)} \tag{22}$$

This leads to

$$f(y(w)) - f(y(0)) = f'(y(\xi w))\left[y(w) - y(0)\right. \tag{23}$$

where $\xi \in [0, 1]$ Since the trajectory of Back propagation Net is given by

$$y(w) = \sum_{i=1}^{n} x_i * w + b \tag{24}$$

where n is the number of neurons connected to the output neuron and b is the bias. $f(x) = \frac{1}{1+e^{-x}}$ $\Re = (0, 1)$ $\frac{\partial f(y(w))}{\partial w} = \frac{df}{dx}\frac{dx}{dw}$

$$= f(x)\left[1 - f(x)\right. \sum_{i=1}^{n} x_i \tag{25}$$

$$\frac{\partial f(y(w))}{\partial w} \leq \sum_{i=1}^{n} x_i \; Hence\, 0 \leq \frac{\partial f(y(w))}{\partial w} \leq \sum_{i=1}^{n} x_i \tag{26}$$

$$f(y(w)) - f(y(0)) \leq \sum_{i=1}^{n} x_i \left[y(w) - y(0)\right.$$

$$f(y(w)) - f(y(0)) \leq L\left[y(w) - y(0)\right. \tag{27}$$

where

$$L = \sum_{i=1}^{n} x_i \tag{28}$$

From the Existence and Uniqueness Theorm, the transient equation is will posed deferential equation with Lipschitz constant

$$\hat{L} = 1 + L\|A\| \tag{29}$$

where $\|A\|$ defined as follows:

$$\|A\| = \frac{\sum_{i=1}^{n} x_i - 1}{\sum_{i=1}^{n} x_i} \tag{30}$$

Hence the perturbation in the trajectory of transient equation in accordance with the changes in the feed forward cloning template is given by

$$|y(w) - y(0)| \leq k\varepsilon, k > 0, \varepsilon < w \tag{31}$$

From Eq. 13, Eq. 15 and let k = 1 we get

$$|f(y(w)) - f(y(0))| \leq \left| w. \sum_{i=1}^{n} x_i \right| \tag{32}$$

This shows the maximum error that occurred in the Back propagation Net if the connection is removed. If the maximum error is less than the tolerance value, the weight value on the trajectory of rough neuron has no effect in the classification process and hence can be removed, and so can decrease the amount of processing and increase the efficiency of the classification process.

4 Rough Neural Model Using Segmentation

The approach used in this system involves three stages as shown in Fig. 1:

- Signature Preprocessing stage
- Feature Extraction
- Rough Neurons in Post Processing

This approach concern on the recognition stage of offline signatures images, it uses grid features and Rough Neurons as a classifier technique.

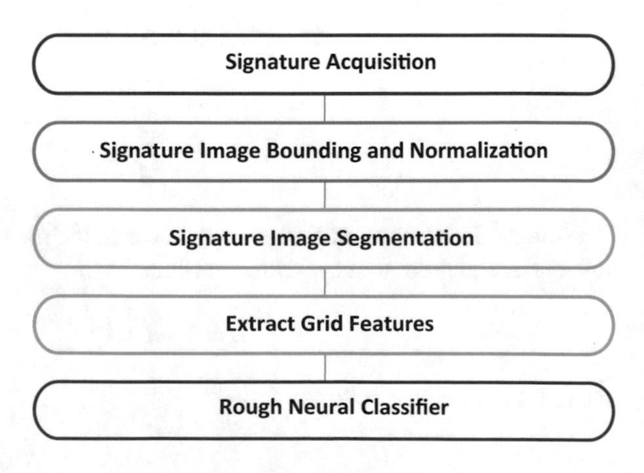

Fig. 1 Signature recognition algorithm using segmentation based RNN

5 Signature Preprocessing Stage

In this stage, our algorithm has to deal with how to avoid the curse of dimensionality and how to improve generalization ability of classifiers. To implement dimensionality reduction, a small but important subset of features is selected and a lower dimensional data preserving the distinguishing characteristics of the original higher dimensional data is extracted. First the signature images are aligned by the centers of their bounding boxes. Next the signature is scaled into a fixed size, in this process the area of Background is reduced by defining a fixed frame 512×467; which from experiments shows features preserving. The size normalization in offline signature verification is important because it establishes a common ground for image comparison. Next the positional information of the signature images is normalized by calculating an angle Θ about the centroid $(\overline{x}, \overline{y})$ such that rotating the signature by Θ brings it back to a uniform baseline. The baseline is the imaginary line about which the signature is assumed to rest [6]. This baseline is calculated from the linear regression equation for y on x which has the following form

$$y = a + bx \tag{33}$$

where b is the slope of the regression line and can be defined as follows:

$$b = \frac{\sum_{i=1}^{n} x_i y_i - \frac{\sum_{i=1}^{n} x_i \sum_{i=1}^{n} y_i}{n}}{\sum_{i=1}^{n} x_i^2 - \frac{(\sum_{i=1}^{n} x_i)^2}{n}} \tag{34}$$

While a is the intercept point of the regression line and the y-axis and can be defined as follows:

$$a = \frac{\sum_{i=1}^{n} y_i - b \sum_{i=1}^{n} x_i}{n} \tag{35}$$

where n is the number of pixels in the signature image. After the baseline is estimated it is rotated along the horizontal line. This is done by calculating the angle Θ between the estimated baseline the horizontal line as follows:

$$\Theta = \tan^{-1}(b) \tag{36}$$

Once the angle is computed, the signature images are signatures images for a particular person has a uniform shape that is identified clearly from the signature images of other trained persons [6].

When features are extracted from an image, their associated spatial information is often lost, as only the frequency of the feature is kept, while its corresponding location within the image is disregarded. It is this frequency information that is used to construct a histogram that represents the signature. Because no spatial information is captured, a potential problem that occurs is that two signatures which have very different appearances may produce similar histograms. The result of this is that these two signatures will produce a similarity score that is not representative of their true similarity.

A method of solving this problem is to use region sampling, which is a technique that provides spatial information for the extracted features by breaking an image up into regions. The concept behind region sampling is that the same region across multiple signatures will capture the same sections of signature. The goal of this is that when corresponding regions are compared, the same sections will also be compared, producing a more accurate similarity.

Once the signature has been normalized, the algorithm uses the uniform grid for regions sampling, since it is simple and the images are size and rotation invariant. In uniform grid: each region is of the same size and shape. This is carried out by placing the grid lines at equally spaced positions along the x-axis of the image; creating the vertical regions. Similarly, the horizontal regions are produced by placing grid lines at equally spaced positions down the y-axis. This forms a 16 × 16 grid region [7] The positions of both the horizontal and vertical grid lines are found by Eq. 38.

$$p_i = \left\lceil i\mathrm{x}\frac{\int}{\mathrm{n}} \right\rceil \quad \mathrm{i} = 1, 2, \ldots, \mathrm{n}\text{-}1 \tag{37}$$

When classifying a signature, one of the key requirements is the extraction of features that uniquely define it. This information allows the signature image to be broken down into components that are more useful for describing its structure, differentiating it from other signatures when compared. These features allow information about the image to be captured in a statistical manner. This statistical information provides a new representation that allows two signature images to be compared with greater ease.

After the signature has been segmented into a 16 × 16 equally regions, the objective is to extract the suitable features from each region to differentiate between them, the extracted features are the **"Width"**, which is the distance between two points in the horizontal projection in a binary signature image and must contain more than three pixels of the image. And **"Height"**, that is the distance between two points in the vertical projection and must contain more than 3 pixels of the image for a binary image. Moreover, the "Aspect ratio," which is defined as width to height of a signature. And the **Area of black pixels**, that is the area of black pixels is obtained by counting the number of black pixels in the binary signature images, separately. Moreover, the **"Normalized area of black pixels"**, that is found by dividing the area of black pixels by the area of signature image (width × height) of the signature. And the "Maximum and Minimum values in vertical projection," this corresponds to the highest and the lowest frequencies of black pixels in the vertical projection. Moreover, the **"Maximum and Minimum values in horizontal projection"**, this corresponds to the highest and the lowest frequencies of black pixels in the horizontal projection. And the **"baseline"**, that is the vertical projection of binary signature image has one peak point and the global baseline corresponds to this point. In addition to these features, **"Center of gravity of a signature image"** is obtained by adding all x, y locations of gray pixels and dividing it by the number of pixels counted. The resulting two numbers (one for x and other for y) is the center of gravity location. The vertical center Cy is given by;

$$C_y = \frac{\sum_{y=1}^{y\max} y \sum_{x=1}^{x\max} b[x,y]}{\sum_{x=1}^{x\max} \sum_{y=1}^{y\max} b[x,y]} \tag{38}$$

While the horizontal center of the signature Cx is given by

$$C_x = \frac{\sum_{x=1}^{x\max} x \sum_{y=1}^{y\max} b[x,y]}{\sum_{x=1}^{x\max} \sum_{y=1}^{y\max} b[x,y]} \tag{39}$$

Moreover, the 'Average of distances of each black pixel' from the bottom left corner of each segment. The vector distance for k th pixel in b th box at location (i, j) is calculated as $d_k^b = \sqrt{(i^2 + j^2)}$. By dividing the sum of distances of all pixels present in a box (region) with their total number , a normalized vector distance γ_b for each box is defined by $\gamma_b = \frac{1}{n_b} \sum_{k=1}^{n_b} d_k^b$ where n_b is number of pixels in b^{th} box.

Moreover, the 'Average of angles of each black pixel' from the bottom left corner of each segment the vector angle for kth pixel in bth box at location (i, j) is calculated as $\theta_k^b = \tan^{-1}\frac{i}{j}$. By dividing the sum of angles of all pixels present in a box (region) with their total number, a normalized vector angle γ_b for each box Is defined by $\gamma_b = \frac{1}{n_b} \sum_{k=1}^{n_b} \theta_k^b$ where n_b is number of pixels in b^{th} Box. Once these above mentioned features have been extracted, the average of them is taken that represent the best choice for all features for the same segmented region. Each segment is represented using this average value, so each signature image can be represented using 16×16 matrix, each value in this matrix is the average value of the features extracted from the corresponding segment [7–9].

6 Rough Neurons in Post Processing

Each segment average value in the 16×16 matrix corresponds to rough neurons in the input pattern. Such that each rough neuron used to represent the maximum value and the minimum value of the corresponding segment regions in the training set of person's signatures.

The proposed system uses a Backpropagation (BP) NNS for classification process. There is a RNN for each of the trained persons. The structure of each RNN consists of input layer, hidden layer, and output layer [9]. Neurons in the input layer are rough neurons; the number of neurons in the input layer is equal to the number of a segmented region in a signature image. The hidden layer consists of crisp neurons; the number of neurons in the hidden layer is approximately double the input layer size. The output layer consists of one crisp neuron that represents the output person corresponding to the current signature input pattern. The computation parameters in the Backpropagation classification algorithm include the learning rate of hidden neurons α_1 and the learning rate of output neuron α_2, the stopping condition during the learning phase include specific numbers of iterations or acceptable error rate.

Backpropagation algorithm uses sigmoid activation function to compute the output, it also uses momentum term to increase the recognition ability. During the test phase; if the output value of the crisp neuron is binary 1, then the input pattern represents that person, if the output value is binary 0, then the input pattern doesn't represent that person [2]. Back-Propagation propagates the input values through the neural network in such way that calculates the output values, compare the obtained output with the desired output and calculate its error, then it backward redistribute the error to the neurons according to their contributions in the error and modifies their synaptic weights [10, 11]. The fitness function of BP utilized is the sum-squared error between the input to the system and the optimized parameters:

$$f(c) = \sum_{i=1}^{n} \left(\left| y(n) - \hat{y}(n) \right| \right)^2 \tag{40}$$

where n represents the number of input/output sample, $y(n)$ represents the desired output and $\hat{y}(n)$ represents the actual output [2, 3].

7 Experimental Results

The signature database includes signature images of ten trained persons, each person has twenty signature images divided as ten signatures used during training phase and the Other used during testing phase. According to our experimental results, the experiments were applied on the set of training images for the trained persons, as depicted in Fig. 2.

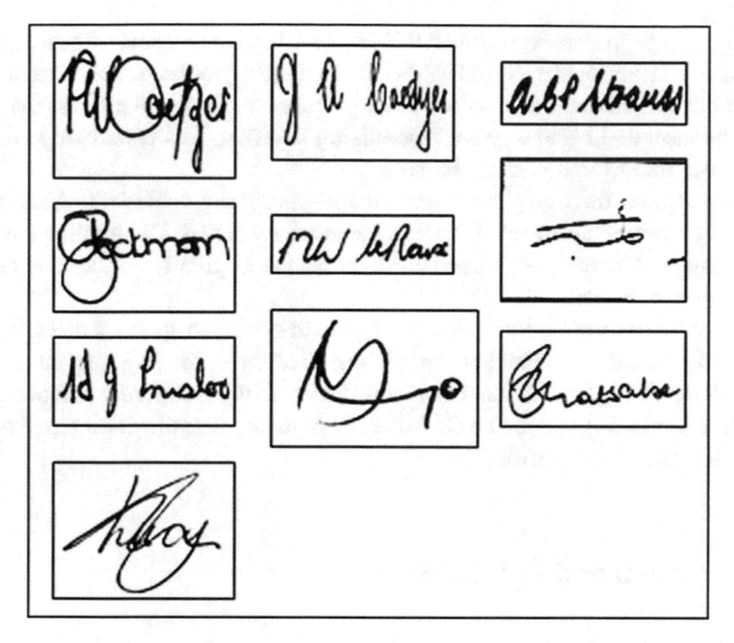

Fig. 2 Sample signature images

Fig. 3 Conventional NN VS Rough NN VS Rough NN with segmentation

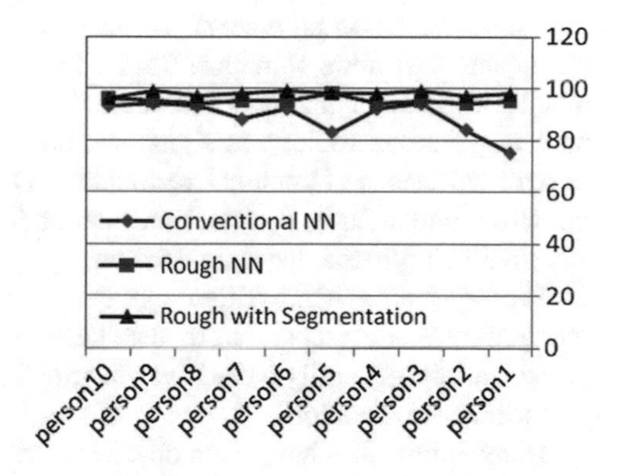

The introduced algorithm is applied with random initial weights and learning rate of hidden neurons α_1 has value of 0.01 and the learning rate of output neuron α_2 has value 0.001, the experiment is repeated five times for each training person due to optimizing the weigh template.

A comparison between conventional NN and RNN and RNN with segmentation for the ten persons is taken into account. As depicted in Fig. 4.3, RNN with segmentation achieves more progressive tests for 10 persons than that of the conventional NN and RNN and shows that the total identification percentage for conventional NN is 89% and for RNN is approximately 95% and for RNN with segmentation is 98% (see Fig. 3).

Then using segmentation with RNN is more effective in terms of mean improvement than Conventional NN and RNN. This chapter proposed a new technique to improve the performance of handwritten signature identification process that is considered an easy and basic way for recognizing persons. This technique is based on grid features and Rough Neural Network.

In the feature extraction phase signature images are segmented into equal number of segments from which a set of local features are extracted. The application of grid features coupled with well defined RNN contributed greatly to the generation of robust signature models.

Uses grid features and rough neural networks where grid features based on segmentation used for reducing the pattern size by extracting the most significant attributes in signatures that can discriminate between different persons classes, Rough neural networks used as a classifier for signature identification based on back-propagation learning algorithm.

8 Conclusion and Future Work

Signature Identification is an important research area in the field of person authentication. The goal of a signature recognition system is to verify the identity of an individual based on an analysis of his or her signature through a process that discriminates a genuine signature from a forgery one. It enables computers to recognize handwritten signatures without human intervention. Signature recognition is a necessary task in society, as signatures have a well established and accepted place as a formal means of personal verification. Due to this, its potential applications are numerous and include, personal verification for access control, banking applications, electronic commerce, legal transactions.

The process of handwritten signature recognition follows the classical pattern recognition model steps; that is, data acquisition, preprocessing, feature extraction, classification (generally called "verification" in the signature verification field), and performance evaluation.

Many approaches have been developed to counterfeit the signature identification problem such as conventional neural networks, Neurofuzzy, Wavelet Neural, hidden Markov models and support vector machines. Although these approaches have made tremendous achievements in the area of person authentication, there is still a scope of work since these techniques have weakness such as lack of semantics, long training time, proneness to overfitting, and large feature vector.

The proposed work aims to provide a Rough Neural Network that is a hybrid technique between modified neural network that contains rough neurons and rough set. It presents a new approach for recognizing handwritten signatures. Rough sets provides mathematical techniques for discovering regularities in data and uncovering patterns in it, which may otherwise be obscured by too many details, it is used for reducing the input pattern size by resolving the variability problem exists in signatures, and to determine which neurons in the input layer considered being rough neurons while Rough neural networks use a combination of rough and conventional neurons. Rough neuron is used as a classifier for signature identification based on back propagation learning algorithm.

In the future we wish to extend the system to deal with more complex types of data and solve inconsistencies found in it using granular computing as an emerging computing paradigm of information processing. Also we wish to implement full biometric system identification such as online signature recognition and other biometrics.

References

1. Al-Shoshan, A.: Handwritten signature verification using image invariants and dynamic features. In: Proceedings of the International Conference on Computer Graphics, Imaging and Visualisation, pp. 173–176 (2006)
2. Saleh, A.: Signature Identification using Evolutionary Rough Neuron. Ain Shams University, Faculty of computer science and Information (2011)
3. Amudha, V., Venkataramani, B., Manikandan, J.: FPGA implementation of isolated digit recognition system using modified back propagation algorithm. In: Proceedings of (ICED) International Conference Electronic Design, pp. 1–6, Dec. 2008
4. Ashwin, G.K.: Data mining tool for semiconductor manufacturing using rough neuro hybrid approach. In: Proceedings of International Conference on Computer Aided Engineering- CAE-2007, IIT Chennai, pp. 13–15, December 2007
5. Kothari, A., Keskar, A.: Rough set approaches to unsupervised neural network based pattern classifier. Adv. Mach. Learn. Data S.-I. Ao (EDS) Anal. **48**, 151–163 (2010)
6. Coetzer, J.: Off-line Signature verification. Ph.D. thesis, University of Stellenbosch (2005)
7. Kisku, D.R., Gupta, P., Sing, J.K.: Offline signature identification by fusion of multiple classifiers using statistical learning theory. Int. J. Secur. Appl. **4**(3) (2010)
8. Zhang, D., Wang, Y.: Filtering image impulse noise based on fuzzy rough neural network. J. Comput. Appl., 2336–2338 (2005)
9. Radwan, E.: Ph.D. thesis. Increasing Cellular Neural Networks Template Robustness by New Artificial Intelligence Techniques. Toin University of Yokohama in Japan, pp. 34–35, Sept. 2006
10. Own, H.S., Al-Mayyan, W., Zedan, H.: Biometric-based authentication system using rough set theory. M. Szczuka. (EDS): Rough Sets Curr. Trends. Comput. **6086**(2010), 560–569 (2010)
11. Impedovo, S., Pirlo, G.: Verification of handwritten signatures: an overview. In: 14th International Conference on Image Analysis and Processing, 2007. ICIAP 2007, pp. 191–196 (2007)
12. Coetzer, J.: Offline signature verification. Chapt. 2, pp. 20–27, April 2005
13. Kothari, A., Keskar, A., Chalasani, R., Srinath, S.: VNIT, Nagpur, Rough neuron based neural classifier. In: Proceedings of International Conference of Emerging Trends in Engineering and Technology, (ICETET), 2008, pp. 624–628, 16–18, July 2008
14. Batista, L., Rivard, D., Sabourin, R., Granger, E., Maupin, P.: State of the art in off-line signature verification (2008)
15. Madasu, V., Lovell, B.C.: An automatic offline signature and forgery detection system. In: Verma, B., Blumenstein, M., (Ed.), Pattern Recognition Technologies and Applications: Recent Advances, PA: Information Science Reference, pp. 63–89 (2008)
16. Lingras, P.: Rough neural networks. In: Proceedings of the 6th International Conference on Information Processing and Management of Uncertainty in Knowledge-based Systems, pp. 1445–1450 (1996)

17. Metawa, N., Elhoseny, M., Kabir Hassan, M., Hassanien, A.: Loan portfolio optimization using genetic algorithm: a case of credit constraints. In: 12th International Computer Engineering Conference (ICENCO), pp. 59–64. IEEE (2016). doi:10.1109/ICENCO.2016.7856446
18. Metawa, N., Kabir Hassan, M., Elhoseny, M.: Genetic algorithm based model for optimizing bank lending decisions, Expert Systems with Applications, vol. 80, 1 Sept 2017, pp. 75–82. ISSN 0957-4174. http://dx.doi.org/10.1016/j.eswa.2017.03.021

A Novel Secure Personal Authentication System with Finger in Face Watermarking Mechanism

Chinta Someswara Rao, K.V.S. Murthy, R. Shiva Shankar and V. Mnssvkr Gupta

Abstract Facial and Finger authentication plays a pivotal role for proving personal verification in any organization, industry, enterprise, etc. In the previous works, authentication systems are developed by using the password, pin number, digital signature, etc., as a single source of identification. But all these systems can be subjected to spoofing attack. In this paper, a novel authentication system is proposed with image-in-image Fast Hadmard Transform (FHT) watermarking and authentication with Singular Value Decomposition (SVD). The proposed system is strong enough from attacks as the authentication is being done using face and finger traits. The proposed work is useful for reducing the size of the database, identification and authentication for bank systems, crime investigations, organizational attendance systems, and for knowing student attendance system, unauthorized copying, etc.

Keywords Face • Finger • Authentication • Personal verification • FHT • SVD

1 Introduction

Authentication is the act of confirming the truth of an attribute of a single piece of data which is claimed true by an entity [1]. In contrast with identification, which refers to the act of stating or otherwise indicating a claim purportedly attesting to a person or thing's identity, authentication is the process of actually confirming that identity. It might involve confirming the identity of a person by validating their credentials like face, finger in biometric systems.

In the previous system authentication can be proved by person's password, pin number, digital signature, etc., but this system remains failed because of spoofing attacks by intruders. To stop these perfect digital copies and modified files, several

C.S. Rao (✉) · K.V.S. Murthy · R.S. Shankar · V.M. Gupta
Department of CSE, S.R.K.R Engineering College,
W.G. District, Bhimavaram 534 204, Andhra Pradesh, India
e-mail: chinta.someswararao@gmail.com

A.E. Hassanien and D.A. Oliva (eds.), *Advances in Soft Computing and Machine Learning in Image Processing*, Studies in Computational Intelligence 730,
https://doi.org/10.1007/978-3-319-63754-9_15

methods such as copy protection and file encryption have been tried and until recently, have failed. These old techniques suffer from major drawbacks. Once an encrypted file has been decoded successfully, it can be copied as many times as for redistribution with no encryption. Thus, there is a strong need for techniques to protect the copyright for the content of owners. Cryptography and digital watermarking are two complementary techniques to protect digital content [2].

Cryptography is the processing of information into an encrypted form for the purpose of secure transmission. Before delivery, the digital content is encrypted by the owner using a secret key. A corresponding decryption key is provided only to a legitimate receiver. The encrypted content is then transmitted via Internet or other public channels, and it will be not possible to pirate without the decryption key. At the receiving end, the receiver decrypts the information using a secret key, which is shared by both the delivering and receiving ends. However, once the encrypted content is decrypted, it has no protection anymore. For example, an adversary can obtain the decryption key by purchasing a legal copy of the media, but then redistribute the decrypted copies of the original [3].

In response to these challenges, in this paper, digital watermarking techniques have been proposed for authentication, storing, and protecting the digital content even after it is decrypted. In digital watermarking, a watermark is embedded into a covertext, resulting in a watermarked signal called stegotext which has no visible difference from the covertext. In a successful watermarking system, watermarks should be embedded in such a way that the watermarked signals are robust to certain distortions caused by either standard data processing in a friendly environment or malicious attacks in an unfriendly environment. In other words, watermarks still can be recovered from the attacked watermarked signal generated by an attacker if the attack is not too much [4, 5].

1.1 Digital Watermarking History and Terminology

1.1.1 History

The idea of communicating secretly is as old as communication itself. Steganographic methods made their record debut a few centuries later in several tales by Herodotus, the father of history. Kautilya's Arthasastra and Lalita Visastra are few famous examples of the Indian literature in which secret writing or steganography have been used. Few other examples of steganography can be found. An important technique was the use of sympathetic inks. Ovid in his Art of Love suggests using milk to write invisibly. Later, chemically affected sympathetic inks were developed. This was used in First and Second World Wars. The origin of steganography is biological and physiological. The term steganography came into use in 1500s after the appearance of Trithemius book on the subject Steganographia. In First World War, for example, German spies used fake orders for cigars to represent various types of British warships-cruisers and destroyers [6, 7].

1.1.2 Terminology

As most of the techniques used to hide or embed data in media share similar principles and basic ideas, this section gives some definitions, to avoid confusion, to clarify and show the difference between the different techniques. The various information hiding techniques can be classified as given in Fig. 1

Steganography stands for the art, science, study, work of communicating in a way, which hides a secret message in the main information. Steganography methods rely generally on the assumption that the existence of the covert data is unknown to unauthorized parties and are mainly used in secret point-to-point communication between trusting parties. In general, the hidden data does not resist manipulation and thus cannot be recovered.
Watermarking as opposed to steganography, in an ideal world can resist to attacks. Thus, even if the existence of the hidden information is known, it should be difficult for an attacker to remove the embedded watermark, even if the algorithmic principle is known.
Data hiding and Data embedding are used in varying context, but they do typically denote either steganography or applications "between" steganography and watermarking applications where the existence of the embedded data are publicly known, but do not need to be protected.
Fingerprinting and labeling are terms that denote special applications of watermarking. They relate to copyright protection applications, where information about originator and recipient of digital data is embedded as watermarks. The individual watermarks, which are unique codes out of a series of codes, are called "fingerprints" or "labels."

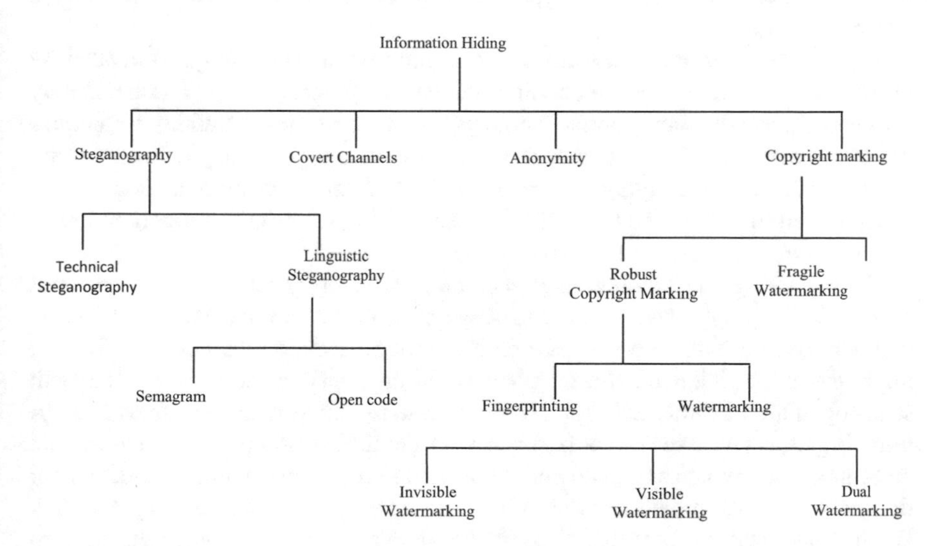

Fig. 1 Information hiding techniques

Bit-stream watermarking is sometimes used for data hiding or watermarking of compressed data, for example, compressed video.
Visible watermarks as the name says, are visual patterns, like logos, which are inserted into or overlaid on images, very similar to visible paper watermarks.
Copy protection attempts to find ways, which limits the access to copyrighted material and/or inhibit the copy process itself.
Copyright protection inserts copyright information into the digital object without the loss of quality. Whenever the copyright of a digital object is in question, this information is extracted to identify the rightful owner. It is also possible to encode the identity of the original buyer along with the identity of the copyright holder, which allows tracing of any unauthorized copies.

1.2 Research Problems and Motivations

A major research problem on digital watermarking is to determine best tradeoffs among the distortion between the covertext and stegotext, the distortion between the stegotext and forgery, the watermark embedding rate, the compression rate and the robustness of the stegotext. Along this direction, some information theoretic results, such as watermarking capacities and watermarking error exponents, have been determined. Watermarking is not a new phenomenon. For nearly one thousand years, watermarks on paper have been used to identify a particular brand (in the case of publishers) and to discourage counterfeiting (in the case of stamps and currency). In the contemporary era, proving authenticity is becoming increasingly important as more of the world's information is stored as readily transferable bits. Digital watermarking is a process, whereby arbitrary information is encoded into an image in such a way that the additional payload is imperceptible to the image observer [8, 9].

Digital watermarking is a form of data hiding or steganography. Motivated by growing concern about the protection of intellectual property on the Internet and by the treat of a ban for encryption technology, the interest of watermarking techniques has been increasing over the recent years [10]. In a digital image watermarking system, information carrying watermark is embedded in an original image. The watermarked image is then transmitted or stored. The received watermarked image is then decoded to resolve the watermark.

Watermarking and cryptography encryption are closely related in the spy craft family. Cryptography scrambles a message so it cannot be understood. A watermarking method hides the message so it cannot be seen, a message in ciphertext might cause suspicion on the recipient while an invisible message created with steganographic methods will not. Note that watermarking is distinct from encryption. Its goal is not to restrict or regulate access to the host signal, but to ensure that embedded data remain inviolate and recoverable [10]. There is little protection for decrypted or descrambled content, which can be redistributed or misappropriated. Digital watermarking is intended by its developers as the solution to the need to provide value added protection on top of data encryption and scrambling for content protection.

1.3 Basic Principles of Watermarking

All watermarking methods share the same generic building blocks, a watermark embedding system and a watermark recovery. Figure 2 shows the generic watermark embedding process. The inputs to the scheme are the watermark, the cover-data and an optional public or secret key. The watermark can be of any nature such as a number, text, or an image. The key may be used to enforce security that is the prevention of unauthorized parties from recovering and manipulating the watermark. All practical systems employ at least one key, or even a combination of several keys [11].

Imperceptibility The modifications caused by watermark embedding should be below the perceptible threshold, which means that some sort of perceptibility criterion should be used not only to design the watermark, but also to quantify the distortion. As a consequence of the required imperceptibility, the individual samples (or pixels, features, etc.) that are used for watermark embedding are only modified by a small amount.
Redundancy To ensure robustness despite the small allowed changes, the watermark information is usually redundantly distributed over many samples (or pixels, features, etc.) of the cover-data, thus providing a global robustness which means that the watermark can usually be recovered from a small fraction of the watermarked data. Obviously watermark recovery is more robust if more of the watermarked data is available in the recovery process.
Keys In general, watermarking systems use one or more cryptographically secure keys to ensure security against manipulation and erasure of the watermark. As soon as a watermark can be read by someone, the same person may easily destroy it because not only the embedding strategy, but also the locations of the watermark are known in this case. These principles apply to watermarking schemes for all kinds of data that can be watermarked, like audio, images, video, formatted text, 3D models, model animation parameters, and others. The generic watermark recovery

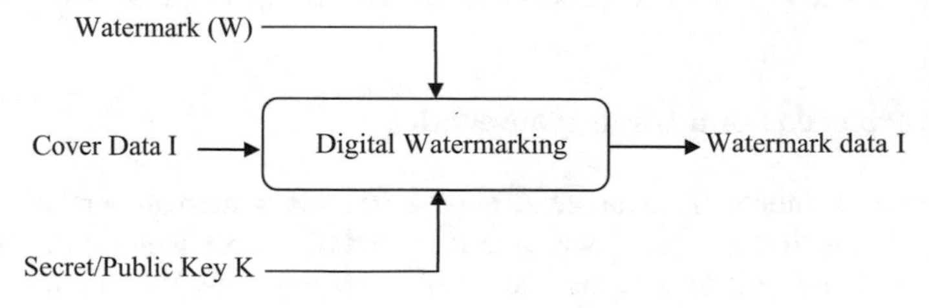

Fig. 2 Generic watermarking schemes

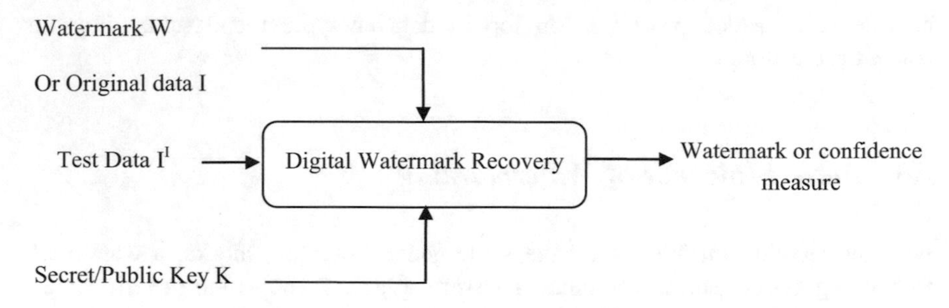

Fig. 3 Generic watermark recovery scheme

process is depicted in Fig. 3. Inputs to the scheme are the watermarked data, the secret or public key and, depending on the method, the original data and/or the original watermark. The output is either the recovered watermark W or some kind of confidence measure.

1.4 Basic Requirements of Digital Watermarking

Depending on the watermarking application and purpose, different requirements arise resulting in various design issues. Watermark imperceptibility is a common requirement and independent of the application purpose. Additional requirements have to be taken into consideration when designing watermarking techniques [12, 13]

1.4.1 Recovery with or Without the Original Data

Depending on the application, the original data is available or is not available to the watermark recovery system. If the original data is available, it is usually advantageous to use it, since systems that use the original data for recovery are typically more robust. However, in applications such as data monitoring, the original data is of no use because the goal is to identify the monitored data. In other applications, such as video watermarking applications, it may be impracticable to use the original data because of the large amount of data that would have to be processed. While most early watermarking techniques require the original data for recovery, there is a clear tendency to devise techniques that do not require the original data set [12, 13].

1.4.2 Verification of a Given Watermark

There are two inherently equivalent approaches for watermark embedding and recovery. In the first approach, one watermark out of a predefined set of admissible watermarks is embedded, and the watermark recovery tests the watermarked data against the admissible set. The output of the watermark recovery is the index of the embedded watermark or a symbol "no watermark found." In the second approach, the embedded watermark is the modulation of a sequence of symbols given to the watermark embedding system. In the detection process, the embedded symbols are extracted through demodulation.

1.4.3 Robustness

Robustness of the watermarked data against modifications and/or malicious attacks is one of the key requirements in watermarking. However, as said before, there are applications where it is less important than for others.

1.4.4 Security Issues and Use of Keys

The conditions for key management differ greatly depending on the application. Obvious examples are public-key watermarking systems versus secret-key systems used for copyright protection. Watermarking methods using the original data set in the recovery process usually feature increased robustness not only toward noise-like distortions, but also distortions in the data geometry since it allows the detection and inversion of geometrical distortions. In many applications, such as data monitoring or tracking, access to the original data is not possible.

2 Cryptography Related to Watermarking

Most content protection mechanisms rely on cryptological (cryptographical or steganographical) means for the provision of functionality. These mechanisms serve one or more of the requirements in definitions for confidentiality through anonymity that are commonly sought for in information security [14].

2.1 Cryptography

Cryptography is the science of using mathematics to encrypt and decrypt data. Cryptography enables to store sensitive information or transmit it across insecure networks, so that it cannot be read by anyone except the intended recipient. While cryptography is the science of securing data, cryptanalysis is the science of analyzing and breaking secure communication.

Classical cryptanalysis involves an interesting combination of analytical reasoning, application of mathematical tools, pattern finding, patience, determination [15]. A related discipline is steganography, which is the science of hiding messages rather than making them unreadable. Steganography is not cryptography; it is a form of coding. It relies on the secrecy of the mechanism used to hide the message. If, for example, encode a secret message by putting each letter as the first letter of the first word of every sentence, its secret until someone knows to look for it, and then it provides no security at all. A message is known as a plaintext or cleartext. The method of disguising the plaintext in such a way as to hide its information is encryption and the encrypted text is also known as a ciphertext. The process of reverting ciphertext back to its original text is decryption.

2.2 Steganography

Steganography is the study of techniques for hiding the existence of a secondary message in the presence of a primary message. The primary message is referred to as the carrier signal or carrier message; the secondary message is referred to as the payload signal or payload message. Classical Steganography can be divided into two areas, technical steganography and linguistic steganography. The classification of the various steganographic techniques is shown in Fig. 4 and described briefly in the following section.

Technical steganography involves the use of technical means to conceal the existence of a message using physical or chemical means. Examples of this type of steganography include invisible inks.

Linguistic steganography itself can be grouped into two categories, open codes and semagrams. The latter category also encompasses visual semagrams. These are

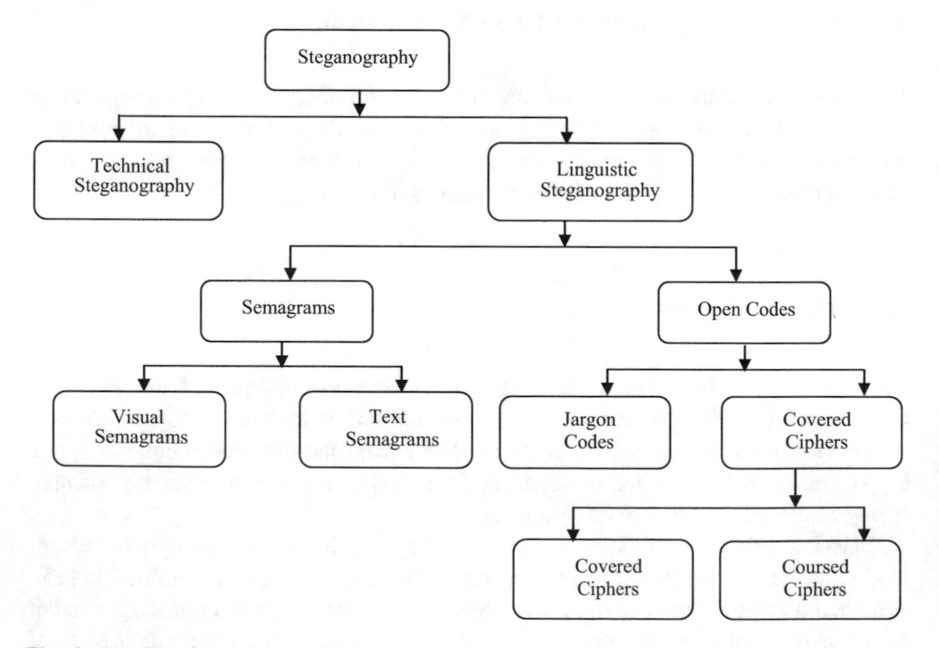

Fig. 4 Classifications of steganographic techniques

physical objects, depictions of objects or other diagrams with an ostensibly innocuous purpose in which a message is encoded. Examples of semagrams include the positioning of figures on a chessboard. Text semagrams are created by modifying the appearance of a text in such a way that the payload message is encoded.

The category of open codes is characterized by embedding the payload signal in the carrier signal in such a way in the carrier signal that the carrier signal itself can be seen as a legitimate message from which an observer may not immediately deduce the existence of the payload signal.

One of the most frequently cited examples is the cue code with which in World War II Japanese diplomats were to be notified of impending conflict. In this code, "HIGASHI NO KAZE AME" ("east wind, rain") signified pending conflict with the United States, while "KITANO KAZE JUMORI" ("north wind, cloudy") indicated no conflict with Russia and "NISHI NO KAZE HARE" ("west wind, clear") with the British Empire. Unlike jargon codes, which lead to atypical language that can be detected by an observer, cue codes are harder to detect provided that their establishment has not been compromised.

2.3 *Watermarking*

Although cryptography and watermarking both describe techniques used for covert communication, cryptography typically relates only to covert point-to-point communication between two parties. Cryptography methods are not robust against attacks or modification of data that might occur during transmission, storage or format conversion. Watermarking, as opposed to cryptography, has an additional requirement of robustness against possible attacks. An ideal cryptographic system would embed a large amount of information securely, with no visible degradation to the cover object. An ideal watermarking system, however, would embed an amount of information that could not be removed or altered without making the cover object entirely unusable. As a side effect of these different requirements, a watermarking system will often trade capacity and perhaps even some security for additional robustness.

The working principle of the watermarking techniques is similar to the cryptography methods. A watermarking system is made up of a watermark embedding system and a watermark recovery system. The system also has a key which could be either a public or a secret key. The key is used to enforce security, which is prevention of unauthorized parties from manipulating or recovering the watermark.

3 Review of the State of the Art

The review of the state of the art is classified into two categories as authentication and watermarking. In authentication category, the literature which is related to biometric authentication is discussed, whereas in second category, the literature which is related to digital image watermarking is discussed.

3.1 Watermarking

A modern proliferation and success of the Internet, together with availability of relatively inexpensive digital recording and storage devices has created an environment in which it became very easy to obtain, replicate, and distribute digital content without any loss in quality. This has become a great concern to the multimedia content publishing industries, because techniques that could be used to protect intellectual property rights for digital media, and prevent unauthorized copying did not exist. While encryption technologies can be used to prevent unauthorized access to digital content, it is clear that encryption has its limitations in protecting intellectual property rights, once content is decrypted, there is nothing to prevent an authorized user from illegally replicating digital content. Some other technology was obviously needed to help establish and prove ownership rights, track content usage, ensure authorized access, facilitate content authentication, and prevent illegal replication. This need attracted attention from the research community and industry leading to a creation of a new information hiding form, called digital watermarking.

In 1994, Van Schyndelem et al. [16] changed the Least Significant Bit (LSB) of an image to embed a watermark. Since then more and more researchers studied digital watermarking problem. This section discusses some of the watermarking techniques that have been implemented so far. Here also the results, advantages including security, robustness, capacity, complexity, and the limitations of some of these schemes are briefly discussed.

3.1.1 Spatial Domain Watermarking

In spatial domain watermarking, each pixel value of the image is modified depending on the watermark signal pattern. Changing the pixel value does affect the image statistics. Due to the constraint that the watermark should be imperceptible, there cannot be much deviation from the original image statistics. This requires that the change introduced in each pixel be less. One disadvantage of spatial domain watermarking is not robust to attacks.

Van Schyndel et al. [16] presents LSB Method. One of the simplest methods to insert data into digital signals in noise free environments is LSB coding. The LSBs of each pixel of the image is first set to zero or one. Zeros and ones are then used to embed information in the LSB plane. For example, a binary black and white image can be embedded in this host image. LSB coding introduces noise of at most one unit, which, in practice, is imperceptible, as long as the host pixel value is not too low. The traditional LSB coding has been used for a long time. Nowadays people do not use LSB watermarking technique.

Van Schydel et al. [17] presents other LSB coding methods that use m-sequences to transparently embed messages with greater security than straightforward LSB coding. A binary sequence is mapped from $\{0, 1\}$ to $\{-1, 1\}$. In one method, the message is embedded in the LSB plane using m-sequences. Unlike the first one changing bit value in the LSB plane, the second adds messages in LSB plane. Disadvantages of these methods are hostile parties know that all pixels are changed with value ± 1, partial knowledge of a m-sequence string enables the recovery of the whole data, and this technique is fragile to all manipulations in the frequency domain. Therefore, this LSB coding method is very vulnerable.

Huang et al. [18] presents a novel and robust color watermarking approach for applications in copy protection and digital archives. This scheme considers chrominance information that can be utilized at information embedding. This work presents an approach for hiding the watermark into Direct Current (DC) components of the color image directly in the spatial domain, followed by a saturation adjustment technique performed in Red Green Blue (RGB) space. The merit of this approach is that it not only provides promising watermarking performance but also is computationally efficient.

3.1.2 Transform Domain Watermarking

In simple terms transform domain means that the image is segmented into multiple frequency bands. Transform domain methods hide messages in significant areas of the cover image which makes them more robust to attacks, such as compression, cropping, and some image processing techniques than the LSB approach. However, while they are more robust to various kinds of signal processing, they remain imperceptible to the human sensory system. Here singular value decomposition-based watermarking scheme is only discussed. Other transform domain techniques are discussed in further sections.

Zhu et al. [19] presents SVD-based watermarking method. SVD deal with the rectangle matrices directly and can extract better-quality watermarks. It takes little time to embed and extract the watermark in large images and this method can avoid some disadvantages such as the distortion caused by the computing error when extracting the watermark in the diagonal direction.

Liang and Qi [20] presents Singular Value Decomposition (SVD)-Discrete Wavelet Transformation (DWT) composite watermarking. Robustness against geometric distortion is one of the crucial important issues in watermarking. Here, watermarking is embedded in high-frequency image by singular value decomposition. This is unlike traditional viewpoint that assumes watermarking should be embedded in low or middle frequency to have good robustness. Experimental evaluation demonstrates that this algorithm is able to withstand a variety of attacks including common geometric attacks.

3.1.3 DCT Domain Watermarking

Discrete Cosine Transformation (DCT)-based watermarking techniques are more robust compared to simple spatial domain watermarking techniques [2, 21]. Such algorithms are robust against simple image processing operations like low-pass filtering, brightness and contrast adjustment, blurring, etc. However, they are difficult to implement and are computationally more expensive. At the same time they are weak against geometric attacks like rotation, scaling, cropping, etc. DCT domain watermarking can be classified into global DCT watermarking and block-based DCT watermarking. One of the first algorithms presented by Cox et al. used global DCT approach to embed a robust watermark in the perceptually significant portion of the human visual system. Embedding in the perceptually significant portion of the image has its own advantages, because most compression schemes remove the perceptually insignificant portion of the image. The main steps of any block-based DCT algorithm is depicted in Algorithm 1.

Algorithm 1: DCT Block-Based Watermarking Algorithm	
1.	Segment the image into nonoverlapping blocks of 8×8
2.	Apply forward DCT to each of these blocks
3.	Apply some block selection criteria (e.g., Human Visual System (HVS))
4.	Apply coefficient selection criteria (e.g., highest)
5.	Embed watermark by modifying the selected coefficients
6.	Apply inverse DCT transform on each block

Most algorithms discussed here are classified based on steps 3 and 4, i.e., the main difference between most algorithms is that they differ either in the block selection criteria or coefficient selection criteria and based on the perceptual modeling strategy incorporated by the watermarking algorithms.

Bors et al. [22] presents image watermarking using DCT domain constraints. This algorithm selects certain blocks in the image based on a Gaussian network classifier. The pixel values of the selected blocks are modified such that their discrete cosine transform coefficients fulfill a constraint imposed by the watermark code. The watermarks embedded by the presented algorithm are resistant to image compression.

Hernandez et al. [23] presents DCT—domain watermarking techniques for still images. The DCT is applied in blocks of 8×8 pixels. The watermark can encode information to track illegal misuses. For flexibility purposes, the original image is not necessary during the ownership verification process, so it must be modeled by noise. As a result of their work, analytical expressions for performance measures such as the probability of error in watermark decoding and probabilities of false alarm in watermark detection are derived.

Lu et al. [24] presents Block DCT-based Robust watermarking using side information extracted by mean filtering. Here, the authors investigate the characteristic of mean filtering to formulize the problems of watermark embedding and extraction. This leads to a nonadditive embedding strategy and a sophisticated blind detection mechanism. With regard to geometric attacks, block-based watermarking and moment normalization are both addressed without relying on pilot signals. The new watermarking scheme has been verified by Stirmark and a new benchmark.

Mohanty et al. [25] presents DCT domain visible watermarking technique for images. In this technique, they describe a visible watermarking scheme that is applied into the host image in the DCT domain. A mathematical model has been developed for that purpose. They also present a modification of the algorithm to make the watermark more robust.

3.1.4 Wavelet Domain Watermarking

This section discusses wavelet-based watermarking algorithms. Here, these algorithms are classified based on their decoder requirements as blind detection or non-blind detection. As mentioned earlier, blind detection does not require the original image for detecting the watermarks, however, non-blind detection requires the original image [26].

DWT Domain Watermarking

In the past few years, wavelet transform has been widely studied in signal processing in general and image compression in particular. In some applications wavelet-based watermarking schemes outperforms DCT-based approaches.

DWT Watermarking

DWT based watermarking schemes follow the same guidelines as DCT-based schemes, i.e., the underlying concept is the same, however, the process to transform the image into its transform domain varies and hence the resulting coefficients are different. Wavelet transforms use wavelet filters to transform the image [27]. There are many available filters, although the most commonly used filters for watermarking are Haar wavelet Filter, Daubechies orthogonal filters, and Daubechies bi-orthogonal filters. Each of these filters decomposes the image into several frequencies. Single-level decomposition gives four frequency representations of the images. These four representations are called the Low Low (LL), Low High (LH), High Low (HL), High High (HH) subbands as shown in Fig. 5.

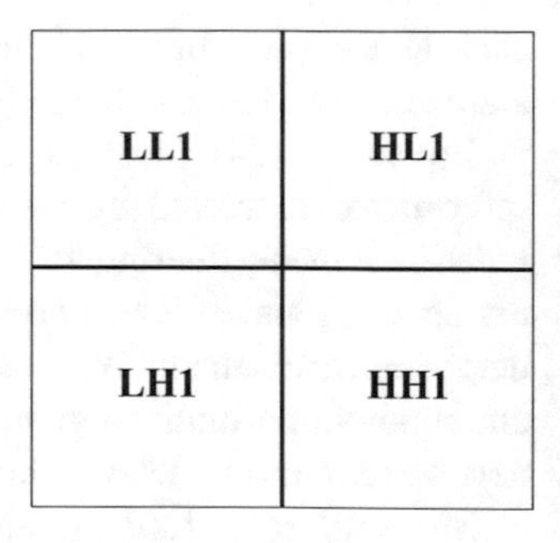

Fig. 5 Single-level decomposition using DWT

DWT-Based Blind Watermark Detection

Lu et al. [28, 29] presents a novel watermarking technique called as "cocktail watermarking". This technique embeds dual watermarks which complement each other. This scheme is resistant to several attacks, and no matter what type of attack is applied, one of the watermarks can be detected. Furthermore, they enhance this technique for image authentication and protection by using the wavelet based just noticeable distortion values. Hence, this technique achieves copyright protection as well as content authentication simultaneously [17, 28].

Zhu et al. [30] presents a multi-resolution watermarking technique for watermarking video and images. The watermark is embedded in all the high-pass bands in a nested manner at multiple resolutions. This technique does not consider the HVS aspect.

Voyatzis and Pitas [31], who presented the "toral automorphism" concept, provides a technique to embed binary logo as a watermark, which can be detected using visual models as well as by statistical means, So in case the image is degraded too much and the logo is not visible, it can be detected statistically using correlation. Watermark embedding is based on a chaotic (mixing) system. Original image is not required for watermark detection. However, the watermark is embedded in spatial domain by modifying the pixel or luminance values. A similar approach is presented for the wavelet domain, where the authors propose a watermarking algorithm based on chaotic encryption.

DWT-Based Non-blind Watermark Detection

This technique requires the original image for detecting the watermark [32]. Most of the techniques found in literature use a smaller image as a watermark and hence cannot use correlation-based detectors for detecting the watermark; as a result they rely on the original image for informed detection. The size of the watermark image normally is smaller compared to the host image.

Xia et a1. [33] presents a wavelet-based non-blind watermarking technique for still images, where watermarks are added to all bands except the approximation band. A multi-resolution-based approach with binary watermarks is presented.

Here, both the watermark logo as well as the host image is decomposed into subbands and later embedded. Watermark is subjectively detected by visual inspection; however, an objective detection is employed by using normalized correlation.

Raval and Rege [34] presents a multiple watermarking technique. The authors argue that if the watermark is embedded in the low-frequency components it is robust against low-pass filtering, lossy compression, and geometric distortions. On the other hand, if the watermark is embedded in high-frequency components, it is robust against contrast and brightness adjustment, gamma correction, histogram equalization, and cropping and vice versa.

Kundur and Hatzinakos [35] presents image fusion watermarking technique. They use salient features of the image to embed the watermark. They use a saliency measure to identify the watermark strength and later embed the watermark additively. Normalized correlation is used to evaluate the robustness of the extracted watermark. Later the authors propose another technique termed as FuseMark, which includes minimum variance fusion for watermark extraction.

Tao and Eskicioglu [36] presents an optimal wavelet-based watermarking technique. They embed binary logo watermark in all the four bands. But they embed the watermarks with variable scaling factor in different bands. The scaling factor is high for thc LL subband but for the other three bands its lower. The quality of the extracted watermark is determined by similarity ratio measurement for objective calculation.

3.1.5 DFT Domain Watermarking

Discrete Fourier Transformation (DFT) domain has been explored by researches because it offers robustness against geometric attacks like rotation, scaling, cropping, translation, etc. This section, discusses some watermarking algorithms based on the DFT domain [37].

Direct Embedding

There are few algorithms that modify these DFT magnitude and phase coefficients to embed watermarks. Ruanaidh et al. proposed a DFT watermarking technique in which the watermark is embedded by modifying the phase information within the DFT. It has been shown that phase-based watermarking is robust against image contrast operation.

Kang et al. [38] presents a new resilient watermarking algorithm. The watermark is embedded in the magnitude coefficients of the Fourier transform re-sampled by log-polar mapping. The technique is however not robust against cropping and shows weak robustness against JPEG compression.

Solachidis and Pitas [39] presents a novel watermarking technique. They embed a circularly symmetric watermark in the magnitude of the DFT domain. Since the watermark is circular in shape with its center at image cenetr, it is robust against geometric rotation attacks. The watermark is centered on the mid frequency region of the DFT magnitude. Neighborhood pixel variance masking is employed to reduce any visible artifacts. The technique is computationally not expensive to recover from rotation.

Ganic and Eskicioglu [40] presents a semi-blind watermarking technique. They embed circular watermarks with one in the lower frequency while the other in the higher frequency. They follow the same argument as that endorsed by embedding watermarks in the low-frequency component, which is robust against one set of attacks, while embedding in the high-frequency components is robust to another set of attacks.

Template-Based Embedding

Pereira and Pun presents robust watermarking algorithm resistant to affine transformations. They introduce the concept of template. A template is a structure which is embedded in the DFT domain to estimate the transformation factor. Once the image undergoes a transformation this template is searched to resynchronize the image, and then use the detector to extract the embedded spread spectrum watermark.

3.1.6 FFT and DHT Domain Watermarking

Pereira et al. [41] presents a watermarking algorithm based on Fast Fourier Transformation (FFT) that is robust against compression attacks. It is a template-based embedding algorithm similar to the one discussed in the previous section. Apart from the template, an informative watermark is embedded to prove ownership. In case the image undergoes a geometric distortion the template is reversed back to its original location and then the watermark is extracted. They employ the concept of log-polar maps and log-log maps to recover the hidden template. This technique is shown to be robust against cropping, print and scan attack.

Falkowski and Lim [42] presents a watermarking technique based on multi-resolution transform and complex Hadamard transform. Initially the multi-resolution Hadamard transform is applied to the image to decompose it into various frequency bands like low–low, low–high, and high–high. The lowest frequency band is then divided into 8 × 8 blocks and 2-Dimensional (D) complex Hadamard transform is applied. Watermark is embedded in this domain by altering the phase component of the most significant image component. The watermark is embedded in the phase component because phase modulation is more robust to noise than amplitude modulation. This scheme is shown to be robust against several attacks like image compression, image scaling, dithering, cropping, and successive watermarking.

Gilani and Skodras [43] presents another watermarking technique, which embeds the watermark by modifying the high-frequency Hadamard coefficients. An image undergoes double frequency transform initially by Haar wavelet transform and later by Hadamard transform. This gives rise to the multi-resolution Hadamard frequency domain. The Hadamard transform concentrates most of the energy in the upper left corner, and hence it is selected to embed watermark information.

3.1.7 Frequency Domain Watermarking

The basic principles of adding or changing components of digital images and other digital documents can be transferred to other value domains [44]. In order to integrate watermark information into frequency components, the document has to be transformed into its frequency components. The LSB method is also applied in the frequency domain by selecting the pixel based on the frequency. Most methods involve the modification of transformed coefficients based on their frequencies due to its imperceptibility. In a frequency-based watermarking scheme, the watermark upon inverse transformation to the spatial domain, is dispersed throughout the image making it very difficult for an attacker to remove the watermark without causing significant damage to the image. In most cases digital watermarks are integrated into the mid-band frequencies.

Shao et al. [45] presents a frequency domain watermarking algorithm for image and video. By using a novel frequency division multiplexing structure, the information watermark and synchronization template are jointly embedded both in the frequency domain, thus the algorithm has resistance to geometric distortions and real-time embedding for bit stream is possible. Detection result is improved by using concatenation coding. Copy attack is combated by introducing low-frequency content information in the watermark generation instead of using a second watermark. Experiments show the presented algorithm is robust and well suited for such applications as copyright protection and broadcast monitoring.

3.1.8 Spread Spectrum Watermarking

Spread spectrum techniques used in digital watermarking is borrowed from the communication field. The basic idea of spread spectrum is to spread the data across a large frequency band. In the case of audio, it is the entire audible spectrum, in the case of images, it is the whole visible spectrum. Spread spectrum is a military technology designed to handle interferences and disturbances. In most cases, signals that represent the information are modulated at low intensity across the source bandwidth. Spread spectrum communication is used in radar, navigation, and communication applications. The information is weaved into the source material using a secret key or an embedding procedure [46].

Xiao and Wang [47] present spread spectrum-based watermarking method. In spread transform dither modulation, the watermark signal is embedded into the spread transformed data, where spread transform is an important factor that affects the performances. Spread transform (also called projection) makes the embedding distortion concentrating on one coefficient spread to multiple coefficients. This leads to some advantages, such as the satisfaction of peak distortion limitations.

3.2 Authentication

The next stage in fingerprint automation occurred at the end of 1994 with the Integrated Automated Fingerprint Identification System (IAFIS) competition. The competition identified and investigated three major challenges:

(1) digital fingerprint acquisition,
(2) local ridge characteristic extraction, and
(3) ridge characteristic pattern matching [48].

The first Automated Fingerprint Identification System (AFIS) was developed by Palm System in 1993.

The year 2000 envisaged the first face recognition vendor test (FRVT, 2000) sponsored by the US Government agencies and the same year paved way for the first research paper on the use of vascular patterns for recognition [49]. During 2003, ICAO (International civil Aviation Organization) adopted blueprints for the integration of biometric identification information into passports and other Machine Readable Travel Documents (MRTDs). Facial recognition was selected as the globally interoperable biometric for machine-assisted identity confirmation with MRTDs

Zhao et al. [50] proposed an adaptive pore model for fingerprint pore extraction. Sweat pores have been recently employed for automated fingerprint recognition, in which the pores are usually extracted by using a computationally expensive skeletonization method or a unitary scale isotropic pore model. In this paper, however, the author shows that real pores are not always isotropic. To accurately and robustly extract pores, they propose an adaptive anisotropic pore model, whose parameters are adjusted adaptively according to the fingerprint ridge direction and period. The fingerprint image is partitioned into blocks and a local pore model is determined for each block. With the local pore model, a matched filter is used to extract the pores within each block. Experiments on a high-resolution (1200 dpi) fingerprint dataset are performed and the results demonstrate that the proposed pore model and pore extraction method can locate pores more accurately and robustly in comparison with other state-of-the-art pore extractors.

Moheb et al. [51] proposed an approach to image extraction and accurate skin detection from web pages. This paper proposes a system to extract images from web pages and then detect the skin color regions of these images. As part of the proposed system, using band object control, they build a Tool bar named "Filter Tool Bar (FTB)" by modifying the Pavel Zolnikov implementation. In the proposed system, they introduce three new methods for extracting images from the web pages (after loading the web page by using the proposed FTB, before loading the web page physically from the local host, and before loading the web page from any server). These methods overcome the drawback of the regular expressions method for extracting images suggested by Ilan Assayag. The second part of the proposed system is concerned with the detection of the skin color regions of the extracted images. So, they studied two famous skin color detection techniques.

Kaur et al. [52] proposed a fingerprint verification system using minutiae extraction technique. Most fingerprint recognition techniques are based on minutiae matching and have been well studied. However, this technology still suffers from problems associated with the handling of poor quality impressions. One problem besetting fingerprint matching is distortion. Distortion changes both geometric position and orientation, and leads to difficulties in establishing a match among multiple impressions acquired from the same finger tip. Marking all the minutiae accurately as well as rejecting false minutiae is another issue still under research. Their work has combined many methods to build a minutia extractor and a minutia matcher. The combination of multiple methods comes from a wide investigation into research papers. Also, some novel changes like segmentation using morphological operations, improved thinning, false minutiae removal methods, minutia marking with special considering the triple branch counting, minutia unification by decomposing a branch into three terminations, and matching in the unified x-y coordinate system after a two-step transformation are used in the work.

Le et al. [53] proposed online fingerprint identification with a fast and distortion tolerant hashing method. National ID card, electronic commerce, and access to computer networks are some scenarios where reliable identification is a must. Existing authentication systems relying on knowledge-based approaches like passwords or token based such as magnetic cards and passports contain serious security risks due to the vulnerability to engineering-social attacks and the easiness of sharing or compromising passwords and PINs. Biometrics such as fingerprint, face, eye retina, and voice offer a more reliable means for authentication. However, due to large biometric database and complicated biometric measures, it is difficult to design both an accurate and fast biometric recognition. Particularly, fast fingerprint indexing is one of the most challenging problems faced in fingerprint authentication system. In this paper, they present a specific contribution by introducing a new robust indexing scheme that is able not only to fasten the fingerprint recognition process but also improve the accuracy of the system.

Ratha et al. [54] proposed an adaptive flow orientation based segmentation or binarization algorithm. In this approach, the orientation field is computed to obtain the ridge directions at each point in the image. To segment the ridges, a 16×16 window oriented along the ridge direction is considered around each pixel. The projection sum along the ridge direction is computed. The centers of the ridges appear as peak points in the projection. The ridge skeleton thus obtained is smoothened by morphological operation. Finally minutiae are detected by locating end points and bifurcations in the thinned binary image.

Jain et al. [55] proposed a Pores and Ridges: Fingerprint Matching Using Level 3 Features. Fingerprint friction ridge details are generally described in a hierarchical order at three levels, namely, level 1 (pattern), level 2 (minutiae points), and level 3 (pores and ridge shape). Although high-resolution sensors ($\sim$1000 dpi) have become commercially available and have made it possible to reliably extract level 3 features, most Automated Fingerprint Identification Systems (AFIS) employ only level 1 and level 2 features. As a result, increasing the scan resolution does not provide any matching performance improvement [17].

Vatsa et al. [56] proposed a combining pores and ridges with minutiae for improved fingerprint verification. This paper presents a fast fingerprint verification algorithm using level-2 minutiae and level-3 pore and ridge features. The proposed algorithm uses a two-stage process to register fingerprint images. In the first stage, Taylor series-based image transformation is used to perform coarse registration, while in the second stage thin plate spline transformation is used for fine registration. A fast feature extraction algorithm is proposed using the Mumford–Shah functional curve evolution to efficiently segment contours and extracts the intricate level-3 pore and ridge features. Further, Delaunay triangulation-based fusion algorithm is proposed to combine level-2 and level-3 information that provides structural stability and robustness to small changes caused due to extraneous noise or nonlinear deformation during image capture. They define eight quantitative measures using level-2 and level-3 topological characteristics to form a feature super vector. A 2n-support vector machine performs the final classification of genuine or impostor cases using the feature super vectors. Experimental results and statistical evaluation show that the feature super vector yields discriminatory information and higher accuracy compared to existing recognition and fusion algorithms.

Uludaga et al. [57] proposed a biometric template selection and update: a case study in fingerprints. Sweat pores have been recently employed for automated fingerprint recognition, in which the pores are usually extracted by using a computationally expensive skeletonization method or a unitary scale isotropic pore model. In this paper, however, real pores are not always isotropic. To accurately and robustly extract pores, they propose an adaptive anisotropic pore model, whose parameters are adjusted adaptively according to the fingerprint ridge direction and period. The fingerprint image is partitioned into blocks and a local pore model is determined for each block. With the local pore model, a matched filter is used to extract the pores within each block. Experiments on a high-resolution (1200 dpi) fingerprint dataset are performed and the results demonstrate that the proposed pore model and pore extraction method can locate pores more accurately and robustly in comparison with other state-of-the-art pore extractors.

Coetzee and Botha [58] proposed a binarization technique based on the use of edges extracted using Marr–Hilderith operator. The resulting edge image is used in conjunction with the original gray-scale image to obtain the binarized image. This is based on the recursive approach of line following and line thinning. Two adaptive windows, the edge window and the gray-scale window are used in each step of the recursive process. To begin with, the pixel with the lowest gray-scale value is chosen and a window is centered on it. The boundary of the window is then examined to determine the next position of the window. The window is successively positioned to trace the ridge boundary and the recursive process terminates when all the ridge pixels have been followed to their respective ends.

Bolle et al. [59] proposed the evaluation techniques for biometrics-based authentication systems. Biometrics-based authentication is becoming popular because of increasing ease-of-use and reliability. Performance evaluation of such systems is an important issue. They endeavor to address two aspects of performance evaluation that have been conventionally neglected. First, the "difficulty" of the data that is used in a study influences the evaluation results. They propose some measures to characterize the data set so that the performance of a given system on different data sets can be compared. Second, conventional studies often have reported the false reject and false accept rates in the form of match score distributions. However, no confidence intervals are computed for these distributions, hence no indication of the significance of the estimates is given. In this paper, they compare the parametric and nonparametric (bootstrap) methods for measuring confidence intervals. They give special attention to false reject rate estimates.

Yuan et al. [60] proposed a real-time fingerprint recognition system based on novel fingerprint matching strategy. In this paper, they present a real-time fingerprint recognition system based on a novel fingerprint minutiae matching algorithm. The system is developed to be applicable to today's embedded systems for fingerprint authentication, in which small area sensors are employed. The system is comprised of fingerprint enhancement and quality control, fingerprint feature extraction, fingerprint matching using a novel matching algorithm, and connection with other identification system. Here, they describe their way to design a more reliable and fast fingerprint recognition system which is based on today's embedded systems in which small area fingerprint sensors are used.

Cui et al. [61] proposed the research of edge detection algorithm for fingerprint images. This paper introduces some edge detection operators and compares their characteristics and performances. At last the experiments show that each algorithm has its advantages and disadvantages, and the suitable algorithm should be selected according to the characteristics of the images detected, so that it can perform perfectly. The Canny operator is not susceptible to the noise interference; it can detect the real weak edge. The advantage is that it uses two different thresholds to detect the strong edge and the weak edge, and the weak edge will be included in the output image only when the weak edge is connected to the strong edge. The Sobel operator has a good performance on the images with gray gradient and high noise, but the location of edges is not very accurate, the edges of the image have more than one pixel. The Binary Image Edge Detection Algorithm is simple, but it can detect the edge of the image accurately, and the processed images are not need to be thinned, it particularly adapts to process various binary images with no noise. So each algorithm has its advantages and disadvantages, and the suitable algorithm should be selected according to the characters of the images detected, then it can perform perfectly.

Li et al. [62] proposed the Image Enhancement Method for Fingerprint Recognition System. In this paper, fingerprint image enhancement method, a refined Gabor filter, is presented. This enhancement method can connect the ridge breaks, ensures the maximal gray values located at the ridge center and has the ability to compensate for the nonlinear deformations. It includes ridge orientation estimation, a Gabor filter processing and a refined Gabor filter processing. The first Gabor filter reduces the noise, provides more accurate distance between the two ridges for the next filter and gets a rough ridge orientation map while the refined Gabor filter with the adjustment parameters significantly enhances the ridge, connects the ridge breaks and ensures the maximal gray values of the image being located at the ridge center. In addition, the algorithm has the ability to compensate for the nonlinear deformations. Furthermore, this method does not result in any spurious ridge structure, which avoids undesired side effects for the subsequent processing and provides a reliable fingerprint image processing for Fingerprint Recognition System. In a word, a refined Gabor filter is applied in fingerprint image processing, then a good quality fingerprint image is achieved, and the performance of Fingerprint Recognition System has been improved.

Mil'shtein et al. [63] proposed a fingerprint recognition algorithm for partial and full fingerprints. In this study, they propose two new algorithms. The first algorithm, called the Spaced Frequency Transformation Algorithm (SFTA), is based on taking the Fast Fourier Transform of the images. The second algorithm, called the Line Scan Algorithm (LSA), was developed to compare partial fingerprints and reduce the time taken to compare full fingerprints. A combination of SFTA and LSA provides a very efficient recognition technique. The most notable advantages of these algorithms are the high accuracy in the case of partial fingerprints. At this time, the major drawback of developed algorithms is lack of pre-classification of examined fingers. Thus, they use minutiae classification scheme to reduce the reference base for given tested finger. When the reference base had shrunk, they apply the LSA and SFTA.

Another paper proposed a novel approach by Ross et al. [64] for minutiae filtering in fingerprint images. Existing structural approaches for minutiae filtering use heuristics and ad hoc rules to eliminate such false positives, where as gray-level approach is based on using raw pixel values and a supervised classifier such as neural networks. They proposed two new techniques for minutiae verification based on nontrivial gray level features. The proposed features intuitively represent the structural properties of the minutiae neighborhood leading to better classification. They use directionally selective steerable wedge filters to differentiate between minutiae and non-minutiae neighborhoods with reasonable accuracy. They also propose a second technique based on Gabor expansions that result in even better discrimination. They present an objective evaluation of both the algorithms. Apart from minutiae verification, the feature description can also be used for minutiae detection and minutiae quality assessment.

Karna et al. [65] proposed normalized cross-correlation-based fingerprint matching. It has been in use for quite some time to perform fingerprint matching based on the number of corresponding minutia pairings. But this technique is not very efficient for recognizing the low-quality fingerprints. To overcome this problem, some researchers suggest the correlation technique which provides better result. Use of correlation-based methods is increasing day-by-day in the field of biometrics as it provides better results. In this paper, they propose normalized cross-correlation technique for fingerprint matching to minimize error rate as well as reduce the computational effort than the minutiae matching method. The Equal Error Rate (ERR) obtained from result till now with minutiae matching method is 3%, while that obtained for the method proposed in this paper is approximately 2% for all types of fingerprints in combined form.

Bazen et al. [66] proposed a correlation-based fingerprint verification system. In this paper, a correlation-based fingerprint verification system is presented. Unlike the traditional minutiae-based systems, this system directly uses the richer gray-scale information of the fingerprints. The correlation-based fingerprint verification system first selects appropriate templates in the primary fingerprint, uses template matching to locate them in the secondary print, and compares the template positions of both fingerprints. Unlike minutiae-based systems, the correlation-based fingerprint verification system is capable of dealing with bad-quality images from which no minutiae can be extracted reliably and with fingerprints that suffer from nonuniform shape distortions. Experiments have shown that the performance of this system at the moment is comparable to the performance of many other fingerprint verification systems.

Lowe [67] proposed an approach to distinctive image features from scale-invariant key points. This paper presents a method for extracting distinctive invariant features from images that can be used to perform reliable matching between different views of an object or scene. The features are invariant to image scale and rotation, and are shown to provide robust matching across a substantial range of affine distortion, change in 3D viewpoint, addition of noise, and change in illumination. The features are highly distinctive, in the sense that a single feature can be correctly matched with high probability against a large database of features from many images. This paper also describes an approach to use these features for object recognition. The recognition proceeds by matching individual features to a database of features from known objects using a fast nearest neighbor algorithm, followed by a Hough transformation to identify clusters belonging to a single object, and finally performing verification through least-squares solution for consistent pores parameters. This approach to recognition can robustly identify objects among clutter and occlusion while achieving near real-time performance.

4 Methodology

Methodology consists of two phases called watermarking and authentication. For watermarking, the FHT mechanism is used which is discussed in Sect. 4.1 in detail. For authentication, the SVD mechanism is used which is discussed in Sect. 4.2 in detail.

4.1 Hadmard Transformation Theory

The Hadamard transform has been used extensively in image processing and image compression [68–70]. In this section, we give a brief review of the Hadamard transform representation of image data, which is used in the watermarking embedding and extraction process. The reason for choosing FHT domain is also discussed. Let $[U]$ represents the original image and $[V]$ the transformed image, the 2D-Hadamard transform is given by Eq. 1.

$$[\mathbf{V}] = \frac{\mathbf{H_n}[\mathbf{U}]\mathbf{H_N}}{\mathbf{N}} \tag{1}$$

where $\boldsymbol{H_n}$ represents an N * N Hadamard matrix, $\boldsymbol{N} = \mathbf{2}^n$, n = 1, 2, 3, …, with element values either +1 or −1.

The advantages of Hadamard transform are that the elements of the transform matrix $\boldsymbol{H}$. are simple: they are binary, real numbers and the rows or columns of $\boldsymbol{H_n}$ are orthogonal. Hence the Hadamard transform matrix has the following property shown in Eq. 2.

$$\mathrm{H_n} = \mathrm{H_n^*} = \mathrm{H^T} = \mathrm{H^{-1}} \tag{2}$$

Since $\boldsymbol{H_n}$ has $\boldsymbol{N}$ orthogonal rows $H_nH_n = \mathrm{NI}$ (I is the identity matrix) and $H_nH_n = NH_nH_n^{-1}$, thus $H_n^{-1} = \frac{H_n}{N}$. The inverse 2D-fast Hadmard transfer (IFHT) is given as Eq. 3.

$$[\mathbf{U}] = \mathbf{H_n^{-1}}[\mathbf{U}]\mathbf{H_n^*} = \frac{\mathbf{H_n}[\mathbf{V}]\mathbf{H_N}}{\mathbf{N}} \tag{3}$$

For N = 2, the Hadmard matrix, $\mathrm{H_1}$, is called as core matrix, which is defined as Eq. 4.

$$H_1 = \begin{bmatrix} 1 & 1 \\ 1 & -1 \end{bmatrix} \tag{4}$$

The Hadamard matrix of the order n is generated in terms of Hadamard matrix of order (n − 1) using Kronecker product ⊗, as Eq. 5 or 6

$$H_n = H_{n-1} \otimes H_1 \tag{5}$$

or

$$H_n = \begin{bmatrix} H_{n-1} & H_{n-1} \\ H_{n-1} & -H_{n-1} \end{bmatrix} \tag{6}$$

Since in our algorithm, the processing is carried out based on the 8 × 8 sub-blocks of the whole image, the third order Hadamard transform matrix H_3 is used. By applying Eq. (5) or (6) we will get the H_3

Equivalently, we can define the Hadamard matrix by its (k, n)-th entry by writing as Eqs. 7 and 8

$$k = \sum_{i=0}^{m-1} k_i 2^i = k_{m-1}2^{m-1} + k_{m-2}2^{m-2} + \ldots . + k_1 2 + k_o \tag{7}$$

and

$$n = \sum_{i=0}^{m-1} n_i 2^i = n_{m-1}2^{m-1} + n_{m-2}2^{m-2} + \ldots . + n_1 2 + n_o \tag{8}$$

where the k_j and n_j are the binary digits (0 or 1) of k and n, respectively. Note that for the element in the top left corner, we define: $k - n - 0$. In this case, we have Eq. 9.

$$(H_m)_{k,n} = \frac{1}{2^{\frac{m}{2}}}(-1) \sum_j k_j n_j \tag{9}$$

The main steps of the watermark embedding process are depicted in Algorithm 2 and the extraction process is provided in Algorithm 3 which are also illustrated in the block diagrams shown in Figs. 6 and 7.

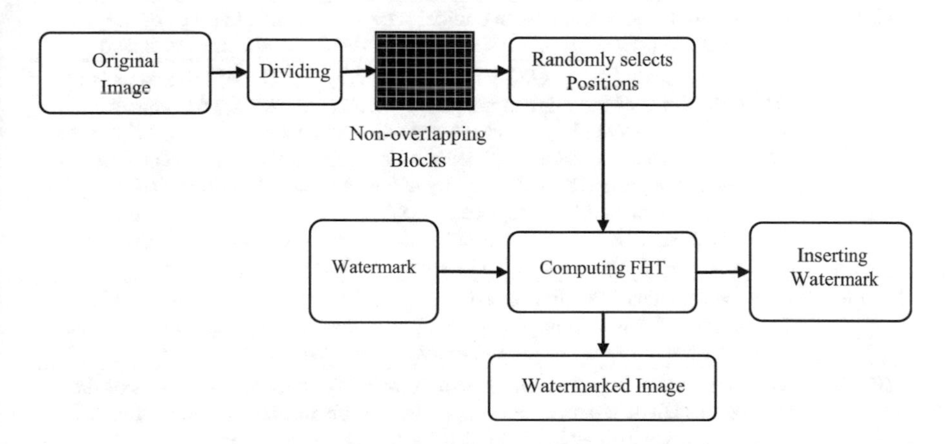

Fig. 6 Image-in-image watermark embedding process

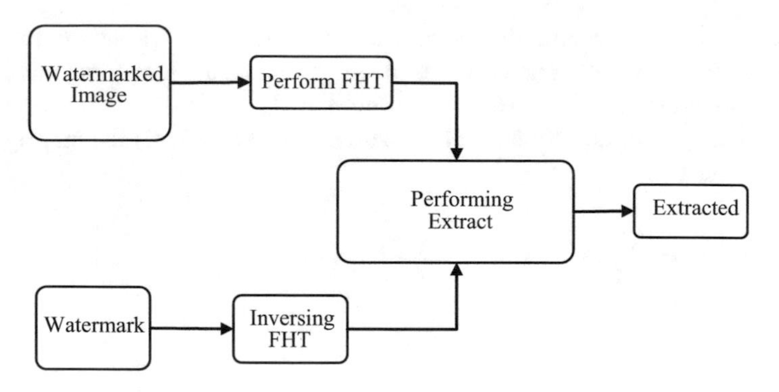

Fig. 7 Image-in-image watermark extraction process

Algorithm 2: Watermark Embedding with FHT

Input: Facial Image (is denoted by $f(x, y)$), Fingerprint (is denoted by $w(x, y)$)
Output: Watermarked Image

STEP 1:	Fingerprint (watermark) $w(x, y)$ is transformed into FHT coefficients. After transformation, 16×16 watermark, Hadamard transform coefficients are obtained. Here the DC component is stored in a key file and the AC components are then selected for embedding
STEP 2:	The facial image $f(x, y)$, is also decomposed into a set of nonoverlapped blocks of $h \times h$, denoted by $f_k(x^1, y^1)$, $k = 0, 1, \ldots, \boldsymbol{K} - 1$, where the subscript k denotes the index of blocks and $\boldsymbol{K}$ denotes the total number of blocks
STEP 3:	This Fast Hadamard Transforms algorithm, pseudo randomly selects the sub-blocks from the original image by using java-sequence random number for watermark insertion
STEP 4:	The seed of this java-sequence and initial state are also stored in the key file. After that, a FHT is performed on each selected sub-blocks of the original image
STEP 5:	Let the watermark FHT coefficients are denoted by m_i. The AC components of FHT coefficients of the original image sub-blocks, before and after inserting watermark are denoted by x_i, and x_i* respectively. Where $i \in (0, n)$, with n the number of the watermarked coefficients which is 16 in this experiment. The watermark strength factor is denoted by α. The embedding formula is $\boldsymbol{X_i^* = \alpha m_i}$ The original coefficient X_i is replaced by X_i* Here α is strength factor and its formula is $\alpha = \beta * mask_1(j, k) * mask_2(j, k)$ Where β is the scaling factor, j and k indicate the positions of the sub-blocks
STEP 6:	After the watermark insertion, a new 8×8 matrix of FHT coefficients is obtained. The Inverse FHT is then applied to the 8×8 matrix using Eq. 1 to obtain the luminance value matrix of the watermarked image sub-block $f_k^1(x^1, y^1)$
STEP 7:	After performing the watermark insertion for all the relevant sub-blocks of the original image, the watermarked image $f^1(x, y)$ is obtained. At the same time, a key file is also generated, which is needed for the decoding process

Algorithm 3: Watermark Extraction with FHT

Input: Watermarked Image (is denoted by $f^{11}(x, y)$)
Output: Facial image and fingerprint are extracted

STEP 1:	Transforming all the relevant sub-blocks, $f_k^{11}(x^1, y^1)$, into the FHT domain, Here obtain all the Hadamard transform coefficients embedded with the watermark
STEP 2:	In each of the sub-block FHT coefficients, the watermark bits are inserted into the bottom right sixteen middle and high frequency components. Let these components are denoted by $X_i{*}^1$, the retrieved watermark FHT coefficients are denoted by m_i^1, where $i \in (0, n)$, and the number of the watermarked coefficients n = 16
STEP 3:	The watermark extraction formula is given as: $m_i^1 = \frac{X_i^{*1}}{\alpha}$ Where α is strength factor and its formula is $\boldsymbol{\alpha = \beta * mask_1(j, k) * mask_2(j, k)}$, Where β is the scaling factor, j and k indicate the positions of the sub-blocks
STEP 4:	The watermark FHT coefficients are extracted from all the sub-blocks of the original image. Along with the DC component stored in the key file, the AC coefficients are rearranged into a 16 × 16 FHT coefficients matrix
STEP 5:	The watermark FHT coefficients are extracted from all the sub-blocks of the original image
STEP 6:	The extracted watermark image, $w(x, y)$, is obtained by Inverse FHT of the 16 × 16 Hadamard coefficients matrix using Eq. 3

4.2 *Theory of Singular Value Decomposition*

Singular Value Decomposition (SVD) is said to be a significant topic in linear algebra by many renowned mathematicians [71]. SVD has many practical and theoretical values; special feature of SVD is that it can be performed on any real (m, n) matrix. Let us say we have a matrix ***A*** with ***m*** rows and ***n*** columns, with rank ***r*** and ***r*** < ***n*** < m. Then the ***A*** can be factorized into three matrices shown in Eq. 10.

$$A = USV^T \tag{10}$$

and it is illustrated as follows:

$$\underset{m\times n}{A} = \underset{m\times m}{U}\;\underset{m\times n}{S}\;\underset{n\times n}{V^T}$$

where Matrix U is an m × m orthogonal matrix shown in Eq. 11

$$U = [u_1, u_2, \ldots, u_r, u_{r+1}, \ldots, u_m] \tag{11}$$

columns vectors u_i, for i = 1, 2, …, m, form an orthonormal set shown in Eq. 12

$$u_i^T u_j = \delta_{ij} = \begin{cases} 1, & i = j \\ 0, & i \neq j \end{cases} \tag{12}$$

and matrix V is an n × n orthogonal matrix shown in Eq. 13

$$V = [v_1, v_2, \ldots, v_r, v_{r+1}, \ldots, v_n] \tag{13}$$

column vectors v_i, for i = 1, 2, …, n, form an orthonormal set shown in Eq. 14

$$v_i^T v_j = \delta_{ij} = \begin{cases} 1, & i=j \\ 0, & i \neq j \end{cases} \tag{14}$$

Here, S is an $m \times n$ diagonal matrix with singular values (SV) on the diagonal. The matrix S can be shown in Eq. 15

$$s = \begin{bmatrix} \sigma_1 & \mathbf{0} & \ldots & \mathbf{0} & \mathbf{0} & \ldots & \mathbf{0} \\ \mathbf{0} & \sigma_2 & \ldots & \mathbf{0} & \mathbf{0} & \ldots & \mathbf{0} \\ \vdots & \vdots & \ddots & \vdots & \vdots & \ddots & \vdots \\ \mathbf{0} & \mathbf{0} & \ldots & \sigma_2 & \mathbf{0} & \ldots & \mathbf{0} \\ \mathbf{0} & \mathbf{0} & \ldots & \sigma_r & \mathbf{0} & \ldots & \mathbf{0} \\ \vdots & \vdots & \ddots & \vdots & \sigma_{r+1} & \ldots & \vdots \\ \mathbf{0} & \mathbf{0} & \ldots & \mathbf{0} & \mathbf{0} & \ldots & \sigma_n \\ \mathbf{0} & \mathbf{0} & \ldots & \mathbf{0} & \mathbf{0} & \ldots & \mathbf{0} \end{bmatrix} \tag{15}$$

For i = 1, 2, …, n, σ_i are called Singular Values (SV) of matrix A. It can be proved that

$$\sigma_1 \geq \sigma_2 \geq \cdots \geq \sigma_r > 0, and\, \sigma_{r+1} = \sigma_{r+2} = \cdots = \sigma_N = 0 \tag{16}$$

For i = 1, 2, …, n, σ_i are called Singular Values (SV) of matrix A. The $\mathrm{v_i}$'s and $\mathrm{u_i}$'s are called right and left singular-vectors of A [1].

4.2.1 Authentication with SVD Approach

In this case, we redefined the matrix A as a set of the training face & finger images. Assume each face &finger image has $\boldsymbol{m} \times \boldsymbol{n} = \boldsymbol{M}$ pixels, and is represented as an $\boldsymbol{M} \times 1$ column vector f, a 'training set' $\boldsymbol{S}$ with $\boldsymbol{N}$ number of face and finger images of known individuals form an $\boldsymbol{M} \times \boldsymbol{N}$ matrix shown in Eq. 17

$$S = \mathrm{f}_1, \mathrm{f}_2, \ldots, \mathrm{f}_N \tag{17}$$

The mean image f of set S, is given by Eq. 18

$$\bar{\mathbf{f}} = \frac{1}{\mathbf{N}} \sum_{\mathbf{i}=1}^{\mathbf{N}} \mathbf{f_i} \tag{18}$$

Subtracting $\bar{f}$ from the original face & finger gives by Eq. 19

$$\mathbf{a_i} = \mathbf{f_i} - \bar{\mathbf{f}}, \mathbf{i} = 1, 2, \ldots, \mathbf{N} \tag{19}$$

This gives another M × N matrix A; given by Eq. 20

$$\mathbf{A} = [\mathbf{a}_1, \mathbf{a}_2, \ldots, \mathbf{a_N}] \tag{20}$$

Since $\{u_1, u_2, \ldots, u_r\}$ form an orthonormal basis for *R*(*A*), the range (column) subspace of matrix A. Since matrix *A* is formed from a training set ***S*** with *N* face & finger images, R(A) is called a 'subspace' in the 'image & finger space' of $\boldsymbol{m} \times \boldsymbol{n}$ pixels, and each u, $\boldsymbol{i} = 1, 2, \ldots, \boldsymbol{r}$, can be called a 'base'.

Let x (= $[x_1, x_2, \ldots, x_r]^T$) be the coordinates (position) of any $m \times n$ face image finger f in the subspace. Then it is the scalar projection of f − f onto the base.

$$x = [u_1, u_2, \ldots, u_r]^T (f - \bar{f}) \tag{21}$$

This coordinate vector x shown in Eq. 21 is used to find which of the training images best describes the image f. That is to find some training face f_i, $\boldsymbol{i} = 1, 2, \ldots, N$, that minimizes the distance and it is shown in Eq. 22.

$$\varepsilon_i = ||x - x_i||_2 = \left[(x - x_i)^T (x - x_i)\right]^{1/2} \tag{22}$$

where x_i is the coordinate vector of f_i, which is the scalar projection of $f_i - \bar{f}$ onto the bases as shown in Eq. 23.

$$x_i = [u_1, u_2, \ldots, u_r]^T (f_i - \bar{f}) \tag{23}$$

A face and finger f is classified as face & finger f_i when the minimum $\boldsymbol{\varepsilon}_i$ is less than some predefined threshold $\boldsymbol{\varepsilon}_0$ Otherwise the face and finger f is classified as "unknown face & finger".

If f is not a finger, its distance to the finger subspace will be greater than 0. Since the vector projection of $f_i - \bar{f}$ onto the face and finger space is given by Eq. 24.

$$f_p = [u_1, u_2, \ldots, u_r] \tag{24}$$

where x is given in Eq. 21.

The distance of f to the face & finger space is the distance between $f_i - \bar{f}$ and the projection $f_{\boldsymbol{p}}$ onto the face and finger space given in Eq. 25.

$$\boldsymbol{\varepsilon}_j = ||(f_i - \bar{f}) - f_p||_2 = \left[(f - \bar{f} - f_p)^T (f - \bar{f} - f_p)\right]^{1/2} \tag{25}$$

If ε_j is greater than some predefined threshold ε_1, then f is not a face and finger image.

The algorithm for face and finger recognition with SVD is explained in the Algorithm 4.

Algorithm 4: Authentication with SVD	
STEP 1:	Compute the mean face and finger $\bar{f}$ of $\boldsymbol{S}$ by Eq. 18
STEP 2:	Form a matrix $\boldsymbol{A}$ as shown in Eq. 20 with the computed $\bar{f}$
STEP 3:	Calculate the SVD of $\boldsymbol{A}$ as shown in Eq. 10
STEP 4:	For each face and finger, compute the coordinate vector x_i. from Eq. 23 Choose a threshold ε_1 that defines the maximum allowable distance from face space Determine a threshold ε_0 that defines the maximum allowable distance from any known face and finger in the training set $\boldsymbol{S}$
STEP 5:	For a new input face and finger image f to be identified, calculate its coordinate vector x from Eq. 21, the vector projection f_p, the distance ε_f to the face and finger space from Eq. 25 If $\varepsilon_f > \varepsilon_1$ the input image is not a face and finger
STEP 6:	If $\varepsilon_f < \varepsilon_1$, compute the distance ε_i to each image in database If all $\varepsilon_i > \varepsilon_0$, the input face and finger classified as unknown face Else if $\varepsilon_i > \varepsilon_0$, classify the input face and finger image as the known individual

5 Training Process

It consists of three phases named as reading, embedding and storing as shown in Fig. 8. The process is depicted in Algorithm 5.

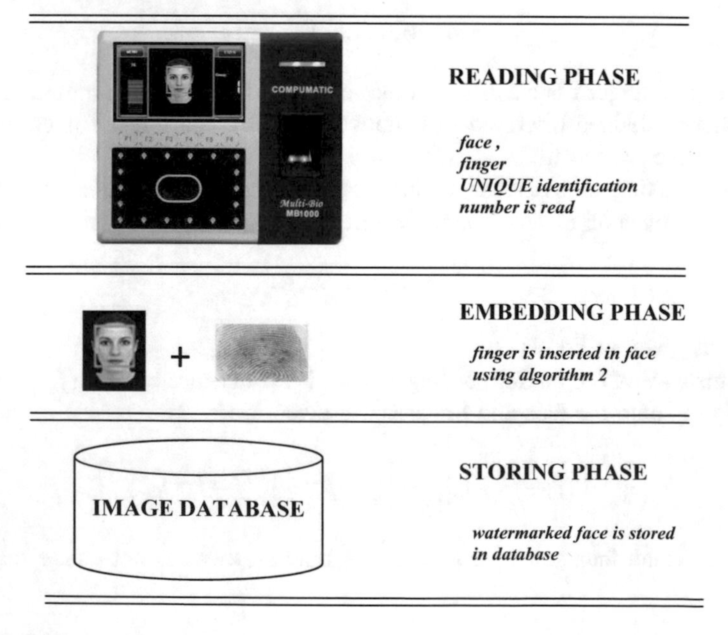

Fig. 8 Training process

Algorithm 5: Training Process	
STEP 1:	The face of a human and fingerprint is read
STEP 2:	Finger is hidden in face with Algorithm 2, the output image is said to watermarked face
STEP 3:	The watermarked face is stored in database using persons's unique identification number

6 Authentication Process

It consists of three phases namely reading, extraction, and authentication as shown in Fig. 9. The process is depicted in Algorithm 6.

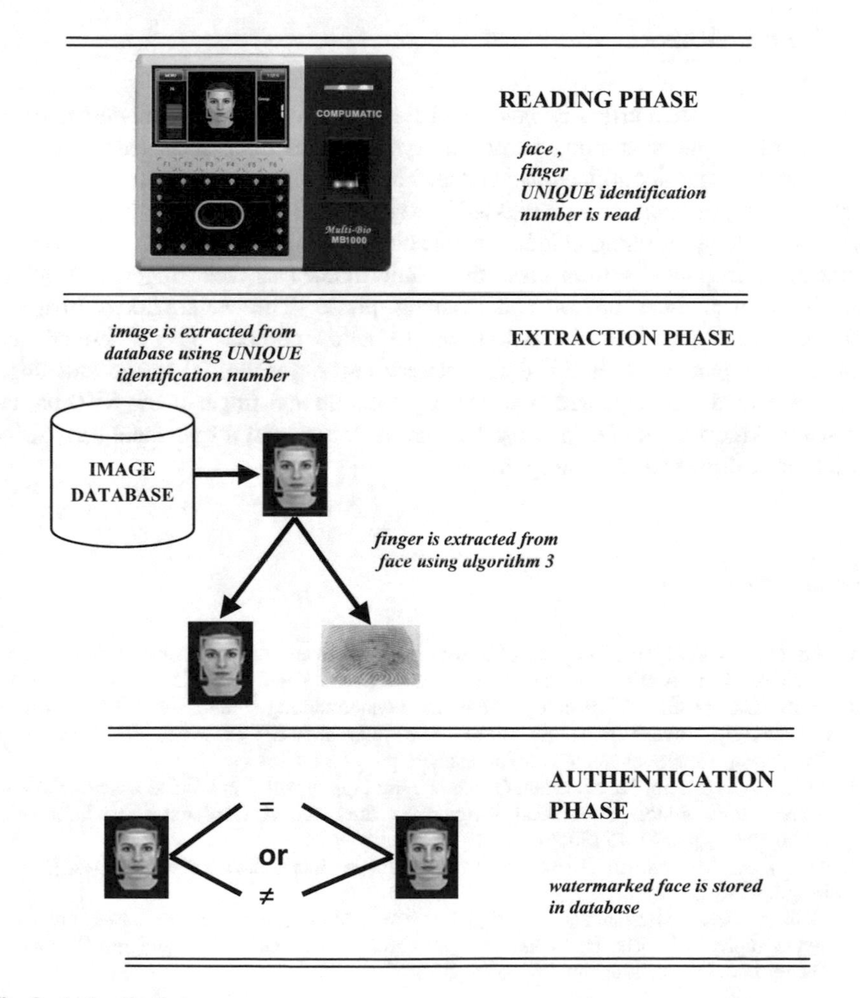

Fig. 9 Authentication process

Algorithm 6: Authentication Process Algorithm	
STEP 1:	The face of a human and fingerprint is read
STEP 2:	The person's unique identification number is read
STEP 3:	The watermarked image is read from database using the person's unique identification number
STEP 4:	The face and fingerprint is separated with Algorithm 3
STEP 5:	The Algorithm 4 is applied to match with extracted data from database

STEP 6:	If the Algorithm 4 gives the positive results then
	authentications is proved
	else
	authentication is failed

7 Conclusions

The proposed system primarily has two phases called training and authentication. In the first phase called training phase, the system takes three traits named as face, finger, and unique identification number. The finger is inserted in face with FHT watermarking mechanism as discussed in Algorithm 2 and that watermarked image is stored in database using unique identification number. In the second phase called authentication phase, system takes three traits named as face, finger, and unique identification number as same in training phase. The watermarked image is retrieved from database with unique identification number. The watermark and image are separated with IFHT as discussed in Algorithm 3. Image and finger which are read are compared with extracted image and finger using SVD as discussed in Algorithm 4. The proposed system is very useful for personal verification in industry, organization, enterprise, etc.

References

1. Wu, M., Chen, J., Zhu, W., Yuan, Z.: Security analysis and enhancements of a multi-factor biometric authentication scheme. Int. J. Electron. Secur. Digit. Forensics 352–365 (2016)
2. Sheth, R.K., Nath, V.V.: Secured digital image watermarking with discrete cosine transform and discrete wavelet transform method. In: International Conference on Advances in Computing, Communication, and Automation, pp. 1–5 (2016)
3. Al-Haj, A., Hussein, N., Abandah, G.: Combining cryptography and digital watermarking for secured transmission of medical images. In: International Conference on Information Management, pp. 40–46 (2016)
4. Al-Haj, A., Mohammad, A.: Crypto-watermarking of transmitted medical images. J. Digit. Imaging 1–3 (2016)
5. Mallick, A.K., Maheshkar, S.: Digital image watermarking scheme based on visual cryptography and SVD. In: International Conference on Frontiers in Intelligent Computing: Theory and Applications, pp. 589–598 (2015)

6. Kumar, S., Dutta, A.: A novel spatial domain technique for digital image watermarking using block entropy. In: International Conference on Recent Trends in Information Technology, pp. 1–4 (2016)
7. Joshi, A.M., Bapna, M., Meena, M.: Blind image watermarking of variable block size for copyright protection. In: International Conference on Recent Cognizance in Wireless Communication and Image Processing, pp. 853–859 (2016)
8. Vellasques, E., Sabourin, R., Granger, E.: A dual-purpose memory approach for dynamic particle swarm optimization of recurrent problems. In: Recent Advances in Computational Intelligence in Defense and Security, pp. 367–389 (2016)
9. Hua, G., Huang, J., Shi, Y.Q., Goh, J., Thing, V.L.: Twenty years of digital audio watermarking—a comprehensive review. Signal Process. 222–242 (2016)
10. Seitz, J.: Digital Watermarking for Digital Media. Information Science Publishing (2004)
11. Bas, P., Furon, T., Cayre, F., Doërr, G., Mathon, B.: A quick tour of watermarking techniques. In: Security in Watermarking, pp. 13–31 (2016)
12. Dragoi, I.C., Coltuc, D.: A simple four-stages reversible watermarking scheme. In: International Symposium on in Signals, Circuits and Systems, pp. 1–4 (2015)
13. Xiang, S., Wang, Y.: Non-integer expansion embedding techniques for reversible image watermarking. J. Adv. Signal Process. 1–2 (2015)
14. Ghosh, S., De, S., Maity, S.P., Rahaman, H.: A novel dual purpose spatial domain algorithm for digital image watermarking and cryptography using Extended Hamming Code. In: International Conference on Electrical Information and Communication Technology, pp. 167–172 (2015)
15. Zhou, Q., Lu, S., Zhang, Z., Sun, J.: Quantum differential cryptanalysis. Quantum Inf. Process. 2101–2109 (2015)
16. Van Schyndel, R.G., Tirkel, A.Z., Osborne, C.F.:A digital watermark. In: International Conference on Image Processing, pp. 86–90 (1994)
17. Van Schyndel, R.G., Tirkel, A.Z., Svalbe, ID.: Key independent watermark detection. In: International Conference on Multimedia Computing and Systems, pp. 580–585 (1999)
18. Huang, P.S., Chiang, C.S., Chang, C.P., Tu. T.M.: Robust spatial watermarking technique for colour images via direct saturation adjustment. In: IEE Proceedings-Vision, Image and Signal Processing, pp. 561–574 (2005)
19. Zhu, X., Zhao, J., Xu, H.: A digital watermarking algorithm and implementation based on improved SVD. In: International Conference on Pattern Recognition, pp. 651–656 (2006)
20. Liang, L., Qi, S.: A new SVD-DWT composite watermarking. In: International Conference on Signal Processing 2006
21. Ali, M., Ahn, C.W., Pant, M.: A robust image watermarking technique using SVD and differential evolution in DCT domain. Optik-Int. J. Light Electron Opt. 428–434 (2014)
22. Bors, A.G., Pitas, I.: Image watermarking using DCT domain constraints. In: International Conference on Image Processing, pp. 231–234 (1996)
23. Hernandez, J.R., Amado, M., Perez-Gonzalez, F.: DCT-domain watermarking techniques for still images: detector performance analysis and a new structure. IEEE Trans. Image Process. 55–68 (2000)
24. Lu, C.S., Chen, J.R., Liao, H.Y., Fan, K.C.: Real-time MPEG2 video watermarking in the VLC domain. In: International Conference on Pattern Recognition, pp. 552–555 (2002)
25. Mohanty, S.P., Ranganathan, N., Balakrishnan, K.: A dual voltage-frequency VLSI chip for image watermarking in DCT domain. IEEE Trans. Circuits Syst. II: Express Briefs 394–398 (2006)
26. Amini, M., Ahmad, M., Swamy, M.: A robust multibit multiplicative watermark decoder using vector-based hidden Markov model in wavelet domain. IEEE Trans. Circuits Syst. Video Technol. (2016)
27. Wei, C., Zhaodan, L.: Robust watermarking algorithm of color image based on DWT-DCT and chaotic system. In: International Conference on Computer Communication and the Internet, pp. 370–373 (2016)
28. Lu, Z.M., Guo, S.Z.: Lossless Information Hiding in Images. Syngress (2016)

29. Lu, C.S., Huang, S.K., Sze, C.J., Liao, H.Y.: Cocktail watermarking for digital image protection. IEEE Trans. Multimedia 209–224 (2000)
30. Hua, L.I., Xi, Z.G., Ting, Z.Y.: A visual model weighted image watermarking method using wavelet decomposition. J. China Inst. 006 (2000)
31. Voyatzis, G., Pitas, I.: Applications of toral automorphisms in image watermarking. In: International Conference on Image Processing, pp. 237–240 (1996)
32. El-Taweel, G.S., Onsi, H.M., Samy, M., Darwish, M.G.: Secure and non-blind watermarking scheme for color images based on DWT. ICGST Int. J. Graph. Vis. Image Process. 15 (2005)
33. Xia, X.-G., Boncelet, C.G., Arce, G.R.: A multiresolution watermark for digital images. In: International Conference on Image Processing, pp. 548–551 (1997)
34. Raval, M.S., Rege, P.P.: Discrete wavelet transform based multiple watermarking scheme. In: International Conference on Convergent Technologies for the Asia-Pacific Region, pp. 935–938 (2003)
35. Kundur, D., Hatzinakos, D.: Digital watermarking using multiresolution wavelet decomposition. In: International Conference on Acoustics, Speech and Signal Processing, pp. 2969–2972 (1998)
36. Tao, P., Eskicioglu, A.M.: A robust multiple watermarking scheme in the discrete wavelet transform domain. Int. Soc. Opt. Photonics 133–144 (2004)
37. Dong, J., Li, J.: A robust zero-watermarking algorithm for encrypted medical images in the DWT-DFT encrypted domain. Innov. Med. Healthc. 197–208 (2016)
38. Kang, X., Huang, J., Shi, Y.Q., Lin, Y.: A DWT-DFT composite watermarking scheme robust to both affine transform and JPEG compression. IEEE Trans. Circuits Syst. Video Technol. 776–786 (2003)
39. Solachidis, V., Pitas, L.: Circularly symmetric watermark embedding in 2-D DFT domain. IEEE Trans. Image Process. 1741–1753 (2001)
40. Ganic, E., Eskicioglu, A.M.: Robust DWT-SVD domain image watermarking: embedding data in all frequencies. In: Workshop on Multimedia and Security, pp. 166–174 (2004)
41. Pereira, S., Pun, T.: Robust template matching for affine resistant image watermarks. IEEE Trans. Image Process. 1123–1129 (2000)
42. Falkowski, B.J., Lim, L.S.: Image watermarking using Hadamard transforms. Electron. Lett. 211–213 (2000)
43. Gilani, S.A., Kostopoulos, I., Skodras, A.N.: Color image-adaptive watermarking. In: International Conference on Digital Signal Processing, pp. 721–724 (2002)
44. Kaur, A., Dutta, M.K., Burget, R., Riha, K.: A heuristic algorithmic approach to challenging robustness of digital audio watermarking using discrete wavelet transform", International Conference on Telecommunications and Signal Processing, pp. 519–522, 2016
45. Shao, Y., Wu, G., Lin, X.: Quantization-based digital watermarking algorithm. J. Tsinghua Univ. 006 (2003)
46. Bose, A., Maity, S.P.: Improved spread spectrum compressive image watermark detection with distortion minimization. In: International Conference on Signal Processing and Communications, pp. 1–5 (2016)
47. Xiao, J., Wang, Y.: Toward a better understanding of DCT coefficients in watermarking. In: Pacific-Asia Workshop on Computational Intelligence and Industrial Application, pp. 206–209 (2008)
48. You, J., Li, W., Zhang, D.: Hierarchical palmprint identification via multiple feature extraction. Pattern Recogn. 847–859 (2002)
49. Holt, S.B.: Finger-print patterns in mongolism. Ann. Hum. Genet. 279–282 (1963)
50. Zhao, Q., Zhang, L., Zhang, D., Luo, N.: Adaptive pore model for fingerprint pore extraction. Proc. IEEE (2008)
51. Girgis, M.R., Mahmoud, T.M., Abd-El-Hafeez, T.: An approach to image extraction and accurate skin detection from web pages. World Acad. Sci. Eng. Technol. 27 (2007)
52. Kaur, M., Singh, M., Girdhar, A., Sandhu, P.S.: Fingerprint verification system using minutiae extraction technique. World Acad. Sci. Eng. Technol. 46 (2008)

53. Le, H., Bui, T.D.: Online fingerprint identification with a fast and distortion tolerant hashing. J. Inf. Assur. Secur. 117–123 (2009)
54. Ratha, N.K., Karu, K., Chen, S., Jain, A.K.: A real-time matching system for large fingerprint databases. Trans. Pattern Anal. Mach. Intell. 799–813 (1996)
55. Jain, A., Chen, Y., Demirkus, M.: Pores and ridges: fingerprint matching using level 3 features. Pattern Recogn. Lett. 2221–2224 (2004)
56. Vatsa, M., Singh, R., Noore, A., Singh, S.K.: Combining pores and ridges with minutiae for improved fingerprint verification. Signal Process. 2676–2685 (2009)
57. Uludaga, U., Ross, A., Jain, A.: Biometric template selection and update: a case study in fingerprints. Pattern Recogn. 1533–1542 (2004)
58. Coetzee, L., Botha, E.C.: Fingerprint recognition in low quality images. Pattern Recogn. (1993)
59. O"Gormann, L., Nickerson, J.V.: An approach to fingerprint filter design. Pattern Recogn. 29–38 (1989)
60. Yuan, W., Lixiu, Y., Fuqiang, Z.: A real time fingerprint recognition system based on novel fingerprint matching strategy. In: International Conference on Electronic Measurement and Instruments (2007)
61. Cui, W., Wu, G., Hua, R., Yang, H.: The research of edge detection algorithm for fingerprint images. Proc. IEEE (2008)
62. Li, S., Wei, M., Tang, H., Zhuang, T., Buonocore, M.H.: Image enhancement method for fingerprint recognition system. In: Annual Conference on Engineering in Medicine and Biology, pp. 3386–3389 (2005)
63. Mil'shtein, S., Pillai, A., Shendye, A., Liessner, C., Baier, M.: Fingerprint recognition algorithms for partial and full fingerprints. Proc. IEEE (2008)
64. Ross, A., Uludag, U., Jain, A.: Biometric template selection and update: a case study in fingerprints. Pattern Recogn. Soc. (2003)
65. Karna, D.K., Agarwal, S., Nikam, S.: Normalized cross-correlation based fingerprint matching. In: Fifth International Conference on Computer Graphics, Imaging and Visualization (2008)
66. Bazen, A.M., Verwaaijen, G.T.B., Gerez, S.H.: A correlation-based fingerprint verification system. In: Workshop on Circuits, Systems and Signal Processing, Veldhoven (2000)
67. Lowe, D.G.: Distinctive image features from scale-invariant key points. Int. J. Comput. Vis. (2004)
68. Schalkoff, R.J.: Digital Image Processing and Computer Vision. Wiley Publications, New York (1989)
69. Saryazdi, S., Nezamabadi-Pour, H.: A blind digital watermark in Hadamard domain. World Acad. Sci. Eng. Technol. 126–129 (2005)
70. Rao, C.S., Murthy, K.V., Gupta, V.M., Raju, G.P., Raju, S.V., Balakrishna, A.: Implementation of object oriented approach for copyright protection using Hadamard transforms. In: International Conference on Computer and Communication Technology, pp. 473–480 (2010)
71. Devi, B.P., Singh, K.M., Roy, S.: A Copyright Protection Scheme for Digital Images Based on Shuffled Singular Value Decomposition and Visual Cryptography, p. 1091. Springer Plus (2016)

Activity Recognition Using Imagery for Smart Home Monitoring

Bradley Schneider and Tanvi Banerjee

Abstract In this chapter, we will describe our comprehensive literature survey on using vision technologies for in-home activity monitoring using computer vision techniques, as well as computational intelligence (CI) approaches. Specifically, through our survey of the body of work, we will address the following questions:

I. What are the challenges of using standard RGB cameras for activity analysis and how are they solved?
II. Why do most existing algorithms perform so poorly in real-world settings?
III. Which the design choices should be considered when deciding between wearable cameras or stationary cameras for activity analysis?
IV. What does CI bring to the vision world as compared to computer vision techniques in the activity analysis domain?

Through our literature survey, as well as based on our own research in both the wearable and non-wearable domain, we share our experiences to enable researchers to make their own design choices as they enter the field of vision-based technologies. We present the hierarchy of the literature survey in Fig. 1.

1 Wearable Vision Sensors

With the shrinking size of computational and sensor hardware, the number of wearable devices available to both researchers and consumers continues to grow. These devices are an attractive option to researchers because of their affordability, availability, and the familiarity with the general public. Perhaps the largest benefit of wearable devices to the scientific community is that techniques which previously required

B. Schneider (✉) · T. Banerjee
Department of Computer Science and Engineering, Wright State University, Fairborn, OH, USA
e-mail: schneider.163@wright.edu

T. Banerjee
e-mail: tanvi.banerjee@wright.edu

A.E. Hassanien and D.A. Oliva (eds.), *Advances in Soft Computing and Machine Learning in Image Processing*, Studies in Computational Intelligence 730,
https://doi.org/10.1007/978-3-319-63754-9_16

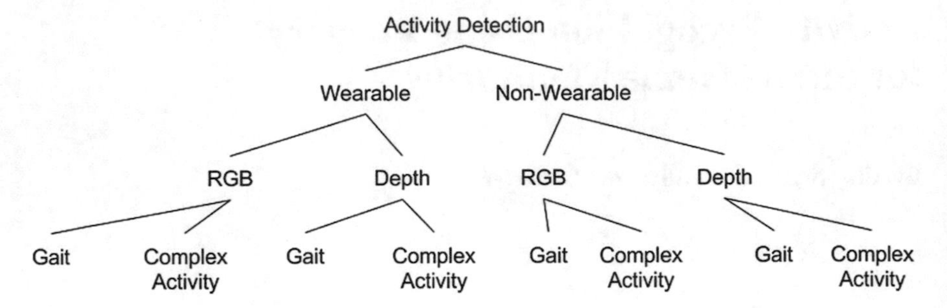

Fig. 1 Hierarchy of common methods for activity detection and analysis using wearable and non-wearable vision sensors

time-consuming preparation of the environment with potentially large equipment, such as multiple cameras [15, 19] or force-measuring treadmills [11], can now be traded for techniques focused on instrumenting the subject of study.

Activity detection through computer vision is classically performed through a third-person viewpoint. Approaches using third-person camera setups can involve multiple, carefully placed cameras [15, 19] to achieve a viewpoint of the subject which gives enough two-dimensional information from multiple perspectives to extract the three-dimensional movement of the subject. As mentioned before, this limits the applicability of these methods to predetermined environments within the field of view of the stationary cameras. Wearable technology allows for the vision sensor to travel with the subject, expanding the environment in which the study may be conducted.

1.1 Wearable Cameras for Gait Analysis

Gait analysis performed with computer vision on wearable devices is a challenging task due to the limited first-person perspective of the device. Consequently, there may be a lack of information available to describe the subjects movement. However, it has the benefit of requiring little configuration and is not prone to accumulated error over time, as wearable accelerometers or positioning devices may be [29, 45]. Approaches of this type rely on the predictability of a wearers movement during locomotion to recognize gait. Patterns of movement in the captured video are recognized and lead to the recognition of the gait activity and extraction of gait parameters.

While vision sensors can provide a convenient solution to the complexity problems other techniques suffer, many published methods of gait analysis with wearable technology focus more on instrumenting the subject with accelerometers or other sensors which can more directly give an accurate indication of movement during locomotion. For this reason, there are fewer methods successfully using computer vision at this point in time.

In an attempt to take advantage of both the accuracy of accelerometers and the convenience of computer vision, Cho et al. describe an approach combining both types of sensors [6]. A multi-resolutional, grid-based optical flow method is applied to video collected from a wearable vision sensor. By examining different regions of the video for flows in different directions, activities such as walking forward or backward, turning, and sitting down are able to be detected from the video. The activities detected by the video are limited to those describing movement in the environment, and two similar activities may not be distinguishable. This approach is based on the perceived direction of movement of the vision sensor through the environment, but the video is not used to estimate the pose of the wearer. Because of the lack of consideration for the physical effects of locomotion on the movement of the camera, this approach is equally suited to estimating vehicular or robotic movement. To improve the detection accuracy, accelerometers were also placed on the subject, contributing information about the estimated pose of the wearer. Taking advantage of well-published information on human joint movement during locomotion to estimate the pose of the subject, rather than just the movement of the subject relative to the environment, can lead to the extraction of more detailed information about the gait. For this purpose, it is necessary to utilize information on human behaviors to construct a model of human motion. Hirasaki et al. document the effect of locomotion on the head and body, demonstrating the translation and rotation during the gait [17]. Unuma et al. modeled human locomotion using Fourier expansions due to the cyclic nature of the activity [40]. These and related works give a basis for building accurate models of human gait for analysis in egocentric video, which enables approaches based entirely on video.

In a purely video-based approach, Watanabe et al. rely on the predictability of limb motion during locomotion to analyze movement and classify gait from a single video source [45]. A camera is attached to the leg, facing downward so that it captures the motion of the environment as the leg moves. A calibration step is performed to document the movement of the camera during the walking activity. Walking samples are taken by a motion capture system and models of various states (slow walk, walk, run). A model is formed based on the waist position, traveling speed, and angular speed at a given time. The behavior is described by up to a fifth-order Fourier transform, which represents the cyclic motion described in the human motion studies [17, 40]. State prediction for the walking state occurs at each sample in time and is based on a likelihood estimation from the captured parameters, and includes an error calculation based on the difference in the expected location of known landmarks in the environment and the actual observed location.

Schneider and Banerjee were able to extract gait parameters from a single wearable camera without requiring the lengthy setup process and without prior knowledge of landmarks in the environment, though the work is performed in a controlled environment [33]. A single head-mounted camera embedded in a pair of glasses provides a first-person view of the environment. As the subject walks, the video mimics the cyclic head motion described in the studies on locomotion [6]. While not directly modeling the subjects pose as Watanabe et al. [45], the cyclic motion is used as a detector of gait. Optical flow extracts the overall motion of each video frame. The

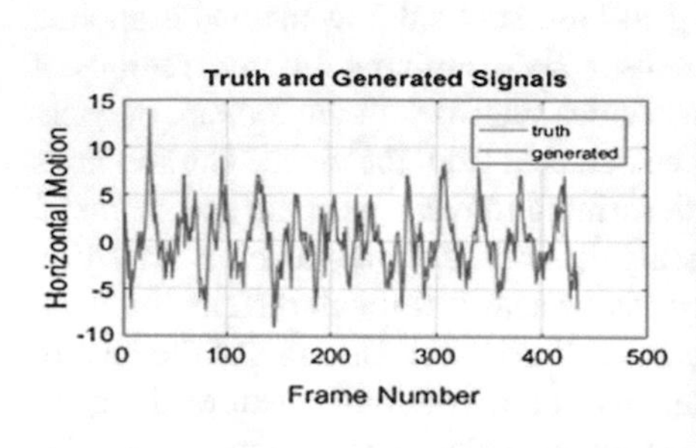

Fig. 2 Waveform of the generated motion vectors over frames of video (*left*), wearable camera sensor embedded in a pair of glasses (*top right*), optical flow vectors overlayed on video collected from the wearable camera (*bottom right*) [33]

series of motion vectors over time constructs a waveform that is analyzed by frequency analysis. Examples of the motion vectors and waveform can be seen in Fig. 2. The detected frequency is indicative of walking speed and number of steps taken. The robustness of this method lies in the ease of use for the subject, the flexibility of the environment, and the computational simplicity of the method. While most methods focus on classifying the activity, this method is focused on determining how the activity is occurring, assuming it has been already been classified.

Gait analysis using only wearable vision sensors remains a relatively immature topic compared to research using third-person vision techniques, or accelerometer and pressure-sensor-based techniques. The combination of studies on the physical movements occurring during human locomotion and wearable vision sensors has shown promising results in both gait classification [6, 45] and extraction of specific parameters [33].

1.2 Wearable Cameras for Complex Activity Analysis

Detection of a variety of other complex activities can also be accomplished using wearable vision sensors. In contrast to the smaller body of knowledge on gait detection, a greater amount of research has focused on detecting activities involving interaction with other people or objects from egocentric video. Since many activities are naturally identified by the objects with which a subject interacts, techniques often rely on recognition of these objects from an egocentric view [1, 29, 30]. Unlike the case of gait, the goal of these techniques is not to model an estimation of the human pose at a given time, but rather to gain knowledge of interactions.

A common challenge to recognizing these activities is consistent detection of objects during interaction. As the subject manipulates an object, it may become difficult to recognize due to either occlusion or the fact that it changes shape. For example, a microwave with its door shut appears very different from a microwave with its door open. These challenges can be overcome by utilizing the ample research published on object detection and tracking.

To detect complex activities from first-person video, Pirsiavash and Ramanan [29] perform object detection using a deformable part-based model method [12] to reliably recognize objects in different positions, both static and during use. This allows for the formation of an object-centric model of activities. Activities are defined as interaction with objects over time, however, so the activity recognition must involve a temporal component. Pirsiavash and Ramanan build a temporal pyramid where each level represents a concatenation of histograms of object scores taken from an increasing number of segments of the entire video. The features of the histogram are fed to linear SVM classifiers.

Matsuo et al. posit that inaccuracies in approaches similar to Pirsiavash and Ramanans [29] may be due to the fact that while the hands are a good natural cue of information that is relevant to the action being performed, the hands do not always present useful information and may not be present in the scene captured by the first-person video during all activities [24]. Instead, an attention-based approach is introduced in order to discriminate between situations in which the hands are important and when they are not. The goal is to build a per-pixel measure of attention to determine regions of high attention. The attention measure is based on existing saliency methods but also takes the egocentric motion of the head into account. The saliency measure is weighted higher if the object is in a region toward which the users head is rotating. For example, if the egocentric motion indicates that the user is turning his head to the right, an object on the right of the frame will have a higher attention measure than one of equal saliency on the left of the frame.

A spatial segmentation of video, rather than temporal segmentation, is performed by Narayan et al. [26]. The segmentation is based on direction and intensity of motion in video captured using a head-mounted camera, rather than the detection of specific objects. An assumption is made that regions with motion of the largest magnitude represent regions significant to the activity, while regions of lower magnitude can be considered background motion. The foreground motions define a motion map, and trajectories of movement are formed based on the optical flow field that is identified, a method described by Wang et al. [43]. Since the camera is worn on the head, an attempt is made to identify motion contributed by head movement in order to recover the actual motion of the objects along their trajectories. The feature descriptors along the trajectories are represented with both bag of words and Fisher vectors and are concatenated for use as input to a nonlinear SVM classifier.

We see from our experience, as well as the literature survey, that while the wearable vision research is sparse, there are several advantages to using such as system that will be discussed later in the chapter (research question III).

2 Non-wearable Vision Sensors

In the domain of non-wearable vision work in activity recognition, there are several studies that demonstrate the validity of these sensors when quantifying and analyzing human activities. In particular, the largest difference in using non-wearables in continuous activity monitoring is the transition from active to passive monitoring: in non-wearable systems, the users require minimal assistance with the sensing devices, whereas the active (wearable) sensors require more input from the users. To that end, non-wearable sensors may be a more appropriate option for activity recognition depending on the goals of the research.

2.1 Standard RGB Cameras for Gait Analysis

Gait analysis has several different applications, from fall risk assessment to human recognition [23]. Most of the research done in this area uses the silhouettes generated from background subtraction of the color images for analysis [16]. Among the earlier body of literature on gait analysis, Bobick and Davis [5] created a series of motion templates using the union of the images in a given image sequence to recognize the activity. In a similar approach, Liu and Sarkar [21] used a simple, but effective technique of averaging the silhouette images within a gait cycle (See Fig. 3). In this case as well as in a similar study by Yu et al. [51], the goal was to identify gaits of different

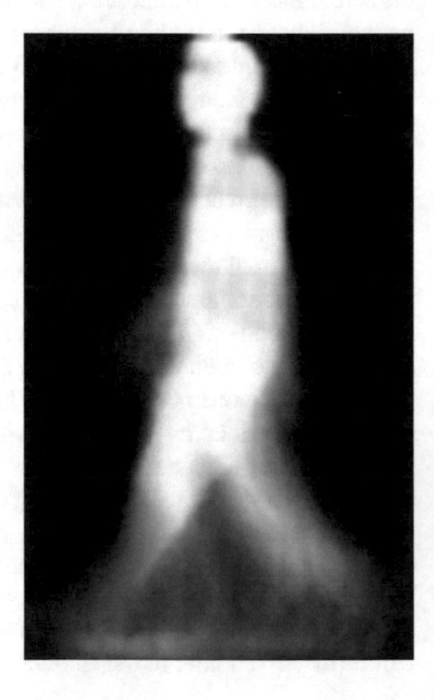

Fig. 3 Sample Gait Energy Image using the CASIA gait dataset [16, 51]. This is an exemplar of the earlier work on image-based feature extraction for gait recognition

individuals for the purpose of human identification. In [7], the authors combined the shape information of the silhouette, with the dynamic motion for gait recognition. Anatomical positions of the shoulder, hip, wrist, and hands were leveraged to divide the human silhouette for further gait analysis. The center of mass was then extracted for the dynamic motion extraction. The results obtained were comparable to the state of the art using the public gait datasets such as CMU MoBo gait and HumanID gait (links provided later). While these studies focus on 2D features, the following studies look at 3D features for gait analysis. Zhao et al. [52] used a stick figure model in conjunction with dynamic time warping to account for the different walking speeds of different individuals. Images of the gait sequences using six cameras were used in combination with Newtons optimization technique to estimate the pose at each frame in the gait sequence. In this study, the characteristics of the different gait sequences were examined but human identification was not discussed.

In all the above studies, the focus of the gait analysis was not for clinical purposes, i.e., extraction of gait parameters for physiological information. Furthermore, the techniques are not precise enough to extract gait parameters of clinical relevance such as step time, step length, stride time, stride length, etc. The following body of work addresses the use of cameras in measuring clinical gait information. In [39], Ugbolue et al. used camera recordings of gait sequences recorded in a specific grid area for measuring parameters like step width and step length. The system strongly corroborated with the gold standard Vicon system, however there were manual components to the gait extraction, i.e., there was manual feature extraction of the gait measures from the video data. In [35], Stone and Skubic computed 3D voxel images by superimposing camera images from two perpendicular viewed cameras. The voxels below a certain height were then identified as the foot region, and dissimilarity based on proximity measures were used to identify individual steps that were then further used for measuring gait speed, as well as stride length and stride time. In a similar approach, Wang et al. [42] used the voxel centroid to measure the gait information, as well as a thresholding technique to extract the feet voxels for measuring the step length for an individual. Both the Vicon and the GaitRite system [46] were used to validate the system for both healthy younger participants, as well as older adults for eldercare applications.

In [20], one of the earlier works using machine learning, Kale et al. leveraged the Markovian dependence of a gait cycle: each gait stance depends on the earlier pose of the human due to the cyclic nature of the activity. Specifically, the width of the silhouette in each frame was used as the feature vector to be fed into a continuous hidden markov model, with individual models trained for each participant. In [22], Lu and Zhang used a combination of shape features from multiple camera views to train a genetic fuzzy support vector machine for human gait recognition. In this study, some more preprocessing steps were discussed in terms of shadow elimination caused by different illumination settings but occlusion caused by partial viewing of the person was not discussed. In [25], Muramatsu et al. used a more rigorous setting with 24 synchronized cameras to generate 3D visual gait hulls to generate view invariant sequences. To achieve this, they proposed a view transformation model which performed well with a participant cohort of 52 subjects.

Fig. 4 (*Top*) An image that shows a frame from a walking sequence. (*Bottom*) The locations of the two feet across time t showing a braided walk pattern [27]

2.2 Standard RGB Cameras for Complex Analysis

In indoor activity recognition research, the classical papers using computer vision used two different approaches: model based and non-model based. Within model based approaches, the two popular approaches involved using either the stick figure model or the contour of the human shape for the activity information. Specifically, in Niyogi and Adelson [27], the authors created a five stick model to look at activities. Simple line fitting algorithms were used to track different activities in this early work. See Fig. 4 for an example of the extracted walk pattern.

As compared to gait recognition, computational intelligence was incorporated with computer vision earlier on for activity recognition. It is not surprising that CI was included in earlier research, given the complexity of activity recognition, as well as the challenges of dynamic motion further compounded by skewness from different camera view angles. In 1992 [48], Yamato et al. used 2D human contours for activity recognition. This study was one of the first to use hidden markov models (HMMs) for sequential activity recognition. However, the authors found that the performance of the classifiers for a particular activity deteriorated if the training data did not include samples from the test data participants. With that in mind, the authors recognized the need for latent variables in activity model building for complex activity recognition. In [31], Rowley and Rehg used optical flow in conjunction with expectation maximization to predict the movement of different body parts for activity recognition. While image segmentation had been tackled earlier, the addition of CI allowed a temporal component to this through prediction the movement of different body parts.

In one of the earliest works in activity recognition using dynamic bayesian networks (DBN), Park and Aggarwal [28] created a very interesting DBN framework to map domain knowledge in activity recognition. In particular, the basic BNs were used to map the relation between different body parts for static pose estimation. By defining a set of finite states for a particular body part and assuming independence between the movements for each part, the DBN structure was simplified for effective activity analysis. Different interactions between two individuals (such as punching, hugging) were analyzed using this method. One key point identified early on in this field was the need to incorporate hierarchical methods [28, 31] for understanding complex activities. A complex activity such as hugging involves two people both moving forward with arms stretched forward, and then stationary, finally moving apart. Both the sequential aspect of such an activity, as well as the hierarchical aspect of complex activities needs to be incorporated in the learning model, either through probabilistic or fuzzy approaches [2]. In addition to these techniques, another interesting approach of understanding activities was implemented by Veeraraghavan et al. [41]. In their study, the authors not only tried to incorporate the temporal change in the complex activities like jogging, or picking up an object, but they also tried to learn the nature of the temporal transformations for each activity. This is useful in not only recognizing activities, but also finding similarities and dissimilarities between people performing the same activity. In recent studies incorporating activity analysis systems in real indoor settings (i.e., non-laboratory), the study by Crispim et al. [9] brings in both CI with computer vision as well as contextual knowledge for detecting instrumental activities of daily living or IADLs. Using a complex systems that detects certain objects in the scene along with an ontology-like apriori of certain activities, complex IADLs are evaluated in the older population with mild cognitive impairment (MCI). One of the crucial needs of computer vision in activity analysis is in healthcare settings for eldercare research. While great strides have been taken toward building systems that can be employed in the real world, there are still challenges in using standard RGB cameras that need to be addressed.

I. Challenges in activity analysis using RGB cameras

The preceding overview has described the benefits of algorithms using standard RGB camera images for activity extraction. However, using RGB cameras has the following drawbacks:

i. Illumination: color images are affected by illumination changes in the indoor settings such as the changes in lighting and sunlight that create the need for shadow elimination and continuous background adaption. Even though the state-of-the-art algorithms try to address this issue, the problem is still challenging.
ii. Camera synchronization: in order to ensure that the activities are identified in different locations in the indoor space, multiple cameras are required. However, in order to integrate the information to create the 3D voxel space, the cameras need to be synchronized to ensure that the image frames from the different cameras are collected at the same precise time point. This is again a challenging task.

iii. Image frame rate: A consequence of (ii), using single or multiple cameras in synchronization can affect the frame rate of the data collection, which in turn can affect the temporal resolution of the activity. In our system at TigerPlace [35, 42], using a two-camera system, we achieved a frame rate of 5 frames per second consistently. However, this may not be sufficient for applications like fall detection where the identification of the fall is critical, but finding the cause of the fall is even more valuable.
iv. Privacy: in order to ensure the feasibility of a system, the privacy of the participants needs to be addressed. This is particularly true in surveillance applications where the data are collected 24×7. In [10], Demiris et al. found that storing just the silhouettes of the image can address this concern. However, if the quality of the stored silhouettes is poor, there will be no way to go back and recover the raw data.

While these concerns can be addressed either through sophisticated algorithms or through a high speed processor, researchers started to transition from standard web cameras to RGBD cameras. This transition was expedited with the introduction of the Microsoft Kinect that allowed the use of a cheap system to extract RGB and D (as well as a microphone, but that is outside the scope of this chapter). This sensor addressed all the concerns described above in the following manner:

i. The depth sensor (near infrared (IR)) is not affected by illumination changes in the scene, as it is immune to variable lighting in the environment.
ii. Since it is a single camera, camera synchronization is not an issue
iii. With a high frame rate of 30 frames per second and an average frame rate of 15 frames per second [2, 36] after considering frame storage, the temporal resolution is sufficient even for fall detection, which is one of the most time-sensitive activity recognition domains in smart home settings.
iv. For applications only storing the raw depth data (i.e., not storing the RGB information), the depth data resemble gray silhouettes hence the privacy is preserved. In our earlier studies, our institution IRB found the data to be blurry enough that human recognition would be impossible thus enabling privacy of depth imagery usage in living spaces as well as hospitals [3]. See Fig. 5 for sample depth images taken inside an apartment at an independent living facility.

2.3 Depth Cameras

A larger body of work is using the depth imagery for activity analysis. As discussed in point (ii) above, acquiring 3D data from depth sensors is more convenient than estimating it from stereo images or using motion capture systems due to the need for a single camera, as opposed to two or more cameras. This was an important cornerstone in computer vision applications, as the information lost in projection from 3D to 2D in the traditional intensity image was partially recovered using the

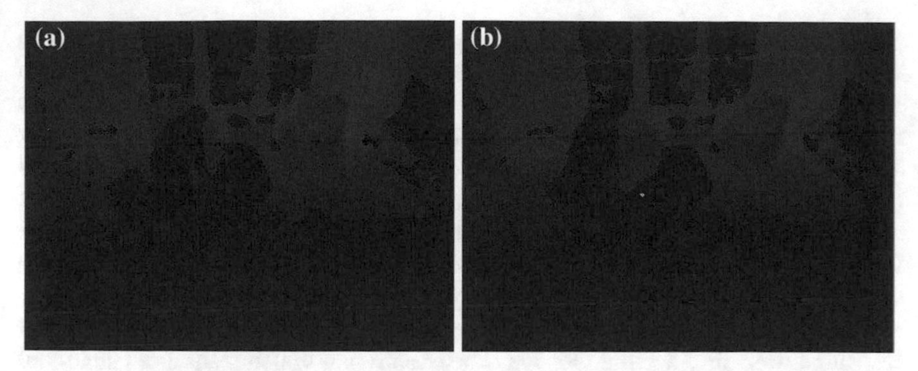

Fig. 5 Sample pseudo-colored depth images from [4] of a resident at TigerPlace. Here, the person is getting up from a recliner (facing away from the camera)

depth component of the RGB-D sensor. Many algorithms have been proposed to address human activity recognition using 3D data.

2.3.1 Depth Cameras for Gait Analysis

In one of the earlier works on gait analysis using the Kinect sensor, Stone and Skubi [36] found that measuring the Kinect and standard web cameras against the Vicon motion-capture system showed that the depth sensor has sufficient resolution to differentiate between different gait parameters and measures comparably against the Vicon system. Gabel et al. [13] did a similar study on a larger population to measure strides, swings, and stances in the gait movement on 23 subjects. The Kinect skeletonization SDK was utilized in the study and the center of mass for the hip joints, shoulder joints, and spine. Regression models were used to measure the gait parameters which were then compared with pressure sensors as ground truth. In later work by Stone et al. [37], a gait model was created using longitudinal depth data collected over several months from the living spaces of older adults living in TigerPlace, an independent living facility. Gait parameters such as stride length, as well as 3D shape parameters like height were incorporated in an individual gait model to identify different residents living in a common space, or to differentiate between the resident and guests for a longitudinal gait study. Gaussian mixture models were used to create the gait model for each individual resident. This was one of the earliest studies with actual systems deployed in the real world for continuous video surveillance applications. Since computer vision techniques were fairly advanced before the advent of the depth sensors, the combination of CI and computer vision began fairly simultaneously. Even in the use of energy images for gait, the researchers used CI techniques for gait recognition. In [18], Hoffman et al. used the depth imagery instead of the color data to obtain the binary silhouettes for the averaging for the energy images. They further used the depth data to compute the gradient depth histogram energy features by computing the histogram of gradients for the depth images, and computing the average of the features over a gait cycle. The dimensions of the features were further reduced using a combination of principal component analysis and linear discriminant analysis. The features were able to differentiate between the gaits from different depth sequences. In more healthcare specific applications, Galna et al. [14] used the Kinect sensor to measure motion in people with Parkinsons disease (PD). In a relatively small cohort size of nine PD patients and 10 control participants, the authors compared the Kinects skeleton model with the Vicon skeletal model and found that the coarser grained activities like sit-to-stand, gait, or stepping were measured comparably between the two measurements but the Kinect was not able to pick up finer motions like hand-clasping or toe tapping. However, the sensors temporal resolution was fairly accurate for the coarser activities which made it a viable tool for tracking changes in the movements of PD patients over time.

2.3.2 Depth Cameras for Complex Activity Analysis

In activity recognition, we see a similar trend with even the earlier works incorporating CI along with the computer vision algorithms. In one of the earliest works by Shotton et al. [34] used random decision forests to predict the labels for a depth pixel for body part labeling as a precursor to activity recognition. In this study, different poses were evaluated to effectively detect the pixel regions with respect to the human body. Not surprisingly, the researchers found that the average precision of the system increased linearly with a corresponding exponential increase in the training data sample. This paper clearly highlights the need for simpler CI techniques given limited training data samples. In [49], Yang and Tian generated feature channels called the eigen joints by taking static and temporal features from the spatial difference in the joint locations across different image frames. Principal component analysis extracted the eigen components of these features that they termed as eigen joints. Using the public datasets like the Cornell human action dataset [38], they compared their algorithm with other techniques, achieving 85% and greater performance accuracies across different activity sequences. In [44], Wang et al. use an interesting approach of fusing short-time fourier transform (STFT) features recursively within each segment of an activity, and combining all the features into a vector defining an actionlet that is a combination of all the STFT features in each segment.This technique was also successfully able to categorize the different activities in the public CMU MoBo dataset [49] and the MSR Action3D datasets [50]. In [47], Xia and Aggarwal expand the idea of spatio-temporal interest points to depth videos or DSTIPs and generate 3D depth cuboids to define activity sequences. One important aspect that is addressed in many of these papers is the need for noise suppression; in standard RGB videos, smoothing filtering techniques usually suffice for noise removal but in the case of depth imagery, the three categories of noise described in [47] are: noise due to the variation in the sensing device that depends on the distance of the objects from the sensor, noise due to the jumping of depth pixel values from foreground to background due to dynamic changes in the scene, and holes present in the depth images present due to reflecting materials, or porous materials present in the scene. After generating the DSTIP cuboid features, support vector machine classification was used to recognize the different activities. For the MSR Action3D dataset, the performance was again compared with the existing algorithms and found to improve the classification performance. In [4], Banerjee et al. demonstrated the use of a simple optical flow based technique in combination with unsupervised clustering techniques to analyze the sedentary behavior of older adults in their homes (See Fig. 5 for sample depth images from the apartment). This work was deployed in the same settings at Stone and Skubi [36] at TigerPlace, an independent living facility for older adults and tested with five residents on a continuous data of around 10 months to map the relation between the amount of time the residents spent sitting in the living spaces, and its mapping with the established fall risk measures.

In a more semantically enhanced method, [2], Banerjee et al. used a knowledge-base framework to incorporate more scene-based features as a means of distinguishing between complex activities of daily living in the eldercare domain. In all of these studies discussed in this section, some common themes are addressed: handling noise in depth imagery, extending the RGB feature framework into the depth imagery framework, and bottlenecks in moving the laboratory tested algorithms into the real-world settings. Of all the studies discussed, only Crispim-Junior [9], Stone [36], and Banerjee [2, 4] discuss results obtained in the real environment, which then begs the question:

II. Why do most existing algorithms perform so poorly in real-world settings?

One critical issue that needs to be addressed is that public datasets like CMUs MoBo [8] or MSR Activity [50] datasets are still far too "clean" to represent dynamic environments, i.e. there is an inherent bias in capturing activity sequences that are not very noisy and show the entire human body from the field of view. This creates a misleading performance evaluation that cannot emulate the performance of these algorithms in actual settings. Collecting and annotating large activity datasets (such as ImageNet [32]) that contain samples of noisy images or video sequences can address this eventually, and is already taking strides towards improvement but is still a long way from creating a gold standard that can be used for assessing algorithms. That said, it is also important to not jump to a dynamic environment directly as that can create too complex a problem for an algorithm to perform. A solution would be to test the algorithm in laboratory settings and once it achieves a decent performance, move it to the real world settings, and not worry too much about incrementally improving its performance in the laboratory settings. Another challenge that the computer vision industry still faces is the increasing gap between cheap computation and expensive communication; memory bandwidth or moving the data from the sensor to the server (where the processing takes place most of the times) is expensive in time as well as cost. One possibility is to have local systems process the data partly, compress and then send to the servers (something we have running at TigerPlace) but the efficiency is still reduced owing to the limitations in the data transfer speeds. Similarly, at the front end, existing computer vision APIs like OpenVX enable acceleration of computer vision applications and works with open-source computer vision

Table 1 Comparison of wearable and non-wearable activity detection

Advantages	Disadvantages
Wearable	
• Usually lower cost	• Incomplete view of subject requires estimation with fewer inputs
• Portable	• Less mature research
• Enables "at-home" usage in many applications	• Duration of activity limited by battery life
	• May physically encumber subject during the activity
Non-wearable	
• Complete view of subject provides ample input	• Not portable
• Usually more controlled and repeatable results due to lower variance in environment	• Complex setup and configuration
• Potentially able to recognize larger variety of parameters	• Installation may be invasive in "at-home" applications

libraries like OpenCV, but is still not at the stage to be used in practical applications requiring various processing units and dedicated hardware to perform specific tasks.

III. What are the design choices one needs to be aware of when deciding between wearable cameras or stationary cameras for activity analysis?

In Table 1, we describe the advantages and disadvantages of using wearable and non-wearable camera systems.

IV. What does CI bring to the vision world as compared to computer vision techniques in the activity analysis domain?

As we observed from the literature survey, we see that the largest body of literature without CI took place in gait analysis using RGB cameras. Even in gait analysis using depth cameras, CI was introduced a lot earlier on since depth images were utilized much later in the computer vision community. One concern raised multiple times is the need for a large amount of training data for supervised applications. As discussed in (I), this is lesser of a concern now with the advent of larger annotated datasets. Another concern for activity classification is the possibility of overfitting using a complex model where simpler algorithms may perform poorer with the training data, but generalize much better compared to complex models using DBNs and HMMs. Another aspect of the comparison between machine learning based approaches and pure computer vision based approaches is the application of the system: machine learning based approaches are predominantly used in the classification domain, whereas computer vision or other signal processing and feature extraction based techniques are more frequently used for parameter learning, such as stride length [36] or walking speed [33] or sitting time [4]. In that sense, a combined computer vision and CI-based approach can be used to first detect a particular activity, such as sitting, and once detected, can be then used to find the variables specific to sedentary behavior. Here, we do see the complementary nature of using a tiered approach for activity analysis. In conclusion, we see that CI-based approaches focus on recognition, and a combined CI and computer vision-based approach can lead to recognition as well as feature learning within the activity.

Acknowledgements This work in part is sponsored by the NIH award #1K01LM012439.

References

1. Bambach, S., Lee, S., Crandall, D.J., Yu, C.: Lending a hand: detecting hands and recognizing activities in complex egocentric interactions. In: 2015 IEEE International Conference on Computer Vision (ICCV), pp. 1949–1957 (2015). doi:10.1109/ICCV.2015.226
2. Banerjee, T., Keller, J.M., Popescu, M., Skubic, M.: Recognizing complex instrumental activities of daily living using scene information and fuzzy logic. Comput. Vis. Image Underst. **140**, 68–82 (2015)
3. Banerjee, T., Rantz, M., Li, M., Popescu, M., Stone, E., Skubic, M., Scott, S.: Monitoring hospital rooms for safety using depth images. AI for Gerontechnology, Arlington, Virginia, USA (2012)
4. Banerjee, T., Yefimova, M., Keller, J., Skubic, M., Woods, D., Rantz, M.: Case studies of older adults sedentary behavior in the primary living area using kinect depth data. J. Ambient Intell, Smart Environ (2016)
5. Bobick, A.F., Davis, J.W.: The recognition of human movement using temporal templates. IEEE Trans. Pattern Anal. Mach. Intell. **23**(3), 257–267 (2001). doi:10.1109/34.910878
6. Cho, Y., Nam, Y., Choi, Y.J., Cho, W.D.: Smartbuckle: human activity recognition using a 3-axis accelerometer and a wearable camera. In: Proceedings of the 2nd International Workshop on Systems and Networking Support for Health Care and Assisted Living Environments, p. 7. ACM (2008)
7. Choudhury, S.D., Tjahjadi, T.: Gait recognition based on shape and motion analysis of silhouette contours. Comput. Vis. Image Underst. **117**(12), 1770–1785 (2013)
8. Collins, R.T., Gross, R., Shi, J.: Silhouette-based human identification from body shape and gait. In: Proceedings of Fifth IEEE International Conference on Automatic Face Gesture Recognition, pp. 366–371 (2002). doi:10.1109/AFGR.2002.1004181
9. Crispim, C.F., Bathrinarayanan, V., Fosty, B., Konig, A., Romdhane, R., Thonnat, M., Bremond, F.: Evaluation of a monitoring system for event recognition of older people. In: 2013 10th IEEE International Conference on Advanced Video and Signal Based Surveillance, pp. 165–170 (2013). doi:10.1109/AVSS.2013.6636634
10. Demiris, G., Oliver, D.P., Giger, J., Skubic, M., Rantz, M.: Older adults' privacy considerations for vision based recognition methods of eldercare applications. Technol. Health Care **17**(1), 41–48 (2009)
11. Dierick, F., Penta, M., Renaut, D., Detrembleur, C.: A force measuring treadmill in clinical gait analysis. Gait Posture **20**(3), 299–303 (2004)
12. Felzenszwalb, P.F., Girshick, R.B., McAllester, D., Ramanan, D.: Object detection with discriminatively trained part-based models. IEEE Trans. Pattern Anal. Mach. Intell. **32**(9), 1627–1645 (2010). doi:10.1109/TPAMI.2009.167
13. Gabel, M., Gilad-Bachrach, R., Renshaw, E., Schuster, A.: Full body gait analysis with kinect. In: 2012 Annual International Conference of the IEEE Engineering in Medicine and Biology Society, pp. 1964–1967 (2012). doi:10.1109/EMBC.2012.6346340

14. Galna, B., Barry, G., Jackson, D., Mhiripiri, D., Olivier, P., Rochester, L.: Accuracy of the microsoft kinect sensor for measuring movement in people with parkinson's disease. Gait Posture **39**(4), 1062–1068 (2014)
15. Gavrila, D.M., Davis, L.S.: 3-d model-based tracking of humans in action: a multi-view approach. In: Proceedings CVPR IEEE Computer Society Conference on Computer Vision and Pattern Recognition, pp. 73–80 (1996). doi:10.1109/CVPR.1996.517056
16. Gonzales, R., Woods, R.: 3 edn. Addison-Wesley Longman Publishing Co. (2007)
17. Hirasaki, E., Moore, S.T., Raphan, T., Cohen, B.: Effects of walking velocity on vertical head and body movements during locomotion. Exp. Brain Res. **127**(2), 117–130 (1999)
18. Hofmann, M., Bachmann, S., Rigoll, G.: 2.5d gait biometrics using the depth gradient histogram energy image. In: 2012 IEEE Fifth International Conference on Biometrics: Theory, Applications and Systems (BTAS), pp. 399–403 (2012). doi:10.1109/BTAS.2012.6374606
19. Kakadiaris, L., Metaxas, D.: Model-based estimation of 3d human motion. IEEE Trans. Pattern Anal. Mach. Intell. **22**(12), 1453–1459 (2000). doi:10.1109/34.895978
20. Kale, A., Rajagopalan, A.N., Cuntoor, N., Kruger, V.: Gait-based recognition of humans using continuous hmms. In: Proceedings of Fifth IEEE International Conference on Automatic Face Gesture Recognition, pp. 336–341 (2002). doi:10.1109/AFGR.2002.1004176
21. Liu, Z., Sarkar, S.: Simplest representation yet for gait recognition: averaged silhouette. In: Proceedings of the 17th International Conference on Pattern Recognition, 2004. ICPR 2004., vol. 4, pp. 211–214 (2004). doi:10.1109/ICPR.2004.1333741
22. Lu, J., Zhang, E.: Gait recognition for human identification based on ica and fuzzy svm through multiple views fusion. Pattern Recogn. Lett. **28**(16), 2401–2411 (2007)
23. Man, J., Bhanu, B.: Individual recognition using gait energy image. IEEE Transactions on Pattern Analysis and Machine Intelligence **28**(2), 316–322 (2006). doi:10.1109/TPAMI.2006.38
24. Matsuo, K., Yamada, K., Ueno, S., Naito, S.: An attention-based activity recognition for egocentric video. In: 2014 IEEE Conference on Computer Vision and Pattern Recognition Workshops, pp. 565–570 (2014). doi:10.1109/CVPRW.2014.87
25. Muramatsu, D., Shiraishi, A., Makihara, Y., Uddin, M.Z., Yagi, Y.: Gait-based person recognition using arbitrary view transformation model. IEEE Trans. Image Process. **24**(1), 140–154 (2015). doi:10.1109/TIP.2014.2371335
26. Narayan, S., Kankanhalli, M.S., Ramakrishnan, K.R.: Action and interaction recognition in first-person videos. In: 2014 IEEE Conference on Computer Vision and Pattern Recognition Workshops, pp. 526–532 (2014). doi:10.1109/CVPRW.2014.82
27. Niyogi, S.A., Adelson, E.H.: Analyzing and recognizing walking figures in xyt. In: 1994 Proceedings of IEEE Conference on Computer Vision and Pattern Recognition, pp. 469–474 (1994). doi:10.1109/CVPR.1994.323868
28. Park, S., Aggarwal, J.K.: A hierarchical bayesian network for event recognition of human actions and interactions. Multimed. Syst. **10**(2), 164–179 (2004)
29. Pirsiavash, H., Ramanan, D.: Detecting activities of daily living in first-person camera views. In: 2012 IEEE Conference on Computer Vision and Pattern Recognition, pp. 2847–2854 (2012). doi:10.1109/CVPR.2012.6248010
30. Rogez, G., Supancic, J.S., Ramanan, D.: Understanding everyday hands in action from rgb-d images. In: 2015 IEEE International Conference on Computer Vision (ICCV), pp. 3889–3897 (2015). doi:10.1109/ICCV.2015.443
31. Rowley, H.A., Rehg, J.M.: Analyzing articulated motion using expectation-maximization. In: Proceedings of IEEE Computer Society Conference on Computer Vision and Pattern Recognition, pp. 935–941 (1997). doi:10.1109/CVPR.1997.609440
32. Russakovsky, O., Deng, J., Su, H., Krause, J., Satheesh, S., Ma, S., Huang, Z., Karpathy, A., Khosla, A., Bernstein, M., et al.: Imagenet large scale visual recognition challenge. Int. J. Comput. Vis. **115**(3), 211–252 (2015)
33. Schneider, B., Banerjee, T.: Preliminary investigation of walking motion using a combination of image and signal processing. In: 2016 International Conference on Computational Science and Computational Intelligence (CSCI) (2016)

34. Shotton, J., Sharp, T., Kipman, A., Fitzgibbon, A., Finocchio, M., Blake, A., Cook, M., Moore, R.: Real-time human pose recognition in parts from single depth images. Commun. ACM **56**(1), 116–124 (2013)
35. Stone, E.E., Anderson, D., Skubic, M., Keller, J.M.: Extracting footfalls from voxel data. In: 2010 Annual International Conference of the IEEE Engineering in Medicine and Biology, pp. 1119–1122 (2010). doi:10.1109/IEMBS.2010.5627102
36. Stone, E.E., Skubic, M.: Passive in-home measurement of stride-to-stride gait variability comparing vision and kinect sensing. In: Engineering in Medicine and Biology Society, EMBC, 2011 Annual International Conference of the IEEE, pp. 6491–6494. IEEE (2011)
37. Stone, E.E., Skubic, M.: Unobtrusive, continuous, in-home gait measurement using the microsoft kinect. IEEE Trans. Biomed. Eng. **60**(10), 2925–2932 (2013). doi:10.1109/TBME.2013.2266341
38. Sung, J., Ponce, C., Selman, B., Saxena, A.: Human activity detection from rgbd images. Plan, activity, and intent recognition, vol. 64 (2011)
39. Ugbolue, U.C., Papi, E., Kaliarntas, K.T., Kerr, A., Earl, L., Pomeroy, V.M., Rowe, P.J.: The evaluation of an inexpensive, 2d, video based gait assessment system for clinical use. Gait Posture **38**(3), 483–489 (2013)
40. Unuma, M., Anjyo, K., Takeuchi, R.: Fourier principles for emotion-based human figure animation. In: Proceedings of the 22nd Annual Conference on Computer Graphics and Interactive Techniques, pp. 91–96. ACM (1995)
41. Veeraraghavan, A., Chellappa, R., Roy-Chowdhury, A.K.: The function space of an activity. In: 2006 IEEE Computer Society Conference on Computer Vision and Pattern Recognition (CVPR'06), vol. 1, pp. 959–968 (2006). doi:10.1109/CVPR.2006.304
42. Wang, F., Stone, E., Skubic, M., Keller, J.M., Abbott, C., Rantz, M.: Toward a passive low-cost in-home gait assessment system for older adults. IEEE J. Biomed. Health Inf. rmatics **17**(2), 346–355 (2013). doi:10.1109/JBHI.2012.2233745
43. Wang, H., Kläser, A., Schmid, C., Liu, C.L.: Dense trajectories and motion boundary descriptors for action recognition. Int. J. Comput. Vis. **103**(1), 60–79 (2013)
44. Wang, J., Liu, Z., Wu, Y., Yuan, J.: Mining actionlet ensemble for action recognition with depth cameras. In: 2012 IEEE Conference on Computer Vision and Pattern Recognition (CVPR), pp. 1290–1297. IEEE (2012)
45. Watanabe, Y., Hatanaka, T., Komuro, T., Ishikawa, M.: Human gait estimation using a wearable camera. In: 2011 IEEE Workshop on Applications of Computer Vision (WACV), pp. 276–281 (2011). doi:10.1109/WACV.2011.5711514
46. Webster, K.E., Wittwer, J.E., Feller, J.A.: Validity of the gaitrite walkway system for the measurement of averaged and individual step parameters of gait. Gait Posture **22**(4), 317–321 (2005)
47. Xia, L., Aggarwal, J.K.: Spatio-temporal depth cuboid similarity feature for activity recognition using depth camera. In: 2013 IEEE Conference on Computer Vision and Pattern Recognition, pp. 2834–2841 (2013). doi:10.1109/CVPR.2013.365
48. Yamato, J., Ohya, J., Ishii, K.: Recognizing human action in time-sequential images using hidden markov model. In: Proceedings 1992 IEEE Computer Society Conference on Computer Vision and Pattern Recognition, pp. 379–385 (1992). doi:10.1109/CVPR.1992.223161
49. Yang, X., Tian, Y.: Effective 3d action recognition using eigenjoints. J. Vis. Commun. Image Represent. **25**(1), 2–11 (2014)
50. Yu, G., Liu, Z., Yuan, J.: Discriminative orderlet mining for real-time recognition of human-object interaction. In: Asian Conference on Computer Vision, pp. 50–65. Springer (2014)
51. Yu, S., Tan, D., Tan, T.: A framework for evaluating the effect of view angle, clothing and carrying condition on gait recognition. In: 18th International Conference on Pattern Recognition (ICPR'06), vol. 4, pp. 441–444 (2006). doi:10.1109/ICPR.2006.67
52. Zhao, G., Liu, G., Li, H., Pietikainen, M.: 3d gait recognition using multiple cameras. In: 7th International Conference on Automatic Face and Gesture Recognition (FGR06), pp. 529–534 (2006). doi:10.1109/FGR.2006.2

Compressive Sensing and Chaos-Based Image Compression Encryption

R. Ponuma and R. Amutha

Abstract Compressive sensing and chaos-based image compression-encryption scheme is proposed. A two-dimensional chaotic map, the sine logistic modulation map is used to generate a chaotic sequence. The chaotic sequence is used to construct two circulant measurement matrices. The sparse representation of the plain image is obtained by employing discrete cosine transform. The transform coefficients are then measured using the two measurement matrices. Two levels of encryption are achieved. The parameters of the chaotic map acts as the key in the first level of encryption. Further, Arnold chaotic map-based scrambling is used to enhance the security of the cipher. Simulation results verify the effectiveness of the algorithm and its robustness against various attacks.

Keywords Chaotic maps · Compressive sensing · Image compression · Image encryption

1 Introduction

The proliferation of information technology has made multimedia communication an important aspect of daily life. Multimedia is used for personal communication, interactive applications, entertainment, education, etc. Multimedia applications extensively use digital image transmission; hence it is necessary to provide secure and fast image transmission. An efficient and secure transmission can be achieved by employing simultaneous compression-encryption techniques. The intrinsic properties of digital images like bulk data capacity, redundancy, and high correlation can be exploited to compress and encrypt the images. Image compression is

R. Ponuma (✉) · R. Amutha
Department of Electronics and Communication Engineering,
SSN College of Engineering, Chennai, India
e-mail: ponumar@ssn.edu.in

R. Amutha
e-mail: amuthar@ssn.edu.in

A.E. Hassanien and D.A. Oliva (eds.), *Advances in Soft Computing and Machine Learning in Image Processing*, Studies in Computational Intelligence 730,
https://doi.org/10.1007/978-3-319-63754-9_17

of paramount importance in applications like surveillance, which involve wireless sensor networks that are resource constrained and requires low bitrate image compression. There are many image compression schemes available in the literature. In [1] the authors proposed a wavelet-based scheme, which extracts and encodes only approximations of the image using fixed-point arithmetic, which considerably improved the lifetime of the sensors. In [2] only shift and add operations together with BinDCT is used to directly extract and encode the DC coefficient and the first three AC coefficients of an 8×8 image block. Traditional encryption schemes are not suitable for image encryption because of the inherent properties of digital images. Chaos-based cryptosystem has gained popularity due to the properties of chaotic systems like periodicity, sensitive dependence on initial conditions and good pseudo-randomness.

Zhou et al. [3] proposed a technique that combined existing one-dimensional chaotic maps like logistic map, sine map, and tent map to generate many new one-dimensional chaotic maps. The new chaotic system had larger chaotic range as well as excellent diffusion and confusion properties. The chaos-based encryption scheme was able to resist brute-force attack. A cryptosystem using chaos and permutation-substitution network was proposed by Belazi et al. [4]. A new one-dimensional chaotic map is first used to diffuse the plain image. Then s-boxes are used to generate the cipher by employing substitution operation. A logistic map is used to again diffuse the cipher, thereby increasing the cryptographic strength. Finally, a permutation function is used to obtain an encrypted image, which can resist statistical attacks. In [5] a double image encryption scheme using logistic map and cellular automata is proposed. The keys for the encryption are generated by the convolution of the logistic maps. The least significant bits of the two images are first combined and then diffused using cellular automata. The performance of the proposed scheme and the randomness of the key are experimentally verified. In [6] a multiple image encryption using logistic map and Fractional Fourier Transform is proposed. The fractional order is used as the key in the encryption process. The performance analysis showed that the proposed scheme is highly secure and requires less bandwidth to transmit the encrypted data. The encryption schemes based on one-dimensional chaos is vulnerable to brute-force attack. Hence hyper-chaos-based image encryption schemes are widely used. In [7] an image encryption based on three-dimensional bit matrix permutation is proposed. A coupled Chen and 3D cat map is used to implement a 3D bit matrix permutation. The plain image is first transformed into a 3D bit matrix and the encryption is done using permutation.

In [8] an image encryption-compression algorithm based on hyper-chaos and Chinese remainder theorem was proposed. Two-dimensional hyper-chaotic system was used to first shuffle the plain image. The Chinese remainder theorem was then used to diffuse and compress the image simultaneously. Guesmi et al. [9] proposed an image encryption algorithm based on DNA masking, Secure Hash Algorithm, and Lorenz chaotic system. The experimental results showed that the entropy of the cipher image was improved significantly, which implies that the cipher is highly random. Also, the proposed scheme has a key space large enough to resist the statistical attacks.

An optical image encryption using three-dimensional chaotic map was proposed by Chen et al. [10]. The joint image scrambling and random encoding was performed in gyrator domains. The key stream was generated using the three-dimensional chaotic map. They demonstrated that the proposed crypto system was robust against noise and occlusion attack. In [11] a lossless symmetric key encryption scheme using Discrete Haar Wavelet transform and Arnold Cat map is proposed. The plain image is transformed using the wavelet transform to obtain the transform coefficients. The chaotic map is used to scramble the transform coefficients to obtain the cipher image. Wu et al. [12] introduced a lossless color image encryption scheme using DWT and hyperchaotic system. A six-dimensional chaotic system generated by coupling two identical Lorenz systems is used. The key for the encryption scheme depends on both the hyperchaotic system and the plain image. The security analysis demonstrates that the proposed algorithm is fast and highly secure.

In [13] hybrid hyper-chaotic system and cellular automata based color image encryption is proposed. A 3D-Arnold Cat map is used to permute the pixels of the plain image. The permutation process reduces the correlation between the adjacent pixels. Key image is created using nonuniform cellular automata. A Chen chaotic system is used to select the key image to encrypt the RGB components of the color image. Information entropy of 7.9991 was achieved for the cipher image, which is very close to the ideal value of 8, it proves that the cipher image is highly random. An encryption scheme for medical images which integrates number theoretic approach and Henon map was proposed in [14]. Modular exponentiation of the primitive roots of the chosen prime in the range of its residual set is employed in the generation of two-dimensional array of keys. The DICOM image is encrypted using a key from the key matrix, whose selection is chaotically controlled by the Henon map. Experimental results verified the strength of the proposed cryptographic system against different statistical attacks.

Compressive Sensing (CS) [15, 16] is a new sampling, reconstruction technique that can perform simultaneous sampling and compression. The signal is sampled linearly at significantly lower rate. The sampling rate is lower than the traditional Nyquist sampling criterion. An optimization technique is used to reconstruct the original signal from the reduced measurements. CS can be integrated into a cryptosystem to provide simultaneous compression and encryption. Rachlin and Baron [17] proved that encryption using the measurement matrix as the key in the sampling process does not achieve Shannon's definition of perfect secrecy. However, CS-based encryption provides computational secrecy as the recovery of the original signal is a NP-hard problem, even though the key is known to the intruder. In [18] the authors review the different types of encryption models that can be implemented using CS. Six different frameworks for encryption using Compressive sensing, chaos and optics are reviewed. Zhou et al. [19] presents a key controlled measurement matrix construction and implements a hybrid image compression-encryption technique. The key controlled scheme reduces the complexity of key generation and distribution. An image compression-encryption scheme is implemented using Cosine Number Transform in [20]. Only modular arithmetic is used for computing the transform as it is defined over algebraic structures in a finite field. The transform coefficients are then compressed and encrypted using a Gaussian measurement matrix.

Zhang et al. [21] introduced a color image encryption scheme using Arnold transform and compressive sensing. The three-color components of the color image are compressed and encrypted simultaneously using compressive sensing. The three-color components are then grouped to form a grayscale image and scrambled using the Arnold transform to enhance the security. In [22] a 2D-compressive sensing-based image compression-encryption scheme is proposed. In two-dimensional compressive sensing the plain image is measured in two directions and then encrypted using a fractional Mellin transform. The fractional Mellin transform is a nonlinear transform, which thwarts the security risks that arise because of the linear property of compressive sensing. Zhou et al. [23] proposed an image cryptosystem using hyper-chaos and 2D compressive sensing to achieve simultaneously compression-encryption. A hyper-chaotic Chen system is used, which increases the key space thereby resisting brute-force attack.

In this paper a compression-encryption algorithm based on 2D compressive sensing and 2D-Sine Logistic Modulation Map (2D-SLMM) is proposed. The measurement matrix is constructed using the chaotic sequence generated by 2D-SLMM. The chaotic sequence is used as the seed to construct a Circulant matrix. Two measurement matrices are used to measure the plain image. The control parameter and the initial values of the 2D-SLMM are used as the security key. The security of the cipher image is further enhanced by scrambling the linear measurements using a Arnold map. The proposed image compression-encryption algorithm is robust against various attacks and has low data volume. Simulation results verify the effectiveness of the proposed scheme.

2 Preliminaries

2.1 Compressive Sensing

Classical data acquisition states that, if a signal is sampled at Nyquist rate, i.e., twice the maximum frequency it can be perfectly reconstructed. In multimedia applications the Nyquist rate can result in large amount of samples, which has to be compressed for storage and faster transmission. Compressive sensing is a new paradigm introduced in [15, 16] that states that if a signal is compressible/sparse in some domain, then it can be perfectly reconstructed by sampling at a rate less than the Nyquist sampling rate. The fundamental requirements in compressive sensing are sparsity, measurement matrix and the reconstruction algorithm. Sparsity indicates that the signal has a concise representation in some basis ψ. If an image is sparse in a basis, it has few large coefficients that capture most of the information and all the other coefficients are small. Sparsity leads to efficient estimation, compression and dimensionality reduction. The image data is naturally sparse in some basis (Discrete Cosine Transform, Wavelet).

A 1D sparse signal with length N can be represented as

$$x = \sum_{i=1}^{N} \alpha_i \psi_i \tag{1}$$

where ψ is an orthonormal basis, and α is the representation of x in ψ. The coefficient vector α is sparse, if the number of nonzero coefficients $K \ll N$. A linear measurement process is used to compute the measurements y with length M ($M \ll N$), by projecting x onto a measurement matrix φ.

$$y = \varphi x = \varphi \psi \alpha = \theta \alpha \tag{2}$$

where φ is an $M \times N$ measurement matrix which is incoherent with basis matrix ψ. The sensor matrix Θ should satisfy the restricted isometry property (RIP), so that the signal x can be reconstructed from the reduced measurements y. Candes [24] presented the Restricted Isometry Property as a sufficient condition on the sensing matrix to perform perfect recovery. In [25] Banderia et al. proved that the determination of RIP for a matrix is NP-hard. An alternative approach for perfect reconstruction is to ensure that the measurement matrix φ is incoherent with the basis ψ. A randomly generated measurement matrix was proved to be incoherent with the sparsifying basis with high probability. The commonly used measurement matrices are Gaussian matrix, Bernoulli matrix, etc. Structurally Random Matrix (SRM) like the circulant and Toeplitz matrix are also proved to be incoherent with ψ.

A 2D signal X of size $N \times N$ can be measured using two measurement matrices. The 2D signal is first transformed using the basis function ψ. The transformed coefficients are measured using the measurement matrix φ_1 to obtain the measurements α_1.

$$\alpha_1 = \varphi_1 \psi^T X \tag{3}$$

where φ_1 is an $M \times N$ measurement matrix and ψ is an $N \times N$ orthogonal basis. The 2D compressed and encrypted measurements Y is obtained by measuring the transposition of α_1 expressed in the ψ domain.

$$\alpha_2 = \psi^T \alpha_1^T = \psi^T X^T \psi \varphi_1^T = \alpha \varphi_1^T \tag{4}$$

where $\alpha = \psi^T X^T \psi$, α_2 is the sparse coefficient of α_1 in the ψ domain. To obtain the $M \times M$ matrix Y from the $N \times N$ matrix X, α_2 is measured with another measurement matrix φ_2

$$Y = \varphi_2 \alpha_2 = \varphi_2 \alpha \varphi_1^T \quad (5)$$

To recover X from Y, the following convex optimization process is necessarily involved

$$\alpha = \arg \min \|\alpha\|_0 \ s.t. \ Y = \varphi_2 \alpha \varphi_1^T \quad (6)$$

The optimization problem can be solved using several reconstruction algorithms. The commonly employed methods are matching pursuit (MP) [26], OMP [27], Sl^0 [28] and so on.

2.2 Two-Dimensional Sine Logistic Modulation Map

The chaotic maps can be classified into one-dimensional chaotic maps and high-dimensional chaotic maps. The one-dimensional chaotic maps usually contain one variable and a few parameters. The 1D-maps like logistic, sine and tent maps have simple structures. The chaotic structure can be determined by using chaotic signal estimation technologies. The encryption scheme using these 1D maps are vulnerable to various cryptographic attacks. Hence, higher dimensional chaotic maps are employed for encryption. Logistic map is simple in structure and it is widely used in image encryption. The logistic map has single control parameter μ and is defined by the following equation:

$$x_{n+1} = \mu x_n (1 - x_n), x_n \in (0.1) \quad (7)$$

when the parameter $\mu \in [3.94, 4]$, the logistic map exhibits good chaotic characters.

The Sine map also has single control parameter μ and is defined by the following equation:

$$x_{n+1} = \mu \sin(\pi x_n), x_n \in (0, 1), \mu > 0 \quad (8)$$

In the logistic and sine map the next iteration value is obtained by a linear transform. Hence, it can be predicted easily. In [29] the two one-dimensional maps are combined to generate a new two-dimensional sine logistic modulation map (2D-SLMM). Hua et al. mathematically represented the 2D-SLMM as

$$\begin{aligned} x_{i+1} &= \gamma (\sin(\pi y_i) + \beta) x_i (1 - x_i) \\ y_{i+1} &= \gamma (\sin(\pi x_{i+1}) + \beta) y_i (1 - y_i) \end{aligned} \quad (9)$$

where γ and β are the control parameters. $\gamma \in [0,1]$ and $\beta \in [0, 3]$. The 2D map has a highly chaotic structure and its values cannot be easily predicted. When parameter $\beta = 3$ and $\gamma \in [0.9, 1]$, 2D-SLMM has wide chaotic range.

2.3 Arnold Scrambling

The Arnold chaotic map is used to scramble the pixel location. The Arnold transform shifts the pixel position using the relation given below:

$$\begin{bmatrix} x' \\ y' \end{bmatrix} = \begin{bmatrix} 1 & 1 \\ 1 & 2 \end{bmatrix} \begin{bmatrix} x \\ y \end{bmatrix} \bmod M \tag{10}$$

where M is the size of the square matrix. (x', y') is the position of the pixels after the scrambling. (x, y) is the original position of the pixels. The cryptographic strength of the cipher image can be improved by repeating the above scrambling process. The number of iteration also acts as one of the keys in the encryption process.

3 Image Compression-Encryption Scheme

The block diagram of the proposed image compression-encryption algorithm based on 2D compressive sensing and chaos is shown in Fig. 1. The image compression-encryption algorithm is as follows:

1. The initial conditions (x_{0i}, y_{0i}) and control parameter (γ_i) are used as the key to generate two chaotic sequences (s_1, s_2) of length $2M$ using the 2D-SLMM.
2. The preceding M elements of the chaotic sequence is neglected to form a sequence of length M.
3. The truncated sequence s_1, s_2 is used as the seed to construct two circulant matrices of dimension $M \times M$.
4. The circulant matrices (φ_1, φ_2) are constructed as given in [19]

$$\begin{aligned} \varphi(i,1) &= \lambda\varphi(i-1,M) \\ \varphi(i,2\colon N) &= \varphi(i-1,1\colon M-1) \end{aligned} \tag{11}$$

   where $2 \leq i \leq M$ and $\lambda > 1$.
5. The plain image is transformed using discrete cosine transform to obtain the transform coefficients α.

$$\alpha = \psi^T X^T \psi \tag{12}$$

6. The coefficients are then measured using φ_1 and φ_2, the measurement Y is given by

$$Y = \varphi_2\, \alpha\, \varphi_1^T \tag{13}$$

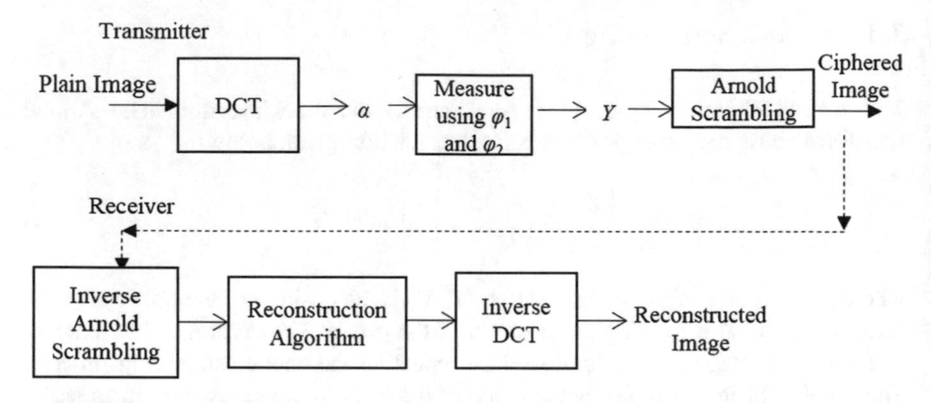

Fig. 1 Block diagram of compression-encryption scheme

7. The measurements Y are scrambled using the Arnold map to obtain the compressed-encrypted image. The scrambling operation is performed for T rounds of iteration.

At the receiver, the cipher obtained can be decrypted using the keys used in the encryption process. The keys consist of the initial and control parameters of the 2D-SLMM, i.e., x_{01}, y_{01}, γ_1, x_{02}, y_{02} and γ_2, parameter λ used in the construction of the circulant matrix and the number of iterations performed in Arnold scrambling. The Smoothed l^0 algorithm is used to recover the transform coefficients. The reconstructed image is obtained from the recovered transform coefficients using the inverse DCT.

4 Simulation Result and Analysis

The test image is taken from USC-SIPI 'Miscellaneous' image dataset. The test image Lena with size 256 × 256, is shown in Fig. 2a. The initial parameters of the 2D-SLMM are $x_{01} = 0.1598$, $y_{01} = 0.8507$, $\gamma_1 = 0.9783$; $x_{02} = 0.9403$, $y_{02} = 0.1722$, $\gamma_2 = 0.9373$. The circulant measurement matrix is generated using the chaotic sequence generated by 2D-SLMM. The image is transformed using DCT and the transform coefficients are measured. The measurements are scrambled using Arnold map to obtain the cipher image as in Fig. 2b. The compression-encryption scheme was analysed with four different images. The Smoothed l^0 algorithm is used to reconstruct the original image. Sl^0 is a fast algorithm to find the sparse solutions of an underdetermined system of linear equations. The algorithm directly tries to minimize the l^0 norm. The reconstructed image using Sl^0 is shown in Fig. 2c.

Fig. 2 **a** Plain image, **b** Cipher image, **c** Reconstructed image

4.1 Entropy Analysis

Entropy is used to measure the randomness of the image. The entropy of the image is given by,

$$H(X) = -\sum_{i=1}^{n} \Pr(x_i) \log_2 \Pr(x_i) \tag{14}$$

where $\Pr(x_i)$ is the probability of x_i. For an image with 256 gray levels the absolute maximum of entropy is 8 bits. The maximum entropy is achieved when the histogram is flat. Hence, the maximum entropy is obtained when the gray levels have equal probability of occurrence. Hence, for a cipher image the entropy value should be close to 8, i.e., in a cipher image all the gray levels must be equally distributed. Table 1, shows the entropy of the plain image and the cipher image. The entropy of

Table 1 Entropy of the plain and cipher image

Test images	Entropy of the plain image	Entropy of the cipher image		
		Proposed scheme	Ref. [4]	Ref. [29]
Airport	6.8303	7.9980	7.9960	7.9969
Barbara	7.6321	7.9971	7.9978	7.9957
Boat	7.4842	7.9974	7.9980	7.9959
Cameraman	7.0097	7.9933	7.9985	7.9964
Couple	6.4207	7.9939	7.9976	7.9980
Elaine	7.5060	7.9974	7.9985	7.9971
House	6.4961	7.9975	7.9981	7.9952
Lena	7.4451	7.9972	7.9963	7.9965
Man	7.5237	7.9983	7.9975	7.9965
Peppers	7.5937	7.9973	7.9985	7.9958
Average		7.9967	7.9977	7.9964

the cipher image is compared with the entropy of the cipher produced by the encryption algorithms proposed in [4, 29]. The entropy of the proposed scheme is higher than that of the values obtained in [29] and lesser than that of the values in [4]. We can infer that the encrypted image has high randomness as the entropy of the cipher image is close to the theoretical value of 8.

4.2 PSNR and SSIM Analysis

Peak signal-to-noise ratio (PSNR) is used for measuring the quality of decrypted digital image. Mathematically PSNR is defined by

$$PSNR = 10\log\frac{255^2}{(1/N^2)\sum_{i=1}^{N}\sum_{j=1}^{N}\left[R(i,j) - I(i,j)\right]^2} \tag{15}$$

where $I(i, j)$ is the plain image and $R(i, j)$ is the reconstructed image. The test image is of size $N \times N$. Structural Similarity Index (SSIM) is used for measuring the similarity between plain image and reconstructed image. The test image shown in Table 2 is of size 256 × 256. The total number of pixels is 65,536 from which approximately 56.25, 25, 6.25% of the pixels are taken as measurements. For a compression ratio of 4:1 the PSNR obtained for the test image is 30.4403 dB. The visual quality of the reconstructed image is also good. Table 2 shows the PSNR and SSIM value for different number of measurements.

Table 2 PSNR and SSIM analysis

Plain Image	Number of Measurements	Cipher Image	Reconstructed Image	PSNR (dB)	SSIM
	36864			34.9370	0.9003
(256 × 256)	16384			30.4403	0.8000
	4096			25.9012	0.6244

4.3 *Pixel Correlation Analysis*

In a natural image, high correlation exists between the neighboring pixels. A good image encryption algorithm should have the ability to break these correlations. Mathematically, data correlation is defined by

$$r_{xy} = \frac{\mathrm{cov}(x, y)}{\sqrt{D(x)\,D(y)}} \tag{16}$$

$$\mathrm{cov}(x,y) = \frac{1}{P}\sum_{i=1}^{P}[x_i - E(x)][y_i - E(y)],\ D(x) = \frac{1}{P}\sum_{i}^{P}[x_i - E(x)]^2,$$

$$E(x) = \frac{1}{P}\sum_{i=1}^{P} x_i$$

2,000 pixels and their corresponding adjacent pixels along the horizontal, vertical and diagonal directions were randomly chosen for the correlation analysis. Table 3, shows the pixel correlation value of the plain and cipher image. The results of the

Table 3 Pixel correlation

Test images	Direction	Plain images	Cipher image		
			Proposed	Ref. [4]	Ref. [29]
Airport	Horizontal	0.9182	−0.0386	−0.0275	0.0094
	Vertical	0.9139	0.0193	−0.0154	−0.0107
	Diagonal	0.8720	0.0136	−0.0187	−0.0007
Barbara	Horizontal	0.9002	0.0051	−0.0052	−0.0187
	Vertical	0.9612	0.0253	−0.0067	0.0006
	Diagonal	0.8817	0.0024	−0.0068	0.0001
Boat	Horizontal	0.9972	0.0002	−0.0100	−0.0295
	Vertical	0.9717	0.0115	−0.0124	−0.0150
	Diagonal	0.9599	0.0122	−0.0185	−0.0224
Cameraman	Horizontal	0.9401	0.0371	−0.0095	−0.0047
	Vertical	0.9646	0.0017	−0.017	−0.0195
	Diagonal	0.9127	−0.0234	−0.0119	0.0279
Couple	Horizontal	0.9437	−0.0068	−0.0251	−0.0236
	Vertical	0.9543	0.0247	−0.0213	−0.0045
	Diagonal	0.9009	−0.0055	−0.0078	0.0016
Elaine	Horizontal	0.9761	0.0200	−0.0232	−0.0066
	Vertical	0.9725	0.0021	−0.042	−0.0019
	Diagonal	0.9679	0.0072	−0.003	0.0007
House	Horizontal	0.961	−0.0079	−0.0095	−0.0339
	Vertical	0.936	−0.0217	−0.0259	0.0180
	Diagonal	0.9363	−0.0108	−0.0094	−0.0001
Lena	Horizontal	0.9396	0.0146	−0.0048	0.0011
	Vertical	0.9639	0.0619	−0.0112	0.0098
	Diagonal	0.9198	−0.0074	−0.0045	−0.0227
Man	Horizontal	0.9796	−0.0266	−0.0155	0.0022
	Vertical	0.9827	−0.0266	−0.0276	−0.0226
	Diagonal	0.9672	0.0039	−0.0157	0.0060
Peppers	Horizontal	0.9769	−0.0055	−0.0056	0.0071
	Vertical	0.9772	−0.0207	−0.0162	−0.0065
	Diagonal	0.9625	−0.0140	−0.0113	−0.0165
Average	Horizontal	–	−0.0008	−0.0135	−0.0097
	Vertical	–	0.0077	−0.0195	−0.0052
	Diagonal	–	−0.0021	−0.0107	−0.0026

plain image are near to one and the results of the cipher image are near to zero, which indicates negligible correlation between the adjacent pixels. Table 3 also shows that the correlation achieved is better than the schemes in [4, 29]. Figure 3 shows the distribution of the pixels in plain and cipher images in the horizontal direction.

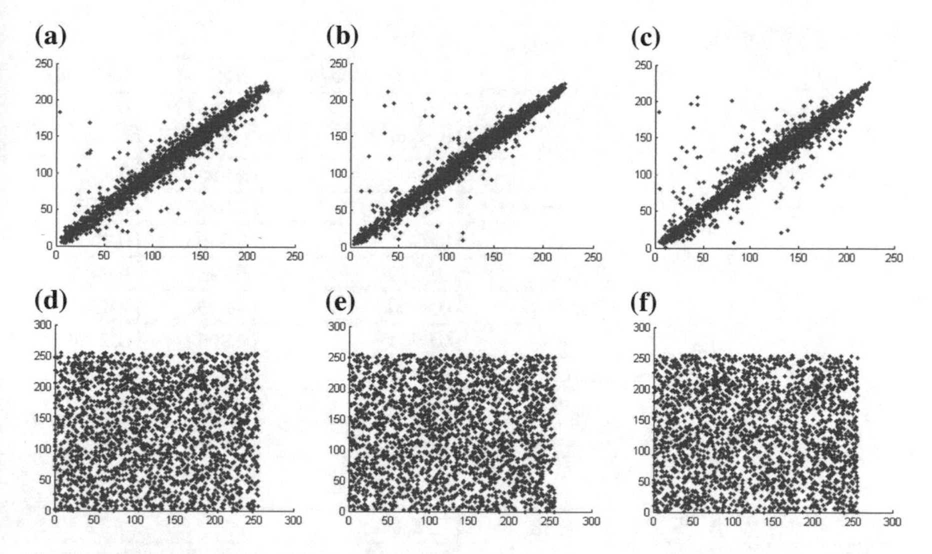

Fig. 3 Scatter plot of the plain image **a** Horizontal direction, **b** Vertical direction, **c** Diagonal direction; Scatter plot of the cipher image in **d** Horizontal direction, **e** Vertical direction, **f** Diagonal direction

4.4 NPCR and UACI Analysis

The one-bit pixel change can lead to considerably distinct cipher image. The sensitivity to this change is measured by the number of pixels change rate (NPCR) and by unified average changing intensity (UACI) computed using (17) and (18).

$$NPCR = \frac{\sum_{i,j} D(i,j)}{N \times N} \times 100\% \tag{17}$$

$$UACI = \frac{1}{N \times N} \frac{\sum_{i,j} |C_1(i,j) - C_2(i,j)|}{L} \times 100\% \tag{18}$$

$C_1(i, j)$ and $C_2(i, j)$ are the values of the pixels in the position (i, j) of the two ciphered-image C_1 and C_2 respectively; L is the number of gray levels. $D(i, j)$ is determined based on the rule

$$D(i,j) = \begin{cases} 0, & C_1(i,j) = C_2(ij) \\ 1, & otherwise \end{cases}$$

The results shown in Table 4 and Table 5 indicates that the algorithm can resist differential attacks as the NPCR and UACI values are close to the theoretical values of 0.9961 and 0.3346 respectively.

Table 4 NPCR of cipher image

Test images	NPCR		
	Proposed scheme	Ref. [4]	Ref. [29]
Airport	0.9961	0.9961	0.9961
Barbara	0.9958	0.9960	0.9964
Boat	0.9962	0.9961	0.9962
Cameraman	0.9964	0.9962	0.9961
Couple	0.9961	0.9963	0.9963
Elaine	0.9959	0.9961	0.9962
House	0.9960	0.9962	0.9961
Lena	0.9961	0.9962	0.9960
Man	0.9962	0.9960	0.9962
Peppers	0.9960	0.9963	0.9962
Average	0.9961	0.9961	0.9961

Table 5 UACI of cipher image

Test images	UACI		
	Proposed scheme	Ref. [4]	Ref. [29]
Airport	0.3392	0.3356	0.3361
Barbara	0.3349	0.3374	0.3367
Boat	0.3334	0.3353	0.3360
Cameraman	0.3366	0.3377	0.3368
Couple	0.3361	0.3380	0.3372
Elaine	0.3331	0.3351	0.3362
House	0.3367	0.3349	0.3372
Lena	0.3336	0.3370	0.3363
Man	0.3390	0.3379	0.3347
Peppers	0.3345	0.3369	0.3373
Average	0.3357	0.3365	0.3364

4.5 Histogram Analysis

An image histogram is the graphical representation of the total distribution of pixel intensities of a digital image. Histogram of the plain images have unique pattern. Hence an attacker can deduce distinctive information about the plain image by statistical analysis of the histogram. To prevent this histogram of encrypted images should have similar distribution and significantly different from the plain image. Figure 4 shows the original image, cipher image and their respective histograms. The histograms of the four test images are distinct. The histogram of the cipher is different from the histogram of the corresponding plain image. Further, the histogram of all the four cipher images is close to each other. Hence an attacker will not be able to deduce any statistical information. The results indicate that the proposed algorithm is robust against statistical attack.

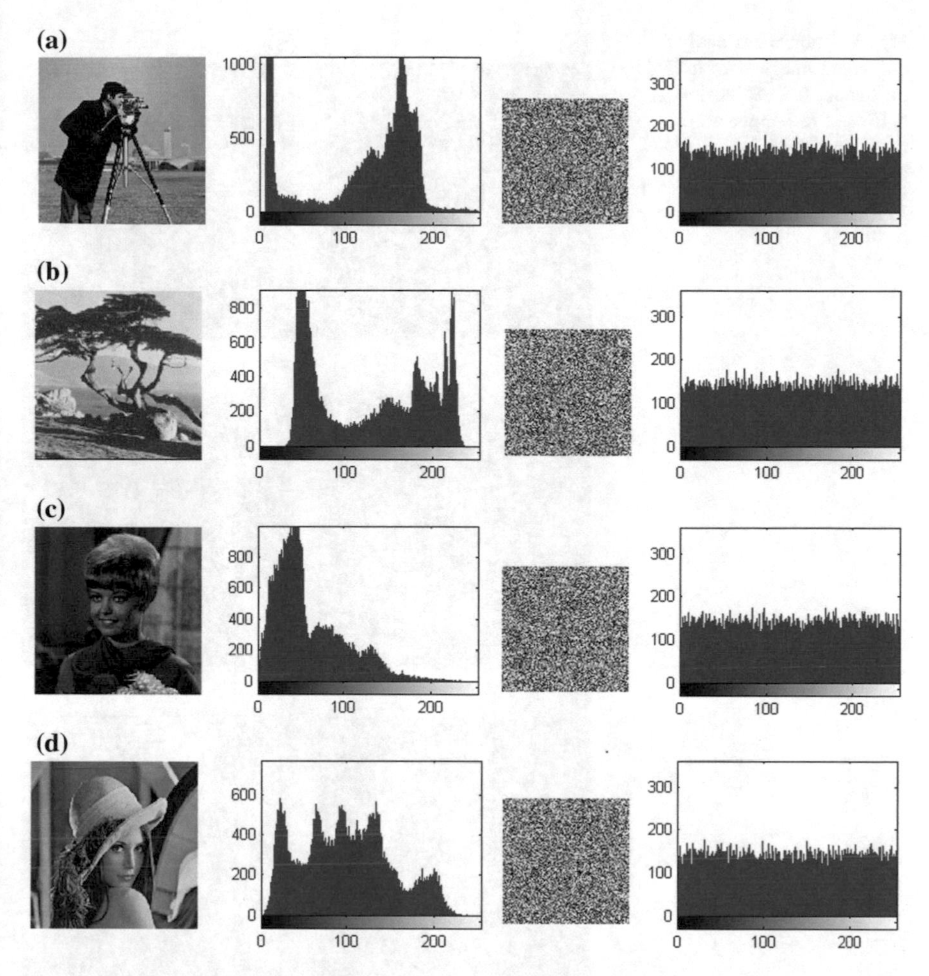

Fig. 4 Histogram of the plain image and corresponding cipher image

4.6 Robustness Analysis

The proposed image compression-encryption algorithms robustness against data loss and noise attack is analysed. Figure 5 shows the robustness analysis of the proposed scheme. It can be seen, even though the cipher images are subjected to different types of noises and data loss, the reconstruction algorithm can still recover the original image. The Cipher images with 25%, 4.8% data loss, 1% salt & pepper noise and 2% speckle noise are shown in Fig. 5a, b, c, d respectively. The reconstructed images with 25%, 4.8% data loss, 1% salt & pepper noise and 2% speckle noise are shown in Fig. 5e, f, g, h respectively.

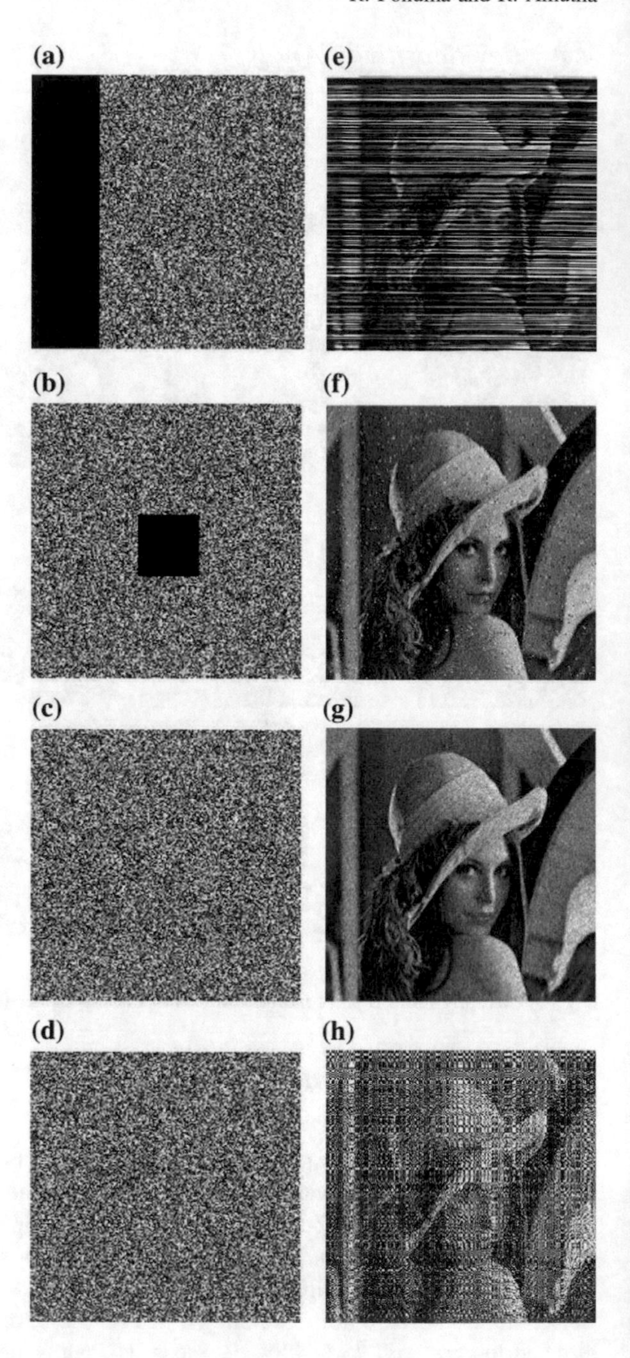

Fig. 5 Robustness analysis: encrypted image with **a** 25% occlusion, **b** 4.8% occlusion, **c** 1% salt & pepper noise, **d** 2% speckle noise; **e**, **f**, **g**, **h** are the corresponding reconstructed images

4.7 Key Sensitivity Analysis

For an efficient encryption scheme the security key should be highly sensitive. A small change in the security key should result in a totally different decrypted image. The cipher is decrypted using the same key as used in encryption and the reconstructed image is shown in Fig. 6a. A small value $\Delta = 10^{-16}$ is added to the correct keys and the cipher image is decrypted. The reconstructed images with incorrect keys are shown in Fig. 6, which shows that the proposed scheme is highly sensitive to even a small change in the key.

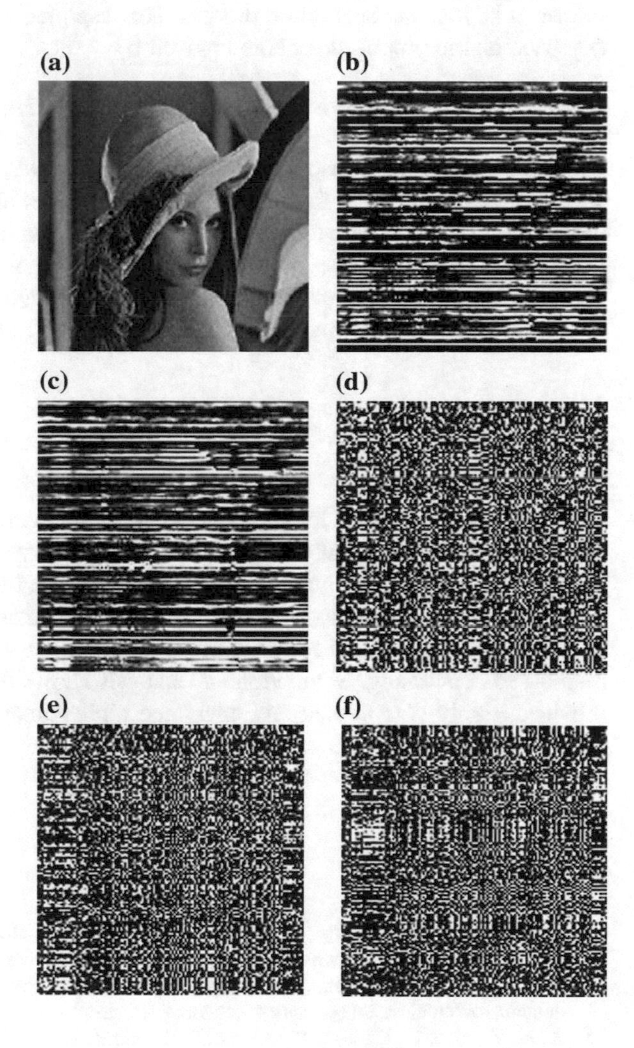

Fig. 6 **a** Reconstructed image with correct keys, Reconstructed image with incorrect key: **b** x_{01}, **c** y_{01}, **d** x_{02}, **e** y_{02}, **f** γ

Table 6 Key space analysis

Proposed scheme	Ref. [4]	Ref. [29]
10^{105}	10^{187}	10^{75}

4.8 Keyspace Analysis

An encryption scheme should have a large keyspace (K_s) to resist the brute-force attack. In the proposed scheme the parameters of the 2D-SLMM (x_{01}, y_{01}, x_{02}, y_{02}, β, γ), the parameter λ used in the generation of the circulant measurement matrix and the number of iterations T performed in the Arnold scrambling process acts as the secret keys in the encryption process. The data precision of the keys is 10^{15}. The keyspace of the proposed scheme is given by

$$K_s = ks_1 \times ks_2 \times ks_3 \times ks_4 \times ks_5 \times ks_6 \times ks_7 \tag{19}$$

where ks_i is the key subspace of the ith key. The keyspace $K_s = 10^{105}$ is large enough to resist the brute-force attack. The comparison of the keyspace achieved by the proposed scheme with that of the encryption techniques in Refs [4, 29] is shown in Table 6. The keyspace of the proposed scheme is better than the keyspace achieved by [29]. The proposed keyspace when compared with the scheme introduced by Belazi et al. is small, but it is large enough to thwart the statistical attacks.

5 Conclusion

In this paper, a compression-encryption algorithm based on compressive sensing and 2D-SLMM chaotic map is proposed. Compressive Sensing is used to simultaneously compress and encrypt the image using two circulant measurement matrices. The construction of the measurement matrix is controlled by the 2D-SLMM. Further the security of the cipher image is enhanced using the Arnold map-based scrambling. Simulation and analysis shows that the proposed algorithm provides highly compressed and encrypted cipher image which is secure against different statistical attacks.

References

1. Phamila, A.V.Y., Amutha, R.: Low complexity energy efficient very low bit-rate image compression scheme for wireless sensor network. Inf. Process. Lett. **113**(18), 672–676 (2013)
2. Phamila, A.V.Y., Amutha, R.: Energy-efficient low bit rate image compression in wavelet domain for wireless image sensor networks. Electron. Lett. **51**(11), 824–826 (2015)
3. Zhou, Y., Bao, L., Chen, C.P.: A new 1D chaotic system for image encryption. Sig. Process. **97**, 172–182 (2014)

4. Belazi, A., El-Latif, A.A.A., Belghith, S.: A novel image encryption scheme based on substitution-permutation network and chaos. Sig. Process. **128**, 155–170 (2016)
5. Hanis, S., Amutha, R.: Double image compression and encryption scheme using logistic mapped convolution and cellular automata. Multimedia Tools Appl. (2017). doi:10.1007/s11042-017-4606-0
6. Deepak, M., Ashwin, V. and Amutha, R.: A new multistage multiple image encryption using a combination of Chaotic Block Cipher and Iterative Fractional Fourier Transform. In: First International Conference on Networks and Soft Computing (ICNSC2014), pp. 360–364 (2014)
7. Zhang, W., Yu, H., Zhao, Y.L., Zhu, Z.L.: Image encryption based on three-dimensional bit matrix permutation. Sig. Process. **118**, 36–50 (2016)
8. Zhu, H., Zhao, C., Zhang, X.: A novel image encryption–compression scheme using hyper-chaos and Chinese remainder theorem. Sig. Process. Image Commun. **28**(6), 670–680 (2013)
9. Guesmi, R., Farah, M.A.B., Kachouri, A., Samet, M.: A novel chaos-based image encryption using DNA sequence operation and Secure Hash Algorithm SHA-2. Nonlinear Dyn. **83**(3), 1123–1136 (2016)
10. Chen, J.X., Zhu, Z.L., Fu, C., Yu, H.: Optical image encryption scheme using 3-D chaotic map based joint image scrambling and random encoding in gyrator domains. Opt. Commun. **341**, 263–270 (2015)
11. Mahesh, M., Srinivasan, D., Kankanala, M., Amutha, R.: Image cryptography using discrete Haar Wavelet transform and Arnold Cat Map. In Communications and Signal Processing (ICCSP), 2015 International Conference, pp. 1849–1855 (2015)
12. Wu, X., Wang, D., Kurths, J., Kan, H.: A novel lossless color image encryption scheme using 2D DWT and 6D hyperchaotic system. Inf. Sci. **349**, 137–153 (2016)
13. Niyat, A.Y., Moattar, M.H., Torshiz, M.N.: Color image encryption based on hybrid hyper-chaotic system and cellular automata. Opt. Lasers Eng. **90**, 225–237 (2017)
14. Chandrasekaran, J. and Thiruvengadam, S.J.: A hybrid chaotic and number theoretic approach for securing DICOM images. Secur. Commun. Netw. (2017)
15. Donoho, D.L.: Compressed sensing. IEEE Trans. Inf. Theory **52**(4), 1289–1306 (2006)
16. Baraniuk, R.G.: Compressive sensing [lecture notes]. IEEE Signal Process. Mag. **24**(4), 118–121 (2007)
17. Rachlin, Y., Baron, D.: The secrecy of compressed sensing measurements. In: Communication, Control, and Computing, 2008 46th Annual Allerton Conference, pp. 813–817 (2008)
18. Zhang, Y., Zhou, J., Chen, F., Zhang, L.Y., Wong, K.W., He, X., Xiao, D.: Embedding cryptographic features in compressive sensing. Neurocomputing **205**, 472–480 (2016)
19. Zhou, N., Zhang, A., Zheng, F., Gong, L.: Novel image compression–encryption hybrid algorithm based on key-controlled measurement matrix in compressive sensing. Opt. Laser Technol. **62**, 152–160 (2014)
20. Ponuma, R., Aarthi, V., Amutha, R.: Cosine Number Transform based hybrid image compression-encryption. In: Wireless Communications, Signal Processing and Networking (WiSPNET), International Conference, pp. 172–176 (2016)
21. Zhang, A., Zhou, N., Gong, L.: Color image encryption algorithm combining compressive sensing with Arnold transform. J. Comput. **8**(11), 2857–2863 (2013)
22. Zhou, N., Li, H., Wang, D., Pan, S., Zhou, Z.: Image compression and encryption scheme based on 2D compressive sensing and fractional Mellin transform. Opt. Commun. **343**, 10–21 (2015)
23. Zhou, N., Pan, S., Cheng, S., Zhou, Z.: Image compression–encryption scheme based on hyper-chaotic system and 2D compressive sensing. Opt. Laser Technol. **82**, 121–133 (2016)
24. Candes, E.J.: The restricted isometry property and its implications for compressed sensing. C. R. Math. **346**(9–10), 589–592 (2008)
25. Bandeira, A.S., Dobriban, E., Mixon, D.G., Sawin, W.F.: Certifying the restricted isometry property is hard. IEEE Trans. Inf. Theory **59**(6), 3448–3450 (2013)

26. Mallat, S.G., Zhang, Z.: Matching pursuits with time-frequency dictionaries. IEEE Trans. Signal Process. **41**(12), 3397–3415 (1993)
27. Liu, E., Temlyakov, V.N.: The orthogonal super greedy algorithm and applications in compressed sensing. IEEE Trans. Inf. Theory **58**(4), 2040–2047 (2012)
28. Mohimani, H., Babaie-Zadeh, M., Jutten, C.: A fast approach for overcomplete sparse decomposition based on smoothed ℓ^0 norm. IEEE Trans. Signal Process. **57**(1), 289–301 (2009)
29. Hua, Z., Zhou, Y., Pun, C.M., Chen, C.P.: 2D sine logistic modulation map for image encryption. Inf. Sci. **297**, 80–94 (2015)

Fingerprint Identification Using Hierarchical Matching and Topological Structures

Meryam Elmouhtadi, Sanaa El fkihi and Driss Aboutajdine

Abstract Fingerprint identification is one of the most popular and efficient biometric techniques used for improving automatic personal identification. In this paper, we will present a new indexing method, based on estimation of singular point considered as an important feature in the fingerprint by using the directional file. On the other hand, a hierarchical Delaunay triangulation is applied on the minutiae around the extracted singular point. Two fingerprints calculated by introducing the barycenter notion to ensure the exact location of the similar triangles is compared. We have performed extensive experiments and comparisons to demonstrate the effectiveness of the proposed approach using a challenging public database (i.e., FVC2000), which contains small area and low-quality fingerprints.

Keywords Fingerprint indexing • Delaunay triangulation • Barycentre
Singular point

1 Introduction

Biometric technologies are already widely disseminated in numerous large-scale nationwide projects [1, 2]. Fingerprint indexing and matching is among the most widely used biometric technologies, it is playing a major role in automated person identification systems deployed to enhance security all over the world. Even

M. Elmouhtadi (✉) · S.E. fkihi · D. Aboutajdine
Faculty of Science, LRIT – CNRST URAC29, University Mohammed V Rabat, Rabat, Morocco
e-mail: meryem.mouhtadi@gmail.com

S.E. fkihi
e-mail: aboutaj@fsr.ac.ma

D. Aboutajdine
e-mail: elfkihi.s@gmail.com

S.E. fkihi
RIITM, ENSIAS, University Mohammed V, Rabat, Morocco

A.E. Hassanien and D.A. Oliva (eds.), *Advances in Soft Computing and Machine Learning in Image Processing*, Studies in Computational Intelligence 730,
https://doi.org/10.1007/978-3-319-63754-9_18

without the advent of environmental noise, two same fingerprint impressions are not guaranteed to be identical due to variability in displacement, rotation, scanned regions, and nonlinear distortion. Displacement, rotation, and disjoint-detected regions are obviously due to the differences in the physical placement of a finger on a scanner.

In the literature, various methods of fingerprint matching exist, such as correlation-based matching, texture descriptor, minutiae points-based matching [3, 4]. Among them, minutiae-based algorithms are the most widely used [5, 6]. Most methods of matching consist of relating two minutiae points. This decreases the performance of systems when one has transformations, then the minutiae in rotation will not aligned and the high-level minutiae can be neglected. Furthermore, the complexity of matching increases more and more considering the number of extracted minutiae, which increases the computing time and reduces the result quality.

In this article, we present a new method based on the extraction of singular point, which represents the central point of the image, using the directional field to locate it, then we will focus on a block of 100 * 100 pixels around the extracted singular point. The selected block is an interval that contains the high-quality minutiae which the probability of losing them still minimum for any type of transformation. In a second step, we will apply the Delaunay triangulation of minutiae points. We apply a process matching first by extracting similar triangles, and second extracting the barycenters of similar triangles extracted in the first, and then reapply the Delaunay triangulation and the extraction of similar triangles. The aim of this method is to ensure the right location of the matched triangles, which means a good matched minutiae points. The strengths of this system are as follows:

- The extraction of the singular point involves the reduction of the minutiae aligned at matching, which implies reducing time and complexity.
- Location of the singular point at the center makes it robust to the loss in case of contour missing. This increases the recognition performance.
- This method is robust to zoom, because it is interested only in the corners of triangles when searching similarity of triangles.
- The change in orientation of the incoming image does not affect the identification results. This approves the performance of using triangulation.
- Considering the barycenters of the similar triangles, ensures that the relevant triangles found are located in the same places in the both compared impression.

The rest of this paper is organized as follows: Sect. 2 analyzes the topological structure of fingerprint and preprocessing steps. Section 3 presents the extraction of the singular point. Section 4 details the proposed hierarchy triangulation Delaunay for matching. Section 5 presents experience and results. Conclusions are drawn in Sect. 6 (Fig. 1).

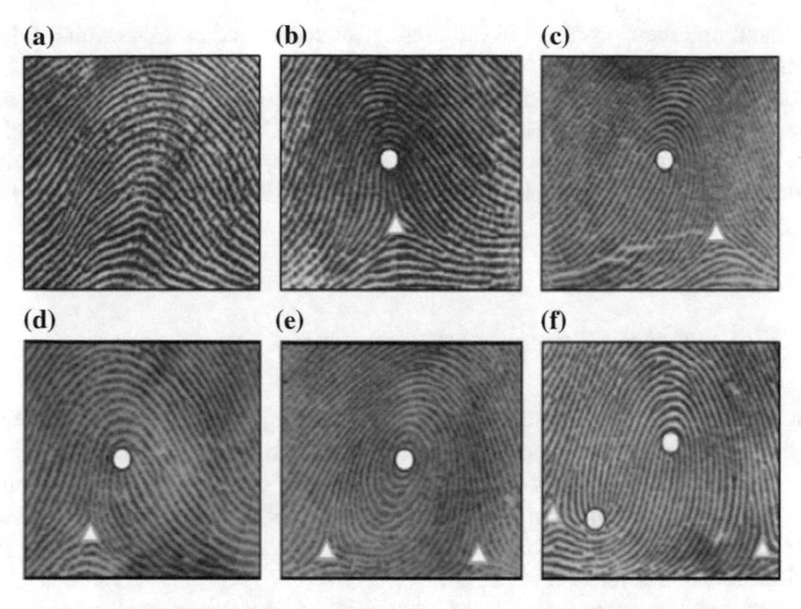

Fig. 1 **a** Arch, **b** tented arch, **c** *left* loop, **d** *right* loop, **e** whorl, and **f** whorl (twin loop)

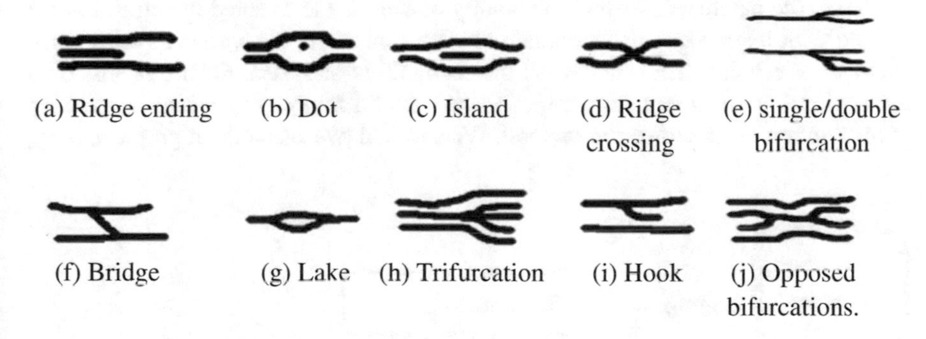

Fig. 2 Different minutiae forms

2 Fingerprinting

2.1 Fingerprint Characteristics

Generally, fingerprint structure is classified into two categories: global characteristics and local ones. The local characteristics called minutiae (see Fig. 2), they represent the form of the ridges intersection, in literature, we find ten forms, but the indexation methods use only two important minutiae's form, defined by end ridge and bifurcation. Because all others shapes are represented as a multiplication of bifurcation or end ridges. These two patterns create the uniqueness of the individual; and their positions in the fingerprint considered as a determining factor of differentiation.

The global characteristics are the singular points (core and delta), obtained at the points when ridges change their orientation which influence largely the directional field. They determine the topological structure and type of fingerprint, allowing a classification of fingerprints into six main classes; defined by Henry [4, 7] (see Fig. 1).

2.2 *Fingerprint Pretreatment*

Typically, the recognition studies are based on the multiple use features found after applying an extraction process. We can classify the fingerprints processing in two major classes (see Fig. 3), preprocessing which is a common step for all approaches. Then the indexing, which includes several process, according to the use of fingerprint extracted characteristics. Finally, we use the indexed database for an identification or verification applications.

Before applying a fingerprint indexing method, the preprocessing step of the fingerprint image is important. The main procedure is to clear the image as to make operations easier and enhance performance.

Due to the nature of skin and the quality of sensor, the scanned fingerprint is not assured with a good quality, it contain always a missing pixels, regions with a bad contrast, or a noise. This is why all the recognition and identification approaches starts with a pretreatment steps which is very useful for keeping a higher accuracy of recognition or identification method. We can find two methods of preprocessing,

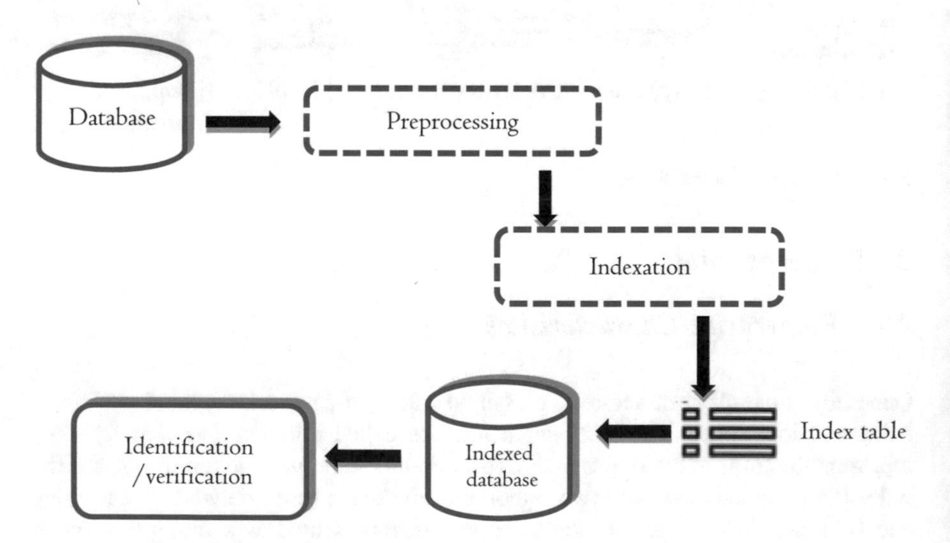

Fig. 3 Fingerprint processing steps

some approaches like the ones [4, 8] are based on the estimation of the ridges orientation noted DF (directional field), followed by binarization and finally skeletonization. Other approaches use only by binarization of the image followed by skeletoization. Both approaches end by filtering the image from noise, and apply the extraction of minutiae.

- **Binarization**: It is important to derive the shape of the ridges and remove the background pixels. Because the true information that could be extracted from a scanned fingerprint must be simply binary; ridges and valleys. The simplest approach is segmentation by adaptive or global thresholding. However, the problem in performing binarization is that not all the fingerprint images have the same contrast characteristics, so a single intensity threshold using a global thresholding cannot applied. Some approaches [9] uses the fact that there is a significant difference in the variation amplitudes of the gray levels along and through the flow ridges. The widely used method is the segmentation by Otsu [10], it consists to maximize the variance interclass, and the more this variance is big, the more the threshold is going to segment the image correctly. This method has showing good results. In figure (Fig. 4), we present the difference between both methods of segmentation:
- **Skeletonization**: In order to facilitate the extraction of the minutiae, this method reduces the thickness of the ridges in a line of one pixel. It does not modify the location and orientation of minutiae compared to the original fingerprint which ensures accurate estimation of minutiae. The approach of Zhang proved important results [11, 12].
- **Minutiae extraction**: Bifurcations and end of ridges are the most significant features in the fingerprint impression. In most approaches, extraction is made by dividing the skeletonized fingerprint into blocks. Alain [5] consider that the bifurcation is identified by a three neighboring pixels, while a termination is pixel with only one neighbor, and bifurcation is a pixel with the tree neighbors. To improve the certainty of minutiae extracted, approaches such as Arcelli [13], have worked by the Cross Number calculated on a blocks of eight pixels as: CN = 0: Isolated pixel, not considered; CN = 1: ridge minutiae; CN = 2: minutiae does not exist. However, the minutiae extraction is not genuine due to image processing and the noise in the fingerprint image. While the error of missing or added minutiae always exist.

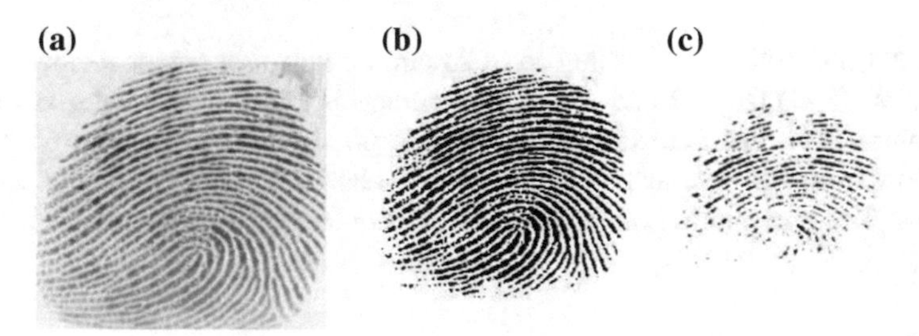

Fig. 4 **a** Original fingerprint. **b** Segmentation with Otsu. **c** Segmentation with global mean

3 Singular Point Extraction

The singular points, cores, and deltas are the most important global characteristics of a fingerprint, they also determine the topological structure. The singular point area defined as a region where the ridge curvature is higher than normal and where the direction of the ridge changes rapidly. A core point defined as the central point of an ogive region, and a delta point is the center of triangular regions where three different direction meet.

In the literature, different approaches were based on the extraction of a core point [14, 15], as a useful point for obtaining the directional registration, classifying fingerprint, matching, and recognition. Most of existing methods are working on directivity ridge estimation to use the directional field images. A number of methods proposed to estimate the fingerprint orientation field [5, 9, 16]. For this, the gradient-based method was adopted by many approaches [15], it has been shown as a very important result for detecting the both singulars points, and robust to very poor image quality and noisy impression. Poincare index is another proposed method and used by many approaches [17, 18] it is considered as very sensitive to noise and the variations of the grayscale level of the input image. However, in this work, we focus our interest for a singular core point location in fact to detect the center of a fingerprint impression as an important region. The performance of core point here is that:

(i) The location of core point at the center makes him a high-level singular point.
(ii) Whatever the finger position on the sensor, we will get the center of the impression in the required image, in contrary to the delta that can be missed in a bad position or in the case of missing parts.

In addition to its applicability by many researchers in the area of fingerprints, estimation of the orientation ridges using a directional filed it used in this work, then we will apply it to our proposed indexation method. Using a gradient approach, we compute the image derivative in the x and y directions to get the gradient vector defined as

$$\begin{pmatrix} G_x I(x,y) \\ G_y I(x,y) \end{pmatrix} = \begin{bmatrix} \frac{\partial I(x,y)}{\partial x} \\ \frac{\partial I(x,y)}{\partial y} \end{bmatrix} \tag{1}$$

Orientation estimated by using least square contour alignment method using a window W (3, 3) (Eqs. 2 and 3). The advantage is that this method gives continuous values [14]. With calculating gradient at pixel level, the orientation of ridge stays orthogonal to average phase angle of changes pixels value indicated by gradients. Then orientation of an image block is determined by averaging the square gradients $\begin{pmatrix} G_{s,x}I(x,y) \\ G_{s,y}I(x,y) \end{pmatrix}$ to eliminate the orientation ambiguity. Figure 5 presents an overview of extracting singular point steps.

$$\begin{pmatrix} G_{s,x}I(x,y) \\ G_{s,y}I(x,y) \end{pmatrix} = \begin{bmatrix} \sum_W G_x^2 - \sum_W G_y^2 \\ 2\sum_W G_x G_y \end{bmatrix} = \begin{bmatrix} G_{xx} - G_{yy} \\ 2G_{xy} \end{bmatrix} \quad (2)$$

Then the gradient orientation $\boldsymbol{\theta}$ estimated as follows:

$$\theta = \frac{1}{2}\nabla\left(G_{xx} - G_{yy}, 2G_{xy}\right) \quad (3)$$

With $\nabla(x, y)$ defined as (Eq. 4):

$$\nabla(x,y) = \begin{cases} \tan^{-1}(x/y)\, x > 0 \\ \tan^{-1}(x/y) + \pi \text{ for } x < 0 \text{ and } y \geq 0 \\ \tan^{-1}(x/y) - \pi x < 0 \text{ and } y \geq 0 \end{cases} \quad (4)$$

As an important region, pixels around the singular point present a high-level point on a fingerprint impression, for this we will focus to extract minutiae around the singular point extracted. Many approach approved that the high number of minutiae extracted can decrease the result matching especially when use minutiae features, because it can generate a bad point and then a false feature. We propose in this work to considering a bloc of (100 * 100) pixels around the singular point extracted, this reduce number of minutiae extracted and help to minimize time and complexity matching (Fig. 6).

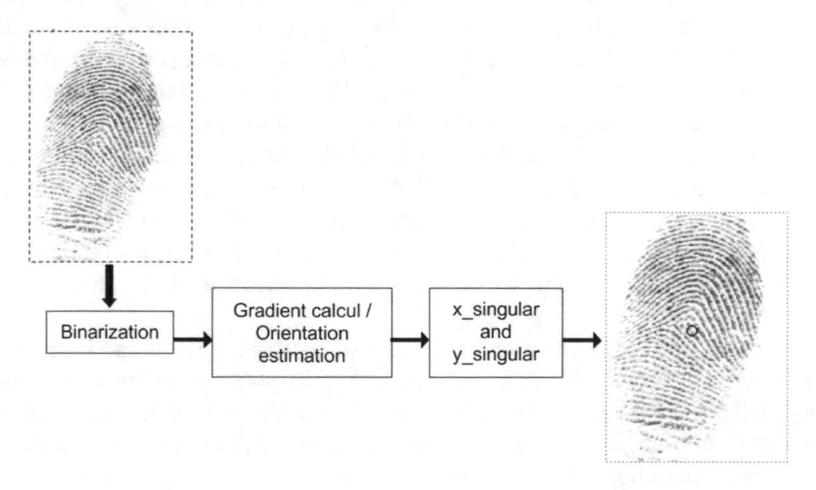

Fig. 5 Overview of singular point extracting

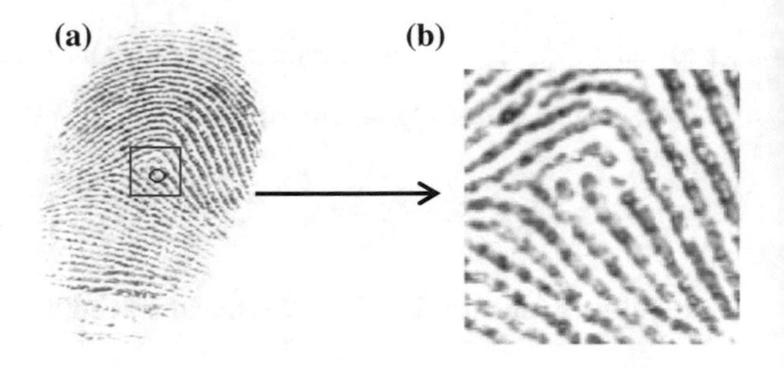

Fig. 6 **a** Original fingerprint with the extracted singular point. **b** The selection bloc of (100 * 100) pixels around singular point

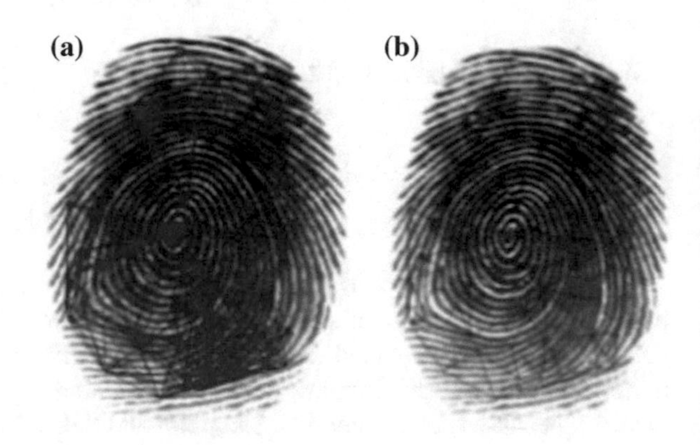

Fig. 7 **a** Delaunay triangulation, **b** triangulation with reducing triplets

4 Hierarchical Delaunay Triangulation

Delaunay triangulation is a matching-based method produces triangles starting from the center point of the fingerprint image, forming an arc with the closest points to cover all the minutiae points. This method provides good results in terms of complexity [7, 19]. Delaunay triangulation [6, 10] consists in generating the largest possible number of triangles formed by the extracted minutiae points.

The main problem is that the number of the generated triangles may be a weakness for these kinds of algorithms. Indeed, as the set of extracted minutiae may contain false ones, so for each false minutia, multiple false triangles will be generated and the result quality will decrease. Thus, some approaches [10, 11] were proposed to reduce the number of the generated triangles, but they were not efficient as it can eliminate useful points (Fig. 7).

Considering two sets of minutiae I (input image) and T (template image). The problem of the based triangles methods that it is possible to find two similar triangles from T and I, but in reality they are not formed by the correspondent minutiae; (i.e., triangle nodes do not have the same position in the two considered impression). Therefore, the challenge is to find similar triangles with the same topological distribution in two similar fingerprints.

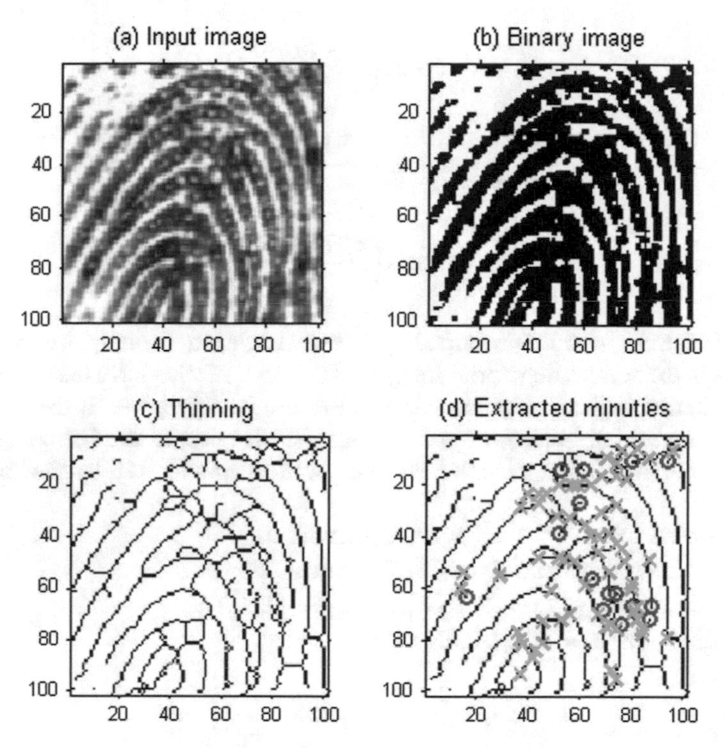

Fig. 8 Preprocessing steps: **a** the input block, **b** binary image, **c** thinned image. **d** Present the extracted minutiae, end ridges presented by circles

In the first of our approach, we apply the preprocessing steps to each fingerprint in order to extract minutiae that will be classified into a vector of ridges and bifurcations. For this, we use the Otsu binarization method for obtaining the binary image. Zhang's skeletonization approach was used for thinning the binary fingerprint. Then, we apply the extracting minutiae process, composed by bifurcation and end of ridges, the Cross Number-based method was applied on a neighborhood of 8 pixels, as mentioned before. Figure 8, show the different steps applied on a extracted singular point region considered as a block of 100 * 100, Next, based on the obtained minutiae vector (ridge and bifurcation), the different possible triangles will be formed applying a Delaunay triangulation (DT) function so as to obtain all possible triplets of the extracted points.

As a comparison and matching method based on triplets minutiae, we define the similarity between an input fingerprint image and the others in a database. In the work, we propose to calculate the three angles of each triangle in the obtained Delaunay triangulation (DT) for the both compared fingerprints (DT), for an input image, and DT2 for database image. Among the existing methods, we use the Akashi theorem (Eqs. 5, 6, and 7) to define the three angles α, β, and γ of each triangle ABC in DT.

$$\alpha = \arccos\left(\frac{b^2 + c^2 - a^2}{2bc}\right) \tag{5}$$

$$\beta = \arccos\left(\frac{a^2 + c^2 - b^2}{2ac}\right) \tag{6}$$

$$\gamma = \arccos\left(\frac{a^2 + b^2 - c^2}{2ab}\right) \tag{7}$$

For each triangle of DT identified by its three angles (α, β, and γ) we look all the triangles in the second fingerprint image; defined in DT2; that have the same angles. This similarity method gives always the best results, cause even if the fingerprint changes orientation, quality, or zoom, since the triangles do not change angles in these cases. The similar triangles will be saved in Similar_DT by the following algorithm:

```
For each triangle Δi(Ai, Bi, Ci) from DT
For each triangle Δj(Aj, Bj, Cj) from DT2
If is member ((Ai, Bi, Ci), (Aj, Bj, Cj)) = (1, 1, 1);
Similar_DT i = = Δi(Ai, Bi, Ci);
End;
End;
End;
Similar_DT = unique (Similar_DT, rows);
```

The problem in this case that we may find similar triangles that are not formed by the same minutiae. Indeed, even if it exists a homothetic transformation between two triangles this does not guarantee that they are formed by the same minutiae points, and not effectively similar because they can be in two different positions in the compared fingerprints. To solve this problem and to take into consideration the location of similar triangles with a similar minutiae, the next step of our method consist to extract the barycenter [20], of each triangle given in Similar_DT (Fig. 9).

Barycenter is calculated by using the mean of A_i, B_i and C_i, nodes of each triangle $\Delta_i(A_i, B_i, C_i)$ given in the extracted similar triangles (Similar_DT). The result saved in P_center (x_center, y_center). To keep the topological structure of minutiae, we apply the Delaunay Triangulation (DT_Similar_DT) of points saved in the vector P_center, and then we measure the similarity between DT_Similar_DT of the entry fingerprint and DT2_Similar_DT2 of the compared fingerprint. This method will ensure improvement the similarity decision of triangles and so the identification of fingerprints. Finally, we define the probability identification (P) of each fingerprint compared to the database images, as follows in Eq. 8:

$$P = \frac{|\text{Similar\_barycenter}|}{|\text{DT\_similar\_DT}|} \tag{8}$$

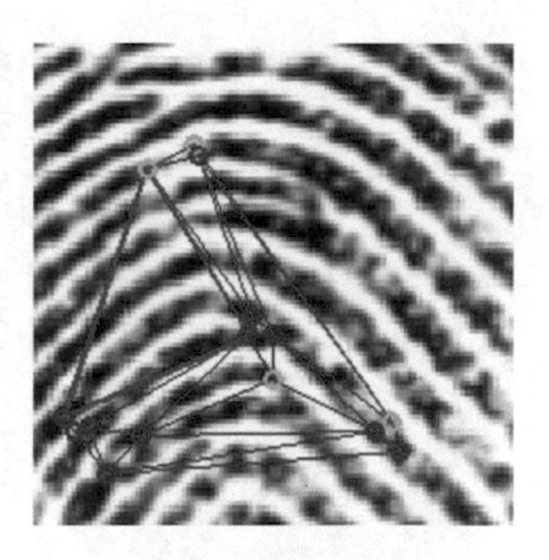

Fig. 9 Delaunay triangulation of minutiae extracted on the chosen block [100 * 100]

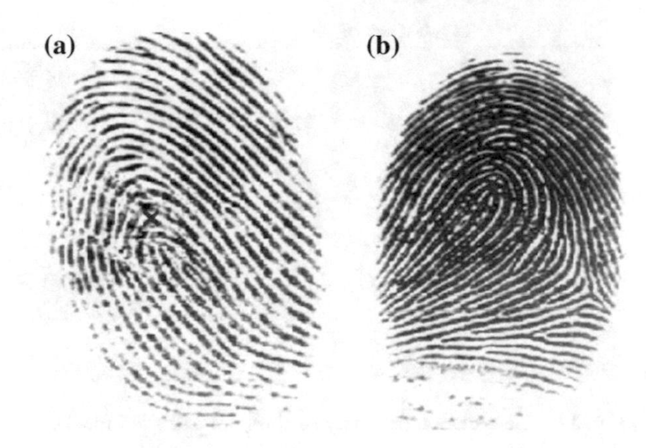

Fig. 10 Singular point detecting in original fingerprints (**a** database image) and (**b** input image)

With:

|DT_similar_DT|: number of similar triangles obtained using Delaunay triangulation of barycenter points.

|Similar_barycenter|: number of similar triangles obtained using Delaunay triangulation of barycenter points.

5 Results and Discussion

To evaluate the performance of the proposed method, we use the benchmarking dataset Db1_a from FVC2002 database available at [21]. FVC2002 DB1 and DB2 contain 880 fingerprint impressions, of various quality, from 110 distinct fingers (i.e., each person is represented by 8 impressions). Three different scanners and the SFinGE synthetic generator were used to collect fingerprints.

In the first, we generate the file of blocs (100 * 100) pixels as shown in Fig. 11, obtained after extraction of the singular point in Fig. 10. Given two different fingerprint **I1** and **I2**, Fig. 12 shows the Delaunay triangle (i.e., DT and DT2) extracted in the both images, the similar triangles founded (i.e., similar_DT and

Fig. 11 Block of (100 * 100) pixels around singular point

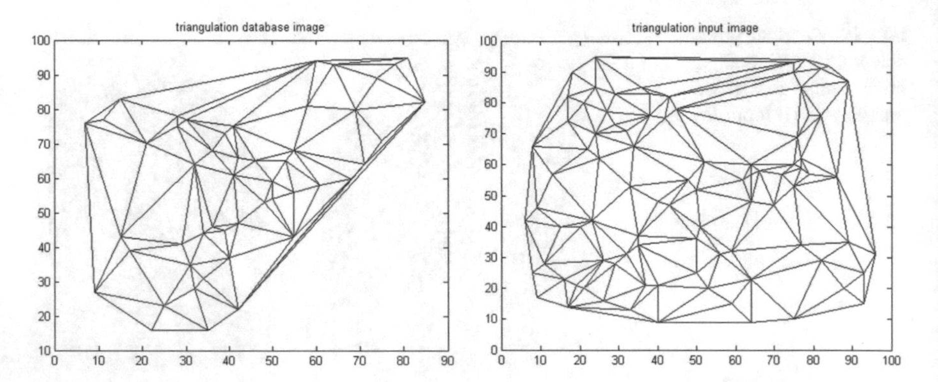

Fig. 12 Triangulation delaunay of the both extracted blocks (i.e., DT and DT2)

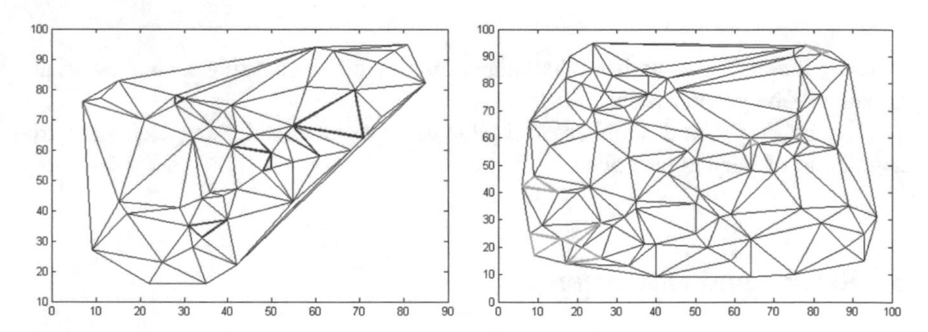

Fig. 13 Similar triangles detected (i.e., similar_DT and similar_DT2)

similar_DT2) as shown in Fig. 13. In Fig. 14, we present the generated barycenter and then the extracted similar triangles formed by barycenter (i.e., DT_similar_DT and DT2_similar_DT2) as mentioned in Figs. 15 and 16. Finally, we calculate the probability P of matching for an input image and others in the database.

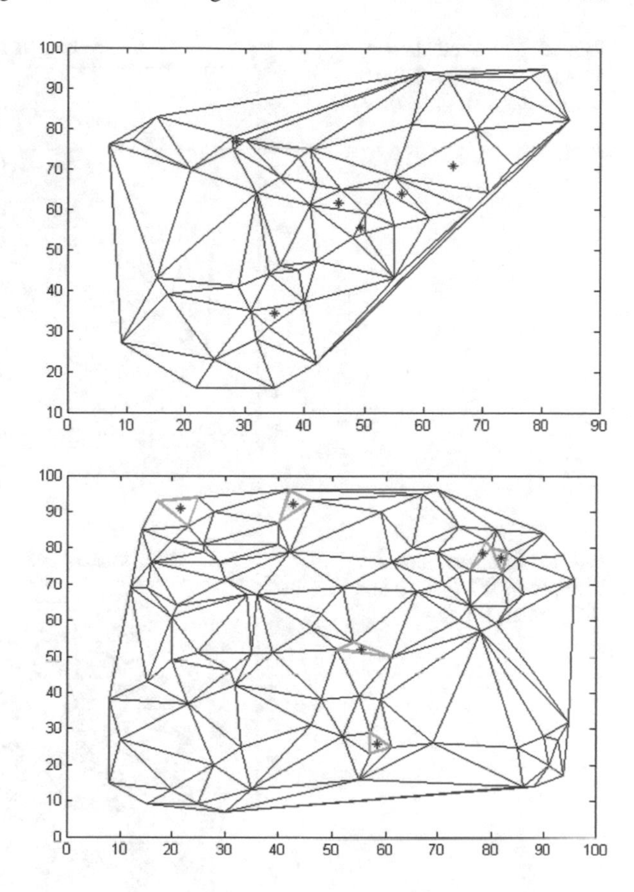

Fig. 14 Generated barycenter in the extracted similar triangles (i.e., similar_DT and similar_DT2)

Comparing a query image (I10) and others selected fingerprint from the database, Table 1 shows that the new approach is more accurate comparing with the other methods; matching using a simple Delaunay triangulation applied on the entry minutiae (SDT), matching with simple Delaunay triangulation around the singular point extracted (SPSDT). We obtain a similarity of 100% only for the query image (I10) by using our proposed method, and for all the others nonsimilar fingerprints we obtain 0% as a similarity rate. In the opposite, in the other methods, there are many false detection rates (rates not equal to 0%) that are determined. They give other relevant images with similarity rates 100%, which decrease the efficiency. The average processing time for each fingerprint is around 0.098 s.

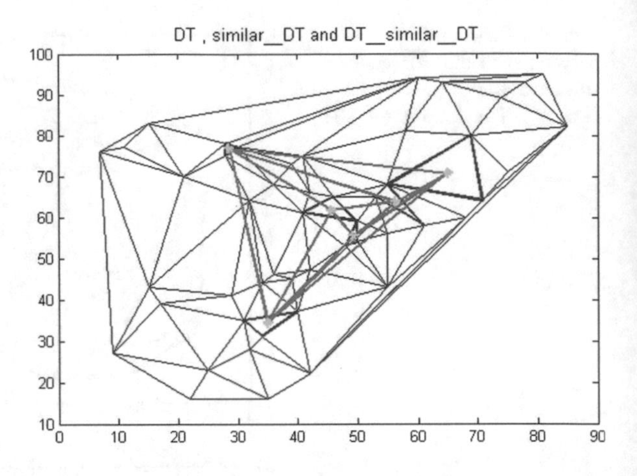

Fig. 15 Extracted similar triangles in DT (*red*) and DT_similar_DT (*green*)

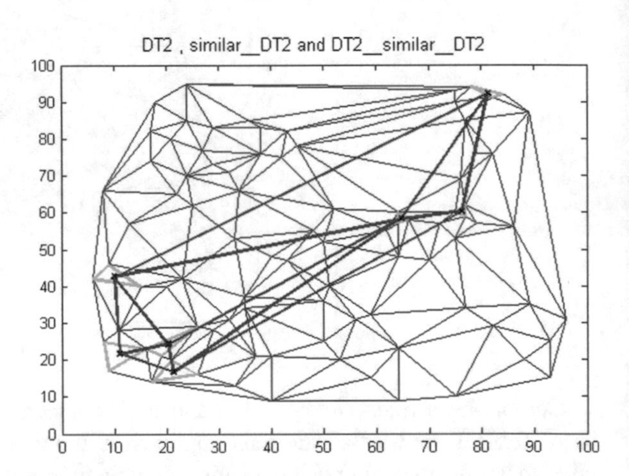

Fig. 16 Extracted similar triangles in DT (*green*) and DT_similar_DT (*red*)

Table 1 Matching results

Compared image	Proposed method (%)	SDT (%)	SPSDT (%)
I1	14	100	30
I2	0	15	0
I3	5	36	9
I4	0	56	0
I5	0	42	3
I6	12	62	0
I7	0	54	12
I8	0	27	0
I9	0	35	0
I10	100	100	100

6 Conclusion

In this paper, we have presented a new fingerprint matching approach based on the extraction of the singular point, and a hierarchical Delaunay triangulation based on triangles features. As an important point, singular point is considered as a high-quality fingerprint pattern with a low probability of loss in case of the missing part on a fingerprint caption. The extraction of singular point helps to approve a high matching performance when considering minutiae on his neighbor. A Delaunay matching method was applied on the minutiae extracted around the singular point. To ensure the best correspondence of similar triangles, we calculate hierarchically similarity of triangles formed by barycenter point of the extracted triangles in the first time. This is to ensure the topological distribution of aligned minutiae. The main experiences approved the efficiency and advantages of proposed method.

References

1. Hassanien, A.E.: Hiding iris data for authentication of digital images using wavelet theory. Int. J. Pattern Recogn. Image Anal. **16**(4), 637–643 (2006)
2. Hassanien, A.E.: A copyright protection using watermarking algorithm. Informatica **17**(2), 187–198 (2006)
3. Kumar, D.A., Begum, T.U.S.: A comparative study on fingerprint matching algorithms for EVM. J. Comput. Sci. Appl. **1**(4), 55–60 (2013)
4. Jain, A., Pankanti, S.: Fingerprint classification and matching. in Handbook for Image and Video Processing. Academic Press (2000)
5. de Boer, J., Bazen, A.M., Gerez, S.H.: Indexing fingerprint databases based on multiple features (2001)
6. Zaeri, N.: Minutiae-based fingerprint extraction and recognition, biometrics. In: Yang, J. (ed.) InTech, (2011). doi:10.5772/17527, http://www.intechopen.com/books/biometrics/minutiae-based-fingerprint-extraction-and-recognition
7. U.S.F.B. of Investigation: The science of fingerprints: classification and uses. United States Department of Justice, Federal Bureau of Investigation (1979)
8. Saleh, A., Wahdan, A., Bahaa, A.: Fingerprint recognition. INTECH Open Access Publisher (2011)
9. Ratha, N.K., Karu, K., Chen, S., Jain, A.K.: A real-time matching system for large fingerprint databases. IEEE Trans. Pattern Anal. Mach. Intell. **18**(8), 799–813 (1996)
10. Sezgin, M., Sankur, B.: Survey over image thresholding techniques and quantitative performance evaluation. J. Electron. Imaging **13**(1), 146–165 (2004)
11. Parker, J.: Algorithms for Image Processing and Computer Vision, ser. IT Pro. Wiley (2010)
12. Tisse, C.-L., Martin, L.,Torres, L., Robert, M.: Systèm automatique de reconnaissance d'empreintes digitales sécurisation de l'authentification sur carte à puce (2001)
13. Muñoz-Briseño, A., Alonso, A.G., Palancar, J.H.: Fingerprint indexing with bad quality areas. Expert Syst. Appl. **40**(5), 1839–1846 (2013)
14. Bazen, A.M., Gerez, S.H.: Systematic methods for the computation of the directional fields and singular points of fingerprints. IEEE Trans. Pattern Anal. Mach. Intell. **24**(7), 905–919 (2002)
15. Zhou, J., Chen, F., Gu, J.: A novel algorithm for detecting singular points from fingerprint images. IEEE Trans. Pattern Anal. Mach. Intell. **31**(7), 1239–1250 (2009)

16. Liu, M., Jiang, X., Kot, A.C.: Efficient fingerprint search based on database clustering". Pattern Recogn. **40**(6), 1793–1803 (2007)
17. Liu, M.: Fingerprint classification based on Adaboost learning from singularity features. Pattern Recogn. **43**(3), 1062–1070 (2010)
18. Awad, A.I., Baba, K.: Singular point detection for efficient fingerprint classification. Int. J. New Comput. Architect. Appl. (IJNCAA) **2**(1), 1–7 (2012)
19. Munoz-Briseno., Alonso, A.G., Palancar, J.H.: Fingerprint indexing with bad quality areas. Expert Syst. Appl. **40**(5), 1839–1846 (2013)
20. Elmouhtadi, M., Aboutajdine, D., El Fkihi, S.: Fingerprint indexing based barycenter triangulation. In: 2015 Third World Conference on Complex Systems (WCCS), Marrakech, pp. 1–6 (2015)
21. Maio, D., Maltoni, D., Cappelli, R., Wayman, J.L., Jain, A.K.: FVC2000: fingerprint verification competition. IEEE Trans. Pattern Anal. Mach. Intell. **24**(3), 402–412 (2002)

A Study of Action Recognition Problems: Dataset and Architectures Perspectives

Bassel S. Chawky, A.S. Elons, A. Ali and Howida A. Shedeed

Abstract Action recognition field has recently grown dramatically due to its importance in many applications like smart surveillance, human–computer interaction, assisting aged citizens or web-video search and retrieval. Many research trials have tackled action recognition as an open problem. Different datasets are built to evaluate architectures variations. In this survey, different action recognition datasets are explored to highlight their ability to evaluate different models. In addition, for each dataset, a usage is proposed based on the content and format of data it includes, the number of classes and challenges it covers. On other hand, another exploration for different architectures is drawn showing the contribution of each of them to handle different action recognition problem challenges and the scientific explanation behind their results. An overall of 21 datasets is covered with 13 architectures that are shallow and deep models.

Keywords Action/activity recognition · Machine learning · Computer vision · Action recognition · Architectures · Shallow models
Deep learning models

B.S. Chawky (✉) · A.S. Elons · A. Ali · H.A. Shedeed
Faculty of Computer and Information Sciences, Scientific Computing Department, Ain Shams University, Cairo, Egypt
e-mail: bassel.safwat@cis.asu.edu.eg

A.S. Elons
e-mail: ahmed.new80@hotmail.com

A. Ali
e-mail: ahmed4a@hotmail.com

H.A. Shedeed
e-mail: dr_howida@cis.asu.edu.eg

A.E. Hassanien and D.A. Oliva (eds.), *Advances in Soft Computing and Machine Learning in Image Processing*, Studies in Computational Intelligence 730,
https://doi.org/10.1007/978-3-319-63754-9_19

1 Introduction

Video action recognition (AR) aims to recognize the set of actions occurring in the input data. For humans, this would be an easy task to do, but for computers, the problem becomes extremely complicated. It consists of extracting the required information from the input data and transforming it into a meaningful representation for computers. There are many challenges in this problem.

- First, the same action can occur in a variety of contexts and backgrounds.
- Second, there are many actions to consider, while each action has intra-class variations caused by viewpoint change, transformation (mainly scale change) occurring on the target object or the change in the speed for the same action.
- Third, the resolution problem, low image resolution might lack some information details, and very high resolution would require a large amount of resources and processing. Similarly, the small frame per second (fps) rate may lack temporal details for the action, while larger fps rates require more processing or at least different handling mechanism.
- Forth, many applications for the action recognition involved object occlusions and illumination changes.
- Fifth, the lack of completely labeled data for training, and thus the need to find new unsupervised algorithms.

The attention received for action/activity recognition either in video or image is due to the large number of vital applications: (1) Video/ Image retrieval by action or behavior: retrieving videos or images based on the actions and activities occurring inside the video is beneficial for the user's search query [1] (2) Videos/ images annotation: a large number of unlabeled videos and images exists online and the speed of acquiring new videos/images is increasing [2]. (3) Automated surveillance: two good examples here are detecting robberies [3] and/or elderly behavior monitoring [4]. (4) Human–computer interaction: computers can assist human by watching them and understanding if they are doing the tasks required correctly and take further decisions based on their actions [5]. (5) Video summarization: understanding the 'verbs' occurring in the video is very helpful for summarizing the videos [6]. This shall be complementary to content understanding. (6) Activity or task recognition: atomic actions units like walking, running, jumping and others are necessary to understand more abstract concepts such as activity or task.

Automatically inferring action in a robust manner remains a challenge despite the great efforts done in this area. It has been tackled in images and videos but the latter got the largest share in this research due to its motion cues. It has been shown the useful applications of AR in still images in the survey [7]. However, it is more challenging and new methodologies are needed.

In this study, different datasets are evaluated based on the challenges of AR they cover. Based on these challenges and the classes they cover, candidate areas of application and usages for each dataset are provided. A total of 21 datasets are discussed with 13 datasets used to discuss action recognition challenges. Moreover, different methodologies, ranging from hand-crafted models to deep architectures, are presented. Then, a discussion is followed showing the models architectures and how each of them succeeded in handling the different AR problems. In effect, a total of 13 methods are covered with 6 shallow models and 7 deep models. Finally, common ideas are summarized in a proposed novel architecture based on the success shown by the deep models.

The chapter is organized as follows. Section 1 provides an introduction and highlights the problems and importance of action recognition. The next section provides a detailed description on the datasets covered in this chapter together with their properties and discussing the pros and cons of each with a suggested usage domain. Section 3 provides different model architectures together with the training and classification algorithms used. In addition, each architecture is commented with the AR challenges it participated in solving. Section 4 presents a review of the results and performances for both hand-crafted features-based models and deep models used in action recognition. Section 5 presents a proposed novel architecture based on the success shown by the deep models. Finally, we draw conclusions in Sect. 6.

2 Action Recognition Datasets

The fact that conducting an accurate expressive evaluation of new algorithms requires owing a dataset of the same complications of real life actions. In this section, we summarize 13 action/activity datasets and discuss their possible usage in action recognition based on their characteristics.

UCF Sports action dataset [8, 9] This dataset is collected from wide range of stock footage websites including BBC Motion gallery and GettyImages. The actions are sport-related and are typically featured on channels like BBC and ESPN. There are 10 human actions: swinging (on the pommel horse and on the floor), diving, kicking, weight-lifting, horse riding, running, skateboarding, swinging (at the high bar), golf-swinging and walking.

The videos are both gray-scaled and colorful with moving and static backgrounds. Different actions classes have similar duration preventing using action length as a factor to classify the action. One or more human can appear in the video, adding a challenge to identify the main action occurring, which is one of the problems in the action recognition field, however, this exists in the low portion of the videos. In addition, the clips show large intra-class variability which encourages researchers to work on unconstrained environments. In this dataset, clips have both static camera and moving the camera (including the camera zooms). This allows training on realistic videos and/or learning robust to transformation models becomes feasible. After all, due to the low number of frames (958 s of duration) models requiring large training set (deep models) are expected to have issues during the learning time, especially for action in complex videos. The dataset might not be a good choice to test motion features [10] because of the low fps, and the fact that each action category has nearly a different domain. In addition, there is a clear issue in the actions with fast motion due to the low frame rate. One more drawback is that the low number of classes in this dataset might not represent the real performance of the model in sport actions as there are much more classes and categories.

Supervised learning can be used in this dataset as each video clip is labeled with a specific action. Annotation of the human body is provided as a bounding box, allowing the evaluation of a human body detection model if used.

Usage of UCF shall be for testing the model against the few sports actions and using a pretrained model would be in the favor of increasing the overall performance of the network.

KTH dataset [11] This dataset has a recorded set of videos of 25 humans performing, for several times, six actions: walking, jogging, running, boxing, hand waving, and hand clapping under 4 scenarios (outdoors, outdoors with scale variation, outdoors with different clothes and indoors).

All the videos are gray-scaled with the static and homogenous background. No objects exist than the main actor, so no background motions noise exists. This makes the dataset an easier example to separate a foreground object from the background. In addition, each sequence contains minimum spatial and temporal (motion) noise compared to more complex datasets with objects interacting or moving in the background. The videos were recorded using a static camera, so camera motion issue is not considered if trained on this dataset. Due to the low resolution of the clips, less computations are required but also less information is provided in the input.

Each video is divided into 4 sequences for the same action and one human doing the action in each sequence allowing supervised training for this dataset as each sequence has a known start/end time. The proposed training/testing is to split the videos into 8 persons for the training set, 8 for the validation set use to optimized the and 9 for the test set.

Usage of KTH shall be for prototype testing mainly and can be considered as a basic dataset for evaluation basic actions in a controlled complex-less environment.

YouTube dataset (UCF11) [12] This dataset is collected from YouTube videos. There are 11 actions in this dataset: basketball shooting, volleyball spiking, trampoline jumping, soccer juggling, horseback riding, cycling, diving, swinging, golf-swinging, tennis-swinging, and walking with a dog. Due to some missing videos and bad annotation in this dataset, it has been updated and is known now as UCF11. It worth noting that this dataset might need further improvements as some videos contain only very few frames like 1 or 2 frames, while a specific file was found to be broken (no frames), namely "v_shooting_24_01.mpg".

The dataset contains colorful videos with backgrounds from real life with both steady and moving backgrounds. This provides background motions in the dataset, which can be viewed as noise regarding the main action being done. The dataset covers different action recognition problems in between: illumination changes, camera movement, background movement, scale change due to the object/camera movement in the z dimension and different viewpoints for the same action introducing high intra-class variance. Due to camera movement, main actors disappear from the scene which introduces a serious problem for hand-crafted models like those relying on tracking the main objects to detect the action occurring.

Each class of action is divided into 25 groups, with a minimum of 4 videos per group. Videos in the same group share common features like the same environment, viewpoint, actors, objects, and light conditions. The suggested training/testing technique is to Leave One Group Out, and taking the average accuracy over all the classes. Each clip is labeled with an action class together with its group and number in the group.

Usage of this dataset would be for training the model on small but challenging sports classes. The complexity of the dataset would provide evaluation against a small number of classes, which can be considered as an indicator whether to move to a larger dataset or not. This dataset is suitable for comparing deep and shallow models.

UCF50 dataset [13] This dataset is collected from YouTube and is an extension to the YouTube Action dataset. There are 50 actions in this dataset.

Backgrounds are indoor and outdoor scenes with colorful videos. Some videos have static backgrounds and others dynamic backgrounds thus, adding a complex environment to deal with. The dataset videos are realistic for different reasons, in between, the variation in camera motion, objects pose and scaling, different view point for the same action, different illuminations, and noisy backgrounds. Therefore, there is high variation between the same action class.

The dataset is grouped into 25 groups with each having more than 4 action clips. Each group usually shares the same features as viewpoint, actors, background. Each video clip is labeled with the corresponding action occurring. The suggested evaluation method is to leave one group out cross validation leading to 25 cross-validations.

Usage of the dataset shall be for testing the model against a wider number of actions. The complexity of the dataset provides a good evaluation metric of the proposed system.

UCF101 dataset [14] This dataset is an extension to UCF50 dataset and contains videos collected from YouTube with 101 action classes. This is one of the largest dataset with respect to number of action category with complex and realistic videos. It has different realistic backgrounds with colored videos. In addition to the higher number of classes, it covers the same AR problems as UCF50 dataset.

The videos are grouped into 25 group with 4 to 7 actions per group sharing similar features. Each video in a group is obtained from longer video, thus actors, point of view, camera motion, illumination conditions, and environment are the same. A proposed evaluation system is to split the train/test data into 3 parts where each sharing videos from the same group in training and testing are not allowed as it this results in fake high performance.

Usage of UCF101 dataset shall be for testing the model against a large number (101) of realistic human actions. This is a very challenging dataset due to the action recognition problems it covers, in addition to the number of classes.

Hollywood2 dataset [15] This dataset is collected from 69 movies and has 12 classes of human actions (answering phone, driving car, Eat, fighting person, get out of the car, hand shaking, hug person, kissing, running, sitting down, sitting up and standing up). Some of the videos are gray-scaled (taken from old movies), while others are colorful. The background is different along the video, thus allowing learning actions from unconstrained environment. Moreover, the dataset clips have high intra-class variability as there are different ways for doing the same action.

There is a total of 3669 clips that are divided into 2517 clips for training/testing action recognition and the rest for scene recognition. Some of the clips contain more than one action, like hugging and kissing. Furthermore, clips contain both automatic and manual action tagging such that for each action, for all the clips, a tagging of 1 or -1 indicating the presence or absence of the action in this clip. For the training set, there are 823 manually checked label videos and 810 automatic generated labels which are considered as labels including noise. The test set contains 884 clips are manually labeled.

One of the important things in this dataset is that there are different viewpoints, scales, and transformations for the same action. In addition, obstacles occur in some of the clip frames, together with different people in the scenes, thus making the dataset a good candidate for daily human actions recognition, i.e., surveillance.

Usage of Hollywood2 action recognition dataset shall be for testing the model performance against real life basic actions in a given video. The number of classes in this dataset is moderate for covering the basic human actions. This dataset can be used also to train a model that provides a short description of a given movie based on action and scenes of the video. It also can be used to train the detection of movie type as containing violence, or detecting action or romantic movies.

HMDB51 dataset [10] This dataset is collected from different sources mostly from movies and partially from public dataset such as Prelinger archive, YouTube, and Google videos. There are 51 classes of human actions, categorized by the authors in five types: general facial actions; facial actions with object manipulation; general body movements; body movements with object interaction; body movements for human interaction. The videos are colorful with challenging backgrounds.

Training and testing sets are constructed such that clips from the same video are on the same sets to break the correlation between the training and testing sets. 3 training/test sets are constructed and the performance is the average of the 3 splits. Each action clip is labeled with the action, and in addition, a metadata describing the properties of the video clip such as view point, video quality (good for videos where finger can be distinguished, medium or bad for the rest of the videos), the number of people involved in the action (ranging from 1 to 3), visible body parts (head, upper or lower body, full body) and camera motion (either in motion or static).

Camera motion due to traveling or shaking hands is present. In addition, a different point of views for the same action and the change of illumination represent high intra-class variability and increases the difficulty of the action recognition for this dataset. This low resolution makes the dataset more challenging.

Usage of the HMDB51 shall be for modeling human actions in a complex environment. The larger number of categories compared to the previous datasets allows the usage of the dataset to train models for a wider range of applications in between, security and surveillance.

J-HMDB dataset [16] This dataset is a subset of 21 actions selected from HMDB51 dataset such that there is one main person in action for each sample. A puppet model is used and manually adjusted on the actor to get the ground truth of the joint position of the actor in the video.

The background and colors are the same as HMDB51. The same illumination, camera motion, obstacles, scaling and different viewpoints exist as in the HMDB51 except that one main actor appears in the video, so it is little less complex.

There are 928 videos in this dataset with detailed annotations: puppet contour, flow of the puppet and the joints position. The training and testing sets are constructed as same as HMDB51 dataset. However, this would be suitable for human detection and segmentation purposes.

Usage of J-HMDB shall be for testing shallow models, especially the models detecting higher features like human pose while detecting the actions.

UIUC SPORT dataset [10] This dataset is a set of images of 8 different sports categories: rowing, polo, snowboarding, sailing, badminton, bocce ball, croquet, and rock climbing.

All images are colorful with different backgrounds in the same category for most images. It worth mentioning that 3 out of 8 categories have similar backgrounds, specifically: bocce, croquet, and polo most of them are having grass in the backgrounds. Each image is labeled with its class together with a level of how challenging is the picture per a human judgment and a level of how close is the foreground to the camera. The foreground (i.e., objects) in the images are of different scales, with some objects being too small to be detected. Due to different camera position, there are different viewpoints for the same action category introducing intra-class variations. Illumination problem also is considered in this dataset as different images have different illumination conditions.

The suggested testing methodology is to use random 70 images for training and 60 images for testing

Usage of UIUC Sport dataset shall be for models dealing with still images, i.e., not considering temporal features. This dataset contains very low number of classes, but due to the challenges this dataset covers, it is suggested to be used to test the model performance regarding these issues. It is suggested also be used to compare the spatial features learned by the models.

MPII Fine-grained Kitchen Activity Dataset [17] This dataset is a recording of 12 subject performing 65 different cooking activities such as cut slices, take from drawer, or pour spice etc.

The videos are colored with indoor static background: kitchen. Data are labeled in two-folds: the first one, each video is manually labeled with an activity category together with the start and end frames for this category. The recording camera is attached to the ceiling; thus, no camera movement present and no scale change are present. Moreover, illumination is the same for all the recording. The dataset has many classes with low inter-class but large intra-class variability.

There is a total of 5609 annotations for the 65 categories. The other fold is annotation of articulated human pose, 5 subjects from different recording (1071 frames) for training and 7 subjects (1277 frames sampled) for testing. Suggested cross validation is to leave one person out and evaluation metrics are multiclass precision (Pr), recall (Rc) and single class average precision (AP) which takes the mean of all the test runs.

Usage of MPII dataset shall be for training models on kitchen activities. It is suggested to be used for training a model for assisting people in cooking recipe or summarizing food-and-cooking videos. Moreover, it is also a good dataset to evaluate what deep models can learn and compare it with different models relying on hand-crafted features.

Kitchen Scene Context based Gesture Recognition dataset (KSCGR) [18] This dataset is a collection of 9 classes performed by 7 actors cooking 5 recipes. The 9 classes are as follow: breaking, mixing, baking, turning, cutting, boiling, seasoning, peeling, none.

The dataset contains colorful image sequence together with depth information for each image (frame) taken by Kinect sensor. The background is an indoor static kitchen. The camera is fixed so there is no camera motion exists. It worth noting that all the dataset videos have the same viewpoint, illumination conditions and the same cooking items are used. There is no scale change of objects.

There are 5 actors for the training and 2 actors for the testing dataset. Each frame is labeled with the action being done.

Usage of KSCGR shall be for testing models on small categories of kitchen activities. It is suggested to be used for testing the ability of a model to learn the cooking action. Modeling human pose and modeling the cooking motion witnessed a lot of efforts using shallow models and would be very competitive with deep learning models and this dataset would be used for this purpose too.

ChaLearn Gesture Recognition 2013 dataset [19] This dataset contains 13, 858 gestures from a lexicon of 20 Italian gesture categories recorded with a Kinect camera, providing the audio, skeletal model, user mask, RGB and depth images. The participants are 81% Italian native speakers while the rests are not Italians but Italian-speakers.

The input information can be used to model human pose which is for the favor of hand-crafted based model. The background is fixed between all the speakers; thus, no background motion noise exists and human segmentation becomes an easy task due to the simplicity of the background. The dataset only focuses on hand gestures as the camera, illumination and viewpoint are fixed. No occlusion exists in the dataset.

There are 956 sequences, divided as follow: 393 (7.754 gestures), validation: 287 (3.362 gestures), and test: 276 (2.742 gestures). Each sequence is labeled with the main gestures categories with the same order of appearance in the video.

Usage of ChaLearn Gesture Recognition shall be training a model to learn the Italian gestures. Moreover, it can be used to evaluate different hand-crafted based models and compare them with the ability of deep models to learn higher features like human pose and gesture occurring from both spatial and temporal dimensions respectively.

Sports-1 M dataset [20] This dataset contains links to 1,133,158 YouTube videos annotated with 487 sports labels. To deal with the dataset, it is recommended to either sample the video and consider a subset of the frames or to resize the frames to 227 × 227 pixels in the spatial resolution. Some videos might be deleted or needing permission to reach them which presents an issue in this dataset.

The videos have different static and backgrounds with moving objects. It worth noting that text might appear in some videos which are considered as a noise for the spatial features, ideas to deal with this problem are to neglect such regions or to consider only the center of the frame. Videos are colorful but there are grayscale videos as well. Sports-1 M dataset consider a wide range of action recognition problem, like different viewpoints, illumination, objects scaling, background noisy movements, occlusions, obstacles, low/high resolutions, and camera motions. Some scenes of the videos are not related to the action being done but to the environment of recoding which represents an additional challenge for the sport recognition. GPU/parallelism is necessary to speed-up the training process.

For each class, there is between 1000–3000 video with around 5% of these videos annotated with more than one class. The dataset is split 70% for the training set, 10% for the validation set and 20% for the test set. It has been reported in [20] that 1755 out of 1 million videos contain a significant fraction of near duplicate frames between the training and testing sets.

Usage of Sports-1 M would be for training model on wide range of sports in realistic environment. It is also recommended to use this dataset to train a rough model against transformation, illumination changes, but also, to learn higher concepts instead of relying on low level features for the classification due to the high intra-class variations.

In this section, we reviewed different action recognition datasets, the dataset classes are listed against both the spatial and temporal resolutions properties. An investigation of the existence of the action recognition problems has been made and a suggested usage of each dataset is summarized in Table 1. Summary of the action recognition problems and number of classes in the reviewed dataset is present in Table 2. Resolution, frame rates and other properties of each dataset content are listed in Table 3. This investigation is very helpful for choosing the datasets when testing a new architecture or a modified one. It is also important for showing the participation of each dataset regarding the challenges of AR, which shall be considered when evaluating a new model. This section can be considered also as a measurement, or an evaluation, of how realistic are these datasets.

Table 1 Recommended usage of reported action recognition dataset

Dataset	Proposed usage
KTH dataset	Evaluating model/prototype for basic human actions (6 actions)
UCF sports action dataset	Evaluating model against small number (10) of realistic sport actions
YouTube (UCF11) dataset.	Evaluating model against small number (11) of realistic sport actions
UCF50 dataset	Evaluating model against moderate number (50) of realistic sport actions
UCF101 dataset	Evaluating model against large number (101) of different realistic human actions
Hollywood2 dataset	Evaluating model against small number (12) of realistic basic human actions
	Description of actions in movies based on action and scenes
	Categorizing movies into types
HMDB51 dataset	Evaluating model against moderate number (51) realistic human actions
J-HMDB dataset	Evaluating model against small number (21) realistic human actions
	Evaluating models detecting human pose
UIUC SPORT dataset	Evaluating model against small number (8) of realistic sport actions
	Evaluate spatial features learned by the model
Fine-grained kitchen activity dataset	Evaluating model against large number (65) of indoor (kitchen) activities
	Train model to assist people in cooking recipes
	Train model to summarize cooking show recipe
	Comparison of deep/shallow models in constrained environment
KSCGR dataset	Evaluating model against small number (9) of indoor (kitchen) activities
	Comparison of deep/shallow models in constrained environment
ChaLearn gesture recognition (2013) dataset	Evaluating model against moderate number (20) of constrained Italian gestures
	Compare deep/shall models in learning higher features like human pose and gesture especially in temporal dimension
Sports-1 M dataset	Evaluating model against very large number (487) of sports actions/activities
	Training a rough model against transformation and illumination changes
	Evaluating model ability to learn higher concepts regarding the sport actions/activities

Table 2 Action recognition problems existence in each dataset in addition to the number of classes

Dataset	Background changes	Background object motion	Scale changes	Camera motion	Different viewpoints for the same action	Speed changes for the same action	Pixel Resolution	FPS	Illumination changes	Occlusions	# actors in most of videos {1, S (Small), M (Many)}	# of classes
KTH dataset	✗	✗	✓	✗	✓	✗	Low	Avg.	✓	✗	1	6
UCF Sports action dataset	✓	✓	✓	✓	✓	✗	High	Low	✓	✗	S	10
YouTube dataset (UCF11)	✓	✓	✓	✓	✓	✓	Low	Avg.	✓	✗	1	11
UCF50	✓	✓	✓	✓	✓	✓	Low	Avg.	✓	✓	1	50
UCF101	✓	✓	✓	✓	✓	✓	Low	Avg.	✓	✓	1	101
Hollywood2 dataset	✓	✓	✓	✗	✓	✓	Low	Avg.	✓	✓	S	12
HMDB51 dataset	✓	✓	✓	✓	✓	✓	Low	Avg.	✓	✓	S	51
J-HMDB dataset	✓	✓	✓	✓	✓	✓	Low	Avg.	✓	✓	1	21
UIUC SPORT dataset (still image)	✓	✗	✓	✗	✓	NA	High	NA	✓	✓	S	8
Fine-grained kitchen activity	✗	✗	✗	✗	✗	✓	High	Avg.	✗	✓	1	12

(continued)

Table 2 (continued)

Dataset	Background changes	Background object motion	Scale changes	Camera motion	Different viewpoints for the same action	Speed changes for the same action	Pixel Resolution	FPS	Illumination changes	Occlusions	# actors in most of videos {1, S (Small), M (Many)}	# of classes
KSCGR	✗	✗	✗	✗	✗	✓	High	Avg.	✗	✓	1	9
ChaLearn gesture recognition (2013)	✗	✗	✗	✗	✗	✓	High	Avg.	✗	✗	1	20
Sports-1 M dataset	✓	✓	✓	✓	✓	✓	N.R.	N.R.	✓	✓	M	487

Table 3 Dataset content properties

Dataset Name	Resolution	FPS	Total clips/frames/duration	Clips per action	Min clip length	Max clip length	Avg clip length	Num. of classes
UCF Sports action	720 × 480 px	10	150 clips	–	2.2 s	14.4 s	–	10
KTH	160 × 120 px	25	–	–	–	–	4 s	6
YouTube	176–320 × 144–262 px	29.9	1600 clips, 2.8 h	–	0.33 s	29.99 s	–	11
UCF50	320–400 × 226–240 px	26.9	6681 videos	100 (videos)	1.06 s	33.84 s	7.44 s	50
UCF101	320 × 240 px	25	13320 clips	–	–	–	–	101
Hollywood2	300–400 × 300–200 px	24	2517 clips, 20.1 h	61–278 (clips)	–	–	–	12
HMDB51	176–592 × 240	30	6766 clips, 6 h	Min: 101 Avg: 132.67 (clips)	0.67 s	35.43 s	3.21 s	51
J-HMDB	176–592 × 240	30	928 clips,	36–55 (clips)	15–40 frames per clip			21
UIUC sport	547–3600 × 350–4158 px	NA	1579 images	137 (Bocce)—250 (Rowing) (images)	NA	NA	NA	8
MPII Fine-grained kitchen activity	1624 × 1224 px	29.4	44 video, 881755 f, >8 h	–	–	–	–	65
KSCGR	640 × 480 px, up to 2048 mm depth	30	–	–	5 to 10 min with 9 k to 18 k frames per video			9
ChaLearn gesture recognition 2013	640 × 480 px, 11 bit for depth	20	956 sequences,	–	each sequence: 1 and 2 min, 8–20 gesture samples, around 1.800 frames			13, 858 gestures
Sports-1 M	176–1280 × 144–720 px	–	–	1000–3000 (video)	–	–	5 min 36 s	487

3 Different Action Recognition Shallow and Deep Models

Different methodologies are exploited to enhance the accuracy of action recognition models. In fact, the common idea to classify an action in a video is by calculating the discriminative video features and classifying based on them. The more discriminative features, the better the results are. As shown in the current section, models/architectures can be divided into 2 main categories. The first category are models implemented by using the careful description of the actions and the second category are models that are less complex but that try to learn the discriminative parts of the actions in a generic way. The first approach will be referred as a shallow model or hand-crafted models and the second approach as deep model. In this section, a discussion of shallow and deep architectures is illustrated in the Sects. 3.1 and 3.2 respectively. For each of these models, we highlight the key points that contribute to solving the action recognition problems.

3.1 Shallow Models

Hand-crafted features are the first trials made in action recognition field to the best of our knowledge [21]. Early action recognition trials were domain specific and required hard work to implement *detectors* and *descriptors* and check which one of their combinations was the best for the given domain. The basic idea is to '*detect*' the important parts in an image (salient regions), and then provide a representation for these parts, by '*describing*' them. The description of these regions is called features and differs based on the detector/descriptors used.

Shallow models are harder to implement as they imply to handle different AR challenges and requires tremendous efforts facing them like the different viewpoints for the same class or objects occlusions. Moreover, for the AR problem, there is a need to deal with the temporal dimension and extract these information in addition to the spatial features as this enhance the overall accuracy.

This section first lists different detectors/descriptors and how to use them to classify the actions. Next, different models are detailed in both the spatial and temporal dimensions. For each of these model a discussion is lead for the action recognition problems it tackles.

Many feature detectors have been developed: 3D-Harris extending the Harris detector for images and 3-D-Hessian extending the Hessian saliency measure for blob detection in images. Between other detectors, there are the Cuboid based on temporal Gabor filters, the Dense Trajectories (DT) and Improved Dense Trajectories (iDT) extracting video clips with regular positions and scales in spatial and temporal space (Fig. 1). The most common descriptors are Cuboid descriptors proposed by Dollar et al. [22] to be used along with the Cuboid detector, HOG/HOF aiming to characterize both the local motion and appearance, HOG3D [23] is based on histograms of 3D gradient orientations (considered as extension of

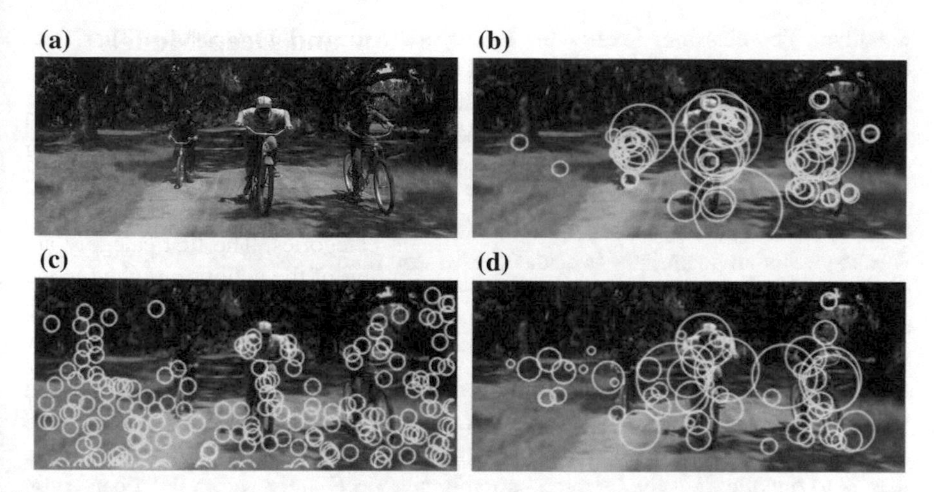

Fig. 1 Different detectors [25] **a** Original image **b** Harris detector **c** Cuboid detector **d** Hessian detector

SIFT features to video sequences), extended SURF (ESURF) which extends the image SURF descriptors [24]. The specific details of each detector/descriptor and their best combination on the different datasets can be found in the evaluation paper done by Wang et al. [25].

Based on these low level local features, the most popular video representation for action recognition is the Bag-Of-Words (BoW) model, also called Bag-of-Visual-Words (BoVW) [26]. The BoW is a general pipeline mainly composed of five phases (1) feature extraction, (2) feature preprocessing, (3) codebook generation, (4) feature encoding, and (5) pooling and normalization (Fig. 2). A detailed study is shown in [26] in which it first provides a comprehensive study of all steps in BoW, then discusses the different local features and the performances of their fusion, and uncovers good practices to produce a state-of-the-art action recognition system in hand-crafted models. The BoW is then used to train a

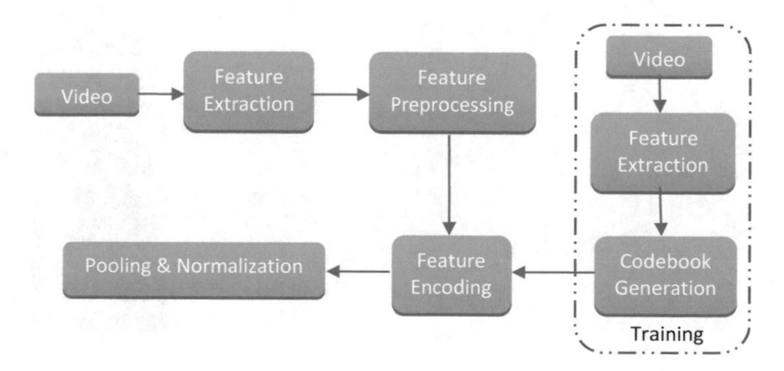

Fig. 2 Bag-of-words general pipeline

classifier. The classifier learns how to recognize the action occurring. Therefore, enhancing the quality and discriminability of BoW is crucial to enhancing the overall accuracy. The most used classifier in shallow models is Support Vector Machine (SVM) and is used in the following researches [25–30]. The standard SVM (with soft margin) was initially proposed by [31]. The core idea of the SVM in classifying the input data, into one of 2 classes, is to find the hyperplane that maximize the space (margin) between its nearest data point from both classes. A one-versus-all technique is usually used for multiclass SVM. The SVM also applies a "kernel trick" by mapping the input from nonlinear space into a higher dimensional feature space that is linear space using a kernel function to enhance the classes separations of the input.

Encouraged by the hierarchical feature learning of deep networks (discussed in the next subsection), a new idea based on hierarchical architectures, that have shown better results than BoW, suggests using a stacked Fisher vectors [28]. The idea is to encode densely sampled sub-volumes via Fisher vectors (FV) encoding, and project them to a low dimensional subspace. These FVs are compressed and encoded by a second FV layer to construct video-level representations. Moreover, another iDT features are calculated over the entire video clip and encoded by a separate FV. A concatenation of features of the FVs and the 2nd layer FV is fed to one-versus-all SVM classifier (Fig. 3). This concatenation improves the accuracy to using the stacked fisher vectors (SFV) alone.

Regarding the features used, the iDT features extracted from the video cuboids are a concatenation of HOG, HOF, and MBH. The HOG is responsible for the spatial features while HOF and MBH capture the temporal information in these cuboids. The cuboids with empty or small number extracted trajectories are removed as they are not salient. These features are reduced and then encoded by a second FV layer to capture and produce higher level features.

The stacked Fisher Vectors (SFV) using costless linear classifier, allowing faster training and therefore, using larger dataset. In addition, normalization is made for

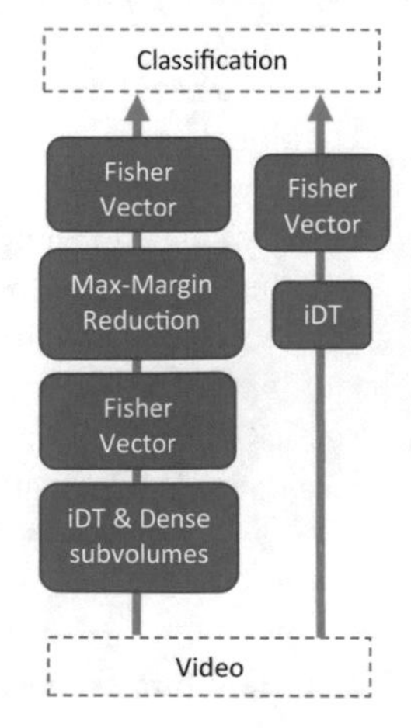

Fig. 3 Stacked Fisher vector architecture

the first and second FV features increasing the model robustness against illumination. It worth noting that by extracting iDT on small video cuboids, it is easier to handle the huge pose and motion variation. The reason is that features collected over large volumes suffer from intra-class variation and deformation issues. The model assists solving different AR problems: scale variance, illumination issues, different viewpoints, occlusion problems in addition to the background change and camera motions challenges.

A well-known idea is to include the non-salient region information in the BoW in a weighted manner as the context convey information that might be needed in certain applications to improve the recognition. Supported by the review [32] which denotes that human visual system (HVS) has the ability to recognize in periphery, a research considers applying this idea by integrating, without weighting, the features of both salient and non-salient regions in the classification step [27]. A Bag-of-Words approach is used with the spatial pyramids and SIFT features for both salient and non-salient regions. This is followed by a Multiple kernel learning (MKL) classifier that learns how to classify based on both salient and non-salient features. It is shown that this idea succeeds in capturing additional information and has a small improvement for the accuracy.

Regarding the features, no temporal features are used as the input was limited for still image. Moreover, the idea can be applied in videos by separating salient and non-salient regions for both spatial and temporal dimensions. The features used are based on SIFT features which are local low level features. SIFT are invariant for objects occlusions, orientation, and uniform scale change. Due to the separation of salient and non-salient features, the model increases the robustness against background changes and viewpoint change challenges.

In Action Recognition system, camera motion features in the background are usually considered noise, and eliminating them is a logical step to increase the accuracy and performance. This was successfully proved in a multiple granularity analysis for action recognition by Ni et al. [33]. The idea behind their work is to include features in BoW in a region around the interaction occurring, thus neglecting the other background motions, and tracking the interaction between human hands and the objects in use as their work was domain specific: kitchen. For the object/hand detection, they used HOG and HSV features. For the action recognition, local motion features used were the dense motion trajectories, where for each trajectory the HOG, Motion Boundary Histogram (MBH), HOF, and trajectory shape (TS) were extracted, and then encoded using k-means. Each video sub-volume is represented as a BoW vector and used to find the likelihood of the action using a Conditional Random Field (CRF). The inferred interaction and hand/object tracking results are used to perform feature pooling and accurate action detection by integrating prior knowledge (interaction status information) and the likelihood (motion features) into a graphical model thus achieving the granulation fusion strategy.

In this fine-coarse granularity approach, the features used are concatenation of spatial (HOG) and temporal (MBH, HOF, TS) features around the interaction (spatial) occurring. The proposed model contributes on handling different AR problems as background motion, scale change, occlusion, camera motion, action speed change and background change. However, this idea can only be applied in specific applications as the focus of the model is the interaction between hand/object. This model is restricted to a specific domain, making recipes in a kitchen. However, this is a general drawback in the shallow models.

Another research study followed the BoW steps, and modeled the *appearance evolution* within the video [29]. In this model, called *videoDarwin,* a new representation for the video is provided. Their contribution was to provide a *video-wide temporal evolution* in which their model learned the appearance and motion features order to predict the action. A combination of HOG, HOF, MBH, and trajectory (TRJ) features is used from the videos followed by a normal fisher vectors for encoding step. Then time varying mean vectors is applied to capture the spatial evolution of the video. This is done by calculating the change of pixels over time by getting the mean of each pixel weighted by the time at a given time. Next, a pooling step is done by ranking function to aggregate these features and represent the video evolution.

Exploring the temporal information of an action by combining the action appearance evolution allowed videoDarwin increasing the accuracy over prior methods as can be seen in the next Sect. 4. This model handles the occlusion problem, speed change, scale change, and background change problems out of the AR problems.

A closely related topic to action recognition is task/plan recognition which can be viewed as set of actions. Instead of channeling information from action system to a plan system and setting grammar or rules to identify the task being done which of course can fail in many scenarios as human actions performing a certain plan can vary. Latent Dirichlet Allocation (LDA) has been investigated in [34] where plans are seen as a distribution of actions thus considering structure relationship between actions that are omitted in BoW approach. Testing dataset contained forty recorded plan executions with RGB-D information, performed in a controlled environment. Different combination of models and number of topics are experienced: ten topics as the documents are built using subsets of ten actions, fifteen topics to check the cases where the differences between left and right hands are distinguishable, and five topics to check the effect of identical poses caused by the lack of position data (cases such as standing and jumping). They proved that by treating Plan Recognition as distributions of actions using LDA model has a potential for real time applications and that extra research will bring more success into this method.

Shallow Models rely on hand-crafted features to provide spatial and temporal features. Due to the amount of details to cover in order to find discriminative feature, shallow model requires good efforts for the design of the model, which implies by practice limiting the model to a specific domain. Recent work attempted to learn in a hierarchical way, and are competing with shallow models as can be seen in the next section.

3.2 Deep Models

Deep Learning is a category of algorithms, for training neural network models with deep architectures, where there are many 'hidden' layers of nonlinear operational units between input and output. The core idea is to learn more abstract concepts from the raw level data where higher levels are more abstract than lower ones. This is encouraged by the way the mammal brain is organized [35].

Although of their computational expensive cost, backpropagation and gradient descent (and its different versions), are the most common algorithms for training the deep networks by minimizing an objective function. Stochastic Gradient Descent (SGD) is an optimization method for the Gradient Descent. The main difference is that it updates the parameters of the object function for each training example instead after the whole dataset, and hence is faster and can be used for online training.

This section lists several deep models. For each model, the architecture is provided followed by a discussion of the spatial features then by the temporal features provided by this model. Each of the spatial/temporal features is linked to the action recognition problems it tackles.

One of the earliest hierarchies is the convolution neural network (CNN) by Yann LeCun et al. in 1998 [36]. This network is composed of convolution, subsampling and fully connected layers, typically between five to seven layers and are hard to train. Convolution is used to decrease the required memory and is applied on small regions to increase the performance. It showed success on the MNISET dataset and was applied for commercial use. In 2012, A. Krizhevsky et al. [37] trained a CNN with 5 convolutional and 3 fully connected layers using SGD, and achieved best results on subsets of ImageNet used in the ILSVRC-2010 and ILSVRC-2012 competitions at the time of the research. They used Rectified Linear Units (ReLUs) as the output of the neurons because it is several times faster than the traditional *tanh*. To improve the generalization, the ReLU is followed by a local normalization scheme. Some of the convolutional layers are followed by a pooling layer, different from other CNN-based models, which was an overlapping max-pooling. This reduced the error rate and made it slightly difficult for overfitting.

The CNN spatial features have many advantages over other models. The connections in each layer of the CNN-based network are limited to a 2D region of the previous layer thus preserving the geometry structure of the objects in the produced features that it learns from the input image. This assisted in using the CNN for objects segmentations. The weights of each learned filter are shared along the layer, thus adding robustness against transformations. To increase the robustness against transformation and illumination changes in the AlexNet model, training examples are augmented by translating, reflecting, and changing the intensity of the RGB in the input images. Without the data-augmentation, the network suffers from overfitting. Decreasing the time required for the model training is one factor that allowed usage of larger dataset. This allowed handling higher level spatial resolution images in addition to tackling different viewpoints and objects occlusion issues.

In these CNN models, no temporal features were generated. However, these spatial features were a baseline used for generating temporal feature. Clearly, in video analysis, the temporal component conveys additional information and shall not be ignored to increase the accuracy and performance.

One of the common methodologies is to use 2 networks with a later fusion, such that one network is trained on the spatial features while the other on the temporal features. This methodology has been applied in [30] using both the *improved trajectory* features and the two-stream convNets architecture. This study involves the advantages of both hand-crafted and deep learned features and outperforms both as it has achieved the state of the art on the datasets used. The convNets used are composed of spatial and temporal networks; the spatial is used to capture appearance features for actions and the temporal is used for learning the motion features. The two-stream convNets network is used to obtain convolutional feature maps, then a normalization step is applied to reduce the effect of the illumination. For the hand-crafted features, the authors used the *improved trajectories* because they are located at high salience regions and they effectively deal with the variations of the motion speed. With this in mind, the normalized convolutional features are sum-pooled around the center of these trajectories. These trajectories constrained sampling produce Trajectory-pooled Deep-convolution Descriptors (TDD). During the training, the input is scaled into multiscale pyramids and fed to the convNet to be able to construct a multiscale TDD which assists handling the AR scale change issues.

The TDD spatial features are convolutional features conveying abstract information. As the TDD model relies on 2 convNets, the spatial net can be pretrained on ImageNet dataset and acquire its benefits. This also assists dealing with both occlusion and different viewpoints AR issues especially when the training phase uses realistic datasets like HMDB51 and UCF101. The convolutional features are sum-pooled around the salient regions calculated by the improved trajectories. The model can then classify the main action and neglect background changes in addition to resisting to the noise resulting from the camera motion.

The TDD temporal features are convolutional features based on stacked optical flow images. Again, limiting the temporal information is not in the essence of deep learning. This can cause loss of temporal information and can suffer from limitations of optical flow features. Issues like static optical flow values and sudden large displacement shall be removed since they are resulting from inaccurate optical flow. It worth noting that the generated temporal features are better than or comparable to the MBH and HOF. Similar to the spatial features in the TDD, the temporal features are sum-pooled around the regions calculated by the improved trajectories. This benefits in having a more robust model against background and camera motion problems.

Deep learning concepts implies that the deeper and wider the network is, the more powerful it is. Consequently, the network learns low level features on the first layers and more abstract features in the latest layers. However, this would require a tremendous effort to train with generalizing. Therefore, increasing the depth and breadth of a network with reduction of the parameters is a common strategy for building new deep architectures.

Besides AlexNet [37] model, GoogLeNet [38] is another commonly used CNN model. It has an overall of 22 layers and uses network in network concept to increase the depth and breadth of the network without a huge increase of the number of parameters to learn [38]. Therefore, these features are more discriminative and provide abstract representation for the input frames, leading to a better handling of the different viewpoint issues and high resolution videos. Moreover, GoogLeNet consider multiscale processing allowing the model to handle the scale change problem.

The research [39] applies 2 different CNN-based models to asses the usage of traditional CNN for learning long temporal relationship. The CNN features are generated by GoogLeNet model as it yielded better features than AlexNet model as shown in the research [39]. The first model uses different pooling techniques for temporally stacked CNN features. Average pooling, fully connected layer and max-pooling have been used with the max-pooling outperforming them. Several variations of the max-pooling architecture are performed with the *Conv Pooling* being the best of them. The *Conv Pooling* performs max-pooling over the final convolutional layer indicating the needs to preserve the spatial information over the time domain. This is followed by fully connected and SoftMax layers (Fig. 4). The video frames are sampled with 1fps causing loss of motion information. Therefore, optical flow features are calculated from 2 adjacent frames sampled at 15 fps and fused with the model. This strategy succeeded in improving video classification and, as shown by the authors, the larger the clips the more accuracy it achieves. However, usage of Optical Flow did not improve the accuracy as the used dataset, Sports 1 M, is very noisy and suffers from AR problems (discussed in Sect. 2).

The second model uses CNN together with a Long Short Term Memory (LSTM) cells to classify videos content/action. The LSTM includes memory cells to store, modify, and access internal state, allowing it to better discover long-range temporal relationships. Specifically, the deep LSTM network consists of five stacked LSTM layers each with 512 memory cells taking as input the output of the final CNN layer at each consecutive video frame. The CNN outputs are processed forward through time and upwards through the five layers of stacked LSTMs (Fig. 5). At each time step, a SoftMax layer predicts the class. The training algorithm of choice is Downpour SGD which is an asynchronous variant of SGD. Like the *Conv Pooling* model, Optical Flow features are used to recover the loss of motion information.

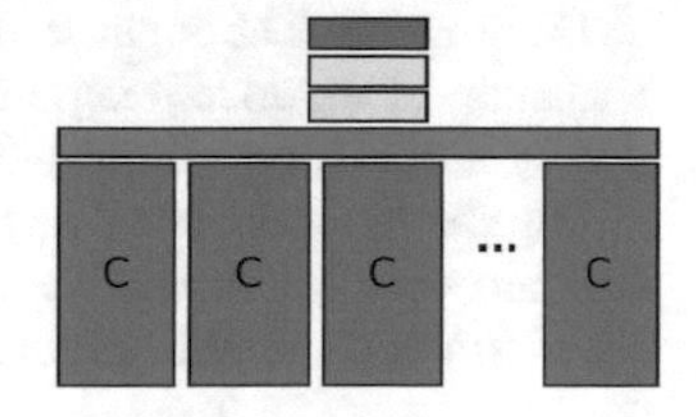

Fig. 4 Conv Pooling architecture used in [39]. Temporally stacked convolution features (denoted by ‘C’) are max-pooled (*blue*) and fed to the fully connected network (*yellow*) followed by a SoftMax layer (*orange*)

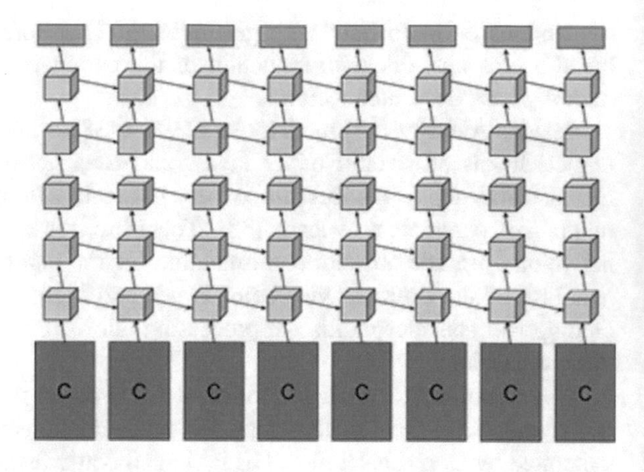

Fig. 5 Deep long short term memory architecture with input as convolutional features and followed by a softmax layer [39]

This is done by a later fusion with the network trained directly on the video frames. Interestingly, it raised the accuracy of the LSTM with 1% indicating that temporal dimension conveys additional information. The deep LSTM network structure is able to discover high level temporal features as it learns the temporal order based on the convolutional features.

Working with videos, spatial and temporal dimensions, recent researchers employ 3D-CNN idea, where features are generated based on cuboids of the input video instead of single images. Each cuboid has $\mathbf{W} \times \mathbf{H} \times \mathbf{T}$ as width, height and time respectively representing part of the video in both spatial and temporal dimensions. This idea was applied with an LSTM network by Yao et al. [40], to provide a description and/or summarization for a video based on the interest temporal segments. The video cuboids are constructed as a concatenation of the histograms of oriented gradients, oriented flow and motion boundary (HoG, HoF, MbH) to extract the 'local' temporal information of the video. Local structure refers to specific time action, short in duration, like standing up or answering the phone. Complementary, there are global temporal information that refers to the order of objects, actions, scenes, and people appearing in the video. Usage of local temporal features instead of the video frames also reduces the computation of the subsequent 3D CNN layers. There are three 3-D convolutional layers that follows, each is followed by ReLU and local max-pooling. Max-pooling is performed to get a vector summarizing the content of the short frame sequences within the video. Regarding the 'global' temporal information, a dynamic weighted sum strategy is used to represent the video. The weights represent the degree of relevance of a given feature vector to the generated words and are updated at each time step of the LSTM (Fig. 6). The input to the 'global' temporal attention mechanism is a concatenation of the local temporal features concatenated with spatial features generated by a 2D GoogLeNet CNN features.

The spatial features of this model take the advantages of GoogLeNet model. They are very discriminative and cover scale issues in addition to high resolution issues among the AR problems. Concatenating these features with the local

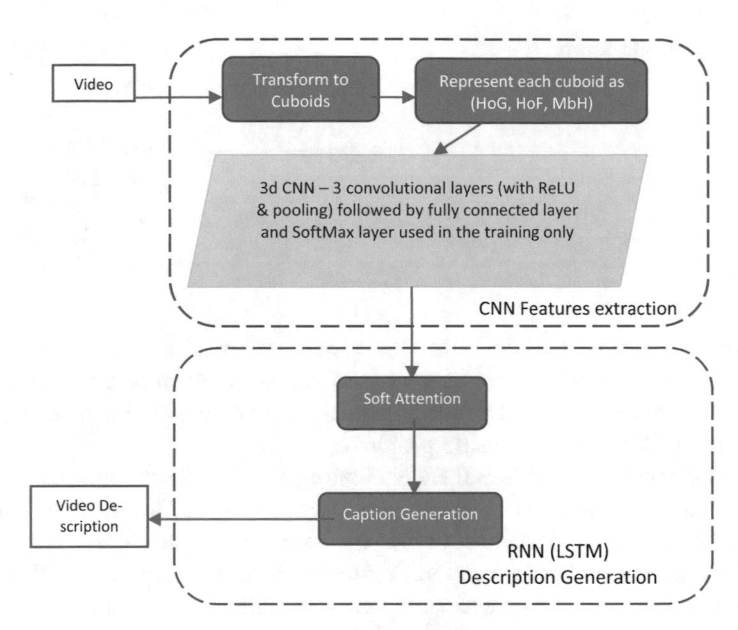

Fig. 6 Encoder-decoder framework. 3D-CNN is used to encode the video followed by LSTM to produce video captions

temporal features generated by the 3D CNN model, increases the model accuracy specifically when higher number of classes is used.

The temporal features in this model are divided into local and global features. The 3D-CNN local temporal features provide robust features against transformations. Equally important, these features capture the salient motion segments of the video. Pooling summarizes the content over short period of time, keeping the important features over the time and discover long term relationships. The global features on its own (with spatial features only) learns the general sequence of actions occurring without keeping track of the action details. Using spatiotemporal features as input adds additional information about the local actions. This results in more accurate action identification like 'Someone is frying a fish in a pot' instead of 'The person is cooking'. Using the hand-crafted features decrease the computations needed, and favors in learning over large datasets. However, this limits the network ability to learn temporal features based on HoG, HoF and MbH features only.

Another deep learning architecture was designed in 2006 with the Deep Belief Networks (DBN) [41]. They are probabilistic generative models that contain many layers of hidden variables and are able to learn complex internal distribution of the input data. The network is constructed of stacking of Restricted Boltzmann Machine (RBM) which is essentially derived from Boltzmann Machine (BM). BM is a type of stochastic recurrent neural network, allowing connection between units of the same layer. On the contrary, RBM does not allow connection between units in the same layer, thus learning becomes quite efficient. Based on this fact, RBMs are used in deep networks like DBN (Fig. 7). Training is done in a greedy wise

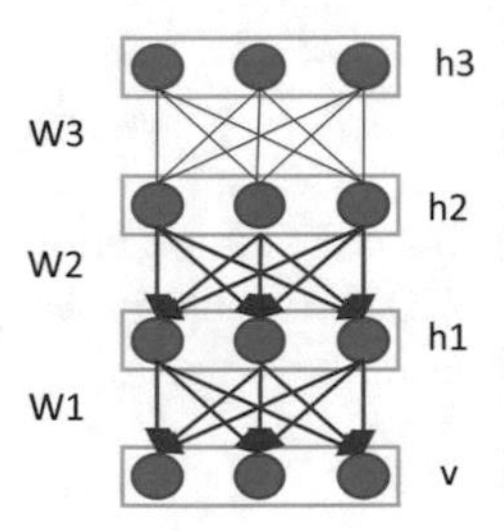

Fig. 7 Deep belief network

algorithm, one layer per time. In this type of networks, each layer captures high-order correlations between the activities of hidden features in the layer below. The DBNs encodes statistical dependencies among the units in the layer below but is challenging for computer vision problems.

Spatially speaking, RBMs and DBNs both ignore the 2D structure of images. It is not practically feasible for computations on large images. One solution is to share its weights among all locations of the image for the same level. This kind of networks does not tackle the temporal information, and if standard DBN is used it would be necessary to use input feature vector containing both spatial and temporal features rather than directly on the raw input data.

A Convolutional Gated Restricted Boltzmann Machine (conv GRBM), (Fig. 8—left) was used in [42] to automatically extract different convolutional features that are also achievable via hand-crafted models like flow field features and edge features. It also provided features helpful for segmenting people from background (it worth noting that the dataset used when person segmentation occurred had uniform background, i.e., KTH dataset). The model was trained on pairs of successive images in image sequence (i.e., video) and learned the transformation occurring. The convGRBM layer is shown in (Fig. 8—right). On top of this layer, is a 3D convolutional layer that captures mid-level spatiotemporal cues. The outputs of this

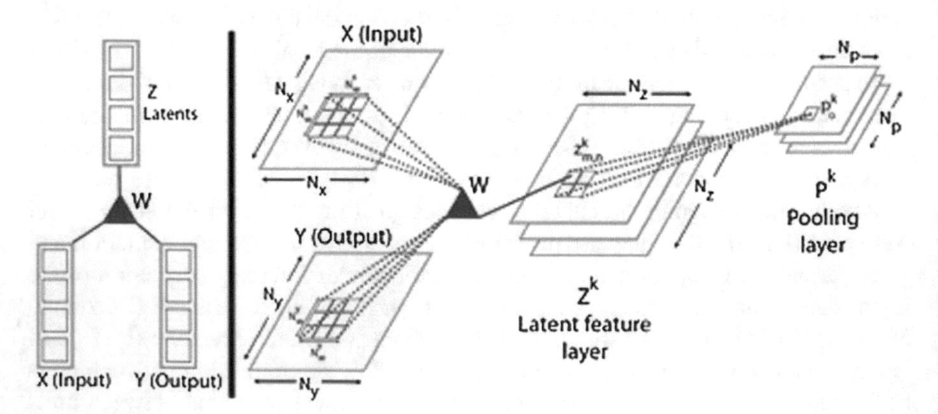

Fig. 8 *Left* A gated RBM. *Right* A convolutional gated RBM using probabilistic max-pooling [42]

layer are 3D feature map, so a max-pooling applied on the temporal dimension is applied to ensure that the mid-level features have consistent vector size, and to consider the important features. Lastly, one or two fully connected layers are used to classify the activity. Regarding the training algorithm, Contrastive Divergence (CD) is used. CD is an approximate to the maximum likelihood algorithm and is commonly used for training RBM models.

Most of the features extracted are motion sensitive but few are static information which represents the context information. The main contribution is the convGRBM layer which is useful as an operator for edge detection beside low level spatial feature generator. Usage of convolution in the GRBM allowed scaling to higher resolutions input video in addition to reducing the occlusion issues.

Regarding the temporal features, the convGRBM layer is able to provide different motion features in between optical flow like features with compact parameters. ConvGRBM layer can be a part of a bigger multistage system or a temporal feature generator as it learns the transformation occurring between 2 input video frames. However, this layer alone is weak against videos including multiple views for the same action in addition to scale change. Fusing this layer features with features of the temporally trained model in a weighted manner may need further researches.

In the exploratory study [43] a network is used with 9 layered locally connected sparse autoencoder. The model can be viewed as sparse deep autoencoder where each feature in the autoencoder connect only small region of the lower layer. The idea that, local filter connects only small region of the lower layer is biological inspired and known as *local receptive fields*. This is different than other CNN models, i.e., [36, 37], as the parameters are not shared across different location of the image. This is a core difference leading to learning more invariances other than translations. One layer is shown in (Fig. 9) and the network replicates this layer. The model has 1 billion trainable parameters trained on a dataset of 10 million 200×200 pixel images. Interesting is that the authors mention that this network as the largest known network is still tiny compared to the human visual cortex which is 10^6 times larger in terms of number of neurons and synapses.

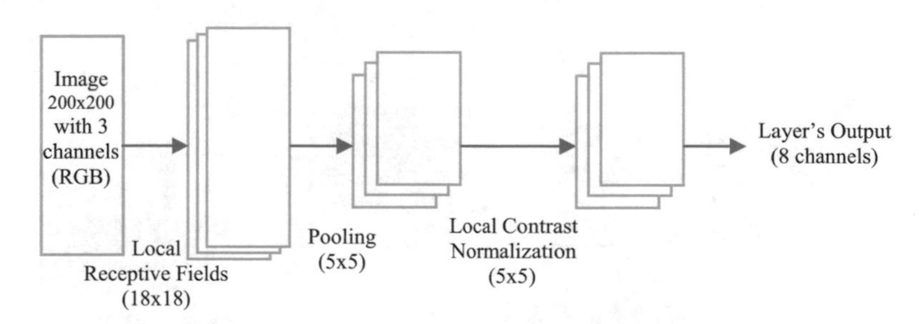

Fig. 9 One layer of sparse deep autoencoder model

This model is designed and uses images for training and results in learning only spatial information. However, the model is very robust against deformation, different position, orientation, scales and viewpoints. Local L2 pooling and local contrast normalization are key factors for this robustness. The network is robust in face recognition, cat and human body detection, and object recognitions and has achieved state-of-the-art results on many datasets as shown in the next section. This model can be used in AR field, with modifications to capture motion features by training additionally on optical flow images as an example. However, as the network is very deep, the need for a large dataset of videos or image sequence represents a challenge to prevent overfitting.

One last model to review is based on Slow Feature Analysis (SFA). SFA extracts the invariant and slowly varying signal from the quickly varying input signal. This is helpful for visual applications to extract high level concepts from the input image, i.e., the objects and their relative positions out of the pixel colors. In [44], SFA features have been investigated following the deep learning concepts: Deep Learning Slow Feature Analysis (DL-SFA) architectures. Two layers of SFA are used. In the first layer, SFA is applied on video cuboids and the learned kernels are convoluted with the video cuboid (3D convolution). The resultant convoluted features are max-pooled and fed to the second layer of SFA (Fig. 10). Stacking the second layer of SFA has shown an improvement of around 7% versus using only one layer. This is an indicator that the hierarchical models have better performance than shallow models. The final feature vector is a combination of first and second layer features.

In this model, the SFA, given a sequence of frames, generates temporal features that describe the action occurring. The kernels learned are convolved, in a 3-D manner, with the spatiotemporal video cuboids to provide feature maps used as input for the next layer. The convolution and max-pooling increased the model translational invariances. The success of this model refers to well extraction of spatiotemporal features instead of separation of the spatial and temporal feature with a later fusion. This study covered several AR problems: scale changes, learning more discriminant features that assist in the different viewpoint of the same action issue, background change and illumination change issues.

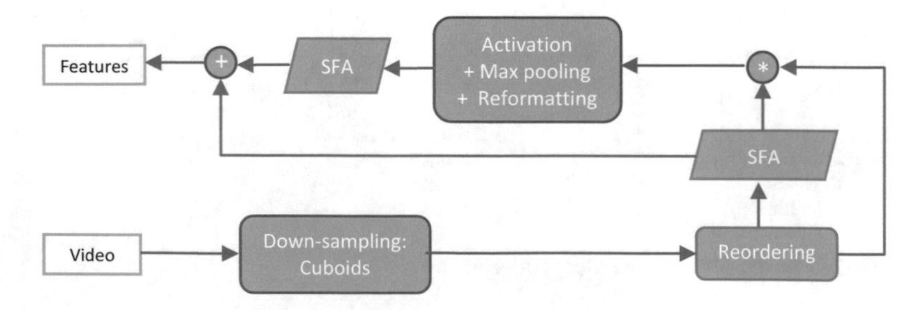

Fig. 10 Deep learning—slow feature analysis architecture

4 Models Accuracy Evaluation

A summary of the results of different shallow and deep models, covered in the selected papers, is presented for each dataset. Based on these results, a discussion is made for the contribution of shallow and deep models in solving the AR problems. Table 4 summarizes the results of different methodologies.

For both KTH and UCF Sports dataset, DL-SFA achieved the best results in the selected papers. Learning more abstract features is therefore for the favor of action recognition. This is an indicator to the necessity to represent the complexity of the actions. The ability of deep models to learn in higher dimensions allow better representation of actions than shallow models.

The Hollywood2 dataset and MPII Fine-grained Kitchen Activity dataset witnessed the best result by shallow model: videoDarwin. This method learning the appearance evolution of the action allowed modeling the action order and proved to be a good discriminative feature for human actions. However, both datasets cover small number of action labels, so more investigation is needed for datasets with higher number of classes. An investigation for a dataset where the same actions are done in different orders is suggested to add new challenges for this method as in real life action can be done in different orders.

As can be seen in the HMDB51 dataset, stacking Fisher vectors yielded the best result of 66.79%. The idea behind this model, is to refine the representation of the hand-crafted features and abstracting the learned information in a hierarchical way is very similar to deep models. This allowed overcoming the variation in shape and motion in the videos along longer duration, and learning global information about the action. Therefore, noise motions caused by camera shaking or background moving objects are reduced. Also, training on different scale cuboids taken from the video is for the favor of training a robust model against scale change.

Regarding the UCF101 dataset, the best result of 91.5% was reached by using a multiscale TDD model. This can be seen as a hybrid model using a combination of both shallow and deep features. The focus to describe, in a deep way, the motion salient regions of a video allowed the model to overcome different AR problems due to neglecting large part of unnecessary features. However, considering the context in a weighted manner is suggested as this might lead to further enhancements the model accuracy.

Consistent with the previous conclusion, deeper hierarchical model is for the favor of inferring much powerful information. The largest known network built was applied on ImageNet Fall 2009 dataset and proved the best results to date: 19.2%. The low accuracy is due to the high number of categories however, it is 15% relative improvements than its best prior result. The model learned invariant features for face recognition and object detection. The authors also showed it is robust against scaling, out of plane rotations, light change, and occlusions. The only drawbacks for this network are the need of high source of computation and the need for a large dataset with various actions classes to allow it modelling the actions.

Table 4 Summary of the results for both hand-crafted features and deep learning models

Dataset	Authors	Results and used methodology
KTH actions dataset [11]	H. Wang et al. (2009)	92.1%—Average accuracy for detector/descriptor combinations: Harris3D + HOF
	G. Taylor (2010)	90.0% convGRBM
	Lin Sun et al. (2014)	**93.1% 2 layers DL-SFA**
UCF sports dataset [8, 9]	**Lin Sun et al. (2014)**	**86.6% 2 layers DL-SFA**
	H. Wang et al. (2009)	85.6% Average accuracy for detector/descriptor combinations: dense sampling + HOG3D
Hollywood2 dataset [15]	H. Wang et al. (2009)	47.4% Mean AP for various detector/descriptor combinations: dense sampling + HOG/HOF
	G. Taylor, (2010)	46.6% convGRBM
	Lin Sun et al. (2014)	48.1% 2 layers DL-SFA
	B. Fernando et al. (2016)	**73.7% using videoDarwin**
HMDB51 dataset [10]	Peng et al. (2014)	61.1% hierarchical representation combining outputs of FV and soft version of SVC and multiple descriptors
	B. Fernando et al.(2016)	63.7% using videoDarwin
	Liming Wang et al. 2015	65.9% using TDD and iDT
	C. Zou et al. (2014)	**66.79% using stacked fisher vectors**
UCF50 dataset [13]	Peng et al. (2014)	92.3% hierarchical representation combining outputs of FV and soft version of SVC and multiple descriptors
UCF101 dataset [14]	Peng et al. (2014)	87.9% hierarchical representation combining outputs of FV and soft version of SVC and multiple descriptors
	J. Yue-Hei Ng et al. (2015)	88.6% LSTM with 30 Frame Unroll (optical flow + image Frames)
	Liming Wang et al. (2015)	**91.5% using TDD and iDT**

(continued)

Table 4 (continued)

Dataset	Authors	Results and used methodology
YouTube dataset [12]	C. Zou et al. (2014)	93.77% using stacked fisher vectors
J-HMDB dataset [16]	C. Zou et al. (2014)	69.03% using stacked fisher vectors
UIUC SPORT dataset [10]	S.F. Dodge, L.J. Karam (2013)	Using a small training size, their method improves upon the classification accuracy of the baseline bag of features approach.
MPII fine-grained kitchen activity dataset [17]	Bingbing Ni et al. (2014)	Prec.: 28.6 Recall: 48.2 AP: 54.3 using interaction tracking
	B. Fernando et al. (2016)	**72% using videoDarwin**
ICPR 2012 kitchen scene context based gesture recognition dataset (KSCGR) [18]	Bingbing Ni et al. (2014)	Mean F-score 0.79 using interaction tracking
ChaLearn gesture recognition dataset [19]	B. Fernando et al. (2016)	Precision: 75.3% Recall: 75.1% F-score: 75.2%
ImageNet [45] Fall 2009 (~9 M images, ~10 K categories)	A. Krizhevsky et al. (2012)	Test error rates: Top-1: 67.4% Top 2: 40.9% using CNN with an additional, sixth convolutional layer over the last pooling layer
	Quoc V. Le et al. (2012)	**19.2% using 9-layered locally connected sparse autoencoder (with unsupervised pre-training)**
ImageNet [45] Fall 2011 (~14 M images, ~22 K categories)	Quoc V. Le et al. (2012)	15.8% (with unsupervised pre-training)
ILSVRC [46] year 2010	A. Krizhevsky et al. (2012)	Test error rates: Top-1: 37.5% Top 2: 17.0% using CNN
ILSVRC [46] year 2012	A. Krizhevsky et al. (2012)	Test error rates: Top-1: 36.7% Top 2: 15.3% using 7 CNNs
Face detection [30]: Dataset sampled from (Wild dataset [47] + ImageNet Fall 2009)	Quoc V. Le et al. (2012)	Best Neuron: 81.7% using 9-layered locally connected sparse autoencoder
Cat Detection [43]: positive example from dataset Zhang et al. [48] negative examples ImageNet dataset	Quoc V. Le et al. (2012)	74.8% using 9-layered locally connected sparse autoencoder
Human Body [43] (subsampled at random from a benchmark dataset Keller et al. [49]	Quoc V. Le et al. (2012)	76.7% using 9-layered locally connected sparse autoencoder
Synthetic video constructed from NORB images [42]	G. Taylor (2010)	Used only to visualize the flow learned by the Hidden Units Learned transformation are richer than Optical Flow features

(continued)

Table 4 (continued)

Dataset	Authors	Results and used methodology
Sports-1 M dataset [20]	J. Yue-Hei Ng et al. (2015)	@Hit 1: 73.1% using LSTM (Image and Optical Flow) @Hit 5: 90.8 using conv pooling (Image and Optical Flow)
Youtube2Text [50]	Li Yao et al. (2015)	BLEU [51] 0.4192 METEOR [52] 0.2960 CIDEr [53] 0.5167 Perplexity 27.55
DVS [54]	Li Yao et al. (2015)	BLEU [51] 0.007 METEOR [52] 0.057 CIDEr [53] 0.061 Perplexity 65.44

More investigation is needed to allow learning the motion dimension in the videos in a discriminant way.

5 A Proposed Architecture

Recent work highlighted the importance of exploring the temporal dimension. In addition, it has been shown the discriminability of the CNN features. In this section, an action recognition hierarchy is suggested. The core idea is to train a Deep Believe Network (DBN) using a sequence of spatial features from CNN network and temporal features extracted from the temporal stacking of CNNs layers (Fig. 11). Each layer of the stacked CNN receives a frame from the successive video frames. This is followed by a Singular Value Decomposition (SVD) that learns the transformation occurring, thus, providing a representation for the temporal dimension of the action occurring. Finally, both spatial and temporal features are fed to a DBN that models the conditional probability distribution of each set of features belonging to each action label. The temporal stacking uses D layers of CNNs with overlapping of L frames to increase the smoothness in the temporal dimension.

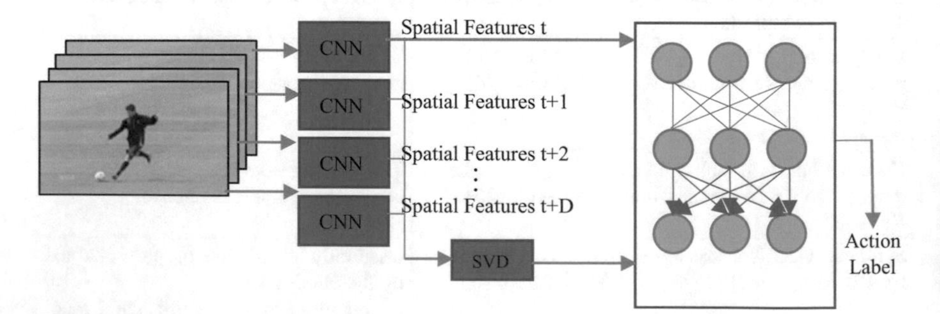

Fig. 11 Proposed hierarchy: stacked CNN with temporal features extraction and DBN network for Human Action Classification

The convolutional features are calculated directly from the input video frames. The reason behind this spatial feature selection is due to its ability to provide abstract representation of the input image. Moreover, its robust against local transformations and illuminations problems.

The second step is to construct a stacking of D features map, generated by the CNN, for D sequential frames. Each stack of features is overlapped by L frames as the start of a given action is not known in prior and to provide smoother representation. For each stack of features, a K dimensional matrix M is factorized by SVD algorithm. The resultant component matrices are considered as temporal features representing the transformation occurring between the input convolutional features for the given action.

Finally, the first convolutional features of the CNN stack are concatenated with the features calculated by the SVD. This concatenation is fed to the DBN to classify the actions. The DBN is powerful in separating similar classes like 'walking' and 'running'. Training a DBN requires a large DBN so the suggested training dataset is Sports-1 M dataset.

This model is expected to handle many challenges. The convolutional features assure robustness against transformation, illumination change, occlusion, and higher number of actors. The SVD provides motion information about the action and increases model robustness against background object motion and camera motion issues. High number of classes challenge is expected to be tackled thanks to the capabilities of DBNs.

6 Conclusion

Action Recognition has a large number of important applications like surveillance and elderly monitoring. To allow building better models, this chapter lists the different challenges in the action recognition problem. The existence of these challenges is investigated in 13 datasets to show their potential for models evaluations. Complementary, 13 shallow and deep models are provided discussing their features from both spatial and temporal perspectives. Hierarchical models proved better accuracies as shown on 21 datasets and can be further enhanced by enhancing the temporal features.

Shallow models are suitable for a constrained application domain while deep models are suitable for large datasets. Moreover, more investigation of the temporal dimension is necessary in action recognition to boost the performance. Based on this review, a proposed deep model based on stacking convolutional features is shown. The idea of stacking CNN features to learn the temporal transformations occurring is expected to provide more discriminative features for the action in the input video and raise the overall performance.

References

1. Shao, L., Jones, S., Li, X.: Efficient search and localization of human actions in video databases. IEEE Trans. Circuits Syst. Video Technol. **24**(3), 504–512 (2014)
2. Wang, F., Xu, D., Lu, W., Xu, H.: Automatic annotation and retrieval for videos. In: Pacific-Rim Symposium on Image and Video Technology, pp. 1030–1040. Springer, Heidelberg (2006)
3. Hung, M.H., Pan, J.S.: A real-time action detection system for surveillance videos using template matching. J. Inf. Hiding Multimedia Signal Process. **6**(6), 1088–1099 (2015)
4. Campo, E., Chan, M.: Detecting abnormal behaviour by real-time monitoring of patients. In: Proceedings of the AAAI-02 Workshop Automation as Caregiver, pp. 8–12 (2002)
5. Mumtaz, M., Habib, H. A.: Evaluation of Activity Recognition Algorithms for Employee Performance Monitoring. Int. J. Comput. Sci. Issues (IJCSI), **9**(5), 203–210 (2012)
6. Regneri, M., Rohrbach, M., Wetzel, D., Thater, S., Schiele, B., Pinkal, M.: Grounding action descriptions in videos. Trans. Assoc. Comput. Linguist. **1**, 25–36 (2013)
7. Guo, G., Lai, A.: A survey on still image based human action recognition. Pattern Recogn. **47** (10), 3343–3361 (2014)
8. Rodriguez, M.: Spatio-temporal maximum average correlation height templates in action recognition and video summarization (2010)
9. Marszalek, M., Laptev, I., Schmid, C.: Actions in context. In: IEEE Conference on Computer Vision and Pattern Recognition, CVPR 2009, pp. 2929–2936. IEEE (2009)
10. Kuehne, H., Jhuang, H., Garrote, E., Poggio, T., Serre, T.: HMDB: a large video database for human motion recognition. In: 2011 International Conference on Computer Vision, pp. 2556–2563. IEEE (2011)
11. Schuldt, C., Laptev, I., Caputo, B.: Recognizing human actions: a local SVM approach. In: Proceedings of the 17th International Conference on Pattern Recognition, ICPR 2004, vol. 3, pp. 32–36. IEEE (2004)
12. Liu, J., Luo, J., Shah, M.: Recognizing realistic actions from videos "in the wild". In: IEEE Conference on Computer Vision and Pattern Recognition, CVPR 2009, pp. 1996–2003. IEEE (2009)
13. Reddy, K.K., Shah, M.: Recognizing 50 human action categories of web videos. Mach. Vis. Appl. **24**(5), 971–981 (2013)
14. Soomro, K., Zamir, A.R., Shah, M.: UCF101: A dataset of 101 human actions classes from videos in the wild (2012). arXiv:1212.0402
15. Li, L.J., Fei-Fei, L.: What, where and who? classifying events by scene and object recognition. In: 2007 IEEE 11th International Conference on Computer Vision, pp. 1–8. IEEE (2007)
16. Jhuang, H., et al.: Towards understanding action recognition. In: Proceedings of the IEEE International Conference on Computer Vision (2013)
17. Rohrbach, M., Amin, S., Andriluka, M., Schiele, B.: A database for fine grained activity detection of cooking activities. In: 2012 IEEE Conference on Computer Vision and Pattern Recognition (CVPR), pp. 1194–1201. IEEE (2012)
18. http://www.murase.m.is.nagoya-u.ac.jp/KSCGR/. Accessed 29 Jan 2013
19. Escalera, S., Gonzàlez, J., Baró, X., Reyes, M., Lopes, O., Guyon, I., Escalante, H.: Multi-modal gesture recognition challenge 2013: dataset and results. In: Proceedings of the 15th ACM on International Conference on Multimodal Interaction, pp. 445–452. ACM (2013)
20. Karpathy, A., Toderici, G., Shetty, S., Leung, T., Sukthankar, R., Fei-Fei, L.: Large-scale Video Classification with Convolutional Neural Networks (2014)
21. Badler, N. I., O'Rourke, J., Platt, S., Morris, M. A.: Human movement understanding: a variety of perspectives. In: AAAI, pp. 53–55 (1980)

22. Dollár, P., Rabaud, V., Cottrell, G., Belongie, S.: Behavior recognition via sparse spatio-temporal features. In: 2005 IEEE International Workshop on Visual Surveillance and Performance Evaluation of Tracking and Surveillance (pp. 65–72). IEEE (2005)
23. Klaser, A., Marszałek, M., Schmid, C. A spatio-temporal descriptor based on 3d-gradients. In: BMVC 2008–19th British Machine Vision Conference, pp. 275–1. British Machine Vision Association (2008)
24. Willems, G., Tuytelaars, T., Van Gool, L.: An efficient dense and scale-invariant spatio-temporal interest point detector. In: European Conference on Computer Vision, pp. 650–663. Springer, Heidelberg (2008)
25. Wang, H., Ullah, M. M., Klaser, A., Laptev, I., Schmid, C.: Evaluation of local spatio-temporal features for action recognition. In: BMVC 2009-British Machine Vision Conference, pp. 124–1. BMVA Press (2009)
26. Peng, X., Wang, L., Wang, X., Qiao, Y.: Bag of visual words and fusion methods for action recognition: Comprehensive study and good practice. Comput. Vis. Image Underst. (2016).
27. Dodge, S. F., Karam, L.J.: Is Bottom-Up Attention Useful for Scene Recognition? (2013). arXiv:1307.5702
28. Peng, X., Zou, C., Qiao, Y., Peng, Q.: Action recognition with stacked fisher vectors. In: European Conference on Computer Vision, pp. 581–595. Springer International Publishing (2014)
29. Fernando, B., Gavves, E., Oramas, J., Ghodrati, A., Tuytelaars, T.: Rank pooling for action recognition (2016)
30. Wang, L., Qiao, Y., Tang, X. Action recognition with trajectory-pooled deep-convolutional descriptors. In: Proceedings of the IEEE Conference on Computer Vision and Pattern Recognition, pp. 4305–4314 (2015)
31. Bottou, L., Vapnik, V.: Local learning algorithms. Neural Comput. **4**(6), 888–900 (1992)
32. Strasburger, H., Rentschler, I., Jüttner, M.: Peripheral vision and pattern recognition: a review. J. Vis. **11**(5), 13–13 (2011)
33. Ni, B., Paramathayalan, V.R., Moulin, P.: Multiple granularity analysis for fine-grained action detection. In: Proceedings of the IEEE Conference on Computer Vision and Pattern Recognition, pp. 756–763 (2014)
34. Freedman, R.G., Jung, H.T., Zilberstein, S.: Plan and activity recognition from a topic modeling perspective. In: ICAPS (2014)
35. Serre, T., Kreiman, G., Kouh, M., Cadieu, C., Knoblich, U., Poggio, T.: A quantitative theory of immediate visual recognition. Prog. Brain Res. **165**, 33–56 (2007)
36. LeCun, Y., Bottou, L., Bengio, Y., Haffner, P.: Gradient-based learning applied to document recognition. Proc. IEEE **86**(11), 2278–2324 (1998)
37. Krizhevsky, A., Sutskever, I., Hinton, G.E.: Imagenet classification with deep convolutional neural networks. In: Advances in Neural Information Processing Systems, pp. 1097–1105 (2012)
38. Szegedy, C., Liu, W., Jia, Y., Sermanet, P., Reed, S., Anguelov, D., Rabinovich, A.: Going deeper with convolutions. In: Proceedings of the IEEE Conference on Computer Vision and Pattern Recognition, pp. 1–9 (2015)
39. Yue-Hei Ng, J., Hausknecht, M., Vijayanarasimhan, S., Vinyals, O., Monga, R., Toderici, G.: Beyond short snippets: Deep networks for video classification. In: Proceedings of the IEEE Conference on Computer Vision and Pattern Recognition, pp. 4694–4702 (2015)
40. Yao, L., Torabi, A., Cho, K., Ballas, N., Pal, C., Larochelle, H., Courville, A.: Describing videos by exploiting temporal structure. In: Proceedings of the IEEE International Conference on Computer Vision, pp. 4507–4515 (2015)
41. Salakhutdinov, R., Hinton, G.E.: Deep boltzmann machines. In: AISTATS, vol. 1, p. 3 (2009)
42. Taylor, G.W., Fergus, R., LeCun, Y., Bregler, C.: Convolutional learning of spatio-temporal features. In: European Conference on Computer Vision, pp. 140–153. Springer, Heidelberg (2010)
43. Le, Q. V.: Building high-level features using large scale unsupervised learning. In: 2013 IEEE International Conference on Acoustics, Speech and Signal Processing pp. 8595–8598 (2013)

44. Sun, L., Jia, K., Chan, T.H., Fang, Y., Wang, G., Yan, S.: DL-SFA: deeply-learned slow feature analysis for action recognition. In: Proceedings of the IEEE Conference on Computer Vision and Pattern Recognition, pp. 2625–2632 (2014)
45. Deng, J., Dong, W., Socher, R., Li, L. J., Li, K., Fei-Fei, L.: Imagenet: A large-scale hierarchical image database. In: IEEE Conference on Computer Vision and Pattern Recognition, CVPR 2009. pp. 248–255. IEEE (2009)
46. Russakovsky, O., Deng, J., Su, H., Krause, J., Satheesh, S., Ma, S., Berg, A.C.: Imagenet large scale visual recognition challenge. Int. J. Comput. Vis. **115**(3), 211–252 (2015)
47. Huang, G.B., Ramesh, M., Berg, T., Learned-Miller, E.: Labeled faces in the wild: A database for studying face recognition in unconstrained environments, vol. 1, no. 2, p. 3, Technical Report 07-49, University of Massachusetts, Amherst (2007)
48. Zhang, W., Sun, J., Tang, X.: Cat head detection-how to effectively exploit shape and texture features. In: European Conference on Computer Vision, pp. 802–816. Springer, Heidelberg (2008)
49. Keller, C. G., Enzweiler, M., Gavrila, D. M.: A new benchmark for stereo-based pedestrian detection. In: 2011 IEEE Intelligent Vehicles Symposium (IV), pp. 691–696. IEEE (2011)
50. Chen, D.L., Dolan, W.B.: Collecting highly parallel data for paraphrase evaluation. In: Proceedings of the 49th Annual Meeting of the Association for Computational Linguistics: Human Language Technologies-Volume 1, pp. 190–200. Association for Computational Linguistics (2011)
51. Papineni, K., Roukos, S., Ward, T., Zhu, W.J.: BLEU: a method for automatic evaluation of machine translation. In: Proceedings of the 40th Annual Meeting on Association for Computational Linguistics, pp. 311–318. Association for Computational Linguistics (2002)
52. Denkowski, M., Lavie, A.: Meteor universal: Language specific translation evaluation for any target language. In: Proceedings of the Ninth Workshop on Statistical Machine Translation (2014)
53. Vedantam, R., Lawrence Zitnick, C., Parikh, D.: Cider: consensus-based image description evaluation. In: Proceedings of the IEEE Conference on Computer Vision and Pattern Recognition, pp. 4566–4575 (2015)
54. Torabi, A., Pal, C., Larochelle, H., Courville, A.: Using descriptive video services to create a large data source for video annotation research (2015). arXiv:1503.01070

Importance of AADHAR-Based Smartcard System's Implementation in Developing Countries

Kamta Nath Mishra

Abstract A smartcard includes fingerprints, iris, face, and palatal patterns where DNA sequence and palatal patterns of the smartcard will be used for identifying a dead person but fingerprints, iris, and face will be used identity verification of a living person. The smartcards are now widely being used as one of the most useful and reliable form of electronic identity verification system. By embedding biometrics in the host, we can formulate a reliable individual identification system as the biometrics possesses. Hence, the conflicts and problems related to the intellectual property rights protection can be potentially prevented. Consequently, it has been decided by governmental institutions in Europe and the U.S. to include digital biometric data in future ID documents. In India, biometric-based unique identification (UID) scheme called AADHAR is being implemented with the objective to issue a unique identification number to all the citizens of the country. This AADHAR number can be used in executing all the money transactions-related activities including all types of purchases, sales, money transfer, hotel bills, hospital expenses, and air tickets, etc. Therefore, the AADHAR-based smartcard system will help the South Asian countries for removing corruption and improving their economies.

Keywords Advanced security system · Biometric information
Human identification · Smartcard and UID

1 Introduction

The smartcards are now widely being used as one of the most useful and reliable form of electronic identity verification system. A smartcard includes fingerprints, iris, face, and palatal patterns where DNA sequence and palatal patterns of the smartcard will be used for identifying a dead person but fingerprints, iris, and face

K.N. Mishra (✉)
Department of CS&E, Birla Institute of Technology (Alld. Campus),
Mesra, Ranchi, India
e-mail: mishrakn@yahoo.com

A.E. Hassanien and D.A. Oliva (eds.), *Advances in Soft Computing and Machine Learning in Image Processing*, Studies in Computational Intelligence 730,
https://doi.org/10.1007/978-3-319-63754-9_20

will be used identity verification of a living person. The Figs. 1 and 2 are showing the structure and security levels of smartcard system [1, 2–4].

Some applications identities can be verified within a population of millions, e.g., thumbprints, palatal patterns, iris images, and face veins, etc. The biometric identification is lower effective error rate and potentially much faster in comparison to a visual comparison of signatures or photo IDs. This has forced the use of biometric technologies for government and industry-based applications [5–8].

To provide the highest level of security in Figs. 1 and 2 the ID verification-based biometric technique is considered to be most important in order to design a secure identification system. Biometric identifiers include digital fingerprints, retinal scans, hand geometry, facial characteristics, and Palatal patterns. Biometric scanning systems do not record the entire imprint of physical features of a person but only that portion, or "template" that should be time-invariant within some statistical limits [9–12]. Since, the body changes with time, the statistical algorithms must be elastic enough to match a stored image with a live scan of the person on later days. This feature will create limitations on the uniqueness of biometric images and we can solve this problem by using multiple images of the same person or we can use the combination biometric images other behavioral features for identifying a person [13, 14, 15].

A Smartcard Identification is a 12 digits string assigned to an entity that identifies an individual uniquely. It is a biometric identification system which ensures that each individual has assigned one and only one Smartcard. It will never generate two or more smartcards for any person. The smartcard is based on 12 digits unique

Fig. 1 Biometric collections for creating smartcard [1–4]

1(a). Finger print	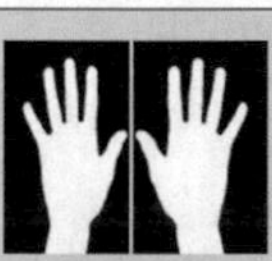
1(b). Iris	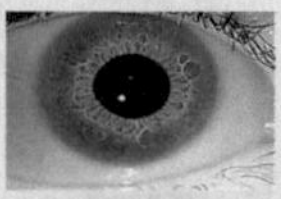
1(c). DNA	
1(d). Face	
1(e). Palatal	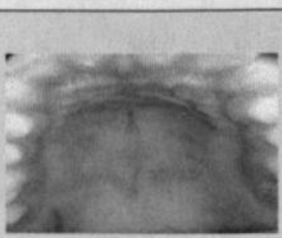

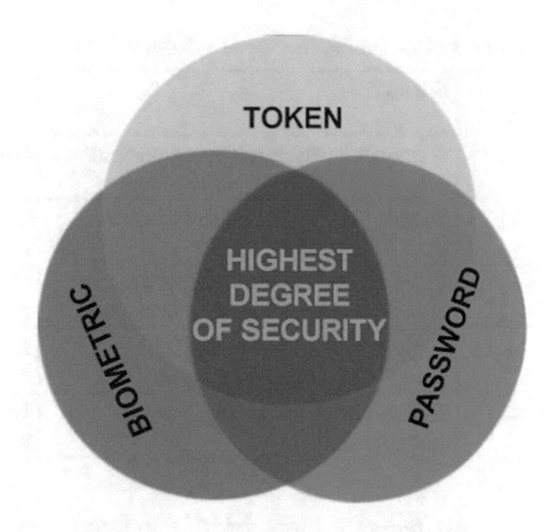

Fig. 2 Highest security level in smartcard-based system [11]

identification number. Smartcards can be used as the proof of address and identity and it can be used everywhere in our country. Each Smartcard number issued will be unique for an individual, and it will remain valid for the lifetime of a human and it will provide the access of banking services, mobile phone connections, advanced security system, etc. [16, 17, 18, 19].

The number generated for identification should be semantic free. The twelve decimal digits will be sufficient to satisfy all the requirements of personal identification in INDIA. The twelve-digit number will support 90–200 billion ID's at the outset. In other words we can say that there should be at least eleven digits available for actual numbering while generating unique ID of a person in INDIA. We should use reserved numbers for the purpose of error detection and corrections in the proposed system [20–23].

We propose a much larger space (80 billion IDs) than the maximum number of IDs needed at any time (estimated to be about no more than 6 billion IDs) for making it harder to speculate a correct valid ID assigned to anybody. This notion is captured by the *density*, which is defined as a ratio of total number of Identities assigned to the total number of identities available. Our target is s to keep this ratio under 0.05 always. The sparseness of using higher number of digits is to confuse the illegal hackers from knowing the exact number of digits used in the smartcard [11, 24].

The check digit is another form of sparseness. But, it will not prevent guessing because it is a publicly disclosed algorithm. The check digit enables error detection at data entry level. The financial transactions for MANREGA scheme may deposit MANREGA wages directly from a Gram Panchayat to a particular resident's account. This transaction could take the form of transfer money from Gram Panchayat UID to resident UID with the specified amount. The several other back end systems currently being used should be replaced by the described technique. This requirement can easily be fulfilled by reserving the value 1 for first digit to entity UIDs.

Table 1 Comparing the security features of a digital card and a smartcard with biometrics

S. no.	Smartcard	Smartcard with biometrics
1.	Automated inspection using readers	All attributes of smartcards with biometric templates
2.	Security markings and materials to help the user identification	Templates are stored on the smartcard for user authentication and enabling the terminal
3.	Cryptographic functionalities on card allows the protection of information	Counter fitting attempts are reduced due to enrollment process that verifies identity
4.	High trusted security-based information shared with users	Extremely high security and excellent user card verification approach
5.	High security and strong user authentication	Enhanced the security level authentication of reader

Most recently biometrics is merged into watermarking technology to improve the credibility of the conventional watermarking techniques. The access control and authenticity verification have been addressed by digital watermarking biometric authentication systems. By fitting biometrics in the host, we can formulate a reliable individual identification system as the biometrics possesses. Hence, the conflicts and problems related to the IPR (intellectual property rights) protection can be potentially prevented [9, 25]. Consequently, Decisions have been taken by government institutions of Europe and the U.S. to include digital biometric data in future ID documents. In India, biometric-based unique identification (UID) scheme called AADHAR is being implemented with the objective to issue a unique identification number to all the citizens of the country (Table 1).

2 Literature Review

Ali and Rameshwar [9] proposed the iris recognition algorithm in 1990's which showed that digital values of iris image can be used for identifying individuals and identical twins. His work included iris codes. In his work, Daugman proposed a methodology which consists of four stages. Here, 1-D Gabor filter is used for iris extraction and calculation of hamming distances. The Daugman's algorithm is not able to deal with noises like reflections on iris [5].

The researcher Pham, D.T. performed segmentation in two steps in his work. In the initial step, the researcher detected pupil center and radius of iris image. The pupil region is extracted through the binary thresh hold values. The edges of pupil are obtained with the help of two virtual orthogonal lines passing through centroid of region. In the next step, Pham [25] detected the edges of iris with the help of average window vector. The feature extraction was performed by Gabor filters [26] and matching through hamming distance. The recognition rate of Pham algorithm on UBIRISv1 came out to be 93.45% [25].

Decker et al. [27] described the segmentation and normalization process for automatic biometric iris recognition system and the algorithm is implemented USING Mat Lab. They used grayscale database images and performed Hough Transform for segmentation. The accuracy of propose systems of [28, 27] were from 78.5 to 85%.

Wang W. et al. focused on feature extraction and iris matching techniques, and they did not say anything about segmentation and normalization approaches [19, 29, 30]. In the area of human computer interaction, an ultimate goal is for machine to understand, communicate with and react to humans in natural ways. Although there are many other avenues to person identification—gait, clothing, hair, voice, and heights are all useful indication of identity of the person, none is as compelling as iris recognition [31, 29, 30]. The Anil K. Jain's study introduced the minutiae, which are local discontinuities in the ridge patterns as discriminating features and showed the uniqueness and permanence of minutiae [32]. According to Jain [32] fingerprint of a person is permanent, i.e., it preserves its characteristics and shape from birth to death. Fingerprint of individual is unique.

The fingerprints, iris, DNA, and palatal patterns-based features can be safely used by researchers, industries, and governments of this world for all types of transactions and ID verifications [29, 33–36].

3 Mathematical Model of Proposed System

Identification of an individual is most important issue for security of any organization. In this research work we actually focus on generating a unique ID and using it for identity verification of an individual. The biometric combinations which we have used in this work for generating unique ID are: Face Image, Iris Image, Finger Prints, Palatal Pattern, and DNA sequencing. The biometric information will helps to identify the individuals. Most people in the world have no physical disability but some time we will have to make the unique identification card of blind person, person who is physical disable. Physically disabled person are also the part of our society and they need security more than the normal person therefore we have designed a new model using the Automata theory for the uniqueness for the each individual person.

Table 2 is reprinting the categories of persons according to their biometric information. In Table 2, category string is an input string of the automata machine, e.g., "YYNYYYY." There are 16 possible categories of strings for a person. Each string of a character of Table 2 has some meaning. These meanings are described in Table 3. Table 3 is consisting of possible combination of string described in Table 2 using biometric information. Table 4 represents the numeric conversion character 'Y' and 'N.'

To mathematical model of our proposed work is modeled using a Non-Deterministic Finite Automata (NDFA) which consists of 5 tuple or quintuple and it is explained using Eq. (1).

Table 2 Categorizing the group of persons on the basis of biometric information

S. no.	Characteristic	Description
1.	YYYYYYY	Normal person
2.	YNYYYYY	Blind (only left eye)
3.	YNYNYYY	Blind (only eye) and handicapped (left hand)
4.	YNYYNYY	Blind (only left eye) and handicapped (right hand)
5.	YNYNNYY	Blind (only left eye) and handicapped (left hand and right hand)
6.	YYNYYYY	Blind (only right eye)
7.	YYNNYYY	Blind (only right eye) and handicapped (only left hand)
8.	YYNYNYY	Blind (only right eye) and handicapped (only right hand)
9.	YYNNNYY	Blind (only right eye) and handicapped (both hands)
10.	YNNYYYY	Blind (both eyes)
11.	YNNNYYY	Blind (both eyes) and handicapped (only left hand)
12.	YNNYNYY	Blind (both eyes) and handicapped (only right hand)
13.	YNNNNYY	Blind (both eyes) and handicapped (both hands)
14.	YYYNYYY	Handicapped (only left hand)
15.	YYYYNYY	Handicapped (only right hand)
16.	YYYNNYY	Handicapped (both hands)

Table 3 Possible strings using biometric information

S. no.	F	L_e	R_e	L_f	R_f	P_l	D
1.	Y	Y	Y	Y	Y	Y	Y
2.	Y	N	Y	Y	Y	Y	Y
3.	Y	N	Y	N	Y	Y	Y
4.	Y	N	Y	Y	N	Y	Y
5.	Y	N	Y	N	N	Y	Y
6.	Y	Y	N	Y	Y	Y	Y
7.	Y	Y	N	N	Y	Y	Y
8.	Y	Y	N	Y	N	Y	Y
9.	Y	Y	N	N	N	Y	Y
10.	Y	N	N	Y	Y	Y	Y
11.	Y	N	N	N	Y	Y	Y
12.	Y	N	N	Y	N	Y	Y
13.	Y	N	N	N	N	Y	Y
14.	Y	Y	Y	N	Y	Y	Y
15.	Y	Y	Y	Y	N	Y	Y
16.	Y	Y	Y	N	N	Y	Y

$$M = \left(\mathrm{Q},\ \mathrm{I},\ Q_0, Q_f,\ \mathrm{T}\ \right) \tag{1}$$

Equation (1) represents a NDFA where M represents automaton machine, 'Q' is the non-empty set of states. 'I' is non-empty set of input symbol. 'Q_0' is initial state of input symbol which belongs to the set of state (Q). The 'Q_F' is the final state and

Table 4 Numerical values for biometric information

S. no.	F	L_e	R_e	L_f	R_f	P_l	D
1.	1	1	1	1	1	1	1
2.	1	0	1	1	1	1	1
3.	1	0	1	0	1	1	1
4.	1	0	1	1	0	1	1
5.	1	0	1	0	0	1	1
6.	1	1	0	1	1	1	1
7.	1	1	0	0	1	1	1
8.	1	1	0	1	0	1	1
9.	1	1	0	0	0	1	1
10.	1	0	0	1	1	1	1
11.	1	0	0	0	1	1	1
12.	1	0	0	1	0	1	1
13.	1	0	0	0	0	1	1
14.	1	1	Y	0	1	1	1
15.	1	1	1	1	0	1	1
16.	1	1	1	0	0	1	1

it also belongs to the set of state (Q) and 'T' is the representation of the transition system which have 2^Q possible moves.

In the process of producing the uniqueness of each person, we need to collect the biometric information which will be represented as the states.

$$Q = \{F, Le, R_e, Lf, Rf, Pl, D\} \tag{2}$$

In Eq. (2), 'Q' is the non-empty set of states which consists of 'F' (face), 'L_e' (left eye), 'R_e' (right eye), 'L_f' (left finger), 'R_f' (right finger), 'P_l' (palatal rugae pattern) and 'D' (DNA sequence).

The combinations 'Y' and 'N' strings can be represented by Eq. (3).

$$I = \{1, 2, 3, 4, 5\} \tag{3}$$

In Eq. (3), '1' represents 'Y,' '2' represents 'NY,' '3' represents 'NNY,' '4' represents 'NNNY' and '5' represents 'NNNNY.' Therefore, we can say that:

$$Q_0 = \{F\} \tag{4}$$

$$Q_f = \{D\} \tag{5}$$

$$T = Q \times I \tag{6}$$

In Table 5 we have shown the methods and combinations of using UID technique for biometric information of an individual. In this technique we have collected five different categories of biometric data (face, palatal, eyes, fingerprints, DNA) of an individual and we compiled these data with the help of Fig. 3.

Table 5 Transaction table

	1	2	3	4	5
F	L_e	R_e	L_f	R_f	P_l
L_e	R_e	L_f	R_f	P_l	–
R_e	L_f	R_f	P_l	–	–
L_f	R_f	Pl	–	–	–
R_f	P_l	–	–	–	–
P_l	D	–	–	–	–
D	–	–	–	–	–

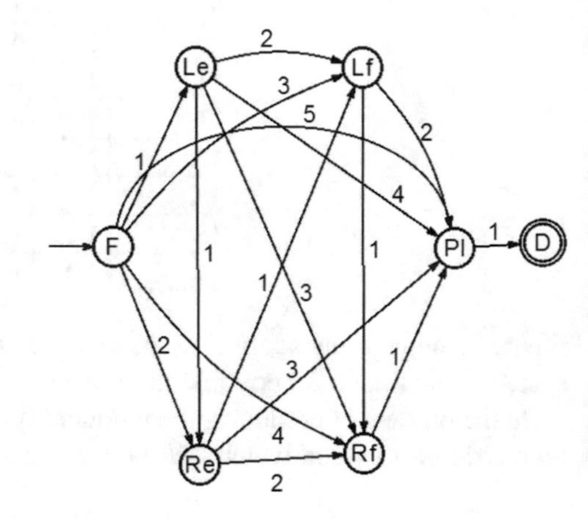

Fig. 3 Transition diagram

4 Proposed Technique

4.1 Algorithms for Code Generation and Enrollment Process

The face, left eye, right eye, left finger, right finger, and palatal patterns are given as input to the proposed code generation algorithm of Fig. 4 and the obtained output is presented in Fig. 5.

The Fig. 4 illustrates the enrollment process of AADHAR-based smartcard technology for overall management of everything related to daily life. In this process the biometric sample of an individual is captured using sensors or cameras and unique features are obtained from the biometric sample image. These biometric templates are stored in the database of a machine or computer system for further matching and usage.

INPUT: *Original Biometrics.*
OUTPUT: *Respective biometric templates*

Algorithm_Code_Generation (Code Generation)
{
1. Input the original biometric sample.
2. Convert the inserted original biometric sample into standard sample (remove noise).
3. Crop standard sample into fixed size during enrollment process. (According to sample of biometrics).
4. Again convert cropped sample into standard sample.
5. Convert those Standard biometric samples into their respected templates.
6: Store those templates in our hard disk or on the machine readable ID card.
}

Fig. 4 Code generation algorithm using face, iris, palatal, and DNA sequences

INPUT: *New biometric sample.*
OUTPUT: *Numeric matching score*

Algorithm_Code_Matching (Code Matching)
{
1. Input a new biometric sample.
2. Convert the inserted new biometric sample into standard sample (remove noise) and crop the image during the enrollment process.
3. Again convert cropped sample into standard sample.
4. Convert those Standard biometric samples into their respective templates and stored as a live templates.
5. Match the live templates with the previously stored templates and generate numeric matching score.
6. Display the results according to numeric score.
}

Fig. 5 Code matching algorithm for identifying a person

4.2 Algorithm for Code Matching

The input is given to code matching algorithm and the outputs obtained from the algorithm are presented by Fig. 5.

The Figs. 6 and 7 illustrate the enrollment and matching processes of biometric sample. The biometric sample is either offline or online obtained for the implementation of matching technique. The unique features are obtained from the biometric feature of the person to create the user's "alive" biometric template. This new template will now be compared with the stored template to verify the identity of a person. The system developers have determined the lower and higher limit values which are required for identity verification using iris image, fingerprints, palatal patterns, face image, and DNA sequences.

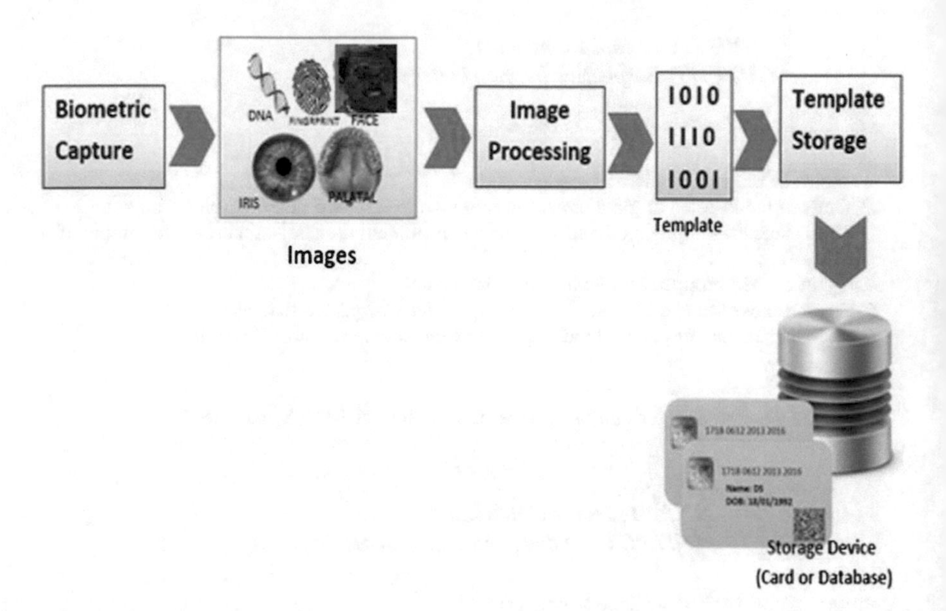

Fig. 6 Enrollment process of biometric sample

Table 6 Selection of biometric in the proposed method

	Finger prints	Face	Iris	Palatal	DNA
Structure of biometrics					
Performance	High	Medium	High	High	High
Acceptability	Medium	High	Low	Low	Low
Accuracy	High	Medium	Medium	High	High
Durability	High	Medium	Medium	High	High
Universality	Medium	High	High	Medium	High
Security level	High	Medium	High	High	High

5 Selection of Biometric Technology

The selection of most appropriate biometric technology will depend on a number of application-oriented factors like accuracy, false acceptance rate, false rejection rate, reliability, and acceptability.

Table 6 shows a comparison of different biometric technologies which are used in our proposed method and their performances are compared with several other biometric features for different factors like performance, acceptability, accuracy, and security level, etc..

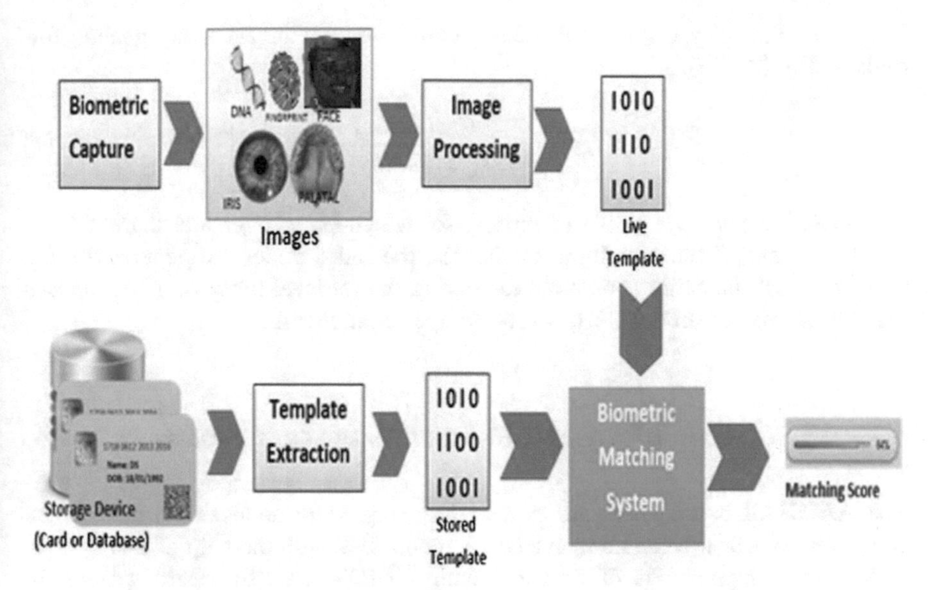

Fig. 7 Matching process of biometric sample

6 Result Analysis and Discussions

To analyze the performance of our AADHAR-based smartcard system we developed a software tool which uses face, iris, palatal, and DNA sequences for generating the unique ID and this unique ID will be used in INDIA and other South Asian countries for removing corruption and bringing prosperity. We created single user (single modularity) and multiuser (multi-modularity)-based system for testing the performance of proposed system. The indexing method used in testing of software tool which is developed using for different biometric features is described as under.

6.1 *Indexing Method*

In this section we have used the iris modality as the case of an example. However, the inferred properties are applicable to the fingerprint modality (as observed in our experiments) and perhaps for other biometric modalities as well.

During identification, the indexing system first computes the index code, Sx, of the input x and then it outputs every enrolled identity 'y,' for which the correlation coefficient between Sy and Sx exceeds a specific threshold value.

The algorithm for obtaining candidate identities can be represented in following steps:

Step I: Let Sx be the index code of the input image, Sy be the index code of an image y from the database, and T be a predefined threshold.

Step II: For all y Compute Pearson's correlation coefficient using general formula of Eq. (7) [37].

$$p(x,y) = \frac{Cov(Sx, Sy)}{[Var(Sx)Var(Sy)]1/2} \tag{7}$$

Step III: *Display the o*utput of those y for which (x, y) is greater than 'T.'

At the time of using multiple modalities, the index codes are generated separately for each modality and combined during the retrieval process. The proposed method of this research work has low storage requirements.

6.2 Applications of AADHAR-Based Smartcard System

The AADHAR-based smartcard system has very wide applications and almost everything of our life can be managed and controlled with the help of this system. The general applications of unique identity (UID)-based smartcard system for managing our day-to-day life-related transactions are presented in Figs. 8 and 9.

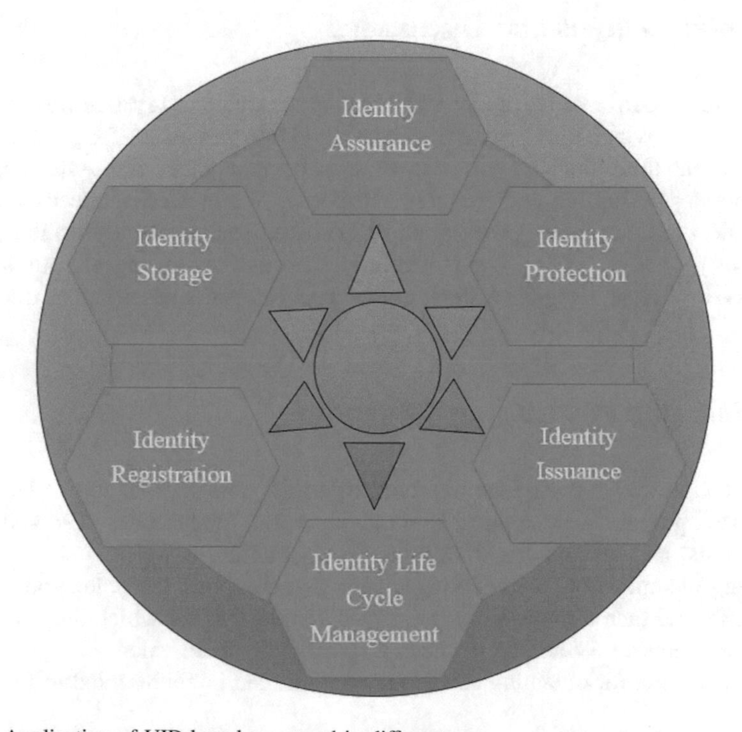

Fig. 8 Application of UID-based smartcard in different area

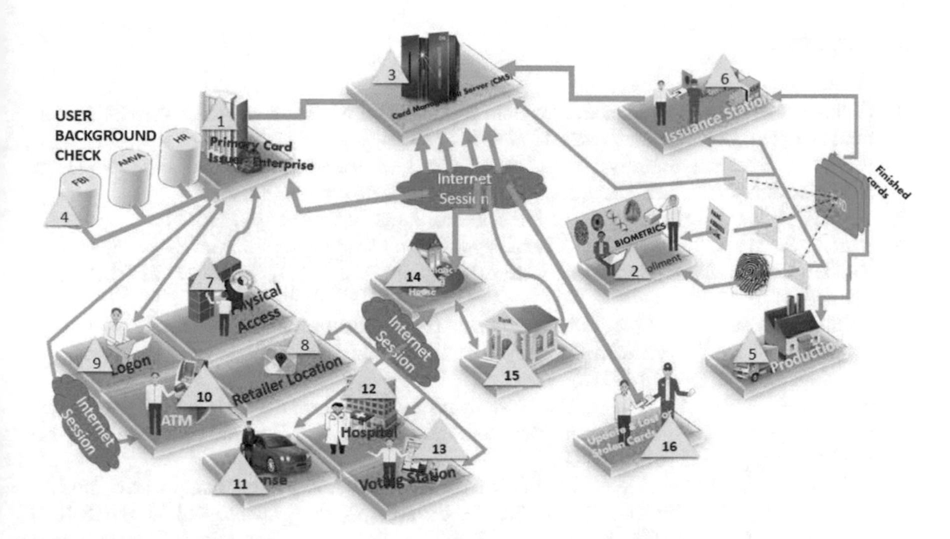

Fig. 9 Different applications of AADHAR-based smartcard system [24, 36, 38]

7 Conclusions

In this research work AADHAR(unique ID)-based smartcard system is presented which uses face, iris (left and right), and palatal features for identifying a living person and DNA sequence with palatal patterns is used for identifying a dead person. The proposed system has very wide applications and it can be used in all types of money transactions which are related to our daily life.

The implementation of AADHAR-based system in our daily life ease many things of our life. It will reduce the corruption level of our society and we can control almost all types of transactions related to the life of a person can be monitored and controlled. It has been decided by governmental institutions in Europe and the U.S. to include digital biometric data in future ID documents [32, 7, 22]. In India, biometric-based UID scheme called AADHAR is started with the goal to issue a unique identification number to all the citizens of INDIA. This AADHAR number can be used in executing all the money transactions-related activities including all types of purchases, sales, money transfer, hotel bills, hospital expenses, and air tickets, etc. Therefore, the AADHAR-based smartcard system will help the South Asian countries in coming out of corruptions and improving their economies.

References

1. Gillberg, J., Ljung, L.: Frequency-domain identification of continuous-time ARMA models from sampled data. Automatic **45**, 1371–1378 (2009)
2. Pillai, J., Shreeramamurthy, T.G.: A public key water making scheme for image authentication. Int. J. Ind. Electron. Electr. Eng. **2**(5), 46–48 (2014)
3. Waleed, J., Jun, H.D., Hameed, S.: Int. J. Secur. Appl. **8**(5), 349–360 (2014)
4. Kandar, S., Maiti, A., Dhara, C.: Visual cryptography scheme for color image using random number with enveloping by digital watermarking. Int. J. Comput. Sci. Issues **2011**(8), 543–549 (2011)
5. Anitha, V., Velusamy, R.L.: Authentication of digital documents using secret key biometric watermarking. In: International Conference on Electrical Engineering and Computer Science, Trivandrum, pp. 243–249 (2012)
6. Nyeem, H., Boles, W., Boyd, C.: Developing a digital image watermarking model. In: Proceedings of IEEE International Conference on Digital Image Computing: Techniques and Applications, pp. 468–473 (2011)
7. Phillips, P.C.B.: The problem of identification in finite parameter continuous time models. J. Econometric **1**(4), 351–362 (1973). http://EconPapers.repec.org/. Accessed 10 March 2015
8. Patel, B.C., Sinha, G.R.: An adaptive K-means clustering algorithm for breast image segmentation. Int. J. Comput. Appl. **10**(4), 35–38 (2010)
9. Ali, M.O., Rameshwar, R.: Digital image watermarking basics and hardware implementation. Int. J. Model. Optim. **2**(1), 19–24 (2012)
10. Karthikeyan, V., Divya, M., Chithra, C.K., PriyaManju, K.: Face verification systems based on integral normalized for. Int. J. Comput. Appl. **66**(9), 39–43 (2013)
11. Hagai, K.: On the unique identification of continuous time autoregressive models from sampled data. IEEE Trans. Signal Proces. **62**(6), 1361–1376 (2014)
12. Hemangi, K., Aniket, Y., Darpan, S., Bhandari, P., Samuya, M., Yadav, A.: Unique id management. Int. J. Comput. Technol. Appl. **3**(2), 520–524 (2012)
13. Inamdar, V., Rege, P., Arya, M.: Offline handwritten signature based blind biometric watermarking and authentication technique using bi-orthogonal wavelet transform. Int. J. Comput. Appl. (0975–8887) **11**, 19–27 (2010)
14. Hemant, K., Nadhamuni, S., Sarma, S.: A UID numbering scheme. White Paper, May 2010
15. Liu, F., Wu, K.: Robust visual cryptography-based watermarking scheme for multiple cover images and multiple owners. IET Inf. Secur. **5**(2), 121–128 (2011)
16. Velusamy, V., Leela, R.L.: Authentication of digital documents using secret key biometric watermarking. Int. J. Commun. Netw. Secur. **1**(4), 5–11 (2012)
17. Corduneanu, C.: Almost Periodic Oscillations and Waves. Springer, New York (2009). doi: 10.1007/978-0-387-09819-7. Accessed 15 Jan 2015
18. Hassanien, E.: Hiding iris data for authentication of digital images using wavelet theory. In: GVIP 05 Conference, Egypt, December 2005
19. Jafri, R., Arabnia, H.R.: A survey of face recognition technique. J. Inf. Proces. Syst. **5**(2), 41–68 (2009)
20. Patel, H., Sharma, V.: Fingerprint recognition by minutiae matching method for evaluating accuracy. Int. J. Eng. Trends Technol. **4**(5) (2013)
21. Prashar, D., Kaur, M.: Human eye iris recognition using discrete 2D reverse bi-orthogonal wavelet. Int. J. Sci. Technol. Res. **3**(8), 266–270 (2014)
22. Johansson, R.: Identification of continuous-time models. IEEE Trans. Signal Proces. **42**(4), 887–897 (1994)
23. Ramya, K.P., Revathi, M.K., Devi, C.R.: PINCODE: protection in provenance conduction over data stream for sensor data. Int. J. Res. Eng. Adv. Technol. **1**(5), 1–7 (2013)
24. Wikipedia: The free encyclopedia, Unique identification authority of India. http://en.wikipedia.org/wiki/Unique_Identification_Authority. 10 Jan 2016

25. Rashid, A.F., Lateef, M., Kaur, B., Aggarwal, O.P., Hamid, S., Gupta, N.: Biometric fingerprint identification is reliable tool are not. J. Indian Acad. Forensic Med. **35**(2), 109–112 (2013)
26. Pham, D.-T.: Estimation of continuous-time autoregressive model from finely sampled data. IEEE Trans. Signal Process. **48**(9), 2576–2584 (2000)
27. Decker, W., Greuel, G.-M., Pfister, G., Schönemann, H.: Singular 3-1-6—a computer algebra system for polynomial computations (2012). http://www.singular.uni-kl.de. Accessed 15 Jan 2015
28. Cox, D.A., Little, J., O'Shea, D.: Ideals, Varieties, and Algorithms: An Introduction to Computational Algebraic Geometry and Commutative Algebra, vol. 10. New York (2007)
29. Ramaswamy, G., Vuda, S., Ramesh, P., Ravi, K.: A novel approach for human identification through fingerprints. Int. J. Comput. Appl. **4**(3), 43–50 (2010)
30. Wong, W., Memon, N.: Secret and public key image watermarking schemes for image authentication and ownership verification. IEEE Trans. Image Proces. **10**(10) (2001)
31. Kekre, H.B., Thepade S.D., Jain, J., Naman, A.: Iris recognition using texture feature extracted from haalet pyramid. Int. J. Comput. Appl. **11**(12), 1–5 (2010)
32. Jain, A.K., Kumar, A.: Biometric of next generation. Book on Second Generation Biometrics, pp. 1–36 (2010)
33. Lokhande, S., Dhongde, V.S.: Fingerprint identification based on neural network. Int. J. Innovative Res. Sci. Eng. Technol. **3**(4), 350–355 (2014)
34. Shakya Kumar, M.: Catching of stolen vehicles with unique identification code using embedded system. Int. J. Electron. Commun. Comput. Technol. **2**(6), 262–266 (2012)
35. Tiwari, S., Singh, A.K., Shukla, V.P.: Statically moments based noise classification using feed forward back propagation neural network. Int. J. Comput. Appl. **18**(2), 36–40 (2011)
36. Venkateswara Rao, M.: Fingerprint analyze. J. Nurs. Health Sci. **3**(1), 14–16 (2014)
37. Marelli, D., Fu, M.: A continuous-time linear system identification method for slowly sampled data. IEEE Trans. Signal Proces. **58**(5), 2521–2533 (2010)
38. Mishra, K.N.: AAdhar based smartcard system for security management in South Asia. Int. Conf. Cont. Comput. Commun. Mater., 1–6 (2016)

Part IV
Object Analysis and Recognition in Digital Images

A Nonlinear Appearance Model for Age Progression

Ali Maina Bukar and Hassan Ugail

Abstract Recently, automatic age progression has gained popularity due to its numerous applications. Among these is the search for missing people, in the UK alone up to 300,000 people are reported missing every year. Although many algorithms have been proposed, most of the methods are affected by image noise, illumination variations, and most importantly facial expressions. To this end we propose to build an age progression framework that utilizes image de-noising and expression normalizing capabilities of kernel principal component analysis (Kernel PCA). Here, Kernel PCA a nonlinear form of PCA that explores higher order correlations between input variables is used to build a model that captures the shape and texture variations of the human face. The extracted facial features are then used to perform age progression via a regression procedure. To evaluate the performance of the framework, rigorous tests are conducted on the FGNET ageing database. Furthermore, the proposed algorithm is used to progress image of Mary Boyle; a 6-year-old that went missing over 39 years ago, she is considered Ireland's youngest missing person. The algorithm presented in this paper could potentially aid, among other applications, the search for missing people worldwide.

Keywords Age progression • Age synthesis • Kernel appearance model
Linear regression • Kernel PCA • Kernel preimage • Mary Boyle

1 Introduction

The human face carries a large amount of information. Faces are used as cues for recognizing identities [1]. Changing facial expressions can reveal underlying emotions [2], and facial structures can even reveal details of developmental disorders [3]. It has been well documented that the shape of the face changes sub-

A.M. Bukar (✉) · H. Ugail
Centre for Visual Computing, University of Bradford, Richmond Road, Bradford BD7 1DP, UK
e-mail: ambukar@student.bradford.ac.uk

A.E. Hassanien and D.A. Oliva (eds.), *Advances in Soft Computing and Machine Learning in Image Processing*, Studies in Computational Intelligence 730,
https://doi.org/10.1007/978-3-319-63754-9_21

stantially from birth to adulthood [4]. During adulthood, while the shape remains relatively constant, there are changes in the facial 'texture' [1]. Consequently, several computing algorithms have been proposed to perform various facial analysis tasks, including but not limited to face recognition, emotion analysis, kinship recognition, and age progression.

In the past two decades, facial age progression has gained more research interest [5]. This is due its numerous practical applications, the most obvious been the search for missing people. Missing person cases are rampant across the globe, in the UK alone, the police record up to 300,000 missing person cases every year [6]; popular examples are the disappearance of Madeleine McCann [7] and Ben Needham [8]. In Ireland, the case of Mary Boyle is considered one of the longest missing person's case. The 6-year-old Irish girl went missing from her grandparent's farm near Ballyshannon, County Donegal, Ireland in March, 1977. The Police have since closed the case, however, the fact that no trace has been found, has left her family most especially her twin sister (Ann Doherty) asking questions, and hoping she will be found someday. Motivated by several missing person cases, this paper introduces a novel approach to facial age progression.

Given an image of a person at a particular age, the aim of age progression is to re-render the facial appearance at a different age, with natural ageing effects. Typically, this procedure is used to age or de-age faces. Starting with an image at age a, the age is altered to b and a new face is generated such that it looks like the person at the target age. Multiple approaches to facial synthesis have been documented in the graphics literature [1, 9].

The approach we have taken to solving this problem is to revisit the classical method of [10], and to improve it by embedding kernel PCA into the model, in order to tackle the problem of image noise, lightening variations, and above all effect of facial expressions. Our contributions include the following:

(1) Development of a nonlinear model that is used to parameterize the face shape and texture. The proposed model is an improvement of the classical active appearance model (AAM) [11], which assumes that faces conform to a multivariate Gaussian distribution. We term this approach Kernel Appearance Model (KAM) and show this model is better suited for the problem of age progression than the conventional AAM.
(2) Formulation of an algebraic procedure of rendering age-progressed faces using the proposed KAM. We further show that our approach is robust to noise, illumination variations, and facial expressions contained within the test images.

2 Related Work

In this section, succinct reviews of age progression and related algorithms are presented.

Age Progression Literature

The conventional method of facial age progression is via artistic sketch produced by forensic artist. Normally, an image is produced manually by hand using the subject's image, their relatives' photos, as well as additional information related to external factors such as life style. A first step towards automation is the use of computer-based sketches; this method is still employed by most police departments around the world. However, this method also requires the experience and knowledge of an artist [12].

In Computer Vision, automatic age progression has been approached through the use of geometric, texture-specific, and appearance-based methods. Geometric techniques define craniofacial growth models [4, 13], by manipulating the facial anthropometric features. An obvious setback of this approach is the fact that it does not take into account the facial texture such as skin tautness as well as wrinkles [2].

Texture models concentrate on manipulating the facial skin, and so wrinkles have been used to generate photo realistic aged faces. Usually this is achieved either by transferring wrinkles from old faces onto young faces, or via independent construction of skin models [1]. The construction of 3D wrinkles has also been reported in the literature [14]. Unfortunately, texture-specific models do not take anthropometric variations into consideration.

Appearance-based approaches model both shape and texture variations. For example, [15] proposed an image-based model that computes average faces for different age groups, then the difference between the target age group and the current age group is computed and added to the subject's image via caricaturing. The same idea has been extended for unconstrained images in [16]. However, AAM is the most popular model used by researchers [10, 12, 17, 18] to achieve appearance-based progression.

Recently, recurrent neural networks [19] have been used to conduct age progression, however this method requires huge amount of labelled data. As observed by Panis et al. [5], age synthesis algorithms are yet to reach the stage of rendering highly accurate predictions. This is attributed to a peculiar problem the aforementioned techniques have in common, i.e. they are affected by facial expressions, as well as poor image quality.

Related Techniques

Active Appearance Model (AAM)

AAM is a statistical technique that captures shape and texture variability from a training dataset [20]. A parameterized model is formed by using PCA to combine shape and texture models. The method provides a means of representing images in terms of a few model parameters giving an avenue for image modelling, interpretation, and description.

To compute an AAM, a shape model is made by labelling facial landmarks for each face in the image dataset, these are then represented as a single vector for each face. Then, PCA is performed on these vectors. Each shape vector can be approximated using a linear equation given by

$$\boldsymbol{s} = \bar{\boldsymbol{s}} + \boldsymbol{P}_s \boldsymbol{b}_S, \tag{1}$$

where $\bar{\boldsymbol{s}}$ is the mean shape, $\boldsymbol{P}_s$ is a matrix containing the eigenvectors of the covariance matrix of the shape vectors, i.e. orthogonal mode of variations, and $\boldsymbol{b}_s$ is a vector of shape parameters [20].

A texture model is constructed by forming a "shape-free-patch" $\boldsymbol{g}$. This involves warping the images to the mean shape $\bar{\boldsymbol{s}}$. Then, another linear equation is formed by performing PCA, thus the image pixels $\boldsymbol{g}$, can be expressed as

$$\boldsymbol{g} = \bar{\boldsymbol{g}} + \boldsymbol{P}_g \boldsymbol{b}_g \tag{2}$$

where $\bar{\boldsymbol{g}}$ is the mean of the normalised texture vector, and $\boldsymbol{P}_g$ and $\boldsymbol{b}_g$ are the eigenvectors and texture parameters respectively. The shape and texture parameters can then be described using the model parameters $\boldsymbol{b}_s$ and $\boldsymbol{b}_g$ to generate novel faces.

Finally, the active appearance model is formed by concatenating the two models in a single vector. A third PCA is then applied to the combined vector in order to decorrelate the shape and texture parameters. Thus, the combined model can be approximated using another linear equation given by

$$\boldsymbol{b} = \boldsymbol{P}\boldsymbol{c}, \tag{3}$$

$\boldsymbol{P}$ is a matrix of orthogonal modes of variation, and $\boldsymbol{c}$ is the vector of AAM parameters. The appearance parameter for a new image, one that has not been used to compute the model, can be computed using,

$$\boldsymbol{c} = \boldsymbol{P}^T \boldsymbol{b}. \tag{4}$$

AAM has proven to be a powerful tool for modelling deformable objects, hence, has been used extensively for diverse applications such as medical image segmentation, object tracking, face recognition, expression recognition, facial age progression [21]. However, PCA which is at the core of the model assumes both shape and texture exhibit linear variation. This assumption is often not true in reality, the method therefore produces sub-optimal performance when the object displays nonlinear variance [22]. Moreover, the texture model computed using PCA assumes a Gaussian distribution even though it is well known that illumination and other external factors invalidate this assumption and therefore can greatly affect its performance [22]. To this end, various improvements have been proposed, these have been discussed in a comprehensive review by Gao et al. [23].

Kernel Machines

Kernel machines have gained popularity in the past two decades due to their computational efficiency when applied to statistical learning theory, signal processing and machine learning in general. The kernel trick, first proposed by Aizeman et al. [24], is a key concept in the development of kernel machines. This approach is used to map data from the original input space X to a higher dimensional (nonlinear) Hilbert space $\mathcal{H}$ given by $\phi\colon \mathrm{X} \to \mathcal{H}$.

The transformation is not done explicitly, it is rather computed from the dot product of the data [25], which in turn is achieved by replacing the inner product operator with a symmetric Hermitian "Kernel" function.

One common application of kernel machines is the kernel principal component analysis (KPCA) that was proposed by Schölkopf et al. [26]. Using kernel methods, [26] generalised PCA into a higher order correlation between input variables. KPCA entails a nonlinear mapping ϕ of data from an input space x into a feature space F and then the computation of conventional PCA in the F space. As stated earlier, the transformation from x to higher dimension is realized using the kernel trick [24].

Given a kernel function $\boldsymbol{K}$ expressed as

$$K(x_i, x_j) = \phi(x_i)^T \phi(x_j), \tag{5}$$

KPCA involves solving the eigenvalue problem

$$\boldsymbol{K}\alpha = N\lambda\alpha, \tag{6}$$

where $\alpha = [\alpha_1, \ldots \alpha_l]$ are set of eigenvectors of $\boldsymbol{K}$, for $\lambda \geq 0$ the set of first k eigenvectors of $\boldsymbol{K}$ can be normalized such that $\boldsymbol{V}^k.\boldsymbol{V}^k = 1$

For the purpose of principal components β_k extraction, the eigenvectors $\boldsymbol{V}^k$ are projected onto the data in F. Assuming x is a test data with image $\phi(\boldsymbol{x})$, its projection is given by

$$\beta_k = \phi(x)^T.\boldsymbol{V}^k = \sum_{i=1}^{l} \alpha_i^k.(\phi(\boldsymbol{x}_i).\phi(\boldsymbol{x})) \tag{7}$$

To ensure the mapped data $\phi(x)$ has a zero mean, centring can be achieved by replacing $\boldsymbol{K}$ with a gram matrix $\widetilde{\boldsymbol{K}}$,

$$\widetilde{\boldsymbol{K}} = \boldsymbol{K} - 1_N\boldsymbol{K} - \boldsymbol{K}1_N + 1_N\boldsymbol{K}1_N \tag{8}$$

where $1_N = \boldsymbol{I}_N - \boldsymbol{J}_N$, $\boldsymbol{I}_N$ is the identity matrix and $\boldsymbol{J}_N$ is an $N \times N$ matrix whose elements are all 1 s.

KPCA has been used in a variety of computer vision applications, and in most cases has proven to outperform the conventional PCA [27]. Furthermore, several research works have shown the power of Kernel PCA in conducting image de-noising [25, 28, 29], illumination normalization, occlusion recovery, as well as facial expression normalization [30].

In this work, we approach the problem of age progression by revisiting the work of Lanitis et al. [10] by integrating KPCA into the conventional appearance model. Thus, creating a Kernel-based Appearance Model (KAM) for extracting facial features. Age progression conducted in kernel space requires a pre-image calculation, this presents us with an avenue for utilising KPCA's image denoising and expression normalization capabilities.

3 Kernel Appearance Model (KAM)

The proposed KAM captures nonlinear shape and texture variabilities from the training dataset. Face shapes are represented by a set of annotated landmarks given by a 2 dimensional vector expressed as

$$\boldsymbol{s} = (x_1, x_2, \ldots, x_n, y_1, y_2, \ldots, y_n)^T. \tag{9}$$

Initially all 2-dimensional vectors representing the face shape are aligned. Then, a nonlinear shape model is built by performing KPCA; $\boldsymbol{s}$ is mapped to the higher dimensional feature space using a kernel method $\boldsymbol{K}_s$.

The mapped data is then centralised using (8),

$$\tilde{\boldsymbol{K}}_s = \boldsymbol{K}_s - 1_N \boldsymbol{K}_s - \boldsymbol{K}_s 1_N + 1_N \boldsymbol{K}_s 1_N. \tag{9}$$

Subsequently, eigen-decomposition is performed using

$$\tilde{\boldsymbol{K}}_s \alpha_s = N \lambda_s \alpha_s, \tag{10}$$

where α_s are the set of eigenvectors and λ_s their corresponding eigenvalues. Consequently, the shape of each face in the training set can be defined by

$$\rho_s = \sum_{i=1}^{N} \alpha_{si}^k \cdot \tilde{\boldsymbol{K}}(\boldsymbol{s}_i, \boldsymbol{s}). \tag{11}$$

To build a nonlinear texture model, first all face images are affine warped to a template shape, this is done to remove unnecessary face size variations. Then, illumination effects are normalized by applying a scaling and an offset to the image pixels $\boldsymbol{g}$ [11]. Subsequently KPCA is applied to $\boldsymbol{g}$ by repeating the procedure described for obtaining nonlinear shape variations. Hence, the texture of each face in the training set can be defined by

$$\tilde{\boldsymbol{K}}_g = \boldsymbol{K}_g - 1_N \boldsymbol{K}_g - \boldsymbol{K}_g 1_N + 1_N \boldsymbol{K}_g 1_N, \tilde{\boldsymbol{K}}_g \alpha_g = N \lambda_g \alpha_g, \rho_g = \sum_{i=1}^{N} \alpha_{gi}^k \cdot \tilde{\boldsymbol{K}}(\boldsymbol{g}_i, \boldsymbol{g}), \tag{12}$$

where ρ_g is a parameter that describes the texture. We can then combine the face shape and texture information into a single appearance model expressed as

$$\rho_a = (\rho_s, \rho_g)^T. \tag{13}$$

Next, conventional PCA is used to reduce the dimension of the combined model. This gives us a linear equation that binds both shape and texture variations into a single parameter $\boldsymbol{f}$,

$$\rho_a = \boldsymbol{Pf}. \tag{14}$$

$\boldsymbol{P}$ is a matrix of eigenvectors associated with both shape and texture. Due to the linear nature of (14), we can reconstruct the shape and texture variations from the appearance parameter $\boldsymbol{f}$,

$$\rho_s = \boldsymbol{P}_{k\_s}\boldsymbol{f},\ \rho_g = \boldsymbol{P}_{k\_g}\boldsymbol{f}, \tag{15}$$

where $\boldsymbol{P}_k = (\boldsymbol{P}_{k_s}, \boldsymbol{P}_{k_g})^T$.

We have now defined a nonlinear extension of the AAM. A single parameter $\boldsymbol{f}$ encodes the shape and texture of the face. Hence, manipulating $\boldsymbol{f}$ results to changes in the face shape and texture. Thus, the derived appearance variable can be used as a parameterised abstraction of the human face.

4 Framework for Age Progression

Since, we now have a nonlinear model that can be used to represent the face, it is then possible to define a linear ageing function that captures the relationship between face appearance and the human ages. This is defined as

$$age = \boldsymbol{f}^T \beta, \tag{16}$$

where β is a vector of regression coeffients.

Once the value of β and $\boldsymbol{f}$ are known, a person's age can be estimated. The reverse is also possible, i.e. for a given value of β the facial appearance at a specific age can be computed. Thus, inverting Eq. (16) is a key part of our age progression framework (shown in Fig. 1). In order to achieve age progression, KAM features $\boldsymbol{f}$ of a face at the current age act as the input to the framework.

Step I: KAM parameters $\hat{f}$ corresponding to the new age *tnew* are computed by inverting the ageing function defined in Eq. (16). First, the Moore–Penrose pseudoinverse † is computed as

$$\hat{f}_{age} = \beta^{\dagger} age_{tnew}. \tag{17}$$

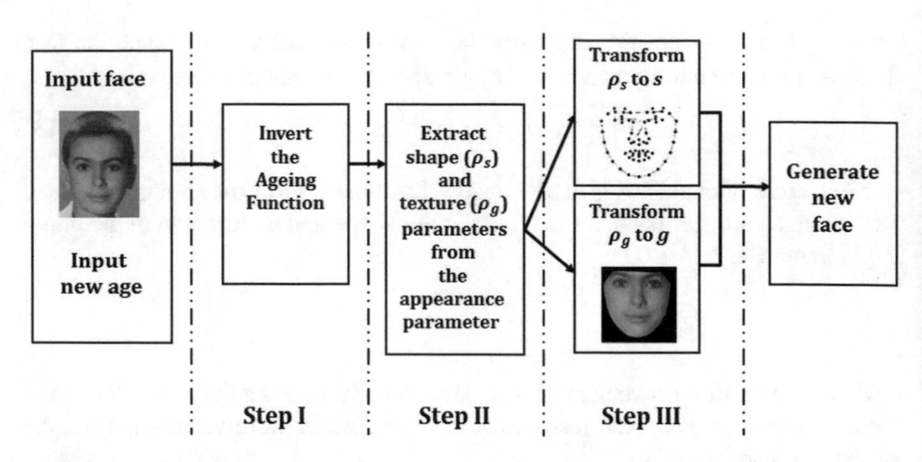

Fig. 1 Age progression framework

Here $\hat{f}_{age}$ computed from (17) is actually a projection, hence two different faces having the same age will also have the same $\hat{f}_{age}$. In linear algebra, two vectors having the same projection have their difference lying in a null space [18]. Hence, one way to conceptualize the problem is to assume that for each individual, the parameter $\widehat{f}$ can be decomposed into two orthogonal component $\hat{f}_{age}$ and f_{id}. Here, we refer to $\hat{f}_{age}$ as the facial features associated with age, while f_{id} is an identity component which lies in the null space, this relationship can be expressed using a simple equation,

$$\hat{f} = \hat{f}_{age} + f_{id}. \tag{18}$$

The procedure of inverting the ageing function and getting a new set of appearance features at a new age can be summarized below.

Algorithm 1: Inversion of Ageing Function

(a) Given the features $\hat{f}_a$ at age a to progress to b
(b) Use Eq. (17) to compute the age component for current age a
(c) Utilize Eq. (18) to calculate the identity component of the facial features
(d) To get the features at age b, use (17) once again to compute the age component for the new age
(e) Sum the results in (c) and (d) to get the appearance parameter for the new age

Step II: Having successfully computed the new appearance parameter $\hat{f}_b$, next we have to separate the fused shape and texture components. This can be achieved via Eq. (15). So this leaves us with ρ_{s_b} and ρ_{g_b} the individual nonlinear shape and texture variations at age b.

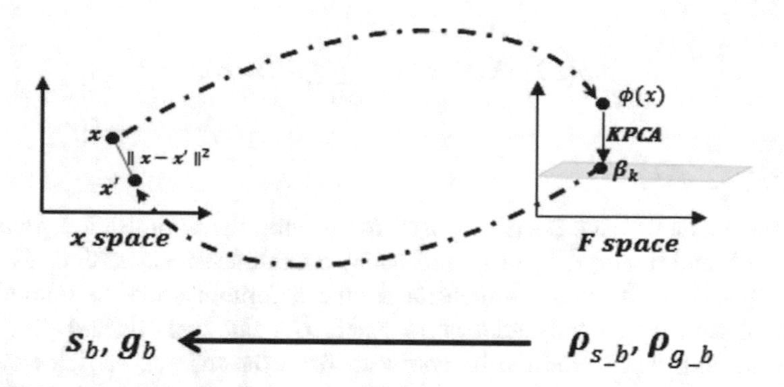

Fig. 2 The preimage (inverse) mapping

Step III: Finally, we have to reverse the nonlinear mapping procedure, so that we get $\boldsymbol{s}_b$ and $\boldsymbol{g}_b$ the facial landmarks and image pixels of the input face at the new age. Easy as it may seem, the problem of mapping back to the initial input space is indeed an ill posed problem also known as the pre-image problem [25, 31]. The pre-image problem involves procedures for finding an approximate inverse map of the feature space in the x space (see Fig. 2).

Given a projection β_k, in F space, its reconstruction in the feature space can be achieved via a projection operator P_n given by,

$$P_n\phi(x) = \sum_{k=1}^{n} \beta_k . V^k, \tag{19}$$

where V^k is a matrix of normalized eigenvectors of kernel $\boldsymbol{K}$. When the vector $P_n\phi(x)$ has no pre-image x' in the input space, solving the pre-image problem will be that of approximating x' via minimising,

$$\rho(x') = \|\phi(x') - P_n\phi(x)\|^2. \tag{20}$$

By discarding terms that are independent of x' we have

$$\rho(x') = \|\phi(x')\|^2 - 2(\phi(x').P_n\phi(x)). \tag{21}$$

Using the kernel trick and substituting (19) into (20) we get

$$\rho(x') = K(x', x') - 2 \sum_{k=1}^{n} \beta_k \sum_{i=1}^{l} \alpha_i^k . K(x', x_i)$$

This can be simplified as

$$\rho(x') = K(x', x') - 2 \sum_{i=1}^{l} \gamma_i . K(x', x_i), \tag{22}$$

where $\gamma_i = \sum_{k=1}^{n} \beta_k \alpha_i^k$.

Several methods have been proposed for solving the optimization problem in (22). Kwok and Tsang [32] proposed using multidimensional scaling (MDS) to embed $P_n\phi(x)$ in the lower dimension x; the algorithm seeks to minimize the pairwise distances in input and feature space. Honeine and Richard [28] used a linear algorithm that learns the inverse map from the training set. However, the most popular approach is the fixed point iteration method proposed by Mika et al. [29], which is guaranteed to produce a preimage that lies within the span of training data. However, it suffers from a number of drawbacks which include; sensitivity to initialization, local minima, and numerical instability. To overcome the first setback re-initialization with different values has been proposed in [33]. In that work Abrahamsen and Hansen also tackled the problem of instability via the use of input space regularization to stop the denominator of the iteration rule from going to zero.

$$\rho(x') = \|\phi(x') - P_n\phi(x)\|^2 + \lambda\|x' - x_o\|, \tag{23}$$

where λ is a non-negative regularization parameter, x_o is an input space value used for the regularization.

With a view to enhancing image reconstruction, centring of the weighting coefficient γ_i has also been suggested in the literature [32–34]. So that,

$$\tilde{\gamma}_i = \gamma_i + \left(\frac{1}{N}\right)\left(1 - \sum_{j=1}^{N} \gamma_j\right). \tag{24}$$

In this work the pre-images of ρ_{snew} and ρ_{gnew} are computed by incorporating (24) into the optimization problem of (23).

Finally, after computing the pre-images; shape $\boldsymbol{s_{new}}$ and texture $\boldsymbol{g_{new}}$, the synthesized face is constructed by warping the new texture to the computed shape. The warping method used here is the piecewise affine method; a simple non parametric warping technique that performs well on local distortions [35].

5 Experiments

Face Database

Experiments were conducted using the FG-NET aging database [36]. This is a publicly available database containing 1002 grey-scale and color images of 82 subjects, with each subject having 12 age separated images. The subjects are multiracial with varying facial expressions and their ages range between 0 and 69 years. Additionally, images contained in the database have varying photographic qualities, i.e. variation in illumination, resolution, and sharpness (see Fig. 3). The database has been distributed to over 4000 researchers across the globe [5], hence it

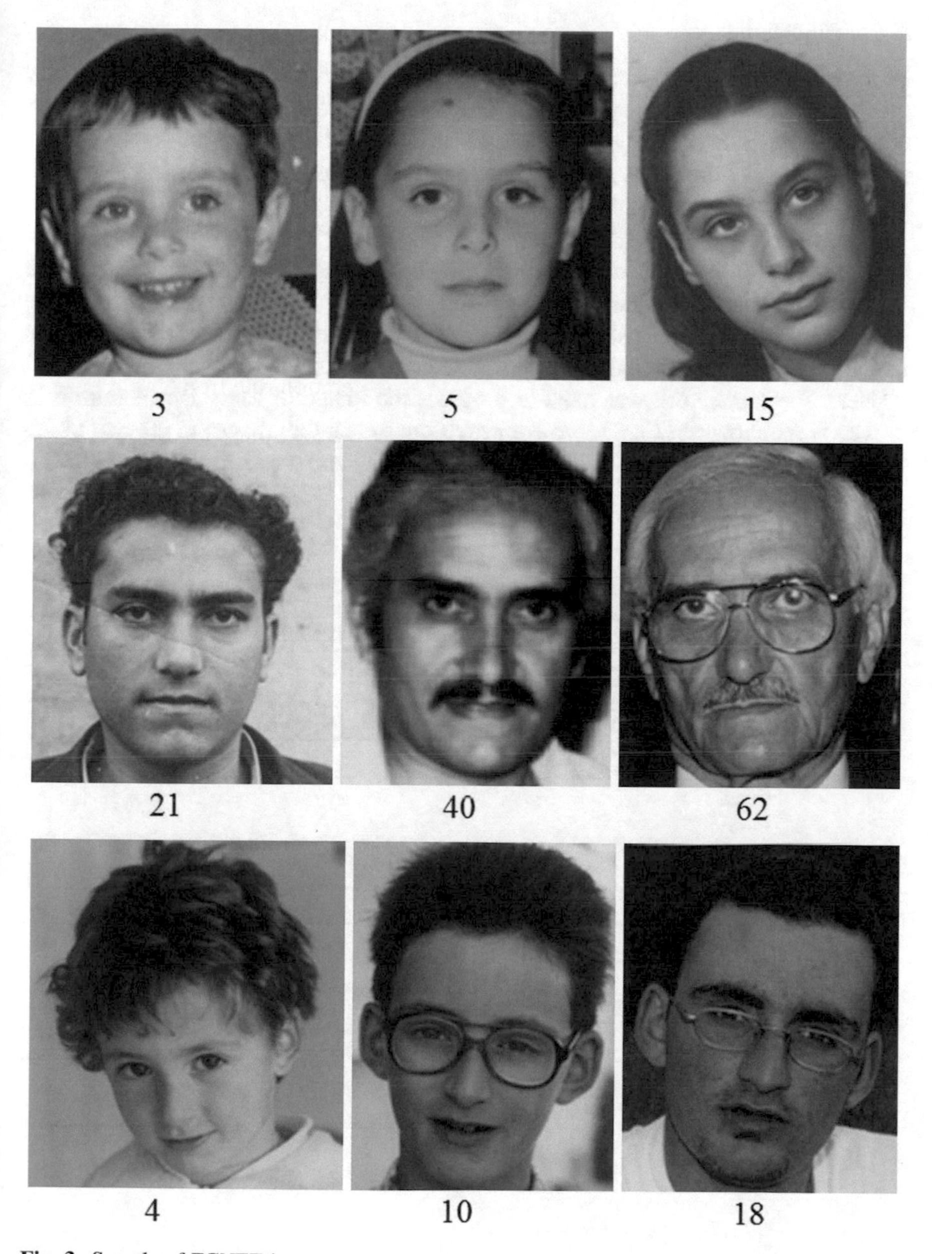

Fig. 3 Sample of FGNET images

is considered one of the most widely used benchmark database for evaluating automatic facial aging algorithms [1]. In our experiments, the face shape was represented by 68 fiducial points, and image pixels (texture) were attained by first cropping all 1002 images to 361 × 300 pixels, in order to reduce image size and therefore computational cost.

Age Progression

We used the kernel functions listed in Table 1 to build three sets of KAMs. Next, we implemented the ageing function defined in Eq. (8) to obtain the regression coefficient β. Subsequently, age progression was achieved by pre-image reconstruction, the optimization rules are presented in Table 2. All tuneable parameters were chosen via cross validation.

A comparison of our technique to that of Lanitis et al. [5], are shown in Fig. 4. Obviously, all 3 versions of the proposed age progression method render realistic images. It is also clear that the de-noising capability of Kernel PCA makes the proposed framework robust to image noise. Furthermore, the synthesized images are less affected by ambient lighting and facial expressions.

Next, the framework was used to progress the image of Mary Boyle from 6 to 45 years. As shown in Fig. 5, we compared our generated images to the real photo of her identical twin sister. Again, our method was compared to that of [10].

Table 1 Kernel methods used for our experiments

Kernel	Expression	Parameter conditions
Gaussian	$\exp\left(\frac{-\lVert x_i - x_j\rVert^2}{2\sigma^2}\right)$	$\sigma > 0$
Sigmoid	$tanh\left(a\left(x_i^T \cdot x_j\right) + r\right)$	$a > 0, r < 0$
Log	$-\log\left(\lVert x_i - x_j\rVert^\beta + 1\right)$	$0 < \beta \leq 2$

Table 2 Kernel preimage iteration rules

Kernels	Gradient of the optimization equation
Gaussian	$\frac{1}{\sigma^2}\sum_{i=1}^{l}\gamma_i \exp(\frac{-\lVert x_* - x_i\rVert^2}{2\sigma^2})(x_* - x_i) + \lambda(x_* - x_o)$
Sigmoid	$ax_*(1 - \tan^2 h(ax_*^T x_* + r)) - \sum_{i=1}^{l}\gamma_i a(1 - \tan^2 h(ax_*^T x_i + r))x_i + \lambda(x_* - x_o)$
Log	$\sum_{i=1}^{l}\gamma_i \frac{\beta}{(\lVert x_* - x_i\rVert^\beta + 1)}(x_* - x_i)^{\beta - 1} + \lambda(x_* - x_o)$

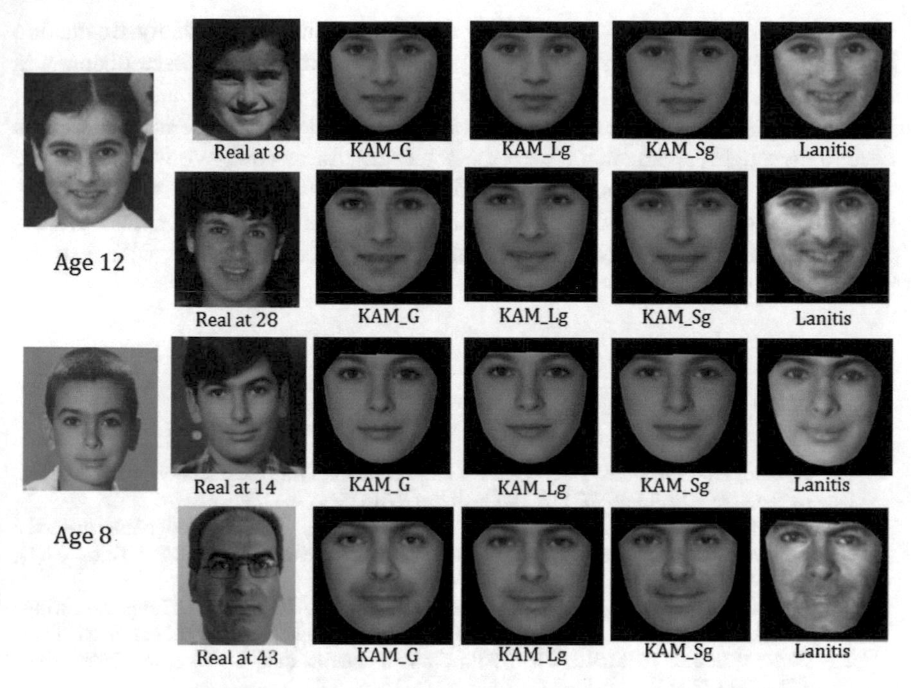

Fig. 4 Age synthesis results: images on the *left most column* are the images to progress, real images and synthesized images at the new age are shown in *each row*. Our methods are denoted KAM_G, KAM_Lg, and KAM_Sg to indicate Gaussian, Logarithm and Sigmoid kernel functions used for the KAM implementation

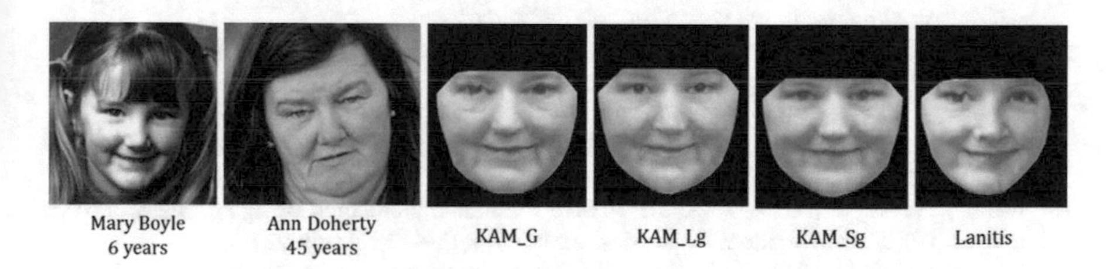

Fig. 5 Age progressed images of Mary Boyle compared to the real image of Ann Doherty

6 Conclusion

In this paper we have proposed a nonlinear variant of the Active Appearance Model (AAM). KAM takes advantage of kernel machines and their ability to explore higher order correlations between input variables. We have shown that KAM when used for age progression generates realistic images despite the effects of image noise, lightening variations and facial expressions. We also showed the application of the proposed framework in real life scenario, i.e. the case of Mary Boyle. We believe that the proposed method has a potential of aiding police investigations in the search for missing people.

However, our work does leave room for future improvements. First, there is a need to thoroughly evaluate the performance of the framework. We will investigate suitable performance metrics for measuring how the framework reserves an individual's identity. Also, with a view to improving the flexibility of the age progression framework, other methods of pre-image computation will also be investigated in the future.

References

1. Fu, Y., Guo, G., Huang, T.: Age synthesis and estimation via faces: a survey. IEEE Trans. Pattern Anal. Mach. Intell. **32**(11), 1955–1976 (2010)
2. Zeng, Z., Pantic, M., Roisman, G., Huang, T.S.: A survey of affect recognition methods: audio, visual, and spontaneous expressions. Pattern Anal. Mach. Intell. IEEE Trans. **31**(1), 39–58 (2009)
3. Zhao, Q., Rosenbaum, K., Okada, K., Zand, D.J., Sze, R., Summar, M., Linguraru, M.G.: Automated down syndrome detection using facial photographs. In: Engineering in Medicine and Biology Society (EMBC), 2013 35th Annual International Conference of the IEEE, pp. 3670–3673 (2013)
4. Ramanathan, N., Chellappa, R.: Modeling age progression in young faces. In: IEEE Computer Society Conference on Computer Vision and Pattern Recognition, vol. 1, pp. 387–394 (2006)
5. Panis, G., Lanitis, A., Tsapatsoulis, N., Cootes, T.F.: Overview of research on facial ageing using the FG-NET ageing database. IET Biometrics **5**(2), 37–46 (2016)
6. Fyfe, N.R., Stevenson, O., Woolnough, P.: Missing persons: the processes and challenges of police investigation. Polic. Soc. **25**(4), 409–425 (2015)
7. Machado, H., Santos, F.: The disappearance of Madeleine McCann: Public drama and trial by media in the Portuguese press. Crime Media Cult. **5**(2), 146–167 (2009)
8. Needham, K.: Ben. Ebury Publishing, London (2013)
9. Haber, J., Terzopoulos, D.: Facial modeling and animation. In: ACM SIGGRAPH 2004 Course Notes, p. 6 (2004)
10. Lanitis, A., Taylor, C., Cootes, T.: Toward automatic simulation of aging effects on face images. IEEE Trans. Pattern Anal. Mach. Intell. **24**(4), 442–455 (2002)
11. Cootes, T.F., Edwards, G.J., Taylor, C.J.: Active appearance models. **23**(6), 681–685 (2001)
12. Patterson, E., Sethuram, A., Albert, M., Ricanek, K.: Comparison of synthetic face aging to age progression by forensic sketch artist. In: International Conference on Visualization, Imaging, and Image Processing, pp. 247–252 (2007)
13. Ramanathan, N., Chellappa, R., Biswas, S.: Age progression in human faces: a survey. J. Vis. Lang. Comput. **15**(1), 3349–3361 (2009)
14. Kono, H., Genda, E.: Wrinkle generation model for 3d facial expression. In: ACM SIGGRAPH 2003 Sketches & Applications, p. 1 (2003)
15. Burt, M., Perrett, D.: Perception of age in adult Caucasian male faces: computer graphic manipulation of shape and colour information. Proc. R. Soc. Lond. Ser. B: Biol. Sci. pp. 137–143 (1995)
16. Kemelmacher-Shlizerman, I., Suwajanakorn, S., Seitz, S.M.: Illumination-aware age progression. In: 2014 IEEE Conference on Computer Vision and Pattern Recognition, pp. 3334–3341, June 2014

17. Geng, X., Zhou, Z.-H., Smith-Miles, K.: Automatic age estimation based on facial aging patterns. IEEE Trans. Pattern Anal. Mach. Intell. **29**(12), 2234–2240 (2007)
18. Bukar, A.M., Ugail, H., Connah, D.: Individualised model of facial age synthesis based on constrained regression. In: 2015 5th International Conference on Image Processing Theory, Tools and Applications (IPTA), pp. 285–290 (2015)
19. Wang, W., Cui, Z., Yan, Y., Feng, J., Yan, S., Shu, X., Sebe, N.: Recurrent face aging. In: Proceedings of the IEEE Conference on Computer Vision and Pattern Recognition (2016)
20. Cootes, T.F., Edwards, G.J., Taylor, C.J.: Active appearance models. In: In Computer Vision —ECCV'98, pp. 484–498 (1998)
21. Lanitis, A.: Evaluating the performance of face-aging algorithms. In: IEEE International Conference on Automatic Face & Gesture Recognition, pp. 1–6 (2008)
22. Christoudias, C.M., Darrell, T.: On modelling nonlinear shape-and-texture appearance manifolds. In: 2005 IEEE Computer Society Conference on Computer Vision and Pattern Recognition, vol. 2, pp. 1067–1074 (2005)
23. Gao, X., Su, Y., Li, X., Tao, D.: A review of active appearance models. IEEE Trans. Syst. Man Cybern. Part C Appl. Rev. **40**(2), 145–158 (2010)
24. Aizerman, M.A., Braverman, E.M., Rozonoer, L.I.: Theoretical foundations of the potential function method in pattern recognition. Autom. Remote Control **25**, 917–936 (1964)
25. Honeine, P., Richard, C.: Preimage problem in kernel-based machine learning. IEEE Signal Proces. Mag. **28**(2), 77–88 (2011)
26. Scholkopf, B., Smola, A., Muller, K.R.: Nonlinear component analysis as a kernel eigenvalue problem. Neural Comput. **10**(5), 1299–1319 (1996)
27. Li, J.-B., Chu, S.-C., Pan, J.-S.: Kernel Learning Algorithms for Face Recognition. Springer, New York (2014)
28. Honeine, P., Richard, C.: A closed-form solution for the pre-image problem in kernel-based machines. J. Signal Proces. Syst. **65**(3), 289–299 (2011)
29. Mika, S., Schölkopf, B., Smola, A., Müller, K., Scholz, M., Rätsch, G.: Kernel PCA and de-noising in feature spaces. Adv. Neural. Inf. Proces. Syst. **11**, 536–542 (1999)
30. Zheng, W., Lai, J., Xie, X., Liang, Y., Yuen, P.C., Zou, Y.: Kernel methods for facial image preprocessing. In: Pattern Recognition, Machine Intelligence and Biometrics, pp. 389–409. Springer (2011)
31. Kabanikhin, S.I.: Definitions and examples of inverse and ill-posed problems. J. Inverse Ill-Posed Probl. **16**(4), 317–357 (2008)
32. Kwok, J.T., Tsang, I.W.: The pre-image problem in kernel methods. IEEE Trans. Neural Netw. **15**(6), 1517–1525 (2004)
33. Abrahamsen, T.J., Hansen, L.K.: Input space regularization stabilizes pre-images for kernel pca de-noising. In: IEEE International Workshop on Machine Learning and Signal Processing. MLSP (2009)
34. Leitner, C., Pernkopf, F.: The Pre-image Problem And Kernel PCA For Speech Enhancement. Springer, Berlin Heidelberg (2011)
35. Glasbey, C.A., Mardia, K.V.: A review of image-warping methods. J. Appl. Stat. **25**(2), 155–171 (1998)
36. FG-NET: The Fg-Net Aging Database (2014)

Efficient Schemes for Playout Latency Reduction in P2P-VoD Systems

Abdulaziz Shehab, Mohamed Elhoseny, Mohamed Abd El Aziz and Aboul Ella Hassanien

Abstract The interest for video delivery systems over the Internet has been gradually growing up last years. It has already become a major application due to client interest of video content and persistent development of network technologies. Recently, Peer-to-Peer (P2P) network plays as an important technology to implement such systems. As a fast growth in population of P2P-VoD system, user behavior is playing an increasingly crucial role in the performance of video system. This chapter proposes an efficient model for P2P-VoD system based on the analysis of the user behavior. The simulation results show that the proposed model can efficiently improvc both scrvcr load and the initial playout latency.

1 Introduction

Multimedia communication systems can improve the level of human–computer interaction (HCI) by providing audio, video, and other media along with traditional media such as text, graphics, and images. Over the past decade, the role of the

A. Shehab · M. Elhoseny (✉)
Faculty of Computers and Information, Mansoura University, Mansoura, Egypt
e-mail: mohamed_elhoseny@mans.edu.eg
URL: http://www.egyptscience.net

A. Shehab
e-mail: Abdulaziz_shehab@mans.edu.eg

M.A. El Aziz
Faculty of Science, Department of Mathematics, Zagazig University, Zagazig, Egypt
e-mail: abd_el_aziz_m@yahoo.com

A.E. Hassanien
Faculty of Computers and Information, Information Technology Department, Cairo University, Cairo, Egypt
e-mail: aboitcairo@gmail.com

M. Elhoseny · M.A. El Aziz · A.E. Hassanien
Scientific Research Group in Egypt (SRGE), Cairo, Egypt

A.E. Hassanien and D.A. Oliva (eds.), *Advances in Soft Computing and Machine Learning in Image Processing*, Studies in Computational Intelligence 730,
https://doi.org/10.1007/978-3-319-63754-9_22

Internet-based video delivery has grown significantly. Video delivery requires high bandwidth that is a significant obstacle in the network infrastructure. Though recent networks have improved bandwidth availability, network interruptions are still an issue for high-quality video delivery [1]. Nowadays, one of the most popular Internet applications is Video on Demand (VoD) which has recently attracted more and more clients over the Internet. In 2014, more than 1 billion unique users visited YouTube each month, over 6 billion hours of video are watched each month, and 100 h of video are uploaded to YouTube every minute [2]. Many years ago, traditional VoD systems rely on client-server architecture where videos are only stored in centralized media servers. The major problem was that the bandwidth of these media servers often turned to be a bottleneck for the whole system. In contrary, each peer signed in a video delivery session in P2P video streaming become a contributor to other neighboring peers [3, 4]. Inside P2P networks, users behavior in watching sessions could help improving the overall performance of video delivery process, such as a higher peer departure has significantly affected other peers [5–7]. Researchers have applied various different paradigms, but certain issues like the playout latency [8–10] and reliability of service have never been near optimal. Recent measurement studies [5, 11–14] have presented new findings and problems about current day P2P media streaming systems. When the service provider provides VoD service, each video title has different popularity and the popularity of a video correlates with the request rate for the video. The popularity of a video may decrease with time since user interest for a video decreases after watching it. It may be also changed due to external factors, such as appearance of new videos and recommendation of videos. Users may have different preferences for each video and have diverse preferable time to watch a video. However, the overall request patterns from all clients follow a uniform curve and constitute daily and weekly access patterns, which are very important parameters for the service provider to consider for their network design.

In this chapter, we have analyzed the user behavior of one of the P2P VoD services. User behavior was analyzed from the real Lancaster Living Lab, which supports high-quality live and on-demand content distribution service. Moreover, a P2P video delivery scheme based on user behavior analysis and chunk-driven philosophy is proposed. The proposed scheme uses two buffers, called watching buffer and service buffer that have valuable chunks for future sharing. The probability that users continuously use the downlink bandwidth all the time is low, and a large part of the downlink bandwidth remains wasted during a session. Therefore, it is effective to deliver highly recommended videos and cache them into a service buffer before the user requests them. Therefore, playout latency could be reduced. Besides, it would significantly minimize server load through minimizing overall number of demanding times. Pre-distributing the video contents prior to user requests according to their interests takes full utilization of available bandwidth and at the same time, provides low latency.

The rest of the chapter is organized as follows: Sect. 2 presents an overview of current P2P streaming systems. Then Sect. 3 presents an analysis for user behavior using a recent data set and presents the main description of the proposed system architecture. Section 4 presents the key schemes of the proposed P2P-VoD model.

Section 5 describes the simulation results. Finally, conclusion and future work are drawn in Sect. 6.

2 Related Work

In the last few years, the preceding works in P2P media delivery have been focused on enhancing certain elements of media streaming metrics using numerous paradigms [15]. Design defects and inefficient peering approach in the earlier works had caused the development of more modern P2P streaming models, most of which can be constructed on chunk-driven philosophy. P2P-based video delivery systems can be classified into three main types: tree-based [16], mesh-based [17], and hybrid systems. Tree-based systems used in many systems, such as o Stream [18] while mesh-based systems used in many implementations, such as pcVoD [19] and Coolstreaming [20]. Some other researches combine both tree-based and mesh-based such as TAG [21].

In fact, many research points could be found in the literature. Nguyen et al. [22] proposed a directory-based P2P VoD architecture. The tracking server acts as a central core that is responsible for monitoring all other devices in the network. It is also responsible for notifying new clients to obtain their services. The media servers store different videos and meets clients demands. In this work, the full dependence on a core server requires it to be strong enough to be capable of meeting global required tasks. Moreover, inter-channel optimizations techniques were not taken into account in their proposal. Xiao et al. [23] proposed a heuristic buffer space concatenation approach. They divide the window into three separate parts for different goals where each part plays different roles for better video quality. Although this work seems good at small scales, many settings are empirical and heuristic. Zhou et al. [24] proposed a hybrid strategy to deliver the video chunks to many peers in a way that enhance playback continuity. The mandatory to have many replications of a video resource in order to support child peers with adequate downloading speed from multiple parent peers is the main problem faced by this approach. Furthermore, if a video resource is not popular, this approach will suffer bandwidth utilization problems. Nguyen et al. [25] focused on collaborating the benefits of different network coding models in a way that might reduce the inherent problems found in unstructured P2P networks. In addition, the authors demonstrate the importance of neighbor selection and shows the evidence of neighbor selection on system performance. The work focuses on the average quality satisfaction of the peer. However, it does not consider the degradation or variation in video quality levels. Chang et al. [26] presented a system dependable on two chunks, the first regards each chunk of equal importance and schedules chunks to be sent in a near-random fashion. The second policy gives each chunk a content-dependent priority, usually in a rate-distortion sense. This work focuses only on live streaming system and ignores many important aspects like deployment of peers inside the overlay and discontinuous playback. Takano and Yoshizawa [27] proposed a scheme that dynamically assigns peers into groups requesting the same media file. Only one of them requests the media from the server and then other clients request the media from the source. Their scheme manages the distributed cache to transfer VoD server load to the groups of clients. Mehbodniya et al. [28] addressed the resource allocation problem for uplink with the aim of improving the spectral efficiency (SE) and reducing the computational complexity. Seyyedi and Akbari [29] proposed hybrid architectures based on Content Distribution Networks (CDN) and P2P networks. They proposed a comparison of the performance of two CDN-P2P architectures: (1) CDN-P2P unconnected mesh in which independent P2P mesh networks are formed under each CDN node and (2) CDN-P2P connected mesh in which CDN nodes and peers join in a single P2P mesh network. Zhang et al. [30] proposed a hybrid strategy that selects the minimum scheduling deadline to achieve optimal delay performance. Zhang et al. [31] proposed both global optimal scheduling scheme and distributed heuristic algorithm to optimize the system throughput. Most of the pre-mentioned works focused on chunk scheduling strategies and overlay structure, ignoring how to utilize available bandwidth capacity to support chunk delivery in a way that minimize the average latency. In this chapter, user behavior was analyzed and an efficient P2P video delivery scheme is proposed. The proposed scheme aims to organize peers into tree-mesh overlay in a way that guarantees low delay. The participating peers are arranged in two layers according to their uplink capacity. Peers are either super-peers or ordinary peers. Super-peers serve as a chunk contributor while ordinary peers serve as a chunk requester.

3 System Architecture

Most of the traditional P2P streaming systems rely on the buffering scheme that use a single-channel overlay. The media data are divided into clips called chunks as a prior step to the streaming over the Internet. Two popular methods are usually followed: One divide the media file into clips according to playing time; the other divide the media file into clips according to the same size. The second method has a problem that different media streams may not synchronize well with the stream to play. To overcome this issue, the media files are divided into basic units, which are usually one second, called a chunk or a slot. This chunk becomes the fundamental unit of data for peers in P2P overlay. The P2P video delivery system comprises four main entities: (1) Web Server, (2) Tracker Server, (3) Media Server, and (4) Peers. They are organized in a tree-mesh overlay. Figure 1 shows the overall architecture of the proposed system model.

- **Web Server**: It informs new participating peers with recent channel list. Peers select a media to watch through browsing the catalog found in the portal.
- **Tracker Server**: It has the job of identifying all necessary information of all online peers, such as providing recent members a list contains information about other joining peer. The tracker keeps track of all joined peers. It sees each peer as an item that assign an IP address, port, bandwidth, playing time and so on. It should track

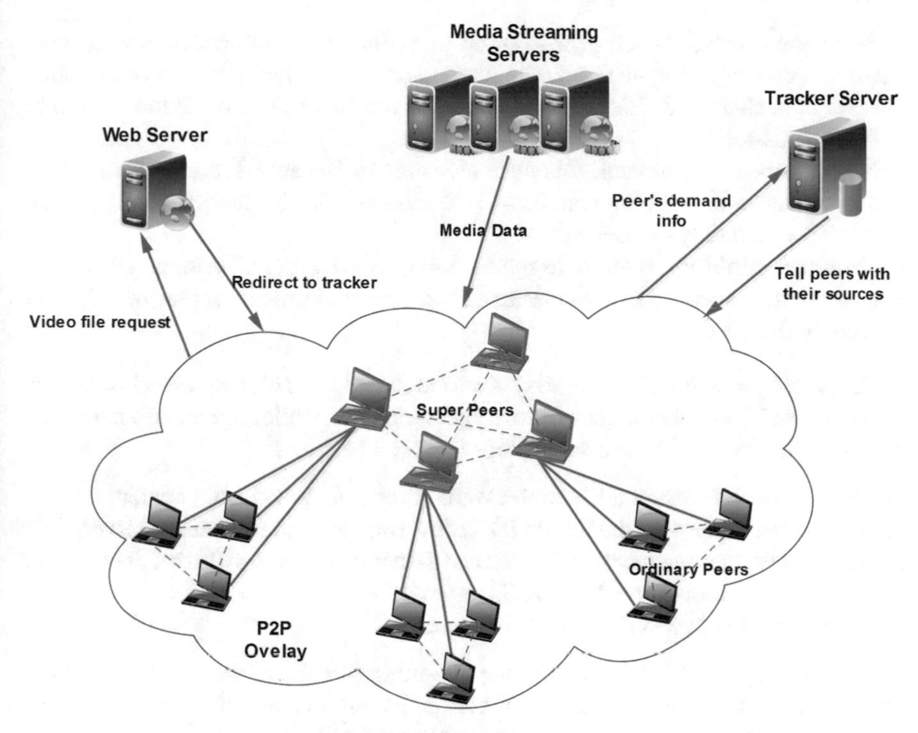

Fig. 1 The architecture of the proposed P2P video delivery system

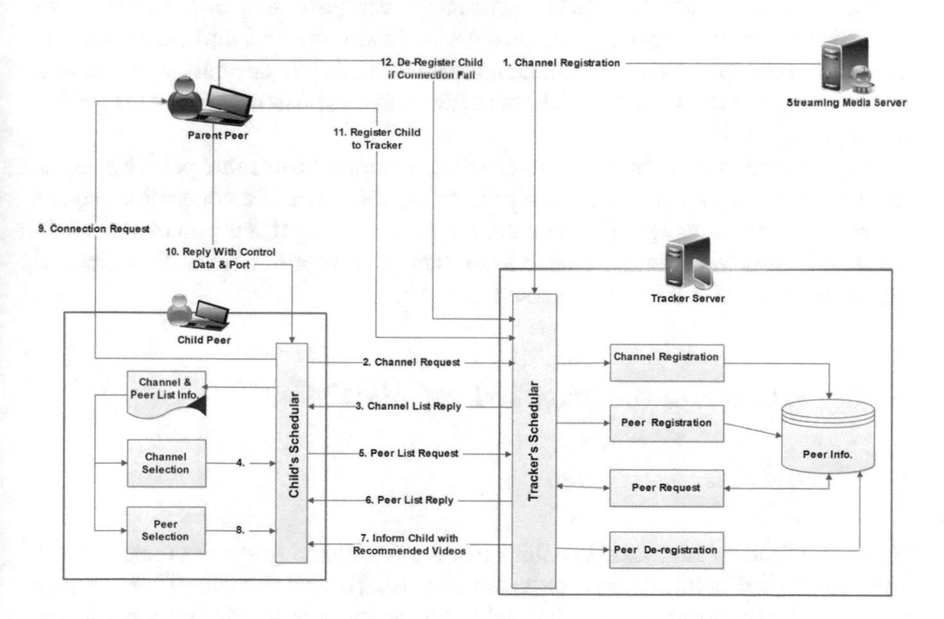

Fig. 2 Connection map between peer, tracker, and media server

the playing time for each peer in order to maintain this information. Each peer sends a message to synchronize its buffer status every minute. This information eases data sharing in P2P overlay and need not be error-free for the system to function correctly.

- **Media Server**: By default, this attribute is set to Distance-Based, which means that the subscriber station connects to the closest BS. This association is permanent when mobility is disabled.
- **Configure Mobility**: It serves as media storage provider for all original video files.
- **Peer**: It serves as a chunk requester either from main media server or neighbor peers in the overlay.

All joined peers in the P2P overlay are arranged into two classes: one class named super-peers and the other named ordinary peers. In fact, while requesting a particular media, peers have to follow a set of steps that are as follows:

1. Peers browse the catalog file to the Web server and select media content.
2. Web server reply with media file ID, redirecting peers to the tracker server.
3. Tracker searches the membership list and then inform peers with a media provider (super-peer, ordinary peer, or media server).
4. Chunks are fetched to the destination.

Schematic diagram that details these steps is shown in Fig. 2. There are four primary entities named streaming media server, tracker, parent peer, and child peer. Channels available are registered at tracker. The tracker scheduler manages many processes, such as channel registration, peer registration, deregistration, and peer requests [32, 33]. On the other hand, peer scheduler maintains channel and peer selection. As soon as these processes are handled, the tracker informs the child with a candidate parent to get the stream. If the connection fails, the parent informs the tracker to deregister the child.

Peer connection establishment starts with a request for candidate peers list. If the list received correctly, peer selection process begins. Then, if a connection request succeeds, the peer receives data and control port. Finally, if the connection established successfully, parent peer have to register child peer at tracker. Flowchart of peer connection establishment is shown in Fig. 3.

4 Key Schemes of the Proposed P2P-VoD Model

4.1 Parent Selection Scheme

One of the fundamental functionalities that a P2P streaming system must have is: from which peer child obtains the video content. To deal with the frequent peer arrivals and departures, a peer regularly updates its peer list during the session. Tracker will next inform the peer with a list containing information of active peers in the session. After receiving an initial list of active peers, the peer will try to make connections to some remote peers on the list, and it starts to deliver video content from its neighbors. Since a minimum guaranteed upload bandwidth equal to the playback rate is required for all connections in the streaming overlay, a child peer has to select parent peers that have sufficient bandwidth [34]. Figure 4 presents the parent selection algorithm. In lines 3–10, tracker server classifies joined peers into super or ordinary peers taking into account their upload capabilities (i.e., super-peers have higher uplink bandwidth than ordinary peers). The peer is considered to be a super-peer if and only if its contribution rate is greater than the requested video bitrate. The contribution rate is calculated by dividing peer upload capacity by the required video bitrate. The higher contribution rate indicates the higher service capability for other peers. After obtaining super-peer list, lines 12–18, the tracker searches their buffer maps according to the required stream in order to find parent list. The tracker informs the requesters with their parent list dynamically according to peer joining time and availability of different parents upload capacity.

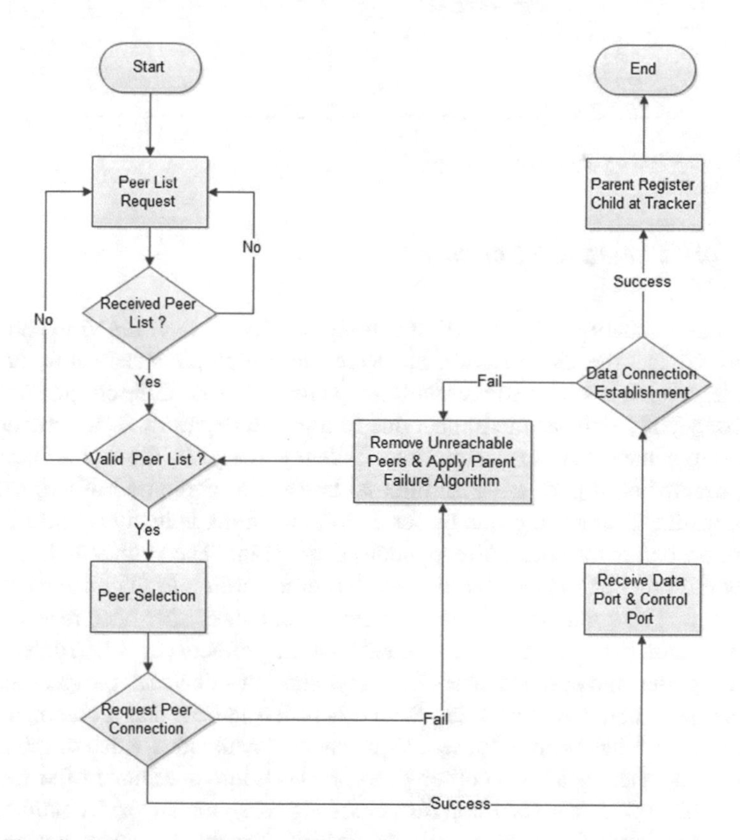

Fig. 3 Flowchart of peer connection establishment

```
Input: P is a list of available peers, r is a video bitrate, sr is a stream of chunks'
sequence number
Output: pr is a list of available parents sorted in a descending manner w.r.t
contribution rate
Algorithm: ParentSelection ( P , r, sr )
1     sp ← intialize an empty super peer list
2     // classify peers
3     For i← 1 to p.length
4         p_i.crate = BW_up(p_i) / r            // contribution rate
5         If ( p_i.crate ≥ 1)                   // the peer could serve at least one child
6             sp.add (p_i)
7         Else
8             Mark p_i as ordinary peer
9         End if
10    End
11    // exclude all super peers that have not the stream
12    For j←1 to sp.length
13        If ( sr ⊆ BufferMap (sp_j)
14            pr.add (sp_j)
15        End
16    End
17    Sort ( pr )
18    Return pr
```

Fig. 4 Parent selection algorithm

4.2 Chunk Selection Scheme

Another basic functionality of P2P streaming system is: how the video stream is delivered. Hence, the video stream is divided into multiple substreams, in which each node can retrieve any sub-stream independently from different parent nodes. This subsequently reduces the impact due to a parent departure or failure, and further helps to achieve better transmission efficiency. Early traditional P2P streaming systems assume each peer setup an internal buffer that composed of two sections: synchronization buffer and cache buffer. Incoming chunk is firstly set into the synchronization buffer for each corresponding sub-stream. The chunks will be stored in sequential order in the synchronization buffer according to their sequence numbers. They will be combined into one stream when blocks have been received from each sub-stream. Each peer uses real-time streaming protocol (RTSP) to deliver these videos. In on-demand systems, users have asynchronous playback progress and cannot exchange content simultaneously with others. To address that problem, a hybrid caching approach has been proposed. Peers cache downloaded video to memory or hard disk, and then relay it to other peers in the future sharing. In the proposed scheme, unlike [27], we go one step further through assigning a service buffer beside currently watching buffer. Multiple videos caching is enabled in a way that each peer could cache chunks of the current channel and, in addition, could cache chunks of other high ranked videos. In general, a video stream is decomposed into K substreams by grouping video chunks according to the chunk selection algorithm illustrated in

Input: *pr* is a list of parent peers, *sr* is the required stream, k is a predefined sub-stream count, p_{new} is a newly joined peer

Output: p_{new} delivered the chunks in *sr*

```
Algorithm: ChunkSelection (pr , k, sr , p_new )
1     If (pr is empty)
2         Send (Server, p_new )
3     Else
4         If( pr.length < k)
5             k← pr.length
6         End if
7         For i← 1 to k
8             n=0
9             For j← sr.Start to sr.End
10                Chunk ← sr[sr.Start + n * k]
11                n←n+1
12            End
13            Send( pr_i, p_new , chunk )
14            sr.Start ← sr.Start + 1
15        End
16    End if
```

Fig. 5 Chunk selection algorithm

Fig. 5. The ith sub-stream contains blocks with sequence numbers (nk + i), where n is a nonnegative incremental integer, and i is a positive integer from1 to k. It implies that a peer can at most receive substreams from k parent peers. An example of single stream decomposition is shown in Fig. 6, in which k = 3 and the stream starts with sequence number 10. In fact, each peer setup an internal buffer that composed of two sections: synchronization buffer and cache buffer as shown in Fig. 7. Incoming chunk is firstly set into the synchronization buffer for each corresponding sub-stream. The chunks will be stored in sequential order in the synchronization buffer according to their sequence numbers. They will be combined into one stream when blocks have been received from each sub-stream. Moreover, the buffer is enlarged to accept three new recommended videos. As shown in Fig. 7, the current buffer has k-substreams $S_1, S_2, \ldots, S_k$ and service buffer has, for example, three related streams R_1, R_2, and R_3 which are ready for usage after currently watched stream.

For chunk selection, the peers follow the algorithm shown in Fig. 5. When a new peer joins the overlay, the tracker informs the requester with a parent list according to aforementioned parent selection algorithm. In line 1–3, if the parent list is empty, the server has the responsibility to serve that peer. In lines 4–6, if the parent list has not enough k parents, the sub-stream count is updated to the available parents in the list. In lines 7–12, the stream in divided into k substreams and requests are made to serving parents.

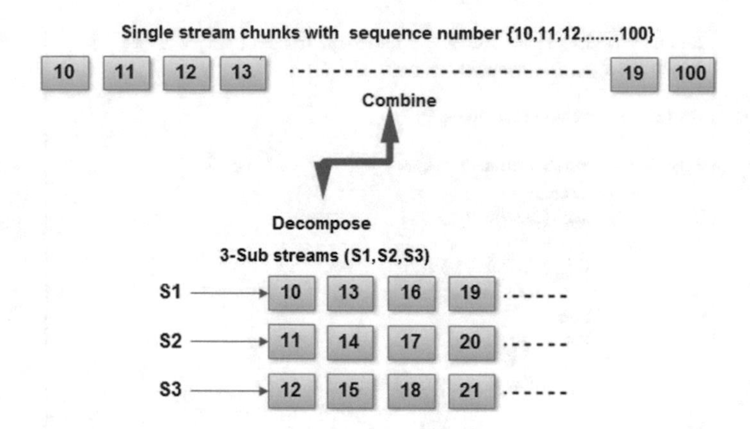

Fig. 6 An example of single stream decomposition

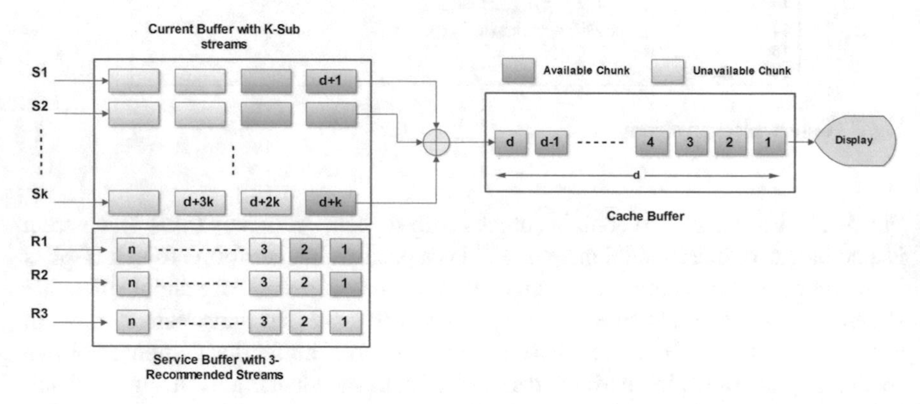

Fig. 7 Proposed structure of the buffer in a node

4.3 Dynamic Prefetching Scheme

Peers in the overlay cache a moving window of the most recent content that they have received. Assume a peer buffers a number of minutes of video in watching buffer. The peer caches the most recent minutes of the video and continuously caches the most recent content as time goes along. This peer can serve any peer requesting the same video be starting at a position within these minutes cached in the buffer. As the amount of cached stream increases, the scalability of service improves. However, the value of the buffer is restricted to: (i) the available cache space at peers, (ii) the available upload capacity at each peer, and (iii) the willingness of peers to serve others for an extra time after finishing watching the video themselves. Tree-based structure has low delay but it is vulnerable to churn, while mesh-based systems are churn resilient but has high delay. Moreover, both structures cannot make full utilization of network bandwidth. The proposed dynamic prefetching scheme implements a tree-mesh mechanism aided by user behavior data gathered by the tracker. It has two

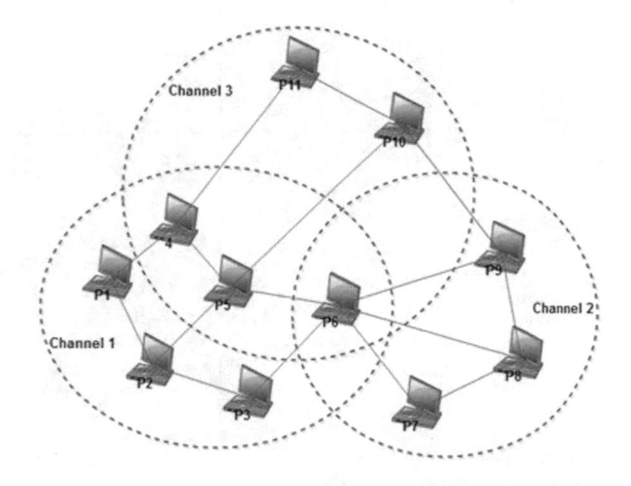

Fig. 8 The proposed multi-channel overlay

buffers when delivering chunks. One catch currently watched stream while the other receive highly related videos. The transmission capacity of the access links has been increased dramatically, and many users will be provided an enormous amount of bandwidth much larger than the playback rate of rich content in the near future. Furthermore, the probability that users continuously use the downlink bandwidth all the time is low, and a large part of the downlink bandwidth remains wasted during a session. Our proposed scheme takes a full utilization of available download bandwidth. Besides, it is always effective to deliver highly recommended videos and cache it into a service buffer before the user requests it. On-demand services always enable clients to watch the whole video from beginning to end. As long as peers cache the initial segment of video content in the buffer, On-demand service can serve other peers that demand it. Caching the initial part of highly recommended videos in a peer service buffer guarantees the chunk availability with low latency. Each peer could assign in multi-channel overlay. It has a main channel that contains currently watching video, and the tracker server could replace it into several service channels. As depicted in Fig. 8, for example, peer P6 main channel is channel 1 and has two service channels: channel 2 and channel 3. Peer P4 (or P5) main channel is channel 1 and service channel is channel 3. P9 main channel is channel 2 and has no service channel.

With the aid of user behavior data, the tracker recommender agent searches for highly ranked videos. Highly candidate videos are chosen through monitoring user requests. Next, the peer would automatically assign a service buffer on the peer side pushing the initial segments of these videos. The proposed scheme aims to reduce server load and waiting time (i.e., minimizing the playout latency). If the content has been pre-distributed and cached in service buffer, chunks are immediately fetched to the decoder and rapidly displayed without the need for requesting from media server. Therefore, by pre-distributing highly ranked videos, media server load could be significantly reduced. A depiction for that is shown in Fig. 9 where the destination peer side has two buffers: watching buffer and service buffer. The watching buffer

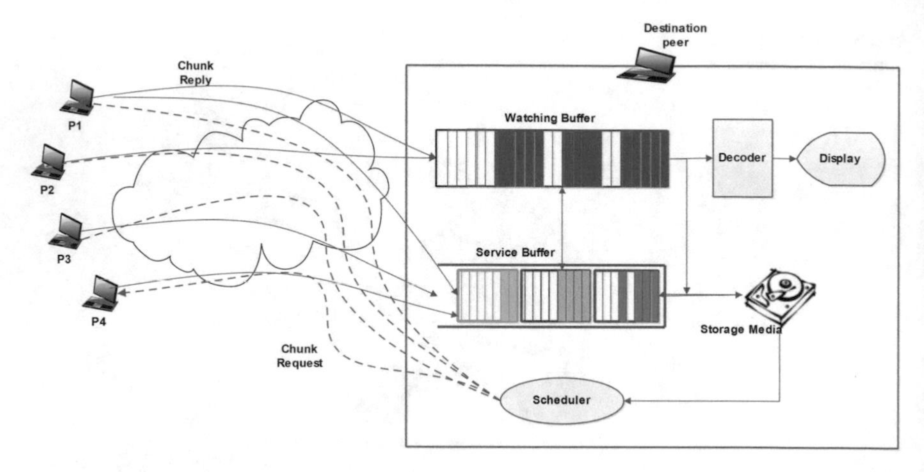

Fig. 9 The proposed hybrid buffering mechanism

Input: P is a list of available peers, C is a currently watched video content, p_{new} is a newly joined peer, b_{Total} is the total buffer size, L_S is the length of initial segment, k is the number of sub-streams, HR is high related (ranked) videos

Output: requester delivered currently watched video and its related videos

Algorithm:

1 $b_{Watching} \leftarrow$ BufferSize[C]
2 $b_{Service} \leftarrow b_{Total} - b_{Watching}$
3 $BW_{Down}(new) \leftarrow BW_{Down}(old) - Bitrate[C]$
4 **While** ($b_{Service}$ is not saturated and $BW_{Down}(new) \geq \min(HR.r_j \ \forall j \in HR)$) **do**:
5 V$\leftarrow$SelectBestRelated(HR, $BW_{Down}(new)$) w.r.t tracker's recommender
6 $L_S \leftarrow \min(L_S, V.length)$
7 $parents \leftarrow$ParentSelection(P ,V.r, V.sr)
8 ChunkSelection($parents$, k, V(1......... L_S), p_{new})
9 $b_{Service} \leftarrow b_{Service}$ - BufferSize[V] // update cache
10 $BW_{Down}(new) \leftarrow BW_{Down}(old) - Bitrate[V]$
11 HR.remove(V)
12 **End**

Fig. 10 Dynamic prefetching algorithm

delivers video chunks from provider peers P1 and P2 while the service buffer delivers three related videos from peers P1, P3, and P4. In fact, Peer (P1), for example, could serve the two buffers at the destination peer as it has chunks of the two requested videos.

Dynamic prefetching algorithm for P2P video delivery is illustrated in Fig. 10. In line 2, service buffer size is estimated by subtracting current watching buffer from the total buffer size. In line 3, the download capacity is affected due to the consumed bitrate of the current watched video. In lines 4–12, while the service buffer is not saturated and the available download bandwidth is sufficient to receive at least one of the recommended (related) videos, the peer start delivering the initial segment of related videos. If the video length is less than, the whole video will be delivered.

Once the initial segment of first high ranked video is delivered, it will be removed from the list and, at the same time, service buffer and download capacity will be dynamically updated. By this way, peers could deliver videos that already cached into service buffer with no waiting time (no latency). Furthermore, in worst cases, if the user does not request any video of those already streamed and cached in service buffer, it still enhance chunk availability through servicing other users need to watch any of these videos.

4.4 Parent Failure Scheme

As in any P2P system, a participant can leave at any time without advanced notice. This action known as peer early departure or churn. Meanwhile, the available bandwidth varies over the time. Therefore, the playback will be interrupted if the available bandwidth becomes less than the required video bit rate (known as bandwidth starvation). Both peer early departures, and bandwidth starvation may interrupt the service. Hence the reconstruction of overlay is required. Chunk buffering has been widely used in combating bandwidth starvation. Buffering a short period of video content and delaying the start of playback is an effective means to deal with the short-term bandwidth starvation. It also allows a peer to locate an alternative parent in case its current parent leaves the system. However, data buffering alone may not be enough to handle peer departures. In parent failure state, multiple secondary parents scheme is used. Every primary peer from k parents has another secondary peer ready to serve once the primary one has been terminated. The child sends a periodic keep alive messages. Once the deadline timer expires, the child searches its parent list informed by the tracker to replace the primary failed peer with a secondary one that has a bandwidth to accommodate one more sub-stream. Finally, the child connects the new parent peer to deliver required chunks. As shown in Fig. 11a, the child has peer 1, 2, and 3 as primary parents and peer 4, 5, and 6 as secondary parents. Once the peer 2 get failed, the child directly removes it from the list and the first secondary one, peer 4, become primary and ready to service as illustrated in Fig. 11b. Based on this way, the client shortens the joining delay and is reconnected quickly without the need to ask the tracker.

For self-failure recovery, the failed peer, x, for example, could get the service, as shown in Fig. 12, through following the steps summarized as follows:

- **Step1**: The peer x reconnect the tracker requesting a video stream
- **Step2**: The tracker reply with new parent list if available
- **Step3**: If the parent list is empty, the tracker redirect x to the server
- **Step4**: The server sends the requested stream to x if the server has enough upload capacity; otherwise x will be rejected.

5 Simulation Results and Discussion

In the following subsections, the performance of both single-channel model and multi-channel model is evaluated. The number of peers typically varies from 50 to 400. All models have been implemented using OPNET Modeler 17.1 [28] as a simulation language tool.

5.1 Simulation Parameters

The proposed models consist of two geographically separated subnets: VoD subnet and peers subnet. The arrival of peer requests follows the Poisson distribution with a mean of λ, where λ ranges from 0 to 100 (requests/min). The peers are assumed to be in heterogeneous bandwidth classes. Additionally, we assume that a peer uplink speed equal to downlink speed. Both subnets are connected to the Internet cloud via DS3 WAN circuits. The Internet (cloud) was configured with a packet discard ratio of 0.001%. Media files originating from a source media server are assumed to vary from 10 to 120 min. It has a 1280 × 720 frame format resolution and a 30 fps

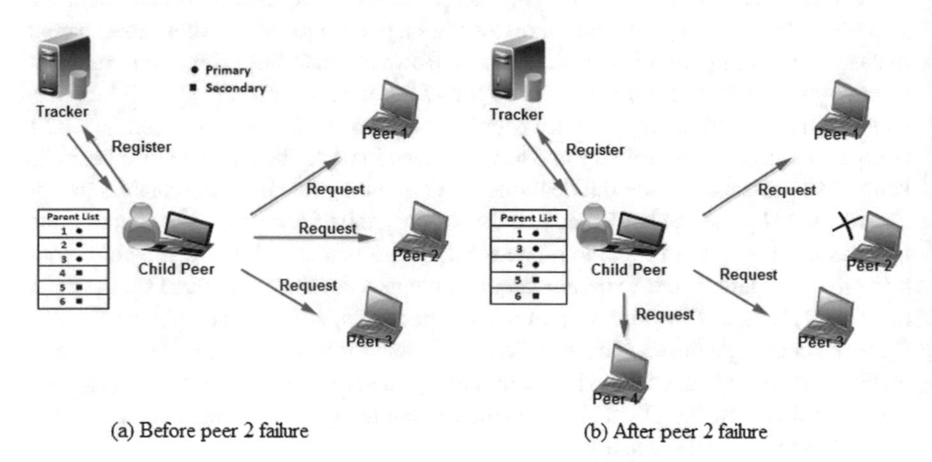

Fig. 11 An example of parent failure state

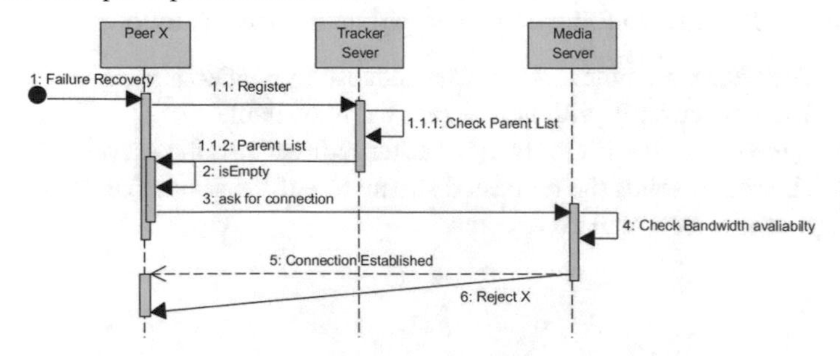

Fig. 12 Self-failure recovery sequence diagram

encoding rate. The length of the initial segment is assumed to be two minutes in our implementation that is long enough to overcome the initial latency.

5.2 Performance Evaluation

To evaluate various aspects of the proposed schemes which are normally difficult to study using real world streams, we use a simulation-based approach to study the same. A simulator gives the ability to experiment with different parameters involved in the system and studies its performance under varying workloads and conditions. Experiments for large scale simulation with a higher number of nodes are difficult to perform using real world implementation for lack of nodes participating in it. Moreover, measuring a publicly deployed P2P VoD system is challenging. This is because all such systems today are proprietary, i.e., their source codes are not publicized and there are no specifications on the communications protocols used in these systems. Till date, there exist no open source codes or specification describing a streaming protocol in detail. Therefore, we simulate both models traditional model which form the base model named Single-Channel as well as the proposed model named Multi-Channel and compare their results. In the proposed model, we simulate all the schemes, i.e., (1) Parent selection scheme (2) Chunk selection scheme (3) Dynamic prefetching scheme and (4) Failure recovery scheme. We compare the results with the recommended values by the ITU-T to provide reliable QoS in IP-Based networks [29]. We consider the most restrictive QoS class of this recommendation, class 0, with limits of 100 ms for the packet transfer delay, 50 ms for the

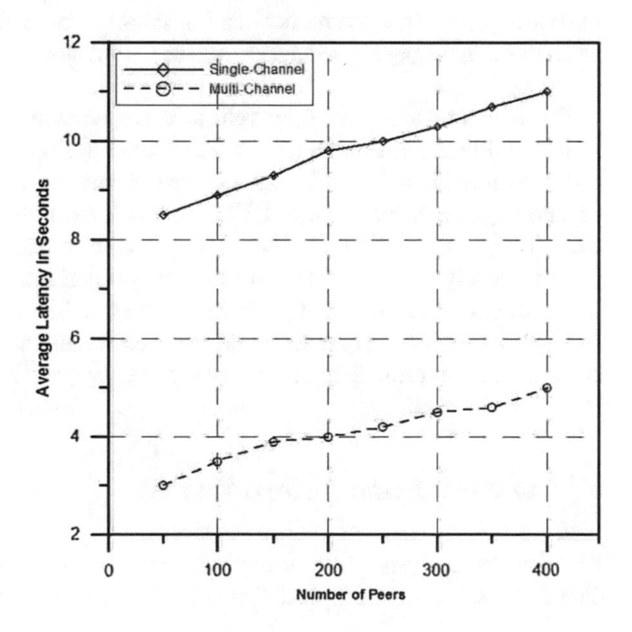

Fig. 13 Average latency versus number of peers

packet delay variation and 0.001 % for the packet loss. In the assessment process, the network bandwidth of each peer is set to 20–100 MB, the memory size for caching data is 300 s, and disk cache size is 200 MB. Figure 13 depicts the average latency in both single-channel and multi-channel models. For single-channel, average latency in seconds varies from approximately 8.5 s for a 50 peers to 11 s for a 400 peers, while multi-channel has considerably lower latency varying from approximately 3 s for a 50 peers to 5 s for a 400 peers. An explanation could be that the existence of hybrid buffer that cache base video and different recommended videos in the overlay, greater are the chances of reduced end to end delay. The proposed model has comparatively lower latency (approximately 5 s) due to the following reasons: (1) It buffers more pieces in service buffer before playback. (2) Due to the systematic nature of data exchange between peers, speeds up the arrival for chunks and hence lower latency. (3) There are at least one or more sources for the content and hence need not request the media server all the time. In fact, these reasons save time as compared to traditional single-channel.

Figure 14 depicts the media server request count during the simulation for both single-channel and multi-channel models. For single-channel, media server request count varies from 750 for a 50 peers to1170 for a 400 peers while multi-channel has considerably lower count varying from 500 for a 50 peers to 531 for a 400 peers. An explanation for this could be that higher availability of multiple videos in P2P overlay, greater are the chances of reduced server request count. The reader can notice that the difference in count between single-channel and multi-channel increases as long as peers increases. The difference is 250 for a 50 peers while it reaches 639 for a 400 peers. As the number of peers gradually increases, we observe that the rise in server request count rate is sharp for single-channel. This effect can be understood considering the impact of rareness of videos force peers to ask the media server. In contrary, the proposed model has less count comparatively. This can be similarly understood that a peer has a higher chance to find new peers with the available pieces of video.

Finally, Fig. 15 depicts jitter rate in average. Sometimes an intermediate piece which could not be fetched, may increase the jitter rate. It can be seen from Fig. 15 that both models perform better with the increasing number of peers, though jitter rate slightly increases after 250 node mark but both, in general, satisfy recommended values mentioned in [14]. The proposed model has a comparatively lower (approximately 0.004%) jitter rate than traditional model. The plausible reason is that in the proposed model, the chunks propagate in an organized fashion from one to other. Moreover, every node could receive the content through multiple parents so chances of an intermediate piece missing are comparatively low.

6 Conclusion and Future Work

Based on the analysis of the current existing P2P video delivery systems, an efficient data distribution model in P2P networks for media streaming has been presented.

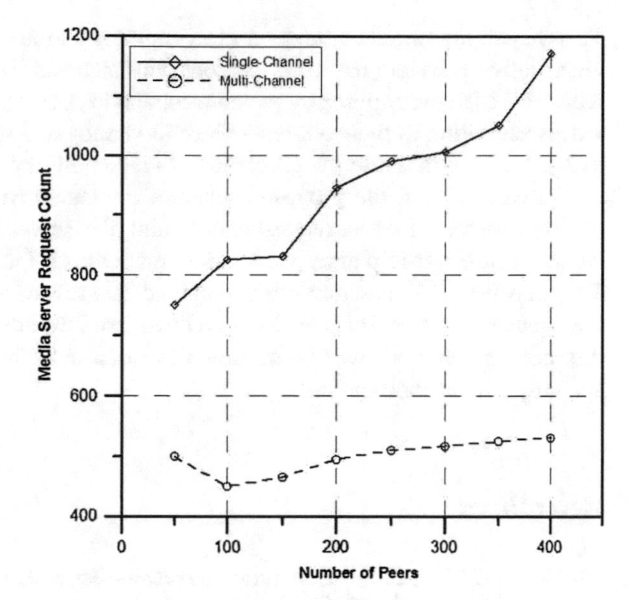

Fig. 14 Media server request count versus number of peers

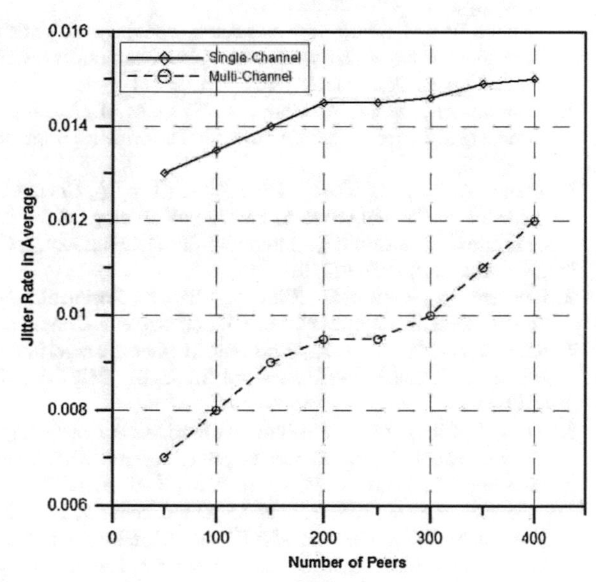

Fig. 15 Jitter rate in average versus number of peers

This chapter presented a strategy and an implementation of efficient schemes in a P2P video delivery system based on user behavior analysis. The proposed schemes make it possible to deliver appropriate media content to peers by prefetching recommended videos according to their interests. Playout latency and server load are two primary metrics used to measure the efficiency of video delivery systems. The experimental results validate that the proposed schemes implemented in multi-channel scenario can improve media server request count (minimize server load) and the initial playout latency. Some participating peers called malicious might not cooperate as desired. They may be selfish and unwilling to upload data to others. How to inhibit and detect malicious peers in P2P networks is an open issue we are going to address. Besides, further research work will be on how to extract more interesting chunks in videos and only deliver these chunks.

References

1. Liu, Y., Guo, Y., Liang, C.: A survey on peer-to-peer video streaming systems. Peer-to-Peer Netw. Appl. **1**(1), 18–28 (2008)
2. Youtube Web Site. https://www.youtube.com/yt/press/statistics. Last accessed on 2 Jan 2015
3. Ramzan, N., Park, H., Izquierdo, E.: Video streaming over P2P networks: challenges and opportunities. Image Commun. **27**, 401–411 (2012)
4. Elhoseny, M., Farouk, A., Zhou, N., Wang, M.M., Abdalla, S., Batle, J.: Dynamic multi hop clustering in a wireless sensor network: Performance improvement. Wirel. Personal Commun. 121 (2007)
5. Zheng, Y., Peng, J., Yu, Q., Huang, D., Chen, Y., Chen, C.: A Measurement study on User Behavior of P2P VoD System. In: The proceedings of the 2nd International Asia Conference on Informatics in Control, Automation and Robotics, vol. 3, CAR 10, pp. 373–376. IEEE Press, Piscataway, NJ, USA (2010)
6. Elhoseny, M., Elminir, H., Riad, A., Yuan, X.: Recent advances of secure clustering protocols in wireless sensor networks. Int. J. Comput. Netw. Commun. Secur. **2**(11), 400–413 (2014)
7. Metawa, N., Hassan, M.K., Elhoseny, M.: Genetic algorithm based model for optimizing bank lending decisions. Expert Syst. Appl. **80**, 75–82 (2017). doi:10.1016/j.eswa.2017.03.021. ISSN 0957-4174
8. Yuan, X., Elhoseny, M., ElMinir, H., Riad, A.: A genetic algorithm-based, dynamic clustering method towards improved wsn longevity. J. Netw. Syst. Manag. **1–26**, 2016 (2016)
9. Elhoseny, M., Yuan, X., El-Minir, H.K., Riad, A.M.: An energy efficient encryption method for secure dynamic WSN. Secur. Commun. Netw. **9**, 2024–2031 (2016)
10. Metawa, N., Elhoseny, M., Kabir Hassan, M., Hassanien, A.: Loan portfolio optimization using genetic algorithm: a case of credit constraints. 12th International Computer Engineering Conference (ICENCO), IEEE, pp. 59–64 (2016). doi:10.1109/ICENCO.2016.7856446
11. Liao, X., Jin, H., Yu, L.: A novel data replication mechanism in P2P VoD system. Future Gener. Comput. Syst. **28**, 930–939 (2012)
12. Ma, K.J., Barto, R., Bhatia, S.: Review: A survey of schemes for internet-based video delivery. J. Netw. Comput. Appl. **34**, 1572–1586 (2011)
13. Hei, X., Liang, C., Liang, J., Liu, Y., Ross, K.: A measurement study of a large-scale P2P IPTV system. IEEE Trans. Multimed. **9**, 1672–1687 (2007)
14. Elhoseny, M., Yuan, X., ElMinir, H., Riad, A.: Extending self-organizing network availability using genetic algorithm. In: International Conference on Computing Communication and Networking Technologies (ICCCNT). IEEE (2014)

15. Purandare, D.: A Framework for Efficient Data Distribution in Peer-to-peer Networks. Ph.D. thesis, Orlando, FL, USA (2008). AAI3335362
16. Lin, W., Shen, W.: Tree-based task scheduling model and dynamic load-balancing algorithm for P2P computing. In: 2010 IEEE 10th International Conference on Computer and Information Technology (CIT), pp. 2903–2907 (2010)
17. Montazeri, A., Akbari, B.: Mesh based P2P video streaming with a distributed incentive mechanism. In: 2011 International Conference on Information Networking (ICOIN), pp. 108–113 (2011)
18. Cui, Y., Li, B., Nahrstedt, K.: Ostream: asynchronous streaming multicast in application-layer overlay Networks. IEEE J. Sel. Areas Commun. **22**, 91–106 (2006)
19. Ying, L., Basu, A.: pcvod: internet peer-to-peer video-on-demand with storage caching on Peers. In: Guercio, A., Arndt, T. (eds.) DMS, pp. 218–223. Knowledge Systems Institute (2005)
20. Li, B., Xie, S., Qu, Y., Keung, G., Lin, C., Liu, J., Zhang, X.: Inside the new coolstreaming: principles, measurements and performance implications. In: INFOCOM 2008. The 27th Conference on Computer Communications. IEEE (2008)
21. Liu, J., Zhou, M.: Tree-assisted gossiping for overlay video distribution. Multimed. Tools Appl. **29**, 211–232 (2006)
22. Nguyen, K., Nguyen, T., Kovchegov, Y.: A P2P Video Delivery Network (P2P-VDN). In: The Proceedings of 18th Internatonal Conference on Computer Communications and Networks (ICCCN), pp. 1–7 (2009)
23. Xiao, X., Shi, Y., Zhang, Q., Shen, J., Gao, Y.: Toward systematical data scheduling for layered streaming in peer-to-peer networks: can we go farther? Parallel Distrib. Syst. IEEE Trans. **21**, 685–697 (2010)
24. Zhou, Y., Chiu, D.-M., Lui, J.: A simple model for Chunk-scheduling strategies in P2P streaming. IEEE/ACM Trans. Netw. **19**, 42–54 (2011)
25. Nguyen, A.T., Li, B., Eliassen, F.: Quality- and context-aware neighbor selection for layered peer-to-peer streaming. In: The Proceedings of the IEEE International Conference on Communications (ICC), pp. 1–6 (2010)
26. Chang, C.-Y., Chou, C.-F., Chen, K.-C.: Content-priority-aware Chunk scheduling over swarm-based P2P live streaming system: from theoretical analysis to practical Design. IEEE J. Emerg. Sel. Topics Circuits Syst. **4**, 57–69 (March 2014)
27. Takano, R., Yoshizawa, Y.: Offloading VoD server organized dynamically distributed cache using P2P delivery. In: The Proceedings of the International Conference on Information Networking (ICOIN), pp. 1–5 (2008)
28. Mehbodniya, A., Peng, W., Adachi, F.: An adaptive multiuser scheduling and chunk allocation algorithm for uplink simo sc-fdma. In: The Proceedings of the IEEE International Conference on Communications (ICC), pp. 2861–2866 (2014)
29. Seyyedi, S., Akbari, B.: Hybrid CDN-P2P architectures for live video streaming: comparative study of connected and unconnected meshes. In: The Proceedings of the International Symposium on Computer Networks and Distributed Systems (CNDS), pp. 175–180 (2011)
30. Zhang, D., Wang, L., Yang, H.: An effective data scheduling algorithm for mesh-based P2P live streaming. In: The Proceedings of the International Conference on Information Science and Technology (ICIST), pp. 1221–1224 (2013)
31. Zhang, M., Xiong, Y., Zhang, Q., Sun, L., Yang, S.: Optimizing the throughput of data-driven peer-to-peer streaming. IEEE Trans. Parallel Distrib. Syst. **20**, 97–110 (2009)
32. Elhoseny, M., Elleithy, K., Elminir, H., Yuan, X., Riad, A.: Dynamic clustering of heterogeneous wireless sensor networks using a genetic algorithm, towards balancing energy exhaustion. Int. J. Sci. Eng. Res. **6**(8) (2015)
33. Riad, A.M., El-minir, H.K., Elhoseny, M.: Secure routing in wireless sensor network: a state of the art. Int. J. Comput. Appl. **67**, 7 (2013)
34. Elhoseny, M., Yuan, X., Yu, Z., Mao, C., El-Minir, H.K., Riad, A.M.: Balancing energy consumption in heterogeneous wireless sensor networks using genetic algorithm. IEEE Commun. Lett. **19**(12), 2194–2197 (2015)

A SetpitextOFF Algorithm-Based Fast Image Projection Analysis

V. Kakulapati and Vijay Pentapati

Abstract This paper proposes a novel image analysis algorithm called setpitextOFF algorithm for image texture interspacing, retrieving the OFF image pixels, run-length pixels block mapping and image fast projection. To explore the Implementation of Image Analysis, we combine the pixels at different space locations with similar retrieving dependencies as a space vector and mapped the space vectors to form interdependency setpi clusters by context building, these setpi clusters were analysed for the proposed setpitextOFF algorithm. Thereafter, we have formulated a double-setpi cluster to regularize the proposed algorithm implementation in a communication channel. Our proposed algorithm used less time to compute with more accuracy in quality performance metrics comparatively. Our proposed research work can be implemented in any digital communication link, for Video, Image and Data analysis.

Keywords Image analysis • Texture • Mapping • Compute Communication link

1 Introduction

In image analysis, it is often necessary to measure an image object [1] with good accuracy. In most of image analysis, an image object is acquired either by using special cameras or by an object recovery [2]. An object recovery utilizes image-processing techniques that require the implementation of special purpose algorithms and use of mathematical tools.

V. Kakulapati (✉) · V. Pentapati
Sreenidhi Institute of Science and Technology, MEC, Yanampet,
Ghatkesar, Hyderabad, Telangana, India
e-mail: vldms@yahoo.com

V. Pentapati
e-mail: vijaya_p51@yahoo.com

A.E. Hassanien and D.A. Oliva (eds.), *Advances in Soft Computing and Machine Learning in Image Processing*, Studies in Computational Intelligence 730,
https://doi.org/10.1007/978-3-319-63754-9_23

Massive integration of image algorithms in all the areas of image-processing demands for image analysis as resource in image-processing systems. Because a stand-alone image-processing system without any object data has no meaning, there was also a need to do analysis about the image between reality and image-processing techniques. This can be achieved by efficient image analysis, or by special image intelligence which is able to recognize the image object by their text in a real environment and implement it into image database resources. Because of this, various image analysis techniques have been developed and image recognition systems are today used in various image-processing applications.

The estimation of different features [3] in an image is a major work in the image analysis, which comes under image illusions, which comes in two-dimensional line drawing patterns. This image illustration have unavoidable benchmarks in image text-position, image pixel-orientation, image object-edge element [4] and image direction-length pattern. In each type of benchmark, illustration needs a model to be presented and analysed.

Recent advances in image analysis provide high resolution images of texture structure, so that the representation of an accurate image illusion model [5] can be measured feasibly. This has resulted in the development of a wide range of new applications for image illusion analysis. Image-based illusion analysis provides valuable information for projection and representation during image rendering problems, both to avoid image benchmarks for projection orientation [6] and localization of the image object of interest. Moreover, comprehensive object-pixel-based analysis [7] has shown a new direction in understanding of the image illusion through its underlying processes, such as image projection analysis and the image restoration that may help to understand the evaluation of image analysis in which image projection analysis plays an important role. [8] Proposed a shape context descriptor, which reduced the coefficient's assignment problem, but the neighbourhood points were not taken into consideration. Under several schemes, [9] proposed to detect shapes in projected images, [10] is proposed to improve the corresponding recovery of the projected image, [11] a different projection settings were used to solve the matching problem [12, 13] are used to project the image data in constraint processing, provides an open-form exists problems under successive projections.

[14] Proposed a hypothesis on feature-integrity theory to display separable features and would be able to detect these in different textures [15, 16]. While the proposed algorithm depends on pixel positions by constraint successive projection in order to attain better accuracy and efficiency for pixel matching.

A vast variety of methods and approaches has emerged in the image projection [1] analysis with increasing complexity. In order to overcome this complexity, we need to implement an algorithm of different methods, their individual responses, their relations, and the methods of measurable parameters, so as to build robust and efficient methods that can be implemented in image projection analysis.

From the algorithm point of view, we have developed a setpitextOFF algorithm to overcome the complexity in image projection analysis and improve the robustness in overcoming the benchmarks quickly. In our proposed algorithm, mid-level features are considered between low-level pixel information and high-level concepts for image visualization. Mid-level features representation helps in robustness in noise and irrelevant image segments with the useful image information. Also, we propose an effective level approach between bottom-up and top-down taps to detect edge-based mid-level features. These features we name as, OFF pixels features provide information about mid-level local edge structures and segments, like from the straight line to strong junctions, curves and parallel lines. Our proposed algorithm can detect the OFF pixel's features by selecting appropriate feature subsets of relevant mid-level information. This relevant mid-level information is obtained from the image taken for projection analysis and from the knowledge of image pattern analysis. The patterns resulted from either cluster of image edge groups or from the skeletonized edges of the image taken for analysis. The key difference between these two allows an efficient and robust OFF pixel edge matching allowing quantifying results to pixel's projection algorithm. The tasks complicated by the OFF pixel algorithm is to find similarities between the shape structure of the mid-level edge of pixel locations, clustering these OFF pixel locations within the image shape similarities, finding the shape prototypes that identify the information about shape score and processing obtained image similarities to discriminate between different OFF edge shapes.

In this paper, we also focus on edge similarity score, proposing a shape matching method under OFF pixel retrieving method with respect to the geometric transformation of image projection. Most methods under these fields represent a node corresponding to connecting OFF pixel edges with the corresponding edge weights to evaluate the similarity values through the cluster graph. The process of evaluating the similarity from node to the pixel edge in a proportionate manner results in repeated random OFF pixel edge detection to improve the image retrieval scores. The proposed OFF pixel retrieving approach includes the match between the mid-level edge similarity with OFF pixel variations between low-level and high-level pixel variations from the dataset taken for analysis and to find the similarity score which contains retrieval information about OFF edge pixel's structure and shape to improve the image retrieval results by analyzing the mutual cluster graph representation yielding improved performance. OFF pixel clusters also identify the corresponding shape cluster prototypes to present a novel OFF pixel clustering approach and project the same possible using the image database as to improve the retrieving task and to obtain novel OFF pixel clustering results.

Section 2 of this paper describes importance and applications image projection analysis, also providing relevant motivational relevancy towards the proposed approach and outlining the focussing points of the proposed analysis. In Sect. 3, proposed setpitextOFF algorithm task development steps are explained, provided the relevant information about the contributed novelty in the paper implementation, explained algorithms of image texture interspacing approach, retrieving the OFF image pixel's method, run-length pixels block mapping and image fast projection. Each algorithm stepwise implementation is explained with a prototype analysis. Section 4 provides an exhaustive evaluation of the image projection analysis implementation and results for the database references, tabulated numerical results predict the performance of proposed image projection analysis. Finally, the proposed analysis algorithm implementation and its scope are concluded.

2 Image Projection Analysis

The most common way to perform image projection is by constructing a database of angled-views. This can be done by taking images of an object through a method of actively acquiring image data, but this method has drawbacks, difficult and time consuming to place labels on the image information, and the projection database is unordered and needs to be reordered to be verifying that there is enough data for proper projection recognition. The amount of data needed for object projection is determined by its projection resolution. Other methods capable of recognizing an object in a scene at different scales and projections have been developed. These types of algorithms can recognize the object up to certain projection angle until features become occluded and unrecognizable from illumination changes.

The projection of an image is the view angle at which an object is observed. Each angle-view includes 2D shape variation creating a different, continuous transition of ordered object projections. The changing angled-views of an object can be captured by an ordered set of object projection analysis methods.

The motivation for image projection analysis is to create a way to find matching images in random projections in unordered images, which can be used to create a 2D model projection structure. The unordering of image projection in the image occurs at different times under varying image angle-views, which are required for capturing 2D image projection.

Another motivation for image projection analysis is for projection recognition of image objects, referring to the design of random views of image object.

Our proposed image projection analysis focuses on the following:

(a) develop a generic and versatile model,
(b) develop methods for image detection, extraction, qualification and analysis,
(c) perform validation of accuracy, robustness and speed of the developed method,
(d) develop specific methods for automated analysis image illustration, and
(e) contribute to the field of image intelligence.

3 A SetpitextOFF Algorithm: Approach and Methodology

Analyzing the disadvantages of the existing methods of image projection analysis, the requirements of the image projection and image illusion, we propose a robust 2D image projection algorithm, developed to integrate image intelligence into the proposed algorithm, using the setpi cluster method.

The task of developing a 2D image projection algorithm for image analysis has two steps as follows:

(a) Design an algorithm which reconstructs the 2D image surfaces in the area of projection analysis.
(b) Add algorithm unlimited modules to maintain the structure of the image area with the previous area, which makes the possibilities to make corrections in the present area, to analyze changes from the last image area presence in order to determine, if area replanning is necessary, and to conduct the image projection further.

In our proposed algorithm, the above steps were addressed. We developed an algorithm to design and implement these approaches for analyzing the acquired image object and projecting it on the 2D image surface.

Among many 2D image projection algorithms, we use the 2D structured image patterns, providing fast object area measurement, by choosing an appropriate projection pattern, to increase the robustness and accuracy in the projected image, of the surface points from setOFF-angulation between surface and image source. The projection angle between the projected and the observed image object has to be taken for two main constraints:

(a) The 2D image projected that the procedure should be dependent of observed position
(b) The algorithm should be fast to project the image object on the 2D surface.

Taking into consideration these two features, we aimed to develop a fast processing algorithm, which can be easily implemented using existing methods. We also develop a pattern mask to producing structured 2D image object.

In this research work, we develop theoretically and implemented a novel purpose, projection algorithm, which allows obtaining the necessary image object 2D reconstruction even for low detail images.

The novelty of the proposed set pixel text OFF algorithm is in the following features:

(1) We used an image texture interspacing the approach on a regular point's pattern to locate the projected 2D image on a grid of setOFF-angulation.
(2) We observed the image object absence during image absence because of OFF pixel's presence in the image object, we propose retrieving the OFF image pixel's method, to find the image object patterns element, which is necessary for 2D image projection, by providing the exact location of OFF pixel's pattern.
(3) Due to image absence, projection algorithm is limited to partial image illusion, which lacks in edge projection to find the image object pattern and their shapes. We apply run-length pixels block mapping procedure to fit the image projection through the image distribution method.
(4) 2D projection in our pattern establish the correlation between the projected and the previous area, the problem is that with low estimation speed of projection, the change in image object is hard to define. To solve this problem, we use the

image fast projection method to correlate the nearest relation between projected and the previous area by pattern structure projection and correlated spacious image analysis.

(5) The initial idea of the proposed algorithm originates from the analysis of the important issue is to have contact over the method to have the similarity between projected and the previous area and to decrease the mapping errors in image object classification. This brings us to develop a setpi clustering algorithm, which is analysed for the proposed image projection algorithm.

3.1 Image Texture Interspacing Approach

Image texture approach is a mathematical realization, very well suited for theoretical reasoning. Image analysis mainly aims to analyze a high quality image versions.

For an observed image i, the image interspacing can be expressed by the following:

$$i = Gx + u, \tag{1}$$

Where G is an interspacing matrix, x is the original image vector and u is the setOFF-angulation vector. Mathematically, a signal $x \in R^V$ can be characterized as a linear arrangement of a few image vectors from the previous area. In the scenario of image interspacing, we have i to interspace over Φ to solve edge artefacts by the following:

$$i_y = ||i - G\Phi\alpha||, \tag{2}$$

And then the reconstructed i, denoted by i_r, is obtained as $i = \Phi\, i_y$. Clearly, iy is expected to be very close to i, so that the predicted area can be close to the projected area. Unfortunately, since i is having the less predicted area, the i_y vector may not predict the i vector, leading to the inaccurate image interspacing of the image texture. In other words, this model can assure i_y being predictive but cannot project as close to i as possible.

In this paper, we propose the concept of image texture interspacing (ITI) to facilitate the discussion of the problem. The IIR of the i is defined as follows:

$$u_x = i_x - i_y, \tag{3}$$

For the given Φ, the image interspacing result depends on the level of IIR setOFF-angulation along u_x, because the image interspace error $u_x = i_r\text{-}i \approx \Phi\, i_y\text{-}\Phi\, i_x = \Phi\, u_x$. The definition of IIR u_x also indicates one way to indicate the projected area from the predicted area that reduces the difference level of u_x (Table 1).

Table 1 Image texture interspacing approach

Algorithm 1 : Image texture interspacing approach
Initialization : i=G x+u,
REPEAT
Step 1 : Find G minimizing Φ with i_r fixed;
Step 2 : Solve iy=||i-G Φ α||,
Step 3 : Solve setOFF-angulation $u_x=i_r-i\approx\Phi\ i_y-\Phi\ i_x=\Phi\ u_x$, ITI index
UNTIL G converges

Table 2 Retrieving the OFF image pixel's method

Algorithm 2 : Retrieving the OFF image pixel's method
Initialization : (λ,μ)
REPEAT
Step 1 : Solve $(\lambda,\mu)=\arg\max_{\lambda,\mu} \log G(u_x|i_x,i_y) + G(i_x,i_y)$;
Step 2 : Solve $S(i_x,i_y)=S(\Phi|\ i_x)S(i_y)=S(\Phi_d|i_x)S(i_y)$,
Step 3 : Solve $S\ u_x=S\ G_i\ X$
UNTIL X_{OFF} converges

3.2 *Retrieving the off Image Pixel's Method*

The basic idea behind ITI is to assume that we can treat the G values as per previous area coefficients. Such idea is essential to retrieve the underlying image pixel during OFF state. Therefore, we might formulate the following maximum retrieving estimation method (MRE)

$$(\lambda, \mu) = \arg\max_{\lambda,\mu} \log G\left(i_x, i_y | u_x\right), \tag{4}$$

Using ITI, we can rewrite the above equation Eq. (4) into the following:

$$(\lambda, \mu) = \arg\max_{\lambda,\mu} \log G\left(u_x | i_x, i_y\right) + G\left(i_x, i_y\right), \tag{5}$$

Where corresponding two terms to image pixel's similarity and prediction distribution, respectively. The first term in Eq. 5 can be characterize by the retrieval model I = R (ON) + S(OFF), namely as follows:

$$R(\Phi) = G\left(u_x | i_x, i_y\right) = 1/\sqrt{2\pi\Phi} \exp\left(-1/2\Phi^2 ||\ u_x - \Phi_d||_2^2\right), \tag{6}$$

where R is ON image pixel retrieval, S is OFF pixel retrieval and Φ_d is the operator Φ dual.

Equation 5 refers to the retrieving the OFF image pixel's model by approximating the second term under the assumption that ITI, we can retrieve S (Φ) into the product of the image object estimators, such method implementation if structured as follows:

$$S(\Phi) = S(i_x, i_y) = S(\Phi|\ i_x)S(i_y) = S(\Phi_d|i_x)S(i_y), \tag{7}$$

where $\Phi_d = \Phi|i_x - \Phi$, defines the deviation between ON and OFF image pixels. Such retrieval-based prediction can be viewed as a level of proposed coding strategy, so Φ_d is approximately independent from $\Phi|i_x$. If we choose to model OFF image pixel's method, by ITI, the model is given by the following:

$$\begin{aligned} S(i_x, i_y) = 1/\sqrt{2\pi}\,\Phi \exp\left(-\Phi_2||u_x||1/\Phi\, x_y\right) * 1/\sqrt{2\pi}\,\Phi_d \\ \exp\left(-\Phi_d^2||\ u_x - \Phi_d||1/\Phi_y\right), \end{aligned} \tag{8}$$

Substituting Eq. (6) and (8) in (5), we obtain the following:

$$\begin{aligned} (\lambda, \mu) = \arg\min ||u_x - \Phi_d||_2^2 + 2\sqrt{2\Phi_x^2}/\Phi_x \\ \sum x||u_x||1 + 2\sqrt{2\Phi_x^2}/\Phi_y \sum x \sum y||u_x - \Phi_d||1, \end{aligned} \tag{9}$$

This is equivalent with Eq. (6) and Eq. (7).

Retrieval settings also allows our approach to generalize Eq. (7) into the following:

$$S\,u_x = S\,G_i X \tag{10}$$

Here S denotes a non-uniform OFF weighting operator in favour of image pixel absence closer to centre of the image object absence. Accordingly, we can extend the Eq. (10) into OFF image pixel retrieving solution

$$X_{OFF} = \left(\sum G_i^T S\, G_i\right) - 1 * \left(G_i^T S\, u_x\right). \tag{11}$$

In our proposed method, we are using a discrete window for S (Table 2).

3.3 Run-Length Pixels Block Mapping

Following the notations used in Algorithm 1 and 2, we establish the connection between ON and OFF pixel in the original image. Now we will take the set of image object coefficients $\delta = \{\delta_x\}$, so called mapping module. Let xi denote the block extracted from x and the mapping location is i; then we evaluate,

$$x_i = BM_i x, \tag{12}$$

where BM_i denotes block mapping operator, we obtain the solution for the original image block mapping as follows:

$$x = \left(\sum_i BM_i^T x_i - 1\right)\left(\sum_i BM_i^T x_i\right), \tag{13}$$

Which gives a mapping of ON and OFF pixels? Each BM operator for the run-length vector Δ, is related to image object coefficients $\{\delta_x\}$ by the following:

$$x_i = \Delta\{\delta_x\}, \tag{14}$$

Substituting Eq. 14 into Eq. 13, we obtain the following:

$$x = G\delta = \left(\sum_i BM_i^T[\Delta\{\delta_x\}] - 1\right)\left(\sum_i BM_i^T[\Delta\{\delta_x\}]\right), \tag{15}$$

Under the BM operator, the run-length model variation can be formulated as follows:

$$BM_i\{\delta_x\} = \arg\min_\delta 1/2 \, || \, x + (\lambda, u)\,\delta||_2^2 + \Delta\,||\{\delta_x\}||1, \tag{16}$$

The key motivations of Δ are for the randomly distributed image object coefficients. Their location certainty is related to block mapping operator, which implies the possibility of pixel mapping in ITI and OFF pixels retrieving. From these two perspectives, run-length pixels blocking the mapping connection is proposed.

3.4 *Image Fast Projection*

By converging into BM operator, we have set of an indices $[1 \le x \le a]$ for $BM_i x$ and set of b indices $[1 \le y \le b]$ for $BM_i y$, where $BM_i x$ denotes the block mapping indices with x and $BM_i y$ denotes the block mapping indices along y.

Let us define the projection constraint, as follows:

$$@p = þ(x,y): \quad \begin{array}{l} \sum a =_1{}^{ax} |\theta\, BM_i x| = 1,\ [1 \le x \le a\,] \\ \sum b =_1{}^{bx} |\theta\, BM = y| = 1,\ [1 \le y \le b\,] \\ \theta\, þ'(x,y) \ge 0,\ \Pi x,y \end{array} \tag{17}$$

Where θ is fast variation operator. The projection constraint @p has been seen as weighted version of G (x, y). The relation algorithm for @p and G (x, y) is given in the following table.

The proposed setpitextOFF algorithm utilizes clusters, instead of the image points. Simplification of clustering is as follows:

$$
\begin{aligned}
Þ'(x,y)(BM_i) &= \arg\min_{Þ \in BM_i} \sum (x,y) \; (\| Þ(i_x,i_y) \| - (\| (\lambda,u)\, \delta * Þ'(i_x,i_y) \|) \\
&= \arg\min_{\Delta x,\delta \in @p} \sum (x,y) \sum a \in BM_i x,\ b \in BM_i y \; (\| \Delta x,\delta (i_x,i_y) \| - (\| Þ'(i_x,i_y)) \\
&= \arg\min_{\Delta x,\delta \in @p} \sum (x,y)\ a \in BM_i x,\ b \in BM_i y \; (\| \Delta x,\delta (i_x,i_y) \| - (\| (\lambda,u)\, \delta * \Delta x,\delta (i_x,i_y) \|)
\end{aligned} \tag{18}
$$

where

$\Delta x,\delta = \sum a \in BM_i x,\ b \in BM_i y\ BM_i\ a,b \,/\, |BM_i x||BM_i y|$.

Therefore, it is natural to define the set pixel clusters as follows:

$$
Þ'(x,y)(BM_i) = \arg\min_{\Delta x,\delta \in @p} \sum (x,y)\ a \in BM_i x,\ b \in BM_i y \; (\| Þ(i_x,i_y) \| - (\| (\lambda,u)\, \delta * \Delta x,\delta (i_x,i_y) \|) \tag{19}
$$

Table 3 Run-length pixels block mapping

Algorithm 3 : Run-length pixels block mapping

```
Initialization : xi=BMi x,
REPEAT
        Step 1 : Denote x using BMi
        Step 2 : Define δ and Δ
        Step 3 : Solve BMi {δx}
        Step 4 : Convert BMi to pixel operator þ by Δx,δ
=
∑ a ϵ BMix, b ϵ BMiy  BMi a,b / |BMix||BMiy|
        Step 5 : Solve þ' (x,y) (BMi) = arg min þ'ϵþ ∑ (x,y)
( || þ (ix,iy) || - ( || (λ,u) δ * þ' (ix,iy) )
        Step 6 : Converts þ' to corresponding þ via:
                For Πx,y
                 If ϵ a, b , so that a ϵ BMix, b ϵ BMiy  ;
then
þ' (x,y) (BMi) Δx,δ ==> þ  (x,y) ;
                End
UNTIL þ converges
```

Table 4 Image fast projection

Algorithm 4 : Image fast projection
Initialization : BM_ix, BM_iy, @p
REPEAT
Step 1 : If $þ \in BM_i$, then $\Delta x, \delta \in$ @p,
where $\Delta x,\delta = \sum a \in BM_ix$, $b \in BM_iy$ BM_i a,b / $|BM_ix||BM_iy|$.
If BMi $\in$ @p then $p = [\, þ (x,y) \,]$ axb $\in BM_i$,
where $þ (x,y) = \Delta x,\delta \mid a \in BM_ix$, $b \in BM_iy$, $1 \le x \le a$, $1 \le y \le b$
Step 2 : Solve $þ' (x,y)$, $(a,b) \in p$
Step 3 : Solve $\sum a =_1{}^{ax} |BM_ix|=1$, x=1,2,3,...i, $(a,b) \in p(x)$
Step 4 : Solve $\sum b =_1{}^{bx} |BM_iy|=1$, y=1,2,3,...i, $(a,b) \in p(y)$
Step 5 : Solve $\Delta x,\delta$, $i \in (|BM_ix||BM_iy|)$, $(a,b) \in p(x,y)$
Step 6 : Augment the constraint indices to p(x), p(y) and p(x,y).
$p'(x) \in \{ (a,b) : \Delta x,\delta \le 0$;
$p'(y) \in \{ b : \sum a =_1{}^{ax} |BM_ix| \ge 1$;
$p'(x,y) \in \{ a : \sum b =_1{}^{bx} |BM_iy| \ge 1$ and $b : \sum a =_1{}^{ax} |BM_ix| \ge 1$;
UNTIL $þ' (x,y) \le 0$, $\Pi x,y$ and $\sum a =_1{}^{ax} |BM_ix| \ge 1$, Πx

From the definition of setpi cluster projection in Eq. 19 and Theorem 4, we can find an approximate solution to @p. The setpi cluster elements of image objects correspond to BM_i, and $(\lambda, u)\ \delta$ are averaged to get the clusters for (BM_i). The fast cluster corresponding projection can result by solving the double-setpi cluster projection by duplicating the $(\lambda, u)\ \delta$ elements with $þ' (x, y)$ (BM_i) in corresponding clusters. The double-setpi cluster projection correspondence is given as follows:

$$Đ\,[þ' (.)] = Đ\,[þ' (\theta_{a,b}, u_a, \delta_b, \lambda_{a,b})] = \sum a =_1{}^{ax} \sum b =_1{}^{bx} |\theta\, BM_ix|\, |\theta\, BM_iy|\ (\|\, þ\, (i_x,i_y) \,\| - (\|\, (\lambda,u)\, \delta * \Delta x,\delta\, (i_x,i_y) \|)\ {}_2{}^2 + \sum a =_1{}^{ax} u_a \left(\sum b =_1{}^{bx} (\|\, þ\, (i_x,i_y) \,\| - 1\,) + \sum b =_1{}^{bx} \delta_b \left(\sum a =_1{}^{ax} (\|\, þ\, (i_x,i_y) \,\| - 1\,) - \sum a =_1{}^{ax} \sum b =_1 \lambda_{a,b}\, \theta_{a,b}\right.\right. \tag{20}$$

The proposed setpitextOFF algorithm guarantees that the inequality constraints of ON and OFF pixels are satisfied with $(\theta_{a,b}, u_a, \delta_b, \lambda_{a,b})$ global variables, thus the proposed image projection analysis is providing a solution to @p with fulfilling the

Table 5 Values of G, Φ and Φ_{ux} for input image ITI index

		Man	Baboon	Peppers	Cameraman	Hill	House	Plant	Straw
x = 512X × 512	G (128 × 128)	29.27	29.33	29.21	29.24	28.26	28.24	28.36	27.26
	Φ(16 × 16)	0.35	0.31	0.32	0.33	0.35	0.36	0.35	0.32
	Φ_{ux}(75)	71	76	79	81	82	83	85	86
x = 256 × 256	G (32 × 32)	21.25	20.25	21.26	23.25	24.26	21.23	23.24	20.12
	Φ (4 × 4)	0.26	0.25	0.23	0.25	0.26	0.27	0.26	0.20
	Φ_{ux}(75)	61	62	65	69	71	75	63	68

(a) Original Input Image : Man

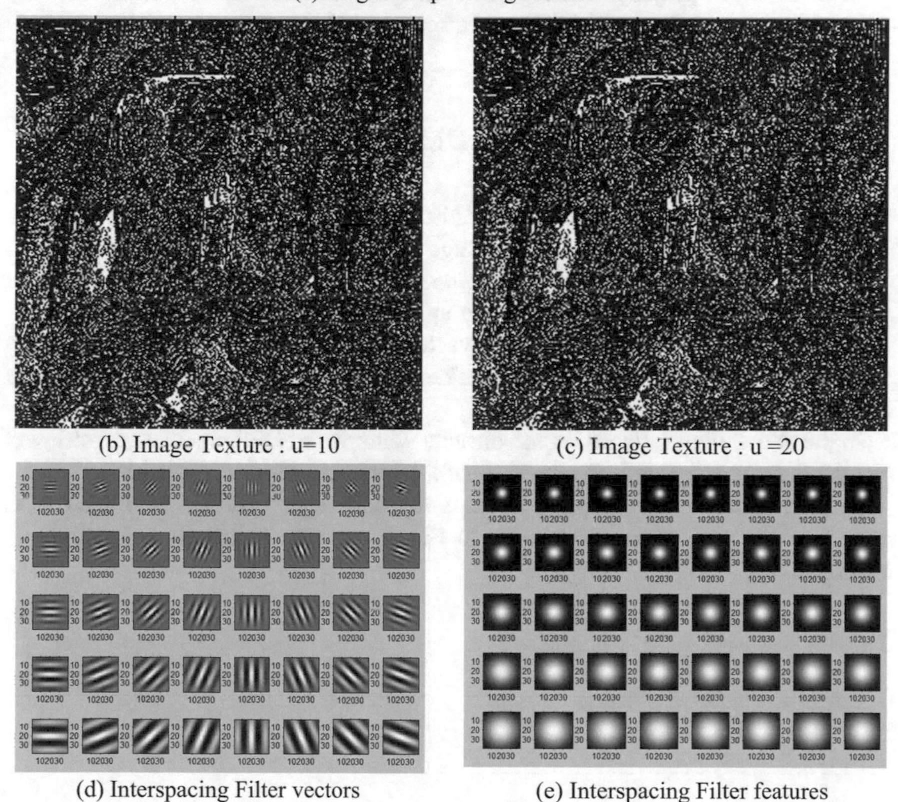

Fig. 1 Input image and its texture details, **a**: Original Input Image: Man, **b**: Image Texture: u = 10, **c**: Image Texture: u = 20, **d**: Interspacing Filter vectors, **e**: Interspacing Filter features, **f**: Texture Interspacing: u = 10, **g**: Texture Interspacing: u = 20

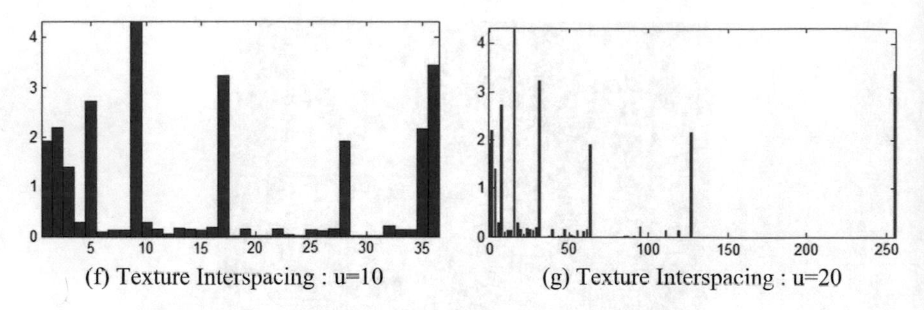

(f) Texture Interspacing : u=10 (g) Texture Interspacing : u=20

Fig. 1 (continued)

block mapping of OFF and ON image object pixels using inter-texture spacing, providing proper interdependency in double-setpi clusters too.

Our proposed algorithms, in Tables 3 and 4, would lead to a good approximation solution to image projection analysis in general. The proposed algorithm is very efficient in computation, reliability and effective with image projection analysis.

3.5 Experiments: Results and Discussions

The performance of the proposed setpitextOFF algorithm is implemented on benchmarks of image retrieval and image projection techniques. Eight different image textures as input images were taken tabulated in Table 5. Figure 1 shows the results of the image texture interspacing approach to input image man.

In the above Fig. 1, the interspacing vectors will retrieve the input and the image features based on the texture details as taken by u value. Figure 2 shows the image OFF AND ON pixels retrieving.

In the above Fig. 2, based on the discrete window the OFF pixels were retrieved and the S ux is estimated. In Fig. 3, block mapping approach is depicted.

In the above Fig. 3, the pixel interspacing and angle variation is plotted with respect to the projection parameters. In Fig. 4, proposed clustering algorithm is picturized.

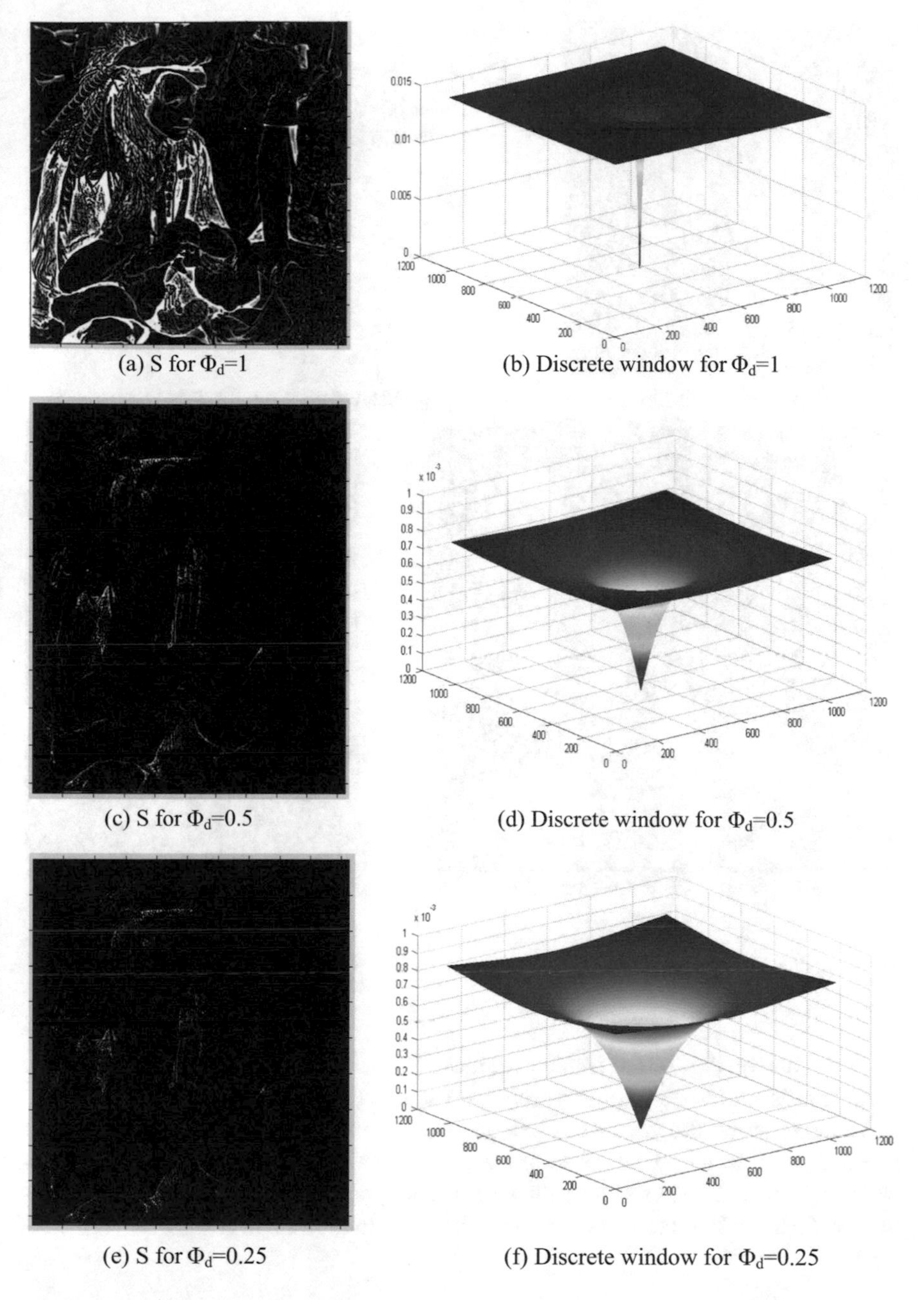

Fig. 2 Retrieving of S image pixels, **a**: S for $\Phi_d = 1$, **b**: Discrete window for $\Phi_d = 1$, **c**: S for $\Phi_d = 0.5$, **d**: Discrete window for $\Phi_d = 0.5$, **e**: S for $\Phi_d = 0.25$, **f**: Discrete window for $\Phi_d = 0.25$, **g**: S for $\Phi_d = 0.125$, **h**: Discrete window for $\Phi_d = 0.125$, **i**: X_{OFF}, **j**: S u_x

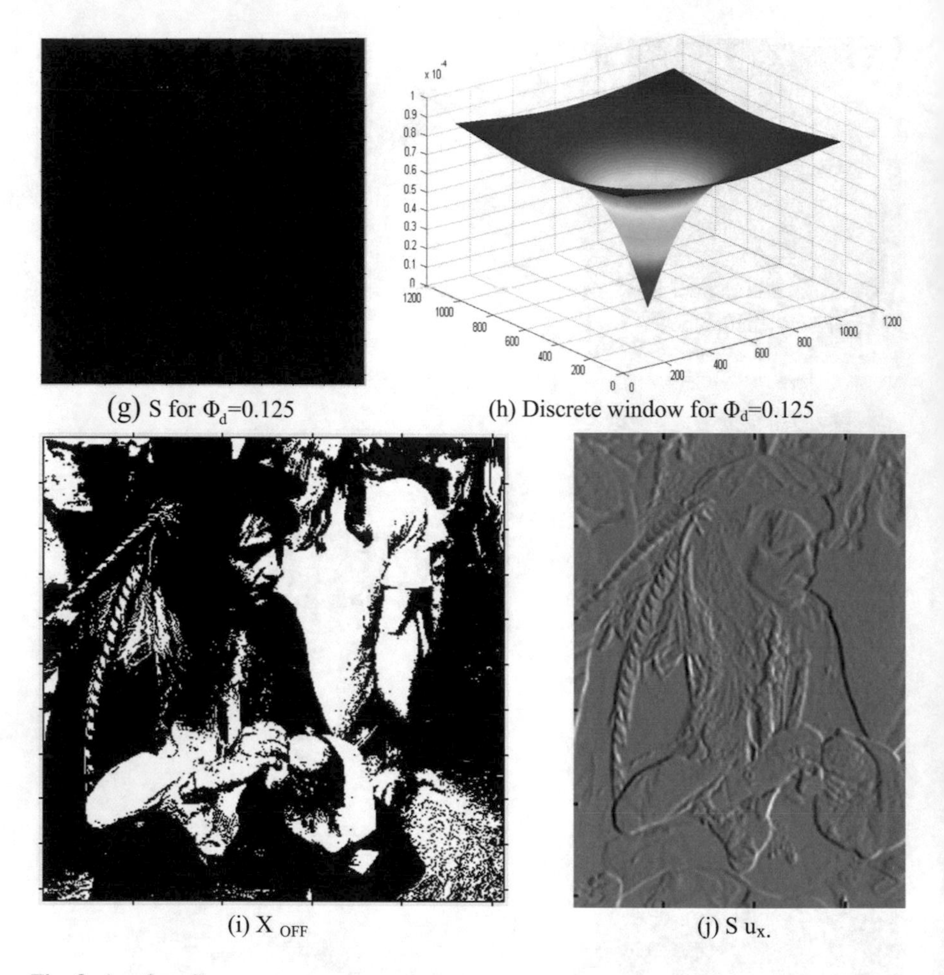

(g) S for Φ_d=0.125 (h) Discrete window for Φ_d=0.125

(i) X $_{OFF}$ (j) S u_x.

Fig. 2 (continued)

In the above Fig. 4, the implementation of clustering algorithm and the image projection is represented through retrieved image (Table 5, Fig. 5).

Table 1, shows the values of G, Φ and Φ_{ux} for the given eight input images. For x = 512 × 512, the values of G are very close to Plant image, the values of Φ is close to House image and the values of Φ_{ux} is close to the straw image. Similarly for x = 256 × 256, the values of G are very close to the Hill image, the values of

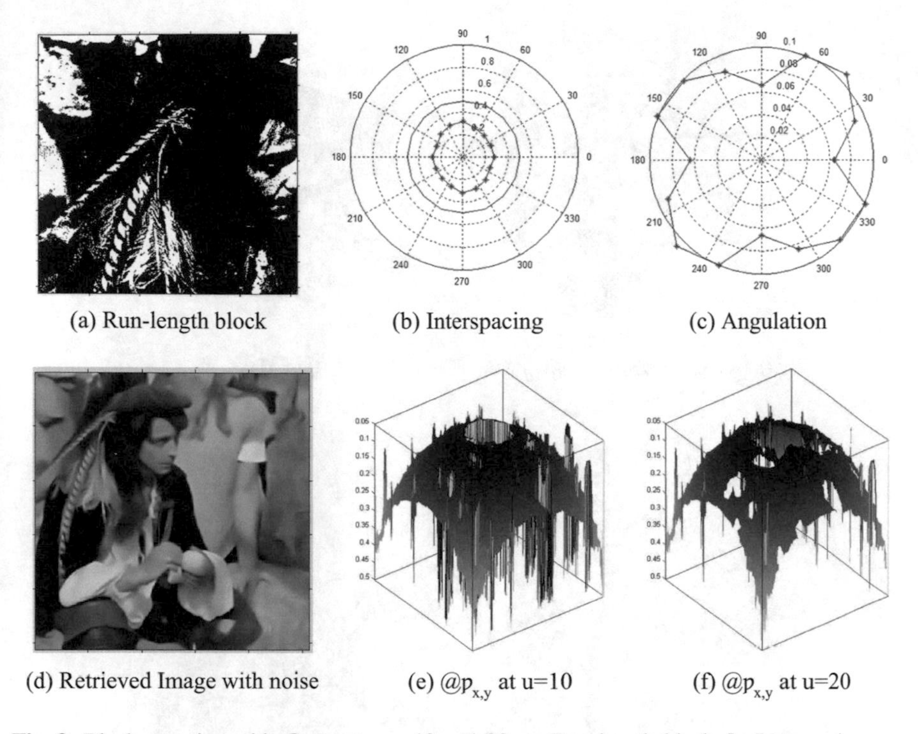

(a) Run-length block (b) Interspacing (c) Angulation

(d) Retrieved Image with noise (e) $@p_{x,y}$ at u=10 (f) $@p_{x,y}$ at u=20

Fig. 3 Block mapping with $@p_{x,y}$ at u = 10 and 20, **a**: Run-length block, **b**: Interspacing, **c**: Angulation, **d**: Retrieved Image with noise, **e**: $@p_{x,y}$ at u = 10, **f**: $@p_{x,y}$ at u = 20

Φ is close to the House image and the values of Φ_{ux} is close to the House image (Table 6).

Table 2, shows the Precision(P) and Recall(R) for the input image taken for analysis, the P value of $R(\Phi)_{OFF}$ is good for the Hill image and for $S(\Phi)_{OFF}$ is good for the Peppers image. Similarly for the remaining images, indication the OFF pixel's retrieval using the proposed method (Table 7).

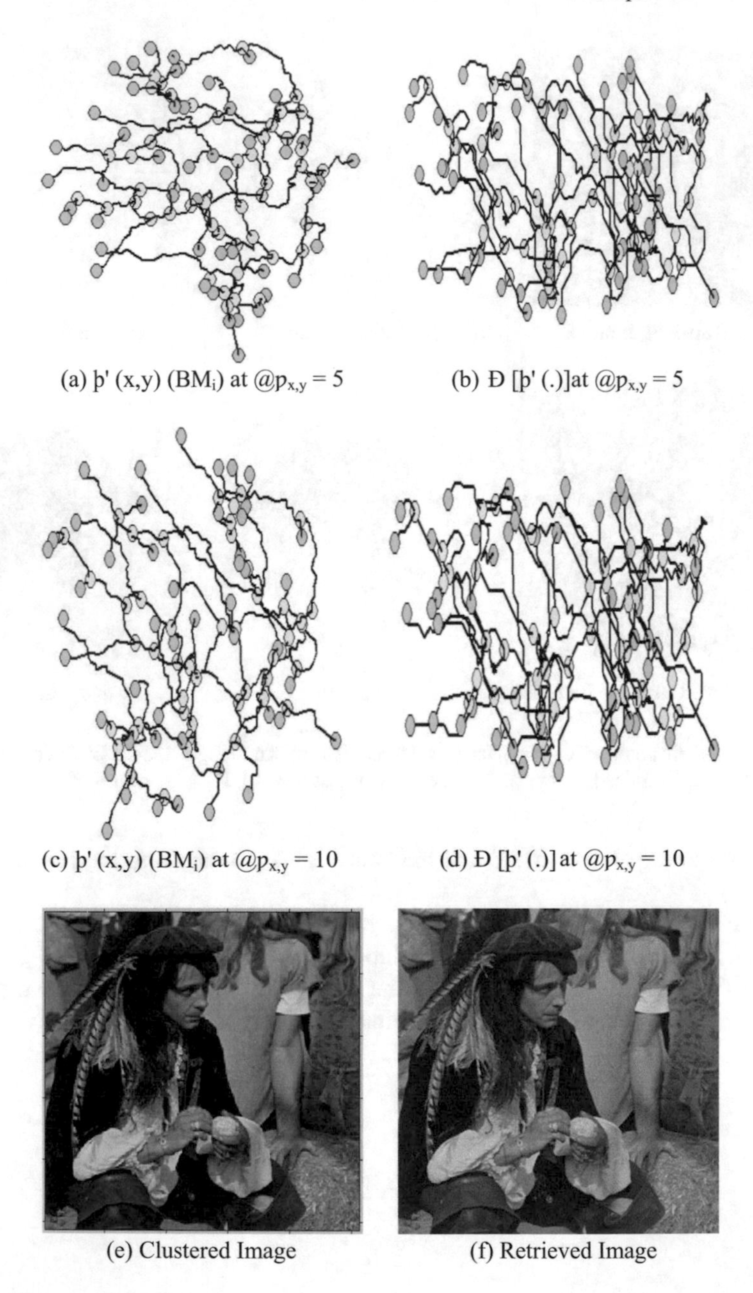

(a) þ' (x,y) (BM_i) at @$p_{x,y} = 5$ (b) Đ [þ' (.)]at @$p_{x,y} = 5$

(c) þ' (x,y) (BM_i) at @$p_{x,y} = 10$ (d) Đ [þ' (.)] at @$p_{x,y} = 10$

(e) Clustered Image (f) Retrieved Image

Fig. 4 Clusters representation

Fig. 5 Shows the eight different types of input images taken for analysis. **a**: Man, **b**: Baboon, **c**: Peppers, **d**: Cameraman, **e**: Hill, **f**: House, **g**: Plant, **h**: Straw

4 Conclusions

In this research paper, we investigated image projection analysis with setpitextOFF algorithm. To better understand the image projection analysis, we introduced the concept of retrieving the OFF pixels and setpi cluster, and it was found from the results that the precision value is mostly near to 50 and recall value is more than 6, indicating the best retrieval technique. To provide the best clustering among the image object classification, two methods were implemented, resulting in the accurate cluster image near to the retrieved image. In addition, the numerical result values of PSNR, depicts the image retrieval for projection analysis is accurate (approximately 50) in image reconstruction. Experimental results on thc input image data (eight) set, demonstrated that the proposed image projection algorithm can significantly outperform other image projection algorithms.

Table 6 Values of Precision (P) and Recall(R) for input image

	Man		Baboon		Peppers		Cameraman		Hill		House		Plant		straw	
	P	R	P	R	P	R	P	R	P	R	P	R	P	R	P	R
$R(\Phi)_{ON}$	50.46	6.23	49.12	6.22	48.25	6.25	49.42	6.24	49.22	6.57	48.26	6.59	47.25	6.24	48.22	6.35
$R(\Phi)_{OFF}$	48.21	6.10	49.25	6.25	48.22	6.35	47.26	6.24	49.27	6.35	48.36	6.57	47.32	6.58	48.22	6.94
$S(\Phi)_{ON}$	50.55	6.24	50.26	6.38	50.36	6.98	50.14	6.87	50.23	6.57	50.33	6.98	50.12	6.24	49.25	6.35
$S(\Phi)_{OFF}$	46.12	6.01	45.23	6.14	46.22	6.25	40.12	6.87	41.12	6.58	43.23	6.24	45.12	6.89	42.11	5.95

Table 7 The values of PSNR (dB) for the projected image with different image object coefficients

		Man	Baboon	Peppers	Cameraman	Hill	House	Plant	straw
δ(512×512)	$@p_x$	45.21	44.21	42.36	42.22	42.36	43.25	45.25	44.23
	$@p_y$	44.21	41.23	42.32	41.11	40.23	41.23	42.23	40.22
	@p(x, y)	51.23	56.23	57.26	52.23	51.23	50.23	57.25	59.22
δ(256×256)	$@p_x$	41.21	45.21	46.21	47.25	48.21	49.21	40.11	41.22
	$@p_y$	42.23	43.34	40.23	40.56	40.22	40.22	40.36	41.2
	@p(x, y)	51.22	52.23	53.25	51.25	54.26	55.23	56.23	55.23

References

1. Besl, P.J., McKay, N.D.: A method for registration of 3-D shapes. IEEE Trans. Pattern Anal. Mach. Intell. **14**(2), 239–256 (1992)
2. Guo, H., Rangarajan, A., Joshi, S., Younes, L.: Non-rigid registration of shapes via diffeomorphic point matching. IEEE Int. Symp. Biomed. Imag., 924–927 (2004)
3. Rangarajan, H., Chui, E.M.: A relationship between spline-based deformable models and weighted graphs in non-rigid matching. In: IEEE International Conference on Computer Vision and Pattern Recognition (CVPR), pp. 897–904 (2001)
4. Belongie, S., Malik, J., Puzicha, J.: Shape matching and object recognition using shape contexts. IEEE Trans. Pattern Anal. Mach. Intell. **24**(4), 509–522 (2002)
5. Yefeng Zheng, D.: Doermann, Robust point matching for nonrigid shapes by preserving local neighborhood structures. IEEE Trans. Pattern Anal. Mach. Intell. **28**(4), 643–649 (2006)
6. Maciel, J., Costeira, J.P.: A global solution to sparse correspondence problems. IEEE Trans. Pattern Anal. Mach. Intell. **25**(2), 187–199 (2003)
7. Medasani, S., Krishnapuram, R., YoungSik, C.: Graph matching by relaxation of fuzzy assignments. IEEE Trans. Fuzzy Syst. **9**(1), 173–182 (2001)
8. Belongie, S., Malik, J., Puzicha, J.: Shape matching and object recognition using shape contexts. IEEE Trans. Pattern Anal. Mach. Intell. **24**(4), 509–522 (2002)
9. Thayananthan, A., Stenger, B., Torr, P.H.S., Cipolla, R.: Shape context and chamfer matching in cluttered scenes. In: IEEE Conference on Computer Vision and Pattern Recognition (CVPR) (2003)
10. Yefeng Zheng, D.: Doermann, Robust point matching for nonrigid shapes by preserving local neighborhood structures. IEEE Trans. Pattern Anal. Mach. Intell. **28**(4), 643–649 (2006)
11. Jiang, H., Drew, M.S., Li, Z.-N.: Matching by linear programming and successive convexification. IEEE Trans. Pattern Anal. Mach. Intell. **29**(6), 959–975 (2007)
12. Stark, H., Yang, Y.: Vector Space Projections: A Numerical Approach to Signal and Image Processing, Neural Nets and Optics. Wiley, New York (1998)
13. Youla, D.C.: Mathematical theory of image restoration by the method of convex projections, Chapter 2. In: Stark, H. (ed.) Image Recovery: Theory and Applications. Academic Press, Orlando, Florida (1987)
14. Treisman, A.: A feature in integration theory of attention. Cogn. Psychol. **12**(1), 97–136 (1980)
15. Julesz, B.: Textons, the elements of texture perception and their interactions. Nature **290** (5802), 91–97 (1981)
16. Julesz, B.: Texton gradients: the texton theory revisited. Biol. Cybern. **54**, 245–251 (1986)

Enhancing Visual Speech Recognition with Lip Protrusion Estimation

Preety Singh

Abstract Visual speech recognition is emerging as an important research area in human–computer interaction. Most of the work done in this area has focused on lip-reading using the frontal view of the speaker or on views available from multiple cameras. However, in absence of views available from different angles, profile information from the speech articulators is lost. This chapter tries to estimate lip protrusion from images available from only the frontal pose of the speaker. With our proposed methodology, an estimated computation of lip profile information from frontal features, increases system efficiency in absence of expensive hardware and without adding to computation overheads. We also show that lip protrusion is a key speech articulator and that other prominent articulators are contained within the centre area of the mouth.

1 Introduction

The use of artificial intelligence to alleviate human effort and time has motivated researchers to look for methods whereby computers can be trained to perform tasks confined to the domain of man. Human–computer interaction is increasingly gaining importance in such a scenario. Understanding and correctly interpreting human inputs is of utmost priority for the machine to perform its task efficiently. An example of extensive use of interaction between man and machine is the field of visual speech recognition. In this system, the machine takes visual inputs from speech articulators (eg. lips, teeth, jaw, etc.) of the speaker and outputs the word being spoken. Visual speech recognition (VSR) has many applications. In speech recognition systems, it has been shown by Sumby and Pollack [44] that visual information from areas in and around the mouth, supplements the audio signal and increases speech recognition accuracy. This is because visual cues are not affected by the presence of noise while the audio signal may be degraded. It has also been used for visual

P. Singh (✉)
The LNM Institute of Information Technology, Jaipur, India
e-mail: prtysingh@gmail.com

A.E. Hassanien and D.A. Oliva (eds.), *Advances in Soft Computing and Machine Learning in Image Processing*, Studies in Computational Intelligence 730,
https://doi.org/10.1007/978-3-319-63754-9_24

text-to-speech conversion [8, 9]. Use of visual speech as a behavioural biometric in a multimodal system is also becoming popular [2, 19, 43].

The inputs to a visual speech recognition system are attributes extracted from the region of mouth, over a sequence of images constituting the speech time frame. These features can be broadly classified into top-down and bottom-up features. While top-down features consist of geometrical and shape parameters of the mouth, mainly the lips [12, 26, 30, 47, 48], bottom-up features comprise of pixel intensities of the region-of-interest (ROI) encompassing the mouth, teeth and tongue [3, 10, 21, 35]. Hybrid features are also used which combine geometrical and pixel intensity information [6, 10, 30].

Features extracted from a particular image sequence of the mouth form a pattern for that particular word and can be used to form a feature vector to train the machine. In most of the researches done for VSR systems, this image sequence is captured from the frontal pose of the speaker [29, 34, 40]. The speaker faces the camera and the movement of the mouth is recorded while he speaks. However, humans do not lip-read in this manner. They prefer lip-reading at a slight angle so that lip protrusion information can also be included in their perception of speech [1, 24, 27]. Considering this, experiments have been performed to study the effect of including lip profile information on visual speech recognition. In [39], research has been done on profile lip reading for recognition of Japanese vowels and words. Five facial points are determined from the profile view of the speaker. Different Euclidean distances and protrusion of the lips are computed from these points. For five Japanese vowels recognition, the accuracy is reported to be 99.6%. For recognition of twenty Japanese words, a recognition accuracy of 85.7% is observed. It is also shown that of all the computed features, the heights of the lips in the profile view and lip protrusion are important features.

In a study by [25], recordings are done on a single speaker uttering 200 sentences from the Resource Management Corpus [36]. Five viewing angles are taken into account using multiple cameras. Shape and appearance parameters are extracted from images forming the video sequences. It is shown through experiments that a full frontal view is not optimal for lip reading but a slightly angled view, at 30°, gives better recognition performance. In an extensive study of lip reading across multiple views [28], Lucey has shown that frontal view gives better results compared to a 90° profile view. While the WER (word error rate) is 27.66% for the frontal pose, it degrades to 38.88% for the full profile view. It has also been shown that a combination of the views greatly enhances performance of the lip-reading system and the WER obtained is 25.36%. While doing a patch analysis, the middle patch of lips, teeth and tongue from the frontal view are shown to be important information providers while it is hypothesized that lip protrusion from the profile view is a prominent attribute.

Iwano et al. [20] have used lip information from side-face images to improve noise robustness for audio-visual speech recognition in mobile environments. Lip contour geometric features and lip-motion velocity features are extracted and used individually or jointly, along with audio features. Experiments are conducted on a database of Japanese connected digit speech contaminated with white noise. It is reported that error reduction rates using side-face images are less compared to frontal-face images, although this cannot be strictly compared as different set of speakers was used for both experiments. In [24], Kumar et al. extract four profile features (upper and lower lip heights as well as protrusions) and three frontal features (lip width and lip heights). Experiments show that the WER for lip reading using frontal features is significantly higher as compared to profile features. Best WER values of 52.5% for profile view and 60.5% for frontal view are reported for a particular speaker.

Goecke et al. [15] performed experiments on a database consisting of sequences covering a range of phonemes and visemes in Australian English with ten native speakers. They have used two video cameras and a head-tracking system. The two images are multiplexed into a single image. The key visual points include the two lip corners and mid-points of upper and lower lips. From these four points, a parameter set consisting of the mouth width, mouth height, protrusions of upper and lower lip are derived. It is shown that the protrusion of the lips and mouth height are important attributes.

Experiments have shown that the cognitive information contained in the puckering of human lips while they speak is lost if only the frontal pose is considered. However, it may be expensive to install multiple cameras to gather information from different angles. Moreover, the dimensionality of the collected information may become very large, leading to problems in storage and handling, apart from being computationally expensive. In this chapter, we try to estimate the lip protrusion of the speaker, as he speaks, from his frontal pose only and without the availablity of any image from any other angular profile of the speaker. Thus, from a two-dimensional aspect, we add a third dimension to enhance the performance of a visual speech recognition system without the use or installation of expensive hardware. However, the estimate is not absolute but with respect to the protrusion of the speaker when his mouth is in the normal closed position.

In real-time implementation, the processing time should be as minimal as possible and our proposed estimation ensures this by being simplistic in its approach. To further reduce processing time, we try to find a subset of features which contribute significantly to the recognition of the class and do not contain redundancy. Diminishing our feature vector will further improve the efficiency of our system. This has been done by applying Correlation-based Feature Selection (CFS), introduced by Mark Hall [17]. The reduced set of features is classified using an ensemble of classifiers to test an unknown input word image sequence. Experiments show that addition of lip protrusion information enhances speech recognition. The experiments have been performed for an isolated word recognition system and are an extension of our work in [41] where we worked with frontal geometrical features only. The main contributions of this chapter are as follows:

- Protrusion of lips, during speech, is estimated from frontal images of the mouth in absence of availability of multiple camera views.
- A subset of visual features is identified which reduces system processing time and increases its efficiency.

- The importance of protrusion towards visual speech recognition accuracy is shown by classifying the reduced feature set using an ensemble of classifiers.
- Inspection of features contained in the feature subset shows that most of the prominent speech articulators are located in the centre area of the mouth.

The chapter is organized as follows: Sect. 2 describes our proposed methodology outlining the extraction of frontal geometrical features, estimation of lip protrusion, feature selection, computation of a near-optimal feature subset and classification. Section 3 presents the experiments, results and analysis of the feature subsets. Section 4 concludes the chapter.

2 Proposed Methodology

The experiments in this research work are based on the top-down approach and utilize the geometrical features extracted from the lips of the speaker. The steps of our proposed methodology are shown in Fig. 1. As the speaker utters a word facing the camera, his video is recorded. From each image frame in the video sequence, the lip contour is detected and segmented from the rest of the image. Various geometrical features, including height and width of lips are determined. Area segments contained within the lip contour, are also computed from the lips. Lip protrusion is estimated from these frontal visual features. To reduce the dimensionality of our feature vector, we apply feature selection. Prominent features, satisfying the conditions of being relevant to the word class but not being redundant, are then identified. Using the prominent features, we train our classifier model, consisting of an ensemble of three classifiers. We test the performance of our features by identifying the word from an unknown input image sequence of the lip movement. The following subsections describe our approach in detail.

2.1 Detection of Lip Contour

The Point Distribution Model (PDM) has been used to extract the lip boundary of the speaker. Segmentation and edge detection are first employed to extract the lip contour in ten training images [16]. An intensity analysis of the images results in determining six *keypoints* on the boundary of the lip. These are the corners of the lips, three points lying on the upper lip arch and the lowest point on the lower lip arch. Other points on the lip contour are interpolated between these keypoints, resulting in 120 such points. Each point is characterized by its $\langle x, y \rangle$ co-ordinates. The mean of all these co-ordinates over the training set yields a mean model of the lip. When this model deforms over an input image, it terminates on detecting a change in intensity. This generates the lip contour in the input image.

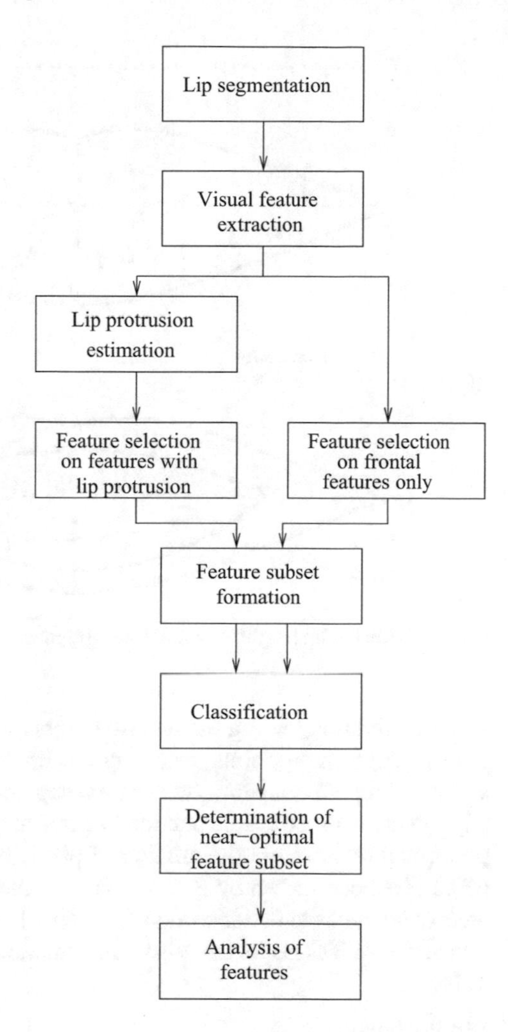

Fig. 1 Proposed methodology

2.2 *Extraction of Visual Features*

From the input two-dimensional image of the mouth of the speaker, we extract high level geometrical features from the lip boundary of the speaker. Geometric parameters are few and easier to derive which is an advantage in a real-time system. They also contain significant speech information. From these features, we estimate the protrusion of the lip of the speaker in each image as discussed in the following subsections.

Frontal Geometrical Features

From the outer boundary of the lip contour, we extract the height of the lip (h) and its width (w). We also divide the area contained within the lip contour into segments.

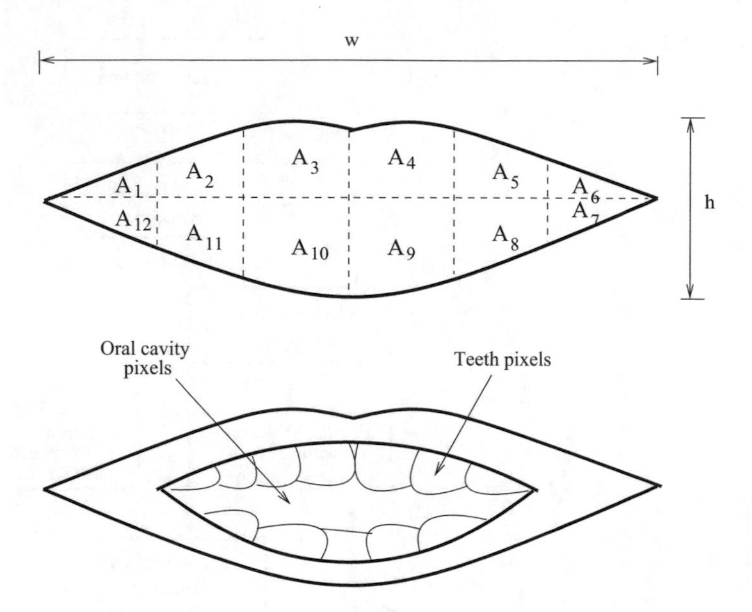

Fig. 2 Extracted lip height, lip width, area segments, teeth pixels and oral cavity pixels

A horizontal line joining the two lip corners is divided into six equal parts. Vertical lines through these points, touching the lip contour at extremities, divide the area within the lip contour into twelve area segments ($A_1 \cdots A_{12}$). Analysis of the intensity values, within the area bounded by the keypoints in the input RGB image, is also performed to ascertain the number of pixels of detectable teeth (Γ) and oral cavity (Θ). It has been shown by Brooke and Summerfield [5] that visibility of tongue and teeth contributes to better recognition. Finn [13] has shown that oral cavity information also provides important visual information. These geometric features are shown in Fig. 2.

Lip Protrusion

We assume that the speaker's lips are in a closed position before he starts speaking. The lip protrusion (if visualized from a side profile) in this frame is assumed to be his normal protrusion. When he speaks, the width (w) of his lips changes as the word is being spoken. Each word consists of a number of *phonemes* [37] and there is a specific mouth shape, the *viseme* [7], associated with it. A phoneme-to-viseme mapping has been proposed by Neti et al. [33]. Intuitively it is known that the protrusion (p) and lip width (w) change relative to each other. If lip width decreases, protrusion is more. If the lips spread out and lip width increases, protrusion is less and may be even less than the normal protrusion. This can be more clearly understood from Fig. 3. We assume that for a particular speaker, the volume of his lips, V, remains constant in all positions of speech utterance. Thus, no matter what position the mouth may be in, volume V will be the same. However, other geometrical features of the lip like, lip height, width, area segments and protrusion would change.

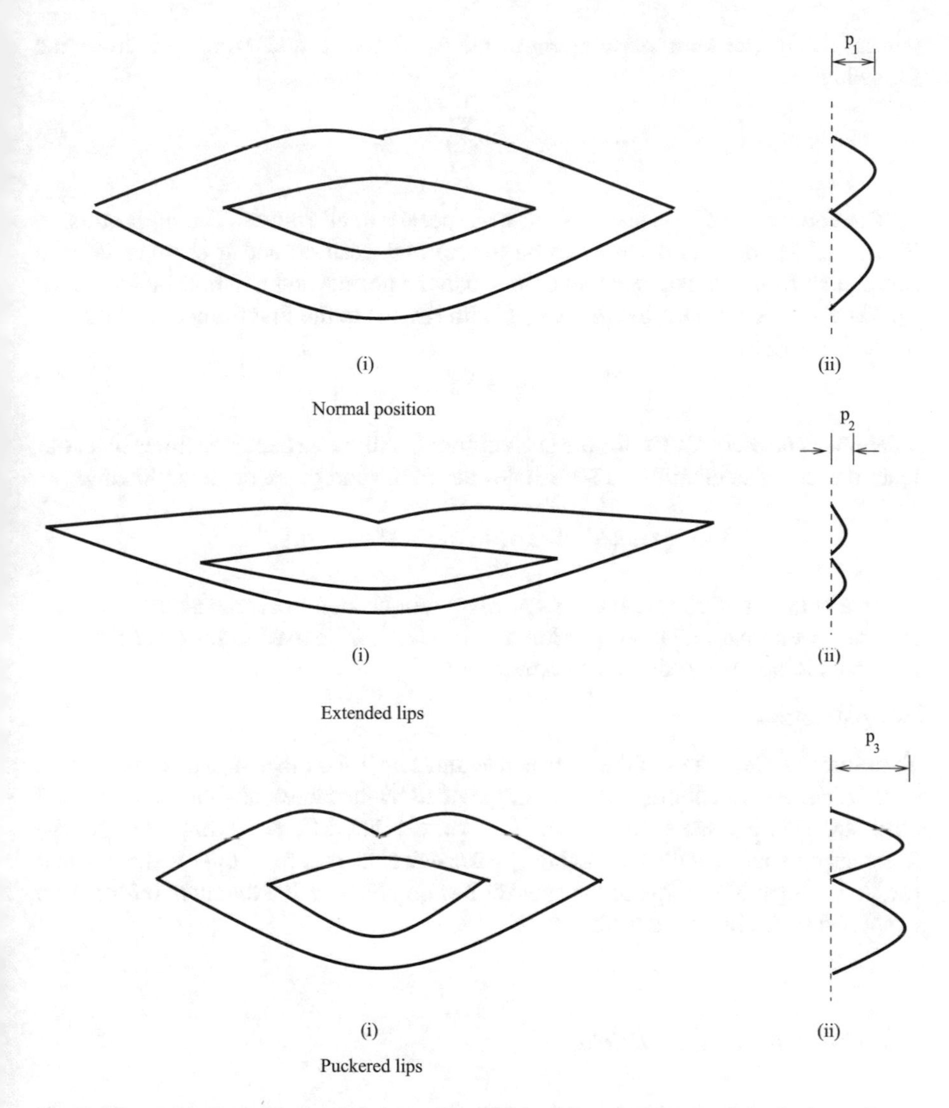

Fig. 3 Changes in lip protrusion relative to lip width in different positions of the lip. In each case (i) shows the frontal view while (ii) depicts the profile view. as can be seen, $p_2 < p_1 < p_3$

Let, the protrusion in each frame be denoted by p_f, where, f is the frame number. We take the protrusion in the first frame of the image sequence, p_1 as the reference value and consider $p_1 = 1$. Let, the sum of the twelve area segments in frame f be A_f. Thus, the volume of the lips of the speaker can be estimated as follows:

$$V = A_1 * p_1 \tag{1}$$

where, A_1 is the sum of area segments in frame 1 and computed from the following:

$$A_f = \sum_{i=1}^{12} A_i \tag{2}$$

The volume V will be constant for that speaker in all frames. The lip protrusion in the first frame is considered to be the normal position and it changes in each subsequent frame. It may be more or less than the normal and will now be a factor of p_1. Thus, for each frame, the protrusion with respect to the first frame, will be given by the following:

$$p_f = V/A_f \tag{3}$$

where A_f is now computed for the current frame. All the extracted features form our feature vector Q, containing 17 attributes per frame and given by the following:

$$Q = \{h \quad w \quad A_1 \cdots A_{12} \quad \Gamma \quad \Theta \quad p\}$$

As a person will speak, these parameters will change in each frame and form a pattern for that particular word. This pattern can now be used to train our machine for recognizing the word in an unknown input.

Normalization

To ensure that the values of the extracted features lie in the range [0, 1], normalization is done. For each attribute, all values are divided by the largest absolute value of that attribute. This is done speaker-wise. It is known that different human beings have different lip shapes and normalizing speaker-wise takes care of this. It ensures that the static shape of the lips of the speaker has no effect on the dynamic information contained in the lip movement.

2.3 *Selection of Features*

We have a total of 17 visual features. This is not a large feature set but our aim is to remove irrelevant and redundant features so that the processing time in real-time implementation can be minimized to the extent possible. For this purpose, we apply feature selection technique, Correlation-based Feature Selection (CFS) [17], which chooses a feature subset based on the individual predictive ability of each feature and the degree of redundancy between features. Its evaluation function is inclined towards subsets which contain features having prominent correlation with the class and minimal correlation with each other. The correlation between the variables is computed using information gain. For ranking of features, we have employed the forward selection method utilizing the greedy algorithm.

Let, N be the complete set of features consisting of all n features and S be the target subset of features, $S \subset N$, where S consists of k relevant features, $k \leq n$. Let, $C = \{C_1, C_2 \cdots C_m\}$ denote the m target classes. Let, $\overline{r_{Cf}}$ represent the mean feature-class correlation ($f \in S$) and $\overline{r_{ff}}$ be the average feature-feature correlation. The feature subset evaluation function, Q_S is given by the following:

$$Q_S = \frac{k\overline{r_{Cf}}}{\sqrt{k + k(k-1)\overline{r_{ff}}}} \tag{4}$$

In this equation, the numerator indicates the predictive measure of feature set towards the class while the denominator indicates the redundancy among features.

To compute symmetrical uncertainty coefficient, *suc*. The probabilistic model of a feature Y can be estimated from the individual probabilities of the values $y \in Y$ from the training data. The entropy of Y is given by the following:

$$H(Y) = -\sum_{y \in Y} p(y) \log_2(p(y)) \tag{5}$$

If a relationship between Y and a second feature X exists and the observed values of Y are partitioned according to values of X, then the entropy of Y with respect to X, represented by $H(Y|X)$, will be less than entropy of Y prior to partitioning.

$$H(Y|X) = -\sum_{x \in X} p(x) \sum_{y \in Y} p(y|x) \log_2(p(y|x)) \tag{6}$$

The decrease in amount of entropy of Y indicates the additional information about Y provided by X. This is the *information gain* and can be computed by:

$$Gain = H(Y) - H(Y|X) \tag{7}$$

Using this value, symmetrical uncertainty coefficient, *suc*, is calculated as follows:

$$suc = 2.0 * \left[\frac{Gain}{H(Y) + H(X)}\right] \tag{8}$$

The value of *suc* can help to compute feature-feature correlation $\overline{r_{ff}}$ and class-feature correlation $\overline{r_{Cf}}$, which then can be used in the evaluation of the merit of a subset.

Formation of Feature Subsets

We form feature subsets starting from the top-ranked feature. This is denoted by Q_1. We add the next ranked feature to this and form Q_2 which then contains the top two ranked features. In this manner, we keep on adding features ranked next in order and obtain feature subsets $Q_1 \cdots Q_{17}$ where Q_k contains the top-ranked k features. It is clear that Q_{17} is equivalent to the complete feature set Q. These feature subsets are now used for classification using an ensemble of classifiers.

2.4 *Classification of Feature Subsets*

An ensemble of classifiers consisting of three classifiers, viz. Naive Bayes, k-Nearest Neighbour and Random Forest are used for classification of the feature subsets. Stacking is used to make a final prediction using a combiner algorithm which utilizes the predictions of other algorithms as inputs [11].

Evaluation Metrics

The classification results of the feature subsets are evaluated using certain metrics, namely, *Precision*, *F–measure* [31] and *ROC Area* [45] are determined. These are defined as follows:

- *Precision* (P): This is computed by taking the ratio of correctly identified speech samples to all speech samples classified as belonging to that class.
- *F–measure* (F): Apart from accuracy and word error rate, F–measure is also a popular metric used in many classification problems [4, 14, 22, 46]. This can be used as a measure to indicate the accuracy of a test [31].
- *ROC Area* (A): The area A under the Receiver Operating Characteristic (ROC) curve can be used to signify the overall performance of a system.

2.5 *Computation of Optimality Factor*

The emphasis of this chapter is on having a subset of features which is small in size but performs reasonably well towards recognition of the spoken word. For determination of this, we compute the *Optimality Factor* of each subset [41, 42]. Let, the complete feature set, Q, contain n features and its F–measure value be F_n. Similarly, let feature subset, Q_k, containing k top features, have an F–measure value denoted by F_k. Let the measure of alteration between the two values be represented by Y_k, where:

$$Y_k = \frac{F_n - F_k}{F_n}$$

Desirable values for Y_k are $Y_k < 0$ and $|Y_k| \rightarrow \infty$. This indicates that the given feature subset exhibits enhanced recognition performance ($F_k > F_n$). All such feature subsets are considered for further analysis. Positive values of Y_k are the indication of degradation in the recognition rate ($F_k < F_n$). However, they maybe considered till a certain threshold. We have taken this threshold as $Y_k \leq 0.1$. If $Y_k = 0$, this shows that the performance of Q_k is similar to Q and there is no modification in the F–measure value. All subsets satisfying the Y_k criteria are short-listed.

Since another important criteria to ascertain a near-optimal subset is the length of the feature vector, we define η_k as the change in the size of the short-listed feature subsets in relation to Q. This is given by the following:

$$\eta_k = \frac{n-k}{n}$$

$\eta_k \rightarrow 1$ will indicate that the feature length has been substantially reduced. However, there should be no compromise in the recognition accuracy. To maintain this balance, Optimality Factor, O_k, of the feature subset is defined, which is computed as follows:

$$O_k = \eta_k * P_k * A_k$$

where, P_k is the precision and A_k is the ROC Area of the feature subset. This ensures the best combination of high evaluation metrics and small feature vector length. Thus, all evaluation metrics are taken into account and the feature subset having the largest value of Optimality Factor is considered.

3 Experiments and Result Analysis

Experiments have been performed on our own audio-visual database. This database has been motivated by a few earlier recorded databases, AVLetters [29], OuluVS [49] and Tulips1 [32]. Twenty subjects, comprising of ten males and ten females, participated in the recording of ten English digits, *zero* to *nine*. Each subject recorded five utterances of each digit, giving a total of 1000 speech samples. To take speaker variablity into account, the recording was done in two sessions. In the first session, three utterances were recorded while in the second recording, three months later, two utterances were recorded.

Recording was done in a relatively isolated laboratory in moderate illumination. The speaker sat facing the camera focused on the lower half of the face of the speaker. The digits were shown to the speakers in a non-repetitive and non-sequential manner. The subjects were requested to speak without movement of their heads and to start and finish each utterance with their mouth in the closed position. Blue colour was applied to the lips of the speakers to aid in easy segmentation of the lips in the HSV space. Blue colour has been used as a marker earlier in [23, 38]. Some sample images from our database are shown in Fig. 4. Frame-grabbing of the video sequence was done at 30 frames per second. The size of each image was 640 × 480.

The video sequence of each utterance is composed of a batch of image frames. The first frame starts with the image of a mouth. Subsequent frames show the mouth in different positions while speaking that particular word. The last frame shows the closed mouth position again. Visual parameters extracted from each frame of the image form a pattern of visual lip movements of that digit, pertaining to that speaker. Thus, each image sequence renders a visual pattern for a word.

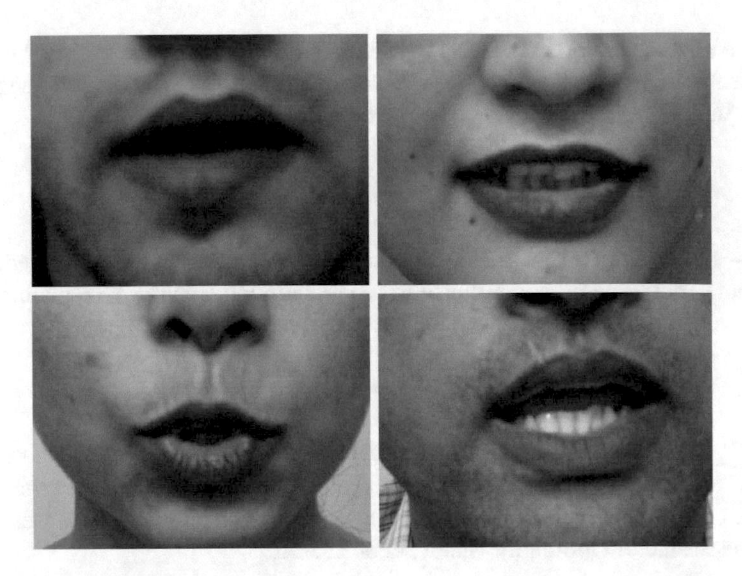

Fig. 4 Images of subjects from recorded database

Correlation-based Feature Selection is applied to rank the normalized features. Feature subsets are then formed and classified using an ensemble of three classifiers: Naive Bayes, k-Nearest Neighbour and Random Forest, available in the popular data-mining tool WEKA [18]. The dataset is divided into a training and testing set. 66% of the speech samples are used for training the classifier model while the remaining speech samples are used for testing. Using the evaluation metrics, the Optimality Factor is computed, as described in Sect. 2.5. A set of most relevant features, having minimal length and reasonable recognition metrics, is thus identified.

To prove that supplemntal information provided by the lip protrusion indeed increases speech recognition performance of the system, we perform the same set of experiments using the frontal features only, i.e. considering all features of Q, except the lip protrusion. Let us denote this set by G. We compare the results of the two feature sets and thus validate our approach.

3.1 Result Analysis

On application of CFS on our feature vector Q, the features are ranked in the following order of prominence:

$$Q = \langle p\ w\ \Gamma\ \Theta\ A_9\ A_7\ A_5\ A_2\ A_{12}\ A_{11}\ A_1\ A_6\ A_4\ h\ A_8\ A_3\ A_{10} \rangle$$

We form feature subsets $Q_1 \cdots Q_{17}$ with these ranked features as explained in Sect. 2.4 and classify them using our ensemble of classifiers. The evaluation

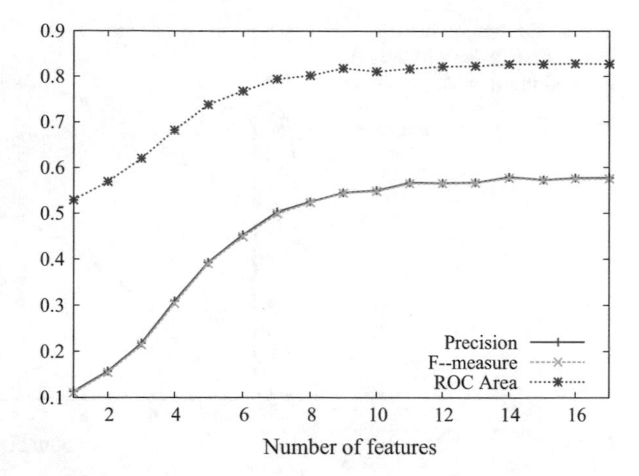

Fig. 5 Evaluation metrics of feature subsets containing lip protrusion

Table 1 Performance of short-listed feature vectors subsets using ensemble of classifiers. Change in size (η_k), F–measure value (Y_k) and Optimality Factor (O_k) are also given. The results for feature vector Q_{17} are also shown

Feature vector	P_k	F_k	A_k	η_k	Y_k	O_k
Q_8	0.525	0.523	0.802	0.529	0.090	0.223
Q_9	0.545	0.544	0.818	0.471	0.054	0.210
Q_{10}	0.550	0.548	0.811	0.412	0.047	0.184
Q_{11}	0.567	0.565	0.817	0.353	0.017	0.163
Q_{12}	0.566	0.565	0.822	0.294	0.017	0.137
Q_{13}	0.567	0.566	0.823	0.235	0.016	0.109
Q_{14}	0.579	0.577	0.827	0.176	–0.003	0.084
Q_{15}	0.573	0.572	0.827	0.117	0.005	0.055
Q_{16}	0.577	0.575	0.828	0.059	0	0.028
Q_{17}	0.577	0.575	0.827	–	–	–

metrics obtained for the subsets are shown in Fig. 5. On application of thresholding on the F–measure values of the feature subsets, we find that only nine feature satisfy the F–measure criteria. These are $Q_8 \cdots Q_{16}$ and their evaluation metrics are shown in Table 1. The metrics for the base vector Q_{17} are also shown. As can be seen, Q_8 has the largest value of Optimality Factor. This implies that Q_8 has the best combination of Precision and ROC Area values along with reduction in feature length as compared to the base vector Q_{17}. Only one feature subset, Q_{14} has a negative value of Y, signifying an increase in F–measure but the increase is negligible. Moreover, its size and other metrics are not comparable to Q_8, as is reflected in its low Optimality Factor. Thus, we can consider Q_8 to be a near-optimal feature subset.

To validate that lip protrusion indeed enhances the system efficiency, we consider the set G which contains all features of Q except the lip protrusion. That is, G consists

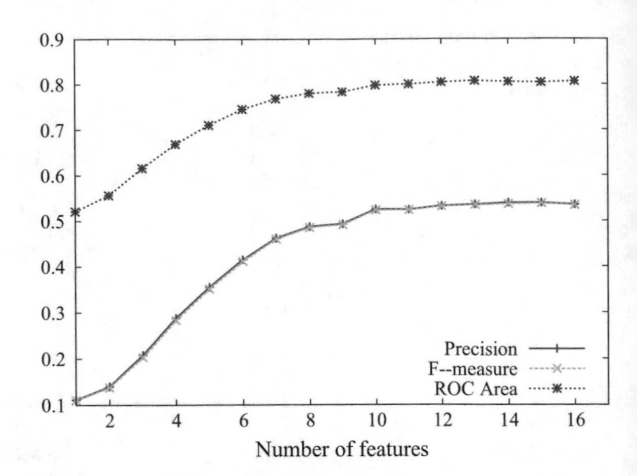

Fig. 6 Evaluation metrics of feature subsets with no lip protrusion information

Table 2 Performance of feature subsets $G_8 \cdots G_{15}$. The results for vector G_{16} are also shown

Feature vector	P_k	F_k	A_k	η_k	Y_k	O_k
G_8	0.487	0.485	0.780	0.500	0.092	0.173
G_9	0.493	0.491	0.783	0.438	0.081	0.158
G_{10}	0.525	0.523	0.798	0.375	0.021	0.138
G_{11}	0.524	0.523	0.800	0.313	0.021	0.119
G_{12}	0.532	0.531	0.804	0.250	0.006	0.094
G_{13}	0.535	0.534	0.807	0.188	0	0.071
G_{14}	0.539	0.536	0.805	0.125	−0.004	0.048
G_{15}	0.539	0.538	0.804	0.063	−0.007	0.024
G_{16}	0.535	0.534	0.806	—	—	—

only the frontal geometrical features. On ranking these features, the order of features is as follows:

$$G = \langle w\ \Gamma\ \Theta\ A_9\ A_7\ A_5\ A_2\ A_{12}\ A_{11}\ A_1\ A_6\ A_4\ h\ A_8\ A_3\ A_{10} \rangle$$

Again feature subsets $G_1 \cdots G_{16}$ are formed using these ranked features. G_{16} is equivalent to the set G as it contains all 16 features. The classification results of feature subsets formed using these frontal features are shown in Fig. 6. On applying thresholding on the F–measure values, the feature subsets that are short-listed are $G_8 \cdots G_{15}$. The evaluation metrics of these subsets along with their base vector G_{16} are shown in Table 2. We observe that G_8 emerges as the most optimal feature subset among these. It has the highest Optimality Factor and a reduced feature set length of eight attributes as compared to 16 of the base vector.

Comparing the results of Q_8 and G_8, it is seen that while the F–measure value of Q_8 is 0.523, that of G_8 is 0.485 only. The Optimality factor increases from 0.173 to 0.223 which is an increase of 28.9%. This indicates an overall improvement in the evaluation metrics and reduction in size of Q_8 as compared to G_8. This will be reflected in enhanced system performance. The visual features contained in both sets are as follows:

$$Q_8 = \langle p \quad w \quad \Gamma \quad \Theta \quad A_9 \quad A_7 \quad A_5 \quad A_2 \rangle$$

$$G_8 = \langle w \quad \Gamma \quad \Theta \quad A_9 \quad A_7 \quad A_5 \quad A_2 \quad A_{12} \rangle$$

Comparing both the feature vectors, it can be argued that addition of lip protrusion information, p, has resulted in better performance of visual speech recognition. It is also seen that while lip width is an important speech articulator, lip height does not contribute much to the system performance. Visible teeth and oral cavity pixels also emerge as prominent attributes [41]. Analysing the features contained in Q_8, it is observed that three of the features (Γ, Θ, A_9) are contained in the centre portion of the mouth. Protrusion and lip width do not have a specific location. Thus, it can be said that most of our prominent speech articulators are contained in the centre region of the mouth [28].

4 Conclusions

This chapter proposes an estimation of the lip protrusion from the frontal pose of the speaker, in absence of other viewing angles or multiple camera inputs. We extract geometrical features from the lips of the speaker and compute the protrusion from this available data. To ensure that our system is suitable for real-time implementation, we reduce the feature vector length by applying Correlation-based Feature Selection. We compute a near-optimal feature subset by analyzing the evaluation metrics of different feature subsets. Through experiments we show that lip profile information is important and enhances visual speech recognition. Lip width, teeth and oral cavity pixels also prove to be important speech articulators. We also perceive that most of the significant speech information lies in the centre region of the mouth. Thus, a simple estimation of the protrusion of the lips results in improving the recognition efficiency of the system. This approach does not require expensive hardware or extensive computing and has a reduced feature set, making it efficient for real-time implementation.

Acknowledgements The author thanks the Department of Science and Technology, Government of India for their support in this research.

References

1. Aizawa, K., Morishima, S., Harashima, H.: An intelligent facial image coding driven by speech and phoneme. In: IEEE International Conference on Acoustics, Speech and Signal Processing, pp. 1795–1798 (1989)
2. Aravabhumi, V., Chenna, R., Reddy, K.: Robust method to identify the speaker using lip motion features. In: International Conference on Mechanical and Electrical Technology (ICMET 2010), pp. 125–129 (2010)
3. Arsic, I., Thiran, J.: Mutual information eigenlips for audio-visual speech recognition. In: 14th European Signal Processing Conference (EUSIPCO) (2006)
4. Batista, F., Caseiro, D., Mamede, N., Trancoso, I.: Recovering punctuation marks for automatic speech recognition. In: Interspeech, pp. 2153–2156 (2007)
5. Brooke, N., Summerfield, A.: Analysis, synthesis, and perception of visible articulatory movements. J. Phon., 63–76 (1983)
6. Chan, M.: Hmm-based audio-visual speech recognition integrating geometric and appearance-based visual features. In: IEEE Fourth Workshop on Multimedia Signal Processing, pp. 9–14 (2001)
7. Chen, T.: Audiovisual speech processing. IEEE Signal Process. Mag., 9–31 (2001)
8. Chen, T., Graf, H., Wang, K.: Lip synchronization using speech-assisted video processing. IEEE Signal Process. Lett. **4**, 57–59 (1995)
9. Cootes, T., Edwards, G., Taylor, C.: Active appearance models. Proc. Eur. Conf. Comput. Vis. **2**, 484–498 (1998)
10. Dupont, S., Luettin, J.: Audio-visual speech modeling for continuous speech recognition. IEEE Trans. Multimed. **2**(3), 141–151 (2000)
11. Dzeroski, S., Zenko, B.: Is combining classifiers with stacking better than selecting the best one? Mach. Learn. **3**, 255–273 (2004)
12. Faruquie, T., Majumdar, A., Rajput, N., Subramaniam, L.: Large vocabulary audio-visual speech recognition using active shape models. Int. Conf. Pattern Recogn. **3**, 106–109 (2000)
13. Finn, K.: An investigation of visible lip information to be used in automated speech recognition. Ph.D. thesis, Washington DC, USA (1986)
14. Florian, R., Ittycheriah, A., Jing, H., Zhang, T.: Named entity recognition through classifier combination. Seventh Conf. Nat. Lang. Learn. **4**, 168–171 (2003)
15. Goecke, R., Millar, J., Zelinsky, A., Zelinsky, E., Ribes, J.: Stereo vision lip-tracking for audio-video speech processing. In: Proceedings IEEE Conference Acoustics, Speech, and Signal Processing (2001)
16. Gupta, D., Singh, P., V.Laxmi, Gaur, M.S.: Comparison of parametric visual features for speech recognition. In: Proceedings of the IEEE International Conference on Network Communication and Computer, pp. 432–435 (2011)
17. Hall, M.: Correlation-based feature subset selection for machine learning. Ph.D. thesis, Hamilton, New Zealand (1998)
18. Hall, M., Frank, E., Holmes, G., Pfahringer, B., Reutemann, P., Witten, I.: The WEKA data mining software: an update. SIGKDD Explor. **11**(1), 10–18 (2009)
19. Ichino, M., Sakano, H., Komatsu, N.: Multimodal biometrics of lip movements and voice using kernel fisher discriminant analysis. In: 9th International Conference on Control, Automation, Robotics and Vision, pp. 1–6 (2006)
20. Iwano, K., Yoshinaga, T., Tamura, S., Furui, S.: Audio-visual speech recognition using lip information extracted from side-face images. EURASIP J. Audio Speech Music Process. **2007** (2007)
21. Jun, H., Hua, Z.: Research on visual speech feature extraction. In: International Conference on Computer Engineering and Technology (ICCET 2009), vol. 2, pp. 499–502 (2009)
22. Kawahara, T., Hasegawa, M.: Automatic indexing of lecture speech by extracting topic-independent discourse markers. In: IEEE International Conference on Acoustics, Speech, and Signal Processing, pp. I.1–I.4 (2002)

23. Kaynak, M., Zhi, Q., Cheok, A., Sengupta, K., Zhang, J., Ko, C.: Analysis of lip geometric features for audio-visual speech recognition. IEEE Trans. Syst. Man Cybern. Part A **34**(4), 564–570 (2004)
24. Kumar, K., Chen, T., Stern, R.: Profile view lip reading. IEEE Int. Conf. Acoust. Speech Signal Process. **4**, 429–432 (2007)
25. Lan, Y., Theobald, B., Harvey, R.: View independent computer lip–reading. In: IEEE International Conference on Multimedia, pp. 432–437 (2012)
26. Lan, Y., Theobald, B., Harvey, R., Ong, E., Bowden, R.: Improving visual features for lip–reading. In: International Conference on Auditory-Visual Speech Processing, pp. 142–147 (2010)
27. Lavagetto, F.: Converting speech into lip movements: a multimedia telephone for hard of hearing people. IEEE Trans. Rehab. Eng. (1995)
28. Lucey, P.: Lipreading across multiple views. Ph.D. thesis (2007)
29. Matthews, I., Cootes, T., Bangham, J., Cox, S., Harvey, R.: Extraction of visual features for lipreading. IEEE Trans. Pattern Anal. Mach. Intell. **24**(2), 198–213 (2002)
30. Matthews, I., Potamianos, G., Neti, C., Luettin, J.: A comparison of model and transform-based visual features for audio-visual LVCSR. In: IEEE International Conference on Multimedia and Expo (ICME 2001), pp. 825–828 (2001)
31. McCowan, I., Moore, D., Dines, J., Gatica-Perez, D., Flynn, M., Wellner, P., Bourlard, H.: On the use of information retrieval measures for speech recognition evaluation. Idiap-rr, IDIAP (2004)
32. Movellan, J.: Visual speech recognition with stochastic networks. In: Tesauro, G., Touretzky, D., Leen, T. (eds.) Advances in Neural Information Processing Systems, vol. 7. MIT Press (1995)
33. Neti, C., Potamianos, G., Luettin, J., Matthews, I., Glotin, H., Vergyri, D., Sison, J., Mashari, A., Zhou, J.: Audio-visual Speech Recognition, Final Workshop 2000 Report. Technical report, The John Hopkins University, Baltimore (2000)
34. Potamianos, G., Neti, C., Gravier, G., Garg, A., Senior, A.: Recent advances in the automatic recognition of audiovisual speech. Proc. IEEE **91**, 1306–1326 (2003)
35. Potamianos, G., Verma, A., Neti, C., Iyengar, G., Basu, S.: A cascade image transform for speaker independent automatic speechreading. In: IEEE International Conference on Multimedia and Expo (II) (ICME 2000), pp. 1097–1100 (2000)
36. Price, P., Fisher, W., Bernstein, J., Pallett, D.: Resource management RM2 2.0. Linguistic Data Consortium, Philadelphia (1993)
37. Rabiner, L.: A tutorial on hidden markov models and selected applications in speech recognition. In: Waibel, A., Lee, K.F. (eds.) Readings in Speech Recognition, pp. 267–296. Morgan Kaufmann Publishers Inc., San Francisco, CA, USA (1990)
38. Saenko, K., Darrell, T., Glass, J.: Articulatory features for robust visual speech recognition. In: 6th International Conference on Multimodal Interfaces, pp. 152–158 (2004)
39. Saitoh, T., Konishi, R.: Profile Lip-Reading for Vowel and Word Recognition. In: Proceedings of the 2010 20th International Conference on Pattern Recognition, pp. 1356–1359 (2010)
40. Saitoh, T., Morishita, K., Konishi, R.: Analysis of efficient lip reading method for various languages. In: International Conference on Pattern Recognition, pp. 1–4 (2008)
41. Singh, P., Laxmi, V., Gaur, M.: Near–optimal geometric feature selection for visual speech recognition. Int. J. Pattern Recogn. Artif. Intell. **27**(8) (2013)
42. Singh, P., Laxmi, V., Gaur, M.S.: Lip peripheral motion for visual surveillance. In: 5th International Conference on Security of Information and Networks, pp. 173–177. ACM (2012)
43. Singh, P., Laxmi, V., Gaur, M.S.: Visual Speech as Behavioural Biometric. Taylor and Francis (2013)
44. Sumby, W.H., Pollack, I.: Visual contribution to speech intelligibility in noise. J. Acoust. Soc. Am. **26**(2), 212–215 (1954)
45. Tan, P.N., Steinbach, M., Kumar, V.: Introduction to Data Mining, 1 edn. Addison-Wesley Longman Publishing Co. Inc. (2005)

46. Yamahata, S., Yamaguchi, Y., Ogawa, A., Masataki, H., Yoshioka, O., Takahashi, S.: Automatic vocabulary adaptation based on semantic similarity and speech recognition confidence measure. In: Interspeech (2012)
47. Zekeriya, S., Gurbuz, S., Tufekci, Z., Patterson, E., Gowdy, J.: Application of affine-invariant fourier descriptors to lipreading for audio-visual speech recognition. In: IEEE International Conference on Acoustics, Speech, and Signal Processing (ICASSP01), pp. 177–180 (2001)
48. Zhang, X., Mersereau, R., Clements, M., Broun, C.: Visual speech feature extraction for improved speech recognition. In: IEEE International Conference on Acoustics, Speech, and Signal Processing (ICASSP), vol. 2, pp. II–1993–II–1996 (2002)
49. Zhao, G., Barnard, M., Pietikäinen, M.: Lipreading with local spatiotemporal descriptors. IEEE Trans. Multimed. **11**(7), 1254–1265 (2009)

Learning-Based Image Scaling Using Neural-Like Structure of Geometric Transformation Paradigm

Roman Tkachenko, Pavlo Tkachenko, Ivan Izonin and Yurij Tsymbal

Abstract In this chapter, it is proposed the solutions of a problem of changing image resolution based on the use of computational intelligence means, which are constructed using the new neuro-paradigm—Geometric Transformations Model. The topologies, the training algorithms, and the usage of neural-like structures of Geometric Transformations Model are described. Two methods of solving a problem of reducing and increasing image resolution are considered: using neural-like structures of Geometric Transformations Model and on the basis of the matrix operator of the weight coefficients of synaptic connections. The influences of the parameters of image preprocessing procedure, as well as the parameters of the neural-like structures of Geometric Transformations Model on the work quality of both methods are investigated. A number of the practical experiments using different quality indicators of synthesized images (PSNR, SSIM, UIQ, MSE) are performed. A comparison of the effectiveness of the developed method with the effectiveness of the existing one is implemented.

1 Introduction

The topological approaches to tasks of image resolution change include the use of finite various coverages by open sets. It gives the possibility to build a variety of feature vectors that characterize an element of certain image topology. The existence of

R. Tkachenko · I. Izonin (✉) · Y. Tsymbal
Lviv Politechnic National University, 12 S. Bandery st., Lviv 79013, Ukraine
e-mail: ivanizonin@gmail.com

R. Tkachenko
e-mail: roman.tkachenko@gmail.com

Y. Tsymbal
e-mail: yurij.tsymbal@gmail.com

P. Tkachenko
Lviv Institute of Banking University, 9 Shevchenko Avenue, Lviv 79005, Ukraine
e-mail: pavlo.tkachenko@gmail.com

A.E. Hassanien and D.A. Oliva (eds.), *Advances in Soft Computing and Machine Learning in Image Processing*, Studies in Computational Intelligence 730,
https://doi.org/10.1007/978-3-319-63754-9_25

significant number of vectors in the feature space, sometimes even large dimension, is a strong argument for use or development of machine learning methods in tasks of digital images processing [1, 2].

This chapter describes the neural network methods for solution of the task of image resolution change. The authors focus is on the task of image resolution increase using machine learning tools. In addition, authors describe the solution of the image resolution decrease using this approach.

The first significant paper in the direction of solving the task of image resolution change based on artificial neural networks is [3]. In the dissertation the author tries to solve the problem of image (fingerprint) resolution increase based on the frame processing by Hopfield network. Extensive development of further research concerning the use of this type of ANN based on the formulation of the problem of image resolution increase as a soft classification problem [4–7] of image frames obtained by remote sensing.

The use of multilayer perceptrons to solve this problem is given in the article [8]. In this paper the task of image resolution change is solved by finding the unknown pixel between two existing ones with the type of ANN. It has used a different approach in [9]. The scheme is compatible using an arbitrary interpolation algorithm and ANN. According to the procedure, the ANN is used to model residual errors between the interpolated image and its corresponding pattern with low resolution.

In [10, 11], it has described another approach to solve the task of image resolution change. The method of using Convolution Neural Network (SRCNN) of a deep learning differs from existing ones primarily to the fact that training occurs directly comparing pairs of low and high image resolution. The results of this comparison are displayed as deep Convolution Neural Network. The low resolution images are given to the input of the method and SRCNN synthesizes high resolution images in the output.

The main disadvantage of existing methods of image scaling based on machine learning, is that they use iterative approaches. This is the reason of obvious drawbacks, including [12, 13]:

- the large computational resources for implementing the training procedures. It limits the solutions of large-dimension tasks;
- the solution depends on the initial random initialization of the neural network;
- the lack of reliability, if activation function of neural network enters to the saturation or inadequate solution getting into the local optima.

We consider an alternative to the existing paradigm of artificial neural networks—geometric transformation model based on fundamentally different spatial and geometric principles [14–16].

The main advantages of this neural-paradigm arising from the basic characteristics of the model are: not iterative training process, orthogonality of solution steps, affinity of training and test procedures. The main features that have practical value for solving this problem include the following ones [13, 15, 16]:

- the high performance in training process, which creates conditions for solution of large-dimension tasks;
- the result repeatability and their mathematical interpretation;
- the ability to solve problems in terms of both large and small training samples;
- the linearity in extrapolation regions;
- the possibility of functioning in automatic mode;
- the significantly better degree of generalization.

1.1 Formulation of Task of Image Resolution Change

In general, each image C can be presented as the result of some continuous function C (i.e., color function):

$$C : \mathrm{N}^{2,+} \rightarrow Color, \tag{1}$$

$$\mathrm{C} = C(\mathrm{N}^{2,+}). \tag{2}$$

Suppose we are given image C from (2) and there is the following:

$$\mathrm{C} = [c_{(i,j)}]_{j \in \overline{1,h}}^{i \in \overline{1,l}}. \tag{3}$$

Then, the task of image resolution (IR) change can be formulated as follows: we must creating new image—C′ in relation to C from (3) such that:

$$\mathrm{C}' = [c_{(i,j)}]_{j \in \overline{1,h'}}^{i \in \overline{1,l'}} \tag{4}$$

with a minimum deterioration in the image quality, where (l', h') determines the matrix dimension, $l' > l \vee h' > h$. In the case of resolution decrease $l' < l \vee h' < h$.

For color images, value $c_{(i,j)}$ should be considered as a vector of the main characteristics of a color palette.

For example, if the image in 24-bit format of RGB palette:

$$c_{(i,j)} = \{c_{(i,j)}^R, c_{(i,j)}^G, c_{(i,j)}^B\}, j \in [1, h], i \in [1, l] \tag{5}$$

where $c_{(i,j)}^R, c_{(i,j)}^G, c_{(i,j)}^B$ values of red, green, and blue components, respectively. Then developed methods might be applied to each of the characteristic separately, and additive operation made in accordance with the principles of building values in color palette, which is chosen.

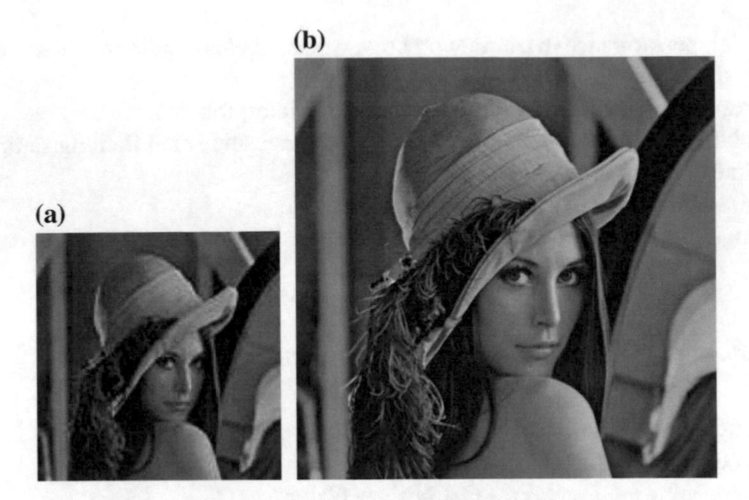

Fig. 1 A pair of 8-bit images: **a** 252 × 252 pixels; **b** 504 × 504 pixels

2 Image Scaling Based on Neural-Like Structure of Geometric Transformation Model

For example couple of two images with low and high resolution are given (Fig. 1). They are used in the training procedure, which is based on applying the neural-like structure of geometric transformation model (NLS GTM) [14–18].

2.1 Image Pre-processing

Let the low-resolution (LR) image is a matrix C, with dimension $(l \times l'), l \in \mathrm{N}$, $l > 0$ and high-resolution (HR) image—matrix $\mathrm{C}^{(m)}$, with dimension $(h \times h'), h \in \mathrm{N}$, $h > 0$ of pixel intensity values. Then:

$$\mathrm{C} = [c_{(i,j)}]_{i,j \in \overline{1,l}}, \ \mathrm{C}^{(m)} = [c^{(m)}_{(i,j)}]_{i,j \in \overline{1,h}} \tag{6}$$

where $c_{(i,j)}$, $c^{(m)}_{(i,j)}$ values of the pixel intensity function in coordinates (i, j) for LR and HR images; $m \in \mathrm{N}, m > 0$ coefficient of image resolution change; $h = l \times m$ variable that determines dimension of the matrix for HR image $\mathrm{C}^{(m)}$.

For implementing the training technology both images are divided into the same frame number $FR_{i,j}$, $FR^{(m)}_{i,j}$—(square regions of image intensity function values) as follows:

$$FR_{i,j} = \begin{pmatrix} c_{k(i-1)+1,k(j-1)+1} & \cdots & c_{k(i-1)+1,kj} \\ \vdots & \vdots & \vdots \\ c_{ki,k(j-1)+1} & \cdots & c_{ki,kj} \end{pmatrix};$$
$$FR_{i,j}^{(m)} = \begin{pmatrix} c^{(m)}_{mk(i-1)+1,mk(j-1)+1} & \cdots & c^{(m)}_{mk(i-1)+1,mkj} \\ \vdots & \vdots & \vdots \\ c^{(m)}_{mki,mk(j-1)+1} & \cdots & c^{(m)}_{mki,mkj} \end{pmatrix}, i,j = \overline{1,n} \tag{7}$$

where $k \in \mathrm{N}, k > 0$ variable that determines dimension of the frame $FR_{i,j}$ for LR image—$\dim(FR_{i,j}) = k \times k$.

The variable n, which determines the number of LR image frames is defined as follows:

$$n = l/k, n \in \mathrm{N}, n > 0. \tag{8}$$

Note that k is multiple to l. If enter the designations as follows:

$$k^{(m)} = mk \tag{9}$$

then dimension of HR image frame will be equal $\dim(FR_{i,j}^{(m)}) = k^{(m)} \times k^{(m)}$. According to this $k^{(m)} \in \mathrm{N}, k^{(m)} > 0$ is a variable that determines dimension of the HR image frame as follows: $FR_{i,j}^{(m)}$. Obviously, $\dim\{FR_{i,j}\} = \dim\{FR_{i,j}^{(m)}\} = n^2$.

Then, according to (9) and determining the h value, values of $k^{(m)}$ is multiple to h value.

Frames coverage is disjunctive. Thus, the matrices (6) can be represented as a set of corresponding frames as follows:

$$\mathrm{C} = [FR_{(i,j)}]_{i,j \in \overline{1,n}}, \; \mathrm{C}^{(m)} = [FR_{(i,j)}^{(m)}]_{i,j \in \overline{1,n}}. \tag{10}$$

Solving the task of image resolution increasing is expressed by equality (9). In the case of image resolution decrease, such equality must be used as follows:

$$k = k^{(m)} \% m \tag{11}$$

where % integer division operation.

One of the few disadvantages of the NLS GTM is input and output data format [13, 19, 20]—(as a table). Therefore, further pre-processing of image pairs before training is as follows:

Each frame $FR_{i,j}$ from C represent as a vector $\mathbf{A}_{j+\frac{l(i-1)}{k}}$, and each frame $FR_{i,j}^{(m)}$ from $\mathrm{C}^{(m)}$ as a vector $\mathbf{A}^{(m)}_{j+\frac{l(i-1)}{k}}$:

$$\mathbf{A}_{j+\frac{l(i-1)}{k}} = \left(c_{ki-k+1,kj-k+1}, \ldots, c_{ki-k+1,kj}, \ldots, c_{ki,kj-k+1}, \ldots, c_{ki,kj}\right), \tag{12}$$

$$\mathbf{A}^{(m)}_{j+\frac{l(i-1)}{k}} = (c^{(m)}_{k^{(m)}i-k^{(m)}+1,k^{(m)}j-k+1}, \ldots, c^{(m)}_{k^{(m)}i-k^{(m)}+1,k^{(m)}j}, \ldots, \\ c^{(m)}_{k^{(m)}i,k^{(m)}j-k+1}, \ldots, c^{(m)}_{k^{(m)}i,k^{(m)}j}) \tag{13}$$

where

$$\dim\left\{\mathbf{A}_{j+\frac{l(i-1)}{k}}\right\} = \dim\left\{\mathbf{A}^{(m)}_{j+\frac{l(i-1)}{k}}\right\} = n^2. \tag{14}$$

Then training set is forming from corresponding sets $\left\{\mathbf{A}_{j+\frac{l(i-1)}{k}}\right\}$ and $\left\{\mathbf{A}^{(m)}_{j+\frac{l(i-1)}{k}}\right\}$ thereby.

Matrix M of the training data for solving the task of image resolution increases using NLS GTM when performed (9), can be formed as follows:

$$M = \begin{pmatrix} \mathbf{A}_1 & \mathbf{A}_1^{(m)} \\ \vdots & \vdots \\ \mathbf{A}_{n^2} & \mathbf{A}_{n^2}^{(m)} \end{pmatrix} \tag{15}$$

where the dimension of the vector $\dim(\mathbf{A}_1)$ defines the number of inputs NLS GTM, and the dimension of the vector $\dim\left(\mathbf{A}_1^{(m)}\right)$—number of outputs neural-like structures of GTM.

Training matrix M for solving the task of image resolution increases using NLS GTM when performed (9), is formed as follows:

$$M = \begin{pmatrix} \mathbf{A}_1^{(m)} & \mathbf{A}_1 \\ \vdots & \vdots \\ \mathbf{A}_{n^2}^{(m)} & \mathbf{A}_{n^2} \end{pmatrix} \tag{16}$$

where dimension of the vector $\dim\left(\mathbf{A}_1^{(m)}\right)$ defines the number of NLS GTM inputs, and dimension of $\dim(\mathbf{A}_1)$—number of neural-like structure outputs.

2.2 *Training of Neural-Like Structure of Geometric Transformation Model*

Matrix M by (15) or by (16) (depending on the task) is fed to the NLS. NLS GTM topologies shown in (Fig. 2). For generalization the training technology descriptions for solving both tasks, matrix (15), or (16) can be represented as follows:

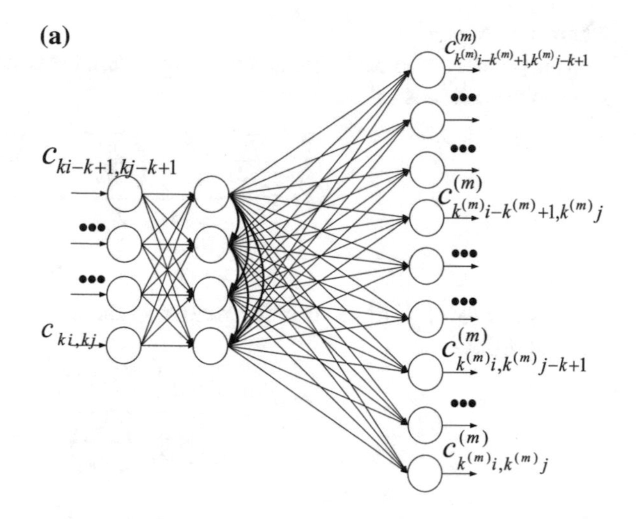

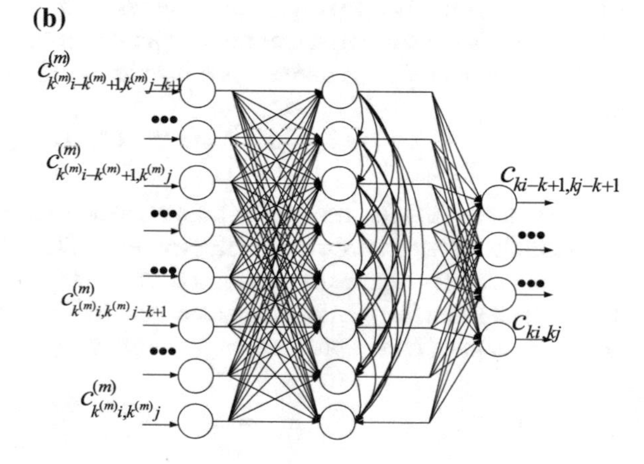

Fig. 2 NLS GTM topologies for solving the task of image resolution change: **a** resolution increase; **b** resolution decrease

$$M = \begin{pmatrix} x_{1,1} & \cdots & x_{1,k^2+(mk)^2} \\ \vdots & \vdots & \vdots \\ x_{n^2,1} & \cdots & x_{n^2,k^2+(mk)^2} \end{pmatrix}. \tag{17}$$

NLS GTM training technology for solving the task of image resolution change involves the following steps [15]. In the first step, the base row $\mathbf{x}_b^{(1)}$, $\mathbf{x}_b^{(1)} = \left(x_{b,1}^{(1)}, \ldots, x_{b,k^2+(mk)^2}^{(1)}\right)$, $1 \leq b \leq n^2$ from the training matrix M is chosen when the sum of its element squares is the maximum.

Each row of initial matrix converted to $\mathbf{x}_N^{(2)}$ as a difference between each row $\mathbf{x}_N^{(1)}$, ($\mathbf{x}_N^{(1)} = \left(x_{N,r}^{(1)}, \ldots, x_{n^2,k^2+(mk)^2}^{(1)}\right)$) from the M, and product of the chosen row $\mathbf{x}_b^{(1)}$ by a coefficient $K_N^{(1)}$:

$$\mathbf{x}_N^{(2)} = \mathbf{x}_N^{(1)} - K_N^{(1)} * \mathbf{x}_b^{(1)} \tag{18}$$

where $1 \leq N \leq n^2$.

The value $K_N^{(1)}$ for each row is determined from the condition of minimum by least squares criterion as follows:

$$K_N^{(1)} = \frac{\sum_{r=1}^{k^2+(km)^2} \left(\mathbf{x}_{N,r}^{(1)} * \mathbf{x}_{b,r}^{(1)} \right)}{\sum_{r=1}^{k^2+km} \left(\mathbf{x}_{b,r}^{(1)} \right)^2}. \quad (19)$$

For each row $\mathbf{x}_N^{(1)}$ of the training matrix M the additional parameter $G_N^{(1)}$ is calculated as follows:

$$G_N^{(1)} = \frac{\sum_{r=1}^{z} \left(\mathbf{x}_{N,r}^{(1)} * \mathbf{x}_{b,r}^{(1)} \right)}{\sum_{r=1}^{z} \left(\mathbf{x}_{b,r}^{(1)} \right)^2} \quad (20)$$

where $z = k^2$—in case (15) or $z = (mk)^2$ in case (16). In fact it is the first component of the numerical characteristic of the first implementation.

The coefficient $K_N^{(1)*}$ is defined as a function $F^{(1)}$ from the parameter $G_N^{(1)}$ as follows:

$$K_N^{(1)*} = F^{(1)} \left(G_N^{(1)} \right). \quad (21)$$

Dependence (21) for the rows of training matrix on the discrete set of key values will reproduced with zero-methodological error [12, 19]. For all other rows from matrix realization—approximately. For the next step, the row $\mathbf{x}_b^{(2)}$ of training matrix M is chosen when the sum of its element squares is the maximum. Note, that value of $\mathbf{x}_b^{(2)}$ are defined on the previous step as follows:

$$\mathbf{x}_b^{(2)} = \mathbf{x}_b^{(1)} - K_N^{(1)*} * \mathbf{x}_b^{(1)}. \quad (22)$$

For q—step of calculations, where $q = \overline{1, k^2 + (km)^2}$, we will have the following:

$$\mathbf{x}_N^{(q+1)} = \mathbf{x}_N^{(q)} - K_N^{(q)*} * \mathbf{x}_b^{(q)}, \quad (23)$$

$$K_N^{(q)*} = F^{(q)} \left(G_N^{(q)} \right), \quad (24)$$

$$G_N^{(q)} = \frac{\sum_{r=1}^{z} \left(\mathbf{x}_{N,r}^{(q)} * \mathbf{x}_{b,r}^{(q)} \right)}{\sum_{r=1}^{z} \left(\mathbf{x}_{b,r}^{(q)} \right)^2}, \quad (25)$$

$$\mathbf{x}_b^{(q+1)} = \mathbf{x}_b^{(q)} - K_N^{(q)*} * \mathbf{x}_b^{(q)}. \quad (26)$$

Expressions (23)–(26) for case of accurate reproduction in nodes are implementing the procedure of Gram-Schmidt orthogonalization [21, 22]. Based on (23)–(26), the rows of initial implementation matrix M can be represented by the ultimate amount (27) in such form as follows:

$$\mathbf{x}_N^{(1)} = \sum_{q=1}^{k^2+(mk)^2} K_N^{(q)^*} * \mathbf{x}_N^{(q)} \tag{27}$$

Based on the results of training using procedures (23)–(26) the set of vectors $\mathbf{x}_N^{(q)}$ and activation functions $F^{(q)}$ are defined.

2.3 Visualization of Geometric Transformations in the Training Mode

For visualization purpose let us consider three-dimensional space of realizations for the function of two variables $y = x_1^{x_2}$.

On each step of training we reduce the dimensionality of hyper body due to orthogonalization, and memorize the information about this reduction in the neural network. We depict the information about objects of modeling with darker color for neural network units. In the first training step all information is contained in inputs and outputs of neural network (Fig. 3).

In the first step of training, geometrical transformations are performed in a space, whose dimensionality is reduced by one (in the given case on the plane). We have only one trained neuron (Fig. 4).

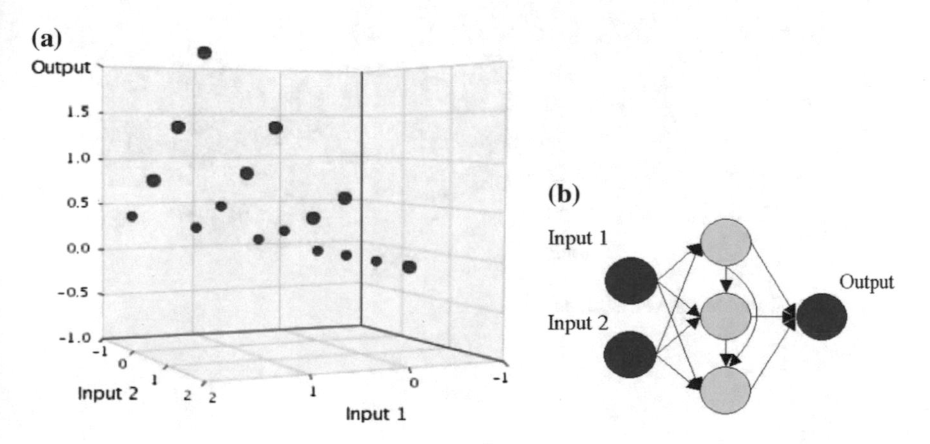

Fig. 3 Initial status

After the second step, the data points are projected on a straight line perpendicular to normal vector. We have trained the second neuron in the hidden layer (Fig. 5).

Finally, as result of performing the third step of transformations, all data points are projected in the coordinate origin (Fig. 6). So we got information about geometrical transformations for regeneration of a training matrix. This information is used for interpolation or extrapolation approach of vectors that did not belong to the training matrix [14, 15].

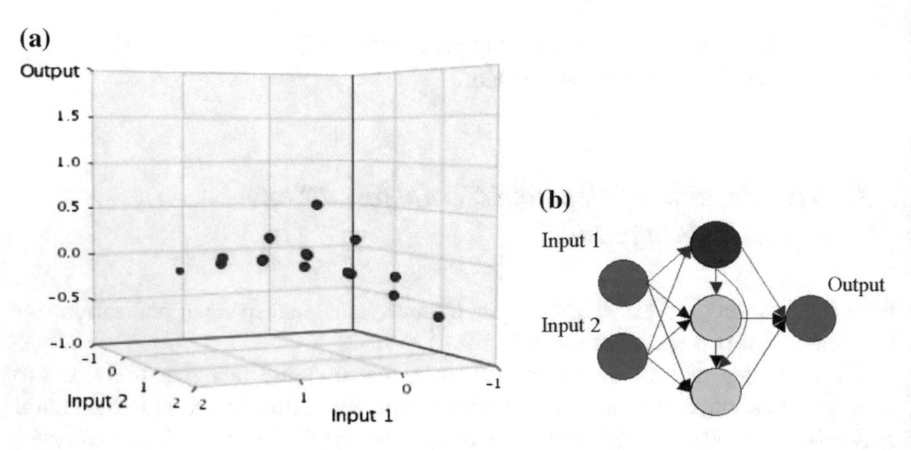

Fig. 4 First step of transformations

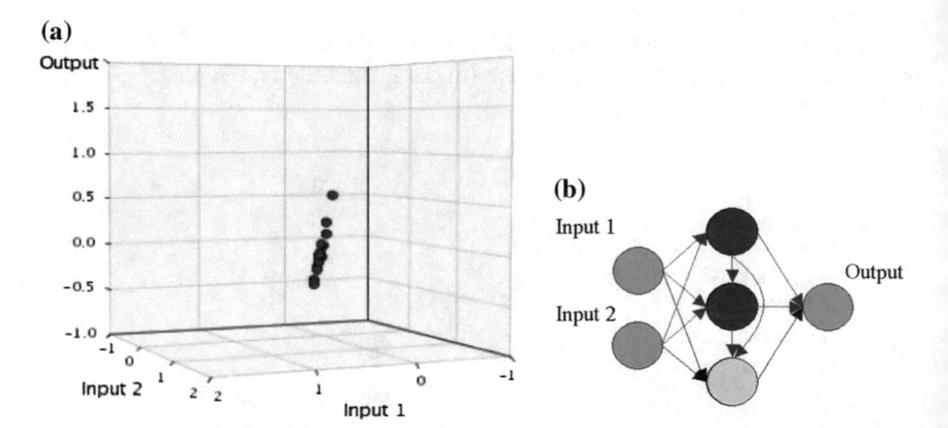

Fig. 5 Second step of transformations

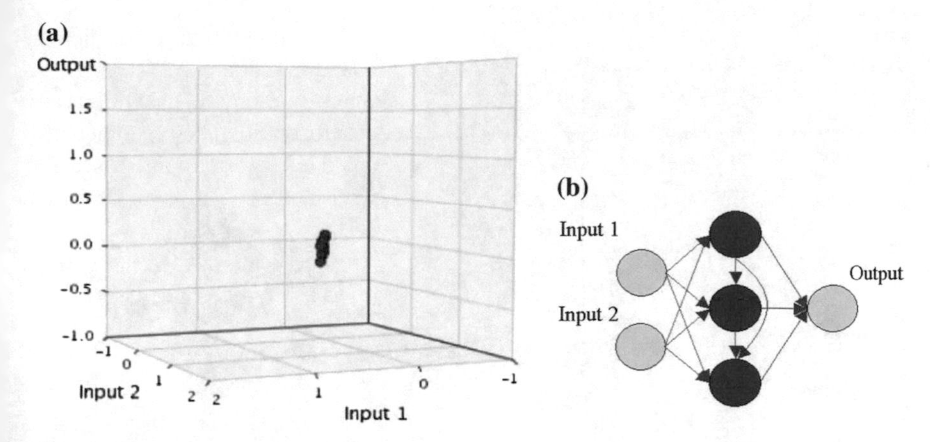

Fig. 6 Third step of transformations

2.4 Using Procedure

In using mode on the input of NLS GTM is served matrix M_t, where $M_t = \begin{pmatrix} \mathbf{A}_1 \\ \vdots \\ \mathbf{A}_{n^2} \end{pmatrix}$, for IR increasing or, $M_t = \begin{pmatrix} \mathbf{A}_1^{(m)} \\ \vdots \\ \mathbf{A}_{n^2}^{(m)} \end{pmatrix}$ in case of IR decrease.

For generalization of using procedure for solving both tasks, matrix M_t can be represented as follows:

$$M_t = \begin{pmatrix} x_{1,1} & \cdots & x_{1,z} \\ \vdots & \vdots & \vdots \\ x_{n^2,1} & \cdots & x_{n^2,z} \end{pmatrix} \tag{28}$$

where z determined in the same way as for (20). Using of NLS GTR in this mode is as follows. For given input vector components $x_N^{(q)}$ of the matrix M_t (28) $G_N^{(1)}$ is calculated according to (20). Values of the coefficients $K_N^{(q)^*}$ are searched according to (19). For $q = 1$, according to (21) the first step of transformation of input vector $\mathbf{x}_N^{(q)}$ is performed.

On the basis of (25) (24) (23) sequential transformations for $q_{\max} = \overline{1, k^2}$ is performed. The main goal of this transformations—search of values, $K_N^{(1)^*}$, $K_N^{(2)^*}$,…, $K_N^{(q_{\max})^*}$.

An unknown output components of each vector from M_t are calculated on the basis of the sum (27).

Rows of newly formed matrix are vectors (12) or (13) of intensity function values from LR or HR image. Therefore, the last step of the procedure is to collect this matrix for the form of C_r or $C_r^{(m)}$.

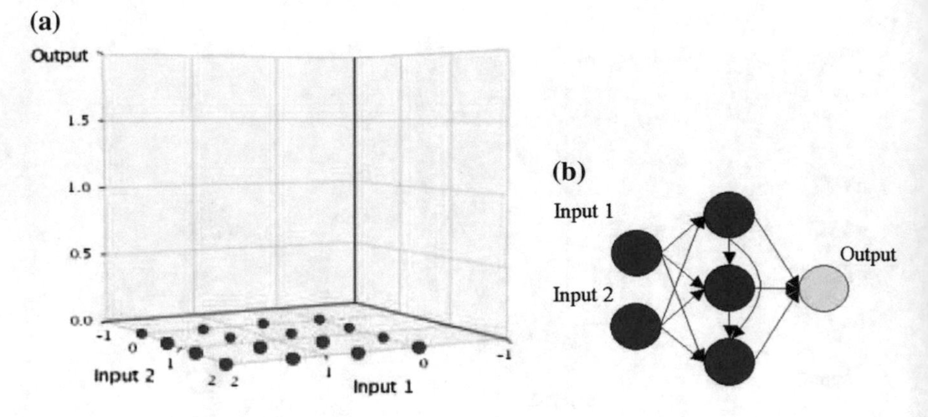

Fig. 7 Initial status

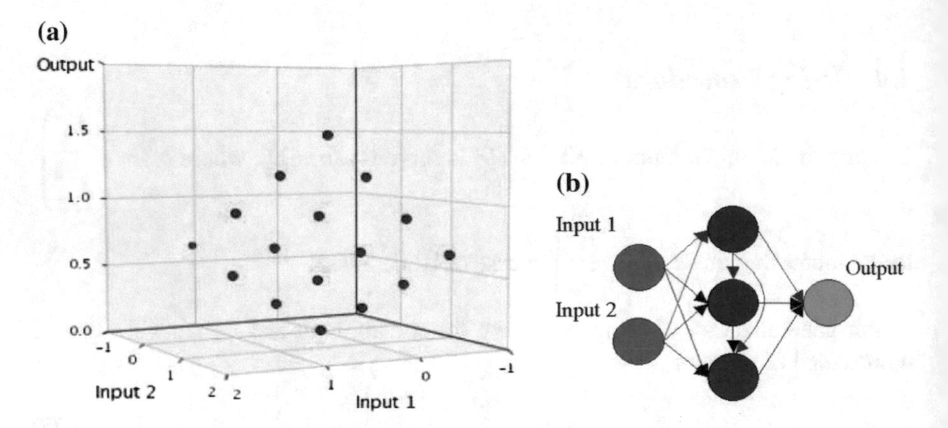

Fig. 8 First step of transformations

This procedure involves inverse representation rows of the resulting matrix from vectors (12) or (13) to form of the corresponding frames $FR_{i,j}$ or $FR_{i,j}^{(m)}$ based on (7). Unknown LR or HR image by collecting of the received frames according $FR_{i,j}$ for C_r or $FR_{i,j}^{(m)}$ for $\mathrm{C}_r^{(m)}$, similar to (10) is formed as follows:

$$\mathrm{C}_r = \left[FR_{i,j}\right]_{i,j=1\ldots n}, \text{ or } \mathrm{C}_r^{(m)} = \left[FR_{i,j}^{(m)}\right]_{i,j=1\ldots n} \tag{29}$$

where C_r—the unknown LR image, $\mathrm{C}_r^{(m)}$—the unknown HR image.

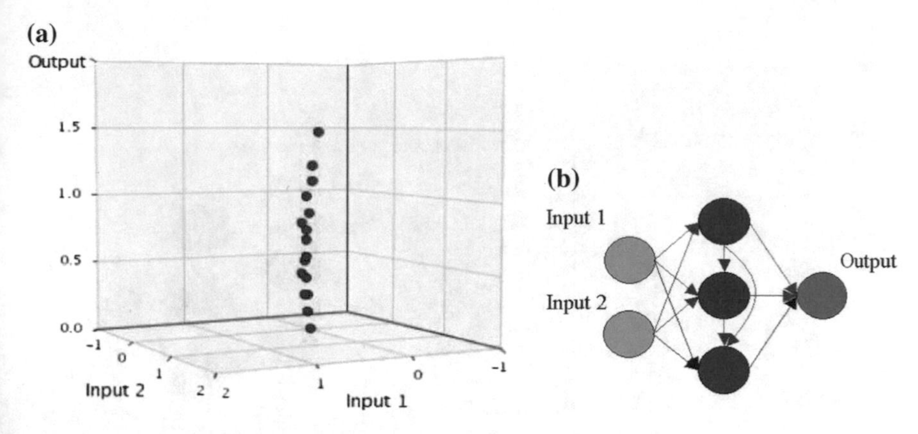

Fig. 9 Second step of transformations

2.5 *Visualization of Geometric Transformations in the Using Mode*

Let us consider previously trained GTM (Figs. 3, 4, 5, and 6). Given input vectors the network should reproduce output (Fig. 7).

Hidden neurons operate sequentially, hyper body information flows from input layer and hidden neurons to output layer (Fig. 8).

Data points on inputs are moving toward coordinate origin. On the output the data points are moving to predicted values. After the second step of operation we have the next (Fig. 9).

After the third step of operation data points are lying around the coordinate origin on inputs and have predicted values on output (Fig. 10). The procedure of use of GTM converges in cases, when the entry vector is real a unit of a matrix of implementation of the object [14, 15].

3 Image Scaling Based on the Matrix Operator of Synaptic Connection Weights

3.1 *Linear Decomposition of Neural-Like Structure of Geometric Transformation Model*

The task of image resolution change through the use of NLS GTM, provides two stages of its use—training and usage. But, for example, the solving of the task of resolution change for the set of same type of images makes inappropriate use of training stage for each of the samples. This is because the statistics of the same

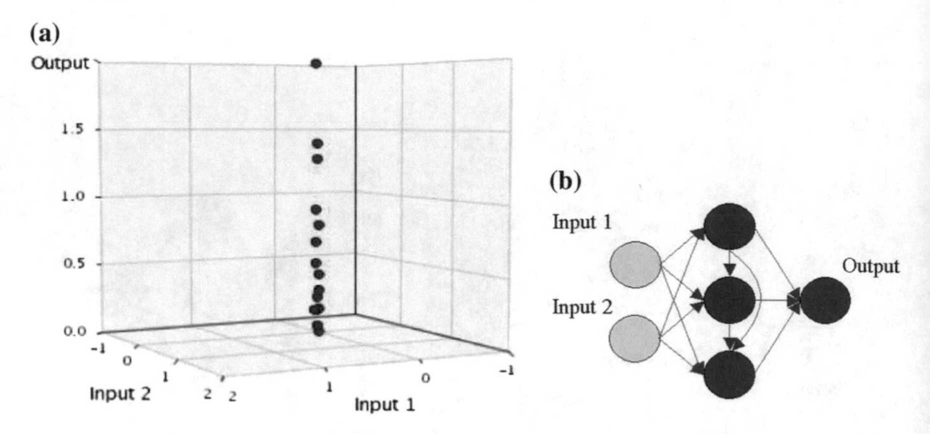

Fig. 10 Third step of transformations

type of images very similar, and the ability to generalization NLS GTM is very high [19, 23]. Therefore, NLS GTM, which is trained for one sample from set can be used to change resolution of another images. The work of NLS GTM in using mode about 2–3 times faster than work in training mode [15]. However, before the using mode we must perform some image pre-processing ((7), then (12) or (13), then (15) or (16)). This is imposes a number of limitations for use the method for solving the series of practical tasks. Therefore, there is a need to increase the performance of the method, in particular by avoiding the use of this pre-processing procedures. For that we use linear structure of NLS GTM, including linear activation function and linear synaptic connections. This allows to apply the principles of linear superposition and move from the initial (Fig. 2) to equivalent topologies of NLS GTM (Fig. 11).

The process of linear decomposition and repetitiveness of the solution, which provides by training technology [16], allows to get the matrix operator of synaptic connection weights from NLS GTM with many outputs. It will provide the ability to solve the task of images resolution change without NLS GTM at using mode. In addition, it will reduce the time for work with the method of image resolution change based on machine learning, in particular, under processing of image set of same type.

3.2 Synthesis of the Matrix Operator of Synaptic Connection Weights from the Trained NLS GTM

To solve the task of images resolution change based on machine learning, weight coefficients of synaptic connection are used. The matrix operator, elements of which are these coefficients is derived from trained NLS GTM. For using this matrix operator to solve the task of images resolution increase, the procedure of obtaining weight coefficients of synaptic connections from the trained neural-like structures of GTM with many outputs is proposed. The main steps of this procedure are as follows [16]:

- we train NLS GTM on a pair of images according to the procedure, which is described above;
- based on trained NLS GTM it is get matrix operator $V^{(m)}$ of response surface coefficients $\alpha_{i,j}$:

$$V^{(m)} = \left[\alpha_{i,j}\right]_{i=\overline{1,k^2+1}}^{j=\overline{1,(km)^2}}. \tag{30}$$

According to (15) the dimension of the matrix $V^{(m)}$ was equal: $\dim(V^{(m)}) = \left(k^2+1\right)(km)^2$.
Matrix $V^{(m)}$ is formed by testing NLS GTM with test signals matrix T. Matrix T can be defined in the form as follows:

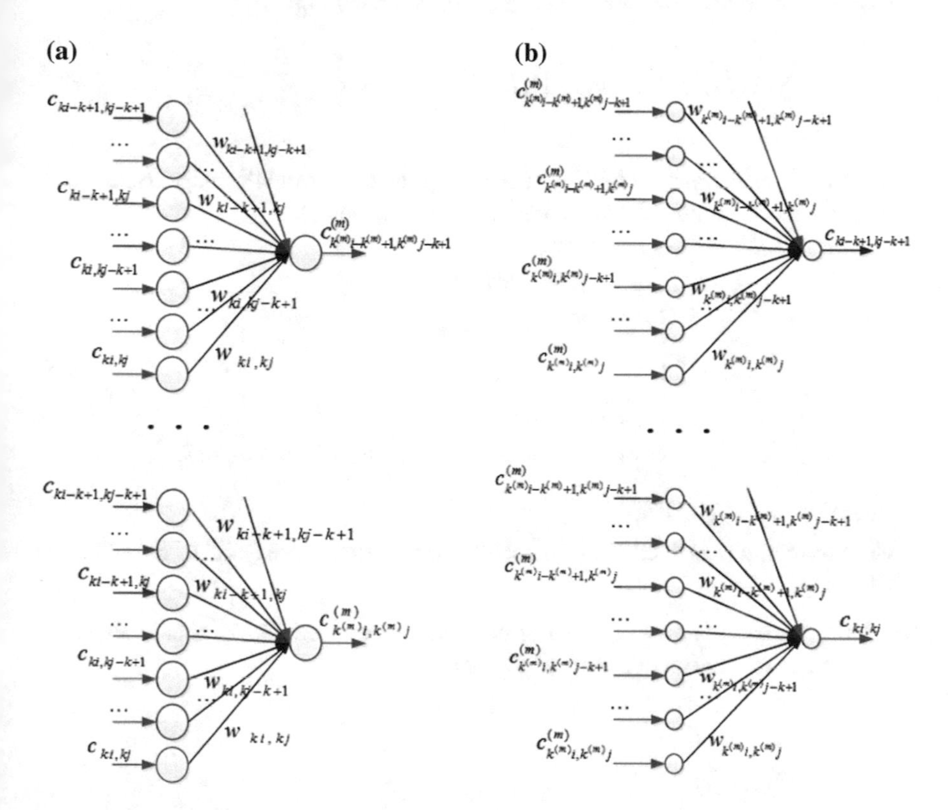

Fig. 11 NLS GTM equivalent topologies for solving the task of image resolution change: **a** increase; **b** decrease

$$T = \begin{pmatrix} \Omega \\ E \end{pmatrix}, \quad \Omega = \left\{ \underbrace{0, \ldots, 0}_{k^2} \right\} \tag{31}$$

where E—an identity matrix, $\dim(E) = k^2 \times k^2$.

- based on the matrix (30), we build new—matrix operator of synaptic connections weights $W^{(m)}$. Each of its j-th column contain a set $(k^2 + 1)$ of synaptic connections weights $\{w_{i,j}\}$ for each unknown pixel from HR image frame of the equivalent topology of linear NLS (Fig. 11b) as follows:

$$W^{(m)} = \left[w_{i,j} \right]_{i=\overline{1,k^2+1}}^{j=\overline{1,(km)^2}} \tag{32}$$

where coefficients from $W^{(m)}$ are calculated as following:

$$\forall i \in \left[1, k^2 + 1\right], \forall j \in \left[1, (km)^2\right] : w_{i,j} = \begin{cases} \alpha_{1,j}, & i = 1; \\ \alpha_{i,j} - \alpha_{1,j}, & i \neq 1. \end{cases} \tag{33}$$

- using the matrix operator $W^{(m)}$ we can obtain the unknown HR image frame $FR_{i,j}^{(m)}$ from respective input LR image frame $FR_{i,j}$, as follows:

$$\forall i, j \in [1, n] :$$
$$FR_{i,j}^{(m)} = \begin{bmatrix} c_{x,y}^{(m)} | c_{x,y}^{(m)} = w_{1,(x-1)mk+y} + \\ + \sum_{a=(i-1)+1}^{ki} \sum_{b=k(j-1)+1}^{kj} \\ c_{a,b} * w_{(a-k(i-1)-1)k+b-k(j-1)+1,(x-1)mk+y} \end{bmatrix}_{x,y=\overline{1,mk}} \tag{34}$$

where $\forall a, b : c_{a,b} \in FR_{i,j}$.

- when applying (34) to all $FR_{i,j}$ we get all appropriate frames $FR_{i,j}^{(m)}$ of the unknown HR image.

Similarly, we can get matrix operator of synaptic connections weights to solve the task of image resolution decrease (Fig. 11b).

4 Results of Practical Experiments

4.1 Investigation of Influence of the Training Sample Dimension Change

To implement the training procedure a pair of images from (Fig. 1) is used. Test samples formed by the most famous images (Fig. 12). They are taken from the base of Signal and Image Processing Institute of Southern California University. Note that all images are scaled to present in this book chapter.

The dimension of the training set to solve various tasks plays an important role when we use means of artificial neural networks [5, 10, 24–26]. Therefore, to study the influence of changing size $FR_{i,j}$ ($\dim(FR_{i,j}) = k \times k$) and appropriate $FR_{i,j}^{(m)}$ ($\dim(FR_{i,j}^{(m)}) = k^{(m)} \times k^{(m)}$) at different m on the quality of received images (Peak signal-to-noise ratio—PSNR [27], Structural SIMilarity—SSIM [28], Universal image quality index—UIQ [29], Mean squared error—MSE [30]), a series of experiments were conducted.

The Task of Image Resolution Increasing. Practical implementation of the method of image resolution increase using neural-like structure of geometric transformations model was conducted by the following parameters of NLIS GTM (Fig. 1a): the number of inputs—$\dim(FR_{i,j})$, hidden layers—1, neuron number in the hidden layer equal $\dim(FR_{i,j})$, the outputs number of NLS determined by condition of solution the task of image resolution increase (9) and equal $\dim(FR_{i,j}^{(m)})$. Linear value of synaptic connections. Coefficients of increase m were 2, 3.

Dependence of HR image's quality values (PSNR and UIQ) from $\dim(FR_{i,j})$ with different m is shown in (Tables 1 and 2).

Evaluation of images quality obtained from the SSIM and MSE are shown in (Table 3) and (Table 4), respectively.

Result of comparing the efficiency of both developed methods has shown full coincidence by all four indexes of quality. This demonstrates that the noise that arises during of decomposition process of NLS GTM is so small that it can be neglected.

As shown in the (Tables 1, 2, 3 and 4) for best results for all test sample on all quality indexes obtained for images, which are divided to frames with the dimension $\dim(FR_{i,j}) = 6 \times 6$. In particular, evaluation of the quality results by PSNR for increasing image resolution with $m = 3$ shown in (Fig. 13) In addition, in (Fig. 14c) a fragment of this image where $\dim(FR_{i,j}^{(m)}) = 18 \times 18$ is demonstrated.

Further increase $\dim(FR_{i,j}^{(m)})$ leads to distortion of test HR images (Fig. 14d), but obviously improves the quality of training image (Fig. 14a).

The results of image resolution increase when $k \leq 4$ is satisfactory. However, among the artifacts that arise during scaling, the presence of visible frame borders significantly reduces image quality at the visual assessment stage (Fig. 14b).

Fig. 12 LR images of test sample

Table 1 The PSNR values of HR images by changing the increase coefficient *m* and $FR_{i,j}$

Image	m = 2					m = 3					
(a)	33.57	34.83	35.42	36.20	36.58	30.61	31.66	31.42	33.12	33.71	34.59
(b)	31.88	32.53	32.77	32.92	32.88	30.16	30.95	30.56	31.15	30.91	30.37
(c)	36.35	38.21	38.82	39.34	39.33	33.54	35.51	35.11	36.12	35.77	34.97
(d)	23.30	23.29	23.28	23.25	23.20	22.69	22.72	22.32	22.56	22.37	22.13
(e)	24.47	24.91	25.12	25.25	25.21	22.88	23.16	22.78	23.26	23.08	22.76
(f)	28.51	29.72	30.10	30.36	30.29	25.68	26.63	25.87	26.81	26.46	25.72
(g)	32.02	33.28	33.66	34.05	33.99	29.43	30.63	30.29	31.07	30.79	30.08
(h)	34.20	35.43	35.88	36.22	36.25	31.80	32.98	32.67	33.47	33.30	32.75
(i)	28.10	28.94	29.18	29.47	29.49	25.60	26.16	25.69	26.38	26.15	25.60
(j)	29.62	30.26	30.49	30.78	30.68	27.04	27.62	27.33	27.80	27.62	27.30
$FR_{i,j}^{(m)}$	4×4	6×6	8×8	12×12	14×14	6×6	9×9	12×12	18×18	21×21	24×24
$FR_{i,j}$	2×2	3×3	4×4	6×6	7×7	2×2	3×3	4×4	6×6	7×7	8×8

Table 2 The SSIM values of HR images by changing the increase coefficient *m* and $FR_{i,j}$

Image	m = 2					m = 3					
(*a*)	0.984	0.987	0.989	0.990	0.990	0.943	0.956	0.947	0.962	0.964	0.965
(*b*)	0.970	0.974	0.975	0.976	0.976	0.936	0.949	0.936	0.948	0.944	0.934
(*c*)	0.990	0.993	0.993	0.993	0.993	0.960	0.977	0.967	0.976	0.973	0.967
(*d*)	0.977	0.982	0.982	0.983	0.982	0.923	0.945	0.923	0.940	0.933	0.915
(*e*)	0.938	0.944	0.946	0.948	0.947	0.800	0.815	0.767	0.824	0.816	0.800
(*f*)	0.977	0.982	0.983	0.984	0.983	0.912	0.932	0.906	0.928	0.920	0.899
(*g*)	0.953	0.967	0.969	0.972	0.973	0.904	0.934	0.925	0.939	0.935	0.923
(*h*)	0.983	0.986	0.987	0.988	0.988	0.951	0.964	0.954	0.965	0.963	0.957
(*i*)	0.967	0.972	0.973	0.974	0.974	0.874	0.890	0.861	0.896	0.885	0.870
(*j*)	0.973	0.977	0.979	0.980	0.980	0.916	0.931	0.916	0.929	0.924	0.913
$FR_{i,j}^{(m)}$	4 × 4	6 × 6	8 × 8	12 × 12	14 × 14	6 × 6	9 × 9	12 × 12	18 × 18	21 × 21	24 × 24
$FR_{i,j}$	2 × 2	3 × 3	4 × 4	6 × 6	7 × 7	2 × 2	3 × 3	4 × 4	6 × 6	7 × 7	8 × 8

Table 3 The MSE values of HR images by changing the increase coefficient *m* and $FR_{i,j}$

Image	m = 2					m = 3					
(*a*)	28.6	21.4	18.7	15.6	14.3	56.5	44.3	46.8	31.7	27.7	22.6
(*b*)	42.2	36.3	34.3	33.2	33.5	62.6	52.3	57.2	49.9	52.8	59.7
(*c*)	15.1	9.8	8.5	7.6	7.6	28.8	18.3	20.0	15.9	17.2	20.7
(*d*)	304.5	304.7	305.3	307.4	311.1	349.7	348.0	380.8	360.5	376.6	398.3
(*e*)	232.4	209.8	200.2	194.2	195.8	335.2	314.1	342.6	307.3	319.6	344.8
(*f*)	91.6	69.4	63.5	59.8	60.8	175.9	141.2	168.3	135.5	146.8	174.3
(*g*)	40.8	30.6	28.0	25.6	26.0	74.2	56.2	60.8	50.9	54.2	63.8
(*h*)	24.7	18.6	16.8	15.5	15.4	43.0	32.7	35.2	29.3	30.4	34.5
(*i*)	100.8	83.0	78.6	73.4	73.2	179.3	157.6	175.6	149.8	157.8	179.1
(*j*)	70.9	61.2	58.1	54.3	55.6	128.6	112.6	120.2	108.0	112.6	121.1
$FR_{i,j}^{(m)}$	4 × 4	6 × 6	8 × 8	12 × 12	14 × 14	6 × 6	9 × 9	12 × 12	18 × 18	21 × 21	24 × 24
$FR_{i,j}$	2 × 2	3 × 3	4 × 4	6 × 6	7 × 7	2 × 2	3 × 3	4 × 4	6 × 6	7 × 7	8 × 8

The results of image resolution increase when $k \geq 7$ shows a high level of noise in some areas of the image (Fig. 14d). Therefore, for practical implementation of the method we recommended use of $k = 6$.

The Task of Image Resolution Decreasing. Practical implementation of method of image resolution decrease using neural-like structure of geometric transformations model (Fig. 2b) and matrix operator of synaptic connections weights for (Fig. 11b) was conducted by the following parameters of NLS GTM: the number of inputs—dim($FR_{i,j}$), hidden layers—1, neuron number of in the hidden layer equal dim($FR_{i,j}^{(m)}$), the number of NLS outputs are determined by condition of solution, the task of image

Table 4 The UIQ values of HR images by changing the increase coefficient m and $FR_{i,j}$

Image	m = 2					m = 3					
(a)	0.76	0.78	0.78	0.79	0.79	0.64	0.67	0.64	0.68	0.69	0.69
(b)	0.68	0.70	0.70	0.71	0.70	0.59	0.61	0.59	0.61	0.60	0.58
(c)	0.77	0.83	0.83	0.83	0.83	0.65	0.72	0.69	0.72	0.70	0.68
(d)	0.75	0.80	0.80	0.81	0.80	0.59	0.66	0.61	0.65	0.63	0.59
(e)	0.71	0.73	0.74	0.75	0.75	0.52	0.54	0.49	0.56	0.55	0.53
(f)	0.72	0.75	0.75	0.75	0.75	0.59	0.62	0.58	0.62	0.60	0.57
(g)	0.69	0.72	0.73	0.73	0.73	0.55	0.59	0.56	0.59	0.57	0.55
(h)	0.82	0.84	0.85	0.85	0.85	0.74	0.77	0.75	0.77	0.76	0.74
(i)	0.77	0.79	0.79	0.80	0.80	0.61	0.63	0.58	0.64	0.62	0.60
(j)	0.74	0.76	0.76	0.77	0.77	0.61	0.63	0.60	0.63	0.62	0.60
$FR_{i,j}^{(m)}$	4 × 4	6 × 6	8 × 8	12 × 12	14 × 14	6 × 6	9 × 9	12 × 12	18 × 18	21 × 21	24 × 24
$FR_{i,j}$	2 × 2	3 × 3	4 × 4	6 × 6	7 × 7	2 × 2	3 × 3	4 × 4	6 × 6	7 × 7	8 × 8

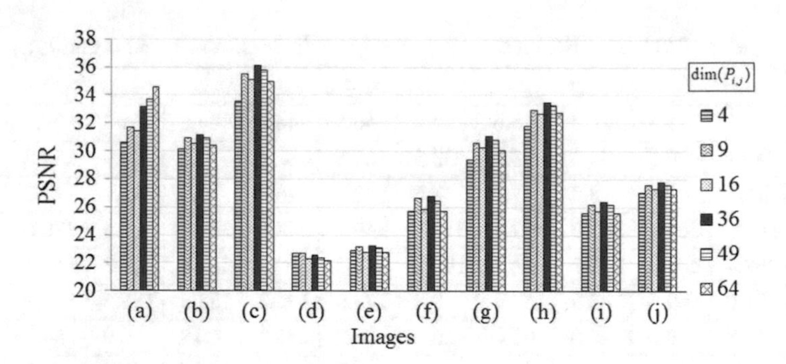

Fig. 13 The results (PSNR values) of image resolution increase if $m = 3$

resolution decrease (11), and equal dim($FR_{i,j}$). Linear value of synaptic connections. Coefficient of decrease m was set to 2, 3, and training process was conducted on the image 1 from the (Fig. 12).

The aim of this experiment is to search for such size frame dim($FR_{i,j}^{(m)}$) = ($k \times k$) that would provide best results based on MSE, UIQ, PSNR, SSIM. SSIM and PSNR values when changing the frame dimensions for input test HR image (h) in the case of a decrease in 2 times are shown in (Figs. 15 and 16) in accordance.

MSE, UIQ show a similar result for this variants of image resolution decrease.

As shown in (Figs. 15 and 16) the best results are obtained in the case of the smallest value k (dim($FR_{i,j}^{(m)}$) = 2 × 2). For visual evaluation of obtained results in (Fig. 17) are shown a set of reduction in twice images when changing k.

As shown in (Fig. 17), when increasing the frame size of dim($FR_{i,j}^{(m)}$), the quality of received images is significantly reduced. The same results were obtained for reducing

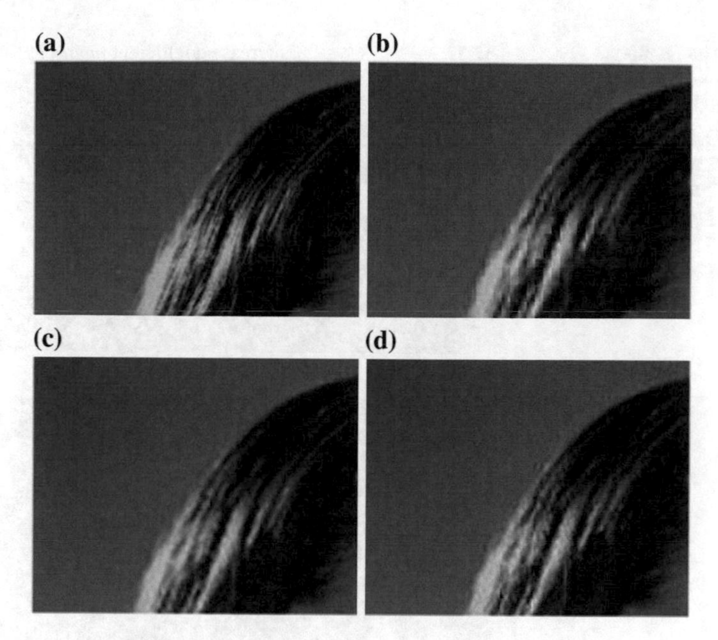

Fig. 14 Fragments of HR images if $m = 3$

Fig. 15 SSIM values by changing $\dim(FR_{i,j}^{(m)})$

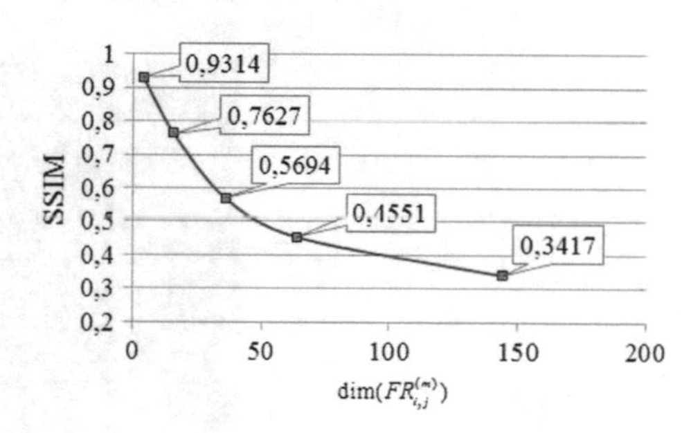

Fig. 16 PSNR values by changing $\dim(FR_{i,j}^{(m)})$

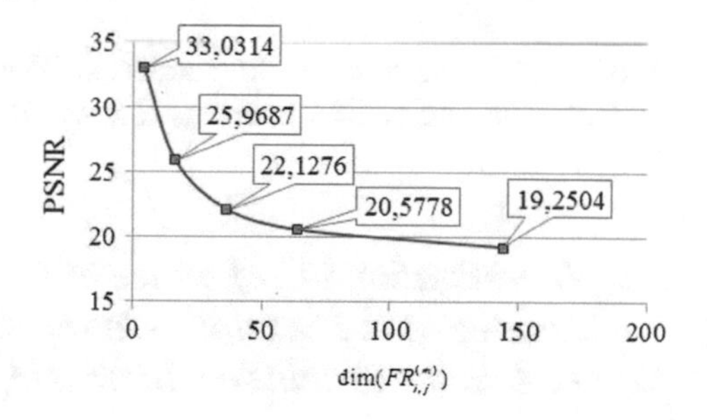

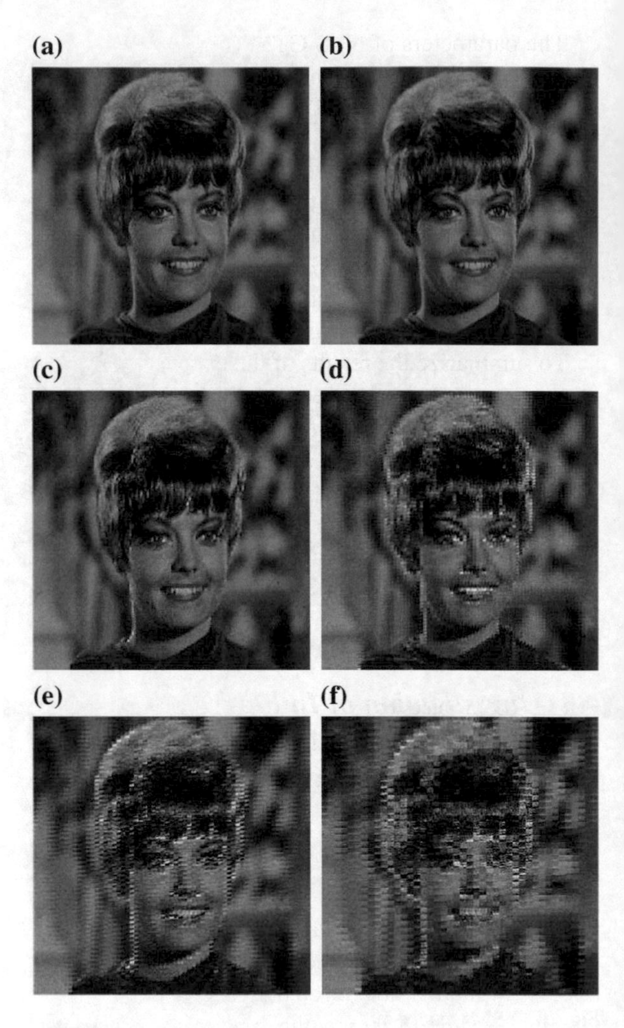

Fig. 17 The results of image resolution decrease twice by changing $\dim(FR_{i,j}^{(m)})$: **a** original image (252×252 pixels); **b** $2 \times 2 - 1 \times 1$; **c** $4 \times 4 - 2 \times 2$; **d** $6 \times 6 - 3 \times 3$; **e** $8 \times 8 - 4 \times 4$; **f** $12 \times 12 - 6 \times 6$

all other test images on all four indexes. Therefore, for practical application of the method, it should be used with the lowest value of input HR image frame size.

4.2 Investigation of Influence of Changing the Nonlinearity Degree of the Synaptic Connections for Solving Task of Images Resolution Increasing

The purpose of this experiment is to determine value of nonlinearity degree of synaptic connections between neurons that would provide sufficient quality of the resulting image and a small number of required computing resources, which are needed for work of NLS GTM. The value of this degree is defined by polynomial order.

The parameters of NLS GTM (Fig. 2a): the number of inputs—$\dim(FR_{i,j}) = 36$, hidden layers—1, neuron number in the hidden layer equal $\dim(FR_{i,j}) = 36$, coefficient of decrease $m = 2$. Therefore, the number of NLS outputs using (9) is equal $\dim(FR_{i,j}^{(m)}) = 144$.

The method is simulated in all test samples by changing of nonlinearity degree of synaptic connections in the range [1–4]. Further increase of this degree makes no sense because its large value significantly increases computing resources, which are needed for work of NLS GTM. Comparative results for test images from (Fig. 12). for the four indexes of quality are shown in (Table 5). Summary results of scaling using PSNR, shown in (Fig. 18).

To summarize the results of the experiment, it can be stated that use of nonlinear synaptic connections between neurons in most test samples did not confirm improved performance of scaling parameters. This occurred through obvious increase in the complexity of the network and by deterioration of its generalization properties.

Obviously, the introduction of nonlinearities significantly increases the computational resources, which are needed for work of NLS GTM. This allows to claim that increasing this parameter makes no sense because solving this task is completely satisfied by linear value of synaptic connections between neurons.

4.3 Investigation of Influence of Changing the Neuron Number in the Hidden Layer for Solving Task of Images Resolution Decreasing

As investigated in the previous paragraph, the training procedure of NLS GTM should take place when $\dim(FR_{i,j}^{(m)}) = 2 \times 2$, $\dim(FR_{i,j}) = 1 \times 1$. In this case, the neuron number in hidden layer was $\dim(FR_{i,j}^{(m)})$, i.e., 4. The next experiment involves searching such NLS parameters, which provide the best results. For this, the neuron number in hidden layer are changing. Since this variable should not be more than the number of NLS inputs, the experiment involves reducing it from 4 to 1.

Here are some (Table 6) simulation results of the method of image resolution decrease in twice for test images (Fig. 12c, e, g).

On (Figs. 19 and 20) results of this experiment are shown for the image (g) from Fig. 12.

Choosing in the hidden layer only one neuron, allows additional increase in the efficiency of the developed method, including the quality of the image scaling with decreased coefficient—2 on all four indexes.

The matrix operator of synaptic connections weights for solution of the task of image resolution decreasing at twice when $\dim(FR_{i,j}^{(m)}) = 2 \times 2$, $\dim(FR_{i,j}) = 1 \times 1$ and one neuron in hidden layer is as follows:

Table 5 The quality indexes of HR images if $m = 2$

Image	MSE				PSNR			
(*a*)	15.58	17.712	19.14	19.476	36.203	35.648	35.311	35.236
(*b*)	33.178	35.811	36.706	37.09	32.922	32.591	32.483	32.438
(*c*)	7.572	8.884	9.547	9.256	39.34	38.64	38.33	38.46
(*d*)	307.45	342.691	343.069	346.411	23.253	22.782	22.777	22.735
(*e*)	194.23	195.908	198.624	199.149	25.248	25.21	25.151	25.139
(*f*)	59.79	62.223	64.709	64.989	30.365	30.191	30.021	30.002
(*g*)	25.569	24.447	25.711	24.615	34.054	34.249	34.03	34.219
(*h*)	15.523	14.979	15.861	15.017	36.221	36.376	36.127	36.365
(*i*)	73.421	74.789	76.865	76.889	29.473	29.392	29.274	29.272
(*j*)	54.33	56.509	57.785	57.984	30.78	30.61	30.513	30.498
Image	UIQ				SSIM			
(*a*)	0.79	0.745	0.736	0.736	0.99	0.985	0.985	0.986
(*b*)	0.706	0.685	0.68	0.679	0.976	0.972	0.972	0.973
(*c*)	0.829	0.784	0.774	0.767	0.993	0.99	0.99	0.99
(*d*)	0.806	0.695	0.682	0.674	0.983	0.973	0.974	0.975
(*e*)	0.747	0.74	0.736	0.735	0.948	0.946	0.945	0.946
(*f*)	0.753	0.722	0.714	0.713	0.984	0.979	0.979	0.98
(*g*)	0.732	0.707	0.696	0.697	0.972	0.977	0.975	0.979
(*h*)	0.849	0.835	0.829	0.829	0.988	0.987	0.987	0.987
(*i*)	0.799	0.788	0.782	0.783	0.974	0.972	0.972	0.972
(*j*)	0.769	0.744	0.737	0.737	0.979	0.976	0.976	0.977
The degree of synapse nonlinearity	1	2	3	4	1	2	3	4

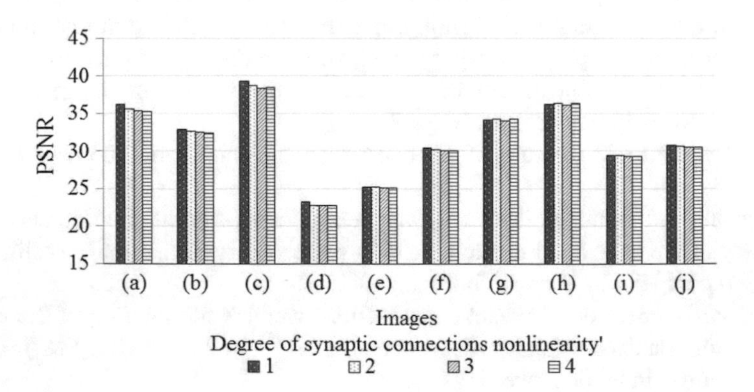

Fig. 18 PSNR values by changing the degree of synaptic connection nonlinearity if $m = 2$

Table 6 The quality indexes of the received images by changing the neuron number in the hidden layer if $m = 2$

Image	Neurons number	SSIM	MSE	UIQ	PSNR
(e)	1	0.99	6.29	0.95	40.14
(c)	1	0.93	61.25	0.93	30.26
(g)	1	0.98	9.26	0.97	38.47
(e)	2	0.97	15.63	0.91	36.19
(c)	2	0.84	166.03	0.83	25.93
(g)	2	0.95	26.15	0.93	33.96
(e)	3	0.96	20.35	0.89	35.05
(c)	3	0.76	281.84	0.75	23.36
(g)	3	0.93	31.12	0.91	33.2
(e)	4	0.96	19.84	0.89	35.16
(c)	4	0.78	255.59	0.77	24.06
(g)	4	0.93	32.35	0.91	33.03

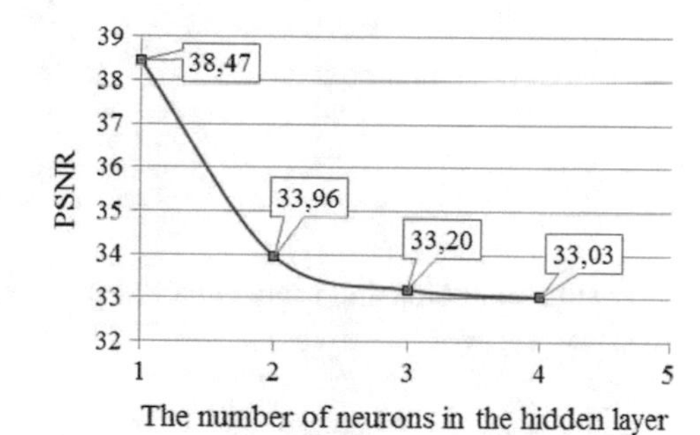

Fig. 19 PSNR values by changing the neuron number in the hidden layer (for image (*g*))

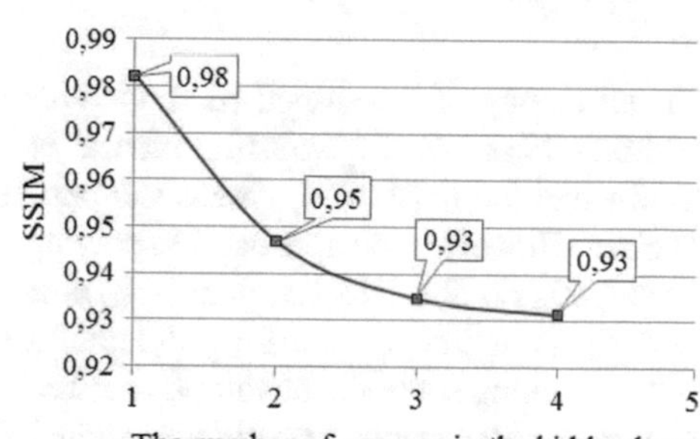

Fig. 20 SSIM values by changing the neuron number in the hidden layer (for image (*g*))

$$W = \begin{bmatrix} 1.7192300387 \\ 0.1976243697 \\ 0.2964612646 \\ 0.2098242399 \\ 0.2844760125 \end{bmatrix} \quad (35)$$

Table 7 The results of image resolution decrease using W

Image	MSE	PSNR	UIQ	SSIM
(*a*)	13.9638	36.6808	0.9503	0.9777
(*b*)	13.4164	36.8544	0.9471	0.964
(*c*)	6.2917	40.1431	0.9517	0.988
(*d*)	259.6678	23.9866	0.9374	0.9717
(*e*)	61.2527	30.2595	0.9279	0.9301
(*f*)	34.5352	32.7482	0.9491	0.9713
(*g*)	15.7173	36.167	0.9165	0.967
(*h*)	9.2588	38.4652	0.9742	0.9821
(*i*)	31.6375	33.1288	0.9535	0.9611
(*j*)	22.2986	34.648	0.9486	0.9664

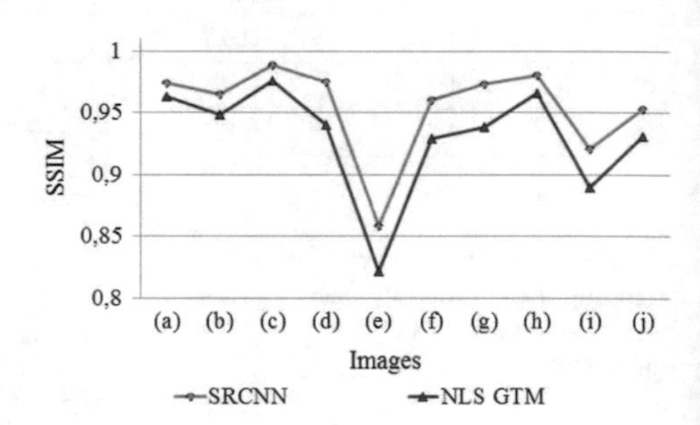

Fig. 21 Comparison of quality (SSIM values) of received HR images if $m = 3$ by different methods

Results of decreasing image resolution at twice using (35) are shown in (Table 7).

5 Comparison of Results

The efficiency of developed methods compared with the efficiency of the known method—based on Convolution Neural Network [10, 11]. It should be noted that known method (SRCNN), shows the best results on high quality of HR images at its class. However, the approach to solving this task is somewhat different from the developed one. SRCNN requires large dimension of the training set (it must contain more than one images pair for good quality of resulting HR images).

The training set of the developed method (as in case NLS GTM as by using matrix operator of synaptic connections weights) formed based on frames from only one pair of images (Fig. 1). Procedure of the practical implementation of the method of image resolution increase was held using $W^{(m)}$ for the following parameters: $m = 3$, $k = 6$. The training model of SRCNN was taken from [11].

As shown in (Fig. 21), SSIM values for two methods are very similar. However, the best results were obtained for synthesis HR images by SRCNN. Nevertheless, developed methods have several significant advantages, including the following:

- the training model of SRCNN method requires a large number of image pairs but the training model of developed method has only one image pair;
- the SRCNN method involves the use of iterative back propagation algorithm and with the same settings can produce different results (i.e., there is a problem of uniqueness of solution). The training process of developed method performs only one iteration and provides unambiguous solution when other things being equal;
- the training procedure of SRCNN method last for about three days. The NLS GTM method due to no iterations offers high performance in a training mode;
- the using mode of the SRCNN method involves to use Convolution neural network. The matrix operator of synaptic connections weights, obtained in the work, avoid to use NLS GTM in using mode. This advantage reduces time for solving the task of image resolution change;
- the SRCNN method designed to implement only the process of image resolution increasing. The NLS GTM-based method makes it possible to perform images resolution change with both coefficients: increase and decrease.

6 Conclusions

In this book chapter, the methods of image scaling based on machine learning are shown. It has chosen the neural-like structure of geometric transformations paradigm for training, because they reduce calculation and time resources which are needed for solving this task. Beside this, NLS GTM enables quick retraining in the automatic mode. To summarize the material it can be stated as follows:

- through the implementation of this approach it is possible to extend the functionality of learning-based methods, including perform scaling process with target coefficients both increase and decrease, that provides the high quality of scaling and reduces the training time;
- the developed methods provide high efficiency of scaling (based on PSNR and SSIM) and are characterized by computing resources decrease, which are needed for implementation of such procedures;
- it is established experimentally that the best value of synapses nonlinearity degree by the quality of the HR images (for the criterion based on PSNR) and computational resources, which are required by NLS GTM for solving this problem is equal to 1;
- it is established experimentally that frame size of the input LR image must be at least 6×6 pixels for synthesis HR image, and no more than 2×2 pixels for image resolution decrease;
- it is established experimentally that the best value of neuron number of hidden layer for synthesis LR image is equal to 1;
- the procedure of obtaining matrix operator of synaptic connections weights from trained NLS GTM with multiple outputs is developed. This allows to implement scaling without reuse of NLS GTM;

- in a practical way it is found that scaling results with use of either NLS GTM or the matrix operator of synaptic connection weights are the same ones. It allows to use this operator to implement changing image resolution methods in an online mode.

References

1. Bodyanskiy, Y., Dolotov, A., Vynokurova, O.: Evolving spiking wavelet-neuro-fuzzy self-learning system. Appl. Soft Comput. **14**, 252–258 (2014)
2. Bodyanskiy, Y., Dolotov, A., Vynokurova, O.: Self-learning cascade spiking neural network for fuzzy clustering based on group method of data handling. J. Autom. Inf. Sci. **45**, 23–33 (2013)
3. Mjolsness, E.: Neural networks, pattern recognition, and fingerprint hallucination: Ph.D. Dissertation, California Institute of Technology, California, (1985)
4. Ling, F., Xiao, F., Xue, H., Wu, S.: Super-resolution land-cover mapping using multiple sub-pixel shifted remotely sensed images. Int. J. Remote Sens. **31**(19), 5023–5040 (2010)
5. Nguyen, Q.M., Atkinson, P.M., Lewis, H.G.: Super-resolution mapping using Hopfield neural network with panchromatic imagery. Int. J. Remote Sens. **32**(21), 6149–6176 (2011)
6. Ling, F., Li, X., Xiao, F., Fang, S., Du, Y.: Object-based subpixel mapping of buildings incorporating the prior shape information from remotely sensed imagery. Int. J. Appl. Earth Obs. Geoinf. **18**, 283–292 (2012)
7. Li, X., Yun, D., Ling, F.: Super-resolution mapping of remotely sensed image based on Hopfield neural network with anisotropic spatial dependence model. IEEE Geosci. Remote Sens. Lett. **11**(10), 1265–1269 (2014)
8. Plaziac, N.: Image interpolation using neural networks. IEEE Trans. Image Process **8**(11), 1647–1651 (1999)
9. Pan, F., Zhang, L.: New image super-resolution scheme based on residual error restoration by neural networks. Opt. Eng. **42**(10), 3038–3046 (2003)
10. Dong, C., Loy, C.C., He, K., Tang, X.: Learning a deep convolutional network for image super resolution. In: Proceedings of the European Conferences on Computer Vision, pp. 184–199. Springer, Zurich, Switzerland (2014)
11. Dong, C., Loy, C.C., He, K., Tang, X.: Image super-resolution using deep convolutional networks. IEEE Trans. Pattern Anal. Mach. Intell. Preprint **14**, (2015)
12. Tkachenko, R., Yurchak, I., Polishchuk, U.: Neurolike networks on the basis of geometrical transformation machine. In: Proceedings of International Scientific and Technical Conference on Perspective Technologies and Methods in MEMS Design, pp. 77–80. Publishing House LPNU, Polyana (2008)
13. Polishchuk, U., Tkachenko, P., Tkachenko, R., Yurchak, I.: Features of the auto-associative neurolike structures of the geometrical transformation machine (GTM). In: Proceedings of 5th International Scientific and Technical Conference on Perspective Technologies and Methods in MEMS Design, pp. 66–67. Publishing House LPNU, Zakarpattya (2009)
14. Tkachenko, R.: Information models of the geometric transformation. In: Proceeding of the Second International Conference on Automatic Control and Information Technology 2013 (ICACIT 13), pp. 48–53. Cracow, Poland (2013)
15. Tkachenko, O., Tkachenko, R.: Neural system based on the geometric transformation model. In: Proceeding of the Second International Conference on Automatic Control and Information Technology 2013 (ICACIT 13), pp. 28–34. Cracow, Poland (2013)
16. Izonin, I., Tkachenko, R., Peleshko, D., Rak, T., Batyuk, D.: Learning-based image super-resolution using weight coefficients of synaptic connections. In: Proceeding of the Xth International Scientific and Technical Conference Computer Sciences and Information Technologies (CSIT), pp. 25–29. Publishing House LPNU, Lviv (2015)

17. Tsmots, I. Tsymbal Y. Doroshenko, A.: Development of a regional energy efficiency control system on the basis of intelligent components. In: XIth International Scientific and Technical Conference Computer Sciences and Information Technologies (CSIT), pp. 18–20. Lviv (2016)
18. Tsmots, I. Skorokhoda, O.: Hardware implementation of the real time neural network components. In: Proceedings of International Scientific and Technical Conference on Perspective Technologies and Methods in MEMS Design, pp. 124–126. Publishing House LPNU, Polyana (2011)
19. Andriyetskyy B., Polishchuk, U., Kulynsky, O.: Geometrical transformation machine in unsupervised learning mode. In: Proceedings of International Scientific and Technical Conference on Perspective Technologies and Methods in MEMS Design, pp. 87–88. Publishing House LPNU, Polyana (2006)
20. Riznyk, V., Riznyk, O., Balych, B., Parubchak, V.: Information encoding method of combinatorial optimization. In: International Conference of Modern Problems of Radio Engineering, Telecommunications, and Computer Science, pp. 357–357. Lviv-Slavsko (2006)
21. Dan, X.: An automatic detection approach for GS orthogonalization. J. Electron. **17**, 79–85 (1995)
22. Yang, X., Hu, X., Liu, Y.: Modified Gram-Schmidt orthogonalization of covariance matrix adaptive beamforming based on data preprocessing. In: IEEE 11th International Conference on Signal Processing, pp. 373–377. Beijing (2012)
23. Veretennikova, N., Pasichnyk, V., Kunanets, N., Gats, B.: E-Science: new paradigms, system integration and scientific research organization. In: Xth International Scientific and Technical Conference Computer Sciences and Information Technologies (CSIT), pp. 76–81. Lviv (2015)
24. Bodyanskiy, Y., Vynokurova, O., Pliss, I., Peleshko D., Rashkevych, Y.: Hybrid generalized additive wavelet-neuro-fuzzy-system and its adaptive learning. In: Dependability Engineering and Complex Systems: Series Advances in Intelligent Systems and Computing, vol. 470, pp. 51–61. Springer International Publishing, Brunow, Poland (2016)
25. Bodyanskiy, Y., Vynokurova, O., Szymaski Z., Kobylin I., Kobylin, O.: Adaptive robust models for identification of nonstationary systems in data stream mining tasks. In: 2016 IEEE First International Conference on Data Stream Mining and Processing (DSMP), pp. 263–268. Publishing House LPNU, Lviv (2016)
26. Bodyanskiy, Y., Vynokurova, O., Pliss, I., Setlak G., Mulesa, P.: Fast learning algorithm for deep evolving GMDH-SVM neural network in data stream mining tasks. In: 2016 IEEE First International Conference on Data Stream Mining and Processing (DSMP), pp. 257–262. Publishing House LPNU, Lviv (2016)
27. Dronyuk, I., Nazarkevych, M., Fedevych, O.: Synthesis of noise-like signal based on AtebfFunctions. In: Vishnevsky V., Kozyrev D. (eds) Distributed Computer and Communication Networks. DCCN 2015. Communications in Computer and Information Science, vol. 601, pp. 132–140 (2016)
28. Wang, Z., Bovik, A., Sheikh, H., Simoncelli, E.P.: Image quality assessment: from error visibility to structural similarity. IEEE Trans. Image Proces. **13**(4), 600–612 (2004)
29. Wang, Z., Bovik, A.: A universal image quality index. IEEE Sign. Proces. Lett. **9**, 81–84 (2002)
30. Khomytska, I., Teslyuk, V.: The method of statistical analysis of the scientific, colloquial, belles-lettres and newspaper styles on the phonological level. Adv. Intell. Syst. Comput. **512**, 149–163 (2016)

2D/3D Object Recognition and Categorization Approaches for Robotic Grasping

Nabila Zrira, Mohamed Hannat, El Houssine Bouyakhf and Haris Ahmad Khan

Abstract Object categorization and manipulation are critical tasks for a robot to operate in the household environment. In this chapter, we propose new methods for visual recognition and categorization. We describe 2D object database and 3D point clouds with 2D/3D local descriptors which we quantify with the k-means clustering algorithm for obtaining the bag of words (BOW). Moreover, we develop a new global descriptor called VFH-Color that combines the original version of Viewpoint Feature Histogram (VFH) descriptor with the color quantization histogram, thus adding the appearance information that improves the recognition rate. The acquired 2D and 3D features are used for training Deep Belief Network (DBN) classifier. Results from our experiments for object recognition and categorization show an average of recognition rate between 91% and 99% which makes it very suitable for robot-assisted tasks.

1 Introduction

In recent years, robots are being deployed in many areas where automation and decision-making skills are required. Robots are not just mechanically advanced but are becoming intelligent as well, and the idea behind these intelligent machines is the creation of systems that imitate the human behavior to be able to perform tasks

N. Zrira (✉) · M. Hannat · E.H. Bouyakhf
LIMIARF Laboratory, Faculty of Sciences Rabat, Mohammed V University Rabat, Rabat, Morocco
e-mail: nabilazrira@gmail.com

M. Hannat
e-mail: mohamedhannat@gmail.com

E.H. Bouyakhf
e-mail: bouyakhf@mtds.com

H. Ahmad Khan
NTNU, Norwegian University of Science and Technology, Gjøvik, Norway
e-mail: haris.a.khan@ntnu.no

A.E. Hassanien and D.A. Oliva (eds.), *Advances in Soft Computing and Machine Learning in Image Processing*, Studies in Computational Intelligence 730,
https://doi.org/10.1007/978-3-319-63754-9_26

which are actually infeasible for humans. The type of tasks for which robots are well adapted includes those that are in unexplored environment such as outer space [61, 64] and undersea [4, 15, 26]. However, the robot tasks are not limited just to complex and difficult problems, but they are covering some industrial [21], medical [17], and domestic applications [20] as well.

A human can search and find an object visually in a cluttered scene. It is a very simple task for human to pick an object and place it in the required place while avoiding obstacles along the path, and without damaging the fragile objects. These simple and trivial tasks for humans become challenging and complex for robots and can overcome their capabilities. The majority of pickup and drop applications through robots are performed in fully known and structured environments. The key question that arises in this context is how robots can perform as well as humans in these tasks when the structure of the environment is varied?

Human vision is extremely robust and can easily classify objects among tens of thousands of possibilities [11] within a fraction of a second [43]. The human system is able to tolerate the tremendous changes in scale, illumination, noise, and viewing angles for object recognition. Contrary to the human vision, the object recognition is a very complex problem and still beyond the capabilities of artificial vision systems. This contrast between vision systems and the human brain for performing visual recognition and classification tasks gave rise to the development of several approaches to visual recognition.

The ability to recognize and manipulate a large variety of objects is critical for mobile robots. Indoor environment often contains several objects on which the robot should make different actions such as "Pick-up the remote control TV!", "Drop it inside the box!". So, how to represent and classify objects to be recognized by robots?

Several techniques have been explored in order to achieve this goal. Recently, appearance-based methods have been successfully applied to the problem of object recognition. These methods typically proceed with two phases. In the first phase, a model is constructed from a set of training images that includes the appearance of the object under different illuminants, scales, and multiple instances. Whereas, in the second phase, the methods try to extract parts from the input image through segmentation or by using the sliding windows over the whole image. The methods then compare extracted parts of the input image with the training set. A popular strategy of appearance-based methods is the bag of words (BoW). BoW is inspired from text-retrieval systems that count how many times a word appears in a document. It aims to represent an image as an orderless set of local regions. In general, local regions are discretized into a visual vocabulary. This method obtains excellent results in image classification [5], image retrieval [66], object detection as well as scene classification [24].

With the advent of new 3D sensors like Microsoft Kinect, 3D perception became a fundamental vision research in mobile robotic applications. The Point Cloud Library (PCL) was developed by Rusu et al. [47] in 2010 and officially published in 2011. This open source library, licensed under Berkeley Software Distribution (BSD) terms, represents a collection of state-of-the-art algorithms and tools that operate with 3D point clouds. Several studies have been made based on PCL detectors and descriptors for 3D object recognition applications. PCL integrates several 3D detectors as well as 3D local and global descriptors. In 3D local descriptors, each point is described by its local geometry. They are developed for specific applications such as object recognition, and local surface categorization. This local category includes Signature of Histograms of OrienTation (SHOT) [57], Point Feature Histograms (PFH) [45], Fast Point Feature Histograms (FPFH) [44], and SHOTCOLOR [58]. On the other hand, the 3D global descriptors describe object geometry and they are not computed for individual points, but for a whole cluster instead. The global descriptors are high-dimensional representations of object geometry. They are usually calculated for subsets of the point clouds that are likely to be objects. The global category encodes only the shape information and includes Viewpoint Feature Histogram (VFH) [46], Clustered Viewpoint Feature Histogram (CVFH) [2], Oriented Unique and Repeatable CVFH (OUR-CVFH) [1], and Ensemble of Shape Functions (ESF) [62].

The ability to recognize objects is highly valuable for performing imperative tasks in mobile robotics. In several works, authors use classification methods such as Support Vector Machines (SVMs) [25, 31, 65], Nearest Neighbor (NN) [38], Artificial Neural Network (ANN) [7], or Hidden Markov Model (HMM) [59] in order to predict the object class. Recently, researchers got interested in deep learning algorithms because they can simultaneously and automatically discover both low-level and high-level features. Deep Belief Network (DBN) is a graphical model consisting of undirected networks at the top hidden layers and directed networks in the lower layers. The learning algorithm uses greedy layer-wise training by stacking restricted Boltzmann machines (RBMs) which contain hidden layer for modeling the probability distribution of visible variables [10].

In this chapter, we present new 2D/3D object recognition and categorization approaches which are based on local and global descriptors. We describe 2D objects and 3D point clouds using 2D/3D local and global descriptors. Then, we train separately these features with Deep Belief Network. We summarize our contributions as follows:

1. We describe an object database with SURF feature points which are quantified with the k-means clustering algorithm to make the 2D bag of words;
2. We describe a point cloud with spin image features which we quantify with the k-means clustering algorithm to generate the 3D bag of words;
3. We propose VFH-Color descriptor that combines both the color information and geometric features extracted from the previous version of VFH descriptor. We extract the color information for point cloud data, and then we use the color quantization technique to obtain the color histogram which is combined with VFH histogram.

This chapter is organized as follows. In Sect. 2, we provide a literature review on the relevant works of 2D/3D object recognition and categorization. Details of the proposed approaches are presented in Sect. 3. The implementation details and the experimental results are presented in Sect. 4. Finally, we conclude the chapter in Sect. 5.

2 State of the Art

2.1 2D Recognition and Categorization

Recently, the approaches that were based on bag of words (BoW), also known as Bag of features produced the promising results on several applications, such as object and scene recognition [13, 33], localization and mapping for mobile robots [19], video retrieval [52], text classification [6], and language modeling for image classification and retrieval [32, 37, 68].

Sivic et al. [51] used Latent Dirichlet Allocation (LDA) and probabilistic Latent Semantic Analysis (pLSA) in order to compute latent concepts in images from the cooccurrences of visual words. The authors aim to generate a consistent vocabulary of visual words that is insensitive to viewpoint changes and illumination. For this reason, they use vector quantized SIFT descriptors which are invariant to translation, rotations, and rescaling of the image.

Csurka et al. [14] developed a generic visual categorization approach for identifying the object content of natural images. In the first step, their approach detects and describes image patches which are clustered with a vector quantization algorithm to generate a vocabulary. The second step constructs a bag of keypoints that counts the number of patches assigned to each cluster. Finally, they use Naive Bayes and SVM to determine image categories.

Fergus et al. [18] suggested an object class recognition method that learns and recognizes object class models from unlabeled and unsegmented cluttered scenes in a scale-invariant manner. The approach exploits a probabilistic model that combines shape, appearance, occlusion, and relative scale, as well as an entropy-based feature detector to select regions and their scale within an image.

Philbin et al. [42] proposed a large-scale object retrieval system with large vocabularies and fast spatial matching. They extract features from each image in a high-dimensional descriptor space which are quantized or clustered to map every feature to a "visual word". This visual word is used to index the images for the search engine.

Wu et al. [63] proposed a new scheme to utilize optimized bag of words models called Semantics Preserving Bag of Words (SPBoW) that aims to map semantically related features to the same visual words. SPBoW computes a distance between identical features as a measurement of the semantic gap and tries to learn a codebook by minimizing this gap.

Larlus et al. [30] combined a bag of words recognition component with spatial regularization based on a random field and a Dirichlet process mixture for category-level object segmentation. The random field (RF) component assures short-range spatial contiguity of the segmentation while a Dirichlet process component assures mid-range spatial contiguity by modeling the image as a composition of blobs. Finally, the bag of words component allows strong intra-class imaging variations and appearance.

Vigo et al. [60] exploited color information in order to improve the bag of words technique. They select highly informative color-based regions for feature extraction.

Then, feature description focuses on shape and can be improved with a color description of the local patches. The experiments show that color information should be used both in the feature detection as well as the feature extraction stages.

Khan et al. [28] suggested integration of spatial information in the bag of visual words. The approach models the global spatial distribution of visual words that consider the interaction among visual words regardless of their spatial distances. The first step consists of computing pair of identical visual words (PIW) that save all the pairs of visual words of the same type. The second step represents a spatial distribution of words as a histogram of orientations of the segments formed by PIW.

2.2 3D Recognition and Categorization

Most of the recent work on 3D object categorization focused on appearance, shapes, and bag of words (BoW) extracted from certain viewing point changes of the 3D objects.

Savarese and Fei-Fei [48] proposed a compact model for representing and learning 3D object categories. Their model aims to solve scale changes and arbitrary rotations problems using appearance and 3D geometric shape. Each object is considered as a linked set of parts that are composed of many local invariant features. Their approach can classify, localize and infer the scale as well as the pose estimation of objects in the given image.

Toldo et al. [56] introduced bag of words (BoW) approach for 3D object categorization. They used spectral clustering to select seed-regions; then computed the geometric features of the object sub-parts. Vector quantization is applied to these features in order to obtain BoW histograms for each mesh. Finally, Support Vector Machine is used to classify different BoW histograms for 3D objects.

Nair and Hinton [40] presented a top-level model of Deep Belief Networks (DBNs) for 3D object recognition. This model is a third-order Boltzmann machine that is trained using a combination of both generative and discriminative gradients. The model performance is evaluated on NORB images where the dimensionality for each stereo-pair image is reduced by using a foveal image. The final representation consists of 8976-dimensional vectors that are learned with a top-level model for Deep Belief Nets (DBNs).

Zhong [67] introduced an approach for 3D point cloud recognition based on a new 3D shape descriptor called Intrinsic Shape Signature (ISS). ISS uses a view-dependent transform encoding for the viewing geometry to facilitate fast pose estimation, and a view-independent representation of the 3D shape in order to match shape patches from different views directly.

Bo et al. [12] introduced a set of kernel features for object recognition. The authors develop kernel descriptors on depth maps that model size, depth edges, and 3D shape. The main match kernel framework defines pixel attributes and designs match kernels in order to measure the similarities of image patches to determine low dimensional match kernels.

Lai et al. [29] built a new RGBD dataset and proposed methods for recognizing RGBD objects. They used SIFT descriptor to extract visual features and spin image descriptor to extract shape features that are used for computing efficient match kernel (EKM). Finally, linear support vector (LiSVM), gaussian kernel support vector machine (kSVM), and random forest (RF) are trained to recognize both the category and the instance of objects.

Mian et al. [39] suggested a 3D object retrieval approach from cluttered scenes based on the repeatability and quality of keypoints. The authors proposed a quality measure to select the best keypoints for extracting local features. They also introduced an automatic scale selection method for extracting scale and multi-scale-invariant features in order to match objects at different unknown scales.

Madry et al. [36] proposed the Global Structure Histogram (GSH) to describe the point cloud information. Their approach encodes the structure of local feature response on a coarse global scale to retain low local variations and keep the advantage of global representativeness. GSH can be instantiated in partial object views and trained using complete or incomplete information about an object.

Tang et al. [55] proposed a Histogram of Oriented Normal Vectors (HONV) feature which is based on local geometric characteristics of an object captured from the depth sensor. They considered that the object category information is presented on its surface. This surface is described by the normal vector at each surface point and the local 3D geometry is presented as a local distribution of the normal vector orientation.

Socher et al. [54] introduced the first convolutional-recursive deep learning model for 3D object recognition. They computed a single CNN layer to extract low-level features from both color and depth images. These representations are provided as input to a set of RNNs with random weights that produce high-quality features. Finally, the concatenation of all the resulting vectors forms the final feature vector for a softmax classifier.

Schwarz et al. [49] developed a meaningful feature set that results from the pre-trained stage of Convolutional Neural Network (CNN). The depth and RGB images are processed independently by CNN and the resulting features are then concatenated to determine the category, instance, and pose of the object.

Eitel et al. [16] presented two separate CNN processing streams for RGBD object recognition. RGB and colorized depth images consist of five convolutional layers and two fully connected layers. Both streams are processed separately through several layers and converge into one fully connected layer and a softmax layer for the classification task.

Alex [3] proposed a new approach for RGBD object classification. Four independent Convolutional Neural Networks (CNNs) are trained, one for each depth data and three for RGB data and then trains these CNNs in a sequence. The decisions of each network are combined to obtain the final classification result.

Ouadiay et al. [41] proposed a new approach for real 3D object recognition and categorization using Deep Belief Networks. First, they extracted 3D keypoints from point clouds using 3D SIFT detector and then they computed SHOT/SHOTCOLOR descriptors. The performance of the approach is evaluated on two datasets: Washington RGBD object dataset and real 3D object dataset.

Madai et al. [35] reinvestigated Deep Convolutional Neural Networks (DCNNs) for RGBD object recognition. They proposed a new method for depth colorization based on surface normals, which colorized the surface normals for every pixel and computed the gradients in a horizontal direction (x-axis) and vertical direction (y-axis) through the Sobel operator. The authors defined two 3D vectors a and b in direction of the z-axis in order to calculate the surface normal n. As n has three dimensions, the authors map each of the three values of the surface normal to a corresponding RGB channels.

3 Our Recognition Pipelines

In this chapter, we suggest new approaches for 2D/3D object recognition and categorization for mobile robotic applications. We introduce two different recognition pipelines, one relies on 2D/3D detectors and descriptors which are quantified with a k-means algorithm to obtain 2D/3D bag of words, while the other one uses our new 3D global descriptor called VFH-Color. Figure 1 summarizes the main steps of 2D/3D bag of words approaches.

1. **Training set**: represents a set of data (images or point clouds) used on our experiment. Training means, creating a dataset with all the objects which we want to recognize.
2. **Keypoint extraction**: is the first step of our approach where keypoints (interest points) are extracted from input data. It reduces the computational complexity by identifying particularly those regions of images which are important for descriptors, in term of high information density.
3. **Keypoint description**: once keypoints are extracted, descriptors are computed on the obtained keypoints and these form a description which is used to represent the data.
4. **Vocabulary**: after the extraction of descriptors, the approach uses the vector quantization technique to cluster descriptors in their feature space. Each cluster is considered as "visual word vocabulary" that represents the specific local pattern shared by the keypoints in this cluster.
5. **Bag of Words**: is a vector containing the (weighted) count or occurrence of each visual word in the data which is used as the feature vector in the recognition and classification tasks.
6. **Classificiation**: all data in training set are represented by their bag of words vectors which represent the input of DBN classifier.

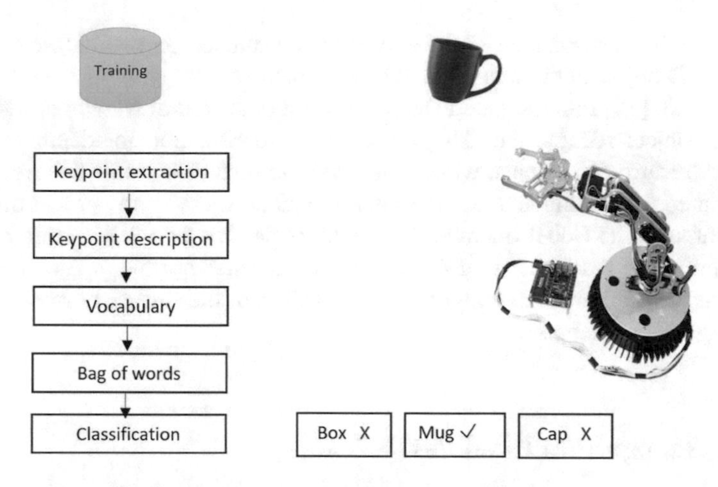

Fig. 1 Overview of 2D/3D bag of words approaches

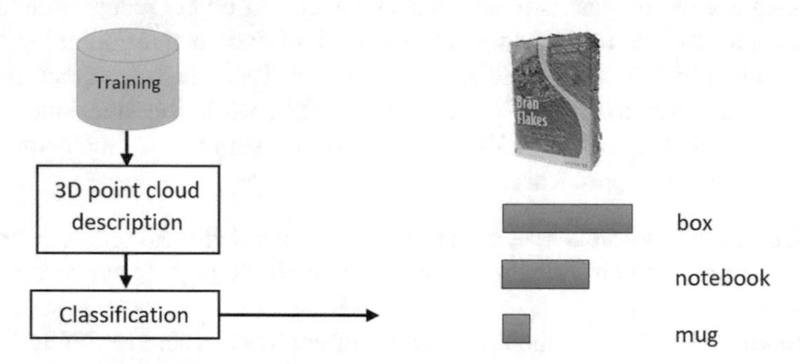

Fig. 2 Overview of 3D global approach

For the global pipeline, we present a new VFH-Color descriptor that combines both the color information and the geometric features extracted from the previous version of VFH descriptor. Figure 2 summarizes the main steps of the global approach.

1. **Training set**: represents a set of point clouds used on our experiment.
2. **3D point description**: extracts the color information for point cloud data, then uses the color quantization technique to obtain the color histogram which is combined with VFH histogram extracted from the previous version of VFH descriptor.
3. **Classificiation**: all point clouds in training set are represented by their VFH-Color features and are provided as the input to DBN classifier.

3.1 *Object Representation*

3.1.1 2D Bag of Words

2D Speeded-Up Robust Features (SURF) detector

Keypoints are important features that are becoming more and more widespread in image analysis. The Speeded-Up Robust Features (SURF) [8, 9] is based on the same steps and principles of SIFT detector [34], but it utilizes a different scheme and provides better results than those obtained with SIFT extractor. SURF is scale- and rotation-invariant keypoint detector that uses a very basic Hessian-matrix approximation because of its good performance in term of accuracy. Gaussian kernels are optimal for scale-space analysis. SURF divides the scale space into levels and octaves where each octave corresponds to a doubling of scale and is divided into uniformly spaced levels. The method builds a pyramid of response maps with various levels within octaves. The keypoints represent the points that are the extrema among eight neighbors in the current level and its 2×9 neighbors in the above and below levels.

2D Speeded-Up Robust Features (SURF) descriptor

SURF descriptor provides a unique and robust description of a featurc that can be generated on the area surrounding a keypoint. SURF descriptor is based on Haar Wavelet responses and can be calculated efficiently with integral images. SURF describes an interesting area with size 20s, then each interest area is divided into 4×4 sub-areas and is described by the values of a wavelet response in the x and y directions. The interest areas are weighted with a Gaussian centered at the keypoint for being robust in deformations and translations. For each sub-area, a vector v is calculated, based on 5×5 samples. The descriptor for keypoint consists of 16 vectors for the sub-areas being concatenated. Finally, the descriptor is normalized, to achieve invariance to contrast variations that will represent themselves as a linear scaling of the descriptor.

Visual vocabulary

Once the keypoint descriptors are obtained, the approach imposes a quantization on the feature space of these descriptors. The standard pipeline to obtain "visual vocabulary" is also called "codebook" which consists of (i) collecting a large sample of a local feature; (ii) quantizing the feature space according to their statistics. Most vector quantization or clustering algorithms are based on hierarchical or iterative square error partitioning methods. Hierarchical methods organize data on groups which can be displayed in the form of the tree. Whereas, square error partitioning algorithms attempt to obtain which maximizes the between cluster scatter or minimizes the within-cluster scatter. In our work, we use a simple k-means clustering algorithm. The "visual words" or "codevector" represent the k cluster centers. A vector quantizer takes a feature vector and maps it to the index of the nearest codevector in the codebook using the Euclidean distance (Fig. 3).

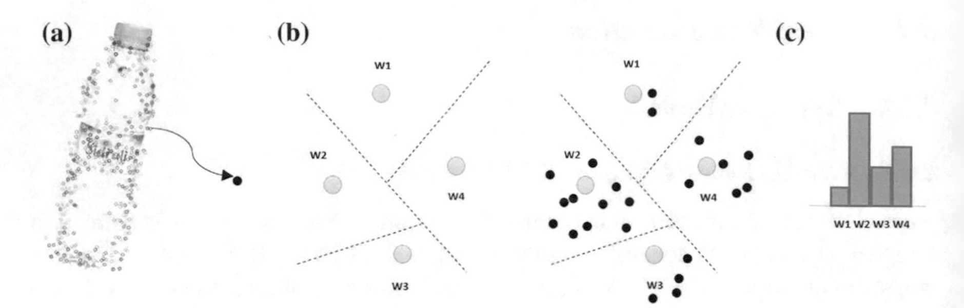

Fig. 3 The schematic illustrates visual vocabulary construction and word assignment. **a** the black dot represents SURF keypoint, the object contains in total 240 SURF keypoints. Next, the approach computes SURF descriptor on each keypoint. **b** Visual words (W1, W2, W3, and W4) denote cluster centers. **c** The sampled features are clustered in order to optimize the space into a discrete number of visual words. A bag of visual words histogram can be used to summarize the entire image. It counts the occurrence of each visual word in the image

Bag of Words

Bag of words is generated by computing the count or occurrence of each visual word in the image which is used as the feature vector in the recognition and classification tasks (Fig. 3).

3.1.2 3D Bag of Words

3D Scale-invariant feature transform (SIFT) detector

Scale-invariant feature transform (SIFT) is an algorithm deployed in the field of computer vision to detect and describe regions in an image and identify similar elements between varying images. This process is called "matching". The algorithm consists of the detected feature points of an image which are used to characterize every point that needs to be recognized by comparing its characteristics with those of the points contained in other images. The general idea of SIFT is to find the keypoints that are invariant to several transformations/changes: rotation, scale, illumination, and viewing angle. The 3D SIFT detector [50] utilizes the Difference-of-Gaussian (DoG) function to extract the extrema points in both spatial and scale dimensions.

Spin image descriptor

The spin image was proposed to describe points of interest by [27]. This descriptor translates the local properties of the surface oriented in a coordinate system fixed and linked to the object. This system is independent of the viewing angle. The spin is defined at a point oriented and designated by its 3D position (p) as well as associated direction (n the normal to the local surface). A 2D local coordinate base is formed using the tangent plane P in the point p, oriented perpendicularly to the normal n, and the line L through p parallel to n. A cylindrical coordinate system (α, β)

of the point p is then deduced. The radial coordinate defining the distance (nonnegative) is perpendicular to L and the elevation coordinate of the defined distance is perpendicular to P (signed positive or negative). The resulting histogram is formed by counting the occurrences of different pairs of discretized distances.

Visual vocabulary

After describing each of the point clouds inside a class with the spin image, we need to make the visual categorization using the probabilistic approach. The method we use consists of applying a quantization operation with the k-means clustering and constructs visual words with the well-known method of the bag of features.

Bag of Words

Instead of considering each feature point a visual word, we consider thanks to the quantization that each of the clusters' center represent a word. The bag of words algorithm consists of computing the number of occurrences of each word in the model database. It is like a probability of the number of words inside the class of objects.

3.1.3 Viewpoint Feature Histogram Color (VFH-Color)

The viewpoint feature histogram (VFH) [46] computes a global descriptor of the point cloud and consists of two components: a surface shape component and a viewpoint direction component. VFH aims to combine the viewpoint direction directly into the relative normal angle calculation in the FPFH descriptor [44]. The viewpoint-dependent component of the descriptor is a histogram of the angles between the vector $(p_c - p_v)$ and each point's normal. This component is binned into a 128-bin histogram. The other component is a simplified point feature histogram (SPFH) estimated for the centroid of the point cloud, and an additional histogram of distances of all points in the cloud to the cloud's centroid. The three angles (α, ϕ, θ) with the distance d between each point and the centroid are binned into a 45-bin histogram. The total length of VFH descriptor is the combination of these two histograms and is equal to 308 bins.

Color quantization is a vector quantization that aims to select K vectors in N dimensional space in order to represent N vectors from that space ($K << N$). In

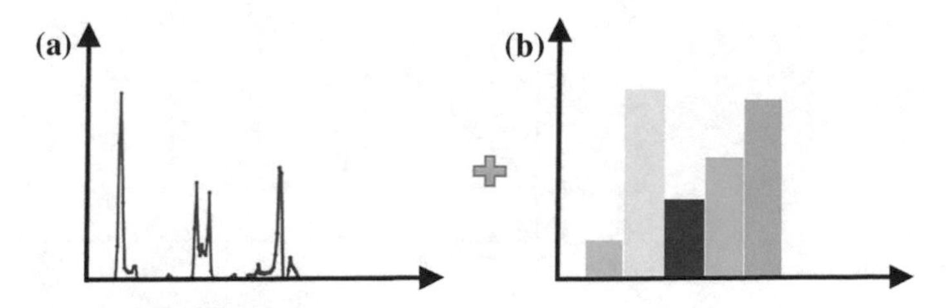

Fig. 4 VFH-Color. **a** VFH descriptor. **b** Color quantization

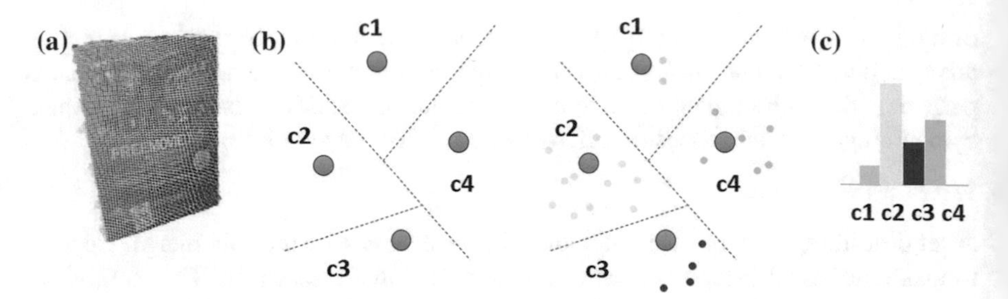

Fig. 5 Color quantization process. **a** point cloud data. **b** codebook (c1, c2, c3, and c4) denote cluster centers. **c** The RGB features are clustered in order to optimize the space. The histogram counts the occurrence of each codebook in the point cloud

general, color quantization is applied to reduce the number of colors in a given image while maintaining the visual appearance of the original image. Color quantization is applied in a three-dimensional space RGB and follows the following steps (Fig. 5):

1. Extract RGB features for each point from the point cloud data;
2. Obtain the matrix of RGB features (number of points×3);
3. Compute k-means algorithm for the RGB matrix in order to generate the codebook (cluster centers);
4. Count the occurrence of each codebook in the point cloud.

The codebook size represents the bins of color quantization histogram. After a set of experiments, we fix the codebook size to 100 bins (see Fig. 6). Therefore, VFH-Color histogram concatenates 308 values of original VFH descriptor and 100 values of color quantization histogram, thus giving the total size of 408 values (Fig. 4).

Fig. 6 The classification performance with respect to the codebook size

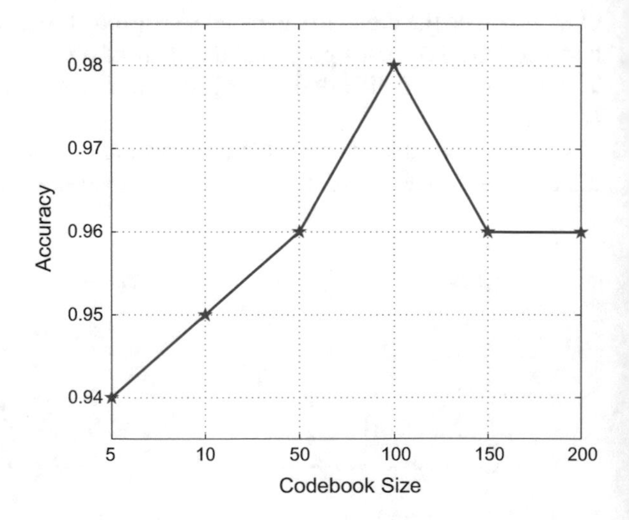

3.2 *Object Classification*

3.2.1 Restricted Boltzmann Machines (RBMs)

Restricted Boltzmann Machines (RBMs) [53] are a specific category of energy based model which include hidden variables. RBMs are restricted in the sense so that no hidden–hidden or variable–variable connections exist. The architecture of a generative RBM is illustrated in Fig. 7.

RBMs are a parameterized generative stochastic neural network which contain stochastic binary units on two layers: the visible layer and the hidden layer.

1. Visible units (the first layer): they contain visible units (x) that correspond to the components of an observation (i.e., 2D/3D features in this case of study);
2. Hidden units (the second layer): they contain hidden units (h) that model dependencies between the components of observations.

The stochastic nature of RBMs results from the fact that the visible and hidden units are stochastic. The units are binary, i.e., $x_i, h_j \in \{0, 1\} \forall\ i$ and j, and the joint probability which characterize the RBM configuration is the Boltzmann distribution

$$p(x, h) = \frac{1}{Z} e^{-E(x,h)} \tag{1}$$

The normalization constant is $Z = \sum_{x,h} e^{-E(x,h)}$ and the energy function of an RBM is defined as

$$E(x, h) = -b^{'} x - c^{'} h - h^{'} W x \tag{2}$$

where:

- W represents the symmetric interaction term between visible units (x) and hidden units (h);
- b and c are vectors that store the visible (input) and hidden biases (respectively).

RBMs are proposed as building blocks of multilayer learning deep architectures called deep belief networks. The idea behind is that the hidden neurons extract pertinent features from the visible neurons. These features can work as the input to another RBM. By stacking RBMs in this way, the model can learn features for a high-level representation.

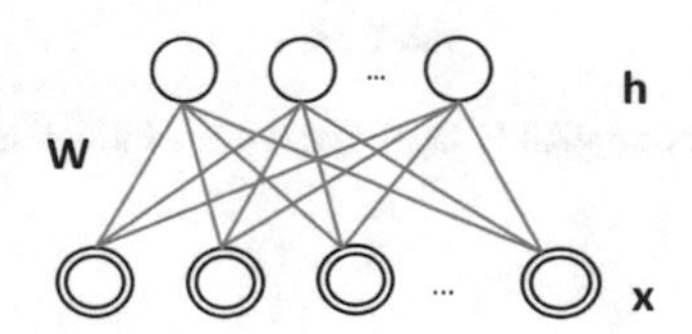

Fig. 7 RBM model. The visible units x and hidden units h are connected through undirected and symmetric connections. There are no intra-layer connections

3.2.2 Deep Belief Network (DBN)

Deep Belief Network (DBN) is the probabilistic generative model with many layers of stochastic and hidden variables. Hinton et al. [23] introduced the motivation for using a deep network versus a single hidden layer (i.e. a DBN vs. an RBM). The power of deep networks is achieved by having more hidden layers. However, one of the major problems for training deep network is how to initialize the weights W between the units of two consecutive layers ($j-1$ and j), and the bias b of the layer j. Random initializations of these parameters can cause poor local minima of the error function resulting in low generalization. For this reason, Hinton *et al.* introduced a DBN architecture based on training sequence of RBMs. DBN train sequentially as many RBMs as the number of hidden layers that constitute its architecture, i.e., for a DBN architecture with l hidden layers, the model has to train l RBMs. For the first RBM, the inputs consist of the DBN's input layer (visible units) and the first hidden layer. For the second RBM, the inputs consist of the hidden unit activations of the previous RBM and the second hidden layer. The same holds for the remaining RBMs to browse through the l layers. After the model performs this layer-wise algorithm, a good initialization of the biases and the hidden weights of the DBN is obtained. At this stage, the model should determine the weights from the last hidden layer for the outputs. To obtain a successfully supervised learning, the model "fine-tunes" the resulting weights of all layers together. Figure 8 illustrates a DBN architecture with one visible layer and three hidden layers.

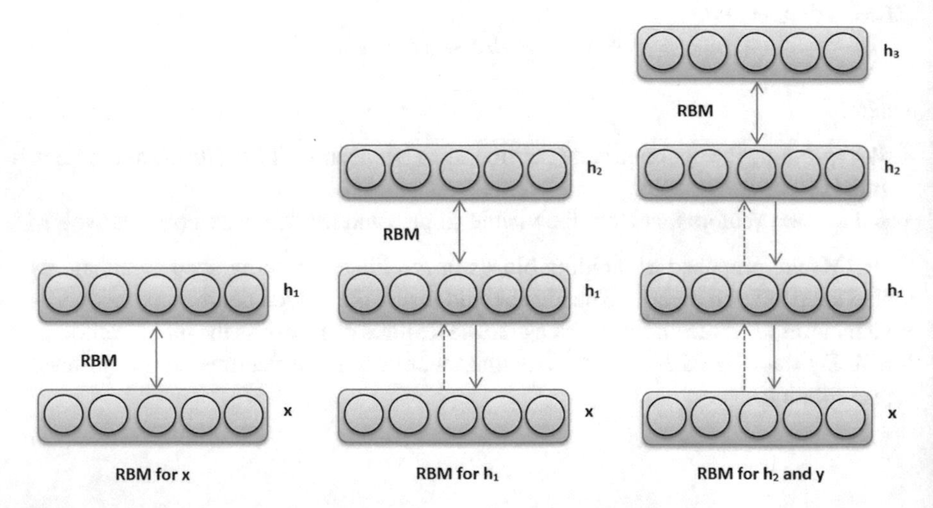

Fig. 8 DBN architecture with one visible layer x and three hidden layers $h1$, $h2$, and $h3$

4 Experimental Results and Discussion

4.1 Datasets

4.1.1 ALOI Dataset

Amsterdam Library of Object Images (ALOI) [22] dataset is an image collection of 1000 small objects recorded for recognition task. 111,250 images are captured by Sony DXC390P 3CCD cameras varying viewing angle, illumination angle and illumination color for each object, and additionally images are captured wide-baseline stereo images (Fig. 9).

4.1.2 Washington RGBD Dataset

Washington RGBD dataset is a large dataset built for 3D object recognition and categorization applications. It is a collection of 300 common household objects which are organized into 51 categories. Each object is placed on a turntable and is captured for one whole rotation in order to obtain all object views using Kinect camera that records synchronized and aligned 640×480 RGB and depth images at 30 Hz [29] (see Fig. 10).

4.2 2D/3D Object Classification

DBN aims to allow each RBM model in the sequence to receive a different representation of the data. In other words, after RBM has been learned, the activity values of its hidden units are used as the training data for learning a higher level RBM. The input layer has a number N of units, equal to the size of sample data x (size of 2D/3D features). The number of units for hidden layers, currently, are predefined according to the experiment. We fixed DBN with three hidden layers $h1$, $h2$ and $h3$. The general DBN characteristics are shown in Table 1.

Fig. 9 The sample images extracted from Amsterdam Library of Object Images (ALOI) dataset

Fig. 10 The sample point clouds extracted from Washington RGBD Dataset

Table 1 DBN characteristics that are used in our experiments

Characteristic	Value
Hidden layers	3
Hidden layer units	600
Learn rates	0.01
Learn rate decays	0.9
Epochs	200
Input layer units	Size of descriptor

4.2.1 2D Bag of Words

Images contain local points or keypoints defined as salient region patches which represent rich local information of the image. We used SURF to automatically detect and describe keypoints from images. Then, we used the vector quantization method in order to cluster the keypoint descriptors in their feature space into a large number of clusters using the k-means clustering algorithm. We test in a set of experiments the impact of the number of clusters on classifier accuracy and we select $k = 1500$ as the size of the codebook (number of visual words) that represents the best accuracy value. We conduct the experiments on ALOI dataset on which we select ten categories: teddy, jam, ball, mug, food_box, towel, shoes, pen, can, and bottle. Figure 9 shows some examples from ALOI dataset which are used in our experiments.

As shown in the confusion matrix (Fig. 12), the classes which are consistently misclassified are the teddy, ball, shoes, can, mug, and bottle which are very similar in appearance (Fig. 11). The results show also that 2D bag of words approach which

Fig. 11 The objects which are missclassified using 2D bag of words classification

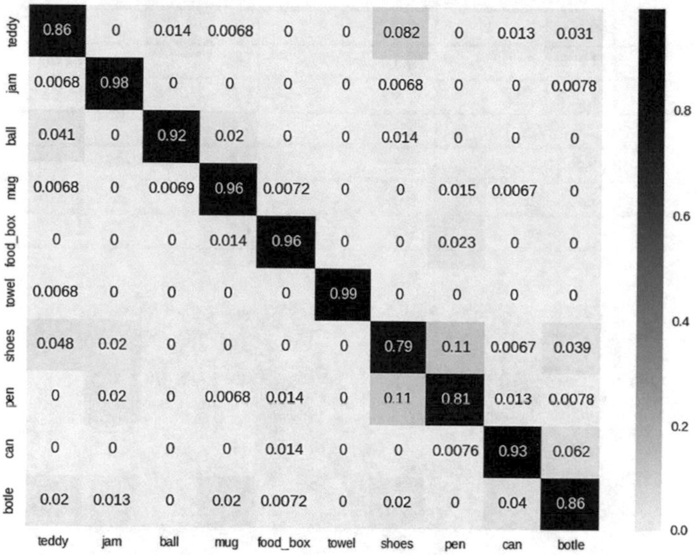

	teddy	jam	ball	mug	food_box	towel	shoes	pen	can	botle
teddy	0.86	0	0.014	0.0068	0	0	0.082	0	0.013	0.031
jam	0.0068	0.98	0	0	0	0	0.0068	0	0	0.0078
ball	0.041	0	0.92	0.02	0	0	0.014	0	0	0
mug	0.0068	0	0.0069	0.96	0.0072	0	0	0.015	0.0067	0
food_box	0	0	0	0.014	0.96	0	0	0.023	0	0
towel	0.0068	0	0	0	0	0.99	0	0	0	0
shoes	0.048	0.02	0	0	0	0	0.79	0.11	0.0067	0.039
pen	0	0.02	0	0.0068	0.014	0	0.11	0.81	0.013	0.0078
can	0	0	0	0	0.014	0	0	0.0076	0.93	0.062
botle	0.02	0.013	0	0.02	0.0072	0	0.02	0	0.04	0.86

Fig. 12 Confusion matrix of 2D bag of words model

is based on SURF features works perfectly with the accuracy rate of 91%. BoW representation encodes only the occurrence of the appearance of the local patches and ignores the object geometry. The lack of geometric features can provide some misclassification especially when the objects are similar in appearance. In Table 2, we report accuracy values for 2D bag of words with both SVM and DBN classifiers. The first row reports the accuracy value of SVM whereas the second row shows the accuracy value of DBN. We notice that the combination of 2D bag of words and DBN outperforms the 2D bag of words with SVM and rises steadily from 88.86% to 90.83%. This result shows the power of deep learning architectures that learn multiple levels of representation depending on the depth of the architecture (Table 5).

4.2.2 3D Bag of Words

After extracting the spin image for the set of point clouds, we constructed a shape dictionary whose size is fixed at k = 250, by clustering all spin image acquired from the whole training set with k-means method. For each bin, a representative local 3D

Table 2 The performance of 2D bag of words

Classes	Metrics			
	Wrong class	f1-score (%)	Recall (%)	Precision (%)
(1)	(3,4,7,9,10)	86	86	87
(2)	(1,7,10)	96	98	95
(3)	(1,4,7)	95	92	98
(4)	(1,3,5,8,9)	95	96	93
(5)	(4,8)	96	96	96
(6)	(1)	100	99	100
(7)	(1,2,8,9,10)	78	79	77
(8)	(2,4,5,7,9,10)	82	81	84
(9)	(5,8,10)	92	93	92
(10)	(1,2,4,5,7,9)	86	86	85
Average	–	91	91	91

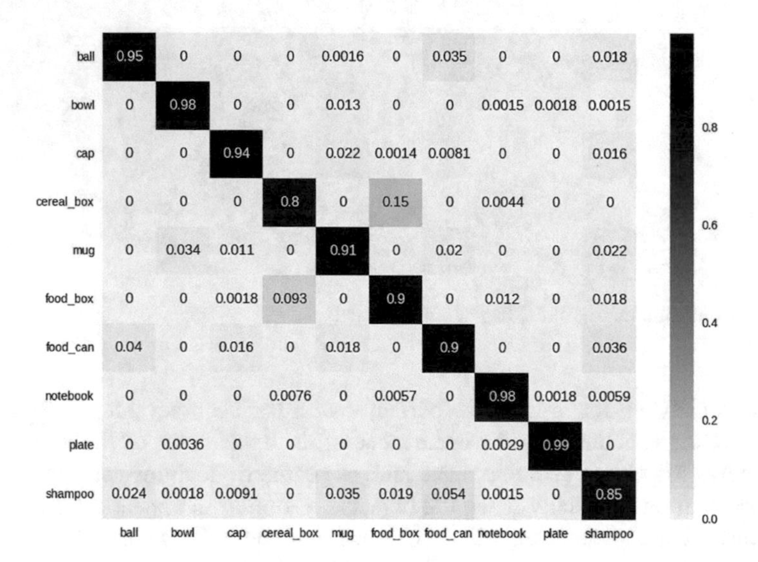

Fig. 13 Confusion matrix of 3D Bag of words

feature description is required. These descriptions are taken from the centroids of each cluster (visual words) determined by k-means clustering on precomputed spin image descriptors.

Figure 13 represents the confusion matrix across all 10 classes. Most model's results are reasonable showing that 3D bag of words can provide high-quality features. The classes that are consistently misclassified are ball-mug-can, bowl-mug-notebook-plate, and food_box-cereal_box-notebook which are very similar in shape.

Table 3 The performance of 3D Bag of Words

Classes	Metrics			
	Wrong class	f1-score (%)	Recall (%)	Precision (%)
(1)	(5,7,10)	94	95	94
(2)	(5,8,9,10)	97	98	96
(3)	(5,6,7,10)	95	94	96
(4)	(6,8)	84	80	89
(5)	(2,3,7,10)	91	91	91
(6)	(4,5,8,10)	87	90	84
(7)	(1,3,5,10)	89	90	88
(8)	(4,6,9,10)	98	98	98
(9)	(2,8)	99	99	100
(10)	(1,2,3,5,6,7,8)	87	85	88
Average	–	92	92	92

Table 3 illustrates the performance metrics of 3D bag of words that encodes only the surface shape of 3D point clouds thanks to the use of spin images descriptor.

4.2.3 Global Pipeline

VFH-Color descriptor combines both the color information and the geometric features extracted from the previous version of VFH descriptor. We extract the color information for point cloud data, then we use the color quantization technique to obtain the color histogram which is combined with VFH histogram. For each point cloud, we extract two types of features: (1) geometric features extracted from Viewpoint Feature Histogram (VFH) (308 bins), and (2) color features extracted from color quantization (100 bins). These features are then combined into a single vector, being 308+100=408 dimensional. Figures 14 and 15 represents the confusion matrix across all 10 classes. Most model's results are very reasonable showing that VFH-Color can provide meaningful features. The classes that are consistently misclassified are mug-cap, cereal_box-food_box, and shampoo-cap-mug-food_can which are very similar in appearance and shape.

Moreover, we evaluate the performance of VFH-Color against the previous version of VFH and SHOTCOLOR. The accuracy using VFH-Color performs 3% better than VFH that models only the geometric features (Tables 4 and 6). This result shows the effectiveness of the approach after adding the color information. We also notice that SHOTCOLOR presents a good accuracy (Table 7 and Figure 16), although this descriptor encounters a problem when it is not able to compute the local reference frame for some point clouds. In this set of experiments, 15% of point clouds from the dataset are not computed with SHOTCOLOR. This problem becomes significant when 3D object recognition is in real time. Indeed, VFH-Color descriptor can

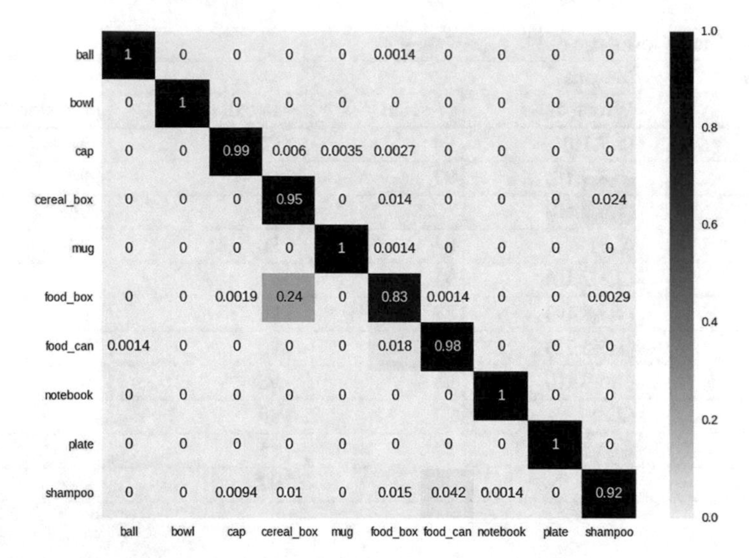

Fig. 14 Confusion matrix of global pipeline using VFH descriptor

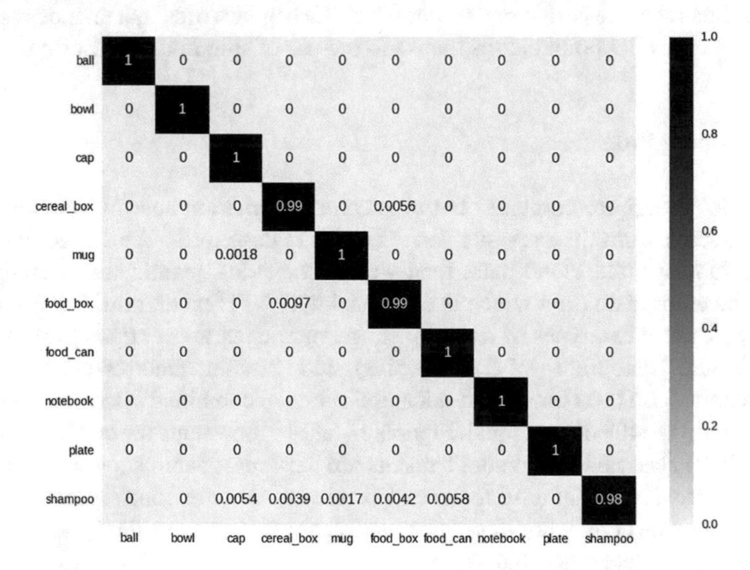

Fig. 15 Confusion Matrix of global pipeline using our VFH-Color descriptor

Table 4 The performance of global pipeline using VFH descriptor

Classes	Metrics			
	Wrong class	f1-score (%)	Recall (%)	Precision (%)
(1)	(6)	100	100	100
(2)	(–)	100	100	100
(3)	(4,5,6)	99	99	99
(4)	(6,10)	86	95	79
(5)	(6)	100	100	100
(6)	(3,4,7,10)	88	83	94
(7)	(1,6)	97	98	96
(8)	(–)	100	100	100
(9)	(–)	100	100	100
(10)	(3,4,6,7,8)	95	92	97
Average	–	95	96	97

Table 5 Accuracy of different proposed approaches using DBN and SVM classifiers

Classifier	BOW2D (%)	BOW3D (%)	VFH (%)	VFH-Color (%)	SHOTCOLOR (%)
SVM	88.86	86.68	95.01	98.34	97.21
DBN	90.83	92.03	96.41	**99.63**	98.63

be used in the real-time applications thanks to its estimation for every point cloud as well as its good recognition rate.

Table 5 shows also that our global pipeline works perfectly with the accuracy rate of 99.63% with DBN architecture that performs the use of SVM classifier. In general, the use of DBN instead of SVM in our approaches increases the accuracy rate thanks to the performance of deep learning algorithms which outperformed the shallow architectures (SVM).

4.3 Comparison to Other Methods

In this subsection, we compare our contributions to the related state-of-the-art approaches. Table 8 shows the main accuracy values and compares our recognition pipelines to the published results [12, 29, 49] and [16, 35]. Lai et al. [29] extract a set of features that captures the shape of the object view using a spin image and another set which captures the visual appearance using SIFT descriptors. These features are extracted separately from both depth and RGB images. A recent work by Schwarz et al. [49] uses both colorizing depth and RGB images that are processed independently by a convolutional neural network. CNN features are then learned using SVM

Table 6 The performance of global pipeline using VFH-Color descriptor

Classes	Metrics			
	Wrong class	f1-score (%)	Recall (%)	Precision (%)
(1)	(–)	100	100	100
(2)	(–)	100	100	100
(3)	(–)	100	100	99
(4)	(6)	99	99	99
(5)	(3)	100	100	100
(6)	(4)	99	99	99
(7)	(–)	100	100	99
(8)	(–)	100	100	100
(9)	(–)	100	100	100
(10)	(3,4,5,6,7)	99	98	100
Average	–	99	99	99

Table 7 The performance of SHOTCOLOR descriptor

Classes	Metrics			
	Wrong class	f1-score (%)	Recall (%)	Precision (%)
(1)	(7)	100	99	100
(2)	(–)	100	100	100
(3)	(–)	100	100	100
(4)	(6,8)	93	92	95
(5)	(–)	100	100	100
(6)	(4,7,10)	95	96	94
(7)	(6)	99	100	99
(8)	(–)	100	100	99
(9)	(–)	100	100	100
(10)	(7)	99	100	99
Average	–	99	99	99

classifier in order to successively determine the category, instance, and pose. The previous approaches [16, 35] used the color-coding depth maps and RGB images for training separately CNN architecture.

In our work, we learn our 3D features using DBN with three hidden layers that model a deep network architecture. The results show also that our global pipeline works perfectly with the accuracy rate of 99.63% thanks to the efficiency of our VFH-Color descriptor and outperforms all methods that are mentioned in the state of the art.

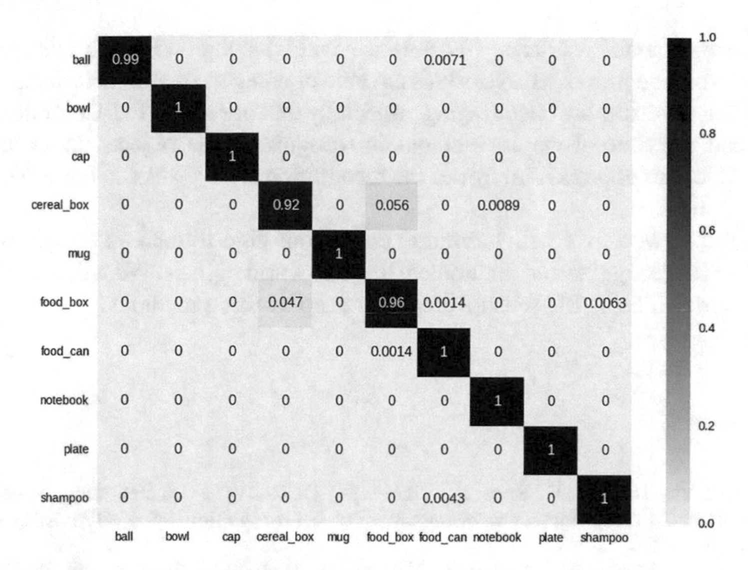

Fig. 16 Confusion matrix of SHOTCOLOR descriptor

Table 8 The comparison of 3D object recognition accuracies and PCL descriptors on the Washington RGBD dataset

Approaches	Accuracy rates (%)
Lai et al. [29]	90.6
Bo et al.[12]	84.5
Eitel et al. [16]	91
Madai et al. [35]	94
Schwarz et al. [49]	94.1
VFH and DBN	96.41
3D BoW and DBN	92
VFH-Color and DBN	**99.63**
SHOTCOLOR and DBN	98.63

5 Conclusion and Future Work

In this chapter, we proposed new approaches for object categorization and recognition in real-world environment. We used the bag of words (BoW) that aims to represent images and point clouds as an orderless of local regions that are discretized into a visual vocabulary. Also, we proposed the VFH-Color descriptor which combined geometric features extracted from Viewpoint Feature Histogram (VFH) descriptor and color information extracted from color quantization method. Then, we learned the 2D and 3D features with Deep Belief Network (DBN) classifier.

The experimental results on ALOI dataset and Washington RGBD dataset clearly ascertain that the proposed algorithms are able of categorizing objects and 3D point clouds. These results are encouraging, especially that our new VFH-Color descriptor performed the state-of-the-art methods in recognizing 3D objects under different views. Also, our approach improved the recognition rates thanks to the use of color information.

In a future work, we will attempt to embed our algorithms in a mobile robot in order for it to recognize and manipulate the real-world objects. We will also develop a new approach using 3D sensors and other deep learning methods.

References

1. Aldoma, A., Tombari, F., Rusu, R., Vincze, M.: OUR-CVFH–oriented, unique and repeatable clustered viewpoint feature histogram for object recognition and 6DOF pose estimation. Springer (2012)
2. Aldoma, A., Vincze, M., Blodow, N., Gossow, D., Gedikli, S., Rusu, R., Bradski, G.: Cad-model recognition and 6dof pose estimation using 3d cues. In: 2011 IEEE International Conference on Computer Vision Workshops (ICCV Workshops, pp. 585–592. IEEE (2011)
3. Alexandre, L.A.: 3d object recognition using convolutional neural networks with transfer learning between input channels. In: Intelligent Autonomous Systems 13, pp. 889–898. Springer (2016)
4. Antonelli, G., Fossen, T.I., Yoerger, D.R.: Underwater robotics. In: Springer Handbook of Robotics, pp. 987–1008. Springer (2008)
5. Avila, S., Thome, N., Cord, M., Valle, E., Araújo, A.D.A.: Bossa: Extended bow formalism for image classification. In: 2011 18th IEEE International Conference on Image Processing, pp. 2909–2912. IEEE (2011)
6. Bai, J., Nie, J.-Y., Paradis, F.: Using language models for text classification. In: Proceedings of the Asia Information Retrieval Symposium, Beijing, China (2004)
7. Basu, J.K., Bhattacharyya, D., Kim, T.-H.: Use of artificial neural network in pattern recognition. Int. J. Softw. Eng. Appl. **4**, 2 (2010)
8. Bay, H., Ess, A., Tuytelaars, T., Van Gool, L.: Speeded-up robust features (surf). Comput. Vis. Image Underst. **110**(3), 346–359 (2008)
9. Bay, H., Tuytelaars, T., Van Gool, L.: Surf: Speeded up robust features. In: Computer vision–ECCV 2006, pp. 404–417. Springer (2006)
10. Bengio, Y.: Learning deep architectures for ai. Foundations and trends®. Mach. Learn. **2**(1), 1–127 (2009)
11. Biederman, I.: Recognition-by-components: a theory of human image understanding. Psychol. Rev. **94**(2), 115 (1987)
12. Bo, L., Ren, X., Fox, D.: Depth kernel descriptors for object recognition. In: 2011 IEEE/RSJ International Conference on Intelligent Robots and Systems (IROS), pp. 821–826. IEEE (2011)
13. Bolovinou, A., Pratikakis, I., Perantonis, S.: Bag of spatio-visual words for context inference in scene classification. Pattern Recogn. **46**(3), 1039–1053 (2013)
14. Csurka, G., Dance, C., Fan, L., Willamowski, J., Bray, C.: Visual categorization with bags of keypoints. In: Workshop on Statistical Learning in Computer Vision, ECCV, vol. 1, Prague, pp. 1–2 (2004)
15. Dunbabin, M., Corke, P., Vasilescu, I., Rus, D.: Data muling over underwater wireless sensor networks using an autonomous underwater vehicle. In: Proceedings 2006 IEEE International Conference on Robotics and Automation, 2006. ICRA 2006, pp. 2091–2098. IEEE (2006)

16. Eitel, A., Springenberg, J.T., Spinello, L., Riedmiller, M., Burgard, W.: Multimodal deep learning for robust rgb-d object recognition. In: 2015 IEEE/RSJ International Conference on Intelligent Robots and Systems (IROS), pp. 681–687. IEEE (2015)
17. Fei, B., Ng, W.S., Chauhan, S., Kwoh, C.K.: The safety issues of medical robotics. Reliab. Eng. Syst. Safety **73**(2), 183–192 (2001)
18. Fergus, R., Perona, P., Zisserman, A.: Object class recognition by unsupervised scale-invariant learning. In: 2003 IEEE Computer Society Conference on Computer Vision and Pattern Recognition, 2003. Proceedings, vol. 2, IEEE, pp. II–264 (2003)
19. Filliat, D.: A visual bag of words method for interactive qualitative localization and mapping. In: 2007 IEEE International Conference on Robotics and Automation, pp. 3921–3926. IEEE (2007)
20. Forlizzi, J., DiSalvo, C.: Service robots in the domestic environment: a study of the roomba vacuum in the home. In: Proceedings of the 1st ACM SIGCHI/SIGART Conference on Human-Robot Interaction, pp. 258–265. ACM (2006)
21. Freund, E.: Fast nonlinear control with arbitrary pole-placement for industrial robots and manipulators. Int. J. Robot. Res. **1**(1), 65–78 (1982)
22. Geusebroek, J.-M., Burghouts, G.J., Smeulders, A.W.: The amsterdam library of object images. Int. J. Comput. Vis. **61**(1), 103–112 (2005)
23. Hinton, G.E., Osindero, S., Teh, Y.-W.: A fast learning algorithm for deep belief nets. Neural Comput. **18**(7), 1527–1554 (2006)
24. Hu, F., Xia, G.-S., Wang, Z., Huang, X., Zhang, L., Sun, H.: Unsupervised feature learning via spectral clustering of multidimensional patches for remotely sensed scene classification. IEEE J. Selected Topics Appl Earth Observ. Remote Sens. **8**, 5 (2015)
25. Janoch, A., Karayev, S., Jia, Y., Barron, J.T., Fritz, M., Saenko, K., Darrell, T.: A category-level 3d object dataset: Putting the kinect to work. In: Consumer Depth Cameras for Computer Vision, pp. 141–165. Springer (2013)
26. Jaulin, L.: Robust set-membership state estimation; application to underwater robotics. Automatica **45**(1), 202–206 (2009)
27. Johnson, A., Hebert, M.: Using spin images for efficient object recognition in cluttered 3d scenes. IEEE Trans. Pattern Anal. Mach. Intell. **21**(5), 433–449 (1999)
28. Khan, R., Barat, C., Muselet, D., Ducottet, C.: Spatial orientations of visual word pairs to improve bag-of-visual-words model. In: Proceedings of the British Machine Vision Conference, pp. 89–1. BMVA Press (2012)
29. Lai, K., Bo, L., Ren, X., Fox, D.: A large-scale hierarchical multi-view rgb-d object dataset. In: 2011 IEEE International Conference on Robotics and Automation (ICRA), pp. 1817–1824. IEEE (2011)
30. Larlus, D., Verbeek, J., Jurie, F.: Category level object segmentation by combining bag-of-words models with dirichlet processes and random fields. Int. J. Comput. Vis. **88**(2), 238–253 (2010)
31. LeCun, Y., Huang, F.J., Bottou, L.: Learning methods for generic object recognition with invariance to pose and lighting. In: Proceedings of the 2004 IEEE Computer Society Conference on Computer Vision and Pattern Recognition, 2004. CVPR 2004, vol. 2, pp. II–97. IEEE (2004)
32. Li, M., Ma, W.-Y., Li, Z., Wu, L.: Visual language modeling for image classification, Feb. 28 2012. US Patent 8,126,274
33. Li, T., Mei, T., Kweon, I.-S., Hua, X.-S.: Contextual bag-of-words for visual categorization. IEEE Trans. Circuits Syst. Video Technol. **21**(4), 381–392 (2011)
34. Lowe, D.G.: Object recognition from local scale-invariant features. In: The proceedings of the Seventh IEEE International Conference on Computer Vision, 1999, vol. 2, pp. 1150–1157. IEEE (1999)
35. Madai-Tahy, L., Otte, S., Hanten, R., Zell, A.: Revisiting deep convolutional neural networks for rgb-d based object recognition. In: International Conference on Artificial Neural Networks, pp. 29–37. Springer (2016)

36. Madry, M., Ek, C.H., Detry, R., Hang, K., Kragic, D.: Improving generalization for 3d object categorization with global structure histograms. In: 2012 IEEE/RSJ International Conference on Intelligent Robots and Systems (IROS), pp. 1379–1386. IEEE (2012)
37. Mc Donald, K.R.: Discrete language models for video retrieval. Ph.D. thesis, Dublin City University (2005)
38. McCann, S., Lowe, D.G.: Local naive bayes nearest neighbor for image classification. In: 2012 IEEE Conference on Computer Vision and Pattern Recognition (CVPR), pp. 3650–3656. IEEE (2012)
39. Mian, A., Bennamoun, M., Owens, R.: On the repeatability and quality of keypoints for local feature-based 3d object retrieval from cluttered scenes. Int. J. Comput. Vis. **89**(2–3), 348–361 (2010)
40. Nair, V., Hinton, G.E.: 3d object recognition with deep belief nets. In: Advances in Neural Information Processing Systems, pp. 1339–1347 (2009)
41. Ouadiay, F.Z., Zrira, N., Bouyakhf, E.H., Himmi, M.M.: 3d object categorization and recognition based on deep belief networks and point clouds. In: Proceedings of the 13th International Conference on Informatics in Control, Automation and Robotics, pp. 311–318 (2016)
42. Philbin, J., Chum, O., Isard, M., Sivic, J., Zisserman, A.: Object retrieval with large vocabularies and fast spatial matching. In: CVPR 2007. IEEE Conference on Computer Vision and Pattern Recognition, 2007, pp. 1–8. IEEE (2007)
43. Potter, M.C.: Short-term conceptual memory for pictures. J. Exp. Psychol: Hum Learn. Mem. **2**(5), 509 (1976)
44. Rusu, R., Blodow, N., Beetz, M.: Fast point feature histograms (fpfh) for 3d registration. In: IEEE International Conference on Robotics and Automation, 2009. ICRA 2009, pp. 3212–3217. IEEE (2009)
45. Rusu, R., Blodow, N., Marton, Z., Beetz, M.: Aligning point cloud views using persistent feature histograms. In: IEEE/RSJ International Conference on Intelligent Robots and Systems, IROS 2008, pp. 3384–3391 (2008)
46. Rusu, R., Bradski, G., Thibaux, R., Hsu, J.: Fast 3d recognition and pose using the viewpoint feature histogram. In: 2010 IEEE/RSJ International Conference on Intelligent Robots and Systems (IROS), pp. 2155–2162. IEEE (2010)
47. Rusu, R., Cousins, S.: 3D is here: point cloud library (PCL). In: IEEE International Conference on Robotics and Automation (ICRA) (Shanghai, China, May 9-13 2011)
48. Savarese, S., Fei-Fei, L.: 3d generic object categorization, localization and pose estimation. In: IEEE 11th International Conference on Computer Vision, 2007. ICCV 2007, pp. 1–8. IEEE (2007)
49. Schwarz, M., Schulz, H., Behnke, S.: Rgb-d object recognition and pose estimation based on pre-trained convolutional neural network features. In: 2015 IEEE International Conference on Robotics and Automation (ICRA), pp. 1329–1335. IEEE (2015)
50. Scovanner, P., Ali, S., Shah, M.: A 3-dimensional sift descriptor and its application to action recognition. In: Proceedings of the 15th International Conference on Multimedia pp. 357–360. ACM (2007)
51. Sivic, J., Russell, B.C., Efros, A.A., Zisserman, A., Freeman, W.T.: Discovering object categories in image collections
52. Sivic, J., Zisserman, A.: Video google: a text retrieval approach to object matching in videos. In: Ninth IEEE International Conference on Computer Vision, Proceedings, pp. 1470–1477. IEEE (2003)
53. Smolensky, P. Information processing in dynamical systems: Foundations of harmony theory
54. Socher, R., Huval, B., Bath, B., Manning, C.D., Ng, A.Y.: Convolutional-recursive deep learning for 3d object classification. In: Advances in Neural Information Processing Systems, pp. 665–673 (2012)
55. Tang, S., Wang, X., Lv, X., Han, T.X., Keller, J., He, Z., Skubic, M., Lao, S.: Histogram of oriented normal vectors for object recognition with a depth sensor. In: Asian Conference on Computer Vision, pp. 525–538. Springer (2012)

56. Toldo, R., Castellani, U., Fusiello, A.: A bag of words approach for 3d object categorization. In: Computer Vision/Computer Graphics CollaborationTechniques, pp. 116–127. Springer (2009)
57. Tombari, F., Salti, S., Stefano, D.L.: Unique signatures of histograms for local surface description. In: Computer Vision–ECCV 2010, pp. 356–369. Springer (2010)
58. Tombari, F., Salti, S., Stefano, L.: A combined texture-shape descriptor for enhanced 3d feature matching. In: 2011 18th IEEE International Conference on Image Processing (ICIP), pp. 809–812. IEEE (2011)
59. Torralba, A., Murphy, K.P., Freeman, W.T., Rubin, M.A.: Context-based vision system for place and object recognition. In: Ninth IEEE International Conference on Computer Vision, 2003. Proceedings, pp. 273–280. IEEE (2003)
60. Vigo, D.A.R., Khan, F.S., Van de Weijer, J., Gevers, T.: The impact of color on bag-of-words based object recognition. In: 2010 20th International Conference on Pattern Recognition (ICPR), pp. 1549–1553. IEEE (2010)
61. Visentin, G., Van Winnendael, M., Putz, P.: Advanced mechatronics in esa's space robotics developments. In: 2001 IEEE/ASME International Conference on Advanced Intelligent Mechatronics, 2001. Proceedings (2001), vol. 2, pp. 1261–1266. IEEE (2001)
62. Wohlkinger, W., Vincze, M.: Ensemble of shape functions for 3d object classification. In: 2011 IEEE International Conference on Robotics and Biomimetics (ROBIO) (2011), pp. 2987–2992. IEEE (2011)
63. Wu, L., Hoi, S.C., Yu, N.: Semantics-preserving bag-of-words models and applications. IEEE Trans. Image Process. **19**(7), 1908–1920 (2010)
64. Yoshida, K.: Achievements in space robotics. IEEE Robot. Automat. Mag. **16**(4), 20–28 (2009)
65. Zhang, H., Berg, A.C., Maire, M., Malik, J.: Svm-knn: discriminative nearest neighbor classification for visual category recognition. In: 2006 IEEE Computer Society Conference on Computer Vision and Pattern Recognition (CVPR 2006), vol. 2, pp. 2126–2136. IEEE (2006)
66. Zheng, L., Wang, S., Liu, Z., Tian, Q.: Packing and padding: Coupled multi-index for accurate image retrieval. In: Proceedings of the IEEE Conference on Computer Vision and Pattern Recognition, pp. 1939–1946 (2014)
67. Zhong, Y.: Intrinsic shape signatures: a shape descriptor for 3d object recognition. In: 2009 IEEE 12th International Conference on Computer Vision Workshops (ICCV Workshops), pp. 689–696. IEEE (2009)
68. Zhu, L., Rao, A.B., Zhang, A.: Theory of keyblock-based image retrieval. ACM Trans. Inf. Syst. (TOIS) **20**(2), 224–257 (2002)

A Distance Function for Comparing Straight-Edge Geometric Figures

Apoorva Honnegowda Roopa and Shrisha Rao

Abstract This chapter defines a distance function that measures the dissimilarity between planar geometric figures formed with straight lines. This function can in turn be used in partial matching of different geometric figures. For a given pair of geometric figures that are graphically isomorphic, one function measures the angular dissimilarity and another function measures the edge-length disproportionality. The distance function is then defined as the convex sum of these two functions. The novelty of the presented function is that it satisfies all properties of a distance function and the computation of the same is done by projecting appropriate features to a cartesian plane. To compute the deviation from the angular similarity property, the Euclidean distance between the given angular pairs and the corresponding points on the $y = x$ line is measured. Further while computing the deviation from the edge-length proportionality property, the best fit line, for the set of edge lengths, which passes through the origin is found, and the Euclidean distance between the given edge-length pairs and the corresponding point on a $y = mx$ line is calculated. Iterative Proportional Fitting Procedure (IPFP) is used to find this best fit line. We demonstrate the behavior of the defined function for some sample pairs of figures.

Keywords Geometric similarity · Iterative Proportional Fitting Procedure · Euclidean distance

2010 Mathematics Subject Classification. 65D10 (primary) · 51K05 (secondary)

1 Introduction

Two geometric figures can be said to be similar if one of the geometric figures can be obtained by either squeezing or enlarging the other. This implies that the considered geometric figures need to have equal number of vertices and edges, matching

---

A. Honnegowda Roopa · S. Rao (✉)
Bangalore, India
e-mail: shrao@ieee.org

A.E. Hassanien and D.A. Oliva (eds.), *Advances in Soft Computing and Machine Learning in Image Processing*, Studies in Computational Intelligence 730,
https://doi.org/10.1007/978-3-319-63754-9_27

corresponding angles, and a fixed proportionality between the corresponding edges. This concept of similarity can be used for partial matching of different geometric figures.

It is well known that geometric shapes and structures are important in determining the behavior of chemical compounds. This is true of smaller molecules [1] as well as larger macromolecules such as DNA and RNA that are studied in bioinformatics [2]. Molecular geometry [3] is, thus, an important aspect of physical and structural chemistry. However, while it is also known that similarity in structures often implies similar observed chemical properties, there is yet no well-defined mathematical approach for comparing geometric shapes, and comparisons are made on an ad hoc basis [4, 5]. Such an approach as proposed here would, thus, allow for a rigorous evaluation of such properties based on the similarity of shapes with molecules with known properties. Similarity in general has wide-ranging applications in many domains [6].

Image similarity and comparisons also play an important role in other domains, such as in models of visual perception and object recognition in humans as well as animals [7, 8], finance and economics [9], computer vision [10], and video analyses [11]. In such contexts also there is much scope for application of this work.

Existing theory in this matter is far from complete. There are heuristic approaches to morphological similarity [12, 13], but no sound mathematical basis for the detection of geometric similarity. Geometric similarity is particularly important in engineering, in comparing a model and its prototype [14, 15], but there, however, does not seem to be a proper universal measure of geometric similarity. The measure in common use in engineering is merely scale-free identity, that all corresponding lengths should be in the same ratio—there is thus no way to properly measure inexact similarity, or to quantitatively state that a figure is more similar to a reference figure, than is some other figure.

Using subgraph isomorphism, alike constituent geometric figures of the original geometric figures can be found and checked for similarity. A simple similarity function can return a boolean value of 1 for similar geometric figures and 0 otherwise. However, such a function would have limited applications. In this chapter, we define instead a distance function that returns a value between 0 (inclusive) and 1. The returned value reflects the dissimilarity between alike planar geometric figures connected with straight lines.

Therefore, the distance function d is defined only when the graphs representing the given geometric figures are isomorphic [16]. The crux of the function is in the measurement of deviations from angular similarity and edge-length proportionality.

The function d is the convex sum of functions α and ρ:

- The function α, which we may call *angular dissimilarity*, measures the deviation from the angular identity between two geometric figures. In order to compute this, angles are projected on a cartesian plane, where the angles of the first geometric figure makes up one axis and the angles of the second geometric figure makes up the other axis. Therefore, a cluster of points in this plane represents corresponding angles of the given geometric figures. If the figures are similar (identical up to scale), the angular similarity property may be said to be satisfied, and the corresponding angle points lie on the $y = x$ line, and the value returned by α is zero. If not, then the deviation from the property is now computed as the distance from the original point to the corresponding point on the $y = x$ line.

- The function ρ, which we may call *edge-length disproportionality*, measures the deviation from edge-length proportionality between geometric figures. In order to compute this, the edge lengths are similarly projected to a cartesian plane, where the edge lengths of the first geometric figure makes up one axis and the edge lengths of the second geometric figure makes up the other axis. The corresponding edge lengths of the given geometric figures are represented as points in this plane. If two figures are proportional (identical up to scale), all corresponding edge-lengths are in a fixed proportion m, all points pass through a line $y = mx$, and the value returned by ρ is zero. In case the edge-lengths are not perfectly proportional, the calculation of ρ comes to finding the best fit line passing through the origin, and measuring the deviation from that line.

The choice of method to find the best fit line needs to consider the fact that the line should pass through the origin. Using the least-squares method of fitting [17] by adding $(0, 0)$ as one of the corresponding edge-length pairs does not give a proper line passing through the origin. This is the reason that the Iterative Proportional Fitting Procedure (IPFP) [18] is used instead. IPFP tries to find a fixed proportion among a set of pairs, thereby giving points on the line passing through origin.

There are many IPFP [19], of which the one used in this chapter is the classical IPFP [20], owing to its simplicity. On obtaining the required points from IPFP, the ratio between any two points gives the values of m, as IPFP creates a fixed proportionality among a set of edge-length pairs. Appendix D explains step by step the IPFP technique used in this chapter. Further, to compute the deviation from the edge-length proportionality, we calculate the Euclidean distance between the original point and the corresponding point on the line $y = mx$. The Euclidean distances of all edge-length pairs are summed up. ρ is computed using this sum and a scaling factor. As the considered geometric figures are alike, the scaling factor is the number of edges in any one of these geometric figures. The need for this scaling factor arises to account for the fact that in a large figure, with a large number of edges, a minor change is less significant in determining overall dissimilarity, than a corresponding change in a smaller figure.

The function d is shown to be a distance function as it satisfies the three properties [21] required: d satisfies the commutativity (Theorem 3.7) and triangular inequality properties (Theorem 3.8) defined over single geometric figures. However, the coincidence axiom is defined over equivalence classes of geometric figures (figures that are alike up to scale). The proofs for these properties are given later in this chapter.

2 The Distance Function

The distance function, represented by d, reflects the degree of dissimilarity between figures.

Let, Γ be the set of straight edge figures for which the distance function is defined then

$$\gamma_i = (V_i, E_i, L_i, \Theta_i) \in \Gamma$$

where V_i denotes the set of vertices, E_i is the set of edges, L_i represents the set of corresponding edge lengths and Θ_i denotes the set of angles that are defined between adjacent edges in terms of radian.

Further, if γ_i and γ_j are said to be "similar", then γ_i and γ_j satisfy the below conditions:

(1) If $g_i = (V_i, E_i)$ is a graph that represents the adjacency of figure γ_i and $g_j = (V_j, E_j)$ is a graph that represents the adjacency of figure γ_j, then graphs g_i and g_j are isomorphic.
(2) All the corresponding angles of γ_i and γ_j are equal, i.e., if $\Theta_i = \{\theta_i(1), \theta_i(2), \ldots, \theta_i(z)\}$ represent the set of angles of γ_i and if $\Theta_j = \{\theta_j(1), \theta_j(2), \ldots, \theta_j(z)\}$ represent the set of corresponding angles of γ_j, then

$$\theta_i(1) = \theta_j(1), \theta_i(2) = \theta_j(2), \ldots, \theta_i(z) = \theta_j(z) \tag{2.1}$$

(3) All the corresponding edge lengths of γ_i and γ_j are proportional, i.e., if $L_i = \{l_i(1), l_i(2), \ldots, l_i(n)\}$ represent the set of edge lengths of γ_i and if $L_j = \{l_j(1), l_j(2), \ldots, l_j(n)\}$ represent the set of corresponding edge lengths of γ_j, then

$$\frac{l_j(1)}{l_i(1)} = \frac{l_j(2)}{l_i(2)} = \ldots = \frac{l_j(z)}{l_i(z)} = m, \text{ a constant.} \tag{2.2}$$

In view of this, the distance function tries to find the extent to which the considered figures deviate from conditions 2 and 3, provided condition 1 is satisfied.

Remark 2.1 A few properties of the d function:

(1) $d : \Gamma \times \Gamma \rightarrow [0, 1)$
(2) $d(\gamma_i, \gamma_i) = 0$
(3) $d(\gamma_i, \gamma_j) = 0$, if and only if $\gamma_i \approx \gamma_j$
where $\approx$ denotes that γ_i and γ_j belong to same equivalence class of figures, i.e., are figures that are identical up to scale.
(4) d satisfies the following:

$$d(\gamma_i, \gamma_j) = \begin{cases} 0 & \text{if } \gamma_i \approx \gamma_j, \\ \lambda \in (0, 1) & \text{otherwise.} \end{cases} \tag{2.3}$$

3 Components of the Distance Function

3.1 Angular Dissimilarity

Let α represent the angular dissimilarity function. Then the function is defined as:

$$\alpha : \Gamma \times \Gamma \to [0, 1) \tag{3.1a}$$

$$\alpha(\gamma_i, \gamma_j) = \begin{cases} \ddagger & \text{if } \delta(g_i, g_j) = 0, \\ \varphi \in (0, 1] & \text{otherwise.} \end{cases} \tag{3.1b}$$

where δ represents the graph isomorphism function.

$$\delta : G \times G \to \{0, 1\}$$

with $G = \{g_1, g_2, \ldots\}$: set of all graphs.

$$\delta(g_i, g_j) = \begin{cases} 1 & \text{if } g_i \approx g_j, \\ 0 & \text{otherwise.} \end{cases} \tag{3.2}$$

In (3.2), the symbol $\approx$ denotes that g_i and g_j satisfy all properties of graph isomorphism.

Assuming $\delta(g_1, g_2) = 1, \alpha(\gamma_i, \gamma_j)$ is computed as follows:

Project each corresponding pair $(\theta_i(u), \theta_j(u))$ into a cartesian plane, wherein the x-axis represents the set Θ_i, while the y-axis represents the set Θ_j. The function α computes the deviation from (2.1). In this cartesian plane, according to (2.1), all corresponding pairs must lie on the line:

$$y = x \tag{3.3}$$

For each point $(\theta_i(u), \theta_j(u))$, calculate the Euclidean distance from its corresponding point on the line (3.3), i.e., $(\theta_i(u), \theta_i(u))$.

$$\begin{aligned} \Lambda_{i,j}(u) &= \sqrt{(\theta_i(u) - \theta_i(u))^2 + (\theta_j(u) - \theta_i(u))^2} \\ &= \sqrt{(\theta_j(u) - \theta_i(u))^2} \\ &= |\theta_j(u) - \theta_i(u)| \end{aligned} \tag{3.4}$$

Therefore,

$$\alpha(\gamma_i, \gamma_j) = \frac{\sum_{u=1}^{n} \Lambda_{i,j}(u)}{1 + \sum_{u=1}^{n} \Lambda_{i,j}(u)} \tag{3.5}$$

Remark 3.1 A few properties of the α function:

(1) $\alpha(\gamma_i, \gamma_j) \geq 0$
(2) $\alpha(\gamma_i, \gamma_i) = 0$
(3) $\alpha(\gamma_i, \gamma_j)$ (which can be equal to 0), where $i \neq j$

Theorem 3.2 $\alpha(\gamma_i, \gamma_j) = \alpha(\gamma_j, \gamma_i)$

Proof We see that the constituents of the α function are commutative:

$$\Lambda_{i,j}(u) = |\theta_j(u) - \theta_i(u)|$$
$$\Lambda_{i,j}(u) = \Lambda_{j,i}(u), \text{ as } |a - b| = |b - a|$$

This follows that $\sum_{u=1}^{n} \Lambda_{i,j}(u) = \sum_{u=1}^{n} \Lambda_{j,i}(u) = e$, a constant
Hence,

$$\begin{aligned}\alpha(\gamma_i, \gamma_j) &= \frac{\sum_{u=1}^{n} \Lambda_{i,j}(u)}{1 + \sum_{u=1}^{n} \Lambda_{i,j}(u)} \\ &= \frac{e}{1+e} \\ &= \frac{\sum_{u=1}^{n} \Lambda_{j,i}(u)}{1 + \sum_{u=1}^{n} \Lambda_{j,i}(u)} \\ &= \alpha(\gamma_j, \gamma_i)\end{aligned}$$

□

Theorem 3.3 $\alpha(\gamma_i, \gamma_k) \leq \alpha(\gamma_i, \gamma_j) + \alpha(\gamma_j, \gamma_k)$

Proof

$$\Lambda_{i,k}(u) = |\theta_k(u) - \theta_i(u)|$$
$$\Lambda_{i,k}(u) \leq \Lambda_{i,j}(u) + \Lambda_{j,k}(u) \qquad \because |c - a| \leq |b - a| + |c - b|$$

Summing the above inequality for $p = 1$ to n, it follows that

$$\sum_{u=1}^{n} \Lambda_{i,k}(u) \leq \sum_{u=1}^{n} \Lambda_{i,j}(u) + \sum_{u=1}^{n} \Lambda_{j,k}(u) \tag{3.6}$$

Let e,f and g represent $\sum_{u=1}^{n} \Lambda_{i,k}(u)$, $\sum_{u=1}^{n} \Lambda_{i,j}(u)$ and $\sum_{u=1}^{n} \Lambda_{j,k}(u)$ respectively. The inequality (3.6) now translates to

$$e \leq f + g \tag{3.7}$$

Assume that the contradiction of Theorem 3.3 is true, i.e.,

$$\alpha(\gamma_i, \gamma_k) > \alpha(\gamma_i, \gamma_j) + \alpha(\gamma_j, \gamma_k) \tag{3.8}$$
$$\frac{e}{1+e} > \frac{f}{1+f} + \frac{g}{1+g}$$

On simplification,

$$e > (f+g) + (2fg + efg) \tag{3.9}$$

As the quantity $(2fg + efg) > 0$, the inequality (3.9) contradicts already proved inequality (3.7). Hence, (3.8) does not hold true, thereby proving Theorem 3.3. □

3.2 *Edge-Length Disproportionality*

Let ρ represent the edge-length disproportionality function. Then the function is defined as:

$$\rho : \Gamma \times \Gamma \to [0, 1) \tag{3.10a}$$

$$\rho(\gamma_i, \gamma_j) = \begin{cases} \ddagger & \text{if } \delta(g_i, g_j) = 0, \\ \tau \in (0, 1] & \text{otherwise.} \end{cases} \tag{3.10b}$$

Assuming $\delta(g_1, g_2) = 1$, $\rho(\gamma_i, \gamma_j)$ is computed as follows:

Project each corresponding pair $(l_i(h), l_j(h))$ into a cartesian plane, wherein the x-axis represents the set L_i, while the y-axis represents the set L_j. The function ρ computes the deviation from (2.2). Consider a part of the same equation.

$$\frac{l_j(h)}{l_i(h)} = m, \text{ a constant} \tag{3.11}$$

In the context of the L_iL_j plane, (3.11) gives the slope of a line that passes through $(0, 0)$ and $(l_i(h), l_j(h))$.

$$\begin{aligned} \text{Slope of a line, } m &= \frac{(y_2 - y_1)}{(x_2 - x_1)} \\ &= \frac{(l_j(h) - 0)}{(l_i(h) - 0)} \\ &= (3.11) \end{aligned}$$

Further extending this concept, it can be seen that in order to satisfy (2.2) all points $(l_i(h), l_j(h))$ should lie on the same line. Therefore, finding edge-length proportionality now boils down to finding for the set of corresponding edge-length pairs the best fit line, which passes though origin.

Let the equation of the required line be as follows:

$$y = mx, \text{ as the line passes through origin.} \tag{3.12}$$

Using IPFP each point $(l_i(h), l_j(h))$ is transformed to $(l'_i(h), l'_j(h))$, which is a point on the line 3.12.

On finding the desired line, the euclidean distance between $(l_i(h), l_j(h))$ and $(l'_i(h), l'_j(h))$ is computed.

$$\Delta_{i,j}(h) = \sqrt{(l_i(h) - l'_i(h))^2 + (l_j(h) - l'_j(h))^2} \tag{3.13}$$

Therefore,

$$\rho(\gamma_i, \gamma_j) = \frac{\sum_{h=1}^{n} \Delta_{i,j}(h)}{n + \sum_{h=1}^{n} \Delta_{i,j}(h)} \tag{3.14}$$

Remark 3.4 A few properties of the ρ function:

(1) $\rho(\gamma_i, \gamma_j) \geq 0$
(2) $\rho(\gamma_i, \gamma_i) = 0$
(3) $\rho(\gamma_i, \gamma_j)$, where $i \neq j$ can be equal to 0.

Theorem 3.5 $\rho(\gamma_i, \gamma_j) = \rho(\gamma_j, \gamma_i)$

Proof The proof is similar to that of Theorem 3.2. □

Theorem 3.6 $\alpha(\gamma_i, \gamma_k) \leq \alpha(\gamma_i, \gamma_j) + \alpha(\gamma_j, \gamma_k)$

Proof The proof is similar to that of Theorem 3.3. □

3.3 Deriving the Function

The function $d(\gamma_i, \gamma_j)$ is the convex sum of $\alpha(\gamma_i, \gamma_j)$ and $\rho(\gamma_i, \gamma_j)$.

$$d : \Gamma \times \Gamma \nrightarrow [0, 1) \tag{3.15a}$$

$$d(\gamma_i, \gamma_j) = \begin{cases} \ddagger & \text{if } \delta(g_i, g_j) = 0, \\ \beta\alpha(\gamma_i, \gamma_j) + (1 - \beta)\rho(\gamma_i, \gamma_j), \text{ where } \beta \in [0, 1] & \text{otherwise.} \end{cases} \tag{3.15b}$$

While computing d using (3.15b) in A, B, C, and D, the value of β is set to 0.5, to equally weight the α and ρ functions. However, other values of $\beta \in [0, 1]$ can be used resulting in similar outcomes for the d function.

Theorem 3.7 $d(\gamma_i, \gamma_j) = d(\gamma_j, \gamma_i)$

Proof According to (3.15b),

$$\begin{aligned} d(\gamma_i, \gamma_j) &= \beta\alpha(\gamma_i, \gamma_j) + (1-\beta)\rho(\gamma_j, \gamma_i), && \text{where } \beta \in [0,1] \\ &= \beta\alpha(\gamma_j, \gamma_i) + (1-\beta)\rho(\gamma_i, \gamma_j), && \text{from Theorems 3.2 and 3.5} \\ &= d(\gamma_j, \gamma_i). \end{aligned}$$

□

Theorem 3.8 $d(\gamma_i, \gamma_k) \leq d(\gamma_i, \gamma_j) + d(\gamma_j, \gamma_k)$

Proof According to (3.15b), $\beta \in [0, 1]$

Multiplying by β both sides of the inequality in Theorem 3.3, we get the following:

$$\beta\alpha(\gamma_i, \gamma_k) \leq \beta\alpha(\gamma_i, \gamma_j) + \beta\alpha(\gamma_j, \gamma_k) \tag{3.16}$$

Multiplying by $(1 - \beta)$ both sides of the inequality in Theorem 3.6, we get the following:

$$(1-\beta)\rho(\gamma_i, \gamma_k) \leq (1-\beta)\rho(\gamma_i, \gamma_j) + (1-\beta)\rho(\gamma_j, \gamma_k) \tag{3.17}$$

Summing up inequalities (3.16) and (3.17), it follows that:

$$\begin{aligned} &\beta\alpha(\gamma_i, \gamma_k) + (1-\beta)\rho(\gamma_i, \gamma_k) \\ &\leq \beta\alpha(\gamma_i, \gamma_j) + (1-\beta)\rho(\gamma_i, \gamma_j) + \beta\alpha(\gamma_j, \gamma_k) + (1-\beta)\rho(\gamma_j, \gamma_k) \end{aligned} \tag{3.18}$$

$d(\gamma_i, \gamma_k) \leq d(\gamma_i, \gamma_j) + d(\gamma_j, \gamma_k)$. □

4 Results

Using the above-discussed method to compute the distance function, d, this section tabulates the results for a few pairs of figures. It can be found that the values of d in Table 1 are reflective of the dissimilarity of considered figures. The same can be said for α and ρ values.

Table 1 Results obtained for a few pairs of figures. See Appendices A, B, C, and D for more details regarding the tabulated results

Figures to be compared		α	ρ	d
1a	1b	0.8073	0.4689	0.6381
4a	4b	0.8073	0.7883	0.7978
7a	7b	0.9281	0.9074	0.9177
10a	10b	0.9201	0.7177	0.8189

Appendix A Computing d for Fig. 1a and b

Computing $\alpha(\gamma_i, \gamma_j)$

$$\Theta_1 = \left\{\frac{\pi}{2}, \frac{3\pi}{4}, \frac{5\pi}{6}, \frac{\pi}{3}, \frac{5\pi}{6}, \frac{3\pi}{4}\right\}$$
$$\Theta_2 = \left\{\frac{2\pi}{3}, \frac{\pi}{3}, \frac{4\pi}{3}, \frac{\pi}{3}, \frac{2\pi}{3}, \frac{2\pi}{3}\right\}$$

Using (3.4), we compute the Euclidean distance.

$$\Lambda_{1,2}(1) = \frac{\pi}{6} \quad \Lambda_{1,2}(2) = \frac{5\pi}{12} \quad \Lambda_{1,2}(3) = \frac{\pi}{2} \quad \Lambda_{1,2}(4) = 0 \quad \Lambda_{1,2}(5) = \frac{\pi}{6} \quad \Lambda_{1,2}(6) = \frac{\pi}{12}$$
$$\sum_{u=1}^{6} \Lambda_{1,2}(u) = \frac{4\pi}{3}$$
$\boldsymbol{\alpha(\gamma_1, \gamma_2)} = 0.8073$, using (3.5)

Figure 2 shows the relation between the corresponding elements of Θ_1 and Θ_2. It also indicates, for each pair $\langle\theta_1(u), \theta_2(u)\rangle$ the corresponding point, $\langle\theta_1(u), \theta_1(u)\rangle$, on the $y = x$ line. Further, Table 2 provides the legend for this figure.

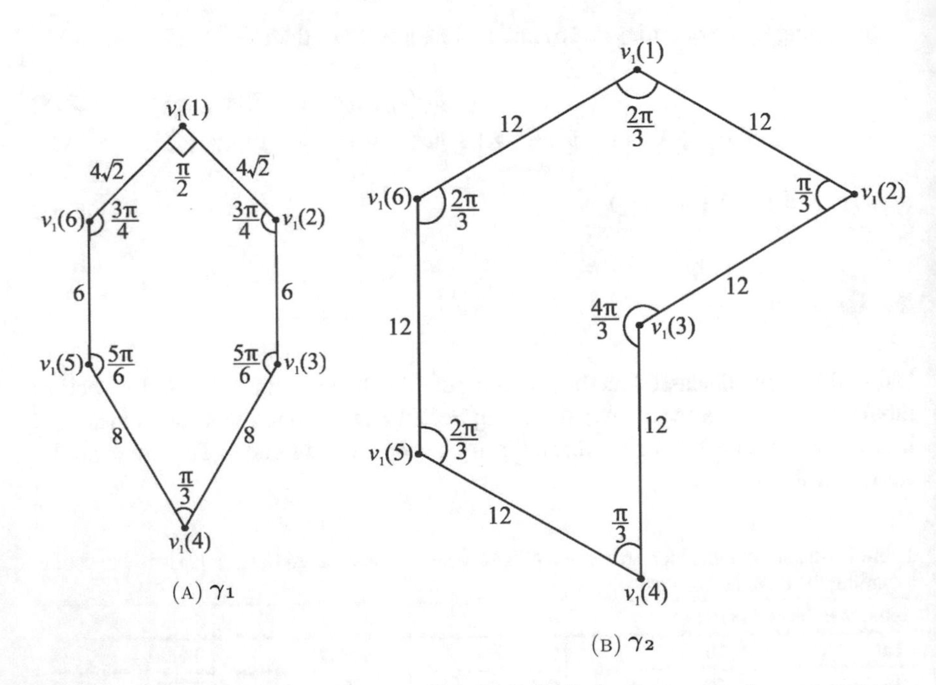

Fig. 1 Planar geometric figures between which the dissimilarity distance is to be computed

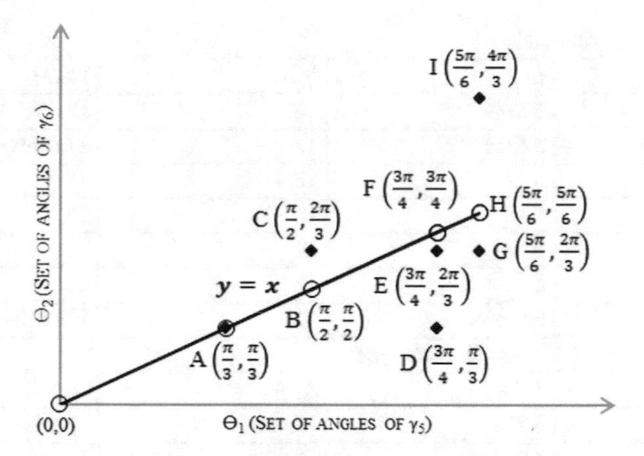

Fig. 2 Solid dots show the existing relation between Θ_1 and Θ_2, whereas the line passing through the hollow dots shows the expected relation between Θ_1 and Θ_2

Table 2 Legend of Fig. 2

$A \in \{\langle\theta_1(4), \theta_2(4)\rangle, \langle\theta_1(4), \theta_1(4)\rangle\}$	$B \in \{\langle\theta_1(1), \theta_1(1)\rangle\}$
$C \in \{\langle\theta_1(1), \theta_2(1)\rangle\}$	$D \in \{\langle\theta_1(2), \theta_2(2)\rangle\}$
$E \in \{\langle\theta_1(6), \theta_2(6)\rangle\}$	$F \in \{\langle\theta_1(6), \theta_1(6)\rangle\}$
$G \in \{\langle\theta_1(5), \theta_2(5)\rangle\}$	$H \in \{\langle\theta_1(3), \theta_1(3)\rangle, \langle\theta_1(5), \theta_1(5)\rangle\}$
$I \in \{\langle\theta_1(3), \theta_2(3)\rangle\}$	

Computing $\rho(\gamma_i, \gamma_j)$

$$L_1 = \{4\sqrt{2}, 4\sqrt{2}, 6, 8, 8, 6\}$$
$$L_2 = \{12, 12, 12, 12, 12, 12\}$$

Table 3 indicates the input and output of IPFP transformation.

Considering any row from the Table 3b, we compute m.

$$m = \frac{11.4208}{6.236} = 1.8314$$

$y = 1.8314x$, equation of the expected line

Figure 3 shows the relation between the corresponding elements of L_1 and L_2. It also indicates, for each pair, $\langle l_1(h), l_2(h)\rangle$, the corresponding point, $\langle l'_1(h), l'_2(h)\rangle$, on line $y = 1.8314x$ line. Further, Table 4 provides the legend for this figure.

Table 3 IPFP Transformation

h	$l_1(h)$	$l_2(h)$	TOTAL
1	5.6569	12	17.6569
2	5.6569	12	17.6569
3	6	12	18
4	8	12	20
5	8	12	20
6	6	12	18
TOTAL	**39.3138**	**72**	**111.3138**

(A) Values on which IPFP is to be performed

h	$l'_1(h)$	$l'_2(h)$	TOTAL
1	6.236	11.4208	17.6569
2	6.236	11.4208	17.6569
3	6.3572	11.6428	18
4	7.0636	12.9364	20
5	7.0636	12.9364	20
6	6.3572	11.6428	18
TOTAL	**39.3138**	**72**	**111.3138**

(B) Values obtained on applying IPFP

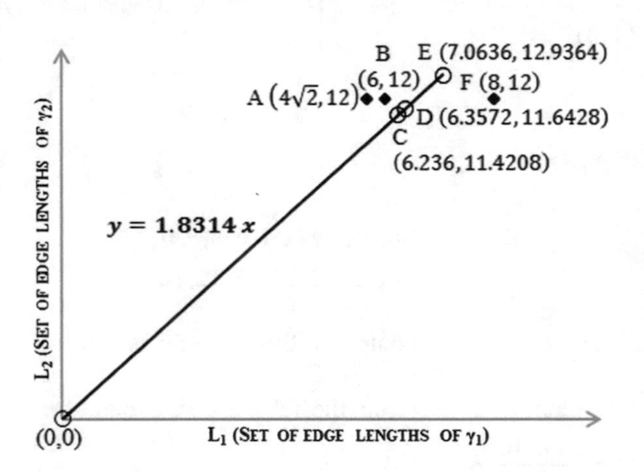

Fig. 3 Solid dots show the existing relation between L_1 and L_2, whereas the line passing through the hollow dots shows the expected relation between L_1 and L_2

Table 4 Legend of Fig. 3

$A \in \{\langle l_1(1), l_2(1)\rangle, \langle l_1(2), l_2(2)\rangle\}$	$B \in \{\langle l_1(3), l_2(3)\rangle, \langle l_1(6), l_2(6)\rangle\}$
$C \in \{\langle l'_1(1), l'_2(1)\rangle, \langle l'_1(2), l'_2(2)\rangle\}$	$D \in \{\langle l'_1(3), l'_2(3)\rangle, \langle l'_1(6), l'_2(6)\rangle\}$
$C \in \{\langle l'_1(4), l'_2(4)\rangle, \langle l'_1(5), l'_2(5)\rangle\}$	$F \in \{\langle l_1(4), l_2(4)\rangle, \langle l_1(5), l_2(5)\rangle\}$

Now, the Euclidean Distance is computed using (3.13)

$$\Delta_{1,2}(1) = 0.8191 \ \Delta_{1,2}(2) = 0.8191 \ \Delta_{1,2}(3) = 0.5052$$
$$\Delta_{1,2}(4) = 1.3243 \ \Delta_{1,2}(5) = 1.3243 \ \Delta_{1,2}(6) = 0.5052$$

$$\sum_{h=1}^{6} \Delta_{1,2}(h) = 5.2971$$

$$\boldsymbol{\rho(\gamma_1, \gamma_2)} = 0.4689, \text{ using (3.14)}$$

Computing $d(\gamma_i, \gamma_j)$

$$\boldsymbol{d(\gamma_1, \gamma_2)} = 0.6381, \text{ with } \beta = 0.5, \text{ using (3.15b)}$$

Appendix B Computing *d* for Fig. 4a and b

Computing $\alpha(\gamma_3, \gamma_4)$

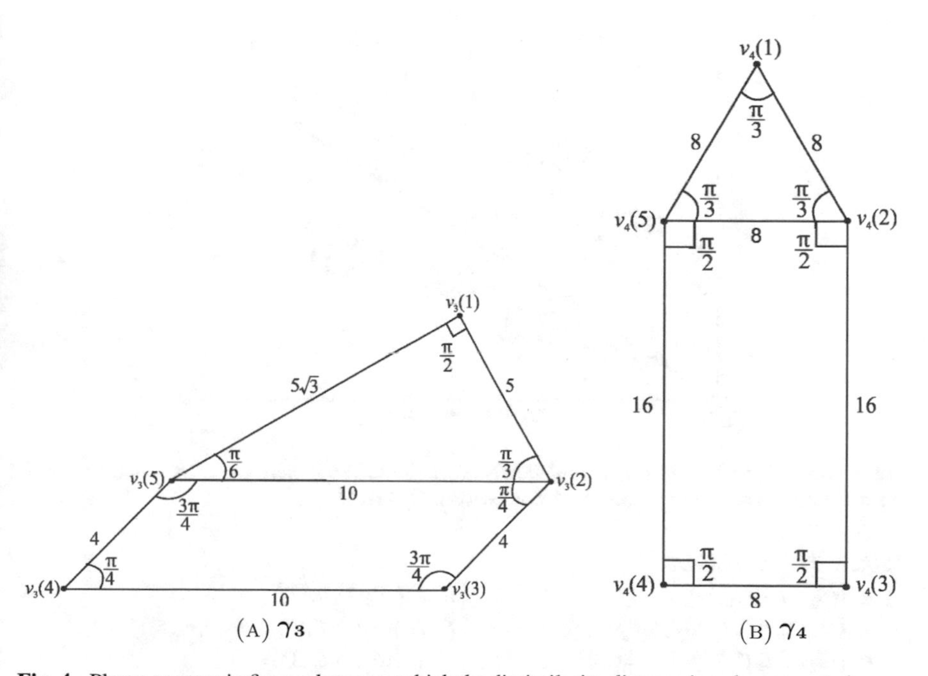

Fig. 4 Planar geometric figures between which the dissimilarity distance is to be computed

$$\Theta_3 = \left\{\frac{\pi}{2}, \frac{\pi}{3}, \frac{3\pi}{4}, \frac{\pi}{4}, \frac{3\pi}{4}, \frac{\pi}{6}\right\}$$

$$\Theta_4 = \left\{\frac{\pi}{3}, \frac{\pi}{3}, \frac{\pi}{2}, \frac{\pi}{2}, \frac{\pi}{2}, \frac{\pi}{3}\right\}$$

Using (3.4) we compute the Euclidean distance.

$$\Lambda_{3,4}(1) = \frac{\pi}{6} \quad \Lambda_{3,4}(2) = 0 \quad \Lambda_{3,4}(3) = \frac{\pi}{4}$$

$$\Lambda_{3,4}(4) = \frac{\pi}{4} \quad \Lambda_{3,4}(5) = \frac{\pi}{4} \quad \Lambda_{3,4}(6) = \frac{\pi}{4}$$

$$\Lambda_{3,4}(7) = \frac{\pi}{6}$$

$$\sum_{u=1}^{7} \Lambda_{3,4}(u) = \frac{4\pi}{3}$$

$\boldsymbol{\alpha(\gamma_3, \gamma_4)} = 0.8073$, using (3.5)

Figure 5 shows the relation between the corresponding elements of Θ_3 and Θ_4. It also indicates, for each pair $\langle\theta_3(u), \theta_4(u)\rangle$ the corresponding point, $\langle\theta_3(u), \theta_3(u)\rangle$, on the $y = x$ line. Further, Table 5 provides the legend for this figure.

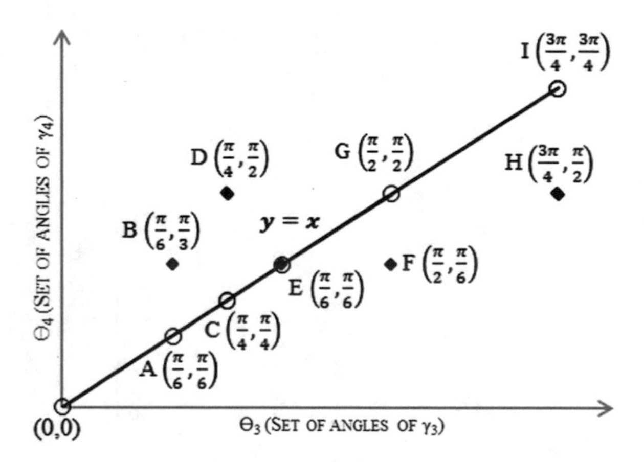

Fig. 5 Solid dots show the existing relation between Θ_3 and Θ_4, whereas the line passing through the hollow dots shows the expected relation between Θ_3 and Θ_4

Table 5 Legend of Fig. 5

$A \in \{\langle\theta_3(7), \theta_3(7)\rangle\}$	$B \in \{\langle\theta_3(7), \theta_4(7)\rangle\}$
$C \in \{\langle\theta_3(3), \theta_3(3)\rangle, \langle\theta_3(5), \theta_3(5)\rangle\}$	$D \in \{\langle\theta_3(3), \theta_4(3)\rangle, \langle\theta_3(5), \theta_4(5)\rangle\}$
$E \in \{\langle\theta_3(2), \theta_4(2)\rangle, \langle\theta_3(2), \theta_3(2)\rangle\}$	$F \in \{\langle\theta_3(1), \theta_4(1)\rangle\}$
$G \in \{\langle\theta_3(1), \theta_3(1)\rangle\}$	$H \in \{\langle\theta_3(4), \theta_4(4)\rangle, \langle\theta_3(6), \theta_4(6)\rangle\}$
$I \in \{\langle\theta_3(4), \theta_3(4)\rangle, \langle\theta_3(6), \theta_3(6)\rangle\}$	

Table 6 IPFP Transformation

h	$l_3(h)$	$l_4(h)$	TOTAL
1	8.6603	8	16.6603
2	5	8	13
3	10	8	18
4	4	16	20
5	10	8	18
6	4	16	20
TOTAL	**41.6603**	**64**	105.6603

(A) Values on which IPFP is to be performed

h	$l'_3(h)$	$l'_4(h)$	TOTAL
1	6.5689	10.0914	16.6603
2	5.1257	7.8743	13
3	7.0971	10.9029	18
4	7.8857	12.1143	20
5	7.0971	10.9029	18
6	7.8857	12.1143	20
TOTAL	**41.6603**	**64**	105.6603

(B) Values obtained on applying IPFP

Computing $\rho(\gamma_3, \gamma_4)$

$$L_3 = \{5\sqrt(3), 5, 10, 4, 10, 4\}$$
$$L_4 = \{8, 8, 8, 16, 8, 16\}$$

Table 6 indicates the input and output of IPFP transformation.

Considering any row from the Table 6b, we compute m

$$m = \frac{10.0914}{6.5689} = 1.5362$$

$y = 1.5362x$, equation of the expected line

Figure 6 shows the relation between the corresponding elements of L_3 and L_4. It also indicates, for each pair $\langle l_3(h), l_4(h)\rangle$ the corresponding point, $\langle l'_3(h), l'_4(h)\rangle$, on the $y = 1.5362x$ line. Further, Table 7 provides the legend for this figure.

Now, the Euclidean Distance is computed using (3.13)

$$\Delta_{3,4}(1) = 2.9577 \ \Delta_{3,4}(2) = 0.1778 \ \Delta_{3,4}(3) = 4.1053$$
$$\Delta_{3,4}(4) = 5.4952 \ \Delta_{3,4}(5) = 4.1053 \ \Delta_{3,4}(6) = 5.4952$$

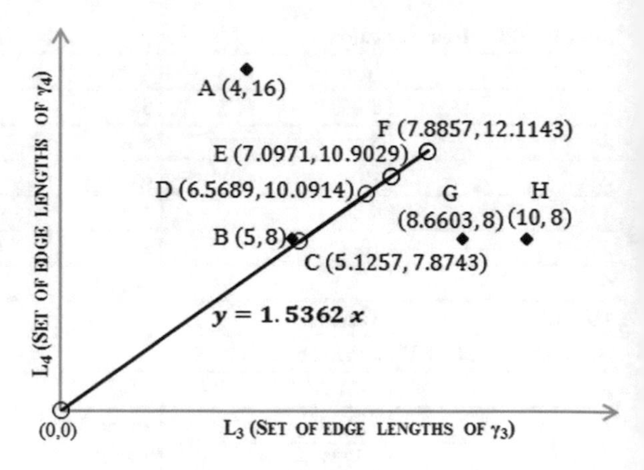

Fig. 6 Solid dots show the existing relation between L_3 and L_4, whereas the line passing through the hollow dots shows the expected relation between L_3 and L_4

Table 7 Legend of Fig. 6

$A \in \{\langle l_3(4), l_4(4)\rangle, \langle l_3(6), l_4(6)\rangle\}$	$B \in \{\langle l_3(2), l_4(2)\rangle\}$
$C \in \{\langle l_3'(2), l_4'(2)\rangle\}$	$D \in \{\langle l_3'(1), l_4'(1)\rangle\}$
$E \in \{\langle l_3'(3), l_4'(3)\rangle, \langle l_3'(5), l_4'(5)\rangle\}$	$F \in \{\langle l_3'(4), l_4'(4)\rangle, \langle l_3'(6), l_4'(6)\rangle\}$
$G \in \{\langle l_3(1), l_4(1)\rangle\}$	$H \in \{\langle l_3(3), l_4(3)\rangle, \langle l_3(5), l_4(5)\rangle\}$

$$\sum_{h=1}^{6} \Delta_{3,4}(h) = 22.3365$$

$$\rho(\gamma_3, \gamma_4) = 0.7883, \text{ using } (3.14)$$

Computing $d(\gamma_3, \gamma_4)$

$$d(\gamma_3, \gamma_4) = 0.7978, \text{ with } \beta = 0.5, \text{ using } (3.15\text{b})$$

Appendix C Computing *d* for Fig. 7a and b

Computing $\alpha(\gamma_5, \gamma_6)$, we get:

$$\Theta_5 = \left\{\frac{\pi}{3}, \frac{\pi}{2}, \frac{\pi}{2}, \frac{\pi}{3}, \frac{\pi}{6}, \frac{\pi}{2}, \frac{5\pi}{6}, \frac{\pi}{4}, \frac{\pi}{4}, \frac{\pi}{2}, \frac{5\pi}{6}, \frac{\pi}{4}, \frac{\pi}{4}, \frac{\pi}{2}, \frac{\pi}{6}, \frac{\pi}{3}, \frac{\pi}{2}, \frac{\pi}{6}, \pi, \frac{\pi}{3}, \frac{\pi}{3}, \frac{\pi}{3}, \frac{2\pi}{3}, \frac{\pi}{6}\right\}$$

$$\Theta_6 = \left\{\frac{\pi}{2}, \frac{\pi}{4}, \frac{2\pi}{3}, \frac{\pi}{2}, \frac{\pi}{4}, \frac{\pi}{4}, \frac{7\pi}{12}, \frac{\pi}{2}, \frac{\pi}{4}, \frac{\pi}{4}, \frac{11\pi}{18}, \frac{\pi}{2}, \frac{\pi}{4}, \frac{\pi}{4}, \frac{5\pi}{9}, \frac{\pi}{2}, \frac{\pi}{4}, \frac{\pi}{4}, \frac{5\pi}{6}, \frac{\pi}{2}, \frac{\pi}{4}, \frac{\pi}{4}, \frac{3\pi}{4}, \frac{\pi}{4}\right\}$$

Using (3.4) we compute the Euclidean distance.

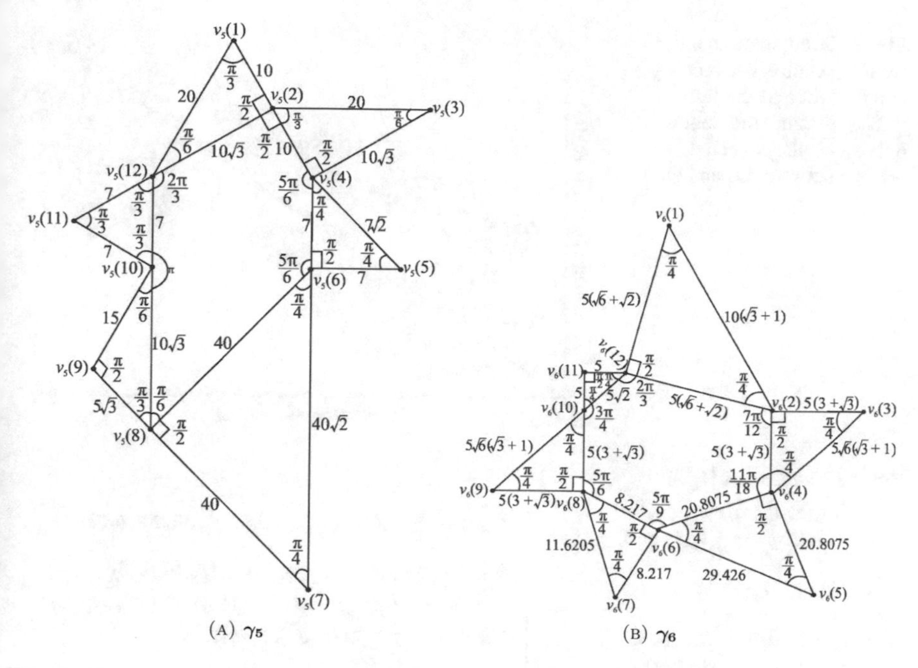

Fig. 7 Planar geometric figures between which the dissimilarity distance is to be computed

$$
\begin{array}{llllll}
\Lambda_{3,4}(1) = \frac{\pi}{6} & \Lambda_{3,4}(2) = \frac{\pi}{4} & \Lambda_{3,4}(3) = \frac{\pi}{6} & \Lambda_{3,4}(4) = \frac{\pi}{6} & \Lambda_{3,4}(5) = \frac{\pi}{12} & \Lambda_{3,4}(6) = \frac{\pi}{4} \\
\Lambda_{3,4}(7) = \frac{\pi}{4} & \Lambda_{3,4}(8) = \frac{\pi}{4} & \Lambda_{3,4}(9) = 0 & \Lambda_{3,4}(10) = \frac{\pi}{4} & \Lambda_{3,4}(11) = \frac{2\pi}{9} & \Lambda_{3,4}(12) = \frac{\pi}{4} \\
\Lambda_{3,4}(13) = 0 & \Lambda_{3,4}(14) = \frac{\pi}{4} & \Lambda_{3,4}(15) = \frac{7\pi}{18} & \Lambda_{3,4}(16) = \frac{\pi}{6} & \Lambda_{3,4}(17) = \frac{\pi}{4} & \Lambda_{3,4}(18) = \frac{\pi}{12} \\
\Lambda_{3,4}(19) = \frac{\pi}{6} & \Lambda_{3,4}(20) = \frac{\pi}{6} & \Lambda_{3,4}(21) = \frac{\pi}{12} & \Lambda_{3,4}(22) = \frac{\pi}{12} & \Lambda_{3,4}(23) = \frac{\pi}{12} & \Lambda_{3,4}(24) = \frac{\pi}{12}
\end{array}
$$

$$\sum_{u=1}^{24} \Lambda_{5,6}(u) = \frac{37\pi}{9}$$

$$\alpha(\gamma_5, \gamma_6) = 0.9281, \text{ using (3.5)}$$

Figure 8 shows the relation between the corresponding elements of Θ_5 and Θ_6. It also indicates, for each pair $\langle \theta_5(u), \theta_6(u) \rangle$ the corresponding point, $\langle \theta_5(u), \theta_5(u) \rangle$, on the $y = x$ line. Further, Table 8 provides the legend for this figure.

Computing $\rho(\gamma_5, \gamma_6)$

$$L_5 = \{20, 10, 10\sqrt{3}, 20, 10\sqrt{3}, 10, 7\sqrt{2}, 7, 7, 40\sqrt{2}, 40, 40, 5\sqrt{3}, 15, 10\sqrt{3}, 7, 7, 7\}$$

$$L_4 = \{5, 5, 5\sqrt{2}, 5(\sqrt{6} + \sqrt{2}), 5\sqrt{2}, 5(\sqrt{6} + \sqrt{2}), 5(3 + \sqrt{3}), 5\sqrt{6}(\sqrt{3} + 1), 5(3 + \sqrt{3}),$$
$$20.8075, 29.426, 20.8075, 8.217, 11.6205, 8.217, 5(3 + \sqrt{3}), 5\sqrt{6}(\sqrt{3} + 1), 5(3 + \sqrt{3})\}$$

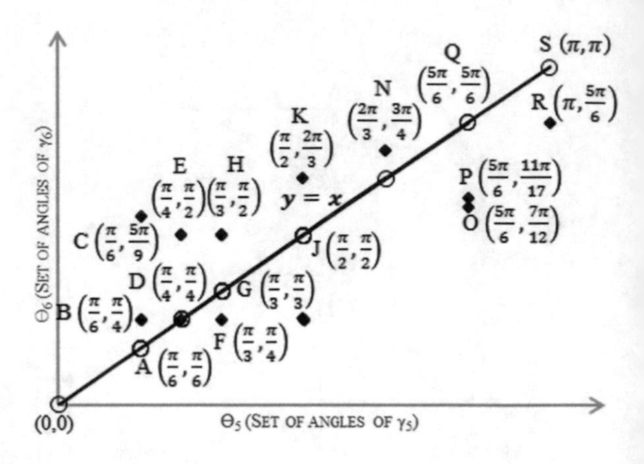

Fig. 8 Solid dots show the existing relation between Θ_5 and Θ_6, whereas the line passing through the hollow dots shows the expected relation between Θ_5 and Θ_6

Table 8 Legend of Fig. 8

$A \in \{\langle\theta_5(5), \theta_5(5)\rangle, \langle\theta_5(15), \theta_5(15)\rangle, \langle\theta_5(18), \theta_5(18)\rangle, \langle\theta_5(24), \theta_5(24)\rangle\}$	$B \in \{\langle\theta_5(5), \theta_6(5)\rangle, \langle\theta_5(18), \theta_6(18)\rangle, \langle\theta_5(24), \theta_6(24)\rangle\}$
$C \in \{\langle\theta_5(15), \theta_6(15)\rangle\}$	$D \in \{\langle\theta_5(8), \theta_5(8)\rangle, \langle\theta_5(9), \theta_6(9)\rangle, \langle\theta_5(9), \theta_5(9)\rangle, \langle\theta_5(12), \theta_5(12)\rangle, \langle\theta_5(13), \theta_6(13)\rangle, \langle\theta_5(13), \theta_5(13)\rangle\}$
$E \in \{\langle\theta_5(8), \theta_6(8)\rangle, \langle\theta_5(12), \theta_6(12)\rangle\}$	$F \in \{\langle\theta_5(21), \theta_6(21)\rangle, \langle\theta_5(22), \theta_6(22)\rangle\}$
$G \in \{\langle\theta_5(1), \theta_5(1)\rangle, \langle\theta_5(4), \theta_5(4)\rangle, \langle\theta_5(16), \theta_5(16)\rangle, \langle\theta_5(20), \theta_5(20)\rangle, \langle\theta_5(21), \theta_5(21)\rangle, \langle\theta_5(22), \theta_5(22)\rangle\}$	$H \in \{\langle\theta_5(1), \theta_6(1)\rangle, \langle\theta_5(4), \theta_6(4)\rangle, \langle\theta_5(16), \theta_6(16)\rangle, \langle\theta_5(20), \theta_6(20)\rangle\}$
$I \in \{\langle\theta_5(2), \theta_6(2)\rangle, \langle\theta_5(6), \theta_6(6)\rangle, \langle\theta_5(10), \theta_6(10)\rangle, \langle\theta_5(14), \theta_6(14)\rangle, \langle\theta_5(17), \theta_6(17)\rangle\}$	$J \in \{\langle\theta_5(2), \theta_5(2)\rangle, \langle\theta_5(3), \theta_5(3)\rangle, \langle\theta_5(6), \theta_5(6)\rangle, \langle\theta_5(10), \theta_5(10)\rangle, \langle\theta_5(14), \theta_5(14)\rangle, \langle\theta_5(17), \theta_5(17)\rangle\}$
$K \in \{\langle\theta_5(3), \theta_6(3)\rangle\}$	$M \in \{\langle\theta_5(23), \theta_5(23)\rangle\}$
$N \in \{\langle\theta_5(23), \theta_6(23)\rangle\}$	$O \in \{\langle\theta_5(7), \theta_6(7)\rangle\}$
$P \in \{\langle\theta_5(11), \theta_6(11)\rangle\}$	$Q \in \{\langle\theta_5(7), \theta_5(7)\rangle, \langle\theta_5(11), \theta_5(11)\rangle\}$
$R \in \{\langle\theta_5(19), \theta_6(19)\rangle\}$	$S \in \{\langle\theta_5(19), \theta_5(19)\rangle\}$

Table 9 indicates the input and output of IPFP transformation.

Considering any row from the Table 9b, we compute m

$$m = \frac{13.0031}{11.9969} = 1.0839$$

$y = 1.0839x$, equation of the expected line

Figure 9 shows the relation between the corresponding elements of L_5 and L_6. It also indicates, for each pair $\langle l_5(h), l_6(h)\rangle$ the corresponding point, $\langle l'_5(h), l'_6(h)\rangle$, on the $y = 1.0839x$ line. Further, Table 10 provides the legend for this figure.

Now, the Euclidean Distance is computed using (3.13)

Table 9 IPFP Transformation

h	$l_5(h)$	$l_6(h)$	TOTAL
1	20	5	25
2	10	5	15
3	17.3205	7.0711	24.3916
4	20	19.3185	39.3185
5	17.3205	27.3205	44.641
6	10	19.3185	29.3185
7	9.8995	23.6603	33.5598
8	7	33.4607	40.4607
9	7	23.6603	30.6603
10	56.5685	20.8075	77.376
11	40	29.426	69.426
12	40	20.8075	60.8075
13	8.6603	8.217	16.8773
14	15	11.6205	26.6205
15	17.3205	8.217	25.5375
16	7	23.6603	30.6603
17	7	33.4607	40.4607
18	7	23.6603	30.6603
TOTAL	**317.0898**	**343.6864**	660.7762

(A) Values on which IPFP is to be performed

h	$l'_5(h)$	$l'_6(h)$	TOTAL
1	11.9969	13.0031	25
2	7.1981	7.8019	15
3	11.7049	12.6867	24.3916
4	18.868	20.4505	39.3185
5	21.4221	23.2189	44.641
6	14.0692	15.2493	29.3185
7	16.1045	17.4553	33.5598
8	19.4161	21.0446	40.4607
9	14.7131	15.9472	30.6603
10	37.1308	40.2452	77.376
11	33.3158	36.1102	69.426
12	29.18	31.6275	60.8075
13	8.099	8.7783	16.8773
14	12.7745	13.846	26.6205
15	12.2548	13.2827	25.5375
16	14.7131	15.9472	30.6603
17	19.4161	21.0446	40.4607
18	14.7131	15.9472	30.6603
TOTAL	**317.0898**	**343.6864**	660.7762

(B) Values obtained on applying IPFP

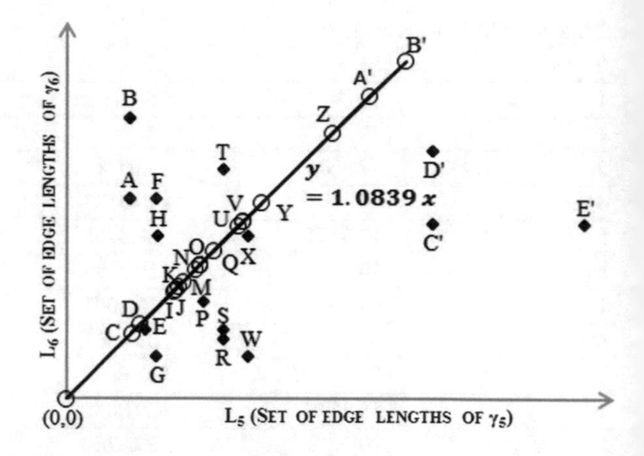

Fig. 9 Solid dots show the existing relation between L_5 and L_6, whereas the line passing through the hollow dots shows the expected relation between L_5 and L_6

$\Delta_{5,6}(1) = 11.3181$ $\Delta_{5,6}(2) = 3.9625$ $\Delta_{5,6}(3) = 7.9417$ $\Delta_{5,6}(4) = 1.6009$
$\Delta_{5,6}(5) = 5.8005$ $\Delta_{5,6}(6) = 5.7547$ $\Delta_{5,6}(7) = 8.7752$ $\Delta_{5,6}(8) = 17.5589$
$\Delta_{5,6}(9) = 10.9079$ $\Delta_{5,6}(10) = 27.4891$ $\Delta_{5,6}(11) = 9.4529$ $\Delta_{5,6}(12) = 15.3018$
$\Delta_{5,6}(13) = 0.7938$ $\Delta_{5,6}(14) = 3.1473$ $\Delta_{5,6}(15) = 7.164$ $\Delta_{5,6}(16) = 10.9079$
$\Delta_{5,6}(17) = 17.5589$ $\Delta_{5,6}(18) = 10.9079$

$$\sum_{h=1}^{18} \Delta_{5,6}(h) = 176.3443$$

$$\rho(\gamma_5, \gamma_6) = 0.9074, \text{ using (3.14)}$$

Computing $d(\gamma_5, \gamma_6)$

$$d(\gamma_5, \gamma_6) = 0.9177, \text{ with } \beta = 0.5, \text{ using (3.15b)}$$

Appendix D IPFP: Step-by-Step

Consider Fig. 10a and b
Computing $\alpha(\gamma_7, \gamma_8)$

Table 10 Legend of Fig. 9

$A = \langle 7, 23.6603\rangle$ $\in \{\langle l_5(9), l_6(9)\rangle, \langle l_5(16), l_6(16)\rangle, \langle l_5(18), l_6(18)\rangle\}$	$B = \langle 7, 33.4607\rangle$ $\in \{\langle l_5(8), l_6(8)\rangle, \langle l_5(17), l_6(17)\rangle\}$
$C = \langle 7.1981, 7.8019\rangle$ $\in \{\langle l'_5(2), l'_6(2)\rangle\}$	$D = \langle 8.099, 8.7783\rangle$ $\in \{\langle l'_5(13), l'_6(13)\rangle\}$
$E = \langle 8.6603, 8.217\rangle$ $\in \{\langle l_5(13), l_6(13)\rangle\}$	$F = \langle 9.8995, 23.6603\rangle$ $\in \{\langle l_5(7), l_6(7)\rangle\}$
$G = \langle 10, 5\rangle$ $\in \{\langle l_5(2), l_6(2)\rangle\}$	$H = \langle 10, 19.3185\rangle$ $\in \{\langle l_5(6), l_6(6)\rangle\}$
$I = \langle 11.7049, 12.6867\rangle$ $\in \{\langle l'_5(3), l'_6(3)\rangle\}$	$J = \langle 11.9969, 13.0031\rangle$ $\in \{\langle l'_5(1), l'_6(1)\rangle\}$
$K = \langle 12.2548, 13.2827\rangle$ $\in \{\langle l'_5(15), l'_6(15)\rangle\}$	$M = \langle 12.7745, 13.846\rangle$ $\in \{\langle l'_5(14), l'_6(14)\rangle\}$
$N = \langle 14.0692, 15.2493\rangle$ $\in \{\langle l'_5(6), l'_6(6)\rangle\}$	$O = \langle 14.7131, 15.9472\rangle$ $\in \{\langle l'_5(9), l'_6(9)\rangle, \langle l'_5(16), l'_6(16)\rangle, \langle l'_5(18), l'_6(18)\rangle\}$
$P = \langle 15, 11.6205\rangle$ $\in \{\langle l_5(14), l_6(14)\rangle\}$	$Q = \langle 16.1045, 17.4553\rangle$ $\in \{\langle l'_5(7), l'_6(7)\rangle\}$
$R = \langle 17.3205, 7.0711\rangle$ $\in \{\langle l_5(3), l_6(3)\rangle\}$	$S = \langle 17.3205, 8.217\rangle$ $\in \{\langle l_5(15), l_6(15)\rangle\}$
$T = \langle 17.3205, 27.3205\rangle$ $\in \{\langle l_5(5), l_6(5)\rangle\}$	$U = \langle 18.868, 20.4505\rangle$ $\in \{\langle l'_5(4), l'_6(4)\rangle\}$
$V = \langle 19.4161, 21.0446\rangle$ $\in \{\langle l'_5(8), l'_6(8)\rangle, \langle l'_5(17), l'_6(17)\rangle\}$	$W = \langle 20, 5\rangle$ $\in \{\langle l_5(1), l_6(1)\rangle\}$
$X = \langle 20, 19.3185\rangle$ $\in \{\langle l_5(4), l_6(4)\rangle\}$	$Y = \langle 21.4221, 23.2189\rangle$ $\in \{\langle l'_5(5), l'_6(5)\rangle\}$
$Z = \langle 29.18, 31.6275\rangle$ $\in \{\langle l'_5(12), l'_6(12)\rangle\}$	$A' = \langle 33.3158, 36.1102\rangle$ $\in \{\langle l'_5(11), l'_6(11)\rangle\}$
$B' = \langle 37.1308, 40.2452\rangle$ $\in \{\langle l'_5(10), l'_6(10)\rangle\}$	$C' = \langle 40, 20.8075\rangle$ $\in \{\langle l_5(12), l_6(12)\rangle\}$
$D' = \langle 40, 29.426\rangle$ $\in \{\langle l_5(11), l_6(11)\rangle\}$	$E' = \langle 56.5685, 20.8075\rangle$ $\in \{\langle l_5(10), l_6(10)\rangle\}$

$$\Theta_7 = \left\{\frac{\pi}{3}, \frac{5\pi}{6}, \frac{\pi}{6}, \frac{2\pi}{3}, \frac{5\pi}{6}, \frac{\pi}{2}, \frac{\pi}{6}, \frac{2\pi}{3}, \frac{\pi}{2}, \frac{2\pi}{3}, \frac{13\pi}{36}, \frac{59\pi}{72}, \frac{13\pi}{72}, \frac{23\pi}{36}, \frac{3\pi}{4}, \frac{11\pi}{18}, \frac{7\pi}{36}, \frac{29\pi}{36}, \frac{7\pi}{18}, \frac{3\pi}{4}, \frac{\pi}{4}, \pi, \frac{\pi}{4}, \frac{\pi}{2}, \frac{2\pi}{3}, \frac{5\pi}{6}, \frac{\pi}{12}, \frac{11\pi}{12}, \frac{\pi}{6}, \frac{5\pi}{6}, \frac{\pi}{2}, \frac{\pi}{2}, \frac{\pi}{2}, \frac{\pi}{2}, \frac{3\pi}{4}, \frac{3\pi}{4}, \frac{\pi}{8}, \frac{7\pi}{8}, \frac{\pi}{4}, \frac{3\pi}{4}\right\}$$

$$\Theta_8 = \left\{\frac{\pi}{2}, \frac{3\pi}{4}, \frac{\pi}{4}, \frac{\pi}{2}, \frac{3\pi}{4}, \frac{3\pi}{4}, \frac{\pi}{8}, \frac{7\pi}{8}, \frac{\pi}{4}, \frac{3\pi}{4}, \frac{\pi}{2}, \frac{3\pi}{4}, \frac{\pi}{4}, \frac{\pi}{2}, \frac{3\pi}{4}, \frac{3\pi}{4}, \frac{\pi}{8}, \frac{7\pi}{8}, \frac{\pi}{4}, \frac{3\pi}{4}, \frac{\pi}{2}, \frac{3\pi}{4}, \frac{\pi}{4}, \frac{\pi}{2}, \frac{3\pi}{4}, \frac{3\pi}{4}, \frac{\pi}{8}, \frac{7\pi}{8}, \frac{\pi}{4}, \frac{3\pi}{4}, \frac{\pi}{2}, \frac{3\pi}{4}, \frac{\pi}{4}, \frac{\pi}{2}, \frac{3\pi}{4}, \frac{3\pi}{4}, \frac{\pi}{8}, \frac{7\pi}{8}, \frac{\pi}{4}, \frac{3\pi}{4}\right\}$$

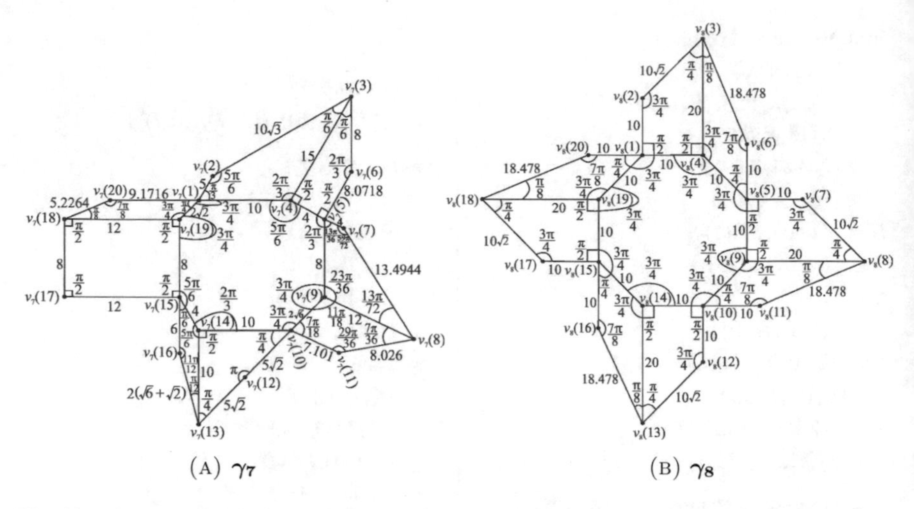

Fig. 10 Planar geometric figures between which the dissimilarity distance is to be computed

Using (3.4) we compute the Euclidean distance.

$$
\begin{array}{lllll}
\Lambda_{7,8}(1)=\frac{\pi}{6} & \Lambda_{7,8}(2)=\frac{\pi}{12} & \Lambda_{7,8}(3)=\frac{\pi}{12} & \Lambda_{7,8}(4)=\frac{\pi}{6} & \Lambda_{7,8}(5)=\frac{\pi}{12}\\
\Lambda_{7,8}(6)=\frac{\pi}{4} & \Lambda_{7,8}(7)=\frac{\pi}{24} & \Lambda_{7,8}(8)=\frac{5\pi}{24} & \Lambda_{7,8}(9)=\frac{\pi}{4} & \Lambda_{7,8}(10)=\frac{\pi}{12}\\
\Lambda_{7,8}(11)=\frac{5\pi}{36} & \Lambda_{7,8}(12)=\frac{5\pi}{72} & \Lambda_{7,8}(13)=\frac{5\pi}{72} & \Lambda_{7,8}(14)=\frac{5\pi}{36} & \Lambda_{7,8}(15)=0\\
\Lambda_{7,8}(16)=\frac{5\pi}{36} & \Lambda_{7,8}(17)=\frac{5\pi}{72} & \Lambda_{7,8}(18)=\frac{5\pi}{72} & \Lambda_{7,8}(19)=\frac{5\pi}{36} & \Lambda_{7,8}(20)=0\\
\Lambda_{7,8}(21)=\frac{\pi}{4} & \Lambda_{7,8}(22)=\frac{\pi}{4} & \Lambda_{7,8}(23)=0 & \Lambda_{7,8}(24)=0 & \Lambda_{7,8}(25)=\frac{\pi}{12}\\
\Lambda_{7,8}(26)=\frac{\pi}{12} & \Lambda_{7,8}(27)=\frac{\pi}{24} & \Lambda_{7,8}(28)=\frac{\pi}{24} & \Lambda_{7,8}(29)=\frac{\pi}{12} & \Lambda_{7,8}(30)=\frac{\pi}{12}\\
\Lambda_{7,8}(31)=0 & \Lambda_{7,8}(32)=\frac{\pi}{4} & \Lambda_{7,8}(33)=\frac{\pi}{4} & \Lambda_{7,8}(34)=0 & \Lambda_{7,8}(35)=0\\
\Lambda_{7,8}(36)=0 & \Lambda_{7,8}(37)=0 & \Lambda_{7,8}(38)=0 & \Lambda_{7,8}(39)=0 & \Lambda_{7,8}(40)=0
\end{array}
$$

$$\sum_{u=1}^{40} \Lambda_{7,8}(u) = \frac{11\pi}{3}$$

$$\boldsymbol{\alpha(\gamma_7, \gamma_8)} = 0.9201, \text{ using (3.5)}$$

Figure 11 shows the relation between the corresponding elements of Θ_7 and Θ_8. It also indicates, for each pair $\langle \theta_7(u), \theta_8(u) \rangle$ the corresponding point, $\langle \theta_7(u), \theta_7(u) \rangle$, on the $y = x$ line. Further, Table 11 provides the legend for this figure.

Computing $\rho(\gamma_7, \gamma_8)$

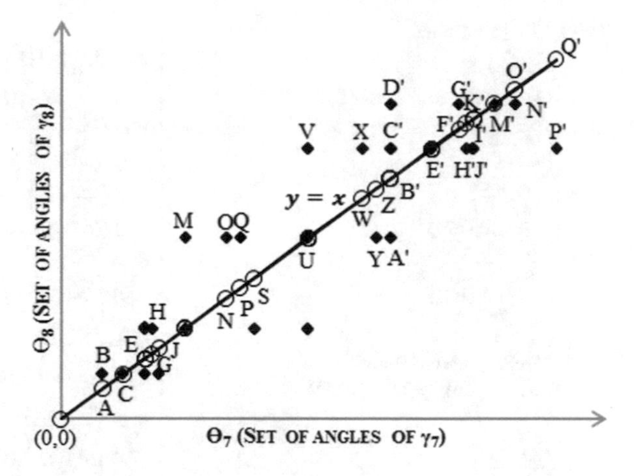

Fig. 11 Solid dots show the existing relation between Θ_7 and Θ_8, whereas the line passing through the hollow dots shows the expected relation between Θ_7 and Θ_8

Table 11 Legend of Fig. 11

$A = \left\langle \frac{\pi}{12}, \frac{\pi}{12} \right\rangle \in \{\langle \theta_7(27), \theta_7(27) \rangle\}$	$B = \left\langle \frac{\pi}{12}, \frac{\pi}{8} \right\rangle \in \{\langle \theta_7(27), \theta_8(27) \rangle\}$
$C = \left\langle \frac{\pi}{8}, \frac{\pi}{8} \right\rangle \in \{\langle \theta_7(37), \theta_8(37) \rangle, \langle \theta_7(37), \theta_7(37) \rangle\}$	$D = \left\langle \frac{\pi}{6}, \frac{\pi}{8} \right\rangle \in \{\langle \theta_7(7), \theta_8(7) \rangle\}$
$E = \left\langle \frac{\pi}{6}, \frac{\pi}{6} \right\rangle \in \{\langle \theta_7(3), \theta_7(3) \rangle, \langle (\theta_7(7), \theta_7(7) \rangle, \langle \theta_7(29), \theta_7(29) \rangle\}$	$F = \left\langle \frac{\pi}{6}, \frac{\pi}{4} \right\rangle \in \{\langle \theta_7(3), \theta_8(3) \rangle, \langle \theta_7(29), \theta_8(29) \rangle\}$
$G = \left\langle \frac{13\pi}{72}, \frac{13\pi}{72} \right\rangle \in \{\langle \theta_7(13), \theta_7(13) \rangle\}$	$H = \left\langle \frac{13\pi}{72}, \frac{\pi}{4} \right\rangle \in \{\langle \theta_7(13), \theta_8(13) \rangle\}$
$I = \left\langle \frac{7\pi}{36}, \frac{\pi}{8} \right\rangle \in \{\langle \theta_7(17), \theta_8(17) \rangle\}$	$J = \left\langle \frac{7\pi}{36}, \frac{7\pi}{36} \right\rangle \in \{\langle \theta_7(17), \theta_7(17) \rangle\}$
$K = \left\langle \frac{\pi}{4}, \frac{\pi}{4} \right\rangle \in \{\langle \theta_7(21), \theta_7(21) \rangle, \langle \theta_7(23), \theta_8(23) \rangle, \langle \theta_7(23), \theta_7(23) \rangle, \langle \theta_7(39), \theta_8(39) \rangle, \langle \theta_7(39), \theta_7(39) \rangle\}$	$M = \left\langle \frac{\pi}{4}, \frac{\pi}{2} \right\rangle \in \{\langle \theta_7(21), \theta_8(21) \rangle\}$
$N = \left\langle \frac{\pi}{3}, \frac{\pi}{3} \right\rangle \in \{\langle \theta_7(1), \theta_7(1) \rangle\}$	$O = \left\langle \frac{\pi}{3}, \frac{\pi}{2} \right\rangle \in \{\langle \theta_7(1), \theta_8(1) \rangle\}$
$P = \left\langle \frac{13\pi}{36}, \frac{13\pi}{36} \right\rangle \in \{\langle \theta_7(11), \theta_7(11) \rangle\}$	$Q = \left\langle \frac{13\pi}{36}, \frac{\pi}{2} \right\rangle \in \{\langle \theta_7(11), \theta_8(11) \rangle\}$
$R = \left\langle \frac{7\pi}{18}, \frac{\pi}{4} \right\rangle \in \{\langle \theta_7(19), \theta_8(19) \rangle\}$	$S = \left\langle \frac{7\pi}{18}, \frac{7\pi}{18} \right\rangle \in \{\langle \theta_7(19), \theta_7(19) \rangle\}$
$T = \left\langle \frac{\pi}{2}, \frac{\pi}{4} \right\rangle \in \{\langle \theta_7(9), \theta_8(9) \rangle, \langle \theta_7(33), \theta_8(33) \rangle\}$	$U = \left\langle \frac{\pi}{2}, \frac{\pi}{2} \right\rangle \in \{\langle \theta_7(6), \theta_7(6) \rangle, \langle \theta_7(9), \theta_7(9) \rangle, \langle \theta_7(24), \theta_8(24) \rangle, \langle \theta_7(24), \theta_7(24) \rangle, \langle \theta_7(31), \theta_8(31) \rangle, \langle \theta_7(31), \theta_7(31) \rangle, \langle \theta_7(32), \theta_7(32) \rangle, \langle \theta_7(33), \theta_7(33) \rangle, \langle \theta_7(34), \theta_8(34) \rangle, \langle \theta_7(34), \theta_7(34) \rangle\}$

(continued)

Table 11 (continued)

$V = \left\langle \frac{\pi}{2}, \frac{3\pi}{4} \right\rangle \in \{\langle \theta_7(6), \theta_8(6) \rangle, \langle \theta_7(32), \theta_8(32) \rangle\}$

$W = \left\langle \frac{11\pi}{18}, \frac{11\pi}{18} \right\rangle \in \{\langle \theta_7(16), \theta_7(16) \rangle\}$

$X = \left\langle \frac{11\pi}{18}, \frac{3\pi}{4} \right\rangle \in \{\langle \theta_7(16), \theta_8(16) \rangle\}$

$Y = \left\langle \frac{23\pi}{36}, \frac{\pi}{2} \right\rangle \in \{\langle \theta_7(14), \theta_8(14) \rangle\}$

$X = \left\langle \frac{11\pi}{18}, \frac{3\pi}{4} \right\rangle \in \{\langle \theta_7(16), \theta_8(16) \rangle\}$

$Y = \left\langle \frac{23\pi}{36}, \frac{\pi}{2} \right\rangle \in \{\langle \theta_7(14), \theta_8(14) \rangle\}$

$Z = \left\langle \frac{23\pi}{36}, \frac{23\pi}{36} \right\rangle \in \{\langle \theta_7(14), \theta_7(14) \rangle\}$

$A' = \left\langle \frac{2\pi}{3}, \frac{\pi}{2} \right\rangle \in \{\langle \theta_7(4), \theta_8(4) \rangle\}$

$B' = \left\langle \frac{2\pi}{3}, \frac{2\pi}{3} \right\rangle \in \{\langle \theta_7(4), \theta_7(4) \rangle, \langle \theta_7(8), \theta_7(8) \rangle, \langle \theta_7(10), \theta_7(10) \rangle, \langle \theta_7(25), \theta_7(25) \rangle\}$

$C' = \left\langle \frac{2\pi}{3}, \frac{3\pi}{4} \right\rangle \in \{\langle \theta_7(10), \theta_8(10) \rangle, \langle \theta_7(25), \theta_8(25) \rangle\}$

$D' = \left\langle \frac{2\pi}{3}, \frac{7\pi}{8} \right\rangle \in \{\langle \theta_7(8), \theta_8(8) \rangle\}$

$E' = \left\langle \frac{3\pi}{4}, \frac{3\pi}{4} \right\rangle \in \{\langle \theta_7(15), \theta_8(15) \rangle, \langle \theta_7(15), \theta_7(15) \rangle, \langle \theta_7(20), \theta_8(20) \rangle, \langle \theta_7(20), \theta_7(20) \rangle, \langle \theta_7(35), \theta_8(35) \rangle, \langle \theta_7(35), \theta_7(35) \rangle, \langle \theta_7(36), \theta_8(36) \rangle, \langle \theta_7(36), \theta_7(36) \rangle, \langle \theta_7(40), \theta_8(40) \rangle, \langle \theta_7(40), \theta_7(40) \rangle\}$

$F' = \left\langle \frac{29\pi}{36}, \frac{29\pi}{36} \right\rangle \in \{\langle \theta_7(18), \theta_7(18) \rangle\}$

$G' = \left\langle \frac{29\pi}{36}, \frac{7\pi}{8} \right\rangle \in \{\langle \theta_7(18), \theta_8(18) \rangle\}$

$H' = \left\langle \frac{59\pi}{72}, \frac{3\pi}{4} \right\rangle \in \{\langle \theta_7(12), \theta_8(12) \rangle\}$

$I' = \left\langle \frac{59\pi}{72}, \frac{59\pi}{72} \right\rangle \in \{\langle \theta_7(12), \theta_7(12) \rangle\}$

$J' = \left\langle \frac{5\pi}{6}, \frac{3\pi}{4} \right\rangle \in \{\langle \theta_7(2), \theta_8(2) \rangle, \langle \theta_7(5), \theta_8(5) \rangle, \langle \theta_7(26), \theta_8(26) \rangle, \langle \theta_7(30), \theta_8(30) \rangle\}$

$K' = \left\langle \frac{5\pi}{6}, \frac{5\pi}{6} \right\rangle \in \{\langle \theta_7(2), \theta_7(2) \rangle, \langle \theta_7(5), \theta_7(5) \rangle, \langle \theta_7(26), \theta_7(26) \rangle, \langle \theta_7(30), \theta_7(30) \rangle\}$

$M' = \left\langle \frac{7\pi}{8}, \frac{7\pi}{8} \right\rangle \in \{\langle \theta_7(38), \theta_8(38) \rangle, \langle \theta_7(38), \theta_7(38) \rangle\}$

$N' = \left\langle \frac{11\pi}{12}, \frac{7\pi}{8} \right\rangle \in \{\langle \theta_7(28), \theta_8(28) \rangle\}$

$O' = \left\langle \frac{11\pi}{12}, \frac{11\pi}{12} \right\rangle \in \{\langle \theta_7(28), \theta_7(28) \rangle\}$

$P' = \left\langle \pi, \frac{3\pi}{4} \right\rangle \in \{\langle \theta_7(22), \theta_8(22) \rangle\}$

$Q' = \langle \pi, \pi \rangle \in \{\langle \theta_7(22), \theta_7(22) \rangle\}$

$$L_7 = \{10, 5, 10\sqrt{3}, 15, 4, 8.0718, 8, 8, 4, 13.4944, 12, 2\sqrt{6}, 7.101, 8.026, 10, 5\sqrt{2}, 5\sqrt{2}, 10, 4, 6, 2(\sqrt{6}+\sqrt{2}), 8, 12, 8, 12, 2\sqrt{2}, 9.1716, 5.2264\}$$

$$L_8 = \{10, 10, 10\sqrt{2}, 20, 10, 10, 18.478, 10, 10, 10\sqrt{2}, 20, 10, 10, 18.478, 10, 10, 10\sqrt{2}, 20, 10, 10, 18.478, 10, 10, 10\sqrt{2}, 20, 10, 10, 18.478\}$$

Tables 12, 13, 14a and b give a detailed explanation of the IPFP transformation used in this chapter.

(1) Table 12 gives the row sum and column sum that will be maintained in row fitting and column fitting respectively.

Table 12 Table whose rows are populated with corresponding elements of sets L_7 and L_8. This table is constructed to compute the Row and Column totals that needs to be maintained in Row Fitting and Column Fitting respectively

h	$l_7(h)$	$l_8(h)$	TOTAL
1	10	10	**20**
2	5	10	**15**
3	17.3205	14.1421	**31.4626**
4	15	20	**35**
5	4	10	**14**
6	8.0718	10	**18.0718**
7	8	18.478	**26.478**
8	8	10	**18**
9	4	10	**14**
10	13.4944	14.1421	**27.6365**
11	12	20	**32**
12	4.899	10	**14.899**
13	7.101	10	**17.101**
14	8.026	18.478	**26.504**
15	10	10	**20**
16	7.0711	10	**17.0711**
17	7.0711	14.1421	**21.2132**
18	10	20	**30**
19	4	10	**14**
20	6	10	**16**
21	7.7274	18.478	**26.2054**
22	8	10	**18**
23	12	10	**22**
24	8	14.1421	**22.1421**
25	12	20	**32**
26	2.8284	10	**12.8284**
27	9.1716	10	**19.1716**
28	5.2264	18.478	**23.7044**
TOTAL	**234.0087**	**370.4805**	**604.4892**

Table 13 Initial table. The table on which the Row fitting of the first iteration

h	$l'_7(h)$	$l'_8(h)$	TOTAL
1	1	1	**2**
2	1	1	**2**
3	1	1	**2**
4	1	1	**2**
5	1	1	**2**
6	1	1	**2**
7	1	1	**2**
8	1	1	**2**
9	1	1	**2**
10	1	1	**2**
11	1	1	**2**
12	1	1	**2**
13	1	1	**2**
14	1	1	**2**
15	1	1	**2**
16	1	1	**2**
17	1	1	**2**
18	1	1	**2**
19	1	1	**2**
20	1	1	**2**
21	1	1	**2**
22	1	1	**2**
23	1	1	**2**
24	1	1	**2**
25	1	1	**2**
26	1	1	**2**
27	1	1	**2**
28	1	1	**2**
TOTAL	**28**	**28**	**56**

Table 14 IPFP First Iteration

h	$l'_7(h)$	$l'_8(h)$	TOTAL
1	10	10	**20**
2	7.5	7.5	**15**
3	15.7313	15.7313	**31.4626**
4	17.5	17.5	**35**
5	7	7	**14**
6	9.0359	9.0359	**18.0718**
7	13.239	13.239	**26.478**
8	9	9	**18**
9	7	7	**14**
10	13.81825	3.818251	**27.6365**
11	16	16	**32**
12	7.4495	7.4995	**14.899**
13	8.5505	8.5505	**17.101**
14	13.2520	13.2520	**26.5040**
15	10	10	**20**
16	8.53555	8.53555	**17.0711**
17	10.6066	10.6066	**21.2132**
18	15	15	**30**
19	7	7	**14**
20	8	8	**16**
21	13.1027	13.1027	**26.2054**
22	9	9	**18**
23	11	11	**22**
24	11.07105	11.07105	**22.1421**
25	16	16	**32**
26	6.4142	6.142	**12.8284**
27	9.5858	9.5858	**19.1716**
28	11.8522	11.8522	**23.7044**
TOTAL	**302.2446**	**302.2446**	**604.4892**

(A) Table obtained on performing row fitting.

h	$l'_7(h)$	$l'_8(h)$	TOTAL
1	7.7424	12.2576	20
2	5.8068	9.1932	15
3	12.1798	19.2829	31.4627
4	13.5491	21.4509	35
5	5.4197	8.5803	14
6	6.9959	11.0759	18.0718
7	10.2501	16.2279	26.478
8	6.9681	11.0319	18

(continued)

Table 14 (continued)

h	$l'_7(h)$	$l'_8(h)$	TOTAL
9	5.4197	8.5803	14
10	10.6986	16.9379	27.6365
11	12.3878	19.6122	32
12	5.7677	9.1313	14.899
13	6.6201	10.4809	17.101
14	10.2602	16.2438	26.504
15	7.7424	12.2576	20
16	6.6085	10.4625	17.071
17	8.212	13.0012	21.2132
18	11.6135	18.3865	30
19	5.4197	8.5803	14
20	6.1939	9.8061	16
21	10.1446	16.0608	26.2054
22	6.9681	11.0319	18
23	8.5166	13.4834	22
24	8.5716	13.5705	22.1421
25	12.3878	19.6122	32
26	4.9661	7.8623	12.8284
27	7.4217	11.7499	19.1716
28	9.1764	14.5280	23.7044
TOTAL	**234.0087**	**370.4805**	604.4892

(B) Table obtained on performing column fitting.

(2) Table 13, is the initial table, on which row fitting of the first iteration is performed.

(3) Table 14a, is the result of row fitting. The value of each cell is obtained as follows:

$$r_{n,o} = \frac{q_{n,o} * s_r(n)}{s_q(n)}$$

where, $r_{n,o}$ represents the value in nth row and oth column of Table 14a, $q_{n,o}$ represents the value in nth row and oth column of Table 13, $s_r(n)$ represents the nth row sum of Table 14a and $s_q(n)$ represents the nth row sum of Table 13.

(4) Table 14b, is the result of column fitting. The value of each cell is obtained as follows:

$$c_{n,o} = \frac{r_{n,o} * s_c(o)}{s_r(o)}$$

where, $c_{n,o}$ represents the value in nth row and oth column of Table 14b, $r_{n,o}$ represents the value in nth row and oth column of Table 14a, $s_c(o)$ represents the zeroth column sum of Table 14b and $s_r(o)$ represents the oth row sum of Table 14a.

The iteration stops after Table 14b, as the column and row sums of this table are equal to that of Table 12, up to 3 decimal places.

Considering any row from the Table 14b, we compute m

$$m = \frac{12.2576}{7.7424} = 1.5832$$

$y = 1.5832x$, equation of the expected line

Figure 12 shows the relation between the corresponding elements of L_7 and L_8. It also indicates, for each pair $\langle l_7(h), l_8(h)\rangle$ the corresponding point, $\langle l'_7(h), l'_8(h)\rangle$, on the $y = 1.5985x$ line. Further, Table 15 provides the legend for this figure.

Now, the Euclidean Distance is computed using (3.13)

$\Delta_{7,8}(1) = 3.1928$ $\Delta_{7,8}(2) = 1.1409$ $\Delta_{7,8}(3) = 7.2701$ $\Delta_{7,8}(4) = 2.0518$
$\Delta_{7,8}(5) = 2.0077$ $\Delta_{7,8}(6) = 1.5215$ $\Delta_{7,8}(7) = 3.1821$ $\Delta_{7,8}(8) = 1.2285$
$\Delta_{7,8}(9) = 2.0077$ $\Delta_{7,8}(10) = 3.9539$ $\Delta_{7,8}(11) = 0.5484$ $\Delta_{7,8}(12) = 1.1803$
$\Delta_{7,8}(13) = 0.6801$ $\Delta_{7,8}(14) = 3.1596$ $\Delta_{7,8}(15) = 3.1928$ $\Delta_{7,8}(16) = 0.6541$
$\Delta_{7,8}(17) = 1.6135$ $\Delta_{7,8}(18) = 2.2819$ $\Delta_{7,8}(19) = 2.0077$ $\Delta_{7,8}(20) = 0.2742$
$\Delta_{7,8}(21) = 3.4184$ $\Delta_{7,8}(22) = 1.4593$ $\Delta_{7,8}(23) = 4.9263$ $\Delta_{7,8}(24) = 0.8084$
$\Delta_{7,8}(25) = 0.5484$ $\Delta_{7,8}(26) = 3.0231$ $\Delta_{7,8}(27) = 2.4748$ $\Delta_{7,8}(28) = 5.5861$

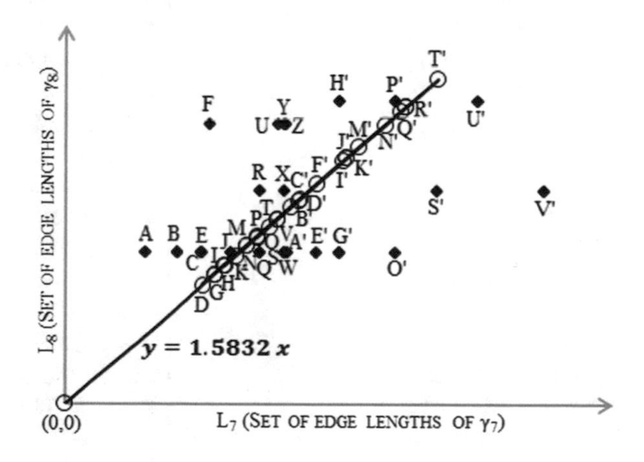

Fig. 12 Solid dots show the existing relation between L_7 and L_8, whereas the line passing through the hollow dots shows the expected relation between L_7 and L_8

Table 15 Legend of Fig. 12

$A = \langle 2.8284, 10\rangle \in \{\langle l_7(26), l_8(26)\rangle\}$	$B = \langle 4, 10\rangle \in \{\langle l_7(5), l_8(5)\rangle, \langle l_7(9), l_8(9)\rangle, \langle l_7(19), l_8(19)\rangle\}$
$C = \langle 4.899, 10\rangle \in \{\langle l_7(12), l_8(12)\rangle\}$	$D = \langle 4.9661, 7.8623\rangle \in \{\langle l'_7(26), l'_8(26)\rangle\}$
$E = \langle 5, 10\rangle \in \{\langle l_7(2), l_8(2)\rangle\}$	$F = \langle 5.2264, 18.478\rangle \in \{\langle l_7(28), l_8(28)\rangle\}$
$G = \langle 5.4197, 8.5803\rangle \in \{\langle l'_7(5), l'_8(5)\rangle, \langle l'_7(9), l'_8(9), \langle l'_7(19), l'_8(19)\rangle\}$	$H = \langle 5.7677, 9.1313\rangle \in \{\langle l'_7(12), l'_8(12)\rangle\}$
$I = \langle 5.8068, 9.1932\rangle \in \{\langle l'_7(2), l'_8(2)\rangle\}$	$J = \langle 6, 10\rangle \in \{\langle l_7(20), l_8(20)\rangle\}$
$K = \langle 6.1939, 9.8061\rangle \in \{\langle l'_7(20), l'_8(20)\rangle\}$	$M = \langle 6.6085, 10.4625\rangle \in \{\langle l'_7(16), l'_8(16)\rangle\}$
$N = \langle 6.6201, 10.4809\rangle \in \{\langle l'_7(13), l'_8(13)\rangle\}$	$O = \langle 6.9681, 11.0319\rangle \in \{\langle l'_7(8), l'_8(8)\rangle, \langle l'_7(22), l'_8(22)\rangle\}$
$P = \langle 6.9959, 11.0759\rangle \in \{\langle l'_7(6), l'_8(6)\rangle\}$	$Q = \langle 7.0711, 10\rangle \in \{\langle l_7(16), l_8(16)\rangle\}$
$R = \langle 7.0711, 14.1421\rangle \in \{\langle l_7(17), l_8(17)\rangle\}$	$S = \langle 7.101, 10\rangle \in \{\langle l_7(13), l_8(13)\rangle\}$
$T = \langle 7.4217, 11.7499\rangle \in \{\langle l'_7(27), l'_8(27)\rangle\}$	$U = \langle 7.7274, 18.4780\rangle \in \{\langle l_7(21), l_8(21)\rangle\}$
$V = \langle 7.7424, 12.2576\rangle \in \{\langle l'_7(1), l'_8(1)\rangle, \langle l'_7(15), l'_8(15)\rangle\}$	$W = \langle 8, 10\rangle \in \{\langle l_7(8), l_8(8)\rangle, \langle l_7(22), l_8(22)\rangle\}$
$X = \langle 8, 14.1421\rangle \in \{\langle l_7(24), l_8(24)\rangle\}$	$Y = \langle 8, 18.478\rangle \in \{\langle l_7(7), l_8(7)\rangle\}$
$Z = \langle 8.0260, 18.4780\rangle \in \{\langle l_7(14), l_8(14)\rangle\}$	$A' = \langle 8.0718, 10\rangle \in \{\langle l_7(6), l_8(6)\rangle\}$
$B' = \langle 8.2120, 13.0012\rangle \in \{\langle l'_7(17), l'_8(17)\rangle\}$	$C' = \langle 8.5166, 13.4834\rangle \in \{\langle l'_7(23), l'_8(23)\rangle\}$
$D' = \langle 8.5716, 13.5705\rangle \in \{\langle l'_7(24), l'_8(24)\rangle\}$	$E' = \langle 8.521, 13.6211\rangle \in \{\langle l'_7(24), l'_8(24)\rangle\}$
$F' = \langle 9.1764, 14.5280\rangle \in \{\langle l'_7(28), l'_8(28)\rangle\}$	$G' = \langle 10, 10\rangle \in \{\langle l_7(1), l_8(1)\rangle, \langle l_7(15), l_8(15)\rangle\}$
$H' = \langle 10, 20\rangle \in \{\langle l_7(18), l_8(18)\rangle\}$	$I' = \langle 10.1446, 16.0608\rangle \in \{\langle l'_7(21), l'_8(21)\rangle\}$
$J' = \langle 10.2501, 16.2279\rangle \in \{\langle l'_7(7), l'_8(7)\rangle\}$	$K' = \langle 10.2602, 16.2438\rangle \in \{\langle l'_7(14), l'_8(14)\rangle\}$
$M' = \langle 10.6986, 16.9379\rangle \in \{\langle l'_7(10), l'_8(10)\rangle\}$	$N' = \langle 11.6135, 18.3865\rangle \in \{\langle l'_7(18), l'_8(18)\rangle\}$
$O' = \langle 12, 10\rangle \in \{\langle l_7(23), l_8(23)\rangle\}$	$P' = \langle 12, 20\rangle \in \{\langle l_7(11), l_8(11)\rangle, \langle l_7(25), l_8(25)\rangle\}$
$Q' = \langle 12.1798, 19.2829\rangle \in \{\langle l'_7(3), l'_8(3)\rangle\}$	$R' = \langle 12.387819.6122\rangle \in \{\langle l'_7(11), l'_8(11)\rangle, \langle l'_7(25), l'_8(25)\rangle\}$
$S' = \langle 13.4944, 14.1421\rangle \in \{\langle l_7(10), l_8(10)\rangle\}$	$T' = \langle 13.549121.4509\rangle \in \{\langle l'_7(4), l'_8(4)\rangle\}$
$U' = \langle 15, 20\rangle \in \{\langle l_7(4), l_8(4)\rangle\}$	$V' = \langle 17.3205, 14.1421\rangle \in \{\langle l_7(3), l_8(3)\rangle\}$

$$\sum_{h=1}^{28} \Delta_{7,8}(h) = 65.6736$$

$$\rho(\gamma_7, \gamma_8) = 0.7011, \text{ using } (3.14)$$

Computing $d(\gamma_7, \gamma_8)$

$$d(\gamma_7, \gamma_8) = 0.8106, \text{ with } \beta = 0.5, \text{ using } (3.15b)$$

References

1. Stephen Stoker, H.: General, organic, and biological chemistry. Cengage Learn. (2009)
2. Bourne, P.E., Gu, J.: Structural bioinformatics, 2 ed., Wiley (2009)
3. Gillespie, R.J., Robinson, E.A.: Models of molecular geometry. Chem. Soc. Rev. **34**, 396–407 (2005)
4. Sen, A., Chebolu, V., Rheingold, A.L.: First structurally characterized geometric isomers of an eight-coordinate complex. structural comparison between cis- and trans-diiodobis(2,5,8-trioxanonane)samarium. Inorg. Chem. **26**(11), 1821–1823 (1987)
5. Stashans, A., Chamba, G., Pinto, H.: Electronic structure, chemical bonding, and geometry of pure and Sr-doped CaCO3. J. Comput. Chem. **29**(3), 343–349 (2008)
6. Zaka, B.: Theory and applications of similarity detection techniques, Ph.D. thesis, Institute for Information Systems and Computer Media (IICM), Graz University of Technology, Graz, Austria, 2 2009
7. Donald, S.B.: The perception of similarity. In: Robert, G.C. (ed.) Avian Visual Cognition, September 2001
8. Michael, J.T., Bülthoff, H.H.: Image-based object recognition in man, monkey and machine. Cognition **67**(1–2), 1–20 (1998). doi:10.1016/S0010-0277(98)00026-2
9. Vermorken, M., Szafarz, A., Pirotte, H.: Sector classification through non-gaussian similarity. Appl. Financ. Econ. **20**(11) (2008). doi:10.1080/09603101003636238
10. Sweeney, C., Kneip, L., Höllerer, T., Turk, M.: Computing similarity transformations from only image correspondences. In: IEEE Conference on Computer Vision and Pattern Recognition (CVPR 2015), pp. 3305–3313, June 2015. doi:10.1109/CVPR.2015.7298951
11. Shiuh-Sheng, Y., Liou, J.-R., Shen, W.-C.: Computational similarity based on chromatic barycenter algorithm. IEEE Trans. Consum. Electron. **42**(2), 216–220 (1996)
12. Komosinski, M., Koczyk, G., Kubiak, M.: On estimating similarity in artificial and real organisms. Theor. Biosci. **120**(3–4), 271–286 (2001)
13. Komosinski, M., Kubiak, M.: Quantitative measure of structural and geometric similarity of 3D morphologies. Complexity **16**(6), 40–52 (2011)
14. Heller, V.: Scale effects in physical hydraulic engineering models. J. Hydraul. Res. **49**(3), 293–306 (2011)
15. Pallett, G.: Geometric similarity–some applications in fluid mechanics. Educ. Train. **3**(2), 36–37 (1961)
16. Ullmann, J.R.: An algorithm for subgraph isomorphism. J. ACM **23**(1), 31–42 (1976)
17. Grewal, B.S., Grewal, J.S.: Higher Engineering Mathematics, 40th edn. Khanna Publishers, New Delhi (2007)
18. Wong, D.W.S.: The reliability of using the iterative proportional fitting procedure, 340–348 (1992)
19. Lahr, M., de Mesnard, L.: Biproportional techniques in input-output analysis: table updating and structural analysis. Econ. Syst. Res. **16**(2), 115–134 (2004)

20. Edwards Deming, W., Stephan, F.F.: On a least squares adjustment of a sampled frequency table when the expected marginal totals are known. Ann. Math. Statist. **11**(4), 427–444 (1940)
21. Rudin, W.: Principles of Mathematical Analysis. McGrawHill Inc., New York (1976)

Rough Set Theory Based on Robust Image Watermarking

Musab Ghadi, Lamri Laouamer, Laurent Nana and Anca Pascu

Abstract Computational intelligence involves convenient adaptation and self-organization concepts, theories, and algorithms, which provide appropriate actions for a complex and changing environment. Fuzzy systems, artificial neural networks, and evolutionary computation are the main computational intelligence approaches used in applications. Rough set theory is one of the important fuzzy systems that have a significant role in extracting rough information from vague and uncertain knowledge. It has a pivotal role in many vague problems linked to image processing, fault diagnosis, intelligent recommendation, and intelligent support decision-making. Image authentication and security are one of the essential demands due to the rapid evolution of tele-image processing systems and to the increase of cyber-attacks on applications relying on such systems. Designing such image authentication and security systems requires the analysis of digital image characteristics which are, in majority, based on uncertain and vague knowledge. Digital watermarking is a well-known solution for image security and authentication. This chapter introduces intelligent systems based on image watermarking and explores the efficiency of rough set theory in designing robust image watermarking with acceptable rate of imperceptibility and robustness against different scenarios of attacks.

M. Ghadi (✉) · L. Laouamer · L. Nana · A. Pascu
Lab-STICC (UMR CNRS 6285), University of Brest, 20 avenue Victor Le Gorgeu, BP 817-CS 93837, 29238 Brest Cedex, France
e-mail: e21409716@etudiant.univ-brest.fr

L. Nana
e-mail: Laurent.Nana@univ-brest.fr

A. Pascu
e-mail: Anca.Pascu@univ-brest.fr

L. Laouamer
Department of Management Information Systems, Qassim University, 6633, Buraidah 51452, Saudi Arabia
e-mail: laoamr@qu.edu.sa

A.E. Hassanien and D.A. Oliva (eds.), *Advances in Soft Computing and Machine Learning in Image Processing*, Studies in Computational Intelligence 730,
https://doi.org/10.1007/978-3-319-63754-9_28

1 Introduction

A collection of methodologies, complementary, and synergistic, which are capable to identify simple algorithms to produce efficient solutions for various problems, are becoming the focus of attention for researchers in different fields. The concepts and paradigms of Computational Intelligence (CI) [1], Soft Computing (SC) [2], and Artificial Intelligence (AI) [3] provide intelligence techniques to develop simple algorithms with efficient solutions for various issues, especially those related to system optimization, pattern recognition, and intelligent control systems. CI has a close relation with AI. The difference between them is that CI employs a subsymbolic knowledge to design a simple algorithm to have an efficient solution for a given problem, while AI uses symbolic knowledge that focuses on finding the best output ignoring the complexity of the proposed algorithm. From the viewpoint of vague and uncertainty concepts, CI is based on numerical and partial set of knowledge (uncertain and incomplete knowledge) that is produced from the given problem, while AI is based on a full knowledge representation decomposed into semantic concepts and logic that are close to the human perspectives. Problems connected with mining, clustering, reduction, and associations are usually solved by CI techniques, while speech recognition, robots, handwriting recognitions, and gaming problems are solved by AI techniques [1, 3].

The SC principle involves all algorithms that are designed to provide a foundation for the conceptions, design, and deployment of intelligent systems. The soft computing methodologies are designed to find efficient solutions for intelligent systems based on uncertain and vague knowledge [2].

The CI combines mainly three techniques including Artificial Neural Networks (ANNs), evolutionary computing, and fuzzy systems. ANNs are inspired by the biological nervous systems, and are learning and adaptive structures used for information processing when it is difficult or not possible to define a set of rules related to a specific problem. The learning task in ANNs is classified into three paradigms: supervised, unsupervised, and reinforcement learning. The Back Propagation ANN algorithm (BPANN) [4] is a supervised learning algorithm that learns by processing the training sets to find approximately optimal network's weight, which is used in turn to enable the algorithm to produce desired output with minimum error. In unsupervised learning methods there is no desired output associated with the training set. It is used usually in clustering and compression applications. The reinforcement learning paradigm is close to supervised learning, except that the change of the network's weight is not related to the error value. Commonly, ANNs-based techniques are applied in classification, frequent pattern, and approximation functions to solve many application problems such as medical diagnosis, image processing, pattern recognition, and data mining.

Evolutionary algorithms are based on the techniques of natural selection and biological evolution. These techniques involve representable and objective functions. Evolutionary algorithms are used when brute-force search is not practical. They are useful for multiparameter optimizations. The Genetic Algorithm (GA) is one of the important kinds of evolutionary algorithms. The fuzzy theory is a heuristic-based approach aimed to introduce efficient solutions based on set of rules and fuzzy membership function. The fuzzy rules deal with incomplete and inexact knowledge such as the concepts of bigger, taller, or faster. Fuzzy set, fuzzy logic, and rough set are the most important techniques in the CI, and they can be combined to give a definition for vagueness and imprecise knowledge in different fields of our life [5–7].

Establishing procedures to preserve digital images security and authentication are significant issues. Many sensitive applications require transmitting a huge amount of digital images in secure way such as medical images analysis, remote sensing images, and objects tracing. Designing the image security and authentication models require considering the major constraints of wireless network including computational complexity, robustness, and fault tolerance against different attacks [8].

Managing these constraints requires an intensive work to deal with image characteristics including the texture/smooth nature, the relationships between the spatial pixels or transformed coefficients, and the structure of image's surface and background. These characteristics, which have significant correlation with the Human Visual System (HVS), are vague and uncertain, since there is no precise meaning or real standard of these characteristics [9].

By demonstrating the texture/smooth nature of digital image, the literature agreed that it was difficult to find a clear definition to the principles of texture and smooth, as well as there is no standard way to discriminate image's regions into texture or smooth. Nevertheless, the problems of texture/smooth analysis are treated through different approaches that attempt to quantify intuitive qualities described by features such as first-order histogram-based features and co-occurrence matrix-based features [10, 11]. These features, which include the uniformity, density, roughness, regularity, linearity, frequency, phase, directionality, coarseness, randomness, fineness, and granulation, are all based on the histogram gray-level intensities. So, the concepts of texture or smooth in image processing are vague and uncertain. They deal with complex patterns that need to be analyzed and characterized using the previously mentioned features.

On the other hand, the relationships between the spatial pixels of digital images or their transformed coefficients by Discrete Wavelet Transform (DWT), Discrete Cosine Transform (DCT), or any other transform processes also have an ambiguity. As an example, the spatial pixel values of neighbored regions have uncertain discernibility due to the disparity of image contrast, so the inability to define exact boundary of spatial pixel values at least for nearby image blocks is considered vague problem, which may lead to degrade the perceptual image quality in case that pixel values are changed. Furthermore, the transformed coefficients of conventional DWT suffer from shift invariant problem due to the variance of wavelet coefficients energy. This problem leads to uncertainty and unpredictability cases to the actual sensitivity to the HVS [12, 13].

However, the mentioned image's characteristics can be analyzed using CI, AI, Image Mining, and Image Coding techniques [14–17]. The associations between the deduced image's characteristics and the techniques mentioned provide efficient solutions for the security and authentication of transmitted images across wireless networks.

CI techniques exhibit many capabilities to adapt and provide multimodal solutions for these complex systems. Although many CI-based models are developed in the fields of fault diagnosis, image classification, recommendation system, and intelligent control system, the employment of these techniques in the field of security and authentication of transmitted multimedia data over the networks is confined, in spite of their significance.

Digital watermarking is one of the solutions which have contributed significantly to the authentication of the transmitted text, images, and videos on the Internet. Many watermarking approaches are proposed in both spatial and frequency domains [18–21], but intelligent techniques based on image watermarking systems have not been widely used. Few works such as those proposed in [22–26] explored the ANNs, GA, and rough set principles in different applications but they did not consider security and authentication of digital images. The key features of rough set theory including the ability to characterize and deal with uncertainty and vague image data, as well as no need to any preliminary or additional information about data like probability in statistics or value of possibility, encourage us to introduce this chapter.

This chapter is organized as follows. Background related to the principles of classical set and rough set is presented in Sect. 2. Section 3 deals with research works based on rough set theory. Some intelligent systems based on image watermarking are then presented in Sect. 4. Section 5 introduces two rough set-based watermarking systems. A comparative study is presented in Sect. 6. The chapter ends with a conclusion and open issues in Sect. 7.

2 Classical Set and Rough Set Principles

The classical set is a primitive notion in mathematics and natural sciences. The set can be defined in such a way that all elements in the universal set are classified definitely into members or nonmembers based on predefined characteristic function. The characteristic function assigns either 0 or 1 for each element in the universe set.

Let U denote the universe set and u denote the general elements, then the characteristic function $F_S(u)$ maps all members in universal set U into set {0, 1}. The mapping process classified the universe elements into crisp sets [27], where the principle of crisp set is defined in such a way that the boundary region of U is empty; this means that all universe set elements are classified definitely either as member or non-member.

The general syntax of characteristic function is mentioned below: $F_S(u)$: $U \rightarrow \{0, 1\}$. Mathematically, the classical sets can be denoted by one of the following expressions:

- List, denoted as $S = \{x_1, x_2, \ldots, x_n\}$
- Formula, denoted as S = {X | X satisfies a given property, for example (X is an even number)}
- Membership function, denoted as

$$F_S(u) = \begin{cases} 1 & \text{if u belongs to S} \\ 0 & \text{if u does not belong to S.} \end{cases}$$

In contrast of crisp set principle, which deals with precise information and knowledge [27], the fuzzy set theory is a mathematical approach to solve the vagueness and uncertainty information about the problem's knowledge. The fuzzy set principle was introduced by Lotfi Zadeh [27] to define a set of elements that is formulated by employing a fuzzy membership function. Any set that is defined by membership function is defined as a fuzzy (imprecise) set not as crisp (precise) set [27].

The membership function expresses the relationship between the value of an element and its degree of membership in a set. The membership function of a fuzzy set S is denoted by F_S, where $F_S : U \rightarrow [0, 1]$. If an element u in the universe set U is a member of the fuzzy set S, then it is member of S with a degree of membership given by $F_S(U) \rightarrow [0, 1]$.

Figure 1 shows the difference between the crisp and fuzzy sets. The crisp set principle is presented by Fig. 1(i), where each element of the three elements {A, B, C} has a crisp value by the characteristic function, while the fuzzy set principle is presented in Fig. 1(ii) in such a way that element B is located on the boundary of crisp set and has a partial membership.

As well, the rough set that was proposed by Pawlak [5] is used efficiently to provide a solution to uncertainty and vagueness knowledge problem. As introduced in [5], the vagueness problem in rough set can be formulated by employing the concept of boundary region of a set rather than partial membership function such as that used in fuzzy set theory [5]. If the boundary region of a set is empty, then it means that the set is crisp (precise); otherwise, the set is rough (imprecise). Nonempty

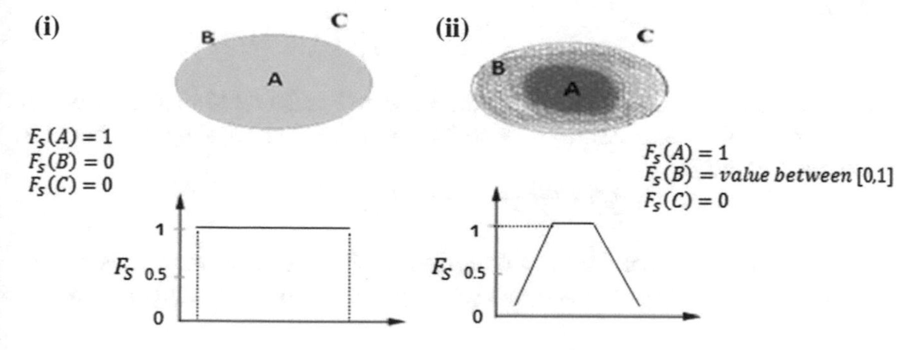

Fig. 1 An example presents the difference between crisp and fuzzy sets

Table 1 An example of information system

Objects	a_1	a_2	a_3	a_4
x_1	Yes	No	Yes	No
x_2	Yes	No	No	Yes
x_3	No	Yes	No	No
x_4	No	No	Yes	Yes
x_5	No	No	No	Yes

boundary region of a set gives indication that our information and knowledge about the problem is vague and uncertain [5].

To describe the rough set problem more precisely, suppose that we have an information system (IS), as presented in Table 1. The IS is expressed by a finite set of objects (U) called (universe) that are described by a finite set of attributes (R). In the sample IS, $U = \{x_1, x_2, x_3, x_4, x_5\}$ and $R = \{a_1, a_2, a_3, a_4\}$.

The representation of lack of knowledge about objects of U can be defined through equivalence relation given by a subset of attributes P, $P \subseteq R$. Let *x* and *y* be arbitrary objects in U, and P an arbitrary nonempty set of R, $P \subseteq R$. Then, objects x and y are denoted to be indiscernible by P, if and only if *x* and *y* have the same vectors of attributes values on all elements in P [7]. The indiscernible relation can denoted as below:

$$\mathrm{ind}(P) = \{(x, y) \in U^2 \mid \forall\, a \in P, a(x) = a(y)\}.$$

The equivalence relation between *x* and *y* is an indiscernible relation by attributes of P, where $(x, y) \in ind(P)$. The equivalence class of an object *x* with respect to P is denoted by $[x]_{ind(P)}$ or $[x]_P$, where $x \in U$ [7, 13]. Based on the IS in Table 1, if $P = \{a_1, a_3\}$, then objects $\{x_3, x_5\}$ are indiscernible.

For a subset $X \subseteq U$, X with respect to P can be characterized by upper and lower approximation sets such as the following:

- The upper approximation of a set *X* with respect to *P* is the set of all objects which can be possibly classified as *X* with respect to *P* (are possibly member of *X* in view of *P*):

$$\overline{AP}(X) = \{x \in U \mid [x]_{ind(P)} \bigcap X \neq \phi\}.$$

- The lower approximation of a set *X* with respect to *P* is the set of all objects, which can with certainty be classified as *X* with respect to *P* (are certainly member of *X* with respect to *P*):

$$\underline{AP}(X) = \{x \in U \mid [x]_{ind(P)} \subseteq X\}.$$

- The boundary region of a set *X* with respect to *P* is the set of all objects, which cannot be classified neither as possibly member of *X* nor as certainly member of *X*:

$$BN_P(X) = \overline{AP}(X) - \underline{AP}(X).$$

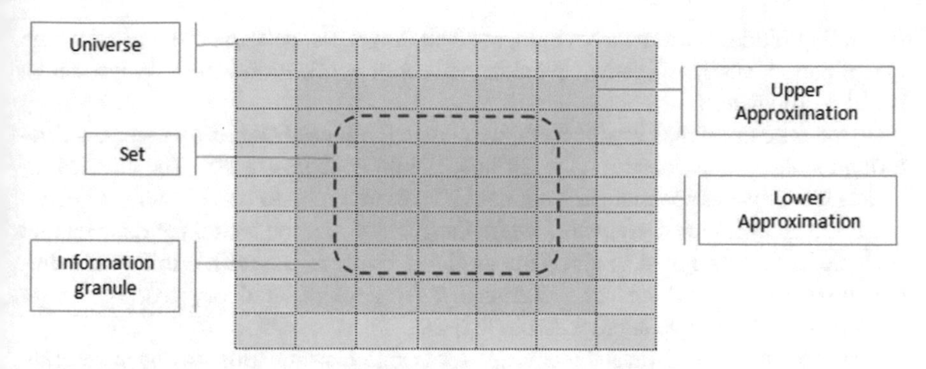

Fig. 2 The elements of rough set theory in terms of approximation sets

Based on these definitions, the set *X* is crisp, if the boundary region of *X* is empty. Otherwise the set *X* is rough. The first case indicates that set *X* is exact with respect to *R*, whereas the second case indicates that set *X* is inexact with respect to *R*. The information granules is another denotation to the equivalence classes of the indiscernibility relation generated by *R*. The granule represents the elementary portion of knowledge that can be recognized due to indiscernibility relation *R*. Then, approximation sets can also be described in terms of granules information [7].

- The upper approximation of set $X(\overline{AP}(X))$ is a union of all granules that have non-empty intersection with the set *X*.
- The lower approximation of a set $X(\underline{AP}(X))$ is a union of all granules that are completely included in the set *X*.
- The boundary region of a set *X* is the difference between the upper and lower approximation sets.

Figure 2 represents the upper and lower approximation sets and the boundary region with means to the information granule.

The roughness metric is usually used to measure the amount of uncertainty of the extracted rough set [7]. For any information system (*IS*) involving set of objects (*U*) and set of attributes (*R*), for any nonempty subset $X \subseteq U$ and attributes $P \subseteq R$, the roughness metric of set *X* with respect to *P* is denoted below:

$$R_P(X) = 1 - \frac{|\underline{AP}(X)|}{|\overline{AP}(X)|},$$

where $X \neq \phi$, |S| denoted the cardinality of the finite set S, and $R_P(X) \in [0, 1]$.

3 Researches Based on Rough Set

In the literature, rough set theory got the attention of the researchers to design efficient algorithms to solve vague and uncertainty problems in different fields such as text processing, fault diagnosis, image processing, and intelligent control system.

Extracting hidden patterns from data, medical diagnosis, pattern recognition, image classification, and intelligent dispatching are set of application whose design can be based on rough set.

In the field of image classification, the classical k-means and fuzzy C-means clustering techniques are used to classify images into specific classes. The classical k-means technique states that each element is assigned only to one cluster, and every cluster must combine k elements, while the fuzzy C-means technique denotes that each element is assigned to all of the available clusters, but with a different membership degree for each cluster. Additionally, using rough set theory to design image classification technique is proposed in [28, 29].

Soft rough fuzzy C-means technique for image classification has been recently proposed in [22]. In this model, the authors aimed to introduce an efficient algorithm to classify MR brain image into three clusters (white matter, gray matter, and the cerebrospinal fluid), where the information and knowledge of this kind of images is vague and uncertain. The fuzzy function in this algorithm expressed the belongingness of each pixel to the cluster centroid without using any weights or thresholds. The upper and lower approximation sets based on rough set theory are computed to find rough regions of images. The contribution of soft rough fuzzy C-means is the capability to classify image's pixels into specified cluster with least error, where the negative region is empty. As well, the use of fuzzy membership function makes it possible to represent the image as soft sets by considering the belongingness of each pixel to the cluster centroid.

The authors in [25] proposed an interesting biological image classification model based on rough set and ANNs. The model defined a set of color image's features, which represents the color intensity of each channel in the processed image, to be the training set of ANNs. Before the training process, the rough inclusion function is applied to define the degree of inclusion of each selected feature, where the value of inclusion function is ranged [0, 1]. Hence, based on resulting inclusion values, the Multi-Layer Perceptron (MLP) algorithm is applied to decide which features are more relevant to classify image with low error rate. The contribution of the proposed model come into view by introducing a simple structure with low computational complexity time for image classification, as well as an efficient approach that may be applied to other kinds of data sets.

The rough set theory is employed also in the field of recommendation systems by determining the most relevant features for creditable classification based on set of quality metrics. This field has a significant role in enhancing the online purchasing system, where designing efficient recommendation system may assist in providing additional and probably unique personalized service for the customers, and increasing the trust and customer loyalty.

The authors in [24] proposed an efficient recommendation model for online purchasing system based on rough set theory and association rules. The model analyzed the customers' transactions on online purchasing system, and used rough set theory to define set of interesting attributes that reflect the preferences of customers. The set of attributes is used to construct a set of rules that are mined by Apriori algorithm [17] and using the predefined minimum support and minimum confidence.

This approach assists to discover the most relevant rules, which after help to match customer preferences prior purchasing process. The contribution of the proposed model is a useful search engine alternative to the common known search engines, which helps the customers to find more relevant things and narrow down the set of choices. Moreover, this model adds a value for the providers by increasing the sales through obtaining more knowledge about customers preferences.

As well, the rough set theory is utilized in applications related to fault diagnosis. The authors in [23] proposed a decision-making scheme using rough set to define a creditable decision table from set of medical data to be a solution for the predictive medical problems. The uncertainty and hierarchical view of medical data are analyzed using rough set and weight functions. The set of attributes based on the rough set principle is used to build set of rules, which after are mined using some performance measurements to generate high credible decision rules. The proposed model is able to analyze the vague medical data and to discover creditable decision rules with reasonable computational complexity.

In the field of text classification, the authors in [26] proposed a model to classify text documents into specific categories based on rough set theory. The motivation of the proposed model come due to the deficiency of exciting text classification systems based on fuzzy set and probability to deal with uncertainty and vague text features. The proposed model extracted a vector of terms from each document to be as a set of features, where each feature is assigned a specific weight using TF/IDF method. The approximation sets based on rough set theory is used to classify the processed documents so as to satisfy user information need.

Attribute reduction and image segmentation are significant processes in the applications of data mining and pattern recognition. Attribute reduction based on rough set theory is proposed in many researches [30–32] to solve the indiscernibility and fuzziness in given information system or dynamic database. In these researches, rough set theory is utilized to remove the redundant/irrelevant attributes and select most informative attributes that preserve the indiscernibility relation and retains the classification ability of a given information system. On the other hand, in the field of image segmentation, the Neutrosophic Sets (NS) theory has been used to reduce and handle the indeterminacy information related to image characteristics such as texture, objects, and edges. In NS theory the image pixels are classified into three sets respectively defined by true, indeterminate, and false membership. Many NS-based image segmentation models are proposed in the literature [33–35]. These models defined efficient segmentation algorithms handling specific indeterminacy information to classify the image pixels into different sets.

4 Intelligent Systems Based on Image Watermarking

This section consists of a state of the art on the application of the intelligent systems in the field of watermarking. It presents some intelligent watermarking systems including the tests and results evaluation.

The authors of [21] proposed an optimized image watermarking system based on genetic algorithm. The system analyzes the processed image with means of HVS characteristics to define texture regions in image, which are more appropriate to embed watermark data robustly. The singular values of Singular Value Decomposition (SVD) transform, which express the contrast of image intensity, are utilized to find the activity factor of each processed image's block using a weight parameter (α). The model selects the high activity factor blocks, which involve a good visual masking effect, to be as input for the watermark embedding process. The embedding process is carried out in the DC coefficients of the transformed DCT image rather than in AC coefficients, where the DC coefficients are more appropriate to embed watermark robustly. DC and AC are terms commonly used in signal processing domain to refer respectively to low- and high-frequency coefficients of DCT. The embedding process as well uses an embedding intensity parameter (β), which controls the degree of image quality. The GA cooperates in this model to optimize the (α and β) parameters, which reflect both the robustness and the perceptual quality of the watermarked image. A fitness function of GA considers the Peak Signal-to-Noise Ratio (PSNR), the Normalized Correlation (NC), and the Structural SIMilarity (SSIM) parameters under several attacking conditions for processed images to find approximately the optimal value of α and β. The proposed watermarking system was tested against additive noise, median filtering, and JPEG loss compression (quality factor = 60) attacks. The PSNR of the proposed model with means of different capacity thresholds ranged 31–46 dB in average, and the experiments result showed that the NC ranged 83–93%.

The authors of [36] proposed an optimized watermarking system based on texture image property and Artificial Bee Colony (ABC). The ABC defined a fitness function used in optimizing the embedding watermark problem through finding optimum embedding parameters (k, p), which are responsible for maintaining the robustness and image quality ratios. The model used two parameters related to the texture feature of processed image blocks. Indeed, the first parameter is related to the difference between the DCT coefficients of neighbored blocks and large difference value expresses more textured areas. The second parameter expresses how much the first parameter can be used for the adjustment of the coefficients. The ABC algorithm employed a fitness function to recombine and rank the solutions using different values of embedding parameters (quality and robustness ratios) under some attack conditions and predefined quality ratio. The result of ABC algorithm defined approximately the optimum embedding parameters, which achieved high perceptual quality and robustness ratios of embedded image. The model used the PSNR and PSNR-JND (Just-Noticeable Difference) metrics to measure the quality between the host image and the watermarked image. The PSNR-JND, which is based on DCT coefficients, is more reliable than other quality measurements metrics. The experiments result proved the efficiency of the proposed tested model in terms of imperceptibility and robustness, where the PSNR in case of Lena image ranged 39.7–46.8 dB with different quality ratios, and the mean value of Bit Correct Rate (BCR) was different for the attack scenarios, but in general it ranged 50–99% for all images.

The authors in [19] proposed an optimized image watermarking system based on GA and SVD transform. First, the singular matrix (S) of the transformed image (USV) is embedded with watermark using a scalar factor $(\alpha) \in [0, 1]$, which is responsible to control the strength of the robustness and the perceptual quality of watermarked image. The model introduced an optimized technique to find approximately optimum value of scale factor using Tournament selection method, which is one of most widely used selection strategies in evolutionary algorithms. The model initially assigns scalar factor (α) to be equal 0.5 and then the fitness value is computed by means of PSNR and NC such as $fitness = robustness - imperceptibility$. The resulting fitness value is considered as reference to the optimization process. Then, the Tournament selection method involves a random selection of two individuals from a population of individuals, which are values between zero and one, to be a parent to produce four chromosomes according to Tournament selection mechanisms. Then, these four values are used in embedding process to find which one gains the minimum fitness value. The one with the minimum fitness is selected for successive generations, till the population evolves toward minimum fitness and then finds approximately the optimal scalar factor (α). The experiments result proved that considering this approach to find scaling factor is efficient in order to obtain high robustness and imperceptibility. In case of Lena image, the PSNR was 47.5 dB, and the NC was 99% against different image processing attacks.

The authors in [15] proposed an optimized digital image watermarking system based on HVS characteristics and Fuzzy Inference System (FIS). The model intended to find approximately best-weighing factors (S1, S2, S3), which are used in the embedding watermark procedure to diminish the conflict between the imperceptibility and robustness of watermarked image. The proposed model uses Matlab packages to compute the set of HVS characteristics from the DCT coefficients of each processed image block. These characteristics include the luminance, texture, edge, and frequency sensitivity, to be the input vector for FIS. The fuzzy inference system uses three inference procedures to find three weighing factors used in the embedding watermark equation. The embedding is done in the center coefficient of each image block DCT to build the watermarked image. The proposed model was tested on Lena and Mandrill images, and the experiments result in case of Lena image showed that the PSNR was 42.3 dB and the NC against different attacks ranged 64–100%.

The authors in [20] aimed to design a robust image watermarking system that resists most geometric attacks based on fuzzy least squares support vector machine (FLS-SVM) and Bessel K form distribution (BKF). The FLS-SVM is a version of the LS-SVM enhanced by reducing the effect of outliers and noises in the data, while the BFK is one of the efficient geometric correction methods. The idea can be organized through two phases: phase 1 involves the embedding watermark process by finding the maximal center region of the host image, where this region typically has least amount of lost data against the rotation and cropping attacks. The watermark is embedded in the low-frequency coefficients of the Quaternion Discrete Fourier Transform (QDFT) of selected region to obtain high robustness and imperceptibility, and the Inverse Quaternion Discrete Fourier Transform (IQDFT) is done to build the watermarked image. In phase 2, the geometric correction on attacked image is applied by BKF and FLS-SVM, where the attacked image is initially converted into grayscale image and the 2QWT (quaternion wavelet transform) is applied on it. The shape and scale parameters of BKF are used to construct the feature vector, where this vector is considered as training data to the FLS-SVM to predict with approximation the best value for rotation angle, scaling factor, and horizontal or vertical distance. Hence, the model will be able to correct the color image. The model is tested against different attacks scenarios on many color images. The experiments result proved the efficiency of the proposed model in terms of imperceptibility and robustness, where the PSNR reached 40 dB, while the Bit Error Rate (BER) ranged 0.003–0.02 against different geometric and non-geometric attacks, except in case of scaling 256×256, where the BER was very high and reached 0.43.

The authors in [16] proposed an optimized image watermarking system based on HVS characteristics and AI techniques. The model can be summarized in three phases: fuzzification phase, where the model calculates the texture and brightness sensitivity characteristics of the DCT coefficients of each image's blocks. These characteristics after that are input into the fuzzy inference system of Mamdani type AND logic. The inference engine phase, where the input parameters are mapped to values between 0 and 1 based on predefined fuzzy inference rules. The result of this phase is a basis used to select some blocks, which are blocks with high texture and high luminance. After that, the centroid method based on BPANN is implemented in the defuzzification phase, where the center value and the eight-neighbor elements for each image block are input into BPANN as a training set to search for optimum weight factor to select approximately most appropriate coefficients to embed watermark bits with good robustness and imperceptibility. The efficient integration between FIS and BPANN in this model provides the ability to optimize intensity factor (α). This factor is used in the embedding equation to balance between the ratio of robustness and imperceptibility. Additionally, the integration between FIS and BPNN introduced a fuzzy crisp set for the value of DCT coefficients that are more appropriate to embed watermark bit. The experiments result proved the efficiency of the proposed system. The PSNR in case of Lena image reached 47 dB, while the NC ranged 73–100% against different attack scenarios.

The authors in [18] dealt with fuzzy inference process of Mamdani type if-then to find a basis to make a decision regarding the amount of watermark bits that can be embedded in host image while maintaining high robustness and imperceptibility. The model based on the moment theory to find the set of orthogonal moments that are embedded by watermark bits by dither modulation. Hence, the FIS uses the moments' parameters and a set of fuzzy linguistic terms to define the fuzzy membership functions, where the parameters of fuzzy membership functions are optimized in terms of robustness and imperceptibility by applying if-then rules and GA. Once the optimized FI parameters are obtained and distilled, the FIS can take a decision on the amount of bits that can be embedded in each moment coefficient independently. The experiments result for the proposed model showed that the PSNR reached 40 dB and the BER in case of Lena image ranged 19–35% against different scales of median filtering, additive white Gaussian noise, and JPEG loss compressing attacks.

5 Rough Set Theory Based on Robust Image Watermarking

This section introduces two proposed watermarking systems based on rough set theory. The first one is in the spatial domain and the second one is in the frequency domain. Each of them aims to solve specific ambiguity in digital image characteristics by defining the upper and lower approximation sets and extracting the rough set, which takes into account the uncertainty and vagueness in the analyzed characteristics.

5.1 Rough Set Theory Based on Robust Image Watermarking in the Spatial Domain

The authors in [12] proposed a robust image watermarking system based on HVS characteristics and rough set theory. The proposed system deals with the vague and uncertainty definition of the textured regions of host image in aims to identify more appropriate regions for embedding watermark with reasonable imperceptibility and robustness ratios. The system took into account two indiscernible HVS characteristics and processed them by rough set theory.

5.1.1 Problem Statement

The proposed watermarking system deals with two problems that are related to the sensitivity of color representations of processed image to the human eyes and the indiscernible effects of DCT coefficients on the perceptual quality of processed image. These problems in case of watermarking system have a close relationship with the principles of HVS in terms of robustness and imperceptibility. The color representation problem deals with the degree of sensitivity of each color space of host image to the human eyes. Many studies in the literature confirmed that analyzing RGB image in means of HVS requires to convert it into another color spaces like YCbCr, which contains the luminance (Y component), the chrominance blue (Cb), and chrominance red (Cr) [37]. The luminance component is very close to the grayscale of the original RGB image, and it expresses the most information in the image, while the chrominance components refer to the color components and they express the details of host image. In means of HVS characteristics, the human eyes are more sensitive to the luminance component and are less sensitive to the Cb component. For designing watermarking system, hiding watermark in Cb component will be more appropriate in terms of imperceptibility and robustness, since the human will not be able to easily see the modification or change in the embedded image. But, the difficulty here is deciding the amount of bits that can be embedded in the Cb component without extreme deficiency in perceptual quality and robustness of watermarked image.

The DCT coefficients ambiguity deals with the DC and AC coefficients of the transformed image by DCT. The literature mentions that the DC coefficient of each image's block expresses the most magnitude information of that block and is used as a good measure to describe the nature of the block (smooth or texture) [15, 17]. These perspectives can be analyzed in terms of HVS and for designing watermarking system. In terms of HVS, the changes in DC coefficients are more sensitive to the human eyes rather than changes in AC coefficients, which define the details of image's information. For designing watermarking system, the literature proved that embedding watermark bits in DC coefficients are more appropriate in terms of robustness than embedding them in AC coefficients [15, 17, 21]. The vague and uncertainty in this case can be described by the amount of bits that can be embedded in the DC coefficients with preservation of the robustness and perceptual quality of the embedded image.

5.1.2 System Model

In order to solve these two ambiguity problems, the proposed watermarking system exploits the capability of rough set theory. Initially, the model builds two information systems related to the nature of host images, which are based on the amount of image's content. Then, rough set theory is applied to define the upper and lower approximation sets and subsequently to extract the rough set, which defines approximately most appropriate blocks to embed watermark in terms of robustness and imperceptibility.

5.1.3 Initialization

The proposed system considers two types of color images: semi-textured and textured images to construct two information systems. Any color image, which is represented by RGB bitmap, is converted to YCbCr components to display luminance component (Y), chrominance blue (Cb), and chrominance red (Cr). The system is designed only by considering the Cb matrix, that is partitioned into non-overlapping 8×8 blocks, based on rough set theory each one called granule. The model analysis each 8×8 block defines two attributes. Attribute (1) defines the average value of pixels in every block in Cb matrix, where each pixel's value ranges between [0–255]. Attribute (2) represents the category value of encoded DC for each block in Cb matrix. This process requires parsing JPEG bitstream file [14] of the Cb matrix. The categories of encoded values for DC coefficients are ranged [0–11]. The structure of system initialization is presented in Fig. 3.

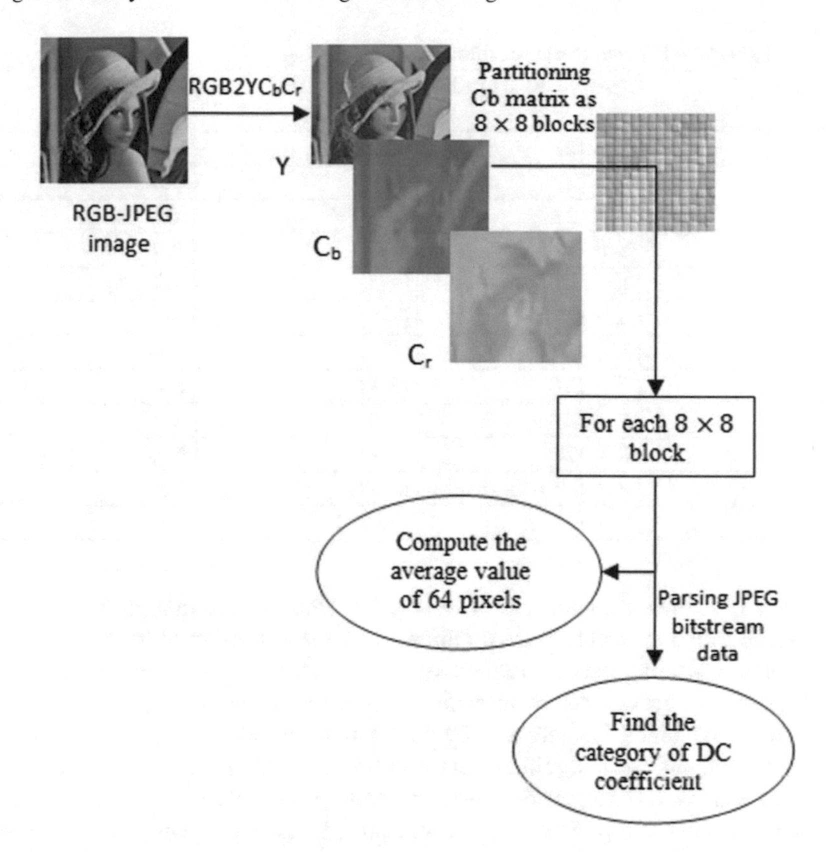

Fig. 3 The structure of system initialization

5.1.4 Construct an Information System for both Semi-textured and Textured Images

The Cb matrix can be represented as an information system $I(U, R)$ that involves a set of objects U and a set of attributes R. The proposed system defines two information systems, which are theoretically well-matched with the watermarking process, and are efficient in terms of robustness and imperceptibility of embedded image.

Each one of these information systems consists of 12 objects, and three attributes. The attributes include the average value of 8×8 pixels corresponding to Cb attribute, the category of DC coefficient that result after parsing JPEG bitstream, and the decision attribute. Defining 12 objects in the information system is arbitrary, the average value of Cb pixels is ranged between [0–255], the category of encoded DC coefficient is ranged [0–11]. The decision attributes express the possibility of the object to hold watermark efficiently (achieve good robustness against different image processing attacks and preserving the perceptual quality of embedded image).

Table 2 Information system of semi-textured images

Class no.	Average value of Cb pixels	Category of DC coefficients	Decision
1	$X \leq 127$	0	Yes
2	$X > 127$	0	Yes
3	$X \leq 127$	[1–3]	Yes
4	$X > 127$	[1–3]	Yes
5	$X \leq 127$	[4–5]	Yes
6	$X > 127$	[4–5]	No
7	$X \leq 127$	[6–7]	No
8	$X > 127$	[6–7]	No
9	$X \leq 127$	[8–9]	No
10	$X > 127$	[8–9]	No
11	$X \leq 127$	[10–11]	No
12	$X > 127$	[10–11]	No

Table 2 illustrates the information system for semi-textured images. The decision of the information system is based on threshold T that corresponds to the class number. By demonstrating the information system in Table 2, we can find that the decision for embedding watermark in semi-textured image blocks depends on $T \leq 5$. This can be explained theoretically by noting that all blocks in any semi-textured images are flat and most significant information content is characterized with low Cb values and low DC categories. Then, embedding watermark through these blocks will become more appropriate to preserve perceptual image quality and to achieve high robustness. In case that some image blocks have DC category in [4–5], this means that these blocks have much information content but it would be significantly low. Therefore, increasing these information by embedding watermark bits will become noticeable by human eyes, and the watermark becomes fragile against different attacks.

On the other hand, Table 3 illustrates the information system for textured images. The decision depends on $T \geq 4$. This can be explained theoretically by noting that all blocks in any textured images are represented by high Cb values and high DC categories, where all blocks have significant information content. Embedding watermark through these blocks will become more appropriate to preserve perceptual image quality and to achieve high robustness. In case that some image blocks have DC category in [1–3], this means that these blocks have low information content. Therefore, increasing this information by embedding watermark bits will become noticeable by human eyes, and the watermark becomes fragile against different attacks.

5.1.5 The Employment of Rough Set Technique

From the information systems illustrated in Tables 2 and 3, we can build a unified information system that deals with any image regardless its nature.

Table 3 Information system of textured images

Class no.	Average value of Cb pixels	Category of DC coefficients	Decision
1	$X \leq 127$	0	No
2	$X > 127$	0	No
3	$X \leq 127$	[1–3]	No
4	$X > 127$	[1–3]	Yes
5	$X \leq 127$	[4–5]	Yes
6	$X > 127$	[4–5]	Yes
7	$X \leq 127$	[6–7]	Yes
8	$X > 127$	[6–7]	Yes
9	$X \leq 127$	[8–9]	Yes
10	$X > 127$	[8–9]	Yes
11	$X \leq 127$	[10–11]	Yes
12	$X > 127$	[10–11]	Yes

The unified information system is illustrated in Table 4, where it expresses the ambiguity in the decision of watermarking process due to the indiscernibility in defining appropriate ranges of Cb and DC category for each candidate block to embed watermark in terms of HVS. By rough set theory, we can extract the upper approximation set ($\overline{AP}(X)$) and lower approximation set ($\underline{AP}(X)$), then we can extract the boundary region (BN) set.

$(\overline{AP}(X)) \rightarrow \{1, 2, 3, 4, 5, 6, 7, 8, 9, 10, 11, 12, 13, 14, 15, 16, 17, 18, 19, 20, 21, 22, 23, 24\}$.

$(\underline{AP}(X)) \rightarrow \{7, 8, 9, 10\}$.

$(BN) \rightarrow \{1, 2, 3, 4, 5, 6, 11, 12, 13, 14, 15, 16, 17, 18, 19, 20, 21, 22, 23, 24\}$.

The representations of upper, lower, and boundary sets for given problem are described in Fig. 4.

Now, the proposed system concerns all of those blocks that are matching the condition for any boundary set element (BN). Based on the results of rough set theory, the selected image's blocks would be defined as the most appropriate blocks to embed watermark in terms of robustness and perceptual quality of embedded image.

5.1.6 Embedding Process

The proposed system used the linear interpolation equation for embedding watermark in host image. This equation gives the ability to control the imperceptibility of embedded image by using a proper interpolation factor t. The pseudocode for embedding watermark process is illustrated by Algorithm 1 and the structure of embedding watermark process is presented in Fig. 5.

Table 4 Unified information system for semi-textured and textured images

Class no.	Average value of Cb pixels	Category of DC coefficients	Decision
1	X ≤ 127	0	Yes
2	X ≤ 127	0	No
3	X > 127	0	Yes
4	X > 127	0	No
5	X ≤ 127	[1–3]	Yes
6	X ≤ 127	[1–3]	No
7	X > 127	[1–3]	Yes
8	X > 127	[1–3]	Yes
9	X ≤ 127	[4–5]	Yes
10	X ≤ 127	[4–5]	Yes
11	X > 127	[4–5]	No
12	X > 127	[4–5]	Yes
13	X ≤ 127	[6–7]	No
14	X ≤ 127	[6–7]	Yes
15	X > 127	[6–7]	No
16	X > 127	[6–7]	Yes
17	X ≤ 127	[8–9]	No
18	X ≤ 127	[8–9]	Yes
19	X > 127	[8–9]	No
20	X > 127	[8–9]	Yes
21	X ≤ 127	[10–11]	No
22	X ≤ 127	[10–11]	Yes
23	X > 127	[10–11]	No
24	X > 127	[10–11]	Yes

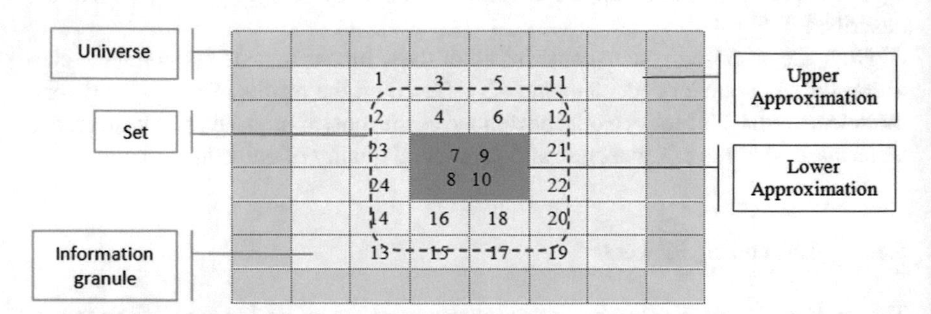

Fig. 4 The representation of upper, lower, and boundary sets for given problem

Algorithm 1 The pseudocode of embedding watermark

1: **Input:** The Cb matrix of host image I and the watermark image w
2: Partitioning Cb matrix into 8×8 blocks (B) where $B = \{1, 2, \ldots, N\}$, N is the total number of 8×8 blocks
3: **for** `i ← 1 to N` **do**
4: $Cb_i \leftarrow$ *Avg value of Cb pixels of* B_i *block*
5: $DC_i \leftarrow$ *Category of encoded DC of* B_i *block*
6: **if** Cb_i and DC_i are matched with the condition of any element in the BN set **then** $B_i^* \leftarrow (1-t)w + tB_i\ ;\ 0<t<1$
7: **loop**
8: $iw \leftarrow$ *combining all* B_i^*

5.1.7 Extraction Process

The proposed system used the inverse linear interpolation equation for extracting watermark from attacked watermarked image, where the watermarked image is usually exposed to different kinds of attacks due to transmission via public networks. The pseudocode for extraction of watermark is illustrated by Algorithm 2 and the structure of extraction watermark process is presented in Fig. 6.

Algorithm 2 The pseudocode of extraction watermark

1: **Input:** The Cb* matrix of attacked watermarked image iw_a and the watermark image w
2: Partitioning Cb* matrix into 8×8 blocks (B*) where $B = \{1, 2, \ldots, N\}$, N is the total number of 8×8 blocks
3: **for** `i ← 1 to N` **do**
4: $Cb_i^* \leftarrow$ *Avg value of Cb pixels of* B_i^* *block*
5: $DC_i^* \leftarrow$ *Category of encoded DC of* B_i^* *block*
6: **if** Cb_i^* and DC_i^* are matched with the condition of any element in BN set **then** $w_i \leftarrow (1/t)w - ((1-t)/t)B_i^*\ ;\ 0<t<1$
7: **loop**
8: $w \leftarrow$ *set of all* w_i

5.1.8 Experiments Result

The proposed system is tested on color JPEG images sized 128×128 and using 8×8 watermark image. All watermarked images are exposed to a variety of geometric and non-geometric attacks using StirMark Benchmark v.4 [38] and Matlab Tool. Figure 7 illustrates a sample of host images, and the watermark image.The PSNR, BER, and Correlation Coefficient (CC) [39–41] are well-known metrics used to evaluate the performance of the proposed system.

The PSNR expresses the imperceptibility of embedded image by measuring the similarity between the original image and embedded image using (1). High PSNR proves that the embedded watermark is imperceptible and did not degrade the quality of original image:

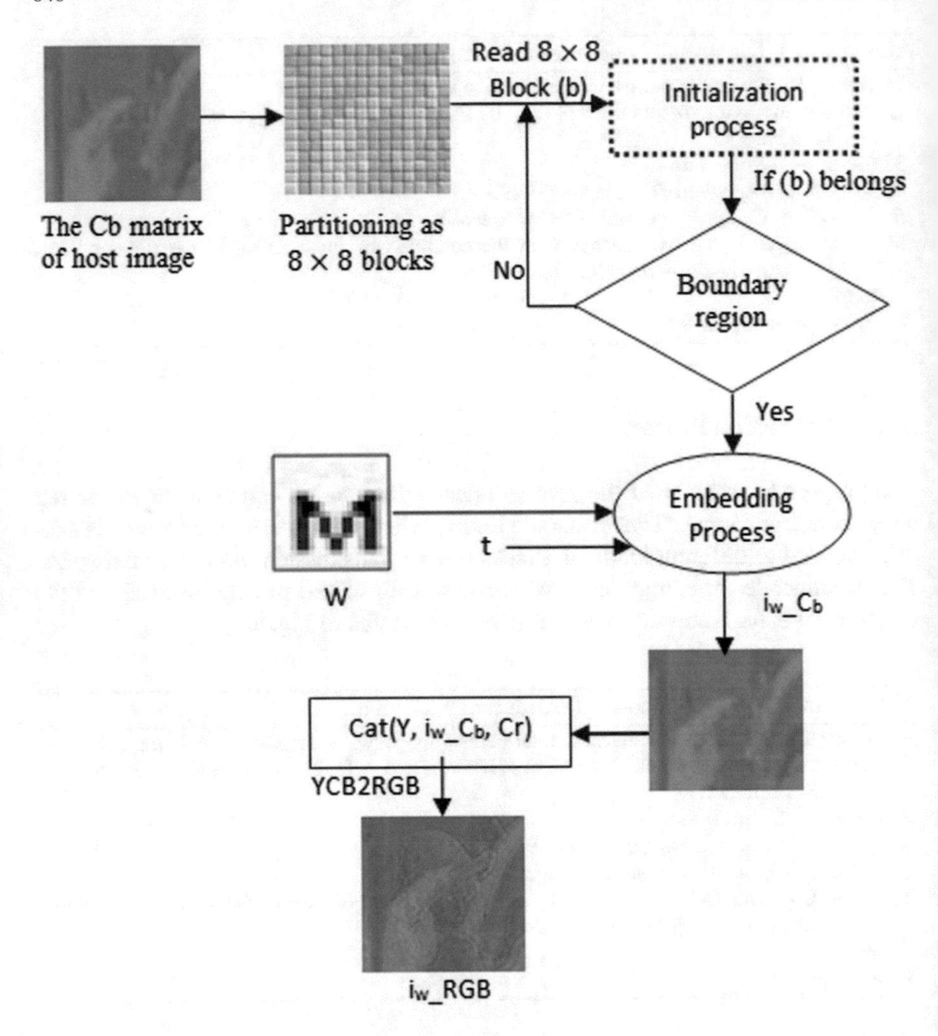

Fig. 5 Structure of embedding process

$$\mathbf{PSNR} = \mathbf{10log_{10}}\left[\frac{\mathbf{255^2}}{\frac{\mathbf{1}}{\mathbf{M \times N}}\sum_{\mathbf{i=1}}^{\mathbf{M}}\sum_{\mathbf{j=1}}^{\mathbf{N}}(\mathbf{A_{ij}} - \mathbf{A'_{ij}})^2}\right], \quad (1)$$

where A_{ij} is the pixel (i, j) in the original image and A'_{ij} is the pixel (i, j) in the watermarked image, $M \times N$ is the size of image.

The CC metric is used to measure the similarity between the original watermark image and the attacked watermark image. High similarity ratio indicates high robustness of proposed watermarking system. The CC ratio ranged $[1, -1]$. If the CC equals -1 this means that two images are completely anti-similar, if the CC equals 1 this means that two images are definitely identical, and if the CC equals 0 this means that two images are completely dissimilar. The CC ratio is computed using (2):

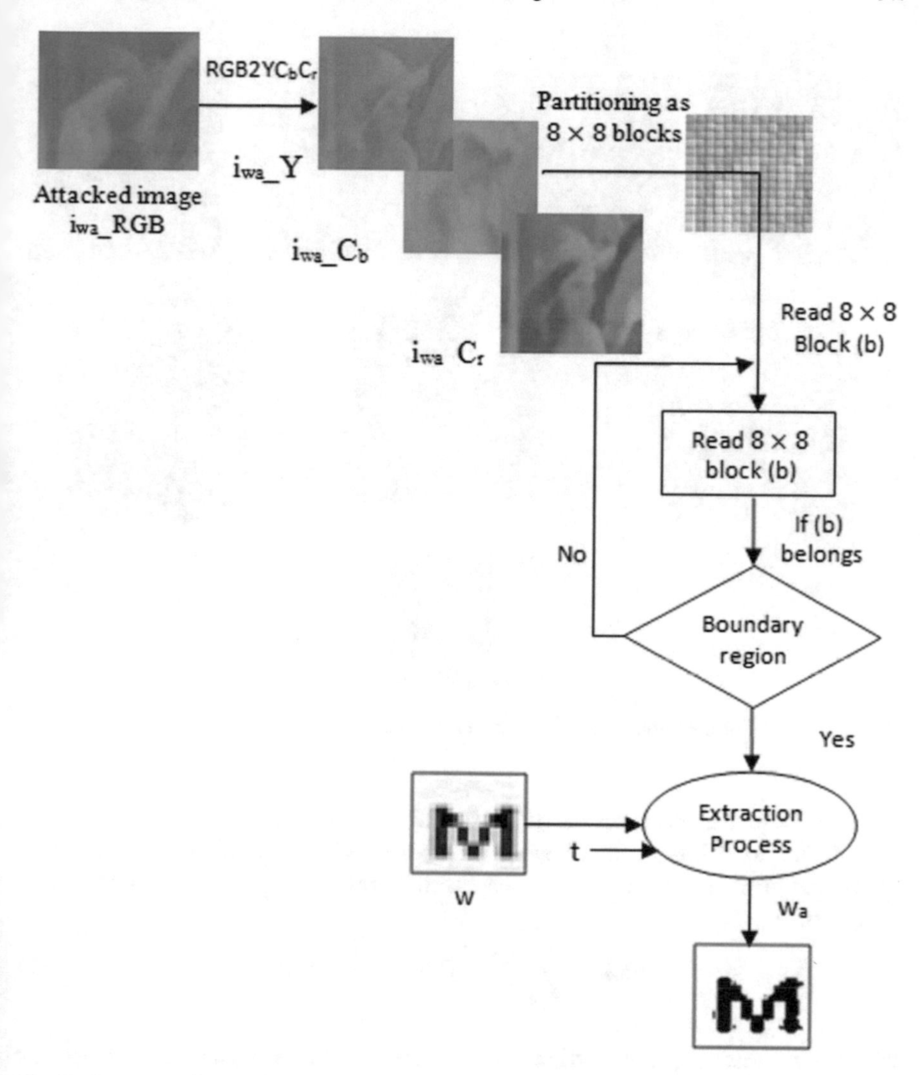

Fig. 6 Structure of extraction process

$$\mathbf{CC} = \frac{\sum_m \sum_n (\mathbf{A}_{mn} - \overline{\mathbf{A}})(\mathbf{B}_{mn} - \overline{\mathbf{B}})}{\sqrt{\left(\sum_m \sum_n \left(\mathbf{A}_{mn} - \overline{\mathbf{A}}\right)^2\right)\left(\sum_m \sum_n \left(\mathbf{B}_{mn} - \overline{\mathbf{B}}\right)^2\right)}}, \quad (2)$$

where A_{mn} is the intensity of *mn* pixel in image *A*, B_{mn} is the intensity of *mn* pixel in image *B*, $\overline{A}$ is the mean intensity of image *A*, and $\overline{B}$ is the mean intensity of image *B*.

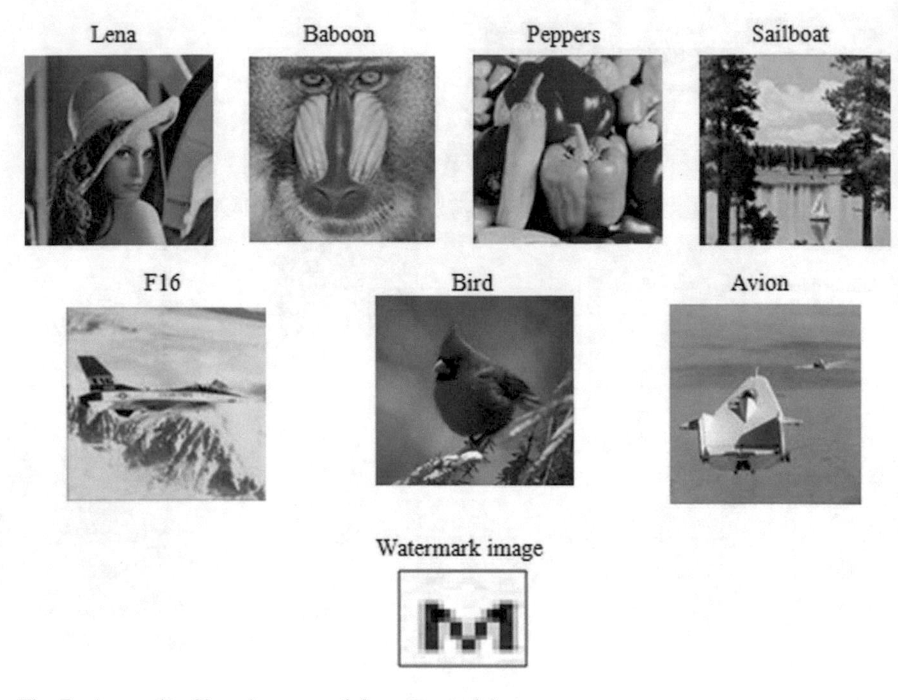

Fig. 7 A sample of host images and the watermark image

The BER metric measures the percentage of erroneous bits in the extracted watermark to the total bits in the original watermark, where low BER expresses high robustness of watermarking system against different attacks. The BER is computed using (3):

$$\mathbf{BER} = \frac{\mathbf{1}}{\mathbf{n}}\left[\sum_{\mathbf{i=1}}^{\mathbf{n}} \mathbf{w(i)} \oplus \overline{\mathbf{w}}\mathbf{(i)}\right] \times \mathbf{100}, \tag{3}$$

where $w(i)$ presents the ith original watermark bit, $\overline{w}(i)$ presents the ith extracted watermark bit, and n is the total number of original watermark bits.

Due to the large number of blocks that satisfy the boundary rough set, the experiments result is displayed in average as shown in Figs. 8, 9, and 10.

The results of PSNR, which computes the imperceptibility between the original (YCbCr) image and the embedded (YCb*Cr) image, are presented in Fig. 8. The PSNR ranges between 38.2–42.0 dB for all processed image. To show the robustness of the proposed watermarking system, the processed images are tested against different attacks including JPEG loss compression (quality factor = 80), Median filtering (9 × 9), Gaussian noise ($\sigma = 10$), and rotation (counterclockwise by 10°, 45°). The attack scenarios on different images and the extracted watermark images are presented in Figs. 9 and 10. The BER and CC are computed between the original watermark and extracted watermark for each attack scenario.

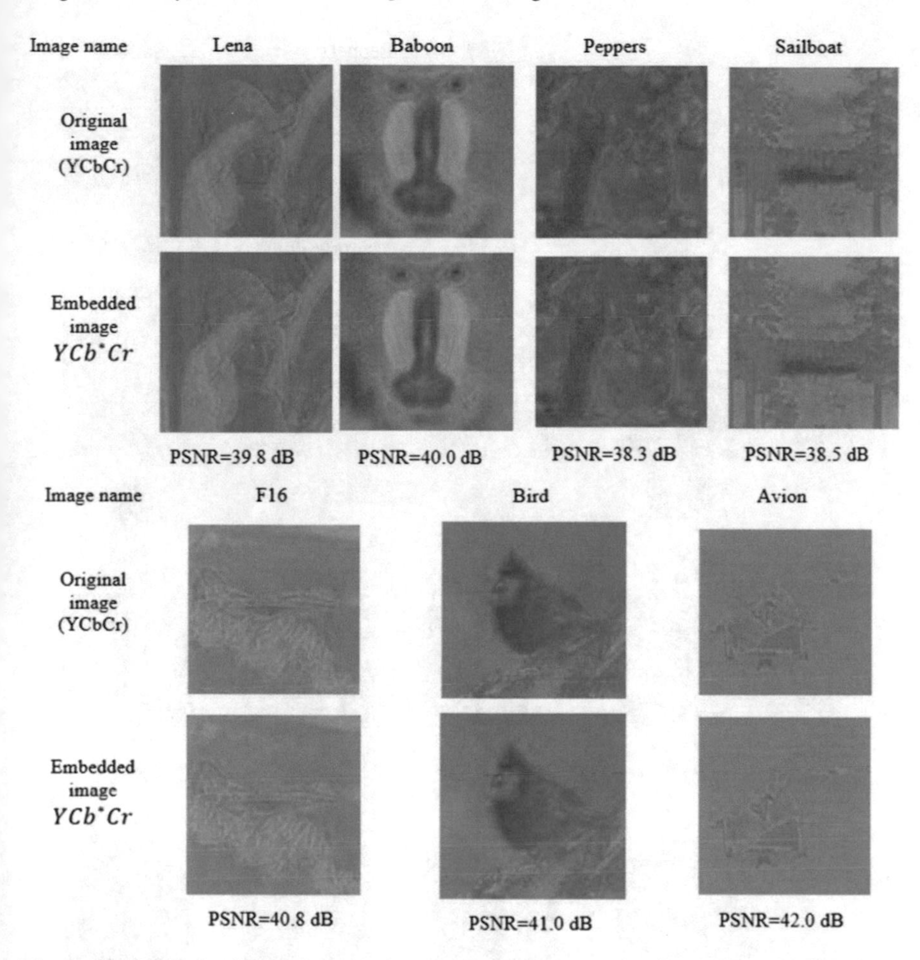

Fig. 8 The PSNR for different attacks (in case of Lena, Baboon, Peppers Sailboat, F16, Bird, and Avion images)

Figure 9 shows the BER and CC ratios between the original watermarks and the extracted attacked watermarks of some processed images, which have much content and appear as textured images.

The CC ratios in most cases reach 1, and the BER ranges 0.18–0.19. Figure 10 also presents the BER and CC ratios between the original watermarks and the extracted attacked watermarks for those images, which do not have much content and appear as semi-textured images.

In most cases the CC ratios reach 1, and the BER ranges 0.19–0.20. The experiments result proves the efficiency of the proposed model and its capability to deal with the color representation and DCT coefficients problems in terms of HVS. This gives us a sense that the proposed rough set-based watermarking technique is very interesting to achieve high robustness and imperceptibility ratios.

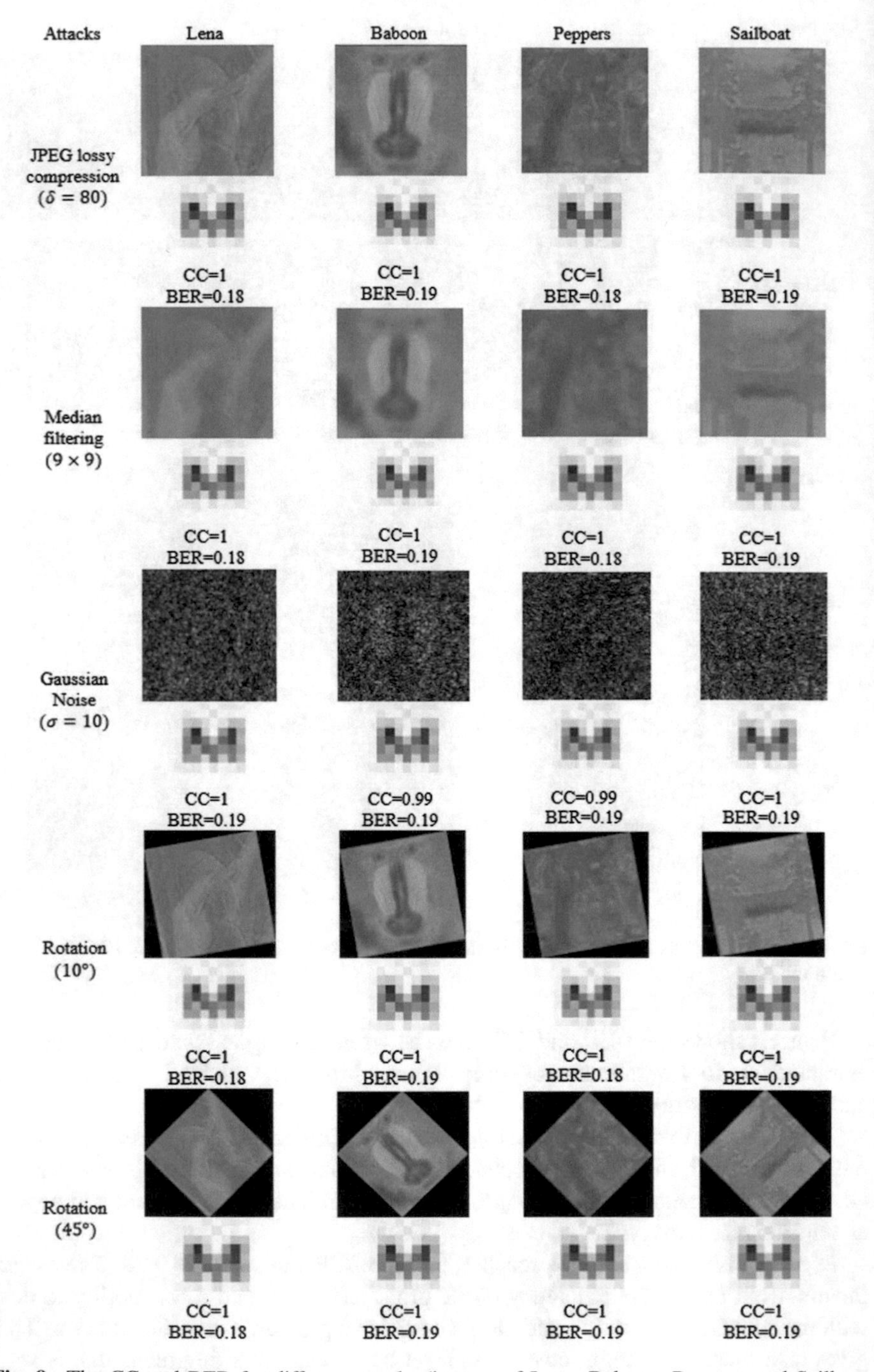

Fig. 9 The CC and BER for different attacks (in case of Lena, Baboon, Peppers, and Sailboat images)

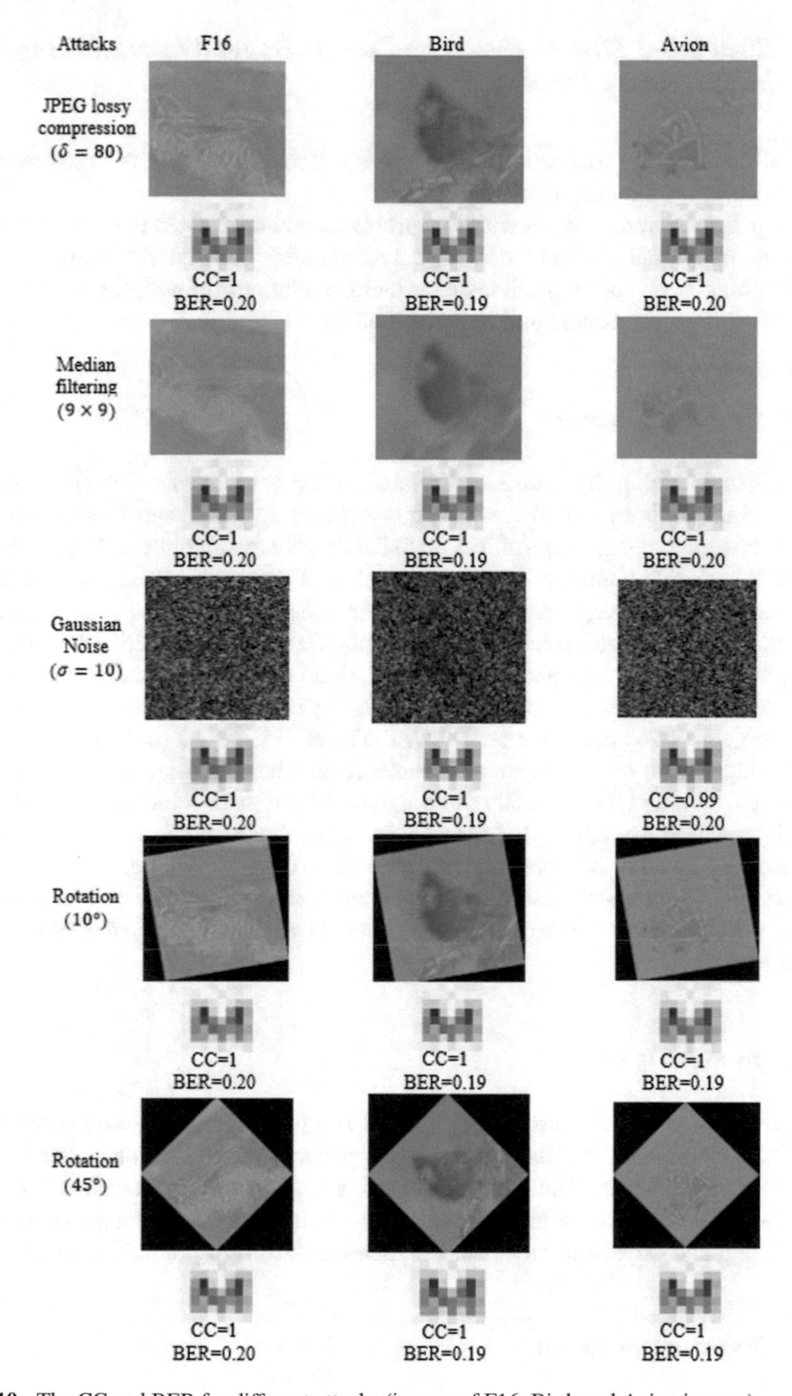

Fig. 10 The CC and BER for different attacks (in case of F16, Bird, and Avion images)

5.2 Rough Set Theory Based on Robust Image Watermarking in Frequency Domain

The authors in [13] proposed a robust image watermarking system based on HVS characteristics and rough set theory.

The proposed system deals with two problems that are related to the boundary of grayscale image's pixels and the statistical redundancy due to the shift invariance in DWT coefficients. These problems have a close relationship with the principles of HVS in terms of robustness and imperceptibility.

5.2.1 Problem Statement

The grayscale ambiguity problem is related to the varying of grayscale values of neighboring pixels in digital image. This variation in pixel values of neighbored regions has imperfect perceptibility due to the deficiency of contrast. Then, it makes it hard to exactly bound the grayscale values of a given image. In terms of HVS, the uncertainty and vague grayscale values may adversely affect image's contrast, then it may weaken the perceptual image quality. Therefore, embedding watermark in vague and uncertain grayscale values could lead to bad perceptual image quality.

Moreover, the statistical redundancy ambiguity occurs due to the shift invariance problem symbolized in conventional DWT. The shift invariant problem symbolizes the variance in the energy of wavelet coefficients whenever the incoming signal is shifted, even though it is basically same signals. The statistical redundancy indicates inability and unpredictability to the actual sensitivity to the HVS. This in turn affects the perceptual quality of embedded image in case of watermarking.

The rough set theory is used in this model to deal with these problems and to design an efficient watermarking system able to ensure the imperceptibility and robustness.

5.2.2 System Model

The proposed watermarking system applied rough set theory on one subband of DWT to approximate its coefficients into upper and lower sets, and after to generate a reference image. The singular value of watermark image is embedded in the singular value of reference image. The pseudocodes of reference image generation, embedding, and extraction processes are illustrated in Algorithms 3, 4, and 5.

5.2.3 Experiments Result

The proposed watermarking system is tested against different geometric and non-geometric attacks; all experiments are conducted on 512×512 grayscale images

Algorithm 3 The pseudocode of reference image generation

1: **Input:** A grayscale host image I
2: Transforming I by 1-level DWT process
3: Select one subband A from the transformed I to generate reference image
4: Find maximum grayscale value of subband A (Max_gray)
5: Find minimum grayscale value of subband A (Min_gray)
6: Decomposing the selected subband A into desired number of granules B_i, where $i \in \{0, 1, \ldots, P\}$ and P is the number of granules
7: $Threshold\ T \leftarrow \frac{max(max(B_i))+min(min(B_i))}{2}$
8: $\underline{AP} \leftarrow \forall B_i < T$
9: $\overline{AP} \leftarrow \forall B_i \geq T$
10: $lowerRE(\underline{RE}) \leftarrow \frac{-e}{2}[(\underline{AP})_{Bi} log_2 (\underline{AP})_{Bi}]$
11: $upperRE(\overline{RE}) \leftarrow \frac{-e}{2}[(\overline{AP})_{Bi} log_2 (\overline{AP})_{Bi}]$
12: Reference image (A^{ref}) $\leftarrow$ inverse of DWT

Algorithm 4 The pseudocode of embedding watermark

1: **Input:** The reference image (A^{ref}), The watermark image (W), lower Approximation ($\underline{AP}$), upper Approximation ($\overline{AP}$), lowerRE($\underline{RE}$), upperRE($\overline{RE}$)
2: Transforming reference image (A^{ref}) and watermark image (W) by SVD then
$A^{ref} \leftarrow U^{ref} S^{ref} V^{ref}$
$W \leftarrow U^W S^W V^W$
3: Compute the watermarked reference image as
$S^{ref*} = S^{ref} + \underline{RE} \times S^w$
$A^{ref*} = U^{ref} \overline{S^{ref*}} V^{ref}$
4: Watermarked image (A_w) $\leftarrow$ IDWT(A^{ref*})

Algorithm 5 The pseudocode of extraction watermark

1: **Input:** The watermarked image (A_w), The original reference image (A^{ref}), The watermark image (W), lowerRE ($\underline{RE}$)
2: Transforming watermarked image (A_w) by 1-level DWT.
3: Extracting the reference watermarked image (A^{ref+})
4: Transforming (A^{ref+}) and (A^{ref}) by SVD then:
$A^{ref+} \leftarrow U^{ref+} S^{ref+} V^{ref+}$
$A^{ref} \leftarrow U^{ref} S^{ref} V^{ref}$
5: Extracting the singular value of watermark as following:
$S^{W*} \leftarrow \frac{S^{ref+} - S^{ref}}{\underline{RE}}$
6: The attacked watermark $W^* \leftarrow U^W S^{W*} V^W$

and using the watermark image sized 64×64. To evaluate the performance of this model in terms of imperceptibility and robustness, the PSNR, NCC, and BER metrics are used. Table 5 presents the PSNR between the original images and watermarked images, while Table 6 presents the NCC between the original watermark and attacked watermark images with different attack scenarios.

Table 5 The PSNR (in case of Barbara, Lena, Pirate, and Living Room images)

Host image	Barbara	Lena	Pirate	Living Room
PSNR (dB)	68.79	52.69	69.52	46.55

Table 6 The NCC with different attacks (in case of Barbara, Lena, Pirate, and Living Room images)

Host image	Barbara	Lena	Pirate	Living Room
Median filtering (13 × 13)	0.57	−0.18	0.86	−0.26
Average filtering (13 × 13)	0.79	−0.19	0.95	−0.26
Rotation (60°)	0.45	−0.21	0.73	0.67
Resizing (512-128-512)	0.60	0.44	0.87	−0.24
Gaussian Noise ($\sigma = 0.1$)	0.38	0.65	0.66	0.91
Cropping (25%)	0.79	0.90	0.88	0.97
JPEG Compression ($\alpha = 50$)	0.55	0.58	0.79	0.91
SaltPepper (50%)	0.38	0.46	0.65	0.74
Motion blure (15 pixels)	0.82	−0.19	0.86	−0.26
Speckel noise (50%)	0.38	0.48	0.67	0.77

6 Comparative Study

This section presents a set of comparative studies between some interesting watermarking systems based on intelligent techniques. Table 7 presents the methodology for each of these works in terms of extraction technique, embedding domain, embedding strength, the sizes of host, and watermark images. The comparisons show the performance of each work in terms of imperceptibility and robustness against different attack scenarios. All comparisons are done on Lena image, and in some cases the sizes of host image or watermark image have been adapted.

Comparing the Imperceptibility Ratios for Related Works in Terms of PSNR

Figure 11 presents a comparison between the intelligent techniques based on watermarking systems to show the achieved imperceptibility between original image and embedded image. For the watermarking system of GA [21] or FIS [15, 18], the PSNR ranged 40–42 dB. In the watermarking system based on rough set using DWT-SVD [13] the PSNR was 52 dB, while in the watermarking system based on rough set in the spatial domain [12] the PSNR was 42 dB.

Table 7 The methodologies of five intelligent techniques based on image watermarking

Proposed system	Extraction technique	Embedding strength	Size of host image	Size of watermark image
Fuzzy inference system based on watermarking in the DCT [15]	Semi-blind	Adaptive	512 × 512	64 × 64
Fuzzy inference system based on watermarking in the orthogonal moment and spatial pixels [18]	Blind	Adaptive	256 × 256	Message bits (100 or 500 bits)
Genetic algorithm based on watermarking in the SVD + DCT [21]	Fragile	Adaptive	256 × 256	Undefined
Rough set-based watermarking in the DWT + SVD [13]	Semi-blind	Adaptive	512 × 512	64 × 64
Rough set-based watermarking in the spatial pixels [12]	Semi-blind	Adaptive	512 × 512	64 × 64

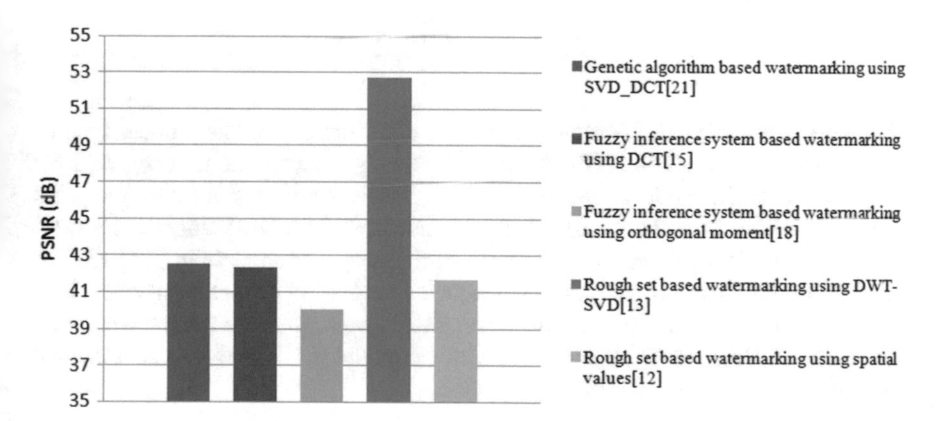

Fig. 11 Comparing the PSNR between different intelligent techniques based on watermarking systems

Comparing the Robustness Ratio for Related Works in Terms of BER and CC

This comparison presents the robustness of rough set-based watermarking systems in terms of BER and CC. Figure 12 presents a comparison between the achieved

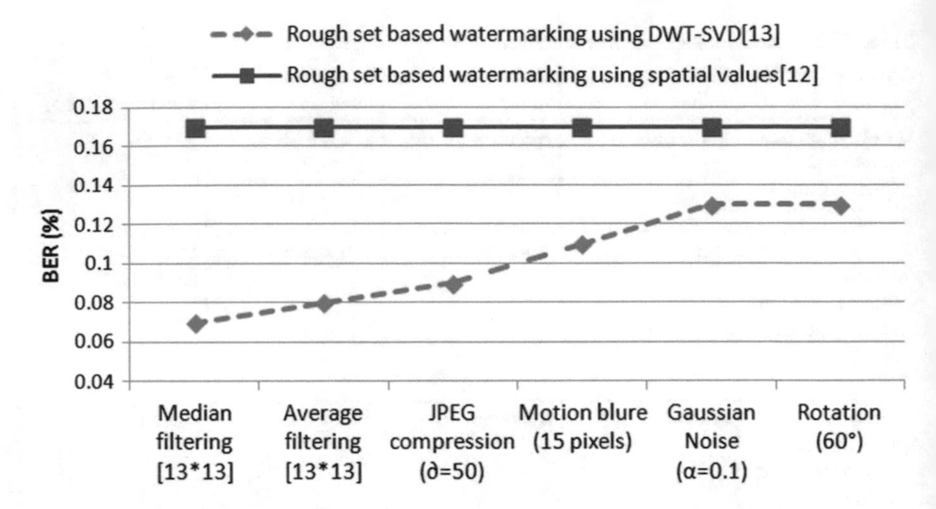

Fig. 12 Comparing the BER between different rough set-based watermarking systems for different attacks

Table 8 Comparing the CC between different intelligent techniques based on watermarking systems for different attacks

Attacks	[15]	[13]	[12]
JPEG compression ($\alpha = 80$)	Na	0.78	1
Gaussian noise ($\sigma = 0.1$)	0.92	0.70	1
Salt and pepper (0.01)	0.64	Na	1
Median filtering (9×9)	0.79	0.87	1
Rotation (10°)	0.75	0.75	1

robustness in models [12, 13] against different attack scenarios. The proposed model in [13], which is based on DWT-SVD, achieved high robustness in terms of BER comparing with the model proposed in [12], which embedded watermark in spatial domain. The BER in [13] ranged 0.07–0.13, while in [12] the BER ranged 0.16–0.17.

As well, Table 8 presents the CC ratio in the proposed models in [12, 13, 15] against different attack scenarios. The proposed model in [12] achieved high similarity between the original watermark and extracted watermark comparing with other models. The CC ratio in [12] reached 1, while in model [13] it ranged 0.70–0.87 and in model [15] it ranged 0.64–92.

7 Conclusion and Open Issues

The rough set theory represents one of important computational intelligence systems that have a significant role in extracting rough information from vague and uncertain knowledge. It is efficient to solve vague problems linked to image processing, fault diagnosis, intelligent recommendation, and intelligent support decision-making. This chapter illustrated the capability of rough set theory to deal with some ambiguity problems in digital images. These problems are associated with image's characteristics, and have a close relation with HVS principles in case of designing an efficient image authentication and security systems. The approximation principle based on rough set theory has been utilized to extract rough information from the vague image's characteristics to suggest an efficient watermarking system in terms of perceptual quality of embedded image and robustness against different image processing attacks.

For future direction, fuzzy equivalence relation and investigating information holding in preference order are two major issues that need to be analyzed from the perspective of rough set theory. The fuzziness in rough set and multi-criteria sorting based on rough set theory are open issues that deal directly with the ambiguity and uncertainty in image's knowledge. In case of watermarking system design, these issues could be analyzed to find other possibilities to minimize the effect of image's ambiguity and uncertainty on the perceptual image quality and the robustness against different attacks.

References

1. Hassanien, A.-E., Abraham, A., Kacprzyk, J., Peters, F.J.: Computational intelligence in multimedia processing: foundation and trends. Stud. Comput. Intell. **96**, 3–49 (2008)
2. Balas, E.V., Jain, L.C., Kovaevi, B.: Soft computing applications. In: Proceedings of the 5th International Workshop Soft Computing Applications. Advances in Intelligent Systems and Computing, vol. 356, pp. 1–724 (2014)
3. Russell, S.: Rationality and intelligence: a brief update. In: Synthese Library, vol. 376, pp. 7–28. Springer International Publishing Switzerland (2016)
4. Karimi, R., Yousefi, F., Ghaedi, M., Dashtian, K.: Back propagation artificial neural network and central composite design modeling of operational parameter impact for sunset yellow and azur (II) adsorption onto MWCNT and MWCNT-Pd-NPs: Isotherm and kinetic study. Chemometr. Intell. Lab. Syst. **159**, 127–137 (2016)
5. Pawlak, Z.: Rough sets. Int. J. Comput. Inf. Sci. **11**, 1–16 (1982)
6. Yao, Y.Y.: A comparative study of fuzzy sets and rough sets. Inf. Sci. **109**, 227–242 (1998)
7. Zhang, Q., Xie, Q., Wang, G.: A survey on rough set theory and its applications. CAAI Trans. Intell. Technol. 1–11 (2016)
8. Ghadi, M., Laouamer, L., Moulahi, T.: Securing data exchange in wireless multimedia sensor networks: perspectives and challenges. Multimedia Tools Appl. **75**, 3425–3451 (2016)
9. Sen, D., Pal, S.K.: Generalized rough sets, entropy, and image ambiguity measures. IEEE Trans. Syst. Man Cybern. **39**, 117–28 (2009)
10. Materka, A., Strzelecki, M.: Texture Analysis Methods—A Review. Technical University of Lodz, Institute of Electronics, COST B11 Report, Brussels (1998)
11. Das, A.: Image Enhancement in Spatial Domain, pp. 43–92. Springer International Publishing, Switzerland (2015)
12. Ghadi, M., Laouamer, L., Nana, L., Pascu, A.: Fuzzy rough set based image watermarking approach. In: Proceedings of the International Conference on Advanced Intelligent Systems and Informatics, vol. 533, pp. 234–245 (2016)
13. Kumar, S., Jaina, N., Fernandes, S.L.: Rough set based effective technique of image watermarking. J. Comput. Sci. 1–17(2016)

14. Ghadi, M., Laouamer, L., Nana, L., Pascu, A.: JPEG bitstream based integrity with lightweight complexity of medical image in WMSNS environment. In: Proceedings of the 7th International ACM Conference on Management of Computational and Collective Intelligence in Digital Ecosystems, pp. 53–58. ACM (2015)
15. Jagadeesh, B., Kumar, R.P., Reddy, C.P.: Fuzzy inference system based robust digital image watermarking technique using discrete cosine transform. Proced. Comput. Sci. **46**, 1618–1625 (2015)
16. Jagadeesh, B., Kumar, R.P., Reddy, C.P.: Robust digital image watermarking based on fuzzy inference system and back propagation neural networks using DCT. Soft Comput. **20**, 3679–3686 (2016)
17. Ghadi, M., Laouamer, L., Nana, L., Pascu, A.: A robust associative watermarking technique based on frequent pattern mining and texture analysis. In: Proceedings of the 8th International ACM Conference on Management of Computational and Collective Intelligence in Digital Ecosystems, pp. 73–81. ACM (2016)
18. Papakostas, G.A., Tsougenis, E.D., Koulouriotis, D.E.: Fuzzy knowledge-based adaptive image watermarking by the method of moments. Complex Intell. Syst. **2**, 205–220 (2016)
19. Lai, C.-C.: Digital watermarking scheme based on singular value decomposition and tiny genetic algorithm. Digital Signal Process. **21**, 522–527 (2011)
20. Wang, C., Wang, X., Zhang, C., Xia, Z.: Geometric correction based color image watermarking using fuzzy least squares support vector machine and Bessel K form distribution. Signal Process. **134**, 197–208 (2017)
21. Han, J., Zhao, X., Qiu, C.: A digital image watermarking method based on host image analysis and genetic algorithm. J. Ambient Intell. Human Comput. **7**, 37–45 (2016)
22. Namburu, A., Samay, K.S., Edara, R.S.: Soft fuzzy rough set-based MR brain image segmentation. Appl. Soft Comput. **51**, 1–11 (2016)
23. Tseng, T.-L., Huang, C.-C., Fraser, K., Ting, H.-W.: Rough set based rule induction in decision making using credible classification and preference from medical application perspective. Comput. Methods Prog. Biomed. **127**, 273–289 (2016)
24. Liao, S.H., Chang, H.K.: A rough set-based association rule approach for a recommendation system for online consumers. Inf. Proces. Manag. **52**, 1142–1160 (2016)
25. Affonso, C., Sassi, J.R., Barreiros, M.R.: Biological image classification using rough-fuzzy artificial neural network. Expert Syst. Appl. **42**, 9482–9488 (2015)
26. Zhang, L., Li, Y., Sun, C., Nadee, W.: Rough set based approach to text classification. In: Proceedings of the International Conferences on Web Intelligence and Intelligent Agent Technology, vol. 3, pp. 245–252 (2013)
27. Zadeh, A.L., Bojadziev, G., Bojadziev, M.: Fuzzy Sets, Fuzzy Logic, and Fuzzy Systems. World Scientific Publishing, USA (1996)
28. Dubey, K.Y., Mushrif, M.M., Mitra, K.: Segmentation of brain MR images using rough set based intuitionistic fuzzy clustering. Biocybern. Biomed. Eng. **36**(2), 413–426 (2016)
29. Phophalia, A., Rajwade, A., Mitra, K.S.: Rough set based image denoising for brain MR images. Signal Process. **103**, 24–35 (2014)
30. Luan, X., Li, Z., Liu, T.: A novel attribute reduction algorithm based on rough set and improved artificial fish swarm algorithm. Neurocomputing **174**, 522–529 (2016)
31. Yang, Y., Chen, D., Wang, H., Tsang, E., Zhang, D.: Fuzzy rough set based incremental attribute reduction from dynamic data with sample arriving. Fuzzy Sets Syst. **312**, 66–86 (2017)
32. Li, H., Li, D., Zhai, Y., Wang, S., Zhang, J.: A novel attribute reduction approach for multi-label data based on rough set theory. Inf. Sci. **367–368**, 827–847 (2016)
33. Sert, E.: A new modified neutrosophic set segmentation approach. Comput. Electr. Eng. 1–17 (2017)
34. Guoa, Y., Sengr, A.: A novel image segmentation algorithm based on neutrosophic similarity clustering. Appl. Soft Comput. **25**, 391–398 (2014)
35. Sengr, A., Guoa, Y.: Color texture image segmentation based on neutrosophic set and wavelet transformation. Comput. Vision Image Underst. **115**, 1134–1144 (2011)

36. Abdelhakim, A., Saleh, H., Nassar, A.: A quality guaranteed robust image watermarking optimization with artificial Bee colony. Expert Syst. Appl. **72**, 1–10 (2016)
37. Khalili, M.: DCT-Arnold chaotic based watermarking using JPEG-YCbCr. Optik **126**, 4367–4371 (2015)
38. Petitcolas, F., Anderson, R., Kuhn, M.: Attacks on copyright marking systems. In: Proceedings of Second Workshop on Information Hiding, Lecture Notes in Computer Science, vol. 1525, pp. 218–238. Springer (1998)
39. Ghadi, M., Laouamer, L., Nana, L., Pascu, A.: A novel zero-watermarking approach of medical images based on Jacobian matrix model. Secur. Commun. Netw. **9**, 5203–5218 (2016)
40. Wang, X., Wang, C., Yang, H., Niu, P.: A robust blind color image watermarking in quaternion fourier transform domain. J. Syst. Softw. **86**, 255–277 (2013)
41. Lu, J., Shen, L., Xu, C., Xu, Y.: Multiplicative noise removal in imaging: an exp-model and its fixed-point proximity algorithm. Appl. Comput. Harmon. Anal. **41**, 518–539 (2015)

CB_pF-IQA: Using Contrast Band-Pass Filtering as Main Axis of Visual Image Quality Assessment

Jesús Jaime Moreno-Escobar, Claudia Lizbeth Martínez-González, Oswaldo Morales-Matamoros and Ricardo Tejeida-Padilla

Abstract Our proposal is to present a Blind and Reference Image Quality Assessment or CBPF-IQA. Thus, the main proposal of this paper is to propose an Interface, which contains not only a Full-Reference Image Quality Assessment (IQA) but also a No-Reference or Blind IQA applying perceptual concepts by means of Contrast Band-Pass Filtering (CBPF). Then, this proposal consists, in contrast, a degraded input image with the filtered versions of several distances by a CBPF, which computes some of the Human Visual System (HVS) variables. If CBPF-IQA detects only one input, it performs a Blind Image Quality Assessment, on the contrary, if CBPF-IQA detects two inputs, it considers that a Reference Image Quality Assessment will be computed. Thus, we first define a Full-Reference IQA and then a No-Reference IQA, which correlation is important when is contrasted with the psychophysical results performed by several observers. CBPF-IQA weights the Peak Signal-to-Noise Ratio by using an algorithm that estimates some properties of the Human Visual System. Then, we compare CB_pF-IQA algorithm not only with the mainstream estimator in IQA and PSNR but also state-of-the-art IQA algorithms, such as Structural SIMilarity (SSIM), Mean Structural SIMilarity (MSSIM), and Visual Information Fidelity (VIF). Our experiments show that the correlation of CBPF-IQA correlated with PSNR is important, but this proposal does not need imperatively the reference image in order to estimate the quality of the recovered image.

1 Introduction and Problem Statement

The evolution of sophisticated Models and applications of Processing of Digital Images gives as a result of extensive and important literature describing several methodologies or algorithms. An important number of these works are dedicated

J.J. Moreno-Escobar (✉) · C.L. Martínez-González · O. Morales-Matamoros · R. Tejeida-Padilla
ESIME Zacatenco, Instituto Politécnico Nacional, Mexico City, Mexico
e-mail: jemoreno@esimez.mx

A.E. Hassanien and D.A. Oliva (eds.), *Advances in Soft Computing and Machine Learning in Image Processing*, Studies in Computational Intelligence 730,
https://doi.org/10.1007/978-3-319-63754-9_29

to methodologies for improving, in some cases, only the image appearance. Nevertheless, we cannot consider that the digital image quality have reach the perfection. We can consider that Natural Images have presumably been distorted during a certain process of codification or representation. Thus, it is important in the representation process of some kind of images improving image quality in order to recognize and obtain a measure of the degree of degradation or quality of a natural digital image.

Today, MSE or Mean Square Error is yet the most used objective metrics, since many other algorithms which evaluate image quality are based on MSE, Peak Signal-to-Noise Ratio (PSNR), for instance. Wang and Bovik [1, 2] consider that MSE is a poor assessment to be employed in systems that predict image quality or in terms of fidelity. So, we want to mention what is probably wrong with respect to MSE estimations. Thus, we would be in conditions to analyze and to propose a different algorithm that introduces some properties of the human eye into the MSE algorithm, maintaining the best properties of the MSE.

Thus, the main proposal of this work is to propose a methodology implemented in an Interface, which contains not only a classical algorithm of Full-Referenced Image Quality Assessment (IQA) but also a No-Referenced or Blind IQA applying some perceptual features of the Human Visual System (HVS) employing a Contrast Band-Pass Filtering (CB_pF-IQA). In the particular case of Blind IQA the main goal is not to present another Blind IQA, our main objective is to propose a blind version of PSNR, since it is the most important metric along the history because it is the most used IQA and its implementations can be obtained without reference.

Then, this proposal lies in contrasting a degraded original or perfect image (an example is depicted in Fig. 1) with the filtered versions of several distances by a CB_pF-IQA, which computes some of HVS characteristics, such as contrast assimilation and sensitivity contrast in terms of intensity.

Fig. 1 Example of original image: *Lena image*

If CB_pF-IQA detects only one input, it considers a No-Referenced Image Quality Assessment, on the contrary if CB_pF-IQA detects two inputs, it considers that a Referenced Image Quality Assessment will be estimated. Thus, we first define the Referenced IQA and No-Reference IQA algorithms, and then we contrast their correlation with the psychophysical results applied on several human observers. CB_pF-IQA modifies the well-known Peak Signal-to-Noise Ratio formula, weighting and estimating the assimilation or contrast the original source at different distances. Finally, we perform a comparison of CB_pF-IQA methodology not only with the mainstream estimator in IQA, PSNR or MSE, but also recent IQA algorithms, such as Structural SIMilarity (SSIM), Mean Structural SIMilarity (MSSIM) or Visual Information Fidelity (VIF), for instance. The result in terms of Blind IQA of our experiments demonstrate that CB_pF-IQA (BIQA) is highly correlated with the answer or response of PSNR, but this proposal does not need mandatory the reference image in order to computes the distortion of the recovered image.

2 Definition of Image Quality Assessment

In this section we a brief review of IQA definition, then, we divide the IQA algorithms in two: Referenced and Non-Referenced approaches, the latter is also known as Blind IQA. Thus, Referenced IQA Metrics can be divided in Bottom-Up and Top-Down Approaches.

Bottom-up approaches for assessing image quality are algorithms those try to emulate well-modeled characteristics of HVS, and the integration of them into the design of algorithms quality evaluation, hopefully, perform similarly as the HVS in the estimation the image quality.

Furthermore, the bottom-up approaches attempt to simulate structural and functional characteristics in HVS which are relevant for the evaluation of image quality assessment. The main goal is to propose algorithms which work alike HVS, at least for attempting assessing of image quality evaluation.

On the contrary, the top-down systems simulate HVS in another way. Top-down algorithms consider HVS as a black box, and only the input-output task is important to considering. A system for assessing image quality from top to bottom can evaluate a little bit different, since it considers the behavior estimation of image quality of an average human observer correctly, namely in some cases this system adapt the results with linear or nonlinear regression.

An important and some times obvious task for the proposition of an algorithm of this type top-down approach is to consider the main challenge of automatic supervised learning, as illustrated in Fig. 2. Thus, HVS is not supervised in order to learn its behavior. The training features are obtained through subjective experiments, where are viewed and evaluated a large number of test images by human observers. Which is way, the main goal is to model the algorithm in a general way giving as a result the minimization of the error between the desired output (subjective assessment) and the model estimation. This is generally a challenge of regression or an approximation function.

By the other hand, no-reference or no-source image quality evaluation is a very difficult task in this field of image quality estimation, but the understanding of the problem is very simple. Somehow, an objective or subjective computational model should assess the quality of any real-world image, without reference to an original image, as HVS does. Thus, this looks like very difficult mission. The quality of an image can be assessed quantitatively without having an methodology of what a good/bad image quality is supposed to be similar. Then, surprisingly, this is a fairly easy assignment for human observers. HSV can easily recognize images with high or good quality when they are contrasted with low or bad quality images, and also the human eye can distinguish what of these two images is good or bad without watching the reference or original image. In addition, humans observers tend to highly correlate with the opinion of other observers. Example of this behavior when the human eye evaluates image quality without comparing the original or reference source, it is very probable that estimates that the image is noisy, fuzzy, or compressed by any image coder, such as JPEG or JPEG2000, for instance. In this way, Figs. 3 and 4 show some examples of JPEG2000 compression, where the recovered images have lower quality than moved and stretched luminance contrast images.

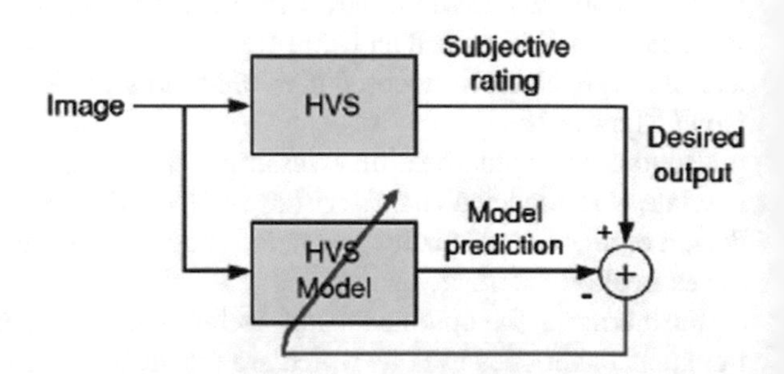

Fig. 2 Learning human visual system

Fig. 3 Baboon image: patches with size = 256 × 256 of recovered images compressed by JPEG2000, PSNR = 32 dB

In future and present of Blind Image Quality Assessment (BIQA), i.e., the objective is to predict the perceptual quality of an image without any prior information of its reference image and distortion type. In this way, we can highlight some research works as following:

1. Wu et al. [3] where they characterize by a new feature fusion scheme and a k-nearest-neighbor (KNN)-based quality prediction model.
2. Li et al. [4] proposed a NR DeBlocked Image Quality (DBIQ) metric by simultaneously evaluating blocking artifacts in smooth regions and blur in textured regions. Their experimental results conducted on the DBID database demonstrate that the proposed metric is effective in evaluating the quality of deblocked images. As an application of this metric is further used for automatic parameter selection in image deblocking algorithms.
3. Lu et al. [5] propose an IQA framework which utilizes minimum amount of structure coefficients to capture the variation of color structure and distortion of degraded image by applying a VPT to remove the visual unperceived coefficients. The difference of the proportion of visual perceived coefficients between distorted and reference image is measured to acquire image quality score.

Some authors have collected and described a survey of existing Blind IQAs by one hand Zhang et al. [6] depict an exhaustive statistical evaluation is conducted to justify the added value of computational saliency in objective image quality assessment, using 20 state-of-the-art saliency models and 12 best-known IQMs. Quantitative results show that the difference in predicting human fixations between saliency models is sufficient to yield a significant difference in performance gain when adding these saliency models to IQMs. By the other hand Kamble et al. [7] describe a survey which includes type of noise and distortions covered, techniques and parameters used by these algorithms, databases on which the algorithms are validated and benchmarking of their performance with each other and also with human visual system.

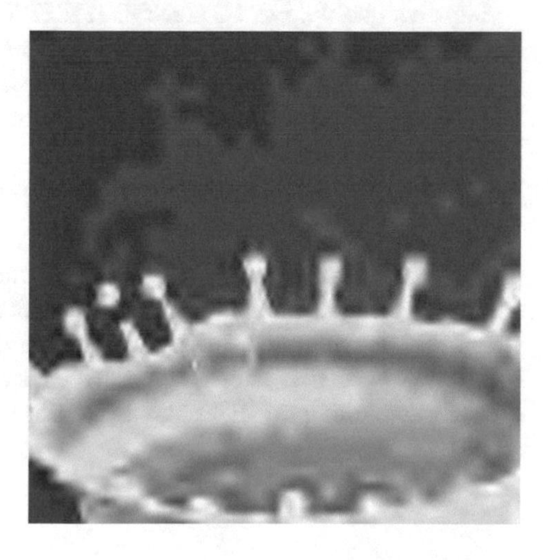

Fig. 4 Splash image: patches with size = 256 × 256 of recovered images compressed by JPEG2000, PSNR = 32 dB

3 MSE Definition

MSE by far is the most important algorithm in the IQA field, which is why we want to define this metric. On one hand, let us define $f(i,j)$ and $\hat{f}(i,j)$ as the couple of images to be compared, which amount of pixels is the size inside each one. Being $f(i,j)$ the original source, considered with best or perfect possible quality or fidelity, and $\hat{f}(i,j)$ a possible degraded estimation of $f(i,j)$, whose quality is subjected to evaluate. On the other hand, let us define first the of the MSE and then the PSNR. Equations 1 and 2 the latter algorithms, respectively.

$$MSE = \frac{1}{l \times m} \sum_{i=1}^{l} \sum_{j=1}^{m} \left[f(i,j) - \hat{f}(i,j) \right]^2 \tag{1}$$

and

$$PSNR = 10 \log_{10} \left(\frac{\alpha^2}{MSE} \right) \tag{2}$$

where α is the maximum value in terms of intensity inside $f(i,j)$, size $= l \times m$. Thus, for images witch contains only one channel, namely 8 bits per pixel (bpp) $\alpha = 2^8 - 1 = 255$. Also, α represents the maximal distortion when the maximal intensity in 8 bpp, 255, is completely degraded, namely a pixel change from 255 to the minimal intensity, 0. Thus, the peak of MSE is $\alpha = 255^2 = 65025$.

For chromatic images, Eq. 2 also defines the estimation of PSNR, but for color images the MSE is separately computed of every component and results of the three channels are averaged.

Both MSE and PSNR are widely employed in the field of image processing, image coding and understanding, because these algorithms have favorable characteristics:

1. Convenient for the purpose of optimizing a certain algorithm that needs to improve quality. For instance in JPEG2000, MSE is employed both in *Optimal Rate Allocation Methodology* [8, 9] and Region of Interest Algorithms [9, 10]. Also, MSE is differentiable and integrable, so its employment could solve these kind of problems in terms of optimization, when it is use along with linear algebra, for instance.
2. By definition MSE compares the square difference of two images, giving as a result a clear meaning of leak of energy.

However, in some cases MSE estimates image quality with a low correlation with quality given by a human observer. A clear example is depicted by Figs. 3 and 4, where both *Baboon* and *Splash* are coded and decoded by JPEG2000 compression with PSNR = 32 dB. Those images have very different visual quality. Then, for this special case, either MSE or PSNR do not correlates with Human Visual System (HVS).

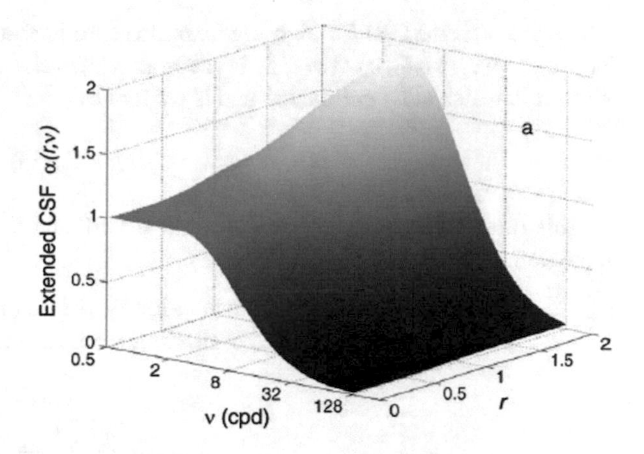

Fig. 5 Representation of the CSF ($\beta_{s,o,i}(r, v)$) for Y channel o illuminate component

4 CB$_p$F-IQA Algorithm

4.1 Contrast Band-Pass Filtering

The *C*ontrast *B*and-*P*ass *F*iltering (CBPF) approximately estimates the image seen by a human observer with a δ separation by filtering some frequencies witch are important o irrelevant for HVS. So, first of all let us define $f(i, j)$ as the mathematical representation of the reference Image and δ as the separation between observer and the screen. Then CBPF estimates a filtered image $\check{f}(i, j)$, when $f(i, j)$ is seen from δ centimeters. CBPF is founded on three main features: frequency of the pixel, spatial scales and surround filtering.

The CBPF methodology decomposes reference image $f(i, j)$ into a set of wavelet planes $\omega(s, o)$ of different spatial scales s (i.e., frequency of the pixel v) and spatial scales as:

$$f(i, j) = \int_{s=1}^{n} \omega(s, o) + c_n \tag{3}$$

where n is the amount of wavelet decompositions, c_n is the plane in the pixel domain and o spatial scale either *v*ertical, *h*orizontal or *d*iagonal.

The filtered image $\check{f}(i, j)$ is recovered by scaling these $\omega(s, o)$ wavelet coefficients employing *Contrast Band-Pass Filtering* function, which is at the same time an approximation Contrast Sensitivity Function (CSF, Fig. 5). The CSF tries to approximate some psychophysical features [11], considering surround filtering information (denoted by r), perceptual frequency denoted by v, which is the gain of frequency either positive or negative depending on δ. Filtered image $\check{f}(i, j)$ is defined by Eq. 4.

$$\check{f}(i, j) = \int_{s=1}^{n} \beta(v, r)\omega(s, o) + c_n \tag{4}$$

where $\beta(\nu, r)$ is the CBPF weighting function reproduce some properties of the HVS. The term $\beta(\nu, r)\ \omega(s, o) \equiv \omega_{s,o;\rho,\delta}$ is the *filtered wavelet coefficients* of image $f(i,j)$ when it is watch at δ centimeters and is written as

$$\beta(\nu, r) = z_{ctr} \cdot C_{\delta}(\dot{s}) + C_{min}(\dot{s}) \tag{5}$$

This function has a shape similar to the e-CSF and the three terms that describe it are defined as

z_{ctr} Nonlinear function and estimation of the central feature contrast relative to its surround contrast, oscillating from zero to one, defined by

$$z_{ctr} = \frac{\left[\frac{\sigma_{cen}}{\sigma_{sur}}\right]^2}{1 + \left[\frac{\sigma_{cen}}{\sigma_{sur}}\right]^2} \tag{6}$$

being σ_{cen} and σ_{sur} the standard deviation of the wavelet coefficients in two concentric rings, which represent a center–surround interaction around each coefficient.

$C_{\delta}(\dot{s})$ Weighting function that approximates to the perceptual e-CSF, emulates some perceptual properties and is defined as a piecewise Gaussian function [12], such as

$$C_{\delta}(\dot{s}) = \begin{cases} e^{-\frac{\dot{s}^2}{2\sigma_1^2}}, \ \dot{s} = s - s_{thr} \leq 0, \\ e^{-\frac{\dot{s}^2}{2\sigma_2^2}}, \ \dot{s} = s - s_{thr} > 0. \end{cases} \tag{7}$$

$C_{min}(\dot{s})$ Term that avoids $\alpha(\nu, r)$ function to be zero and is defined by

$$C_{min}(\dot{s}) = \begin{cases} \frac{1}{2}\ e^{-\frac{\dot{s}^2}{2\sigma_1^2}}, \ \dot{s} = s - s_{thr} \leq 0, \\ \frac{1}{2}, \quad \dot{s} = s - s_{thr} > 0. \end{cases} \tag{8}$$

taking $\sigma_1 = 2$ and $\sigma_2 = 2\sigma_1$. Both $C_{min}(\dot{s})$ and $C_{\delta}(\dot{s})$ depend on the factor s_{thr}, which is the scale associated to 4cpd when an image is observed from the distance δ with a pixel size SZ_p and one visual degree, whose expression is defined by Eq. 9. Where s_{thr} value is associated to the e-CSF maximum value.

$$s_{thr} = \log_2\left(\frac{\delta \tan(1^{\circ})}{4\ SZ_p}\right) \tag{9}$$

Fig. 6 Perceptual images obtained by CBPF with $\delta = 30$ cm

Fig. 7 Perceptual images obtained by CBPF with $\delta = 100$ cm

Figures 6, 7, and 8 depicts three examples of CBPF images of *Lenna* (Fig. 1), calculated by Eq. 4 for a 19 inch monitor with 1280 pixels of resolution in the horizontal, at $\delta = \{30, 100, 200\}$ centimeters. Those figures can show that the higher distance the more distorted the image is.

4.2 General Methodology

Algorithm 1 shows the main methodology of this proposal. Thus, CB_pF-IQA Algorithm estimates the referenced visual quality of the distorted image $\hat{f}(i,j)$ regarding $f(i,j)$ the original reference image, if it exists, otherwise CB_pF-IQA estimates a blind

Fig. 8 Perceptual images obtained by CBPF with $\delta = 200$ cm

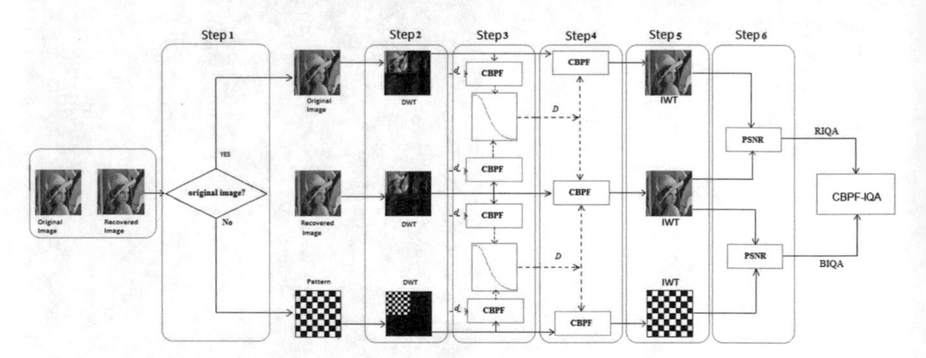

Fig. 9 General explanation of the CB_pF-IQA algorithm, which contains both RIQA and BIQA subprocess

visual image quality. Both algorithms need the definition of the Observational Distance d given by the observer, so if d is not defined, we estimate the distance d from the actual observer by means of 3D/stereoscopic methodology, Algorithm 3. The main algorithm is also presented in Fig. 9.

Then a full-reference image quality metric is performed, there is an reference image $f(i,j)$ and a recovered presumably distorted version $\hat{f}(i,j) = \theta[f(i,j)]$ that is contrasted against $f(i,j)$. It is important to mention θ is the algorithm that distorts the reference image and henceforth we refer the Full-reference image quality algorithm in CB_pF-IQA as RIQA. Otherwise, in the no-referenced image quality issue we refer CB_pF-IQA as BIQA. Furthermore, it is important to mention that BIQA only processes a degraded version of $\hat{f}(i,j)$. Thus from Fig. 10a and b, we compare $\hat{f}(i,j)$ against a repetitive pattern $Y([0,1;1,0])$. Then, we perform the same algorithm in RIQA.

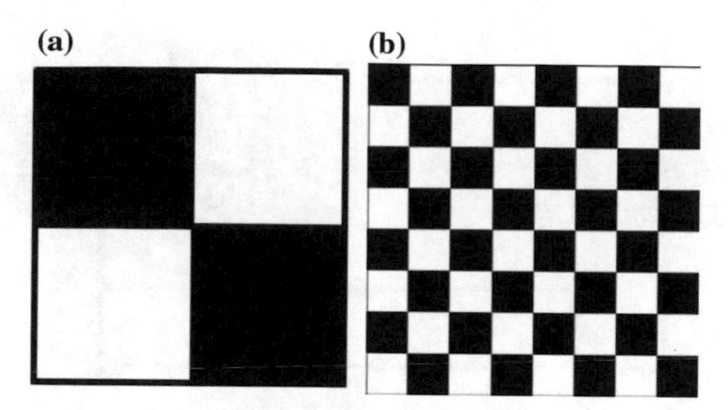

Fig. 10 **a** Primary pattern [0,1;1,0] or Y. **b** Sixteenth pattern or Y^{16}

Algorithm 1: CB_pF-IQA: Framework to assess the quality of a digital image.

Input: $f(i,j), \hat{f}(i,j)$, and δ
Output: *ImageQuality*
1 **if** *d does not exist* **then**
2   d=Compute Observational Distance by means of 3D/stereoscopic approach in Algorithm 3.
3 **if** $f(i,j)$ *exists* **then**
4   ImageQuality = Algorithm $2(f(i,j), \hat{f}(i,j), \delta)$, Referenced-IQA
5 **else**
6   Estimate $f(i,j)$ from a pattern of the same size of $\hat{f}(i,j)$, Fig. 10
7   ImageQuality = Algorithm $2(f(i,j), \hat{f}(i,j), \delta)$, Blind-IQA

Since both $f(i,j)$ and $\hat{f}(i,j)$ are observed at the same time at an observational distance δ, if the similarity between $f(i,j)$ and $\hat{f}(i,j)$ appcars to be better perceived is because δ tends to 0. In contrast, if the observer judges $f(i,j)$ and $\hat{f}(i,j)$ when δ tends to ∞ the correlation between reference and distorted image would be the same. As any algorithm we need to approximate the $\delta = \infty$, namely where similarity is so big that the observer confuse both images, we propose a nonlinear regression for approximating ∞ to $\delta = \Delta$.

Either Reference Assessment or Blind Assessment, our proposal is based on Algorithm 2.

Algorithm 2: Algorithm for estimating Visual Image Quality Assessment.

Input: $f(i,j), \hat{f}(i,j)$, and δ
Output: *ImageQuality*
1 Direct Wavelet Transformation of images $f(i,j)$ and $\hat{f}(i,j)$
2 Estimation of Distance Δ (Eq. 10), The distance where the observer cannot distinguish any difference in terms of quality between $f(i,j)$ and $\hat{f}(i,j)$
3 Compute $f_p(i,j)$ and $\hat{f}_p(i,j)$, namely, contrast band-pass filtered wavelet coefficients at a distance Δ. Where $\omega_{(s,o;\rho,\Delta)} = \text{CBPF}(f(i,j), \Delta)$ and $\hat{\omega}_{(s,o;\rho,\Delta)} = \text{CBPF}(\hat{f}(i,j), \Delta)$
4 Inverse Wavelet Transformation of $\omega_{(s,o;\rho,\Delta)}$ and $\hat{\omega}_{(s,o;\rho,\Delta)}$ obtaining the contrast band-pass filtered images $f_\rho(i,j)$ and $\hat{f}_\rho(i,j)$, rerspectively.
5 *ImageQuality* =PSNR between contrast band-pass filtered images $f_\rho(i,j)$ and $\hat{f}_\rho(i,j)$.

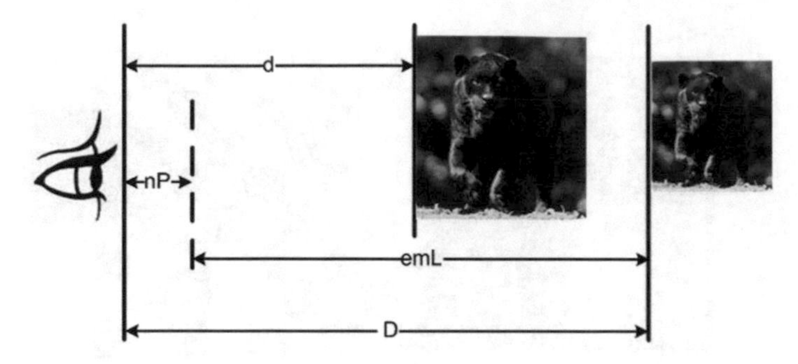

Fig. 11 Portrayal of distances employed by the CB_pF-IQA algorithm. D and $n\mathcal{P}$ graphical representation

Fig. 12 Inside an $\varepsilon\mathcal{R}$ chart

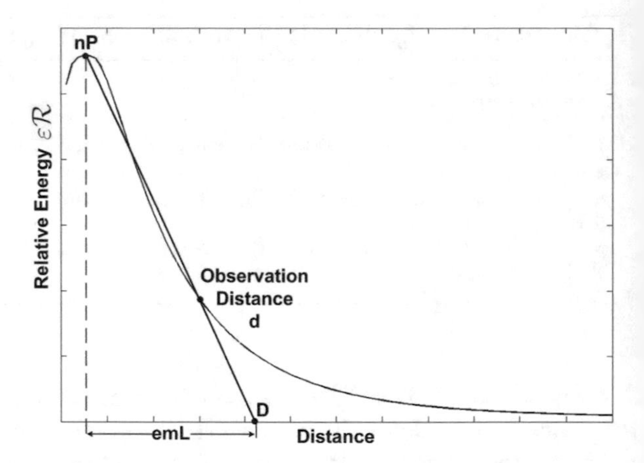

$n\mathcal{P}$ and $\varepsilon m\mathcal{L}$ are two features involved in the evaluation of Distance $\varDelta$. Equation 10 show the estimation of $\varDelta$, besides these two parameters it is important to know or estimate also δ in order to figure out the $n\mathcal{P}$ and $\varepsilon m\mathcal{L}$ distances. Furthermore Figs. 11 and 12 depict the Wavelet Energy Loss or $\varepsilon\mathcal{R}$, which shows not only the behavior of the relative energy but also the significance of $\varDelta$, $n\mathcal{P}$ and $\varepsilon m\mathcal{L}$ inside an $\varepsilon\mathcal{R}$ chart (a).

$$\mathcal{D} = n\mathcal{P} + \varepsilon m\mathcal{L} \tag{10}$$

Furthermore Fig. 12 also show that the pinnacle inside the function is $n\mathcal{P}$, which is describe for the eye specialist as Near Point, which is between 15 to 20 cm for an adult. Thereby, $n\mathcal{P}$ also can be defined as the distance where human eye can evaluate a pair of images $f(i,j)$ and $\hat{f}(i,j)$. From this point $n\mathcal{P}$, fewer the differences are perceived by the observer, until these differences disappear in the ∞. We find $\varDelta$ by projecting the points $(n\mathcal{P}, \varepsilon\mathcal{R}\,(n\mathcal{P}))$ and $(d, \varepsilon\mathcal{R}\,(d))$ to $(\varDelta, 0)$.

4.3 Estimation of the Observational Distance δ

Estimation of the Observational Distance δ is based on Algorithm 3, which divided in six steps and is described as follows:

Step 1: Camera calibration by means of Function *Stereo Calibration*. Calibration Results are stored in a structure, which is defined as *stereoParams*.
Step 2: Ones both left and right cameras are calibrated, we take two images I_l and I_r.
Step 3: With the parameters defined in *stereoParams*, we calibrate both I_l and I_r images using *undistortImage* function, giving as a result Ic_l and Ic_r.
Step 4: In both Ic_l and Ic_r images we estimate two human characteristics: face and eyes detection. This detection is made by means of the function *vision. Cascade-ObjectDetector*. If in both Ic_l and Ic_r images are detected faces, then we detect eyes. This procedure increases the probability to find the head of the observer in the stereo-pair.
Step 5: We estimate the center of the heads located both in Ic_l and Ic_r images.
Step 6: Finally, we estimated the distance δ in centimeters between cameras and the observer, using the function *triangulate*.

It is important to mention that all functions employed in this methodology are toolboxes of MatLab R2015a.

Algorithm 3: Estimation of the Observational Distance δ.

Input: void
Output: δ
1 Camera Calibration
2 Taking stereo-pair I_l and I_r
3 Calibration of the stereo-pair, Ic_l and Ic_r
4 Head detection in the stereo-pair
5 Center of detected Heads in original stereo-pair
6 Estimation of distance δ between cameras and the observer

5 Experimental Results

5.1 Referenced Image Quality Assessment

MSE, PSNR [13], SSIM, VIFP [14], MSSIM [15], VSNR [16], VIF [17], UQI [18], IFC [19], NQM [20], WSNR [21] and SNR are compared against the performance of CB$_p$F-IQA for JPEG2000 compression distortion. We chose for evaluating these assessments the implementation provided in [22], since it is based on the parameters proposed by the author of each indicator.

Table 1 KROCC of RIQA and other quality assessment algorithms on multiple image databases using JPEG2000 distortion. The higher the KROCC the more accurate image assessment. Bold and italicized entries represent the best and the second best performers in the database, respectively. The last column shows the KROCC average of all image databases

Metrics	Image database				
	TID2008	LIVE	CSIQ	IVC	All
Images	100	228	150	50	528
IFC	0.7905	0.7936	0.7667	0.7788	0.7824
MSE	0.6382	0.8249	0.7708	0.7262	0.7400
MSSIM	*0.8656*	*0.8818*	0.8335	*0.7821*	*0.8408*
NQM	0.8034	0.8574	0.8242	0.6801	0.7913
PSNR	0.6382	0.8249	0.7708	0.7262	0.7400
SNR	0.5767	0.8055	0.7665	0.6538	0.7006
SSIM	0.8573	0.8597	0.7592	0.6916	0.7919
UQI	0.7415	0.7893	0.6995	0.6061	0.6602
VIF	0.8515	0.8590	0.8301	0.7903	0.8327
VIFP	0.8215	0.8547	*0.8447*	0.7229	0.8110
VSNR	0.8042	0.8472	0.7117	0.6949	0.7645
WSNR	0.8152	0.8402	0.8362	0.7656	0.8143
RIQA	**0.8718**	**0.8837**	**0.8682**	**0.7981**	**0.8555**

Table 1 shows the performance of RIQA and the other 12 image quality assessments across the set of images from TID2008, LIVE, CSIQ, and IVC image databases employing Kendall Rank-Order Correlation Coefficient (KROCC) for testing the distortion produced by a JPEG2000 compression.

Thus, for JPEG2000 compression distortion, RIQA is getting the best results in all databases. RIQA correlates in 0.8837 for a database of 228 images of the LIVE database. On the average, RIQA algorithm is also correlates in 0.8555, using KROCC. Furthermore, JPEG2000 compression distortion, MSSIM is the second best indicator not only for TID2008, LIVE, and IVC image databases but also on the average, since VIFP occupies second place for CSIQ image database. Thus, the correlation between the opinion of observers and the results of MSSIM is 0.0143 less than the ones of RIQA. So in general, we can conclude that PSNR can be improved its performance in 11.5% if it includes four steps of filtering, RIQA.

Table 2 shows the performance of RIQA and the other twelve image quality assessments across the set of images from TID2008, LIVE, CSIQ and IVC image databases employing KROCC for testing the distortion produced by a JPEG compression.

Table 2 also shows an average performances for the 534 images of the cited image databases. Bold and Italicized represent the best and the second best performance assessment, respectively. It is appropriate to say that RIQA is the best performer both in each image database and average of them. MSSIM is the second best-ranked

Table 2 KROCC of RIQA and other quality assessment algorithms on multiple image databases using JPEG distortion. The higher the KROCC the more accurate image assessment. Bold and italicized entries represent the best and the secondbest performers in the database, respectively. The last column shows the KROCC average of all image databases

Metrics	Image database				
	TID2008	LIVE	CSIQ	IVC	All
Images	100	234	150	50	534
MSE	0.7308	0.7816	0.6961	0.5187	0.6818
PSNR	0.7308	0.7816	0.6961	0.5187	0.6818
SSIM	0.7334	0.8287	0.7529	0.6303	0.7363
MSSIM	*0.7580*	*0.8435*	0.8097	0.7797	*0.7977*
VSNR	0.7344	0.8149	0.7117	0.5827	0.7109
VIF	0.7195	0.8268	*0.8287*	0.7911	0.7915
VIFP	0.7004	0.8140	0.8188	0.6763	0.7524
UQI	0.5445	0.7718	0.6990	0.6254	0.6602
IFC	0.5909	0.7767	0.7644	*0.8158*	0.7369
NQM	0.7142	0.8269	0.7907	0.6664	0.7495
WSNR	0.7300	0.8181	0.8020	0.6959	0.7615
SNR	0.6035	0.7735	0.6942	0.4481	0.6298
CB_pF-IQA	**0.7616**	**0.8457**	**0.8473**	**0.8335**	**0.8220**

metrics not only in all databases but also on the average, except for the CSIQ database, where VIF has this place. RIQA is better 0.0243 than MSSIM and improves the performance of PSNR or MSE by 0.1402 for JPEG compression degradation.

5.2 *Blind Image Quality Assessment*

Some metrics estimate Quality as PSNR does, but some metrics estimates degradation, MSE, for instance. It is important to mention that BIQA estimates the degradation. This degradation tends to zero means that the overall quality is getting better. We already check the behavior of RIQA, so in this section we develop comparisons for verifying the performance BIQA by comparing significance performance of different compress versions of the image *Baboon*. BIQA is a metric that gives decibels as PSNR does, so instead of employing a Non-Parametric Correlation, we use a parametric correlation coefficient, i.e., Pearson correlation coefficient in order to better compare the results between BIQA and PSNR.

Figure 13 depicts three JPEG2000 distorted versions of the image *Lenna* with 0.05 (Fig. 13a), 0.50 (Fig. 13b) and 1.00 (Fig. 13c) bits per pixel. PSNR estimates 23.41, 32.74 and 34.96 dB, respectively. While CB_pF-IQA computes 48.42, 36.56 and 35.95 dB, respectively. Thus, both PSNR and CB_pF-IQA estimate that image at 1.00 bpp has lower distortion. When this experiment is extended computing the

Fig. 13 JPEG2000 Distorted versions of color image *Lenna* at different bit rates expressed in bits per pixel (bpp). **a** High distortion, **b** medium distortion and **c** low distortion

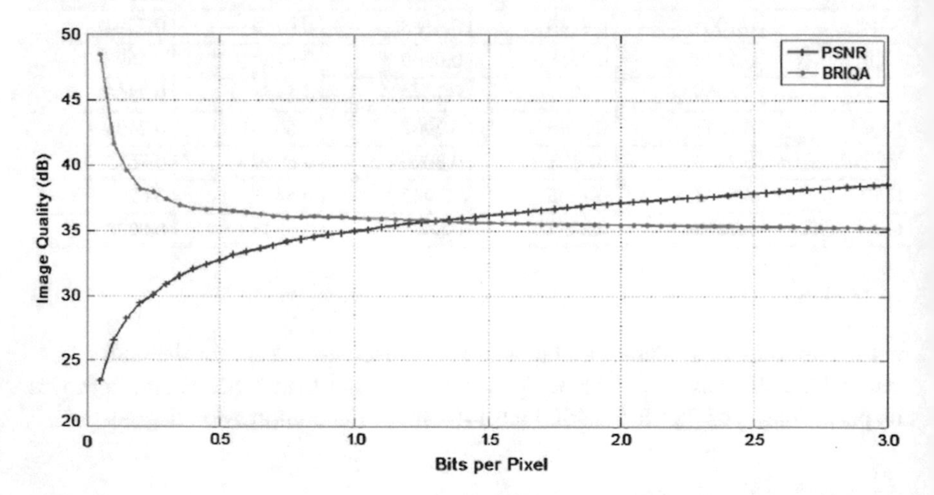

Fig. 14 Comparison of PSNR and CB_pF-IQA for the JPEG2000 distorted versions of image *Lenna*

JPEG2000 distorted versions from 0.05 to 3.00 bpp (increments of 0.05 bpp, depicted at Fig. 14), we found that the correlation between PSNR and CB_pF-IQA is 99.32 image *Lenna* for every 10,000 estimation CB_pF-IQA misses only in 68 assessments.

Figure 15a, b, and c depict three JPEG2000 compression of the image *Baboon* with 0.05, 0.50, and 1.00 bits per pixel, respectively. Thereby PSNR estimates 18.55 dB for Fig. 15a, 23.05 dB for Fig. 15b, and 25.11 dB Fig. 15c. While BIQA computes 43.49, 30.07 and 28.71 dB, respectively. Thus for the 0.05 bpp (Fig. 15a), higher distortion is estimated both PSNR and BIQA.

Figure 16 depicts multiple JPEG2000 decoded images from 0.05 to 3.00 bpp, the increments of varies every 0.05 bpp. With the later data we can found that PSNR and BIQA between them is 0.9695, namely, for image *Baboon* for every 1000 tests BIQA estimates in a wrong way only 30 assessments.

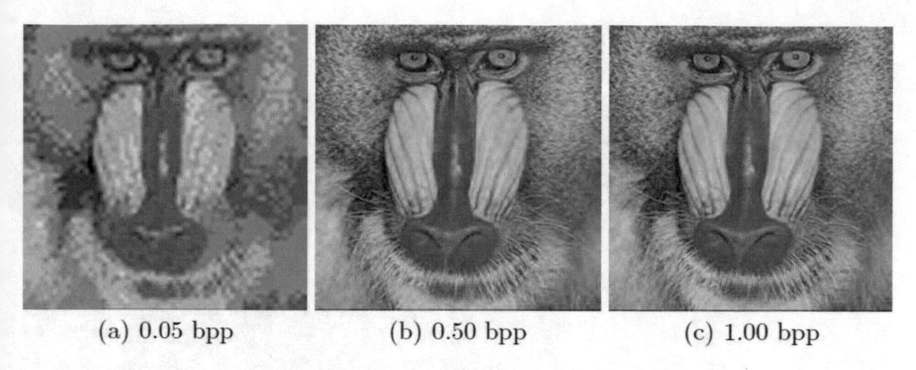

Fig. 15 JPEG2000 Distorted versions of color image *Baboon* at different bit rates expressed in bits per pixel (bpp). **a** High distortion, **b** medium distortion and **c** low distortion

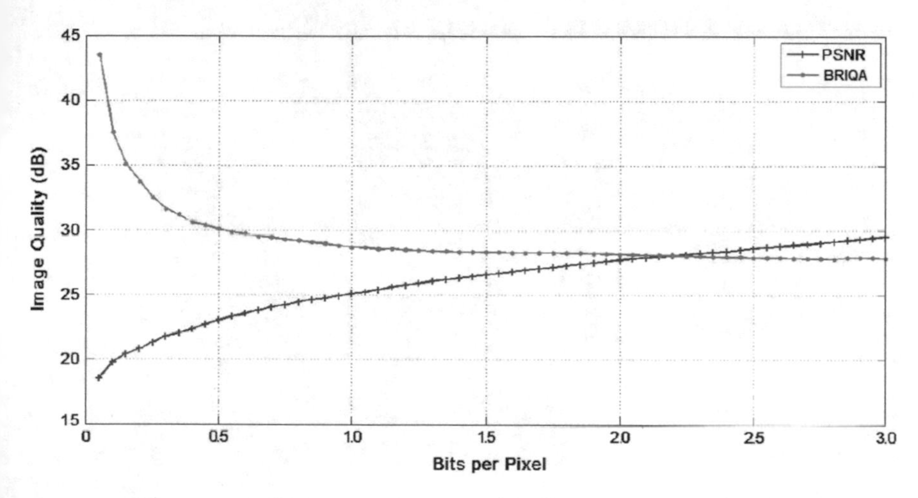

Fig. 16 Comparison of PSNR and CB_pF-IQA for the JPEG2000 distorted versions of image *Baboon*

5.3 CB_pF-*IQA Interface*

In Fig. 17 the graphic interface is shown that allows upload pictures and calculate their quality by using methods with and without reference which are selectable via a drop-down menu. The observer can also select the type of distance δ used the code which is the distance from the screen to the face of the observer.

In the Fig. 18, it shows that selecting the metric referenced by the drop-down menu, you can load the original images (without compression) and distorted (noisy) using the buttons to load image which display a window that lets you explore folders and select the image. Pressing the button Calculate the Legend RIQA method allowing the return a numeric result associated with the image quality is applied (Fig. 19).

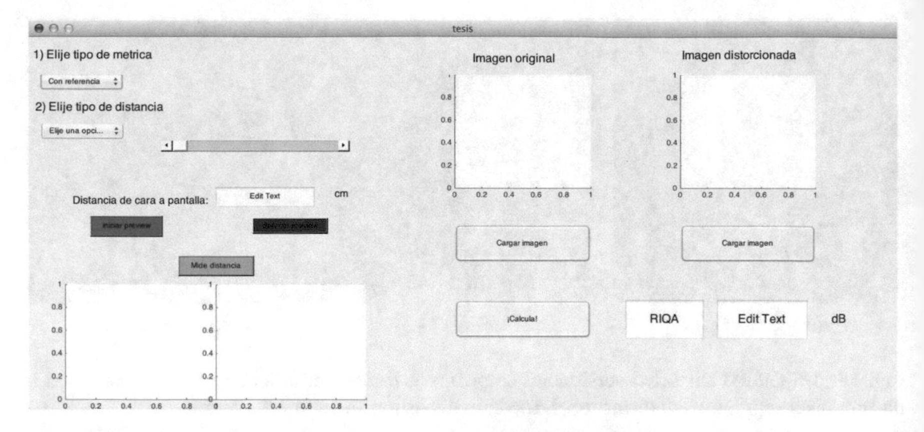

Fig. 17 Estimation of distance: static distance CB_pF-IQA graphic interface

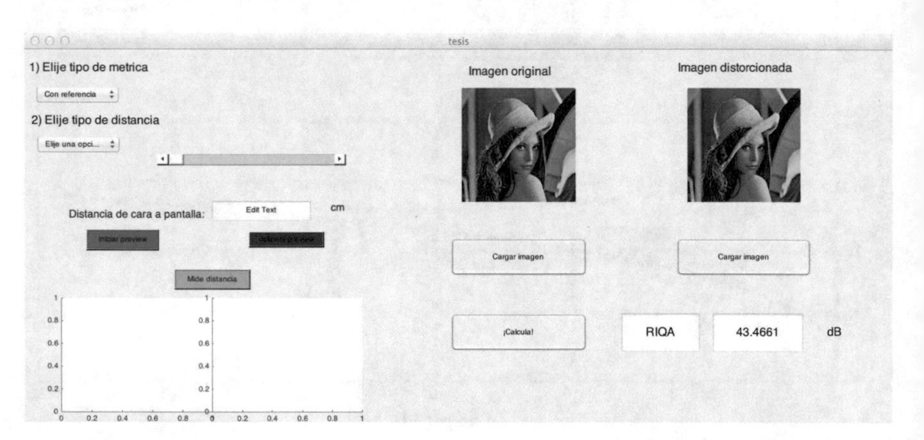

Fig. 18 Estimation of distance: static distance. Referenced image quality assessment

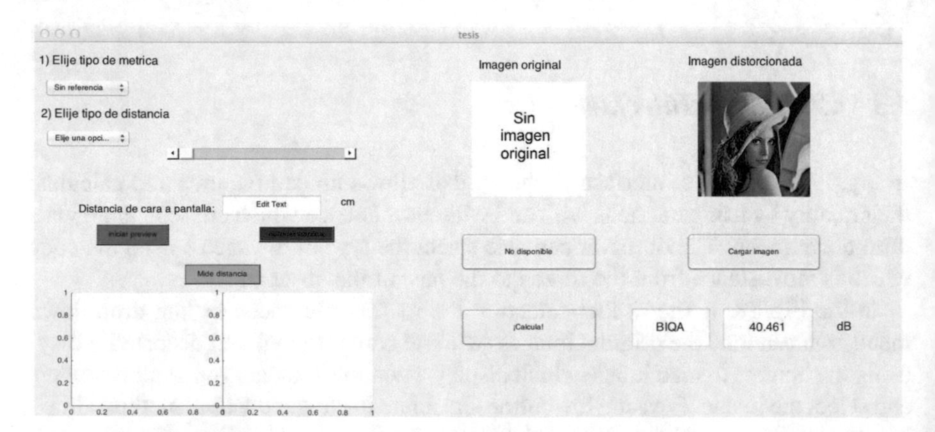

Fig. 19 Estimation of distance: static distance. Blind image quality assessment

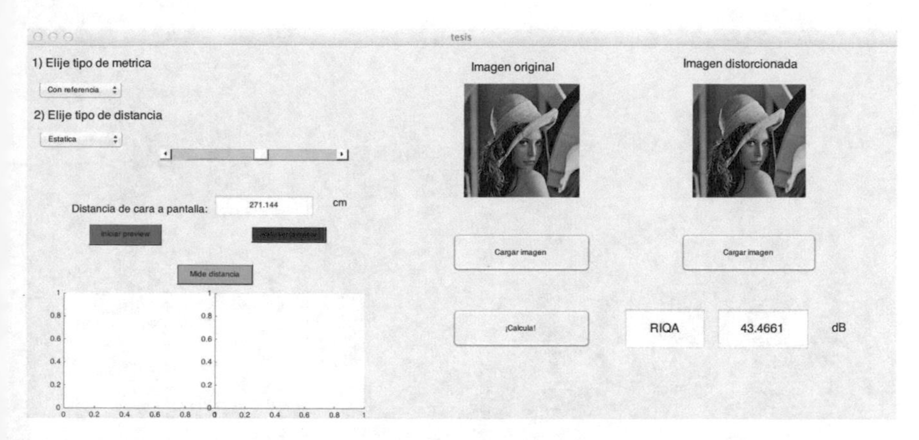

Fig. 20 Estimation of distance: dynamic distance. Selecting the distance

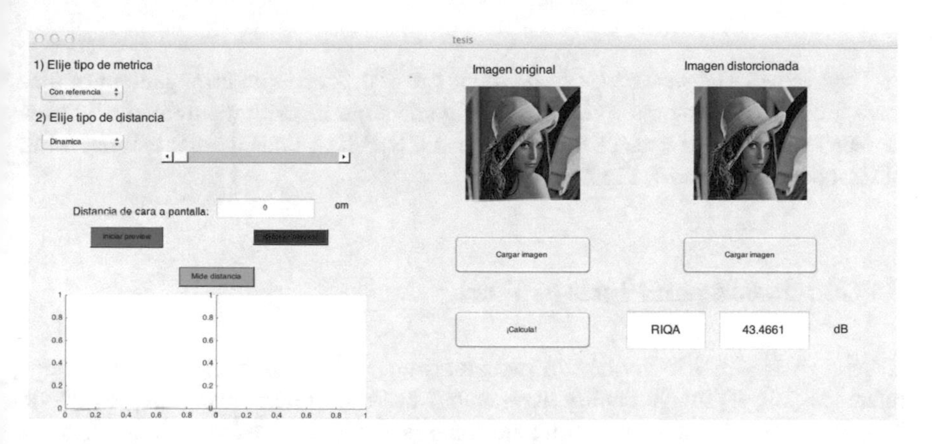

Fig. 21 Estimation of distance: dynamic distance. Taking stereo-pair $I(r, l)$, i.e., I_r and I_l images

Figure 18 shows that selecting the metric without reference automatically displays a window with the caption: *No original image* and the button to load the original image now is disabled.

By selecting the metric without reference *Calculate* button, the algorithm automatically switches to a method without reference to return a numeric value associated with the image quality, namely BIQA.

When selecting the type of distance in static mode, it is possible to move the slider that lets you change the value of the distance used in the algorithms metrics with and without reference. By selecting the distance of dynamic type green buttons (enabled *preview*), show preview (blue) stop preview (red) are enabled allowing handling. Otherwise slider is disabled, Fig. 20.

When the *Preview button* is pressed stereo-cameras transmit the images to the computer and it is displayed on the screen, Fig. 21.

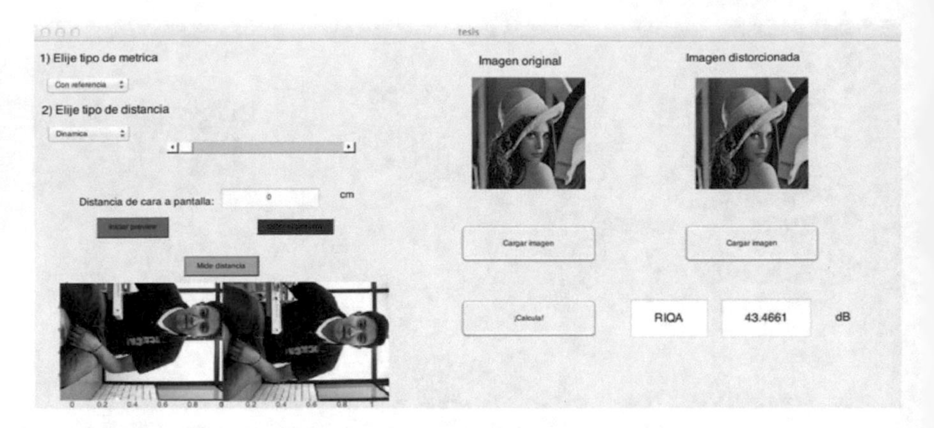

Fig. 22 Estimation of distance: dynamic distance. Dynamic estimation of distance δ

Thus, When *Measure distance button* is pressed, it takes an arrangement of pictures from stereo-cameras. With this arrangement our algorithm automatically tries to detect the face and eyes of the observer if CB_pF-IQA finds them it estimates the observational distance δ, Fig. 22.

6 Conclusions and Future Work

CB_pF-IQA is a metric divided in two algorithms full-reference (RIQA) and non-reference (BIQA) image quality assessments based on filtered weighting of PSNR by using a model that tries to simulate some features of the Human Visual System (CBPF model). Both proposed metrics in CB_pF-IQA are based on five steps.

When we compared RIQA Image Quality Assessment against several state-of-the-art metrics our experiments gave us as a result that RIQA was the best-ranked image quality algorithm in the well-known image databases such as TID2008, LIVE, CSIQ and IVC, JPEG2000 compression algorithm is used as a method of distorting the cited image databases. Thus, it is 2.5 and 1.5% better than the second best performing method, MSSIM. On average, RIQA improves the results of PSNR in 14% and 11.5% for MSE.

In the Blind Image Quality Assessment, BIQA assessment correlates almost perfect for JPEG2000 distortions, since difference between BIQA and PSNR, on the average is only 0.0187.

Combine both RIQA and BIQA in the same interface was the main contribution of this work. Thus, a expert or non-expert in the quality images assessment field can perform its own experiments. These experiments could include dynamic quality estimations or static ones. As a future work of this paper could be to include a set of quality images estimators including RIQA and BIQA.

Acknowledgements This work is supported by National Polytechnic Institute of Mexico (Instituto Politécnico Nacional, México) by means of Project No. SIP-20171179, the Academic Secretary and the Committee of Operation and Promotion of Academic Activities (COFAA) and National Council of Science and Technology of Mexico (CONACyT).
It is important to mention that Sects. 4 and 5 are part of the degree thesis supported by Eduardo García and Yasser Sánchez.

References

1. Wang, Z., Bovik, A.: Mean squared error: Love it or leave it? a new look at signal fidelity measures. Signal Proces. Mag. IEEE **26**(1), 98–117 (2009)
2. Wang, Z., Bovik, A.C.: Modern Image Quality Assessment, 1st edn. Synthesis Lectures on Image, Video, & Multimedia Processing. Morgan & Claypool Publishers (2006)
3. Wu, Q., Li, H., Meng, F., Ngan, K.N., Luo, B., Huang, C., Zeng, B.: Blind image quality assessment based on multichannel feature fusion and label transfer. IEEE Trans. Circ. Syst. Video Technol. **26**(3), 425–440 (2016)
4. Li, L., Zhou, Y., Lin, W., Wu, J., Zhang, X., Chen, B.: No-reference quality assessment of deblocked images. Neurocomputing **177**, 572–584 (2016). http://www.sciencedirect.com/science/article/pii/S092523121501869X
5. Lu, W., Xu, T., Ren, Y., He, L.: On combining visual perception and color structure based image quality assessment. Neurocomputing **212**, 128–134 (2016), Chinese Conference on Computer Vision 2015 (CCCV 2015). http://www.sciencedirect.com/science/article/pii/S092523121630697X
6. Zhang, W., Borji, A., Wang, Z., Callet, P.L., Liu, H.: The application of visual saliency models in objective image quality assessment: A statistical evaluation. IEEE Trans. Neural Netw. Learn. Syst. **27**(6), 1266–1278 (2016)
7. Kamble, V., Bhurchandi, K.: No-reference image quality assessment algorithms: a survey. Optik—Int. J. Light Electron Opt. **126**(1112), 1090–1097 (2015). http://www.sciencedirect.com/science/article/pii/S003040261500145X
8. Auli-Llinas, F., Serra-Sagrista, J.: Low complexity JPEG2000 rate control through reverse subband scanning order and coding passes concatenation. IEEE Signal Proces. Lett. **14**(4), 251–254 (2007)
9. Taubman, D.S., Marcellin, M.W.: *JPEG2000*: Image Compression Fundamentals, Standards and Practice, ser. Kluwer Academic Publishers (2002). ISBN: 0-7923-7519-X
10. Bartrina-Rapesta, J., Auli-Llinas, F., Serra-Sagrista, J., Monteagudo-Pereira, J.: JPEG2000 arbitrary ROI coding through rate-distortion optimization techniques. In: Data Compression Conference, 25-27 2008, pp. 292 –301
11. Mullen, K.: The contrast sensitivity of human color vision to red-green and blue-yellow chromatic gratings. J. Physiol. **359**, 381–400 (1985)
12. Mullen, K.T.: The contrast sensitivity of human colour vision to red-green and blue-yellow chromatic gratings. J. Physiol. **359**, 381–400 (1985)
13. Huynh-Thu, Q., Ghanbari, M.: Scope of validity of PSNR in image/video quality assessment. Electron. Lett. **44**(13), 800–801 (2008)
14. Sheikh, H., Bovik, A.: Image information and visual quality. IEEE Trans. Image Proces. **15**(2), 430–444 (2006)
15. Wang, Z., Simoncelli, E., Bovik, A.: Multiscale structural similarity for image quality assessment. In: Conference Record of the Thirty-Seventh Asilomar Conference on Signals, Systems and Computers, vol. 2, pp. 1398–1402 (2003)
16. Chandler, D., Hemami, S.: Vsnr: a wavelet-based visual signal-to-noise ratio for natural images. IEEE Trans. Image Proces. **16**(9), 2284–2298 (2007)

17. Wang, Z., Bovik, A., Sheikh, H., Simoncelli, E.: Image quality assessment: from error visibility to structural similarity. IEEE Trans. Image Proces. **13**(4), 600–612 (2004)
18. Wang, Z., Bovik, A.: A universal image quality index. IEEE Signal Proces. Lett. **9**, 81–84 (2002)
19. Sheikh, R., Bovik, A., de Veciana, G.: An information fidelity criterion for image quality assessment using natural scene statistics. IEEE Trans. Image Proces. **14**, 2117–2128 (2005)
20. Damera-Venkata, N., Kite, T., Geisler, W., Evans, B., Bovik, A.: Image quality assessment based on a degradation model. IEEE Trans. Image Proces. **9**, 636–650 (2000)
21. Mitsa, T., Varkur, K.: Evaluation of contrast sensitivity functions for formulation of quality measures incorporated in halftoning algorithms. IEEE Int. Conf. Acoust. Speech Signal Proces. **5**, 301–304 (1993)
22. C.U.V.C. Laboratory: MeTriXMuX Visual Quality Assessment Package. Cornell University Visual Communications Laboratory (2010). http://foulard.ece.cornell.edu/gaubatz/metrix_mux/

Digital Image Watermarking Performance Improvement Using Bio-Inspired Algorithms

Mohamed Issa

Abstract Copyrights protection and ownership of multimedia is a vital task nowadays used in a lot of fields such as broadcasting media. Hence digital media watermarking techniques were developed to embed a watermark image into the original media (image or videos). The watermarking techniques aim to improve the robustness of watermarked image against attacks and increase the impeccability of significant regions of the media. This chapter focuses on explaining the rule of using metaheuristic algorithms for optimizing the robustness and impeccability of image watermarking techniques. This will be discussed through two watermarking techniques one used genetic algorithm optimization and the other use cuckoo search optimization approach.

Keywords Digital watermarking · Bio-inspired algorithms
Genetic algorithm and cuckoo search optimization

1 Introduction

Digital multimedia has been used and distributed profusely in the past few years. Protection of it is a vital issue. Two essential approaches are used for protecting copyrighted data, first cryptography key based approach and the second is watermarking. Cryptography methods guarantee secure transmission of data but if the data de-encrypted then the redistribution and its spread are not under control. The solution of cryptography methods' limitations is using digital watermarking. In other words, it is the operation of hiding data information in an embedded signal to the carrier signal and the core target of using watermarking is identifying ownership and copyrighting. Digital watermarking can be used as distribution ownership proof

M. Issa (✉)
Faculty of Engineering, Computer and Systems Department,
member at Scientific Research Group in Egypt (SRGE),
Zagazig University, Zagazig, Egypt
e-mail: mamohamedali@eng.zu.edu.eg

A.E. Hassanien and D.A. Oliva (eds.), *Advances in Soft Computing and Machine Learning in Image Processing*, Studies in Computational Intelligence 730,
https://doi.org/10.1007/978-3-319-63754-9_30

in business application. Also, modern media player devices such as DVD and Divx are developed so that it can detect the distorted watermark data and refuse to play. Digital watermarking is applied to various domains including fingerprinting secret communications, authentication, copyrights control, and digital signatures. Carrier signal may be text, video, audio, and images so the watermark may be the same kind of it and may be different kinds of watermark are embedded on the carrier signal. Watermarking is used in a lot of application such as copyright protection, authentication of videos, social networks content management, and watermarking broadcasting media. Watermarking process consist of three stages, embedding, attack, and detection. As shown in Fig. 1 the carrier signal (host signal) and the embedded signal (watermark) are input to the embedding function to generate watermarked signal (S) and transmitted or stored (red row). Through sending, it attacks a lot of modifications may distort the original watermarked image and propose different one from the original (SE). The last stage is detecting and retrieving the original signal from distorted one. If the embedded data is not modified during the attacks, it will be retrieved and this case called robustness watermarking. Otherwise, the detection fails to detect the original signals and called fragile watermarking. An example of watermarked image is as shown in Fig. 2.

This chapter focuses on image watermarking. There are two important objectives in image watermarking design are imperceptibility and robustness. Imperceptibility represents indistinguishably difference between original carrier signal and the watermarked one. Robustness represents resistance of watermark against class of transformation of marked image signal.

The digital watermarking can be categorized depending on a lot of kinds as shown in Fig. 3. The first categorization is the domain (spatial of frequency based). In spatial domain, the watermark image is embedded into the carrier image by modifying the pixel values [1, 2]. These techniques required computational efforts and less robust against image processing attacks. For frequency domain techniques, the spatial domain representation is transformed and the frequency coefficients are modified to merge the watermark image. Four categories of transform watermarking techniques are singular value decomposition (SVD) [3], discrete cosine transform (DCT) [4, 5], discrete wavelet transform (DWT) [6, 7], and discrete fourier transform (DFT) [8, 9]. These techniques are more robust against signals attacks and more image imperceptibility but these are more computational consumption than

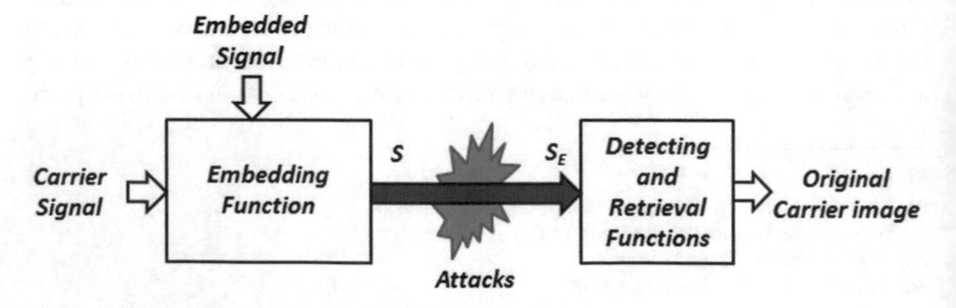

Fig. 1 Watermarking stages (Embedding, Attack and Retrieving)

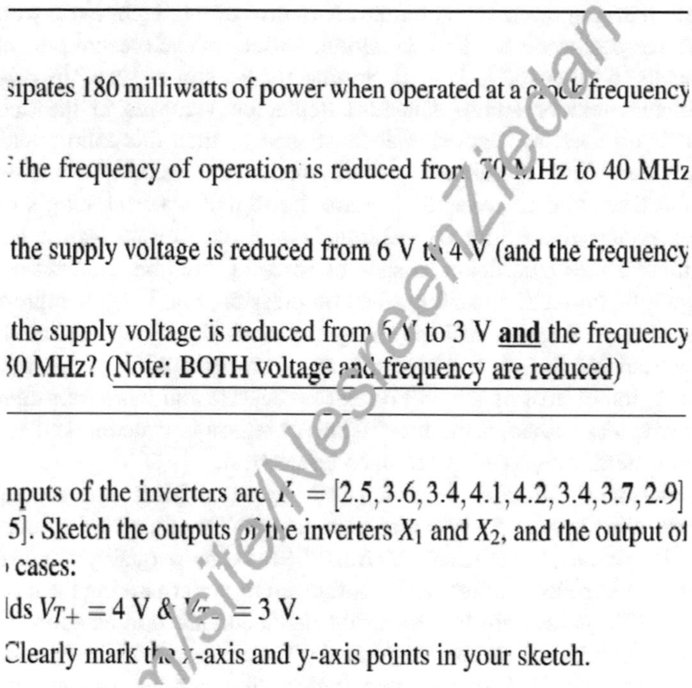

sipates 180 milliwatts of power when operated at a clock frequency

the frequency of operation is reduced from 70 MHz to 40 MHz

the supply voltage is reduced from 6 V to 4 V (and the frequency

the supply voltage is reduced from 6 V to 3 V **and** the frequency
30 MHz? (Note: BOTH voltage and frequency are reduced)

nputs of the inverters are $X_1 = [2.5, 3.6, 3.4, 4.1, 4.2, 3.4, 3.7, 2.9]$
5]. Sketch the outputs of the inverters X_1 and X_2, and the output of
cases:
lds $V_{T+} = 4$ V & $V_{T-} = 3$ V.
Clearly mark the x-axis and y-axis points in your sketch.

Fig. 2 Watermark image example

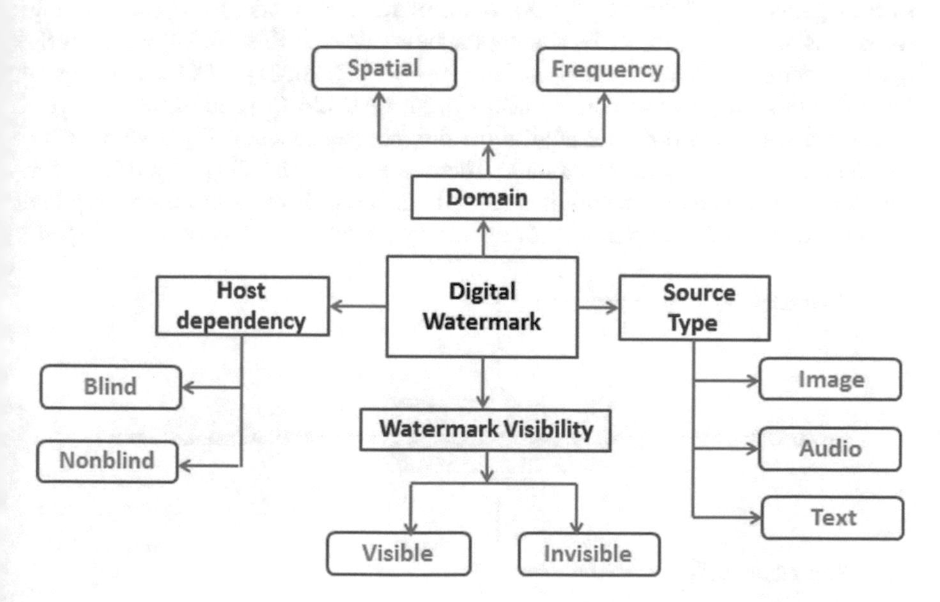

Fig. 3 Digital watermarking categorization

spatial watermarking domain techniques. A hybrid of two techniques were merged to improve the performance of watermarking based on each technique compensate the drawbacks of other [6, 7, 10–12]. Second for the source type, the original data may be image, text or audio. The host dependency means in the detection of watermark if no need for the original (host) source then this kind is called blind otherwise is non-blind. Watermark visibility means the visibility of the watermark.

The robustness and imperceptibility are the digital watermarking's objectives. Robustness means the ability of watermarked data to resist against attacks and modifications. These modifications such as resizing, rotation, contrast adjustment, filtering, gamma correction, scaling, file compression, and lossy compression. The non-perceptibility concerns with covering the most significant part of the image. There is a tradeoff between robustness and non-perceptibility, where the most significant region of host image will be less perceptual and more robustness to hide the watermark. In contrast, if the less important region is watermarked so it is easy to hide the watermark and so it becomes less robust.

The watermark' strength is controlled by modifying the scaling factors (SFs) where the larger the SF of watermark, the quality of the carrier image will be decreased. In contrast, for smaller watermark' SF the more quality of carrier image. Robustness and imperceptibility can be optimized by watermarking the most suitable regions and SF's values are chosen using metaheuristic optimization techniques. Metaheuristic techniques are stochastic methods that try to find the optimal solution using heuristic strategy inspired from nature like particle swarm optimization (PSO) [13], firefly optimization algorithm [14] ant colony optimization (ACO) [15] and bee colony optimization (BCO) [16] or from physical systems like simulated annealing [17], firefly algorithm optimization (FAO) [14] or evolutionary systems such as genetic algorithm (GA) [18]. A lot of research works for optimizing the robustness and impeccability by finding the best values of SFs. [4, 19] were developed for merging GA in watermarking and for PSO [12, 20, 21]. ACO also was used for optimizing image watermarking [22, 23], for FAO [24–26] and BCO [27, 28].

The watermarking does not modify the file (source) functionality it works with the data structure non-perceptibility. There are an embedding algorithms for embedding watermark on the original data (I) may be with secret key for encryption and another extraction algorithms for extracting the watermark as shown in Fig. 4.

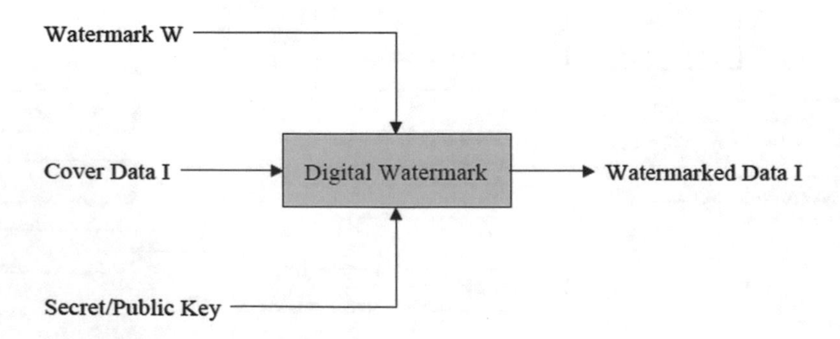

Fig. 4 Inputs and outputs of watermarking operation

The rest of chapter is organized as follows: Sect. 2 presents a techniques for SVD watermarking using GA [29] that explain the rule of using GA for optimizing the watermarking problem. Section 3 show using cuckoo search (CS) optimization techniques for optimizing a watermarking approach hybrid of redundant discrete wavelet transform (RDWT) and SVD. Section 4 proposed a technique used hybrid of chaotic and DNA coding to improve the robustness and non-perceptibility.

2 SVD-GA Watermarking [29]

This work proposed the development of SVD image watermarking techniques using GA to get the maximum robustness without degradation of transparency [29]. The idea is based mainly on modification of singular values of the host images (gray level) to generate invisible watermark image by finding the optimum values of multiple SFs. Figure 5 shows the general flow chart of GA. The populations of GA represent a possible solution by containing a set of SFs and initialized with random values. The robustness of this technique was measured against various attacks such as sharpening, resizing, rotation, and average filtering. Figure 6 summarizes briefly the steps of SVD-GA watermarking technique. The fitness evaluation is computed depending on the robustness (CORRW) and the transparency (CORRI). For CORRW is computed as the correlation between original watermark and extracted watermark. CORRI is computed as the correlation between the watermark image and host image. Equation (1) is the objective function that is used to evaluate the fitness where t is the number of attacks methods and i is the solution number. Tables 1, 2 and 3 show the result of correlations the GA's parameters: Mutation Rate (MR), Cross Rate (CR), and Population Size (PS). The results of $CORR_W$ were studied against the four attacks methods. The results at MR = 0.04 are the optimal result that produce fitness value 142.4. While the fitness at MR values 0.02 and 0.06 are 127.9 and 130.2 respectively. Tables 4, 5 and 6 propose the same test but with variation of PS and the bold values are the optimal with fitness 187.3. The fitness values at PS' values 50 and 100 are 142.4 and 164.3 respectively. As shown in Table 2 with increasing the population size of GA, the fitness value also increases. Various variables of SFs were generating using GA ranging from 1 up to 32 and the values of SFs ranging from −9.3 to 1713. The multiple SFs show the improvement of robustness and transparency over using constant single SF.

$$Fitness_i = \left[\frac{1}{t^{-1} \sum_{i=1}^{t} CORR_W(W, W_i^*)} - CORR_I(I, I_W) \right]^{-1} \quad (1)$$

Figures 7 and 8 show the result of resistance of the proposed technique against various attacks.

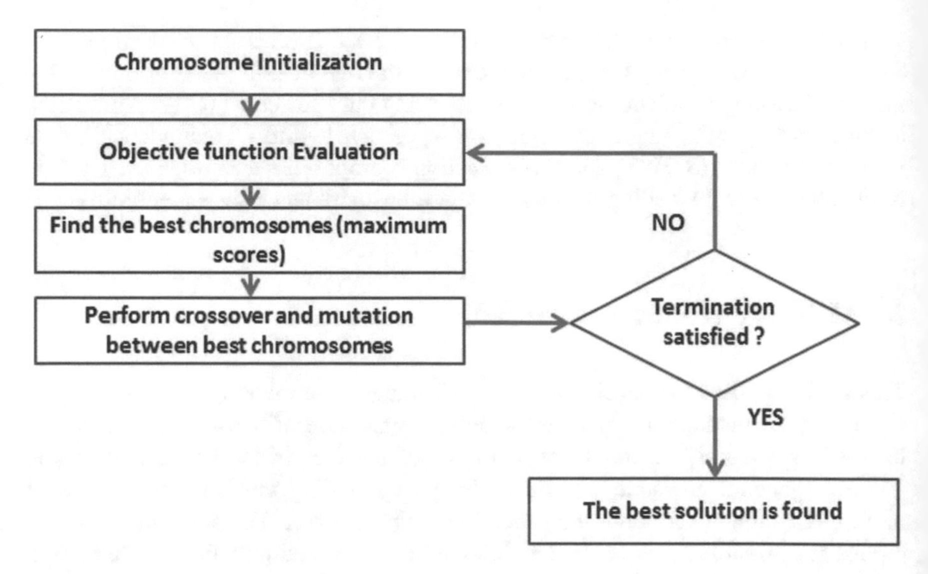

Fig. 5 General flow chart of genetic algorithm

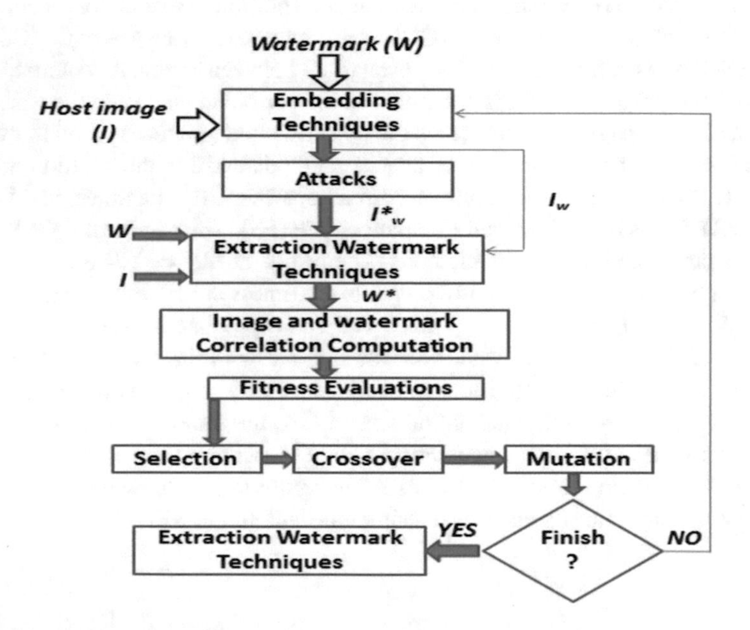

Fig. 6 SVD-GA image watermarking block diagram [29]

Table 1 Results of correlations against GA's parameters (CR = 0.7, PS = 50)

MR	$CORR_I$		$CORR_W$	
			Rotation	
	Avg	Stddev	Avg	Stddev
0.02	0.9993	0.00013	0.9956	0.00083
0.04	**0.9994**	**0.00018**	**0.9958**	**0.00102**
0.06	0.9993	0.00014	0.9952	0.00071

Table 2 Results of correlations against GA's parameters (CR = 0.7, PS = 50)

MR	$CORR_W$			
	Sharpening		Average filtering	
	Avg	Stddev	Avg	Stddev
0.02	0.9834	0.00150	0.9970	0.00035
0.04	**0.9863**	**0.00286**	**0.9969**	**0.00035**
0.06	0.9839	0.00229	0.9972	0.00064

Table 3 Results of correlations against GA's parameters (CR = 0.7, PS = 50)

MR	$CORR_W$	
	Resizing	
	Avg	Stddev
0.02	0.9956	0.00046
0.04	**0.9956**	**0.00049**
0.06	0.9959	0.00076

Table 4 Results of correlations against GA's parameters (CR = 0.7, MR = 0.04)

PS	$CORR_I$		$CORR_W$	
			Rotation	
	Avg	Stddev	Avg	Stddev
50	0.9994	0.00018	0.9958	0.00102
100	0.9994	0.0002	0.9960	0.00052
150	**0.9996**	**0.00021**	**0.9965**	**0.00068**

Table 5 Results of correlations against GA's parameters (CR = 0.7, MR = 0.04)

PS	$CORR_W$			
	Sharpening		Average filtering	
	Avg	Stddev	Avg	Stddev
50	0.9956	0.00049	0.9863	0.00286
100	0.9967	0.00028	0.9877	0.00158
150	**0.9965**	**0.00058**	**0.9896**	**0.00287**

Table 6 Results of correlations against GA's parameters (CR = 0.7, MR = 0.04)

PS	$CORR_W$	
	Average filtering	
	Avg	Stddev
50	0.9996	0.00035
100	0.9977	0.00021
150	**0.9976**	**0.00042**

Fig. 7 Watermarking proposed technique against attacks [29]

Fig. 8 Watermarking proposed technique against attacks [29]

3 Hybrid of RDWT and SVD Watermarking Technique Using Cuckoo Search [30]

This work proposed a technique for finding the SFs in image watermarking using CS optimization algorithm [30]. CS succeeds in solving a lot of real problems in many fields [31–35]. Figure 9 shows the general flow chart of CS optimization algorithm. The watermarking approach that was used in this work is a (RDWT) and SVD [36]. The SFs values used were 0.05 and 0.005 for low- and high-frequency sub-bands in order. The constant SFs may be suitable for some set of images but not for others. Hence, the role of using CS optimization was selecting the best value of SF that improves the robustness and imperceptibly. The host image is divided into four blocks (LL, HL, LH, and HH), so the solution of CS contains four values of SF for the four blocks. The main step of technique as follows:

1. The population is initialized with values ranges from zero to one and blocks of watermark are multiplied with the SFs and are embedded into the corresponding blocks of host image.
2. The watermarked image is tested for different attacks and the watermark are extracted from the distorted watermark using the extracted techniques.

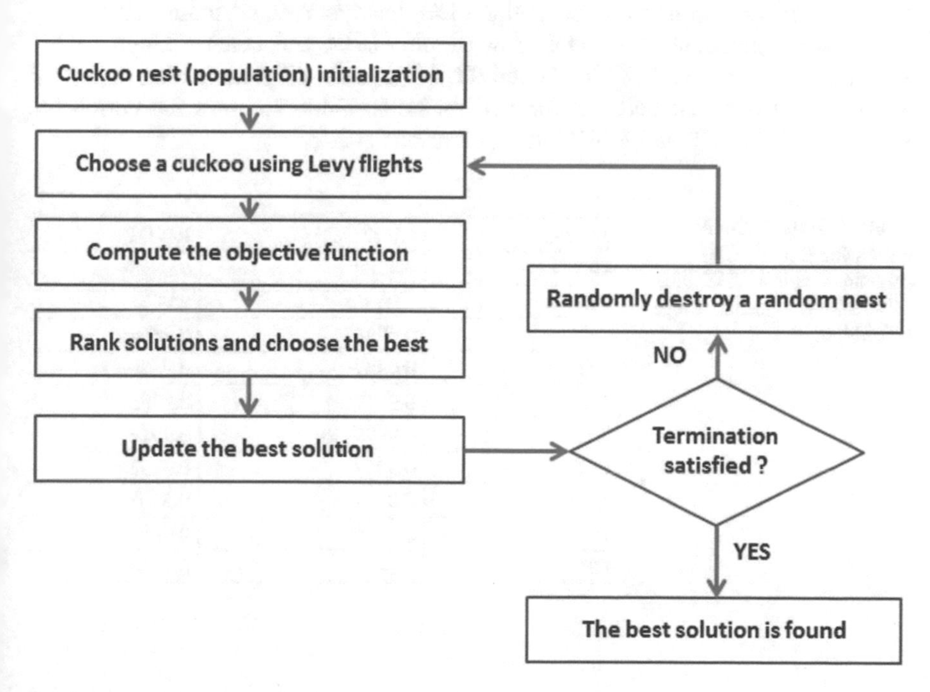

Fig. 9 General flowchart of CS optimization algorithms

3. The fitness of each population is computed according to Eq. (2).

$$Minimize\,Fitness_i = \left[\frac{N}{\sum_{i=1}^{N} NC(W, W_i^*)} - NC(I, I_W) \right] \quad (2)$$

N represents the number of attacks and NC is normalized correlation function. IW and I are the watermarked image and host image in order. W * i and W are the extracted watermark and embedded watermark before attacks in order.
4. A number of solutions that have fitness smaller than certain threshold are chosen for the reproduction process to start the new loop if the termination condition is not satisfied.
5. The smallest fitness value is the best SF value.

The technique was tested with the "Lena", "baboon" and "perpper" images. The perceptual of the watermarked image is tested by evaluating peak-to-signal noise ratio (PSNR). The technique succeeded to provide watermark image non-distinguishable from the original image. In addition, the techniques used various attacks methods such as: gamma correction (GC), cropping (CR), median filtering (MF), rotation (RO), translation (TR), histogram equalization (HE), JPEG compression, gaussian noise (GN), row flipping (RF), and column flipping (CF). A comparison between RWDT-CS and RWDT was done by comparing the PSNR and no large different between the two methods. Table 7 shows the correlation values of RWDT-CS and RWDT for the various attacks.

Table 7 Experimental results for RWDT-CS compared with RWDT form 10 attacks for "Lena" image benchmark

Attacks	RWDT-CS	RWDT
GC	0.9988	0.9896
MF	0.9951	0.9821
CR	0.9936	0.9864
RO	0.9987	0.9828
TR	0.9959	0.9796
HE	0.9979	0.9912
GN	0.9917	0.9792
JPEG	0.9978	0.9903
RF	0.9989	0.9842
CF	0.9987	0.9848

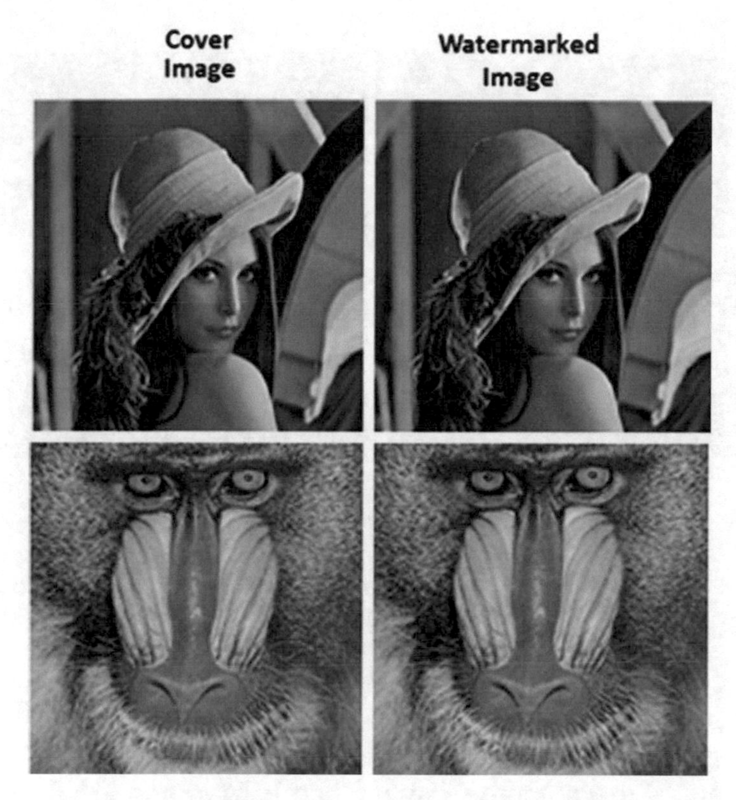

Fig. 10 Difference between cover and watermarked image of RDWT-SV-CS [30]

Figure 10 shows two images before watermarking and after. The PNSR for the first image is 36.39 dB and for the second is 36.28 dB. Figure 11 shows the robustness of the watermarked image "Lena" after various attacks: (a) gamma correction, (b) median filtering, (c) cropping, (d) rotation, (e) translation, (f) histogram equalization, (g) Gaussian noise, (h) JPEG compression, (i) row flipping, (j) column flipping.

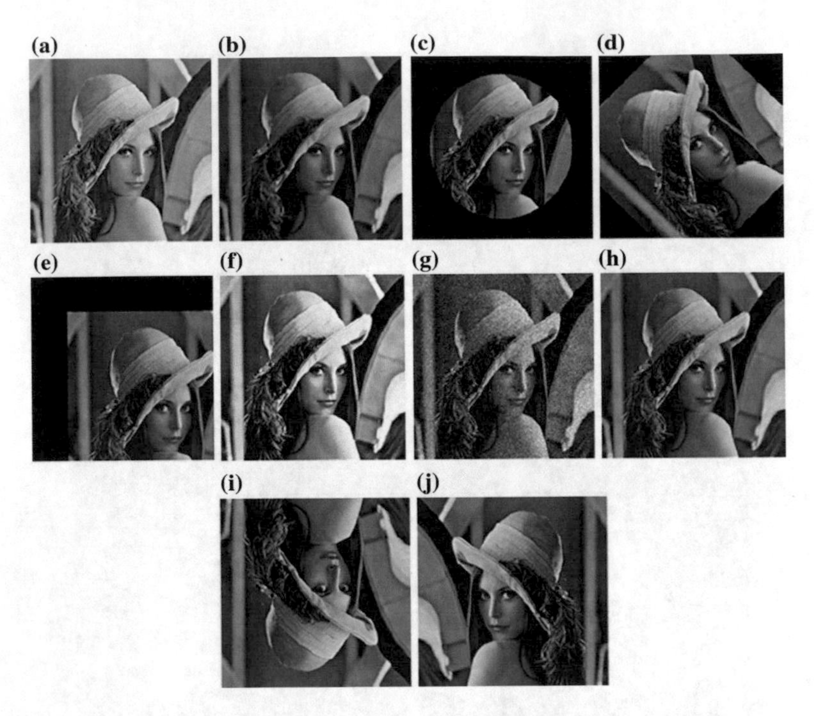

Fig. 11 Watermarked image "Lena" using RDWT-SVD-CS after various attacks

4 Image Watermarking Using Chaotic Map and DNA Coding [37]

The recent significant digital watermarking research implies DNA sequences coding with jointing with chaotic map. In this section, a proposed technique architecture of image watermarking that increases the security of the copyright of cover image using Chaotic map and DNA-based information security is proposed.

This architecture has the feature of higher security than the previous work that is due to using two logistic chaotic maps and embedding the image watermarking into the least significant bit of cover image. Embedding the DNA coding increase the Embedding Rate (ER) but not significantly, however the main gain of using DNA coding is accelerating the development of DNA-based image watermarking. The watermark embedding process is summarized in Fig. 12. Embedding the DNA with chaotic maps is as shown in Fig. 13.

The logistic chaotic map was used is $Xi + 1 = M\ Xi\ (1 - Xi)$, Two logistic maps were used where M1 and M2 are in the range (3.9–4) and the X0(1) and X0(2) are in the range [0–1].

1. Initially, generating the relating key.
2. Iterating the logistic maps and obtaining the chaotic orbits.

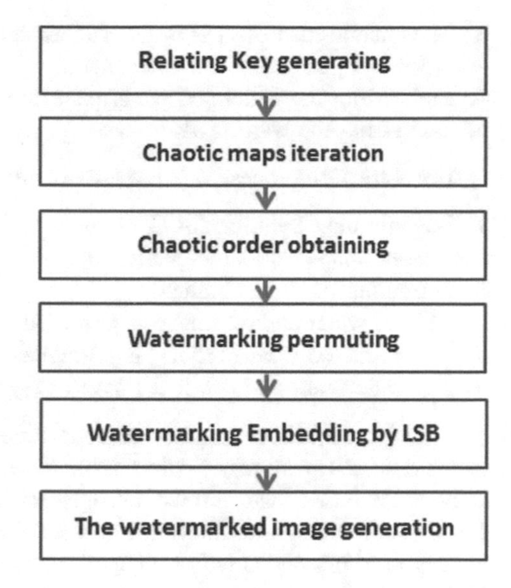

Fig. 12 Embedding process using chaotic maps [37]

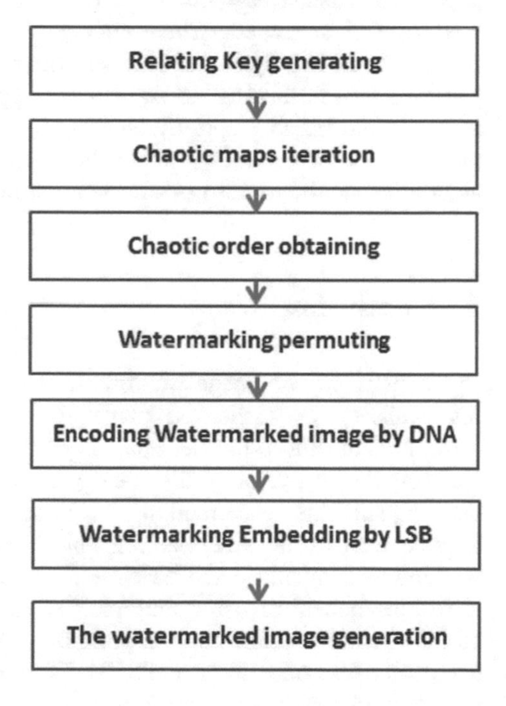

Fig. 13 Embedding process using chaotic maps and DNA [37]

3. The chaotic map orbit is ordered in ascending order and randomly generate the order that permuting the watermark.
4. Embedding the generated watermark into the cover image using LSB.
5. Outputting the watermark image.

The extraction process is the reverse of embedding process and stated as follows:

1. Logistic map iterating for Q times.
2. Chaotic maps orbits are generating.
3. Permuting order is obtained.
4. LSB of watermarked image is extracted.
5. The extracted watermarking is permuted.
6. The watermark and original image are generated.

A DNA sequence consists of four different bases, namely A (adenine), C (cytosine), G (guanine), T (thymine). Base pairs, which form between specific nucleotides bases (also termed nitrogenous bases) are the building blocks of the DNA double helix and contribute to the folded structure of both DNA and RNA, namely A with T and C with G [12]. For the binary bit, 0 with 1 are complementary. So 00 with 11 are complementary, and 01 with 10 are also complementary. In this paper, we consider A = 00, T = 11, C = 01, and G = 10 to convert binary message to DNA sequences. There are eight DNA coding methods to convert binary message to DNA sequences. The performance of this work is briefly without DNA encoding the PSNR is 50 dB and ER is 1 bpp. But using DNA encoding the PSNR is 43 dB and ER is 2 bpp. Embedding DNA with chaotic maps present improvement in non-perceptibility and robustness. As a future work the bio-inspired algorithms can be used to obtain more improvement on the watermarking.

5 Conclusion

The control key of image watermarking is SF that affect on the robustness and impeccability of watermarked image. So, the rule of using metaheuristic algorithms for optimizing the robustness and impeccability of image watermarking through choosing the best value of SF was explained. Two image watermarking techniques that explain the watermarking optimization was discussed that proves the improvement of the watermarking objectives. The first is optimizing SVD watermarking for gray image and invisible watermark using genetic algorithm. This work shows the benefit of using multiple SFs instead of one SF. The second is using cuckoo search optimization algorithm for optimizing a hybrid watermarking approach based on RDWT and SVD. The correlations between original image and extracted original image were shown. In addition to the correlation between original watermarked image and the watermarked image after attacks. Besides, the effect of attacks operation on the results of the two techniques was also explained.

References

1. Liu, J.-C., Chen, S.-Y.: Fast two-layer image watermarking without referring to the original image and watermark. Image Vis. Comput. **19**(14), 1083–1097 (2001)
2. Liu, R., Tan, T.: An SVD-based watermarking scheme for protecting rightful ownership. IEEE Trans. Multimedia **4**(1), 121–128 (2002)
3. Nikolaidis, N., Pitas, I.: Robust image watermarking in the spatial domain. Sig. Process. **66** (3), 385–403 (1998)
4. Phadikar, A., Maity, S.P., Verma, B.: Region based QIM digital watermarking scheme for image database in DCT domain. Comput. Electr. Eng. **37**, 339–355 (2011)
5. Wu, X., Sun, W.: Robust copyright protection scheme for digital images using overlapping DCT and SVD. Appl. Soft Comput. **13**(2), 1170–1182 (2013)
6. Ouhsain, M., Hamza, A.B.: Image watermarking scheme using nonnegative matrix factorization and wavelet transform. Expert Syst. Appl. **36**(2), 2123–2129 (2009)
7. Ganic, E., Eskicioglu, A.M.: Robust DWT-SVD domain image watermarking: embedding data in all frequencies. In: Proceedings of the ACM Multimedia and Security Workshop, pp. 166–174 (2004)
8. Rawat, S., Raman, B.: A blind watermarking algorithm based on fractional fourier transform and visual cryptography. Sig. Process. **92**(6), 1480–1491 (2012)
9. Lu, W., Lu, H., Chung, F.-L.: Feature based robust watermarking using image normalization. Comput. Electr. Eng. **36**, 2–18 (2010)
10. Song, C., Sudirman, S., Merabti, M.: A robust region-adaptive dual image watermarking technique. J. Vis. Commun. Image Represent. **23**, 549–568 (2012)
11. Rastegar, S., Namazi, F., Yaghmaie, K., Aliabadian, A.: Hybrid watermarking algorithm based on singular value decomposition and radon transform. Int. J. Electr. Commun. **65**, 658–663 (2011)
12. Run, R.-S., Horng, S.-J., Lai, J.-L., Kao, T.-W., Chen, R.-J.: An improved SVD-based watermarking technique for copyright protection. Expert Syst. Appl. **39**, 673–689 (2012)
13. Eberhart, R., Kennedy, J.: A new optimizer using particle swarm theory. In: Proceedings of the Sixth International Symposium on Micro Machine and Human Science, 1995. MHS'95. IEEE (1995)
14. Yang, X.-S.: Firefly algorithm, stochastic test functions and design optimisation. Int. J. Bio-Inspired Comput. **2**(2), 78–84 (2010)
15. Dorigo, M.: Optimization, learning and natural algorithms. Ph.D. Thesis, Politecnico di Milano, Italy (1992)
16. Karaboga, D.: An idea based on honey bee swarm for numerical optimization. Volume 200. Technical report-tr06, Erciyes University, Engineering Faculty, Computer Engineering Department (2005)
17. Kirkpatrick, S., Gelatt, C.D., Vecchi, M.P.: Optimization by simulated annealing. Science **220** (4598), 671–680 (1983)
18. Aryanezhad, M.B., Hemati, M.: A new genetic algorithm for solving nonconvex nonlinear programming problems. Appl. Math. Comput. **199**(1), 186–194 (2008)
19. Papakostas, G.A., Tsougenis, E.D., Koulouriotis, D.E.: Moment based local image watermarking via genetic optimization. Appl. Math. Comput. **227**, 222–236 (2014)
20. Vahedi, E., Zoroofi, R.A., Shiva, M.: Toward a new wavelet-based watermarking approach for color images using bio-inspired optimization principles. Digit. Signal Process. **22**, 153–162 (2012)
21. Tsai, H.-H., Jhuang, Y.-J., Lai, Y.-S.: An SVD-based image watermarking in wavelet domain using SVR and PSO. Appl. Soft Comput. **12**(8), 2242–2453 (2012)
22. Al-Qaheri, Hameed, Mustafi, Abhijit, Banerjee, Soumya: Digital watermarking using ant colony optimization in fractional Fourier domain. J. Inf. Hiding Multimed. Signal Process. **1**(3), 179–189 (2010)

23. Loukhaoukha, K., Chouinard J.-Y., Taieb, M.H.: Optimal image watermarking algorithm based on LWT-SVD via multi-objective ant colony optimization. J. Inf. Hiding Multimed. Signal Process. **2**(4), 303–319 (2011)
24. Mishra, A., et al.: Optimized gray-scale image watermarking using DWT–SVD and Firefly algorithm. Expert Syst. Appl. **41**(17), 7858–7867 (2014)
25. Dey, N., et al.: Firefly algorithm for optimization of scaling factors during embedding of manifold medical information: an application in ophthalmology imaging. J. Med. Imaging Health Inf. **4**(3), 384–394 (2014)
26. Ali, M., Ahn, C.W.: Comments on "Optimized gray-scale image watermarking using DWT-SVD and firefly algorithm". Expert Syst. Appl. **42**(5), 2392–2394 (2015)
27. Ali, M., et al.: An image watermarking scheme in wavelet domain with optimized compensation of singular value decomposition via artificial bee colony. Inf. Sci. **301**, 44–60 (2015)
28. Mohammadi, F.G., Saniee Abadeh, M.: Image steganalysis using a bee colony based feature selection algorithm. Eng. Appl. Artif. Intell. **31**, 35–43 (2014)
29. Aslantas, Veysel: A singular-value decomposition-based image watermarking using genetic algorithm. AEU-Int. J. Electr. Commun. **62**(5), 386–394 (2008)
30. Ali, M., Ahn, C.W., Pant, M.: Cuckoo search algorithm for the selection of optimal scaling factors in image watermarking. In: Proceedings of the Third International Conference on Soft Computing for Problem Solving. Springer, India (2014)
31. Bhargava, V., Fateen, S.E.K., Bonilla-Petriciolet, A.: Cuckoo search: a new nature-inspired optimization method for phase equilibrium calculations. Fluid Phase Equilib. **337**, 191–200 (2013)
32. Bulatović, R.R., Đorđević, S.R., Đorđević, V.S.: Cuckoo search algorithm: a metaheuristic approach to solving the problem of optimum synthesis of a six-bar double dwell linkage. Mech. Mach. Theory **61**, 1–13 (2013)
33. Yildiz, A.R.: Cuckoo search algorithm for the selection of optimal machining parameters in milling operations. Int. J. Adv. Manuf. Technol. **64**, 55–61 (2013)
34. Valian, E., Tavakoli, S., Mohanna, S., Haghi, A.: Improved cuckoo search for reliability optimization problems. Comput. Ind. Eng. **64**, 459–468 (2013)
35. Moravej, Z., Akhlaghi, A.: A novel approach based on cuckoo search for DG allocation in distribution network. Electr. Power Energy Syst. **44**, 672–679 (2013)
36. Makbol, N.M., Khoo, B.E.: Robust blind image watermarking scheme based on redundant discrete wavelet transform and singular value decomposition. Int. J. Electron. Commun. (AEÜ) **65**, 658–663 (2012)
37. Wang, B., et al.: Image watermarking using chaotic map and DNA coding. Optik-Int. J. Light Electr. Opt. **126**(24), 4846–4851 (2015)

Image Reconstruction Using Novel Two-Dimensional Fourier Transform

S. Kala, S. Nalesh, Babita R. Jose and Jimson Mathew

Abstract Reconstruction of a signal from its subset is used in various contexts in the field of signal processing. Image reconstruction is one such example which finds widespread application in face recognition, medical imaging, computer vision etc. Image reconstruction is computationally complex, and efficient implementations need to exploit the parallelism inherent in this operation. Discrete Fourier Transform (DFT) is a widely used technique for image reconstruction. Fast Fourier Transform (FFT) algorithms are used to compute DFTs efficiently. In this paper we propose a novel two dimensional (2D) Fast Fourier Transform technique for efficient reconstruction of a 2D image. The algorithm first applies 1D FFT based on radix-4^n along the rows of the image followed by same FFT operation along columns, to obtain a 2D FFT. Radix-4^n technique used here provides significant savings in memory required in the intermediate stages and considerable improvement in latency. The proposed FFT algorithm can be easily extended to three dimensional and higher dimensional

---

S. Kala · B.R. Jose
Division of Electronics, Cochin University of Science and Technology, Kochi, India
e-mail: kalas@cusat.ac.in

B.R. Jose
e-mail: babitajose@cusat.ac.in

S. Nalesh
CAD Lab, Indian Institute of Science, Bangalore, India
e-mail: nalesh@cadl.iisc.ernet.in

J. Mathew (✉)
Department of Computer Science and Engineering, Indian Institute of Technology Patna, Patna, India
e-mail: jimson@iitp.ac.in

A.E. Hassanien and D.A. Oliva (eds.), *Advances in Soft Computing and Machine Learning in Image Processing*, Studies in Computational Intelligence 730,
https://doi.org/10.1007/978-3-319-63754-9_31

FFTs. Simulated results for image reconstruction based on this technique are presented in the paper. 64 point FFT based on radix-4^3 has been implemented using 130nm CMOS technology and operates at a maximum clock frequency of 350 MHz.

1 Introduction

Discrete Fourier Transform (DFT) plays key role in variety of digital signal processing applications. DFTs are generally not computed directly owing to high computational complexity. Several different techniques are available for computing DFT [1]. Fast Fourier Transform (FFT) is one of the most widely used algorithm for efficient DFT computations. FFTs are used for applications such as spectrum sensing in Cognitive Radio, OFDM, Image processing, Radar Signal Processing, etc. In most of these applications, FFT is the most compute intensive and time-consuming operation. Multi-dimensional FFT is frequently used in various applications like medical imaging, radar data processing, etc. [2]. Reconstruction of images is an important operation in most of these applications and involves complex computations in real time. Two-dimensional (2D) FFTs when used for reconstructing image from raw data is computationally intensive and needs efficient implementations for catering to real-time applications. These applications require large memory sizes to support large size images. Therefore it is necessary to have an FFT architecture which optimizes the memory but supports various image sizes, and provides the required throughput.

Large variety of algorithms and architectures for FFT are proposed in literature over the last few decades. FFT architectures are mainly classified into memory based and pipelined architectures [3]. Pipeline-based FFTs can provide higher throughput and gives better hardware utilization. Single-path Delay Feedback (SDF), Single-path Delay Commutator (SDC), Multi-path Delay Commutator (MDC) are the various existing pipelined architectures [3]. Delay feedback-based approaches are more efficient in terms of its memory requirements since the output of a butterfly operation and input shares the same storage. SDF architectures are memory efficient, while MDC provides high throughput [4]. Various SDF pipelined architectures [5, 6] based on the radix-2 (radix-2^2, radix-2^3, radix-2^4, radix-2^5) and radix-4 are used in FFT/IFFT computation. Optimal choice of FFT architecture depends on the application.

Several different architectures for two-dimensional FFT are available in the literature. Software solutions for 2D FFT like FFTW [7], Spiral [8], gives high performance, but consume more power and are not suitable for embedded system applications. Various hardware solutions such as FFT processor ASICs and FPGA implementations are available [9, 10]. In [11], a technique for optimizing the energy consumption of 2D FFT is presented, which uses an FPGA-based one-dimensional FFT kernel. A 2D FFT architecture for large-sized input data based on a 2D decomposition algorithm is proposed in [12]. A 2D FFT based on the novel loop unrolling technique is presented in [13]. Most of these architectures for FFT are not scalable for catering to higher FFT sizes. In [14], a stream architecture for computing 2D

FFT of non-power of two FFT size is presented. This architecture is based on the row-column decomposition algorithm for 2D FFT and can compute variable size of FFTs.

FFT algorithms are compute intensive and it is challenging to choose a suitable solution. Several solutions based on the ASIC, DSP, configurable processors, FPGA and GPP are available. There exist several design trade-offs between different solutions and a suitable solution has to be considered which will meet our requirement. Generally hardware implementations provide better performance per watt and are more suited for real-time embedded applications. DSP chips are flexible but are not suited to meet high throughput. ASICs are excellent in performance and power and we can tune the hardware specific to our application which results in high application specific performance. FPGAs provides reconfigurability at the expense of power. In this work we present an ASIC-based implementation for the 2D FFT.

A 2D FFT is typically computed from one-dimensional (1D) FFT. Computation of 2D FFT consists of 2N one-dimensional FFT computations. Thus improving the efficiency of 1D FFT computation will influence the resulting 2D FFT. Since image processing requires computation of large size FFTs, higher radix algorithms are more efficient. We use radix-4^3 algorithm based on the novel radix-4 butterfly unit as the building block in our work. This paper proposes a new 2D FFT architecture for implementing the algorithm over 64×64 complex points. We implement a radix-4^3 block such that, the outputs generated are already reordered, which results in savings in intermediate memory and reduces latency. Our method uses parallel unrolled architecture of radix-4^3 for implementing 2D FFT, which is an extension of our previous work on 1D FFT [15].

The paper is organized as follows. An overview of FFT algorithms is given in Sect. 2. Section 3 discusses the proposed technique which uses radix-4^3 algorithm to get a 2D Fourier Transform. Section 4 provides MATLAB simulation results and ASIC implementation details. Section 5 summarizes the paper.

2 FFT Algorithms

Among various FFT algorithms, Cooley–Tukey is the most popular because of its reduced complexity, i.e., O(Nlog_2N) instead of O(N^2) when compared to DFT [1]. The formula for finding DFT of a sequence $x(n)$ is given in Eq. (1).

$$X(k) = \sum_{n=0}^{N-1} x(n) W_N^{nk} \quad k = 0, 1, 2, \ldots N-1 \tag{1}$$

$$W_N = e^{-2\pi i/N} \tag{2}$$

where W_N, the Twiddle Factor, denotes the Nth primitive root of unity, with its exponent evaluated to modulo N and is introduced by the Eq. (2). Inverse DFT of $X(k)$ is given in Eq. (3) and follows the same computational algorithm as that of DFT. So the discussion here is restricted to DFT.

$$x(n) = \frac{1}{N}\sum_{k=0}^{N-1} X(k)W_N^{-nk} \quad n = 0, 1, 2, \ldots N-1 \tag{3}$$

The Cooley–Tukey algorithm uses a divide-and-conquer approach to reduce the number of computations to arrive at the same result of DFT. The size of an FFT decomposition is generally referred as 'radix'. Radix-k FFT is efficient for computing an N point DFT, if N=k^M, where k and M are integers. Based on different decompositions, different algorithms are used to compute FFT.

2.1 *Radix-2 FFT*

Cooley–Tukey-based FFT algorithms can be computed using Decimation-in-Time (DIT) or Decimation-in-Frequency (DIF). In the case of DIT algorithms, the input sequence is stored in bit-reversed order and the output sequence will be in normal order. But in the case of DIF algorithms, the input sequences are in natural order and the output sequences are obtained in bit-reversed order. For deriving DIF algorithm from the DFT formula given in Eq. (1), X(k) is split into odd and even samples as given in Eq. (4) [1].

$$\begin{aligned} X(2k) &= \sum_{n=0}^{(N/2)-1} [x(n) + x(n+\frac{N}{2})]W_{N/2}^{nk} \quad k = 0, 1, 2, \ldots \frac{N}{2}-1 \\ X(2k+1) &= \sum_{n=0}^{(N/2)-1} [x(n) - x(n+\frac{N}{2})]W_N^{nk}W_{N/2}^{nk} \quad k = 0, 1, 2, \ldots \frac{N}{2}-1 \end{aligned} \tag{4}$$

This procedure can be repeated and for realizing an N point FFT, log_2N stages of decimation is required. Butterfly computations can be done 'in-place' [1] which means that once butterfly operation is done on a pair of data, the results are stored in the same locations as that of input data.

2.2 *Radix-4 FFT*

If the number of input data sequence N is a power of 4, instead of using radix-2, more efficient way of FFT computation is radix-4 FFT algorithm. For realizing an N point FFT, log_4N stages are required. Here, an N point DFT sequence X(k) in Eq. (1) is broken into four smaller DFTs, X(4k), X(4k + 1), X(4k + 2) and X(4k + 3), where $k = 0, 1, 2, \ldots \frac{N}{4} - 1$. The radix-4 DIF FFT is obtained as follows:

$$X(4k) = \sum_{n=0}^{(N/4)-1} [x(n) + x(n + \frac{N}{4}) + x(n + \frac{N}{2}) + x(n + 3\frac{N}{4})]W_{N/4}^{nk}$$

$$X(4k+1) = \sum_{n=0}^{(N/4)-1} [x(n) - jx(n + \frac{N}{4}) - x(n + \frac{N}{2}) + jx(n + 3\frac{N}{4})]W_N^n W_{N/4}^{nk}$$

$$X(4k+2) = \sum_{n=0}^{(N/4)-1} [x(n) - x(n + \frac{N}{4}) + x(n + \frac{N}{2}) - x(n + 3\frac{N}{4})]W_N^{2n} W_{N/4}^{nk}$$

$$X(4k+3) = \sum_{n=0}^{(N/4)-1} [x(n) + jx(n + \frac{N}{4}) - x(n + \frac{N}{2}) - jx(n + 3\frac{N}{4})]W_N^{3n} W_{N/4}^{nk} \tag{5}$$

Radix-4 FFT computation results in digit reversed output where each digit is with regard to a number scheme of base 4 [16].

3 Proposed Technique

A novel, radix-4^n based 2D FFT reconstruction algorithm which uses linear decomposition technique is proposed here. For reconstructing images of size 64×64, a 64-point 1D FFT is required for which we use radix-4^3 algorithm. Similarly for an image of size 256×256, radix-4^4s algorithm can be used. Combinations of different values of n in radix-4^n can be used for various image sizes. In this work, we present a 2D FFT, which is built from two radix-4^3 1D FFT.

3.1 2D Fourier Transform

A two-dimensional DFT of size $N \times N$, with inputs $x(i_1, i_2)$ is given as,

$$y(k_1, k_2) = \sum_{i_1=0}^{N-1} \sum_{i_2=0}^{N-1} x(i_1, i_2) W_N^{k_1 i_1 + k_2 i_2} \quad where \quad k_1, k_2 = 0, 1, 2, \ldots N-1 \tag{6}$$

Using two one-dimensional DFTs, a 2D DFT can be performed based on Row-Column decomposition algorithm, as given in Eq. (7).

$$X(k_1, i_2) = \sum_{i_1=0}^{N-1} x(i_1, i_2) W_N^{k_1 i_1} \quad where \quad k_1 = 0, 1, 2, \ldots N-1 \tag{7}$$

$$Y(k_1, k_2) = \sum_{i_2=0}^{N-1} X(k_1, i_2) W_N^{k_2 i_2} \quad where \quad k_2 = 0, 1, 2, \ldots N-1 \tag{8}$$

3.2 *Radix-4³ Algorithm*

Radix-4^3 algorithm which is equivalent to radix-64 is derived using a four-dimensional index mapping of n and k to the basic DFT formula given in Eq. (9). From the Discrete Fourier Transform formula [1],

$$X(k) = \sum_{n=0}^{N-1} x(n) W_N^{nk} \quad k = 0, 1, 2, \ldots N-1 \tag{9}$$

$$n = n_1 + \frac{N}{64}n_2 + \frac{N}{16}n_3 + \frac{N}{4}n_4 \quad k = 64k_1 + 16k_2 + 4k_3 + k_4 \tag{10}$$

Apply Eq. (10) to the DFT formula in (9),

$$X(64k_1 + 16k_2 + 4k_3 + k_4) = \sum_{n_1=0}^{\frac{N}{64}-1} \sum_{n_2=0}^{3} \sum_{n_3=0}^{3} \sum_{n_4=0}^{3} x(n_1 + \frac{N}{64}n_2 + \frac{N}{16}n_3 + \frac{N}{4}n_4) W_N^{nk} \tag{11}$$

Decomposing the twiddle factors,

$$W_N^{nk} = W_{\frac{N}{64}}^{n_1 k_1} W_N^{n_1(16k_2+4k_3+k_4)} W_{64}^{n_2(4k_3+k_4)} W_{16}^{n_3 k_4} (-j)^{(n_2 k_2 + n_3 k_3 + n_4 k_4)} \tag{12}$$

Substituting (12) in (11) and expanding the summation with index n_4,

$$X(64k_1 + 16k_2 + 4k_3 + k_4) = \sum_{n_1=0}^{\frac{N}{64}-1} \sum_{n_2=0}^{3} \sum_{n_3=0}^{3} [B(n_1 + \frac{N}{64}n_2 + \frac{N}{16}n_3)](-j)^{(n_2 k_2 + n_3 k_3)} W_{\frac{N}{64}}^{n_1 k_1} \times W_N^{n_1(16k_2+4k_3+k_4)} W_{64}^{n_2(4k_3+k_4)} W_{16}^{n_3(k_4)} \tag{13}$$

The first butterfly unit, B is given by

$$
\begin{aligned}
B = x(n_1 + \frac{N}{64}n_2 + \frac{N}{16}n_3)(-j)^{k_4} \\
x(n_1 + \frac{N}{64}n_2 + \frac{N}{16}n_3 + \frac{N}{4}) + \\
(-1)^{k_4}x(n_1 + \frac{N}{64}n_2 + \frac{N}{16}n_3 + \frac{N}{2}) + \\
(j)^{k_4}x(n_1 + \frac{N}{64}n_2 + \frac{N}{16}n_3 + \frac{3N}{4})
\end{aligned}
\tag{14}
$$

Expanding the summation in (13) with index n_3,

$$
\begin{aligned}
X(64k_1 + 16k_2 + 4k_3 + k_4) = \\
\sum_{n_1=0}^{\frac{N}{64}-1}\sum_{n_2=0}^{3}[H(n_1 + \frac{N}{64}n_2)] \times \\
(-j)^{(n_2k_2}W_{\frac{N}{64}}^{n_1k_1}W_N^{n_1(16k_2+4k_3+k_4)}W_{64}^{n_2(4k_3+k_4)}W_{64}^{n_3(k_4)}
\end{aligned}
\tag{15}
$$

The second butterfly unit, H is given by

$$
\begin{aligned}
H = B(n_1 + \frac{N}{64}n_2) + \\
(-j)^{k_3}[B(n_1 + \frac{N}{64}n_2)]W_{16}^{k_4} + \\
(-1)^{k_3}[B(n_1 + \frac{N}{64}n_2 + \frac{N}{8})]W_8^{k_4} + \\
(j)^{k_3}[B(n_1 + \frac{N}{64}n_2 + \frac{3N}{16})]W_{16}^{k_4}W_8^{k_4}
\end{aligned}
\tag{16}
$$

Expanding the summation in (15) with index n_2,

$$
\begin{aligned}
X(64k_1 + 16k_2 + 4k_3 + k_4) = \\
\sum_{n_1=0}^{\frac{N}{64}-1}[T(n_1)W_N^{n_1(16k_2+4k_3+k_4)}]W_{\frac{N}{64}}^{n_1k_1}
\end{aligned}
\tag{17}
$$

The third butterfly unit, T is given by

$$
\begin{aligned}
T = H(n_1) + (-j)^{k_2}[H(n_1 + \\
\frac{N}{64})]W_{16}^{k_3}W_{64}^{k_4} + \\
(-1)^{k_2}[H(n_1 + \frac{N}{32})]W_{16}^{k_3}W_8^{k_4} + \\
(j)^{k_2}[H(n_1 + \frac{3N}{64})]W_{64}^{3(4k_3+k_4)}
\end{aligned}
\tag{18}
$$

Equations (14), (16) and (18) are the three butterfly units in radix-4^3 algorithm. Each of these butterfly operations are based on radix-4 butterfly and hence the name radix-4^3 (R4^3).

3.3 Proposed 2D Fourier Transform

In this section, we present an architecture for 2D FFT using radix-4^3 algorithm. 2D DFT for size 64×64, with inputs $x(i_1, i_2)$ can be expressed as,

$$y(k_1, k_2) = \sum_{i_2=0}^{63} \left[\sum_{i_1=0}^{63} x(i_1, i_2) W_{64}^{k_1 i_1} \right] W_{64}^{k_2 i_2} \quad where \quad k_1, k_2 = 0, 1, 2, \ldots 63 \tag{19}$$

We can develop this 2D FFT by cascading two radix-4^3 (R4^3) blocks. Each radix-4^3 has three stages of novel radix-4 (R4) butterfly units [15]. Signal Flow Graph (SFG) for radix-4^3 algorithm with three stages of radix-4 units is shown in Fig. 1. Most of the FFT architectures follow conventional Cooley-Tukey algorithm with only one radix-4 butterfly unit in each stage [17]. The outputs from each stage of radix-4 need to be reordered before applying to the next stage. Similarly if two R4^3 engines are cascaded, the intermediate outputs also have to be reordered. Since the subsequent blocks will require the inputs to be reordered before processing, intermediate RAMs are required between every cascaded stage in this implementation. This reordering needs intermediate memory and memory size grows as FFT length increases. For example for a 64×64 2D FFT, a 4096 (64×64) deep memory is required between the two blocks.

In our proposed architecture we use a parallel unrolled implementation for radix-4^3. The order of output to be generated at each stage is controlled using *mode select* signals. With this architecture, for a given set of inputs, at each stage only those outputs required for the next stage can be computed without losing any performance, so that most of the intermediate buffers can be avoided. The resulting implementation is thus fully optimized in terms of memory and latency. Block diagram for the proposed two-dimensional FFT is given in Fig. 2. A one-dimensional FFT using radix-4^3 is performed row wise and then another one-dimensional FFT using second radix-4^3 block is performed column wise, to get the two-dimensional FFT. The first processor performs sixty-four 64-point FFT operations which gives 4K intermediate values. The second FFT processor performs sixty-four 64-point FFT operation on these outputs and gives the final 64×64 point FFT. Input memory consists of two sets of 64 RAMs, each of 32 bit wide, which function in ping-pong fashion.

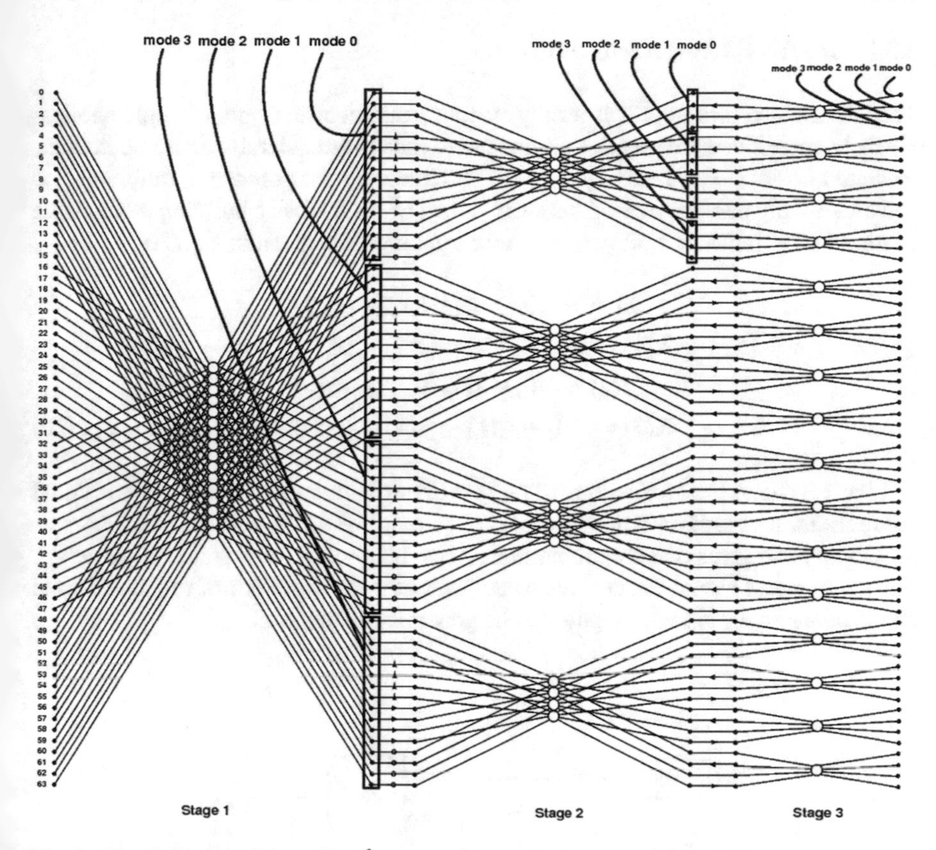

Fig. 1 Signal flow graph for radix-4^3

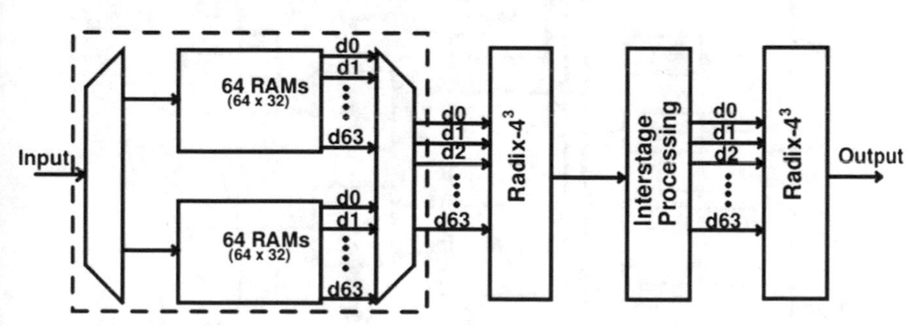

Fig. 2 Proposed two-dimensional FFT using radix-4^3 Blocks

3.3.1 Radix-4^3 ($R4^3$) Architecture

The $R4^3$ architecture in [17] is a fully systolic architecture for giga sample applications. As mentioned already this architecture leads to considerable increase in intermediate buffer size and latency. The $R4^3$ architecture we propose is fully unrolled and uses a fully parallel radix-4 butterfly units [15] as the basic building block. Here a radix-4 (R4) butterfly unit performs four operations as given in Eq. (20).

$$
\begin{aligned}
X(0) &= x(0) + x(1) + x(2) + x(3) \\
X(1) &= x(0) - jx(1) - [x(2) - jx(3)] \\
X(2) &= x(0) - x(1) + x(2) - x(3) \\
X(3) &= x(0) + jx(1) - [x(2) + jx(3)]
\end{aligned} \tag{20}
$$

Figure 3 shows the architecture of radix-4 butterfly unit used in this work. A nodal representation of radix-4 butterfly unit is shown in Fig. 4. Radix-4 unit has four parallel inputs and gives one output based on a two bit control input called *mode select*. The *mode select* signal decides the generation of one output out of the four. Based on the *mode select* signal, outputs can be generated in any order.

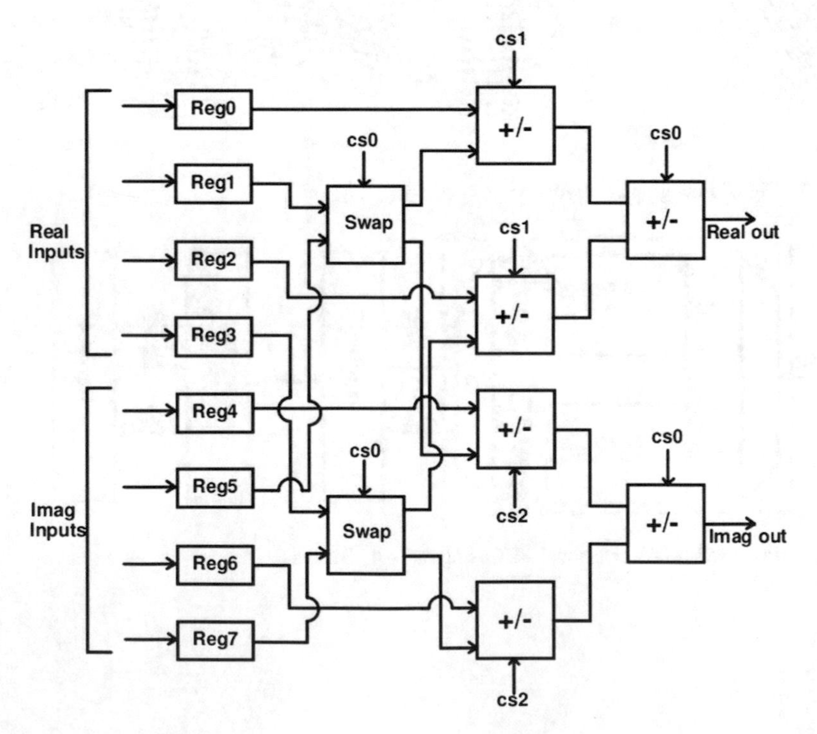

Fig. 3 Radix-4 butterfly unit

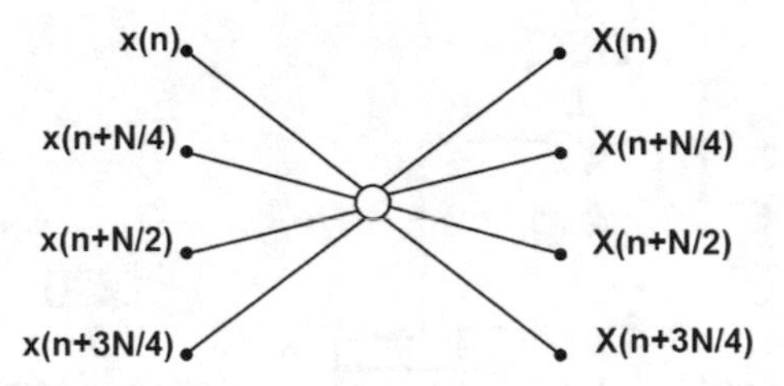

Fig. 4 Radix-4 butterfly node

The parallel unrolled radix-4^3 architecture is shown in Fig. 5 [18]. The intermediate memory requirement is significantly reduced in [15] using mode selection in parallel R4 butterfly unit. Radix-4^3 block consists of 16 radix-4 (R4) units in the first stage, 4 R4 units in the second stage and one R4 unit in the last stage. All R4 units in the same stage are controlled by the same *mode select* input. Six bit wide *mode select* input is needed for all three stages. The sequence of *mode select* input, controls the order of output to be produced by the $R4^3$ block. There are 16 twiddle factor (TF) ROMs in first stage, for storing the twiddle factors denoted as `W16`. Each ROM stores four TF values. In second stage only four ROMS with 16 TF values (`W64`) are required. Thc radix-4^3 architecture presented in [18] reduces intermediate buffering and improves latency, without any degradation in performance or area.

3.3.2 Mode Selection in $R4^3$ Block

In the SFG shown in Fig. 1, the first 16 outputs which are grouped together, are the first output of all 16 butterflies. This is obtained by configuring *mode select* of all radix-4 units in the first stage as *mode 0*. These outputs are multiplied with the corresponding twiddle factors. Similar operation is performed in the second stage. Here, four radix-4 units are required to process the 16 outputs that are obtained from first stage. Again, first four outputs are generated by configuring *mode select* of all four radix-4 units as *mode 0*, keeping the first stage radix-4 units in *mode 0* itself. Now using this four outputs, we can generate the output required in final stage with another mode selection in final stage. For obtaining the next set of four outputs, keeping the first stage radix-4 units in *mode 0*, change the mode of second stage to *mode 1*. This procedure is repeated to get all 64 outputs. Mode select signals in different stages of radix-4^3 block are given in Table 1. Final outputs in the table are in digit reversed order. However using an appropriate sequence of mode select, any required output ordering can be achieved.

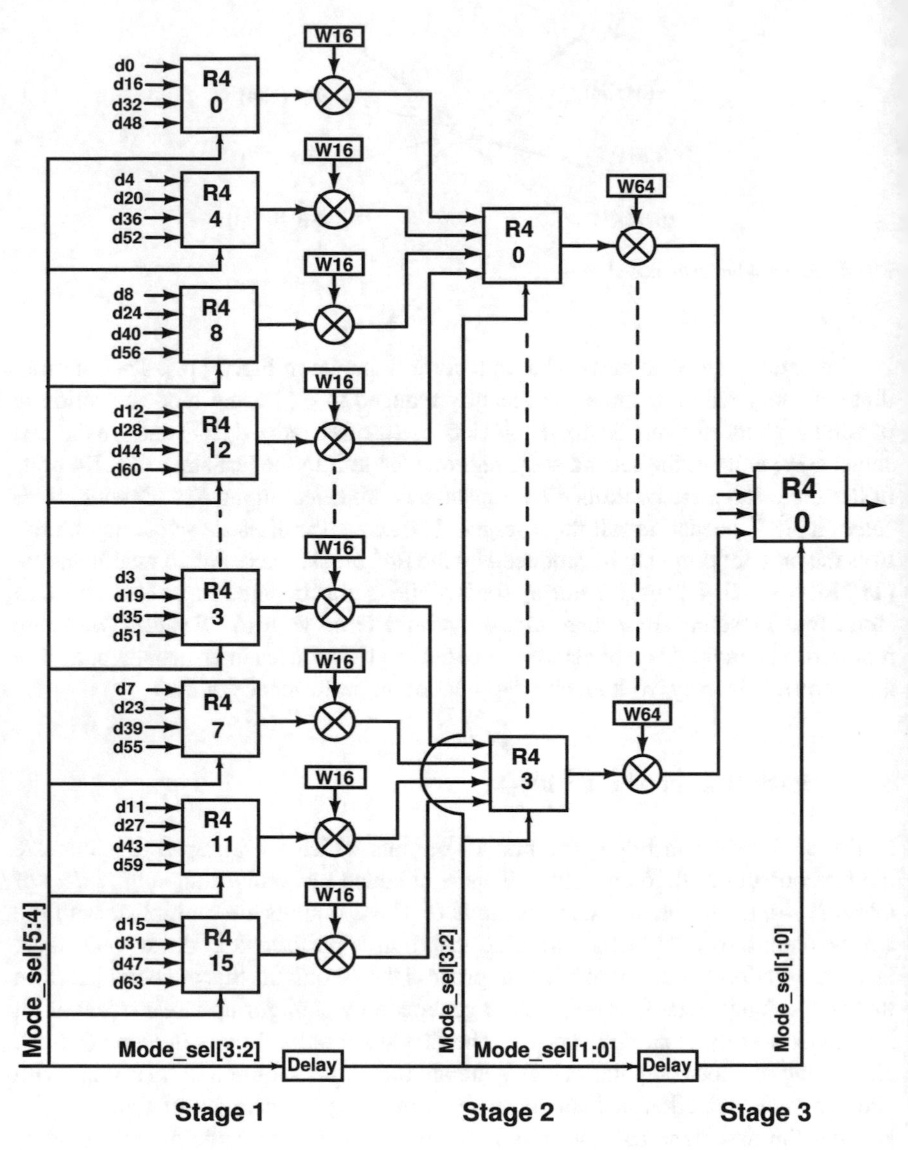

Fig. 5 Radix-4^3 block

Table 1 Mode selection in different stages of $R4^3$ block

Stage 1	Stage 2	Stage 3	Output generated
00	00	00	0
		01	16
		10	32
		11	48
00	01	00	1
		01	17
		10	33
		11	49
:	:	:	:
:	:	:	:
11	11	00	15
		01	31
		10	47
		11	63

4 Implementation and Results

The proposed 2D FFT algorithm has been implemented in MATLAB and outputs are compared with outputs from the inbuilt 2D FFT function of Matlab for validation. Two variants of the algorithm has been implemented. One uses single precision floating point for data representation while the second variant uses fixed point representation. Floating point representation gives accurate results but is not efficient in terms of hardware complexity and power. Hence fixed point implementations are preferred. Selection of word length is a major design issue for fixed point implementations. Word length has to be chosen to meet the Signal-to-Noise Ratio (SNR) demanded by the application. Selection of bit length varies in different architectures based on the required FFT size and SNR [4, 19–21]. Maharatna et al. [22] uses 16 bits for implementing 64-point FFT and in [20] 10 bits are chosen for implementing 128 point FFT.

For choosing the appropriate word length, Signal-to-Noise Ratio (SNR in dB) for various word lengths has been analyzed. Outputs from fixed point implementation with varying word length has been compared with output from inbuilt FFT function. For fixed point implementation, in order to avoid overflow at all stages, inputs at each butterfly node are scaled down by 0.25. Hence floating point FFT outputs also have to be scaled down by the FFT size before performing the comparison. The result of this analysis is shown in Fig. 6. Based on this analysis, we have selected 16 bits as word length. 16 bit word length gives SNR of 59.9 dB, which is sufficient for image reconstruction. For real and imaginary parts of data words, out of 16 bits, 1 bit is used for sign and 15 bits for fraction, giving a range of -1 to $+1$. For representing

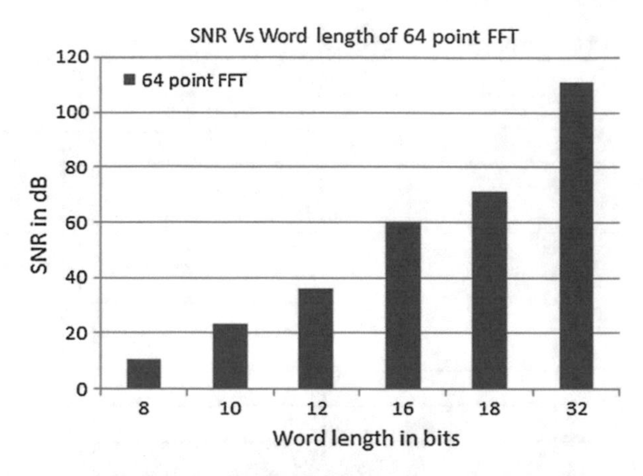

Fig. 6 SNR versus word length for 64-point FFT using radix-4^3 algorithm

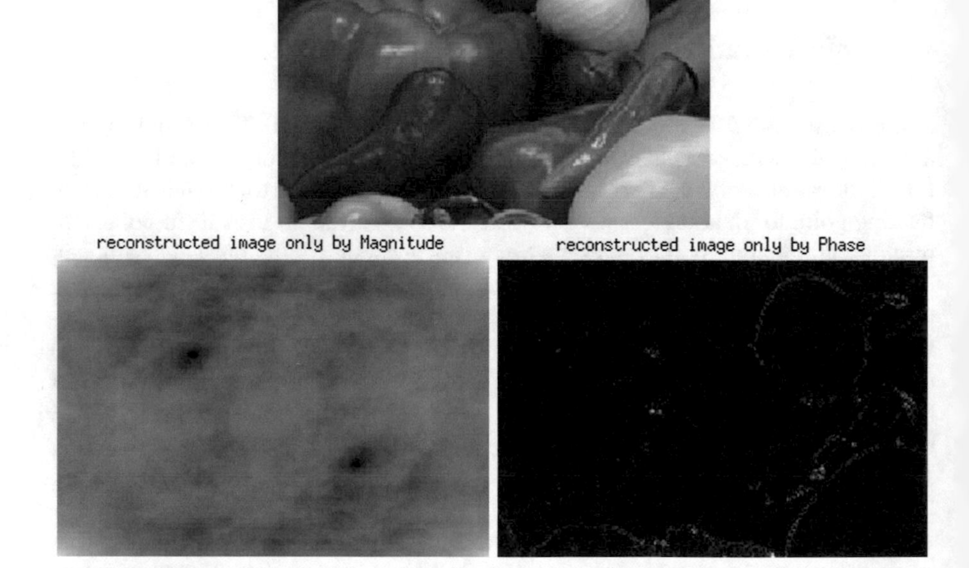

Fig. 7 Reconstructed image

twiddle factors, a 16 bit word with 1 bit for integer, 1 bit for sign and 14 fractional bits is used.

Using fixed point FFT implementation, we have performed image reconstruction in MATLAB. Reconstruction from both magnitude and phase of the image were done separately. The simulated results are shown in Fig. 7.

Table 2 Hardware complexity of FFT architectures

	Complex multipliers	Complex adders	Memory	Control
R2SDF	$2log_4N - 1$	$4log_4N$	$N - 1$	Simple
R4SDF	$log_4N - 1$	$8log_4N$	$N - 1$	Medium
$R2^2$SDF	$log_4N - 1$	$4log_4N$	$N - 1$	Simple
$R2^3$SDF	$2(log_8N - 1)$	$6log_8N$	$N - 1$	Simple
R2MDC	$2log_4N - 1$	$4log_4N$	$3N/2 - 2$	Simple
R4MDC	$3(log_4N - 1)$	$8log_4N$	$5N/2 - 4$	Simple
$R2^2$MDC	$log_2N - 2$	$2log_2N$	$3N/2 - 2$	Simple
R4SDC	$log_4N - 1$	log_4N	$2N - 2$	Complex
$R4^3$ Systolic	$log_4N - 1$	$3log_4N$	$(N/3)log_4N$	Simple

We have implemented a 64-point FFT based on the radix-4^3 in RTL using Verilog. Table 2 compares hardware complexity of various FFT architectures from the literature [17, 23, 24]. For large size FFTs, radix-4^3 systolic is efficient in terms of adders and multipliers. Although the proposed parallel architecture of radix-4^3 uses more adders and multipliers when compared to other architectures, the size of intermediate buffers is lower. For very large FFTs, area savings in adders and multipliers are offset by the area required for these additional buffers [15].

The implemented RTL is simulated using Modelsim and simulated outputs are fed to MATLAB for comparing with the fixed point implementation. The test bench structure is shown in Fig. 8. RTL has been synthesized with Synopsys Design Compiler using Faraday 130 nm standard cell library which is tailored for UMC's 130 nm 1.2 V/3.3 V 1P8M HS Logic Process. 130 nm technology is chosen here for comparison purposes. Gate density for this library is 250,000 gates/mm^2. For RAMs, we used "Flop based asynchronous single port RAM" from Synopsys DesignWare Library.

After synthesis, the DUT in test bench structure shown in Fig. 8 is replaced with synthesized Verilog netlist and Post Synthesis simulation is done using Modelsim. Modelsim generates the 'activity factor' associated with each net in the netlist into a VCD (Value Change Dump) file. This along with the standard cell library is given as input to Synopsys PrimeTime power simulator to get the power dissipation for the design. Physical Synthesis (Floorplanning, Placement and Routing) of the FFT Processor is done using Cadence SoC Encounter. The layout of radix-4^3 FFT processor is shown in Fig. 9. Logic synthesis results for radix-4^3 FFT architecture are given in Table 3.

Table 4 shows comparison of radix-4^3 FFT with existing architectures. Compared to other implementations, the proposed radix-4^3 FFT has the lowest latency. When compared to [17], the proposed FFT implementation operates at half the frequency and has less area but power dissipation is more than double. But [17] uses only single R4 butterfly unit in each stage and needs large intermediate RAMs making it unsuitable for large size FFTs. Maharatna et al. [22] and Lin et al. [25] are targeted

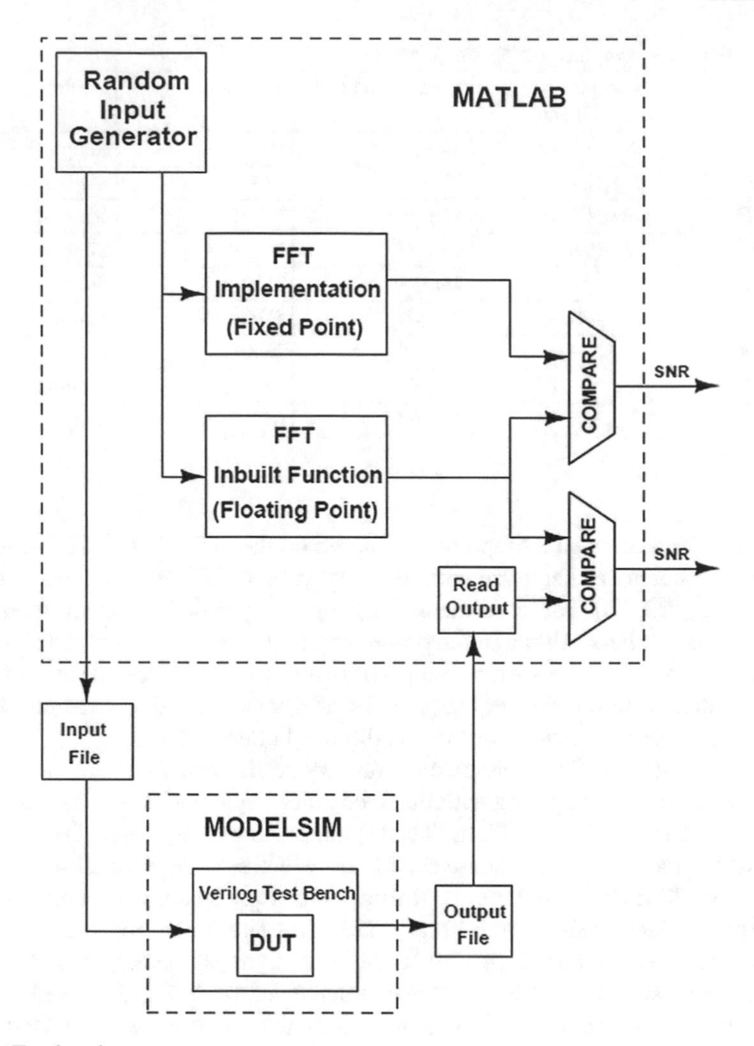

Fig. 8 Test bench structure

at lower operating frequencies and gives lower power dissipation compared to the proposed architecture. However by targeting an operating frequency of 20 MHz and by reducing the pipeline stages, comparable power consumption of 2.27 mW can be obtained from our implementation.

Latency Calculation: Total latency of radix-4^3 FFT includes 64 cycles for input memory block and 19 cycles for radix-4^3 block. Latency of radix-4^3 includes four cycles for each of the three R4 engines, three cycles each for the two complex multipliers and one cycle for input register. The total latency is 83 cycles for a single radix-4^3 FFT. Because of its pipelined nature, the proposed FFT block can produce one output in each clock cycle. 64 clock cycles are needed for computation of 64-points

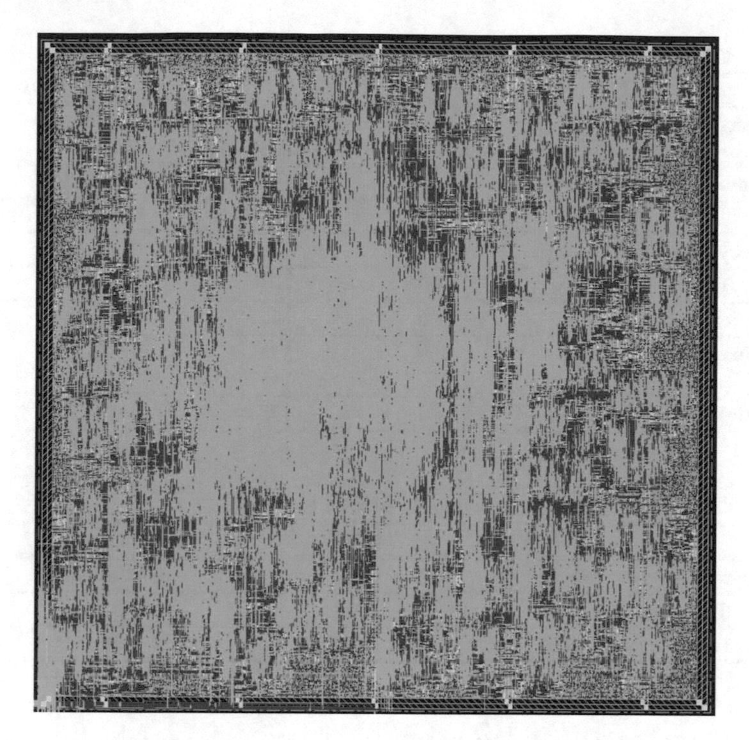

Fig. 9 Layout of Radix-4^3 FFT

Table 3 Synthesis results for radix-4^3 FFT

FFT size	64 point
Frequency (MHz)	350
Throughput (MSPS)	350
Latency (cycles)	83
Area (mm^2)	2.12
Power (W)	0.175

FFT. The ping-pong structure provided at the input memory block allows feeding next set of inputs while the current inputs are processed by the radix-4^3 block. Thus the proposed radix-4^3 FFT supports a throughput of one word per cycle and effective computation time of 64 cycles per 64-point FFT.

For a 64×64 FFT, input memory block requires 4096 cycles, two radix-4^3 block requires 19 cycles each and intermediate registers take 65 cycles. Total latency of 64×64 point FFT is 4199 clock cycles. Effective computation time is 4096 cycles per 64×64 FFT. Our architecture has an operating frequency of 350 MHz. Computation time for various 2D FFT architectures are compared in Table 5. Table shows

Table 4 Comparison of 64-point FFTs

	[22]	[25]	[17]	Our work
Word length (bits)	16	16	14	16
Algorithm	R 2	R 8	R 4^3	R 4^3
Process (μm)	0.25	0.13	0.13	0.13
Frequency (MHz)	20	20	604.5	350
Latency (μs)	3.2	3.6	0.33	0.091
Area (mm^2)	13.5	1.66	5.09	2.12
Normalized area (mm^2)	3.65	1.66	5.09	2.12
Power (mW)	41	22.36	78.2	175

Table 5 Comparison of computation time for 64×64 FFTs

	[26]	[14]	Our work
Clock cycles	4516	–	4199
Clock (MHz)	100	263	350
Computation time (μs)	45.16	15.6	11.9

Table 6 Comparison of scalability of various 2D FFTs

Reference	Architecture/Kernel	Scalability
[11]	Xilinx FPGA FFT kernel, Row-Column algorithm	No
[13]	ILUT-based radix-2, Row-Column algorithm	No
[12]	Xilinx FPGA FFT kernel, 2D decomposition algorithm	No
[14]	Non power of two, Row-Column algorithm	Yes
[27]	Spiral	No
Our work	Radix-4^3 parallel, Row-Column algorithm	Yes

that our architecture gives significant reduction in computation time compared to other implementations.

The architecture that we propose here is a scalable architecture. Higher order 2D FFTs can be developed from the basic radix-4^n architecture. Comparison of scalability of various 2D FFTs are given in Table 6. Even though [14] is a scalable architecture, intermediate memory requirements are high, compared to the proposed architecture.

5 Conclusion and Future Work

In this paper, we present a novel technique for computing two dimensional Fast Fourier Transform. The proposed method uses radix-4^n as the building block. We present a 64×64 FFT architecture using two radix-4^3 blocks, which shows considerable reduction in size of intermediate buffers when compared to existing architectures. Comparing hardware complexity of various FFT architectures, it is observed that the proposed radix-4^n algorithm is suitable for computing large FFT sizes. A fixed point implementation of the proposed algorithm has been done in MATLAB. For choosing the right word length, Signal-to-Noise Ratio (SNR) analysis for various word lengths has been performed and word length of 16 bits has been chosen as per the application requirement. Original image is reconstructed from phase and magnitude separately, using MATLAB simulations. Radix-4^3 FFT architecture, which is the building block of our 2D FFT, has been implemented in RTL using Verilog and simulated using Modelsim. The RTL has been synthesized with Synopsys Design Compiler using Faraday 130 nm standard cell library which is tailored for UMC's 130 nm 1.2 V/3.3 V 1P8M HS Logic Process. Our architecture gives significant reduction in latency when compared with existing architectures. For larger image sizes, the proposed architecture can be scaled-up easily while providing significant reduction in intermediate buffers. Future work includes hardware implementation of the two-dimensional FFT architecture presented in this paper. A reconfigurable FFT which supports various image sizes using the proposed architecture can also be developed.

References

1. Proakis, J.G., Manolakis, D.G.: Digital Signal Processing Principles, Algorithms and Applications. Prentice-Hall (1996)
2. Yu, C.-L., Chakrabarti, C., Park, S., Narayanan, V.: Bandwidth-intensive FPGA architecture for multi-dimensional DFT. In: IEEE International Conference on Acoustics, Speech, Signal Processing, pp. 1486–1489 (2010)
3. Gold, B., Rabiner, L.R.: Theory and Application of Digital Signal Processing. Prentice-Hall (1975)
4. Corts, A., Vlez, I., Zalbide, I., Irizar, A., Sevillano, J.F.: An FFT core for DVB-T/DVB-H receivers. VLSI Design (2008)
5. He, S., Torkelson, M.: A new approach to pipeline FFT processor. In: 10th International Parallel Processing Symposium, IPPS '96, pp. 766–770 (1996)
6. He, S., Torkelson, M.: Designing pipeline FFT processor for OFDM (de)modulation. Signals Syst. Electron. (1998)
7. Frigo, M., Johnson, S.: FFTW: an adaptive software of the FFT. In: IEEE International Conference on Acoustics, Speech, and Signal Processing, pp. 1381–1384 (1998)
8. Pschel, M., et al.: SPIRAL: code generation for DSP transforms. Proc. IEEE **93**(2), 232–275 (2005)
9. D'Alberto, P., et al.: Generating FPGA accelerated DFT libraries. In: IEEE Symposium on Field-Programmable Custom Computing Machines (FCCM), pp. 173–184 (2007)

10. Dillon, T.: Two virtex-II FPGAs deliver fastest, cheapest, best high-performance image processing system. Xilinx Xcell J. **41**, 70–73 (2001)
11. Chen, R., Prasanna, V.K.: Energy optimizations for FPGA-based 2-D FFT architecture. In: IEEE High Performance Extreme Computing Conference (HPEC) (2014)
12. Yu, C.-L., Kim, J.-S., Deng, L., Kestur, S., Narayanan, V., Chakrabarti, C.: FPGA architecture for 2D discrete Fourier Transform based on 2D decomposition for large-sized data. Signal Process. Syst. **64**(1), 109–122 (2011)
13. Kee, H., Bhattacharyya, S.S., Petersen, N., Kornerup, J.: Resource-efficient acceleration of 2-Dimensional Fast Fourier Transform computations on FPGAs. In: International Conference on Distributed Smart Cameras, Como, Italy (2009)
14. Wang, W., Duan, B., Zhang, C., Zhang, P., Sun, N.: Accelerating 2D FFT with non-power-of-two problem size on FPGA. In: International Conference on Reconfigurable Computing (2010)
15. Kala, S., Nalesh, S., Arka, M., Nandy, S.K., Narayan, R.: High throughput, low latency, memory optimized 64K point FFT architecture using novel radix-4 butterfly unit. In: IEEE International Symposium on Circuits and Systems, ISCAS, pp. 3034–3037 (2013)
16. Chidambaram, R.: A scalable and high-performance FFT processor, optimized for UWB-OFDM. M.S. thesis, Delft University of Technology (2005)
17. Babionitakis, K., Chouliaras, V.A., Manolopoulos, K., Nakos, K., Reisis, D., Vlassopoulos, N.: Fully systolic FFT architecture for Giga-sample applications. J. Signal Process. Syst. **58**, 281–299 (2010)
18. Kala, S., Nalesh, S., Nandy, S.K., Narayan, R.: Scalable and energy efficient, dynamically reconfigurable Fast Fourier Transform architecture. J. Low Power Electron. **11**(3), 426–435 (2015)
19. Li, N., ASIC FFT Processor for MB-OFDM UWB System: M.Sc. thesis, Delft University of Technology (2008)
20. Lin, Y.N., Liu, H.Y., Lee, C.Y.: A 1-GS/s FFT/IFFT processor for UWB applications. IEEE J. Solid State Circuits **40**(8) (2005)
21. Lin, Y.-W., Liu, H.-Y., Lee, C.-Y.: A dynamic scaling FFT processor for DVB-T applications. IEEE J. Solid State Circuits **39**(11) (2004)
22. Maharatna, K., Grass, E., Jagdhold, U.: A 64-point Fourier Transform chip for high-speed wireless LAN application using OFDM. IEEE J. Solid State Circuits **39**(3), 484–493 (2004)
23. He, S., Torkelson, M.: Designing pipeline FFT processor for OFDM (de)modulation. Signals Syst. Electron. (1998)
24. Lin, C.-T., Yu, Y.-C.: Cost-effective pipeline FFT/IFFT VLSI architecture for DVB-H system. In: National Symposium on Telecommunications (2007)
25. Lin, C.T., Yu, Y.C., Van, L.D.: A low power 64-point FFT/IFFT design for IEEE 802.11a WLAN application. In: IEEE International Symposium on Circuits and Systems, ISCAS, pp. 4523–4526 (2006)
26. Rodrguez-Ramos, J.M., Magdaleno Castell, E., Domnguez Conde, C., Rodrguez Valido, M., Marichal-Hernndez, J.G.: 2D-FFT implementation on FPGA for wavefront phase recovery from the CAFADIS camera. Proc. SPIE (2008)
27. Akn, B., Milder, P.A., Franchetti, F., Hoe, J.C.: Memory bandwidth efficient two-dimensional Fast Fourier Transform algorithm and implementation for large problem sizes. In: IEEE 20th International Symposium on Field-Programmable Custom Computing Machines (2012)

Printed in the United States
By Bookmasters